PERSPECTIVES FOR THE INTERACTING BOSON MODEL

ON THE OCCASION OF ITS 20TH ANNIVERSARY

PERSPECTIVES FOR THE INTERACTING BOSON MODEL

ON THE OCCASION OF ITS 20TH ANNIVERSARY

Padova, Italy 13 – 17 June 1994

Editors

R. F. Casten
Brookhaven National Laboratory

A. Vitturi
University of Padova

A. B. Balantekin
University of Wisconsin

B. R. Barrett
University of Arizona

J. N. Ginocchio
Los Alamos National Laboratory

G. Maino
ENEA, Bologna

T. Otsuka
University of Tokyo

World Scientific
Singapore • New Jersey • London • Hong Kong

Published by

World Scientific Publishing Co. Pte. Ltd.
P O Box 128, Farrer Road, Singapore 9128
USA office: Suite 1B, 1060 Main Street, River Edge, NJ 07661
UK office: 73 Lynton Mead, Totteridge, London N20 8DH

PERSPECTIVES FOR THE INTERACTING BOSON MODEL

Copyright © 1994 by World Scientific Publishing Co. Pte. Ltd.

All rights reserved. This book, or parts thereof, may not be reproduced in any form or by any means, electronic or mechanical, including photocopying, recording or any information storage and retrieval system now known or to be invented, without written permission from the Publisher.

For photocopying of material in this volume, please pay a copying fee through the Copyright Clearance Center, Inc., 27 Congress Street, Salem, MA 01970, USA.

ISBN: 981-02-2071-5

Printed in Singapore by Utopia Press.

PREFACE

On June 13-17, 1994, the International Conference on Perspectives for the Interacting Boson Model on the Occasion of Its 20th Anniversary was held in Padova, Italy. This volume contains the Proceedings of that Conference.

The meeting was attended by about 130 scientists, representing 29 countries and 6 continents. The Conference presented an exciting array of new research results and fostered active formal and informal discussions, demonstrating the continuing vitality of the IBM model and its application to a number of new areas. The Conference also featured talks on related algebraic approaches to fields as diverse as molecules, scattering phenomena, and baryon structure. We are grateful to all those on the International Advisory Committee for their many suggestions for speakers. The Conference commenced with welcome addresses by M. Mammi, Deputy-Rector of the Università degli Studi di Padova, and A. Bettini, Vice-President of the Istituto Nazionale di Fisica Nucleare. This was followed by a special Ceremonial Session with talks by F. Iachello and I. Talmi that offered a scientific perspective and historical comments as well as recent research results.

The first four days of the Conference were held in the beautiful Palazzo del Bo, the main building of the Università degli Studi di Padova. The first day's sessions took place in the Aula Magna, while subsequent sessions were held in the Sala dell' Archivio Antico. The last day of the Conference was held at the nearby Laboratori Nazionali di Legnaro and featured, besides the formal talks, a day-long Poster Session and a tour of the impressive experimental facilities there.

We would like to take this occasion to thank a number of people and organizations. We are grateful to the Universita degli Studi di Padova both for providing the superb facilities and ambiance for the Conference and for financial support. We are also grateful for support from the Dipartimento di Fisica "G. Galilei" as well as for the venue for the Opening Reception on Sunday, June 12. We appreciate the support of the Laboratori Nazionali di Legnaro in conjunction with the sessions held there on Friday. We would like to acknowledge with gratitude the generous support of the Istituto Nazionale di Fisica Nucleare, whose substantial grant was essential to the success of the Conference. We gratefully acknowledge grants from the International Science Foundation for support of travel and lodging for scientists from the former Soviet Union. Finally, we thank the Physics Department of Brookhaven National Laboratory for infrastructure and technical services in support of the Conference.

We owe a great debt of gratitude to the Conference Secretary, Jackie Mooney, whose work on behalf of the Conference, on both sides of the Atlantic, was invaluable. Her expertise and dedication are well-recognized in the nuclear physics community and were crucial in developing and arranging the Conference. We also are grateful for the substantial help of Laura Salvadori and Ines Campo, secretaries in the Dipartimento di Fisica "G. Galilei" whose help before and during the Conference greatly helped smooth its operation.

                            R. F. Casten
                            A. Vitturi
                            A. B. Balantekin
                            B. R. Barrett
                            J. N. Ginocchio
                            G. Maino
                            T. Otsuka

Advisory Committee

J. M. Arias (Sevilla)
A. Arima (RIKEN)
P. von Brentano (Köln)
D. M. Brink (Trento)
A.E.L. Dieperink (Groningen)
J. Dobes (Prague)
A. Faessler (Tübingen)
H. Feshbach (MIT)
A. Frank (Mexico)
H. B. Geyer (Stellenbosch)
K. Heyde (Gent)
S. T. Hsieh (Taiwan)
M. Kirson (Weizmann)
F. Iachello (Yale)
P. O. Lipas (Jyvaskyla)
G. Molnar (Budapest)
I. Morrison (Melbourne)
V. Paar (Zagreb)
S. Pittel (Bartol)
H.-Z. Sun (Beijing)
I. Talmi (Weizmann)
M. Vergnes (Orsay)
D. D. Warner (Daresbury)
N. V. Zamfir (Bucharest)
P. Van Isacker (GANIL)
N. Yoshinaga (Saitama)

Advisory Committee

J. M. Arias (Sevilla)
A. Arima (RIKEN)
F. von Brentano (Köln)
D. M. Brink (Trento)
A. E. L. Dieperink (Groningen)
J. Dobeš (Prague)
J. Eisenstein (Tübingen)
H. Feshbach (MIT)
A. Frank (Mexico)
H. B. Geyer (Stellenbosch)
K. Heyde (Gent)
S. T. Hsieh (Taiwan)
M. Kirson (Weizmann)
F. Iachello (Yale)
R. O. Lipas (Jyväskylä)
J. Molnár (Budapest)
I. Morrison (Melbourne)
V. Paar (Zagreb)
S. Pittel (Bartol)
H. Z. Sun (Beijing)
I. Talmi (Weizmann)
M. Vergnes (Orsay)
D. D. Warner (Daresbury)
N. V. Zamfir (Bucharest)
P. Van Isacker (GANIL)
N. Yoshinaga (Saitama)

Contents

Preface	v
Perpectives for the Interacting Boson Model *F. Iachello*	1
From Independent Nucleons to Interacting Bosons *Igal Talmi*	11
Robust Predictions of the Interacting Boson Model *R.F. Casten*	21
Interacting Boson Model for O(6) Nuclei *Takaharu Otsuka, Takahiro Mizusaki, and Ka-Hae Kim*	33
How Far Can a Dynamical Symmetry Approach be Pursued? *J. Jolie*	45
The Scissors Mode and Related Modes Revisited in the IBM and other Models *A. Richter*	59
From Semiclassical to Microscopic Descriptions of Scissors Mode *N. Lo Iudice*	79
Mixed-Symmetry States in O(6) Nuclei *G. Molnár, T. Belgya, B. Fazekas, D.P. Diprete, R.A. Gatenby, S.W. Yates, and T. Otsuka*	87
Asymptotic Evaluation of F-Spin Content and M1 Sum Rule *A.F. Daillo, B.R. Barrett, and D.E. Davis*	95
Particle–Hole Excitations in the Interacting Boson Model *K. Heyde*	103
Formulation and Application of Multiparticle – Multihole Configurations within the Proton–Neutron Interacting Boson Model *B.R. Barrett and A.F. Barfield*	117
Partial Dynamical Symmetry in the Interacting Boson Model *Amiram Leviatan*	129

1/N Expansion in Boson Models: Present Status and Future Applications
    Serdar Kuyucak   143

From Seniority to Collectivity
    Akito Arima   157

Quadrupole "Phonon" Excitations in Even Xenon Model
    P. von Brentano, O. Vogel, N. Pietralla, A. Gelberg,
    and I. Wiedenhöver   165

The Pair–Coupling Model
    David J. Rowe   177

Boson Mappings and Phenomenological Boson Models
    Hendrik B. Geyer, Peter Navrátil, and Jacek Dobaczewski   189

Finite Boson Mappings of Fermion Systems
    Calvin W. Johnson and Joseph N. Ginocchio   201

Favored Pairs and Renormalization Effects from Non-Collective Pairs
    Naotaka Yoshinaga   213

Broken Pairs in the Interacting Boson Model: Description of High Spin States
    D. Vretenar, G. Bonsignori, and M. Savoia   225

Testing the IBM Plus Broken Pairs Model at High Spin: An Experimental Viewpoint
    C.J. Lister, P. Chowdhury, and D. Vretenar   239

Boson Mappings and Microscopy of the Interacting Boson Model
    J. Dobeš and P. Navrátil   249

Microscopic Generalization of Standard Interacting Boson Model in the Restricted Dynamics Approach
    J.A. Castilho Alcaras, J. Tambergs, J. Ruža, T. Krasta,
    and O. Katkevičius   257

Symmetries in Odd-Mass Nuclei and Their Applications
    P. van Isacker   265

Low-Lying States in Odd-A Ce Isotopes: Experimental Results and Description by the IBFM-1 Model
    Jean Gizon   275

Intrinsic Frame Description of Composed Boson–Fermion Systems
  C.E. Alonso, J.M. Arias, and A. Vitturi — 283

The Boson–Fermion Interaction in the Generalised Holstein–Primakoff
Boson Expansion
  J. Dukelsky and S. Pittel — 291

Interacting Boson–Fermion Model for Strongly Deformed Nuclei
  R.V. Jolos and A. Gelberg — 299

Soft Mode Dynamics in Transitional Nuclei
  V.G. Zelevinsky — 307

Reflection Asymmetric Shapes in Interacting Boson–Boson and
Boson–Fermion Systems
  *C.E. Alonso, J.M. Arias, A. Frank, H.M. Sofia, S.M. Lenzi,
  and A. Vitturi* — 319

E1 and E3 Transition Rates in the sdf-IBA
  N.V. Zamfir — 327

sdg Interacting Boson Model: Some Analytical and Numerical Aspects
  Y.D. Devi and V.K.B. Kota — 335

Superdeformation and IBM
  Michio Honma and Takaharu Otsuka — 343

Multi–Fermion Dynamical Supersymmetries in Superdeformed Nuclei
  J.A. Cizewski — 351

Contribution of the Two–Phonon Configurations to the Wave Function of
Low–Lying States in Deformed Nuclei
  V.G. Soloviev — 359

New Challenges from Radioactive Beams
  David D. Warner — 373

Global Systematics of Unique Parity Quasibands in Odd–A Nuclei
  D. Bucurescu, G. Cata-Danil, M. Ivascu, L. Stroe, and C.A. Ur — 389

Electric Dipole Excitations in Rare Earth Nuclei
  *A. Zilges, P. von Brentano, R.-D. Herzberg, U. Kneissl, J. Margraf,
  and H.H. Pitz* — 393

IBA Calculation of the Effective Gamma Deformation in Barium Nuclei
   O. Vogel, A. Gelberg, P. von Brentano, and P van Isacker   397

Anomaly in ^{196}Pt: A Clean Test?
   M.K. Harder and B. Krusche   401

Evolution of the Triaxial Asymmetry at Variation of the Symmetric
Quadrupole Deformation
   W. Andrejtscheff, P. Petrov, and N.V. Zamfir   405

The Revival of the L–S Coupling Scheme at Superdeformation
   K. Sugawara–Tanabe   409

On the Reconciliation of Microscopic and Fitted Boson g–Factors
   E.D. Davis and P. Navrátil   413

IBM Approach to the Rotational Damping
   Takahiro Mizusaki, Takaharu Otsuka, and P. von Brentano   417

Algebraic and Geometric Approaches to the Collective Enhancement of
Nuclear level Densities
   A. Mengoni, G. Maino, A. Ventura, and Y. Nakajima   421

F-Spin Multiplets and M1 Transitions in the $A=100$ Mass Region
   A. Gelberg and T. Otsuka   425

Applications of IBM-3 to the $Z \sim N \sim 40$ Nuclei
   Michiaki Sugita   431

Sum Rules in the Proton–Neutron Interacting Boson Model
   C. de Coster and K. Heyde   435

Exploring the Validity of the $Z=38$ and $Z=50$ Proton Closed Shells in
Even–Even Mo Isotopes, and Odd–A Tc Isotopes
   Haydeh-Dejbakhsh and S. Shlomo   439

Supersymmetry in the Pt–Au Region
   S.M. Fischer, A. Aprahamian, X. Wu, J.X. Saladin,
   and M.P. Metlay   443

New Experimental Systematics in the $A \sim 100$ Region
   T. Borello-Lewin, L.B. Horodynski-Matsushigue, J.L.M. Duarte,
   L.C. Gomes, and G.M. Ukita   449

IBFM Calculation of Xe and Cs Isotopes
  N. Yoshida, A. Gelberg, T. Otsuka, and H. Sagawa — 453

Single-Nucleon Transfer Reactions in the IBFFM
  P.E. Garrett and D.G. Burke — 457

Surface Plasmons of Metal Clusters in the Interacting Boson Model
  E. Lipparini and A. Ventura — 461

Effective E1-Operator in IBFM
  D.N. Dojnikov and V.M. Mikhajlov — 465

Structure of Odd-Odd Sb Nuclei
  Z.S. Dombrádi, T. Fényes, Z. Gácsi, J. Gulyás, S. Brant, and V. Paar — 469

Rotational Bands in the sp(6) $\supset$ u(3) Algebra of the Fermion Dynamical Symmetry Model
  Sudha R. Swaminathan and K.T. Hecht — 473

On Possible Equidistance of Some Groups of Levels in Deformed Nuclei at Excitation Energies up to 5 MeV
  A.V. Voinov — 477

The NT-Dependence of the IBM-3 Hamiltonian
  J.P. Elliott, J.A. Evans, G.L. Long, and V.-S. Lac — 483

Nuclear Collective Motion with Isospin
  Joseph N. Ginocchio and Amiram Leviatan — 495

Multistep Processes in Medium Energy Scattering
  R. Bijker — 507

The Algebraic Scattering Theory and its Application to Heavy-Ions Reactions
  Rubens Lichtenthäler Filho — 521

Interacting Boson Techniques in Cluster Studies
  J. Cseh, G. Lévai, and W. Scheid — 529

Interacting Bosons in Molecular Structure
  Alejandro Frank and Renato Lemus — 537

Description of Nuclear Structure Effects in Subbarrier Fusion by the
Interacting Boson Model
    A.B. Balantekin   545

QCD and the Nuclear Physics IBM
    Yuval Ne'eman and Djordje Sijacki   559

q-Deformed Vibron Model for Diatomic Molecules
    R.N. Alvarez, Dennis Bonatsos, and Yu F. Smirnov   575

Chaos in the Collective Dynamics of Nuclei
    Y. Alhassid   591

Baryon Mapping of Constituent Quark Models
    S. Pittel, J.M. Arias, J. Dukelsky, and A. Frank   605

Weak Interactions in the Interacting Boson–Fermion Model
    Giuseppe Maino   617

Twenty Years of IBA and Transient Fields
    N. Benczer-Koller, G. Kumbartzki, A. Mountford, T. Vass,
    M. Satteson, N. Matt, R. Tanczyn, C.L. Lister, and P. Chowdhury   631

Magnetic Moments in Transitional Nuclei: Probing F-Spin Symmetry
and Supersymmetry
    Andrew E. Stuchbery   639

G-Factor Measurements in Some Stable Even–Even Heavy Nuclei as a
Severe Test of IBM and Other Theoretical Models
    F. Brandolini   647

Consistent Treatment of M1 Observables Within the IBA
    A Wolf   657

Transitional Odd–Odd Nuclei in IBFM-2
    A. Ventura, G. Maino, A.M. Bizzeti-Sona, P. Blasi,
    and A.A. Stefanini   665

Structure of Odd–Odd Ga and As Nuclei, Dynamical and
Supersymmetries
    T. Fényes, A. Algora, Z.S. Podolyák, D. Sohler, J. Timár,
    V. Paar, S. Brant, Lj. Šimičič   673

IBFFM Calculations in ^{126}Cs
  N. Blasi   683

Conference Summary

Concluding remarks
  Herman Fleshbach   693

Contributed Abstracts   701

Scientific Program   739

List of Participants   745

Author Index   755

PERSPECTIVES FOR THE INTERACTING BOSON MODEL

F. Iachello

*Center for Theoretical Physics, Sloane Physics Laboratory,
Yale University, New Haven, CT 06520-8120*

Abstract

Some aspects of the interacting boson model that require further study are briefly indicated. In particular, weak interactions in heavy nuclei are discussed. The possible occurrence of a new collective mode in medium mass and heavy nuclei is suggested.

1 Introduction

The purpose of this Meeting is to discuss recent developments in the interacting boson model and indicate perspectives for future work. The interacting boson model introduced[1] in 1974 in order to describe collective low-lying states in medium and heavy-mass nuclei has grown in the ensuing years into a model describing most aspects of nuclear structure and reactions[2]. Several of these aspects will be discussed at this Conference and I refer to the following articles for detailed expositions.

The interacting boson model has two main aspects, one related to the description of properties of nuclei (the model in the proper sense of the word, or IBM)and the other related to its microscopic foundations (the interacting boson approximation, or IBA). The cornerstones of the IBM are its algebraic structure and dynamic symmetries. The latter concept, introduced in the 60's for applications to particle physics, has had its most important and useful applications in nuclear physics in the context of the interacting boson model (and related models). The original IBM, or IBM-1 as it is called today, has three possible dynamic symmetries characterized by the algebraic structure

$$U(6) \begin{array}{c} \nearrow \\ \longrightarrow \\ \searrow \end{array} \begin{array}{l} U(5) \supset O(5) \supset O(3) \supset O(2) \ , \\ SU(3) \supset O(3) \supset O(2) \ , \\ O(6) \supset O(5) \supset O(3) \supset O(2) \ . \end{array} \quad (1)$$

These three dynamic symmetries have formed the basis upon which the classification scheme for complex nuclei has been built. An important result found in the years which have elapsed since the introduction of the IBM has been that nuclei indeed display all three symmetries, in some cases even to a better degree of accuracy that it was thought. Some examples will be discussed at this conference.

The cornerstone of the second aspect, the interacting boson approximation, is the concept of correlated pairs. The original IBA, or IBA-1 as it is called today, introduced only nucleon pairs coupled to angular momentum, J=0, S-pairs, and angular momentum, J=2, D-pairs,

$$S^\dagger = \sum_j \alpha_j S_j^\dagger, \qquad S_j^\dagger = (a_j^\dagger \times a_j^\dagger)^{(0)};$$
$$D_\mu^\dagger = \sum_{jj'} \beta_{jj'} D_{jj',\mu}^\dagger, \qquad D_{jj',\mu}^\dagger = (a_j^\dagger \times a_{j'}^\dagger)_\mu^{(2)}. \qquad (2)$$

Later, separate proton (π) and neutron (ν) pairs were introduced, and pairs with larger values of the angular momentum (G pairs, $\cdots$) included.

In view of the success of the IBM (and IBA) in describing many properties of nuclei, one may inquire what open problems, if any, still remain in this field of physics. The answer to this question is that, despite all the work that has gone into this subject in the last 20 years, still many open problems remain, one of which I will discuss here and others that I will only mention in passing. In general terms, one can see two main perspectives for the interacting boson model in future years:
(i) the study of phenomena which have been either overlooked or only marginally investigated;
(ii) the search for other collective degrees of freedom in addition to those of s and d (proton and neutron) bosons.

2 Aspects That Need Further Study

In the last 20 years, strong and electromagnetic processes in nuclei have been extensively investigated through transfer reactions and γ-ray spectroscopy. Indeed, the latter process has become the dominant way of studying nuclear structure. On the contrary, weak processes (β^- decay, β^+ decay, e capture, ν capture) have been poorly investigated. These processes are important for astrophysics, neutrino detection, detection of WIMPs, double β-decay, e-μ universality, $\cdots$, and require further study.

I will divide the discussion of weak interactions in heavy nuclei into three parts, depending on the nature of the decaying nucleus:
A. Odd-odd $\rightarrow$ Even-even
B. Even-even $\rightarrow$ Odd-odd
C. Even-odd $\rightarrow$ Odd-even
Weak interactions are characterized by the ft-value,

$$ft = \frac{6163}{<M_F>^2 + (\frac{G_A}{G_V})^2 <M_{GT}>^2} sec \quad . \qquad (3)$$

The quantities appearing in (3) are: the matrix elements of the Fermi interaction, $<M_F>^2$, with selection rules $\Delta J = 0$, $\Delta T = 0$; the matrix elements of the Gamow-Teller interaction, $<M_{GT}>^2$, with selection rules $\Delta J = 0, \pm 1$ ($0 \not\rightarrow 0$) , $\Delta T = 0, 1$; and the ratio of the axial vector to vector coupling constants, G_A/G_V. In free space, this ratio is $(G_A/G_V)^2 = 1.59 \pm 0.02$. Wilkinson[3] has analysed this ratio in light nuclei and finds a value of ~ 1.27. Although the question of why the ratio $(G_A/G_V)^2$ is quenched in nuclei is an important one, I will not discuss it here and only address, by way of examples, the problem of how β-decay can be used to study nuclear structure.

A. Odd-odd (1^+) $\overset{GT}{\to}$ Even-even ($0^+, 1^+, 2^+$).

I will begin the discussion with the Gamow-Teller decays of odd-odd nuclei with $J^\pi = 1^+$. These decays provide considerable information on the structure of 0^+ and 2^+ states in even-even nuclei. The analysis of β-decay requires, as a minimum, the introduction of IBM-2, since it involves both proton and neutron degrees of freedom. The basic constituents of IBM-2 are π and ν bosons forming an F-spin doublet

$$F = \frac{1}{2} \quad : \quad \begin{pmatrix} \pi \\ \nu \end{pmatrix} . \tag{4}$$

Consider the spectrum of a nucleus with U(5) symmetry, Fig.1. In addition to the states with

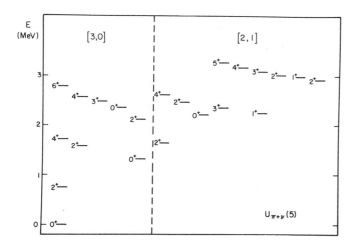

FIG. 1. Schematic representation of the spectrum of an even-even nucleus in the U(5) limit of IBM-2.

totally symmetric wave functions, and Young tableaux $[N_\pi + N_\nu, 0]$, there are also states with mixed symmetry $[N_\pi + N_\nu - 1, 1], \cdots$. In general, one has all states originating from the product

$$[N_\pi] \otimes [N_\nu] = \oplus \sum_{k=0}^{N_\nu} [N_\pi + N_\nu - k, k]; \quad N_\pi \geq N_\nu . \tag{5}$$

While the information on states with maximum F-spin (totally symmetric states) is nowadays good, the information on mixed symmetry states is very scant and β-decay may help elucidating it[4]. To be specific, consider, for example, the decay

$$^{100}_{43}Tc_{57}(1^+) \to ^{100}_{44}Ru_{56} \quad . \tag{6}$$

In the U(5) limit, the initial wave function can be written as $s_\pi^3 s_\nu^3(\pi g_{9/2}\nu g_{7/2})1^+$. In a Gamow-Teller β^--decay, a neutron changes into a proton, $\nu g_{7/2} \to \pi g_{9/2}$, thus leaving in the final state, after a recoupling transformation, the configurations $(\pi g_{9/2} \pi g_{9/2})0^+$, $(\pi g_{9/2} \pi g_{9/2})2^+$. These configurations are components of the boson states s_π and d_π respectively.

This simple argument suggests that, in the case in which the protons and neutrons are particle-like, the Gamow-Teller transition operator from odd-odd to even-even can be written, in the boson-fermion space, as

$$\hat{T}^{(GT)} = \left[\sum_j \alpha_j s_\pi^\dagger \times \left[\tilde{a}_\pi^{(j)} \times \tilde{a}_\pi^{(j)}\right]^{(0)} + h.c. \right.$$
$$+ \sum_{jj''} \beta_{jj''} d_\pi^\dagger \times \left[\tilde{a}_\pi^{(j'')} \times \tilde{a}_\pi^{(j)}\right]^{(2)} + h.c. \Big]^{(0)}$$
$$\left. \times \sum_{jj'} \eta_{jj'} \left[P^{\dagger(j)}_\pi \times \tilde{P}^{(j')}_\nu\right]^{(1)} \right. , \tag{7}$$

with appropriate modifications, if protons and/or neutrons are hole-like. Here the notation is as in Ref. 5. If the mapping to the boson-fermion space is appropriate, GT transitions from odd-odd to even-even nuclei, will test, in the U(5) limit, those components of the wave function which contain either no d-bosons, or only one d_π boson. In this symmetry limit only the ground 0^+ state will be populated since it has wave function $s_\pi^{N_\pi} s_\nu^{N_\nu}$ and only two 2^+ states will be populated, the symmetric and antisymmetric combinations (when $N_\pi = N_\nu$)

$$|\psi_S> = \frac{1}{\sqrt{2}}(d_\pi s_\pi^{N_\pi-1} s_\nu^{N_\nu} + s_\pi^{N_\pi} d_\nu s_\nu^{N_\nu-1}) ,$$
$$|\psi_A> = \frac{1}{\sqrt{2}}(d_\pi s_\pi^{N_\pi-1} s_\nu^{N_\nu} - s_\pi^{N_\pi} d_\nu s_\nu^{N_\nu-1}). \tag{8}$$

(If $N_\pi \neq N_\nu$ the two pieces are multiplied by factors $\sqrt{N_\pi/N}$ and $\sqrt{N_\nu/N}$). When the U(5) symmetry is broken, the GT matrix elements will be proportional to

$$<M_{GT}>^2 \propto |<\psi_f 0^+ | s_\pi^{N_\pi} s_\nu^{N_\nu} >|^2 ,$$
$$<M_{GT}>^2 \propto |<\psi_f 2^+ | s_\pi^{N_\pi-1} d_\pi s_\nu^{N_\nu} >|^2 \quad . \tag{9}$$

An analysis of the observed situation is shown in Table I. One should note that, above the energy needed to break a pair ($\approx 2MeV$ in this region), β-decay populates the broken pair states $s_\pi^{N_\pi-1} s_\nu^{N_\nu} (\pi g_{9/2} \pi g_{9/2}) 0^+$ and $s_\pi^{N_\pi-1} s_\nu^{N_\nu} (\pi g_{9/2} \pi g_{9/2}) 2^+$. This fact makes the analysis of states above $2\,MeV$ more difficult and requires a calculation in which broken pair states are explicitly included.

Table I. Gamow-Teller matrix elements[a] in the decay $^{100}Tc \to ^{100}Ru$.

State	E(MeV)	logft	$<M_{GT}>^2 \times (10^3)$ exp	$<M_{GT}>^2 \times (10^3)$ calc
0^+	0	4.6	110	110[b]
0^+	1.130	5.0	44	13
0^+	1.741	6.3	2	1
0^+	2.052	5.1	35	⎫
(0)?	2.387	5.4	17	⎬ Broken pairs
(0,1,2)?	2.838	4.9	55	⎭

State	E(MeV)	logft	$<M_{GT}>^2 \times (10^5)$ exp	$<M_{GT}>^2 \times (10^5)$ calc
2^+	0.539	6.5	139	139[b]
2^+	1.362	7.1	35	11
$(1,2)^+$	1.865	6.5	139	113
2?	2.099	6.9	55	⎫
2?	2.240	7.3	20	⎬ Broken pairs
(1,2)	2.660	5.9	∼500	⎭

[a] Using $\overline{(G_A/G_V)^2} = 1.40$; [b] Normalized to experiment. From IBM-2 calculations of Giannatiempo and Sona[4].

B. Even-even $(0^+) \stackrel{GT}{\to}$ Odd-odd (1^+).

As a second example, consider the Gamow-Teller decays of even-even nuclei. These decays provide information on the structure of 1^+ states in odd-odd nuclei. In order to illustrate some new features which may occur here, consider the case in which nuclei are described within the framework of IBM-4. The basic constituents of IBM-4 are four bosons[6] arranged into an isospin triplet and singlet

$$T=1, S=0: \begin{pmatrix} \pi \\ \delta \\ \nu \end{pmatrix} ; \qquad T=0, S=1: \quad (\theta) \quad . \tag{10}$$

The spectrum of states, in the U(5) limit of IBM-4, is shown in Fig. 2. In IBM-4, one expects superallowed GT transitions with log ft values of about 3 and $<M_{GT}>^2 \sim 1$. These transitions are collective transitions, in the sense that they correspond to transitions in which a π boson is changed into a θ boson and a θ boson into a ν boson (β^+ -decays). The transitions are $\nu \to \theta$

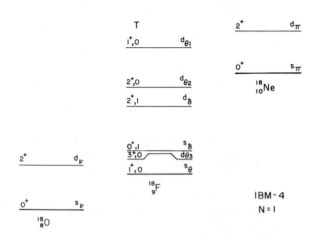

FIG. 2. Schematic representation of the spectrum of states in the U(5) limit of IBM-4. This spectrum comprises two even-even nuclei and the odd-odd nucleus inbetween.

and $\theta \to \pi$, in β^- -decay. The π, θ, ν bosons are correlated pairs of two protons, one proton - one neutron and two neutrons respectively. They are a mixture of two particle states, Eq.(2). The Gamow-Teller operator can be written in this case as

$$\hat{T}^{(GT)} = \frac{g_A}{\sqrt{4\pi}} (s_\pi^\dagger s_\theta + s_\theta^\dagger s_\pi) , \quad \beta^+ - transitions , \qquad (11)$$

and similar expressions for $\theta \to \nu$, etc. Superallowed transitions are observed in light nuclei, an example of which is shown in Fig.3. An important question is the extent to which such transitions occur in heavy nuclei. A scan of the available data shows that there are few cases in which the measured log ft is of order 3. An example is shown in Fig. 4. The particularly low value of log ft = 3.3 suggests that the answer to the question posed above is positive. Collective (pn) pairs represent new collective modes which are of considerable interest. They are a generalization of Wigner SU(4) symmetry to heavy nuclei. They are expected to be particularly low in energy in odd-odd N=Z nuclei. This suggests that β -decay in the region of ^{100}Sn be dominated by this new mode. Perspectives here for the interacting boson model are thus: (i) a search for superallowed decays in heavy nuclei, especially in regions with N=Z; (ii) an extension of IBM-2 to IBM-3 and IBM-4, and (iii) the calculation of β -decay strengths in nuclei far from stability.

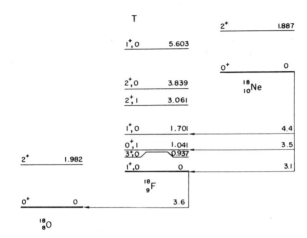

FIG. 3. Partial level scheme of mass A=18. The log ft values are shown next to the decay branch. Energies are in MeV.

The decays even-even to odd-odd are also of relevance to the study of double β-decay since the matrix elements of the double β-decay operator can be thought as the succession of two β-decays, even-even → odd-odd → even-even. The collective θ-mode may thus play an important role in these decays.

C. Even-odd $\stackrel{GT}{\rightarrow}$ Odd-even.

The study of Gamow-Teller decays of even-odd nuclei gives information on the structure of single-particle states in odd-even nuclei. This study has in part been done and some results will be presented at this conference[7]. The GT operator inducing these transitions has the form[5]

$$\hat{T}^{(GT)} = \sum_{jj'} \eta_{jj'} \left[P_\pi^{t(j)} \times \tilde{P}_\nu^{(j')} \right]^{(1)} . \qquad (12)$$

It is thus simpler than those of parts A and B. The study of β-decay in odd-even nuclei has implications also for other fields of physics, such as the study of the s- and r-processes in astrophysics and the detection of neutrinos based on neutrino absorption in $^{127}_{53}I_{74}$.

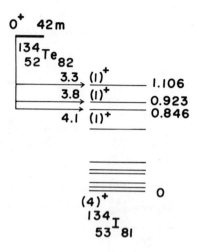

FIG. 4. β-decay of $^{134}_{52}Te_{82}$.

The preliminary calculations of β-decay in odd-even nuclei[5] indicate a quenching of the GT strength relative to that calculated in IBM-2, of the order 3.5. An understanding of this quenching is of crucial importance for the study of weak interactions in nuclei. Equally important is the question of whether the quenching occurs only in odd-even nuclei or also in even-even and odd-odd nuclei.

3 Conclusions

The brief remarks made in Parts A-B-C of the preceding section suggest that the study of weak interactions in heavy nuclei is one of the perspectives for the interacting boson model in the next few years. The way to proceed in this direction is: (i) map the weak operators into the boson-fermion space, (ii) analyze the existing data (some of these data are old and need more accurate measurement) to understand the extent to which the mapped operators describe β-decay into collective states, and the extent of quenching of the axial vector coupling constant in heavy nuclei; (iii) calculate β-decay strengths in nuclei far from stability, especially on the N=Z side, and search for superallowed decays.

In addition to weak interactions, other areas need further study. Among these other areas particularly important are: (i) the study of broken pair states in even-even and even-odd nuclei; this area is of current experimental interest especially in connection with the physics of states with large angular momentum; (ii) the study of odd-odd nuclei. The study of these areas has just begun and needs further work. Both are related to the understanding of residual nucleon-nucleon interactions in heavy nuclei, n-n and p-p in broken pair states and p-n in odd-odd nuclei.

In summary, in spite of the considerable progress which as been made in the interacting boson model in the last 20 years, still many open problems remain. The situation in this respect is similar to what has been in molecular and atomic physics for some time and to what has become in particle physics in recent years. We do have in place a framework within which nuclear structure and reactions can be studied, but we need to go beyond this framework. This type of physics, requires precision and depth to fully understand the observed phenomena and discover new ones.

Acknowledgement

This work was supported in part by Department of Energy Grant DE-FG02-91ER40608.

References

[1] F. Iachello, Proc. Int. Conf. on Nucl. Structure and Spectroscopy, H.P. Blok and A.E.L. Dieperink, eds., *Scholar's Press Amsterdam* (1974), p.163-181.

[2] For reviews see, F. Iachello and A. Arima, "The Interacting Boson Model" *Cambridge University Press*, Cambridge (1987); F. Iachello and P. van Isacker, "Interacting Boson-Fermion Model", *Cambridge University Press*, Cambridge (1991); D. Bonatsos, "Interacting Boson Model of Nuclear Structure", Clarendon, Oxford; (1988). I. Talmi, "Simple Models of Complex Nuclei: The Shell Model and Interacting Boson Model", *Harwood Press*, Chur (1993).

[3] D.H. Wilkinson, *Nucl. Phys.* **209**, 470 (1973).

[4] A. Giannatempo and P. Sona, private communication.

[5] F. Dellagiacoma and F. Iachello, *Phys. Lett.* **218B**, 399 (1989).

[6] J.P. Elliott and A.P. White, *Phys. Lett.* **97B**, 169 (1980); J.P. Elliott and J.A. Evans, *Phys. Lett.* **101B**, 216 (1981);

[7] G. Maino, these Proceedings.

FROM INDEPENDENT NUCLEONS TO INTERACTING BOSONS

IGAL TALMI
Department of Particle Physics, The Weizmann Institute of Science
Rehovot 76100, Israel

ABSTRACT

The shell model basis of the interacting boson model is briefly reviewed. Its history is sketched, its principles are described and some problems are discussed.

Twenty years ago, September 9-13, 1974, in Amsterdam, there was an international conference on Nuclear Structure and Spectroscopy. In that Conference, Franco Iachello gave a talk on the forerunner of IBA, the model with d-bosons only. This talk was the first exposure of the boson model to the community of nuclear physicists. I gave the summary talk of that conference and would like to quote what I said about the model[1].

"The model described by Iachello, on which he and Arima have been working is refreshingly different. Most important, it gives in certain cases excellent agreement with experiment. Ground state bands are described in terms of elementary excitations which are considered as J=2 Bosons. I agree with Soloviev that there are many difficulties with this model. It is not clear, for instance, why J=4 Bosons do not also play an important role. It is not clear how good is the Boson description of a system with a rather small number of Fermions. Nevertheless, it was nice to hear that Iachello is going to work on the model with enthusiasm and determination. If the agreement with experiment will be established even better, we may expect to eventually understand why the model works so well. We may then be able to understand the transition from spherical nuclei to highly deformed ones".

What impressed me about the model was not the use of seniority for d-bosons which Arima and Iachello have used. I was mostly impressed by the simplicity of the model, which was in refreshing contrast to other boson models, and with the good agreement with experiment. Sometimes it is difficult to find how many adjustable parameters are included implicitly in a given model. Here, however, everything was explicit and clear.

I was sufficiently intrigued by the model so I tried to formulate it in the space of nucleons. The d-bosons could have been replaced by irreducible (single nucleon) tensor operators with rank $L = 2$. As in the formalism of generalized seniority, they had to satisfy certain commutation relations. The difficulty was that a commutator of two such operators is another single nucleon operator. Hence, the single nucleon operators had to satisfy many conditions which were not operator conditions but held only if applied to the ground state. Much of the physics was hidden in the ground state and its properties determined to a large extent the nature of the spectrum.

Only in 1975 when Arima and Iachello introduced the s,d boson model[2] the relation between the boson model and the shell model became clearer. I worked on this approach with Franco and then we collaborated with Arima and Otsuka (spelled then Ohtsuka) who developed the same ideas. In our approach, the s-bosons were viewed as corresponding to nucleon pairs with $J = 0$ (S pairs) and d-bosons as corresponding to $J = 2$ (D pairs) of generalized seniority. Such a correspondence implied that there should be two kinds of bosons, proton s and d-bosons and neutron s and d-bosons leading to IBA-2[3].

It is perhaps worth while to restate the fact that IBA-1 *need not* have a shell model basis and may be considered as an approximation to the collective model. The latter has been formulated in the d-boson formalism and if U(6) symmetry is imposed on it (a la Jolos et.al[4]) eigenstates are linear combinations of a limited boson number $n \leq N$. The introduction of s-bosons simplifies the formalism and allows a better way to look at the various symmetries of the model. A clear demonstration of such a situation is offered by the s,p-boson description of molecular spectra introduced by Iachello[5] where N is conserved and yet there is no microscopic basis of that model. In their first publication, Arima and Iachello considered N as just characterizing the irreducible representation to which the states belong. In fact, in the 1976 Varenna Summer School[6] I showed a fit between experimental and calculated levels of ^{170}Er which Franco sent me (Fig.20 of ref.6). In that fit, $N = 5$ was adopted whereas the value based on the shell model interpretation is $N = 17$. In the other direction, several authors have shown how the collective model emerges as the limit of the boson model for large values of N^{7-9}. It could be expected that IBA-1 would have been embraced by the fathers of the collective model. Ironically, this did not happen and instead, connections with the shell model, where fermions are the building blocks, were studied in detail.

It is also perhaps worth while to restate the fact that IBA-1 *cannot* have a direct shell model basis. Some authors consider pairs of valence identical nucleons as such a basis. States of identical nucleons, however, have lowest possible space and spin symmetry characterized by maximum value of isospin. Collective low lying spectra occur only in nuclei with both valence protons and neutrons. Ground states of such nuclei have the lowest possible isospins due to the interaction of valence protons and neutrons. In such nuclei, states with maximum isospin of valence nucleons, allegedly described by IBA-1, lie several MeV above ground states (due to symmetry energy).

In the shell model basis of the boson model, states constructed from s- and d-bosons correspond to states constructed from S and D pairs of identical nucleons. If several nucleon configurations are mixed, the S pair creation operator is defined by[10]

$$S^+ = \sum_j \alpha_j S_j^+ = \sum_j \alpha_j \sum_m (-1)^{j-m} a^+_{jm} a^+_{j,-m} \quad (1)$$

In the quasi-spin scheme, with SU(2) symmetry, the $J = 0$ pair creation operator with all α_j coefficients equal. In this case there is no special $J = 2$ pair and all $J = 2$ states with $v = 2$ have constant spacings independent of number of nucleons, with the $J = 0$ ground state. In the general case, in actual nuclei, the coefficients α_j are not equal. Then there is a special pair creation operator with $J = 2$

$$D_M^+ = \sum_{j \leq j'} \beta_{jj'} \sum_{mm'} (jmj'm'|jj'J = 2, M) a^+_{jm} a^+_{j'm'} \quad (2)$$

which leads to $J = 2$ states with special properties.

If the operator S^+ in Eq.(1) satisfies the two conditions

$$HS^+|0> = [H, S^+]|0> = V_0 S|0> \quad [[H, S^+], S^+] = W(S^+)^2 \quad (3)$$

then a set of eigenstates of the shell model Hamiltonian H is given by

$$H(S^+)^N|0> = (NV_0 + WN(N-1)/2)(S^+)^N|0> \quad (4)$$

If the D_M^+ operator in Eq.(2) satisfies the two conditions

$$HD_\mu^+|0> = V_2 D_\mu^+|0> \quad [[H, S^+], D_\mu^+] = WS^+ D_\mu^+ \quad (5)$$

then a set of eigenstates with $J = 2$ of H is given by

$$H(S^+)^{N-1} D_\mu^+|0> = ((N-1)V_0 + V_2 + WN(N-1)/2)(S^+)^{N-1} D_\mu^+|0> \quad (6)$$

The height of the special states in Eq.(6) above the ground states in Eq.(4) is constant, independent of N, as observed experimentally in semi-magic nuclei.

All states of valence protons and neutrons may be constructed by coupling all states of the valence protons with all states of the valence neutrons. In most nuclei under consideration, valence protons and neutrons are in different major shells. In ground configurations of such nuclei all states have definite isospins, $T = (N - Z)/2$. If satisfactory mapping is achieved between states and operators of identical nucleons and boson states and operators, mapping of states of valence protons and neutrons is straightforward. The corresponding boson states are obtained by coupling proton boson states with neutron boson states. The shell model Hamiltonian of the valence nucleons may be written as

$$H = H_\pi + H_\nu + V_{\pi\nu} = H_\pi + H_\nu + \kappa(Q_\pi . Q_\nu) \tag{7}$$

The proton neutron interaction in Eq.(7) was approximated by a quadrupole term. We take the single nucleon operators Q in Eq.(7) to be equal to those which transform S pairs into D pairs[11],

$$D_\mu^+ = [Q_\mu, S^+] \tag{8}$$

If the commutator of Q and D^+ is a linear combination of S^+ and D^+ then the quadrupole interaction has non-vanishing matrix elements only within the space of states obtained by coupling S pairs and D pairs. The shell model Hamiltonian may then be diagonalized, even if approximately, in this limited S-D space.

Thus, if H_π and Q_π boson operators can be found with the same matrix elements as between corresponding proton states and H_ν and Q_ν boson operators can be found with the same matrix elements as between corresponding neutron states, the boson model can replace the shell model in the space considered. A very simple example of such a situation for identical nucleons is given by mapping the states in Eq.(4) and Eq.(6) onto boson states

$$(s^+)^N |0> \quad \text{and} \quad (s^+)^{N-1} d_\mu^+ |0> \tag{9}$$

The corresponding eigenvalues in Eq.(4) and Eq.(6) are obtained by applying to the states in Eq.(9) the boson Hamiltonian[12]

$$V_0 s^+ s + V_2 \sum_\mu d_\mu^+ d_\mu + (W/2)(s^+)^2 s^2 + W s^+ s \sum_\mu d_\mu^+ d_\mu \tag{10}$$

If more than one D pair is used in the construction of nucleon states, the mapping becomes more complicated.

It should be made clear that the states in Eq.(4) and Eq.(6) are fermion states in which the Pauli principle is strictly obeyed. Still, they correspond to the simple boson states in Eq.(9), where creation operators and their hermitean conjugates obey the standard Bose commutation relations. The eigenvalues of the latter states for the boson Hamiltonian in Eq.(10) are equal to the eigenvalues of the fermion states for shell model Hamiltonians with generalized seniority eigenstates.

There are shell model spaces in which the mapping of states and operators is simple and exact. In the Ginocchio models[13] there is O(6) or SU(3) symmetry. Hence, there is exact correspondence between generators and states of irreducible representations of these groups in the spaces of nucleons and s,d bosons. In the Ginocchio models it is possible to replace exactly the diagonalization of a submatrix of a certain shell model Hamiltonian by diagonalization of a simple boson Hamiltonian. In the case of actual nuclei, states of valence identical nucleons do not possess O(6) nor SU(3) symmetry and the mapping is more complicated.

The natural extension of the states in Eq.(9) are states constructed with several d-boson operators

$$(s^+)^{N-n_d}(d^+)^{n_d}_{\gamma JM}|0> \qquad (11)$$

In Eq.(11) the n_d boson creation operators are coupled to a state with definite values of J and M and γ is an additional quantum number which may be necessary to distinguish between several such states. Boson states in Eq.(11) with different values of n_d are orthogonal. States of nucleons constructed from different numbers of D pairs are not necessarily orthogonal. In order to establish the correspondence with boson states in Eq.(11) they have to be diagonalized. This may be carried out by a simple projection in the seniority scheme or in the quasi-spin scheme. In these cases the boson states in Eq.(11) correspond to nucleon states with seniority quantum numbers $v = 2n_d$. These states of (identical) fermions are exact eigenstates of the pairing interaction. In the general case the situation is more complicated and no simple prescription is available for the process of orthogonalization.

The nucleon states in Eq.(4) vanish for $N > \Omega = \Sigma(2j+1)/2$ and the states in Eq.(6) for $N > \Omega - 1$. On the other hand, the boson states in Eq.(11) exist for any value of N. Such boson states which do not correspond to nucleon states may be called spurious states. States of more d-bosons may become spurious for smaller values of N. The seniority quantum number v of nucleon states cannot exceed Ω. Also in the general case, all nucleon states with $N > \Omega$ contain S-pairs and cannot survive the projection in the orthogonalization procedure[14]. Therefore, beyond the middle of the major shell ($N > \Omega$) nucleon hole states are mapped onto boson states. Even with this restriction of N there may be spurious boson states. It is expected that they lie higher in energy and it is hoped that they do not affect much the low lying levels. In the case of the O(6) Ginnochio model there are no spurious boson states for $N \leq \Omega$ whereas in the SU(3) model, for $N > \Omega/3$ there are spurious states which span bases of certain irreducible representations of SU(3).

The shell model basis of the boson model leads to IBA-2, a model with proton s- and d-bosons and neutron s- and d-bosons. The question arises how does IBA-1 with its beautiful symmetries emerge from IBA-2. The simplest way to achieve this goal is to consider a boson Hamiltonian which is fully symmetric in proton and neutron bosons. Such Hamiltonian should have equal single proton boson and single neutron boson energies and equal interactions for proton-proton, neutron-neutron and proton-neutron bosons. The Hamiltonian should commute with operators that change a proton boson into a neutron boson and vice versa. Such operators are components of the F-spin vector[3]

$$F^+ = s_\pi^+ s_\nu + \sum_\kappa d_{\pi\kappa}^+ d_{\nu\kappa}^+ \qquad F^- = s_\nu^+ s_\pi + \sum_\kappa d_{\nu\kappa}^+ d_{\pi\kappa}$$

$$F^0 = [F^+, F^-]/2 = (s_\pi^+ s_\pi - s_\nu^+ s_\nu + \sum_\kappa d_{\pi\kappa}^+ d_{\pi\kappa} - \sum_\kappa d_{\nu\kappa}^+ d_{\nu\kappa})/2 = (N_\pi - N_\nu)/2$$
(12)

If H commutes with the operators in Eq.(12) then its eigenstates can be characterized by the eigenvalues of $\mathbf{F}^2$ given by $F(F+1)$ and $F^0 = (N_\pi - N_\nu)/2$. States with $F = N/2 = (N_\pi + N_\nu)/2$ are fully symmetric in proton and neutron bosons. They are completely equivalent to IBA-1 states and within the space of these states the IBA-2 Hamiltonian is fully equivalent to an IBA-1 Hamiltonian. Any IBA-2 Hamiltonian has other eigenstates which do not have $F = N/2$ and thus are absent from IBA-1 spectra. An interesting one of these is the 1^+ collective state which has been observed in many deformed nuclei.

If we take the coefficient of the single boson term $\epsilon = \epsilon_d - \epsilon_s$ equal to $V_2 - V_0$ of the shell model, it is a good approximation to put $\epsilon_\pi = \epsilon_\nu$. There are, however, big differences between the shell model effective interaction between identical nucleons and the proton-neutron interaction. The $T = 1$ effective interaction leads to states with seniority or generalized seniority as good quantum numbers. On the other hand, the proton-neutron interaction, including $T = 0$ effective interactions, breaks seniority in a major way[12]. This feature of the latter is characteristic of a quadrupole-quadrupole interaction as in Eq.(7). In some applications the interaction between identical nucleons is approximated by a single nucleon term. The boson Hamiltonian in Eq.(7), apart from terms which contribute only to binding energies, may then be written as

$$H = \epsilon(n_{d\pi} + n_{d\nu}) + \kappa(Q_\pi \cdot Q_\nu) \qquad (13)$$

The expression in Eq.(13) does not appear as having eigenstates with good F-spins. In various applications a Majorana term has been added to H in Eq.(13). Such a term has eigenvalues proportional to $(N/2)(N/2+1) - F(F+1)$. If its coefficient is positive, it pushes states with $F < N/2$ higher and low lying states tend to become states with $F = N/2$.

The origin of the Majorana term, if indeed necessary, is unclear. More precisely, its part which affects antisymmetric $J = 2$ states of two bosons has no direct origin in the shell model. It could emerge from the renormalization of the effective interaction due to the very drastic reduction of shell model states mapped onto boson states. Instead of 10^{14} states with $J = 0, 2, 4, \ldots$ due to couplings of valence protons and neutrons in some rare earth nuclei, there are only several thousands of states mapped onto IBA-2 states. Surely, neglecting all other states, in particular states constructed of pairs with $J = 4, 6, \ldots$, could lead to drastic renormalization of the effective interactions. The difficulty is to find a reliable method to calculate these renormalization effects. We should recall the situation of effective interaction in the shell model. In spite of intensive efforts during the last forty years no reliable way is available to calculate them from the interaction between free protons and neutrons. So far, they have been determined from energies of complex nuclei.

There is another term in the boson Hamiltonian which may exhibit effects of renormalization. The single d-boson energy ϵ should be equal, according to the shell model interpretation, to $V_2 - V_0$. In most IBA-1 analyses of collective spectra much smaller values have been used. As a result, the calculated levels are rather close to those obtained in the cases of dynamical symmetry. Also in some IBA-2 fits to such spectra rather low values of have been adopted. In an attempt to see whether this is really necessary, IBA-2 calculations have been carried out where ϵ was taken to be equal to the experimental values of $V_2 - V_0$. Such a prescription gave good results for nuclei which in IBA-1 have O(6)-like spectra. Both energy levels and E2 transitions were very well reproduced[15]. In nuclei with SU(3)-like spectra it turned out that to obtain good agreement with electromagnetic transitions, lower values of ϵ had to be used[16]. Perhaps this is due to the fact that in strongly deformed nuclei the pairs with $J = 4, J = 6$ etc. play a more important role than in O(6)-like nuclei.

Calculations of energy levels and electromagnetic transition probabilities using IBA-1 give very good agreement with experiment[17]. It is possible to ask whether this fact constitutes a proof that IBA-2 wave functions of these states have definite values of F-spin, $F = N/2$. We can look at levels of ^{178}Hf and their wave functions as calculated in IBA-2[18]. In the $J = 0$ ground state the weight of states with maximum F-spin, $F = 15/2$, is only 82.1% ($F = 13/2$ states amount to 16.4%). In the second $J = 0$ state (of the β-band) the corresponding weights are 63.6% and 29.6% respectively. Thus, F-spin seems to be badly broken and the success of IBA-1 in the case of ^{178}Hf cannot be attributed to F-spin being a good quantum number of IBA-2 states. If we consider only states with maximum F-spin, we limit the IBA-2 Hamiltonian to its submatrix defined by $F = 15/2$. This submatrix is an IBA-1 Hamiltonian[18,19] which can be directly diagonalized. The ground state energy obtained this way is, as expected, higher by .3 MeV than the lowest exact IBA-2 eigenvalue. This shift in energy is rather large but it turned out to be almost the same for all low lying levels. Spacings between levels calculated from the IBA-1 Hamiltonian are rather equal to those obtained in IBA-2. It seems that for these spacings IBA-1 with its symmetries can serve as a good model for the more realistic calculation with IBA-2.

Attempts have been made to determine F-spin admixtures by using measured rates of M1 transitions. Matrix elements of the M1 operator and hence, rates of corresponding transitions, are very sensitive to small admixtures to the wave functions. This holds in particular at the jj-coupling limit. It is therefore difficult to distinguish F-spin admixtures from other ones. In IBA, as in the collective model, the quadrupole degree of freedom plays the important role. Hence, rates of E2 transitions offer the real test of wave functions obtained in the boson model.

The interacting boson model is rather simple and yet it is very versatile. It can successfully describe various kinds of collective motion in nuclei[20]. A most attractive way to derive the collective model is to start with nucleon pairs, proceed to IBA-2 and then to IBA-1 which naturally leads to the collective model. At the same time, the s- and d-bosons are defined in a frame of reference fixed in space. This makes possible the connection with the (spherical) shell model. States of identical nucleons, well described by generalized seniority, correspond to states of s- and d-bosons. States of proton and neutron bosons are strongly coupled by the quadrupole interaction which corresponds to the effective interaction between valence protons and neutrons in actual nuclei.

The success of the interacting boson model can be attributed to its ability to incorporate the bosons equivalents of the two basic ingredients of the effective interaction between nucleons in the shell model. The first is the $T = 1$ effective interaction between identical nucleons which has eigenstates with good seniority or generalized seniority. The second one is the strong and attractive $T = 0$ effective interaction which breaks seniority in a major way like the quadrupole interaction between protons and neutrons. The former tends to keep the spherical shape of semi-magic nuclei whereas the latter leads to lowering of 0-2 spacings, collective spectra and nuclear deformation. The variety of nuclear spectra arises from the competition between these interactions and this is very clearly described by the interacting boson model.

References
1. Proc. Int. Conf. *Nucl. Structure and Spectroscopy*, H.P. Blok and A.E.L. Dieperink Eds., (Scholars Press, Amsterdam, 1974).
2. A. Arima and F. Iachello, *Phys. Rev. Lett.* **35** (1975) 1069.
3. A. Arima, T. Otsuka, F. Iachello and I. Talmi, *Phys. Lett.* **66B** (1977) 205.
   T. Otsuka, A. Arima, F. Iachello and I. Talmi, *Phys. Lett.* **76B** (1978) 141.
4. D. Janssen, R.V. Jolos and F. Dönau, *Nucl. Phys.* **A224** (1974) 93.
5. F. Iachello, *Chem. Phys. Lett.* **78** (1981) 581.
6. I. Talmi in *Elementary Modes of Excitation in Nuclei*, A. Bohr and R.A. Broglia Eds. (North-Holland, Amsterdam, 1977) p. 352.
7. A.E.L. Dieperink, O. Scholten and F. Iachello, *Phys. Rev. Lett.* **44** (1980) 1747.
8. J.N. Ginocchio and M.W. Kirson, *Nucl. Phys.* **A350**, (1980) 31.
9. R. Hatch and S. Levit, *Phys. Rev.* **C25** (1982) 614.
10. I. Talmi, *Nucl. Phys.* **A172** (1971) 1.
11. S. Shlomo and I. Talmi, *Nucl. Phys* **A198** (1972) 81.
12. I. Talmi in *From Nuclei to Particles* (Proc. 1980 Varenna Summer School), A.Molinari Ed. (North-Holland, Amsterdam, 1981) p.172.
13. J.N. Ginocchio, *Ann. Phys. (NY)* **126** (1980) 234 where earlier references are given.
14. I. Talmi, *Phys. Rev.* **C25** (1982) 3189.
15. A. Novoselsky and I. Talmi, *Phys. Lett.* **172B** (1986) 139.
16. A. Novoselsky, *Nucl. Phys.* **A483** (1988) 282.
17. R.F. Casten, *Nuclear Structure from a Simple Perspective* (Oxford Univer-

sity Press, 1990).
18. A. Novoselsky and I. Talmi, *Phys. Lett.* **160B** (1980) 13.
19. O. Scholten, Ph.D. Thesis (Groningen, 1980) unpublished.
    H. Harter, A. Gelberg, and P. von Brentano, *Phys. Lett.* **157 B** (1985) 1.
20. *Algebraic Approaches to Nuclear Structure, Interacting Boson and Fermion Models*, R.F. Casten Ed., (Harwood Academic Publishers, 1993).

ROBUST PREDICTIONS OF THE INTERACTING BOSON MODEL

R. F. Casten
Brookhaven National Laboratory, Upton, NY, 11973, USA
Institut für Kernphysik, University of Köln, Köln, Germany

ABSTRACT

While most recognized for its symmetries and algebraic structure, the IBA model has other less-well-known but equally intrinsic properties which give *unavoidable,* parameter-free predictions. These predictions concern central aspects of low-energy nuclear collective structure. This paper outlines these "robust" predictions and compares them with the data.

1. Introduction

The Interacting Boson Approximation Model (IBA) [1,2] has become one of the standard approaches to nuclear structure. While the dynamical symmetries and algebraic structure of the IBA are its best known characteristics, there are other aspects that are equally intrinsic to the model. Many of these are inherent, automatic, or ineluctable predictions of the IBA in the sense that they are parameter free (or nearly so with any reasonable set of parameters): *were the data different it would be virtually impossible to alter the model to accommodate them.* They reflect the intrinsic structure of the model and relate to very basic aspects of collectivity in nuclei. We call these predictions "robust". They comprise a set of tests of the basic relevance of the model to low-energy nuclear structure. This review will discuss these robust properties of the IBA and their comparison with the data. Our discussion deals with the IBA-1 and much of it is based on ref. 3.

2. Robust Predictions

2.1. $B(E2:0_1^+ \rightarrow 2_1^+)$ Values-Mean Field Collectivity

An excellent measure of the mean field structure of nuclei is the B(E2) value to the first 2^+ state. The IBA makes a specific prediction here of a strong increase in $B(E2:0_1^+ \rightarrow 2_1^+)$ values in deformed nuclei toward mid-shell, arising from the explicit inclusion of valence nucleon number, virtually independent of any parameter variations. Two aspects of finite N_B are involved, namely the finite value itself and its partition into n_s s-bosons and n_d d-bosons. For deformed nuclei both $<n_s>$ and $<n_d>$ increase with N_B for low lying levels and each factor in the (dominant) $(s^\dagger d + d^\dagger s)$ term in the E2 operator gives a power of N_B, leading to an approximate dependence on N_B^2. In particular, in the SU(3) limit, $B(E2:0_1^+ \rightarrow 2_1^+) \propto (2N_B+3)N_B$. This prediction is shown on the right of fig. 1, in agreement with the data shown on the left.

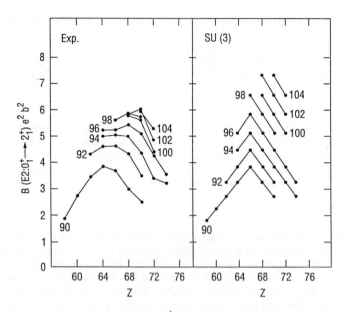

Fig. 1. Empirical and calculated $B(E2:0_1^+ \to 2_1^+)$ values. The calculations are for the SU(3) limit with constant effective charge.

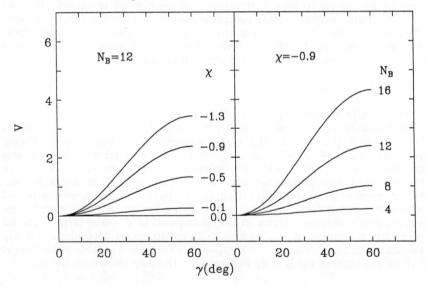

Fig. 2. The IBA potential $V(\gamma)$ vs. γ. Based on ref. 6. Left side, several values of χ with fixed N_B. Right side, several N_B values with constant χ.

2.2. Axial Asymmetry
a) γ-Softness vs. Rigidity at Low Spin

Consider an IBA-1 Hamiltonian which is satisfactory for nearly all calculations, namely

$$H = \varepsilon n_d + \kappa Q \cdot Q \tag{1}$$

with $\quad Q = (s^\dagger d + d^\dagger s) + \chi(d^\dagger d)^2 \tag{2}$

We note that $T(E2)=e_B Q$, where e_B is a boson effective charge. The same χ is used in H and T(E2) which defines the Consistent Q Formalism or CQF [ref. 5]. The three dynamical symmetries of the IBA-1 are reproduced in terms of eqs. 1,2, with $\kappa=0$, $\chi=0$ for U(5), $\varepsilon=0$, $\chi=0$ for O(6), and $\varepsilon=0$, $\chi=-\sqrt{7}/2=-1.32$ for SU(3). For simplicity and ease of understanding, we phrase much of the ensuing discussion in terms of eqs. 1,2 although most if not all the results apply to more general IBA-1 Hamiltonians.

It is possible to associate specific potentials with different IBA Hamiltonians through the intrinsic state formalism [6]. The γ-dependence of the potential enters through the value of χ. Several cases are illustrated in fig. 2. The IBA potential either has no minimum [the O(6) extreme of a γ-flat potential ($\chi=0$) with $\gamma_{ave}\sim 30°$] or a minimum at $0°$. There are no minima for $\gamma \neq 0°$. For IBA potentials with a minimum at $\gamma=0°$, the steepness increases with N_B. Hence, strict axial symmetry can only occur in the infinite N_B limit, and finite γ values arise only from zero point motion in a γ-soft potential. Thus, the IBA-1 corresponds to the opposite picture of the origin of axial asymmetry as the rigid triaxial Davydov model.

Axial asymmetry plays several related roles in the IBA. It is possible to test each. Perhaps the most basic is the nature of empirical axial asymmetry itself: soft or rigid? The best empirical signature for this is, fortunately, quite easy to measure: it relies on energy staggering of γ-band levels. In a γ-soft potential, the quasi γ-band levels cluster in energy couplets as 2^+, $(3^+,4^+)$, $(5^+,6^+)$, ... whereas in a γ-rigid potential the clustering goes as $(2^+,3^+)$, $(4^+,5^+)$, ... A sensitive signature is therefore the "staggering index"

$$S(\gamma) = \frac{[E(4^+_\gamma)-E(3^+_\gamma)] - [E(3^+_\gamma)-E(2^+_\gamma)]}{E(2^+_1)} - 0.33 \tag{3}$$

$S(\gamma)$ is defined so that the symmetric rotor limit corresponds to $S(\gamma)=0$. *Negative* $S(\gamma)$ values correspond to axial asymmetry which is *soft* in origin, and *positive* values to *γ-rigid* motion: the magnitude of $S(\gamma)$, of course, depends on the value of γ at the potential minimum for γ-rigid shapes and on the average value of γ and the steepness of the potential for the γ-soft case. Empirical $S(\gamma)$ values for all even-even nuclei from Z=30 to the actinides are shown in fig. 3 for comparison with both the IBA and Davydov models [8]. Small negative $S(\gamma)$ values correspond to nearly symmetric nuclei with steep

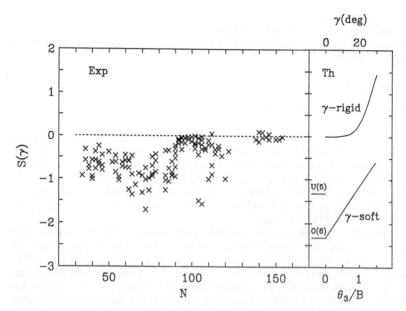

Fig. 3. Left: Empirical values of the staggering index S(γ), defined in eq. 3. Data from ref. 7. Right: Predictions of S(γ) for the γ-rigid Davydov model [8] as a function of γ (top scale) and for the IBA in the U(5) and O(6) limits (short horizontal lines) and for O(6) as a function of θ$_3$/B. This parameter specifies the contribution of a cubic term that adds a weak component with a minimum at γ=30° to the γ-flat O(6) potential. Based on ref. 9.

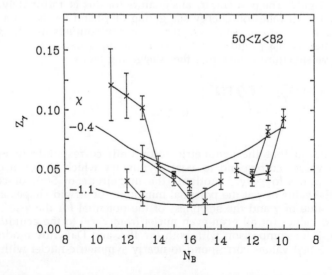

Fig. 4. Empirical Z$_\gamma$ values for rare-earth nuclei. From ref. 10. The IBA curves are for two χ values that bound those used in typical calculations.

potentials and minima at $\gamma=0°$ and, hence, rather small γ_{ave} values. More negative $S(\gamma)$ values reflect potentials that are softer in γ: U(5) (the vibrator) gives $S(\gamma)=-1.33$ (a spherical vibrator potential is γ-flat) and O(6) has $S(\gamma)=-2.33$. Unless other mechanisms not included in the figure can lead to negative $S(\gamma)$ values, it seems that, at least at low spin, nuclear axial asymmetry in even-even nuclides arises from γ-softness, in accord with the IBA view.

b) Rotation-Vibration Interaction

It is a familiar notion that the rotation-vibration interaction, or, equivalently, K band mixing, in deformed nuclei minimizes near mid-shell. This is qualitatively attributed to a distinction between rotational and vibrational energy scales near mid-shell. However, these scales barely change across well deformed regions. Even if the behavior of the rotation-vibration interaction is related to energy scales, this interpretation is hardly quantitative. It is interesting therefore that the IBA automatically embodies the observed phenomenology as an inherent, robust, prediction and that it arises essentially as an effect of finite boson number and of specific properties of the dynamical symmetries in the IBA.

For the γ and ground bands, the K-band mixing amplitude is proportional to a band mixing parameter Z_γ. The empirical values of Z_γ for rare-earth nuclei extracted from the slopes in Mikhailov plots are shown in fig. 4. Using the CQF, calculated values of Z_γ can also be obtained from predicted $B(E2:J_\gamma \rightarrow J'_g)$ values. The results are shown in fig. 4 for two *fixed* values of χ that encompass the range encountered in realistic calculations. Though the Hamiltonian parameters remain constant, Z_γ varies parabolically as a result of the interplay of symmetry structure and N_B effects. γ-g band mixing arises largely from γ-β mixing, which occurs even in the SU(3) limit. γ-β mixing decreases rapidly with increasing N_B, and is transmitted to γ-g band mixing via $\Delta K=0$ β-g mixing which is the dominant mixing when SU(3) is broken [11]. In realistic calculations χ (which is usually fixed by fitting a $\gamma \rightarrow g$ B(E2) value) will vary somewhat for different nuclei and the steepness of the parabolas will increase, giving improved agreement with the data.

c) γ values for deformed nuclei

We have discussed the origin of axial asymmetry and its systematics and found that the IBA automatically mirrors the data. A third key aspect of axial asymmetry is its actual magnitude. As suggested on the left in fig. 2, the effective value of γ (arising from zero point motion) in the IBA is directly correlated with the parameter χ. Indeed, a γ-χ correlation can be obtained by comparing observables in the IBA and in the Davydov model. This is done in fig. 5 using predicted values of $E(2^+_\gamma)/E(2^+_g)$ for several N_B values. Similar γ-χ correlations result if other observables are used instead [12]. The figure shows a result which is perhaps surprising, namely that γ_{eff} does *not* equal 0°, *even* in SU(3). Only as $N_B \rightarrow \infty$ does $\gamma_{eff} \rightarrow 0°$. The reason is simply (see fig. 2) that the potential has finite steepness for finite N_B. Since actual N_B values are ~14-20, the IBA inherently predicts an actual numerical value for the minimum γ_{eff} values near mid-shell, namely ~8°. [Note that, by using $E(2^+_\gamma)$

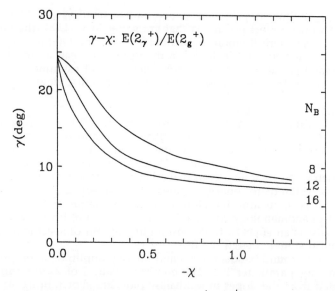

Fig. 5. γ–χ relationship obtained by calculating $E(2_\gamma^+)/E(2_g^+)$ in the Davydov and IBA models (for several N_B values) and equating the values of γ in the former to those of χ in the latter that give the same values of this energy ratio. Note that γ_{ave} does *not* go to zero even in the SU(3) limit but remains finite at ~7-10° in accord with the data (see fig. 6). From ref. 3.

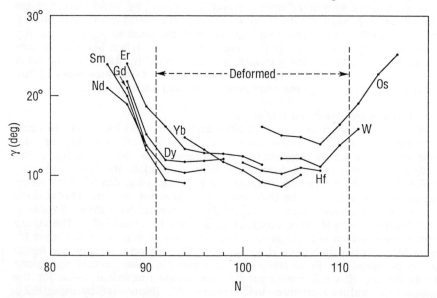

Fig. 6. Empirical values of γ_{eff} for some rare-earth nuclei obtained from $E(2_\gamma^+)/E(2_g^+)$ values. Data from ref. 7. Compare fig. 5.

values to extract γ, the γ values in fig. 5 are inherently connected to the γ band. In ref. 13 it was shown that $γ_g$ does indeed tend to zero even for finite N_B in the IBA: this is not inconsistent with the present result.]

Empirical γ values have traditionally been obtained from the same energy ratio used in fig. 5. These empirical values are shown in fig. 6 for the rare earth region where it is indeed seen that most well-deformed nuclei have γ values remarkably close to the IBA prediction of ~8-10°.

2.3 Nature of the Lowest K = 0 Intrinsic Excitation: The "β" Band

For decades the lowest K=0 excitation in deformed nuclei has been considered a β-vibration (an axially symmetric quadrupole oscillation in the β, or deformation, degree of freedom). Theoretical efforts to understand this mode have been fraught with difficulty; empirically, it behaves erratically.

Interestingly, the IBA makes a number of robust predictions here as well (fig. 7). Moreover, when compared to the data, they suggest a rather different interpretation of the "β" vibration than heretofore accepted. In the CQF deformed nuclei can be characterized by a single parameter χ. The $ε_d$ term in eq. 1 is either absent or N_B is so large that the Q•Q term dominates. Thus, IBA predictions (aside from an overall energy scale) can be expressed in terms of χ and N_B and shown in contour plots, as in fig. 7. The results can be schematically summarized very succinctly (for convenience, we continue to use the label "β" for the lowest K=0 mode):

a) $E_{"β"}$ ~ (1.2 ↔ 1.8) $E_γ$.
b) B(E2:"β"→g) << B(E2:γ→g): "β"→g transitions are *not* collective.
c) B(E2:"β"→γ) ~ B(E2:γ→g): "β"→γ transitions, which are forbidden in the geometrical picture of independent K=0 and 2 vibrations, are predicted to be nearly as collective as γ→g transitions.

Let us inspect the data for each of these. Figure 8 shows the empirical ratios of "β" and γ-band energies. Remarkably, for deformed nuclei (solid dots), the ratio is nearly always between 0.8 and 1.8. Since this ratio can take on a large range of values--and does for non-deformed nuclei (as seen in fig. 8)--the agreement of the IBA predictions with the data is excellent.

Figure 9 shows the data for the ("β"→g)/(γ→g) B(E2) ratio shown in fig. 7. The experimental values are nearly all <0.2, and most are <0.1. Though this exceeds the predictions of fig. 7, the qualitative point that "β"→g E2 matrix elements are much less than γ→g values is indeed observed.

The third robust IBA prediction (fig. 7) concerning the lowest K=0 excitation is that there should be collective E2 matrix elements *from* the "β" band to the γ band comparable in magnitude to γ→g matrix elements. The data were long thought to disagree with this unconventional description, but that turned out to be largely an artifact of the rather low transition energies and the $E_γ^5$ dependence of E2 transition intensities. However, when sufficiently sensitive experiments were carried out, using the unique crystal spectrometers of the ILL, with their very-high energy resolution and dynamic range of observable intensities, collective β↔γ transitions were indeed found.

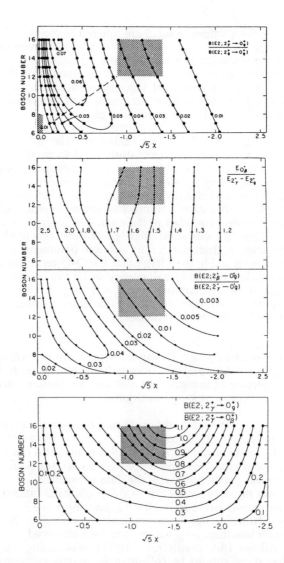

Fig. 7. Contour plots of various observables against χ and N_B calculated with the Hamiltonian $H = -\kappa Q \cdot Q$ in the CQF. From ref. 5. Note the extra factor of $\sqrt{5}$ on the abscissa (retained here for historical reasons): thus SU(3) corresponds to $\sqrt{5}\,\chi = -2.958$ and typical deformed nuclei have $\sqrt{5}\,\chi \sim -1.1$.

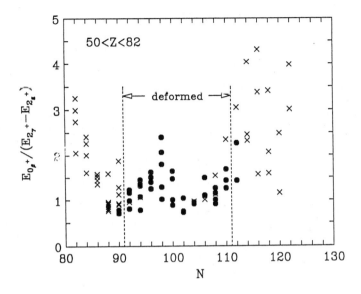

Fig. 8. Empirical energy ratios of β to γ-band energies. Data from ref. 7. Deformed nuclei $[E(4_1^+)/E(2_1^+) > 3.0]$ are indicated by solid dots.

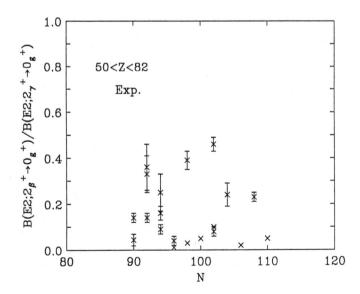

Fig. 9. Empirical energy ratios of the indicated B(E2) ratio. The IBA (see fig. 7) predicts that this ratio is <<1, in agreement with the data.

The best known example [14] is ^{168}Er but similar results characterize [15] the nuclei ^{152}Sm, ^{154}Gd, ^{158}Gd, and ^{166}Er. In ^{168}Er, the ratio of reduced intrinsic matrix elements $<2_\gamma^+||E2||0_{\beta''}^+>/<2_\gamma^+||E2||0_g^+> = 0.37$ which is somewhat smaller than the IBA prediction of ~ unity but which is an order-of-magnitude or more larger than commonly expected, is of collective magnitude, and is at variance with traditional concepts of β band structure.

These results suggest a different view of the "β" vibration (see fig. 10). We are faced here with a collective mode whose energy is related to, and slightly higher than, the γ-band energy, which has collective E2 transitions to the γ band, and which does *not* decay collectively to the ground band. Its ground band B(E2) values are also highly erratic. These results hardly suggest a collective quadrupole excitation of the ground state. Rather, they point to a phonon excitation *built on* the γ band. Whether this can be thought of as a K=0 double γ vibration or not is difficult to decide. What does seem clear is that the "β" vibration is largely a kind of 2-phonon excitation involving a quadrupole phonon excitation based on the γ vibration.

2.4 Universal Anharmonic Vibrator (AHV) Behavior

Recently a remarkable universal empirical feature of the evolution of nuclear structure was disclosed [16] by studying the correlation between two of the simplest observables in even-even nuclei, $E(4_1^+)$ and $E(2_1^+)$. Specifically, they exhibit a universal linear relation for all collective, non-rotational nuclei from Z=38 to Z=82. The data (fig. 11a) lie along a compact straight line

$$E(4_1^+) = 2.0\, E(2_1^+) + 156(10) \text{ keV} \tag{4}$$

That is, despite the widely varying internal structure of these nuclei, as evidenced by the variations in the approximate $R_{4/2} \equiv E(4_1^+)/E(2_1^+)$ scale along the top, the data satisfy the equation (eq. 4) for an AHV with *constant anharmonicity*. How this can be is a major challenge to microscopic theory. Pending that, it is interesting to see what the IBA predicts and if it can provide clues to the origin of this behavior. We have carried out an extensive set of IBA calculations spanning the entire symmetry triangle of the IBA, with the Hamiltonian of eq. 1 with constant κ. The results shown in fig. 11b are as surprising as the data. Regardless of the IBA parameters, that is, regardless of the choice of the parameters ε or χ or of the boson number, the predictions fall along a compact straight line with slope of 2.0 in complete agreement with the data. This straight line results specifically from the constancy of κ, that is, of the strength of the bosonic quadrupole interaction in the IBA. This suggests that the empirical phenomenology may reflect particular constraints on the strength of the fermionic quadrupole interaction that are compatible with constant κ in the boson Hamiltonian. This conclusion may provide clues to a microscopic understanding of the startling empirical result and it may provide interesting constraints on fermion-boson mappings.

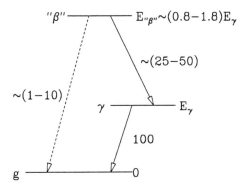

Fig. 10. Highly schematic summary of energies and transitions pertaining to the β and γ bands of deformed nuclei. See text and figs. 7, 8 and 9.

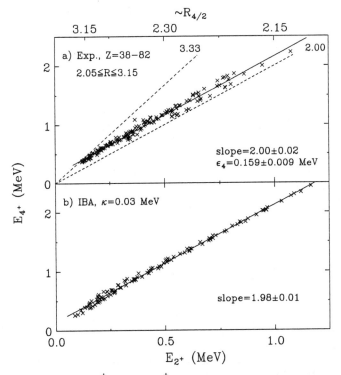

Fig. 11. Correlation of $E(4_1^+)$ against $E(2_1^+)$. a) Empirical values for all even-even nuclei from Z=38-82 with $R_{4/2} \equiv E(4_1^+)/E(2_1^+)$ values between 2.05 and 3.15. An approximate $R_{4/2}$ scale is given at the top. b) IBA calculations spanning the entire IBA symmetry triangle. The straight lines are least squares fit to the data and calculations. The fitted slopes are also given. From refs. 16.

3. Summary

We have shown that the IBA contains a large number of inherent, automatic, intrinsic, essentially parameter-free predictions, some of which were unexpected when first proposed, some of which are at variance with other models, and some of which (e.g., the β, γ relationship) are beyond the scope of a number of other models, but all of which are in agreement with the data. Some of these predictions have their origins in the character of the IBA symmetries, others reflect the effects of, and variations in, finite boson number across series of nuclei, some are reflections of the nature of symmetry-breaking mechanisms in the IBA, and some show the interplay of one or more of these features. All are inherent and robust, none can be reversed by parameter changes, and all reflect real nuclei.

4. Acknowledgements

I am grateful to S. Pittel, F. Iachello, A. Arima, W. Nazarewicz, B.R. Mottelson, and I. Hamamoto for very useful discussions which motivated this work, as well as to my collaborators in the studies comprising this overview, especially D.D. Warner, P. von Brentano and N.V. Zamfir. Research supported in part by Cont. No. DE-AC02-76CH00016 with the USDOE and by the BMFT under Cont. No. 06-OK-602-I.

5. References

1. A. Arima and F. Iachello, Phys. Rev. Lett. **35**, 1069 (1975).
2. F. Iachello, A. Arima, Phys. Rev. Lett. **40**, 385 (1978); F. Iachello, A. Arima, The Interacting Boson Model (Cambridge Univ. Press, Cambridge, 1987).
3. R.F. Casten, P. von Brentano, N.V. Zamfir, Phys. Rev. **C49**, 1940 (1994).
4. S. Raman et al., At. Data and Nucl. Data Tables **36**, 1 (1987).
5. D.D. Warner and R.F. Casten, Phys. Rev. Lett. **48**, 1385 (1982).
6. J.N. Ginocchio and M.W. Kirson, Nucl. Phys. **A350**, 31 (1980).
7. M. Sakai, At. Data and Nucl. Data Tables **31**, 399 (1984).
8. A.S. Davydov and G.F. Filippov, Nucl. Phys. **8**, 237 (1958).
9. N.V. Zamfir and R.F. Casten, Phys. Lett. **260B**, 265 (1991).
10. R.F. Casten, D.D. Warner, A. Aprahamian, Phys. Rev. **C28**, 894 (1983).
11. R.F. Casten and D.D. Warner, Phys. Rev. Lett. **48**, 666 (1982).
12. R.F. Casten, A. Aprahamian, D.D. Warner, Phys. Rev. **C29**, 356 (1984).
13. O. Castanos, A. Frank, P. Van Isacker, Phys. Rev. Lett. **52**, 263 (1984).
14. D.D. Warner, R.F. Casten, W.F. Davidson, Phys. Rev. Lett. **45**, 1761 (1980).
15. See R.F. Casten, D.D. Warner, Rev. Mod. Phys. **60**, 389 (1988); R.F. Casten, et al., Algebraic Approaches to Nuclear Structure, ed. R.F. Casten (Harwood Academic Publications, Langhorne, PA, 1993), Chap. 3.
16. R.F. Casten, N.V. Zamfir and D.S. Brenner, Phys. Rev. Lett. **71**, 227 (1993), and to be published.

INTERACTING BOSON MODEL FOR O(6) NUCLEI

TAKAHARU OTSUKA, TAKAHIRO MIZUSAKI and KA-HAE KIM
Department of Physics, University of Tokyo,
Hongo, Bunkyo-ku, Tokyo 113, Japan

ABSTRACT

The O(6) limit of the Interacting Boson Model is studied from two viewpoints. It is shown, in terms of microscopic realistic IBM calculations, that the O(6) limit does exist in real nuclei, by taking examples of Xe and Ba isotopes. It is presented that the O(6) limit has the multiphonon structure built on a γ-unstable deformed ground state.

1. Introduction

The Interacting Boson Model (IBM) proposed twenty years ago by Arima and Iachello [1] has been extremely successful in various senses, as will be demonstrated in this anniversary conference. Its advanced version including the proton and neutron degrees of freedom, which is referred to as IBM-2, has been proposed three years later (*i.e.*, 1977) [2,3,4]. The IBM-2 has a direct relation to the microscopic nuclear structure, *i.e.*, the shell model [3,4]. Although there have been many important achievements of IBM, the microscopic calculations, which mean IBM calculations based on the shell model, have remained rather undeveloped. In the first half of this talk we shall discuss recent microscopic calculations of IBM-2 for the Xe-Ba region. A part of this work has been published [5], and more detailed report will be made [6].

The second part is for a newly proposed picture of the O(6) limit [7]: the multiphonon structure. The O(6) eigenstates of $\sigma = N$ are described as multiphonon states built upon the γ-unstable deformed ground state, in contrast to the presently accepted picture, γ-unstable *rotor*.

2. γ-soft nuclei in a microscopic IBM-2

2.1. Hamiltonian

We present results of microscopic IBM-2 calculations for Te, Xe and Ba isotopes. The method we shall use is the OAI mapping [4] as explained in detail in Ref. [8]. The doubly-closed shells of Z=50 and N=82 are assumed for protons and neutrons, respectively. Since the neutron number of the nuclei to be considered is between 66 and 82, valence neutrons are treated in terms of holes, whereas valence protons are particles. The nucleon Hamiltonian is assumed to consist of the single-particle energies,

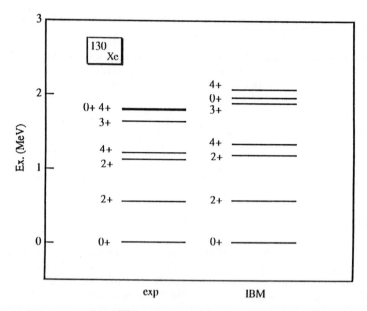

Figure 1: Energy levels of ^{130}Xe. The left part is experiment, and the right is the result of the present microscopic IBM-2 calculation.

the monopole, quadrupole and hexadecupole pairing interactions between identical nucleons, and the quadrupole-quadrupole interaction between identical nucleons and that between proton and neutron. Although this is still a model Hamiltonian in comparison with more fundamental interactions, for instance, the G-matrix, we should mention that this Hamiltonian appears to be quite realistic in particular with respect to features related to the low-lying quadrupole collective states. For instance, the single-particle energies are not degenerate, and the pairing interaction and the quadrupole-quadrupole interaction compete in determining the equilibrium shape.

All parameters such as the single-particle energies and interaction strengths are determined by the spectra of relevant single-closed nuclei except the strength of the proton-neutron interaction. The single-particle energies are determined so as to fit observed levels of one-quasiparticle states of relevant single-closed nuclei, for instance, Sn isotopes for neutrons. The proton (neutron) single-particle energies are basically constant, but include small variation (less than 1 MeV) as a function of the proton (neutron) number.

The strengths of the interactions are kept constant. Those for the interactions between identical nucleons are determined so that levels of 2_1^+ and 4_1^+ of single-closed

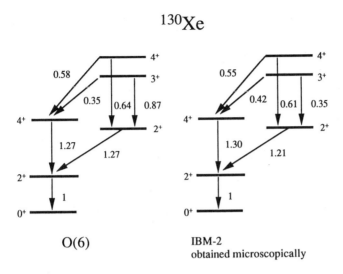

Figure 2: Relative B(E2) values of ^{130}Xe. The left part is obtained in the O(6) limit, while the right is the result of the present microscopic IBM-2 calculation. All B(E2) values are normalized by the $2_1^+ \to 0_1^+$ transition.

nuclei are reproduced well. Those levels are calculated in terms of one-broken pair approximation [6,9]. The proton-neutron interaction strength is determined so as to reproduce the 2_1^+ level of ^{132}Te. Note that ^{132}Te is the minimal proton-neutron open-shell nucleus; two valence protons and two neutron holes. Thus, we determine all the parameters in the nucleon Hamiltonian, before we carry out microscopic IBM-2 calculations.

2.2. Energy levels and E2 decays

Figure 1 shows the energy levels obtained by the present calculation in comparison with experiments for ^{130}Xe as an example [10]. There are non-vanishing matrix elements connecting SD pair states to other states in general. Such coupling effects are taken into account by the renormalization of the boson Hamiltonian. The renormalization is actually carried out in terms of the perturbation [5,6].

The agreement of the present microscopic IBM-2 calculation to experiment [10] is quite good in Fig. 1. The ordering of levels demonstrates the O(6) pattern quite clearly, in particular the low-lying 2_2^+ state. The E2 transition matrix elements also show a clear O(6) pattern [5,6]. Figure 2 shows the relative B(E2) values between

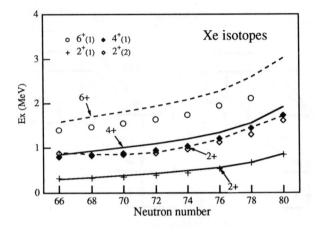

Figure 3: Energy levels of various Xe isotopes. The levels of 2_1^+, 2_2^+, 4_1^+ and 6_1^+ are shown. The neutron number varies from 66 to 80.

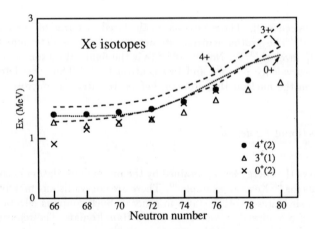

Figure 4: Energy levels of various Xe isotopes. The levels of 0_2^+, 3_1^+ and 4_2^+ are shown. The neutron number varies from 66 to 80.

low-lying states in comparison to the O(6) limit prediction. One sees an excellent correspondence between the O(6) and the present patterns. Note that nothing has been adjusted in the present microscopic calculation so as to obtain such a pattern. We also report the overlaps between the corresponding states in the left and right parts of Fig. 1. The overlaps are 0.954, 0.970, 0.937, 0.981 and 0.844 for the 0_1^+, 2_1^+, 2_2^+, 4_1^+ and 3_1^+, respectively. One finds quite large values, confirming the O(6) structure.

The energy levels and E2 matrix elements are calculated for other Xe isotopes. Figures 3 and 4 show systematic behaviors of low-lying energy levels as functions of the neutron number for Xe isotopes. A salient agreement to experiment can be seen again.

Figure 5 shows the B(E2) value of the $0_1^+ \to 2_1^+$ transition for various Xe isotopes as a function of the neutron number. The present results are obtained with the proton charge 1.6e and the neutron charge 1.1e, which are fixed so that the $0_1^+ \to 2_1^+$ transitions of single-closed nuclei are reproduced [11]. The present B(E2) values are in a good agreement with experiments [11]. Figure 5 also indicates that the present calculation gives a better description than the dynamical symmetry limits.

2.3. "Shape" parameters χ_π and χ_ν.

In the IBM-2 description, the nuclear shape, prolate, oblate, γ-soft, etc. is determined by the "*shape*" parameters, χ_π and χ_ν, which are contained in the boson quadrupole operators as explained in Refs. [5,8]. Figure 2 shows χ_π and χ_ν for Te, Xe and Ba isotopes. These parameters, χ_π and χ_ν, are calculated microscopically, and it turns out that they have opposite signs with similar magnitudes for most of the Xe and Ba isotopes with the neutron number around 70~78, as indicated in Fig. 6. The relation $\chi_\pi \sim -\chi_\nu$ is a key feature in understanding the structure of γ-soft nuclei, because this relation is the indispensable condition of the appearance of the (approximate) O(6) dynamical symmetry in the IBM-2 calculation [5].

The relation $\chi_\pi \sim -\chi_\nu$ holds to a good extent for many Xe and Ba isotopes. In fact, this relation holds much better in the present calculation than in a phenomenological fitting by Puddu *et al.*[12] It has been discussed by Casten and von Brentano [13] that the O(6) region can be extended in the Xe-Ba region rather widely. The present calculation clearly shows that this empirical observation is quite reasonable from the microscopic viewpoint. The value of the χ parameter is determined by shell effects and Pauli effects. In order to understand this, we consider an extreme case that there is a sub-shell closure due to the separation of the $h_{11/2}$ single-orbit shell and the shell of the other natural-parity orbits, and that neutrons (*i.e.*, neutron valence particles) occupy the natural-parity shell first and the $h_{11/2}$ shell next. The neutron number dependence of χ_ν can then be seen analytically [8]. Up to the neutron number 70, the natural-parity shell is being occupied, and χ_ν increases gradually from negative to positive values as the neutron number increases [8]. After the neutron number 70, the $h_{11/2}$ shell is being occupied, and the χ_ν parameter then decreases from positive

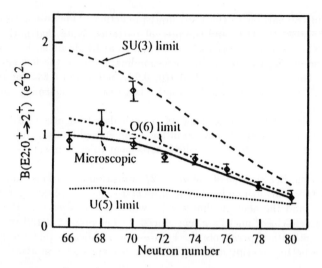

Figure 5: B(E2 ; $0_1^+ \to 2_1^+$) values of various Xe isotopes. The neutron number varies from 66 to 80. The predictions by the dynamical symmetry limits are also shown.

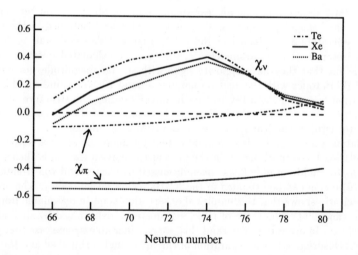

Figure 6: Shape parameters χ_π and χ_ν for Te-Xe-Ba isotopes obtained in the present microscopic IBM-2 calculation.

to negative. This produces a bump in the positive side around the neutron number 70. This schematic consideration is somewhat far from reality, and χ_ν stays rather constant forming a plateau around the neutron number 70.

The results in Fig. 6 shows also that the χ_π parameter is positive for Te isotopes. Hence the relation $\chi_\pi \sim -\chi_\nu$ does not hold and the O(6) structure is not anticipated for Te isotopes.

We are currently working on improvements on this respect, for instance by using more realistic nucleon-nucleon interactions [6]. We mention here that the so-called Pauli blocking renormalization introduced in a microscopic calculation by Egmond and Allaart [14] is absent in the present work, and its necessity remains open.

3. Multiphonon structure of O(6) nuclei

The O(6) dynamical symmetry [15] of the Interacting Boson Model (IBM) [1] has been used for the description of quite a few nuclei, especially in the Xe-Ba [13] and Pt [16] regions. Besides such success in phenomenological description, the O(6) has attracted much interest regarding its interpretation in terms of a more intuitive picture. The commonly accepted picture [17,18,19] has been the γ-unstable rotor of Wilet and Jean [20]. Its relation to the rigid triaxial rotor of Davidov and Filippov [21] has been also discussed [22-27]. In this section, we shall present a completely different picture of O(6) limit. This is a multiphonon description with a strong ground state correlation, where the phonons are built upon a γ unstable ground state. The aim of this work is to show that the phonon description arises in a natural way from the basic properties of the O(6) Hamiltonian.

We focus upon the $\sigma = N$ eigenstates of O(6), where σ and N denote, respectively, the O(6) quantum number and the total boson number (*i.e.*, SU(6) quantum number) [1,15]. The states with $\sigma < N$ are situated at higher energies for usual boson Hamiltonians, and can be described in a similar way, as stated at the end of this article.

The possible use of the O(6) quadrupole operator in the classification of the O(6) eigenstates has been shown in Ref. [28]. It was shown[28] that low-lying states can be constructed by successive operations of the quadrupole operator. Such construction of low-lying collective states are of great significance, particularly in extension to nuclei away from the symmetry limit. However, the origin of the multiphonon structure was not discussed in Ref. [28]. In the following discussions, we shall present how the multiphonon structure emerges in the O(6) limit.

The Hamiltonian we shall consider is

$$H = -\kappa(Q \cdot Q), \quad (1)$$

where κ denotes the strength parameter, the symbol ($\cdot$) means a scalar product, and

$$Q = d^\dagger s + s^\dagger \tilde{d}, \quad (2)$$

with $\tilde{d}$ being the modified annihilation operator ($\tilde{d}_m = (-1)^m d_{-m}$). This Hamiltonian is a linear combination of quadratic Casimir operators of O(6), O(5) and O(3) [15,1], and manifests the feature of the quadrupole collectivity of O(6). By this Hamiltonian, we do not loose the generality of the following discussions. We shall comment on this point later. The strength κ is supposed to be positive, and hence eq.(1) means an attractive quadrupole-quadrupole interaction.

We now construct the ground state for the Hamiltonian in eq.(1). This Hamiltonian can be rewritten as

$$H = -\kappa[\sqrt{5}\{[d^\dagger d^\dagger]^{(0)}ss + s^\dagger s^\dagger[\tilde{d}\tilde{d}]^{(0)}\} + 2(d^\dagger \cdot \tilde{d})s^\dagger s + (d^\dagger \cdot \tilde{d}) + 5s^\dagger s], \quad (3)$$

where $[\ ]^{(L)}$ means the coupling to an angular momentum L. Here, on the right hand side (RHS), the first two terms are those of the monopole pairing, the third is a monopole-monopole interaction, and the remaining terms are single-particle energies, because $(d^\dagger \cdot \tilde{d})$ is nothing but the d-boson number operator. Therefore, the ground state should be of the form

$$|0_1^+\rangle = \sum_n c_n \{[d^\dagger d^\dagger]^{(0)}\}^n (s^\dagger)^{N-2n}|0\rangle, \quad (4)$$

where $|0\rangle$ is the boson vacuum, and the c_n's stand for amplitudes. Here the RHS should be normalized.

We shall first show the commutation relation

$$\begin{aligned}[Q_M, Q_{M'}] &= d_M^\dagger \tilde{d}_{M'} - d_{M'}^\dagger \tilde{d}_M \\ &\equiv R_{M,M'}.\end{aligned} \quad (5)$$

The $R_{M,M'}$ operator in eq.(5) can be expressed through $[d^\dagger \tilde{d}]^{(1)}$ and $[d^\dagger \tilde{d}]^{(3)}$ operators. We mention that

$$R_{M,M'}|0_1^+\rangle = 0, \quad (6)$$

because $[R_{M,M'}, [d^\dagger d^\dagger]^{(0)}]$ is identically zero. This relation plays a key role in the following procedure.

The commutation relation with the Hamiltonian then becomes

$$[H, Q_M] = 4\kappa Q_M - 2\kappa \sum_m (-1)^m Q_{-m} R_{m,M}. \quad (7)$$

This results in

$$[H, Q_M]|0_1^+\rangle = 4\kappa Q_M|0_1^+\rangle, \quad (8)$$

which means that $Q_M|0_1^+\rangle$ is an eigenstate with the excitation energy 4κ. The state $Q_M|0_1^+\rangle$ is nothing but the first 2^+ state, as seen later.

We proceed to another illustrative example. The states with double Q's can be treated as

$$HQ_M Q_N|0_1^+\rangle$$
$$= \{8\kappa Q_M Q_N + 2\kappa Q_N Q_M - 2\kappa(-)^N \delta_{M,-N}(Q \cdot Q) - \kappa Q_M Q_N(Q \cdot Q)\}|0_1^+\rangle, \quad (9)$$

by using the relation

$$[\sum_m (-1)^m Q_{-m} R_{m,M},\ Q_N] = (-)^N \delta_{M,-N}(Q \cdot Q) - Q_N Q_M , \tag{10}$$

which arises from the double commutator $[[H, Q_M], Q_N]$. For the states $[QQ]^{(L)}|0_1^+\rangle$ with $L = 4$ or 2, the excitation energy turns out to be 10κ, whereas it vanishes for $L = 0$. The latter is natural, because $[QQ]^{(0)}|0_1^+\rangle \propto |0_1^+\rangle$. In other words, the double action of the Q operator produces eigenstates of $L = 4$ and 2, which are the first 4^+ and second 2^+ states, respectively, as seen later also.

We shall now consider the general cases. In eq.(9), there are two important features; (i) both $Q_M Q_N$ and $Q_N Q_M$ appear on the right hand side (RHS), (ii) the third term on the RHS produces non-vanishing effects only for two Q's coupled to $L = 0$ because of $\sum_{M,N}(2M2N|LM + N)(-)^N \delta_{M,-N} = \sqrt{5}\delta_{L,0}$. Considering these points we construct a state as

$$|\Psi\rangle = \sum C(\{M_1, M_2, \cdots, M_n\}) S\{Q_{M_1} Q_{M_2} \cdots Q_{M_n}\}|0_1^+\rangle, \tag{11}$$

where S implies a symmetrizer with respect to $M_1, M_2, \cdots, M_n$, and the C's mean amplitudes. By choosing proper C's, the state $|\Psi\rangle$ can have a good angular momentum, and one can introduce a set of $|\Psi\rangle$'s so that different $|\Psi\rangle$'s are orthogonal to each other. Here, we impose a condition on $|\Psi\rangle$ that any pair of two Q's is not coupled to angular momentum $L = 0$. Therefore, in the case of $n = 2$, only the total angular momenta $L = 4$ and 2 (and their linear combinations) are allowed in eq.(11).

We then consider $H|\Psi\rangle$. The first term of the RHS of eq.(7) yields $4\kappa Q_{M_i}$ from the same Q_{M_i} at the same place. This keeps the state unchanged. On the other hand, R_{m,M_i} of the second term must form a commutation relation with one of the Q operators further right, because of eq.(6). Using eq.(10), one obtains from $[R_{m,M_i}, Q_{M_j}]$,

$$+ 2\kappa Q_{M_1} \cdots Q_{M_j} \cdots Q_{M_i} \cdots Q_{M_n}|0_1^+\rangle, \tag{12}$$

where S and the C's are omitted for brevity. Note that Q_{M_i} and Q_{M_j} are interchanged with a factor 2κ in eq.(12) due to the double commutation discussed above. The first term on the RHS of eq.(10) does not contribute because no pair of the Q's is coupled to $L = 0$, as required in the construction of the state $|\Psi\rangle$. Thus, one ends up with

$$H|\Psi\rangle = \{4\kappa n + 2\kappa \frac{1}{2}n(n-1)\}|\Psi\rangle + E(0_1^+)|\Psi\rangle, \tag{13}$$

for all states constructed according to eq.(11). Here $E(0_1^+)$ is the energy of the ground state. Table 1 shows the energy levels of some low-lying states, highlighting several characteristic features.

We would like to mention several points; (i) the energy level is determined only by n, i.e., the number of the Q's, (ii) the energy level can be expressed by n and $\frac{1}{2}n(n-1)$ which can be viewed as a one phonon energy and its anharmonicity, (iii)

Table 1: Classification scheme of lowest O(6) eigenstates (of $\sigma = N$) in terms of phonon quanta (n) and excitation energies (E_x) normalized by 4κ.

phonon quanta (n)	$E_x/(4\kappa)$	angular momenta of eigenstates
0	0	0 (ground state)
1	1	2
2	2.5	4, 2
3	4.5	6, 4, 3, 0
4	7	8, 6, 5, 4, 2
.	.	....

the symmetrizer in eq.(11) produces only phonon-like states, (iv) two Q's coupled to $L = 0$ is forbidden. The first three points strongly suggest that the phonon structure dominate the present system. It is evident that the Q operator with the symmetrization plays the role of the phonon operator. Note that n stands for the number of the phonon quanta.

The fourth point is due to the strong ground state correlation, which can be seen in the structure of the Q operator; the $d^\dagger s$ term of eq.(2) is the usual "phonon creation" operator as is in the U(5) limit of IBM [29]. The second term, $s^\dagger d$, corresponds to the so-called backward amplitude in the random phase approximation, and annihilates $L = 0$ pairs of the d bosons (i.e., $[d^\dagger d^\dagger]^{(0)}$) when it is acting on $|0_1^+\rangle$. We need this second term with the equal strength as the first term, in order to make up the present scheme.

The symmetrization and the elimination of two Q's coupled to $L = 0$ in eq.(11) imply that the states constructed in eq.(11) can be classified in terms of the τ quantum number of O(5) as a matter of mathematics. In fact, also from the comparison between energy levels of eq.(13) and those of the O(6) limit, one finds that the states of n in eq.(11) are nothing but the states of $\tau = n$ in the O(6) limit with the excitation energy rewritten as $\kappa\tau(\tau + 3)$. Thus, it turns out that all the states of $\sigma = N$ are created by eq.(11).

It is of interest that one can obtain the present ground state exactly from the γ-unstable intrinsic states with the integration over the γ variable [18,19], while one can extract the ground state in a good approximation from the rigid-triaxial intrinsic state of $\gamma = 30°$ for smaller boson numbers [26,27]. Clearly the ground state is characterized also as a γ-unstable or triaxial state, and then it is most likely that the phonons introduced in this note preserve the γ-softness to a good extent. This point should be better clarified in the future. We would like to point out that the present result does not contradict the γ-unstable or triaxial nature of the O(6) system as a whole. We should stress, on the other hand, that the excitation mechanism is indeed of the phonon nature.

There are higher-lying states with $\sigma < N$ in the O(6) spectrum [15,1]. The lowest state of a given $\sigma(< N)$ is a 0^+ state, which contain $(N - \sigma)/2$ boson pairs with a specific structure. This pair is monopole, and is referred to usually as the P_6 pair [15,1,30]. The P_6 pair is not included in the ground state in eq.(4). This lowest state of $\sigma(< N)$ can be decomposed into a sector created solely by the P_6 pairs and the rest [30]. In other words, this state is created by $(N - \sigma)/2$ times successive actions of the P_6 pair-creation operator on the rest part. This rest part has a similar structure to the ground state in eq.(4), but consists of $\sigma(< N)$ bosons. The phonon operator, Q, commutes with the P_6 pair operators, and acts only to the rest part. Thus, the Q operator produces phonon excitations without disturbing the P_6 pair sector. To be more precise, the phonon operator conserves the σ quantum numbers, and the phonon excitation occurs within a subspace belonging to the given σ. Thus, one can construct all the states of an O(6) nucleus in terms of the multiphonon excitation and the P_6 boson pairing mode [30].

We have chosen the Hamiltonian in eq.(1). There are three independent terms in the general O(6) Hamiltonian [15,1]. Besides the present term in eq.(1), one of them is the total angular momentum, which does not change the wave function and yields the trivial variations of the energies. The third term can be the pairing interaction for the P_6 boson pairs. This interaction shifts all the levels of a given σ by the same amount. It does not change relative energies for the states belonging to the same σ. By including this interaction, the wave functions are not changed either. Thus, the above discussions based on the Hamiltonian in eq.(1) are quite general for the O(6) limit.

4. Summary

We reported recent results of the microscopic IBM-2 calculation on Xe-Ba isotopes. The present results exhibit good agreement with experiments, and simultaneously demonstrate excellent similarities to the O(6) predictions. Thus, the O(6) limit has been justified from a realistic shell model through the OAI mapping. The O(6) symmetry is shown to appear in a wide range of Xe and Ba isotopes.

We have presented that the low-lying ($\sigma = N$) O(6) states are multiphonon states built upon the γ unstable ground state, where the ground state correlation is dominant and a rather large number of d bosons are contained reflecting a strong deformation. The energies are represented in terms of phonon quanta and two-phonon anharmonicity. This consequence appears to be different from the usual picture of O(6) as a γ unstable "rotor", although the triaxial nature is inherent in this multiphonon picture through the ground state. The $\sigma(< N)$ states are constructed as a product of the P_6 pair sector and the present multiphonon states.

5. Acknowledgements

The authors appreciate the valuable discussions with Professor P. von Brentano and Professor A. Gelberg. The authors acknowledge partial support by the International Joint Research Projects of the Japan Society for the Promotion of Sciences, by Deutsche Forschungsgemeinschaft under contract no. Br 799/5-1, and by the JSPS-

DFG cooperation agreement. This work is supported in part also by Grant-in-Aid for Scientific Research on Priority Areas (No.05243102) by the Ministry of Education, Science and Culture.

6. References

1. F. Iachello and A. Arima, *The Interacting Boson Model* (Cambridge Univ. Press, New York, 1987), and references therein.
2. A. Arima, T. Otsuka, F. Iachello and I. Talmi, *Phys. Lett.* **66B** (1977) 205.
3. T. Otsuka, A. Arima, F. Iachello and I. Talmi, *Phys. Lett.* **76B** (1978) 139.
4. T. Otsuka, A. Arima and F. Iachello, *Nucl. Phys.* **A309** (1978) 1.
5. T. Otsuka, *Nucl. Phys.* **A557** (1993) 531c.
6. T. Mizusaki and T. Otsuka, in preparation.
7. T. Otsuka and K.H. Kim, submitted.
8. T. Otsuka, *Algebraic Approaches to Nuclear Structure: Interacting Boson and Fermion Models*, edited by R.F. Casten, (Harwood Academic Pub., 1993), Chapter 4.
9. K. Allaart et al., *Phys. Rep.* **169** (1988) 209.
10. M. Sakai, *At. Data Nucl. Data Tables* **31** (1984) 399.
11. S. Raman et al., *At. Data Nucl. Data Tables* **42** (1989) 1; S. Raman et al., *At. Data Nucl. Data Tables* **36** (1987) 1.
12. G. Puddu, O. Scholten and T. Otsuka, *Nucl. Phys.* **A348** (1980) 109.
13. R.F. Casten and P. von Brentano, *Phys. Lett.* **152B** (1985) 22.
14. A. van Egmond and K. Allaart, *Nucl. Phys.* **A425** (1984) 275.
15. A. Arima and F. Iachello, *Ann. Phys.* (NY) **123**, 468 (1979).
16. J. A. Cizewski, R. F. Casten, G. J. Smith, M. L. Stelts, W. R. Kane, H. G. Borner, and W. F. Davidson, *Phys. Rev. Lett.* **40** (1978) 167.
17. J. Meyer-Ter-Vehn, *Phys. Lett.* **B84** (1979) 10.
18. J. N. Ginocchio and M. W. Kirson, *Nucl. Phys.* **A350** (1980) 31.
19. A. E. L. Dieperink and O. Scholten, *Nucl. Phys.* **A346** (1980) 125.
20. L. Wilets and M. Jean, *Phys. Rev.* **102** (1956) 788.
21. A. S. Davydov and G. F. Filippov, *Nucl. Phys.* **8** (1958) 237.
22. R. F. Casten, A. Aprahamian and D. D. Warner, *Phys. Rev.* **C29** (1984) 356.
23. O. Castanos, A. Frank and P. Van Isacker, *Phys. Rev. Lett.* **52** (1984) 263.
24. J. Dobeš, *Phys. Lett.* **B158** (1985) 97.
25. J. P. Elliott, J. A. Evans and P. Van Isacker, *Phys. Rev. Lett.* **57** (1986) 1124.
26. T. Otsuka and M. Sugita, *Phys. Rev. Lett.* **59** (1987) 1541.
27. M. Sugita, T. Otsuka and A. Gelberg, *Nucl. Phys.* **A493** (1989) 350.
28. G. Siems, U. Neuneyer, I. Wiedenhöver, S. Albers, M. Eschenauer, R. Wirowski, A. Gelberg, P. von Brentano and T. Otsuka, *Phys. Lett.* **B320** (1994) 1.
29. A. Arima and F. Iachello, *Ann. Phys.* (NY) **99** (1976) 253.
30. A. Gelberg, T. Otsuka and P. von Brentano, to be published (1994).

HOW FAR CAN A DYNAMICAL SYMMETRY APPROACH BE PURSUED?

J. JOLIE
*Institut de Physique, Université de Fribourg,
Pérolles, CH-1700 Fribourg, Switzerland*

ABSTRACT

After reviewing recent experimental studies of nuclei corresponding to the symmetry limits of IBM-1, future applications of dynamical symmetries are discussed. They are illustrated by showing the effects of the O(5) group on shape-coexistence. An attempt to quantify the dynamical symmetry content is made by introducing the concept of wentropy. Wentropy is then applied to two level mixing, shape coexistence and the Casten triangle.

1. Introduction.

Physics has always been a quest to simplify our perception of nature in order to understand its underlying laws. For this quest, symmetries are often excellent guides into unknown territory. What we expect from them is twofold. First, to enable us to neglect a large part of the physical problem by showing that it is similar to the solution obtained by solving a part of the problem times a general solution treating the symmetry (e.g. reducing central force problems to the particular radial solution times spherical harmonics). Those symmetries we call physical symmetries or invariances. Second, to provide analytically solvable simplifications (dynamical symmetry limits) which greatly enhance our understanding. Hamiltonians, describing a quantal system, can exhibit such a dynamical symmetry when the physical system is such that it fulfils certain constraints: it is expressed in terms of constants of motion. If the operators describing these constants of motion all commute we speak of a dynamical symmetry limit. In that case the hamiltonian consists of a combination of Casimir operators of a group chain. From a theoretical point of view these limits are appealing because one can obtain a elegant analytic description of complex systems. In this sense the present search for dynamical symmetries in quantal systems is analogous to the one of physicists of previous centuries for solvable problems. This was brought into a new light by the fact that it was shown that only they correspond to integrable systems when going over to the classical limit[1].

In the field of nuclear physics the Interacting Boson Model (IBM)[2] has clearly shown the utility of dynamical symmetries. We will show that the dynamical symmetry approach of the IBM is well established for systems having a relatively 'simple' structure. Then we will focus on how it can be used to describe excitations in which the atomic nucleus exhibits a more complex structure, i.e. where it leads to a high, but not yet statistical, level density. Finally we introduce the concept of wentropy which, among other things, allows the measurement of the dynamical symmetry content of a particular physical system.

2. Recent experimental evidence for dynamical symmetry limits.

Most of the atomic nuclei we now consider to be typical examples of the dynamical symmetry limits of the IBM have been extensively studied after the IBM was proposed as a new nuclear model. Such studies firmly established the validity of the symmetry approach for the lowest lying levels in some even-even nuclei. Here, I want to report on some experiments on these nuclei testing the theoretical predictions for higher lying states. These tests have only been performed in recent years. Most of them have become possible due to the elaboration of the GRID technique[3], which allows to combine standard (n,γ) spectroscopy with a lifetime measuring technique. Exception made for the first example, they also clearly illustrate that extensions of the original symmetry limits are often needed at higher excitation energies. These can often make full use of the dynamical symmetry concept, as will be shown in paragraph 3.

The γ-unstable nucleus ^{196}Pt is known to be a very good example of the O(6) limit of the IBM. The low-lying excited levels and the E2 transitions between them follow nicely the selection rules up to rather high energy. Only the non-vanishing quadrupole moment of the first excited state is in contradiction with O(6). This has led to some attempts to describe this nucleus as a strongly anharmonic vibrator. In essence this is possible due to the fact that the O(5) symmetry is conserved in the whole U(5) to O(6) transitional region[4] and that the O(5) symmetry is a dominant factor in many experimental observables. This is illustrated in Fig. 1 and will be discussed in more detail in paragraph 3. In order to fully establish the existence of the O(6) dynamical symmetry it is needed to study high-lying states that do not belong to the maximal O(6) represensentation σ = N. The lowest σ = N-2 state is a 0$^+$ state at 1402 keV which was identified on the basis of its decay properties although absolute transition rates were unknown. In this case these rates are crucial for determining the nature of this state and as such the dynamical symmetry. In particular if the B(E2) value to the lowest 2$^+$ state vanishes we have a manifestation of the Δσ = 0 selection rule and an O(6) symmetry. If the B(E2) value is 0.46 e^2b^2 we have a two to one phonon transition and the U(5) symmetry is present. Finally, for intermediate values we only have the O(5) symmetry. With GRID a lower limit of B(E2,0->2)<0.034e^2b^2 could be measured[5]. This clearly establishes the O(6) character of ^{196}Pt.

The existence of atomic nuclei corresponding to the SU(3) limit is less clear. It seems that the SU(3) symmetry is too stringent, due to the fact that it implies degeneracies between states since the K quantum

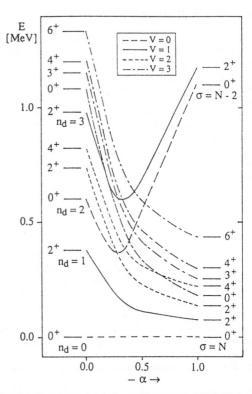

Figure 1: The figure shows how the lowest states of the U(5) conserve their O(5) symmetry during the U(5)-O(6) transition. The transition was described varying α using eq(14) with a= -25, b=10, c= 300, d= 0 and e = 6 (all in keV).

number is a missing label. To overcome this, the SU(3) symmetry has to be broken. Candidates for broken SU(3) symmetry are found from Gd up to the Hf isotopes. There exist two standard ways to break this symmetry. We will give examples of both here. The first way is to allow deviations of the exact symmetry by adding non-SU(3) interactions to the hamiltonian and transition operators. This approach was used in a recent very complete study of ^{156}Gd including GRID measurements[6]. This nucleus was long time considered a good example of SU(3) due to the near degeneracy of the beta and gamma band, but failed to fulfil the stringent symmetry criteria that forbid transitions between these bands and the ground state band. In the study performed in ref 6 it was shown that a small mixture of the U(5) limit into the hamiltonian and the use of an E2 transition operator:

$$T^{(E2)} = T^{(E2)}_{SU(3)} + a\, T^{(E2)}_{U(5)} = (s^\dagger \tilde{d} + d^\dagger s - \frac{\sqrt{7}}{2} d^\dagger \tilde{d})^{(2)} + a\, (d^\dagger \tilde{d})^{(2)} \qquad (1)$$

which is dominated by the U(5) part (a equaled -1.9) gives a remarkable description of the absolute B(E2) values (see Table 1). This result is in contrast to former perturbations on the

J_f\J_i	gsb J_f+2	β-band J_f+2	β-band J_f	β-band J_f-2	γ-band J_f+2	γ-band J_f+1	γ-band J_f	γ-band J_f-1	γ-band J_f-2
0	187|187	0.63|0.97	-	-	4.4|4.2	-	-	-	-
2	258|263	1.2|1.2	3.3|1.4	(7.8)|6.0	1.6|1.9	7.3|7.3	7.1|6.8	-	-
4	295|281	-	1.8|1.0	3.6|3.6	-	5.9|5.7	9.2|8.1	5.6|7.8	0.6|0.4
6	320|282	-	-	-	-	-	-	-	-
8	315|273	-	-	-	-	-	-	-	-

Table 1: Comparison between theoretical and experimental absolute B(E2) values in W.u. (shown as exp.|th.) for transitions towards the ground state band (gsb) in ^{156}Gd. The wavefunctions used were obtained using H (keV) = -7.7 C$_2$[SU(3)] + 57 C$_1$[U(5)] + 13.2 C$_2$[SO(3)] following ref 6.

transition operator that favored 0 < a < +1.3 .

In many rotational nuclei a proper description of the observed states cannot be obtained in the conventional sd IBM-1 but needs to incorporate g-bosons. In another GRID experiment on ^{168}Er the double γ vibration was studied[7]. This vibration is predicted by many models to lead to a low-lying rotational band with K$^\pi$ = 4$^+$. In ^{168}Er such a rotational band with the J$^\pi$ = 4$^+$ bandhead at 2055 keV was considered to be the most probable candidate for a two phonon state. Using the GRID technique the lifetime of this state was measured and 0.014 e^{2}b^2< B(E2;4$^+$->2$^+$) <0.041 e^{2}b^2 could be deduced for the decay to the gamma band[7]. This collective B(E2) value nicely confirmed the two-phonon character of the K$^\pi$ = 4$^+$ band. Of all models the sdg-IBM yielded the best prediction of both the ratio of the energies of the K$^\pi$ = 4$^+$ and K$^\pi$ = 2$^+$ bands and the measured decay probability. Thus, concerning the SU(3) limit we conclude that although the exact symmetry of the IBM is not realised a very good description of deformed nuclei can be obtained by hamiltonians having a broken SU(3) symmetry.

Having found strong evidence for the exact realisation of the O(6) symmetry and no evidence for the exact SU(3) symmetry, the question arises whether there exist atomic nuclei exhibiting the U(5) dynamical symmetry. Oncemore, recent studies[8,9,10] of 110,112,114Cd have extended the experimental knowledge on the Cd isotopes proposed in ref 2 as typical U(5) nuclei. The situation for the Cd istopes is particularly complicated due to the presence of low-lying collective 2particle-2hole (2p-2h) excitations. To theoretically analyse the results in such a complicated situation we have assumed a very simplified structure to describe the Cd isotopes. Consequences of this choice will be discussed in detail in the following paragraph. We assume, in first a instance, that the normal and intruder states globally do not interact strongly.

Global in contrast to local, following earlier work, refers here to their gross structure such as reflected by the number of states with a given spin and their excitation energies[11]. Assuming for the normal states, which form only a part of the observed states, due to the presence of intruders, the U(5) symmetry we have first performed a systematic fit of these 'well-behaved' states using the U(5) limit[9,10,11]. The results of these fits for 110,112,114Cd are given in Fig. 2. The agreement in energies, between the U(5) limit and, in particular, states belonging to the ground-state and quasi-gamma band can be extended to high high spin and very high energy [9,10]. This obsevation is so striking that it hardly can be by sheer chance. Having described 'normal' beha-

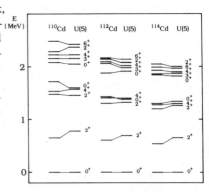

Figure 2: Comparison between the normal states in the Cd isotopes and the U(5) limit.

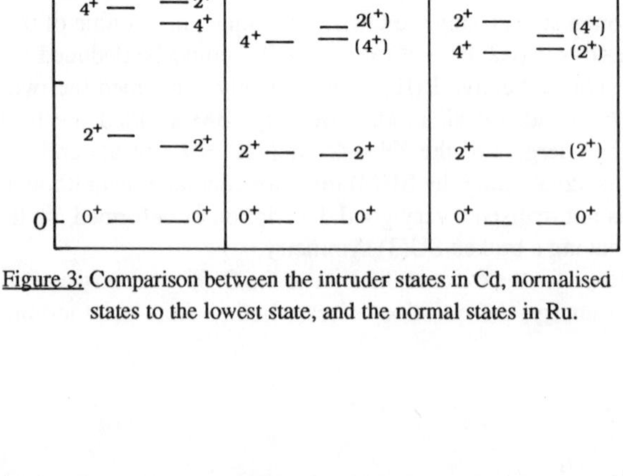

ving states, it is possible to study the intruder states that were left over. They are shown in Fig. 3 after renormalising the energies with respect to the lowest of these states. Now, if the intruder picture has sense, these remaining states might resemble nuclei having six proton holes as those should show rather similar collectivity. A direct comparison to the low-lying excited states in the isotopes 106,108,110Ru is given in Fig. 3. Within the experimental knowledge it is possible to establish a one-to-one correspondence between the states. This is quite surprising as we compare only experimental quantities. The correspondence is especially good for the 'yrast' states.

Figure 3: Comparison between the intruder states in Cd, normalised states to the lowest state, and the normal states in Ru.

Like most similarities in physics, one can formalise the similarity observed in Fig. 3 by its underlying basis: no difference for the nuclear structure is observed whether the proton bosons are made of holes or particles. This similarity can be formalised by assigning a new spin quantum number, called Intruder-spin (I-spin) to the bosons resulting in intruder analog states[12,13]. The levels shown in Fig. 3 can be described by the O(6) limit of the IBM as shown in ref. 9,10,11. Combining both symmetries, U(5) for the normal and O(6) for the intruder states all levels in ^{110}Cd and ^{112}Cd can be described up to high excitation energy. This is a very satisfying situation in view of the simplicity of the model. In ^{114}Cd, where an extensive set of absolute B(E2) values is available, the situation is more complex[8]. This is probably due to the presence of rotational features in the intruder configuration. Those permit the strong disturbance of the normal states as will be explained below.

Before proceeding we would like to mention that the study of dynamical symmetries and supersymmetries in odd-A and odd-odd nuclei, as well the dynamical symmetries of the IBM-2, has to be pursued in the view of the recent results on the symmetries of IBM-1. Also in this field some recent results have illustrated the power of the dynamical symmetry approach. Due to lack of space and time the interested reader is referred to ref 14-20 for some recent developments in this field.

Manifestations of dynamical symmetries associated with limits of the IBM-1 have been investigated in great detail at higher excitation energies. From these studies we conclude that a) the realisation of an O(6) symmetry in atomic nuclei is well above all doubt, b) although the SU(3) symmetry is a good starting point for a IBM-analysis of atomic nuclei there is only evidence for broken SU(3) symmetry. However, this breaking can be established easily in the model either using the consistent-Q formalism towards O(6) nuclei or via U(5) inducing terms or either by introducing g-bosons. It is noteworthy that the E2 transition operator needs to be heavily broken by the perturbations. c) For the U(5) symmetry the situation is more complicated due to the presence of, well understood, intruder excitations. Recently, strong evidence for the global survival of vibrational states even up to high energies has been obtained in the light Cd isotopes. Nevertheless, this evidence needs to be confirmed by more stringent tests. In general, from the results obtained in this paragraph we conclude that the usefulness of dynamical symmetries lies not only in providing an as complete as possible description of a particular many body system, but also in providing a good framework to describe nuclei where the dynamical symmetry is broken.

3. An example of new implications of dynamical symmetries: the U(5)-O(6) model of shape-coexistence.

In the applications to ^{156}Gd and the Cd isotopes a second rôle of dynamical symmetries, not highlighted in the previous paragraph, has been determining: Imposing symmetry constraints greatly restricts the number of free parameters that occur in enlarged IBM models when new degrees of freedom need to be implemented. These extensions of the IBM often destroy the original beauty of the model as they lead to a large set of parameters. The approach of imposing symmetry constraints was also used succesfully in other extensions of the IBM and led to some quite spectacular advances, notably the description of the scissors mode[21] and the introduction of supersymmetry into nuclear physics for the description of odd-A[22] and odd-odd nuclei[23]. Although imposing symmetries might seem to be an awful approach from a microscopic point of view, for the study of complex nuclei this becomes a condituo sine qua non. Although microscopic theories often cannot treat complicated many-body systems in sufficient detail they do have an important rôle in guiding more phenomenological approaches, as we pursue here.

In this paragraph we will show that imposing dynamical symmetries to new complex problems still yields surprises. We will concentrate on a result obtained in the previous paragraph: the Cd isotopes can be very well described by a U(5)-O(6) model that assumes that the normal states resemble very closely U(5) nuclei, while the intruder configurations are dominated by the O(6) symmetry. In view of this, it is worthwhile to recall the group chains associated to the U(5) and O(6) limit of the IBM-1, given in Fig.4.

$$U(5) \supset O(5) \supset O(3)$$
$$| \quad | \quad |$$
$$\{n_d\} \quad (v) \quad L$$

$$U(6) \supset$$
$$|$$
$$[N]$$

$$O(6) \supset O(5) \supset O(3)$$
$$| \quad | \quad |$$
$$<\sigma> \quad (\tau) \quad L$$

Figure 4: The U(5) and O(6) group chains of the IBM. Also given are the associated quantum numbers labelling the irreps.

The question we want to address here is what happens when both symmetries are present within the same atomic nucleus. In our analysis of the light Cd isotopes such a situation appears: the normal configuration made from N s and d bosons approaches the U(5) limit while the intruder configuration with its N+2 bosons is close to the O(6) limit. When this happens, the symmetries lead to consequences that might contradict many-a-physicists common sense. Notably, the rule-of-thumb that the closer two states were before mixing, the stronger they interact is highly affected.

In fact, effects due to the common group O(5) of both group chains become dominant.

To see this we first notice that, up to first order, both configurations can only interact (mix) with a hamiltonian that contains two boson creation operators and which is hermitian and a scalar. Thus the interaction or mixing hamiltonian in IBM-1 has to read like:

$$H_{mix} = \alpha \, (s^\dagger s^\dagger + s \, s)^{(0)} + \beta \, (d^\dagger d^\dagger + \tilde{d} \, \tilde{d})^{(0)} \qquad (2)$$

The mixing hamiltonian (2) has the following tensorial character in the U(5) basis:

$$H_{mix} = \alpha \, T^{[2]\{0\}(0)0} + \beta \, T^{[2]\{2\}(0)0} . \qquad (3)$$

From (3) we see that the mixing hamiltonian is not only a scalar under the group O(3) but also a scalar under the group O(5). This has some important consequences[24]. A first observation is that because the mixing hamiltonian is an O(5) scalar we need to have that for two mixing states $v = \tau$. Since the lowest states have the lowest values for v and therefore do not exhibit a strong N dependence we can refer to Fig. 1 in discussing the consequences of this rule. It is clear that effects due to selective mixing show up already for the first intruder states (such as was found in the numerical calculations of ref. 9). If we compare states that can mix one notices that (3) prohibits the mixing of the second normal 2^+ level and the first intruder 2^+ state. The same holds for the second normal 4^+ state and the first intruder 4^+ state. On the otherhand the 0^+ states can mix heavily. The second result generalises the $\Delta v = 0$ rule to a much broader region of application. Since the O(5) symmetry is conserved over the whole U(5) to O(6) transitional region[4], as illustrated by Fig 1, the selective mixing holds for all states in the normal and intruder configuration if both can be described with a hamiltonian that has no contribution of $C_2[SU(3)]$. As intruder states appear near closed shells where such contributions are the smallest, atomic nuclei exhibiting shape coexistence between two structures with good O(5) symmetry are not so unlikely. In those nuclei effects due to this more general rule, valid outside the exact limits, should show up. Using the U(5)-O(6) model in its strict sense (only dynamical symmetries) more results follow[24,25]. In particular, the following relation holds for the interaction between normal states and intruder states belonging to the irrep σ = N:

$$<[N+2]<N+2>(v)L|H_{mix}|[N]\{n_d\}(v)L> = F(N,n_d,v) \left(\alpha + \frac{\beta}{\sqrt{5}} \right), \qquad (4)$$

with $F(N,n_d,v)$ a purely geometric factor. Equation (4) shows that the mixing interaction does not distinguish between the relative strenght of the s and d boson part. The geometric factor $F(N,n_d,v)$ has no dependence on the spin L. This leads to stability of some B(E2) ratios, that are governed by the O(5) symmetry[4], even when there is mixing between normal and intruder states[24]. To illustrate that the selective mixing stays present in standard IBM-2 mixing calculations we have plotted in Fig.5 the normal components of the wavefunctions obtained in the semi-microscopic calculation[27] for $^{110-114}$Cd for the second and third 0^+ and 2^+ states, versus the energy difference:

$$\Delta E = E(J_3^+) - E(J_2^+) \qquad (5)$$

between these states. One clearly observes how the O(5) selective mixing is still present in the IBM-2 results. Finally the U(5)-O(6) model has permitted for the first time to obtain an analytic description of shape coexistence[25,26]. This allows to attack the problem starting from the local properties[11] whom are dominated by the nature of the wavefunctions and their mixing instead of the standard procedure of first describing the global properties.

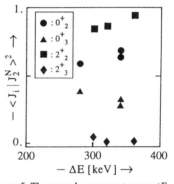

Figure 5: The normal component versus ΔE for the lowest mixed states in $^{110-114}$Cd.

In this paragraph we have illustrated how the dynamical symmetry approach helps to understand the complicated structures that atomic nuclei can exhibit. In particular, we have focussed ourselves on shape coexistence and shown how the O(5) symmetry can dominate the interaction of normal and intruder states. This allows understanding of some pecularities that appear in the semi-microscopic IBM-2 calculations of ref. 27.

4. Wavefunction entropy: a measure of perturbations and symmetry breaking.

In the previous paragraphs we have talked a lot about dynamical symmetries and their breaking. Here, we investigate whether it is possible to measure the amount of order or disorder in the wavefunction with respect to a given first order approach. This is important because the amount of symmetry breaking will stay a field of research for the coming years as the symmetries are now well studied and because of its relation to quantal chaos[1]. The question we want to address resembles the one of order (symmetries) and disorder (no symmetries) and therefore we will refer to these terms.

To quantify the degree of order we first observe that the order is generally related to the projection of wavefunctions $|\Psi(i)\rangle$ onto a basis $|\Phi(j)\rangle$:

$$|\Psi(i)\rangle = \sum_{j=1}^{n} a_{ij}|\Phi(j)\rangle. \qquad (6)$$

What we want to measure is how the components of the wavefunction a_{ij} are distributed for given i. If they are random with respect to $|\Phi(j)\rangle$ we will say there is disorder, if this is not so we talk of order. In order to measure this distribution we use an analogy to the calculation of entropy in statistical mechanics. In statistical physics the task is to apply the law of probability to a system having a complicated structure, by considering the system as being in a macro state made up of an ensemble of micro states that are occupied following probabilistic rules. Likewise, the final wavefunction $|\Psi(i)\rangle$ can be considered to be a macro state which is made out of microstates $|\Phi(j)\rangle$. In our case the occupation probability of the microstate j in the macrostate i is given by:

$$p_{ij} = a_{ij}^2. \qquad (7)$$

In statistical physics the numbers n_j of particles in the micro states j need to fulfil:

$$N = \sum_{j=1}^{n} n_j, \qquad (8)$$

when distributing the N particles into the microstates. In our case a similar constraint is implied by the normalisation of the wavefunctions:

$$1 = \sum_{j=1}^{n} a_{ij}^2 = \sum_{j=1}^{n} p_{ij}. \qquad (9)$$

An additional complication occurs in our case due to the fact that the macrostates need not only to be normalised but also to be orthogonalised.

In contrast to a statistical treatment, we will here try to obtain information on the distribution law instead of implementing a probability law to the occupation probabilities p_{ij}. To measure this distribution, analogous to the Gibbs-Boltzmann approach, a concept analogous to entropy is used[28]. The wavefunction entropy is measured by the relation:

$$W = -\sum_{j=1}^{n} p_{ij}\ln(p_{ij}). \qquad (10)$$

We will denote W by the name wentropy, for wavefunction entropy. This name avoids confusion as, in contrast to entropy, wentropy is a dimensionless entity. Wentropy is closely related to the entropy of the response of a system to a probe as introduced by Iachello and Levine[29]. The main difference to earlier applications is that wentropy is measured with respect to a complete basis and not only with respect to a symmetry.

Although wentropy is not so well founded as entropy it will turn out to be useful for nuclear physics. It is important to note that the wentropy is always with respect to a given unperturbed basis, one should therefore state the wentropy with respect to....

5. Application of wentropy to two level mixing, shape coexistence and the Casten triangle.

As illustrative examples of the possible use of wentropy we will discuss three examples[28]. The examples are chosen in increasing order of complexity and all do not refer to the same kind of wentropy. In the first case some properties of wentropy are illustrated in the instructive case of two-level mixing. Then we apply wentropy to an actual case of multistate mixing: shape coexistence in the Cd isotopes. To illustrate the fact that wentropy allows model independent analysis of the effect of perturbations we apply it both to the results of the previously mentioned U(5)-O(6) model and the microscopically founded IBM-2 calculations[27] for ^{112}Cd. These two examples measure the wentropy of the perturbed, or mixed, wavefunctions versus the unperturbed wavefunctions and are thus of very general nature. In the third example we address another question by measuring the U(5)-wentropy of atomic nuclei as represented by the Casten triangle.

In order to illustrate some properties of wentropy, we consider first the mixing between two states having wavefunctions |Φ(j)> that before mixing fulfil:

$$H |\Phi(j)> = E_j |\Phi(j)> \quad j=1,2 . \tag{11}$$

Under the effect of a residual interaction V, the final wavefunctions |Ψ(i)> fulfil:

$$(H+V) |\Psi(i)> = E_i |\Psi(i)> \quad i=1,2 . \tag{12}$$

The solution of (12) in terms of those of (11) can be written as:

$$|\Psi(i)> = \sum_{j=1}^{2} a_{ij} |\Phi(j)> . \tag{13}$$

We have trivially $W(|\Psi(1)>) = W(|\Psi(2)>)$. Using the orthonormality imposed on the wavefunctions, the wentropy of the state |Ψ(1)> is shown as a function of a_{11} in Fig 6. The behavior shown in this figure clearly illustrates the utility of wentropy as a measure for the behavior of mixing amplitudes. First we note that the wentropy is zero if a_{11} is either 0 or 1. This means that the system stays regular in the basis spanned by the solutions of H. The maximum wentropy obtained from Fig. 6 corresponds to $W = 0.6931 = \ln(2)$ and is obtained at $a_{11} = 0.7071$. Thus the wentropy is maximal when the mixing is maximal. Generalising this result for n level mixing the maximal wentropy is equal to $W_{max} = \ln(n)$. As wentropy is always defined to an unperturbed system, i.e. the basis states |Φ(j)>, it is clear that the wentropy ratio W/W_{max} measures how good an approximation this basis is for the sytem under investigation.

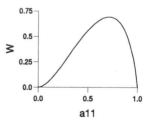

Figure 6: Wentropy as a function of a_{11} in the case of two-level mixing.

To investigate this in more detail we calculate the wentropy in the full mixing calculations[9,27] for ^{112}Cd. The mixing that is allowed concerns the lowest four normal and intruder states. As in the previous example we calculate the wentropy of the mixed wavefunctions with respect to the unperturbed states. The results for the wentropy ratios W/W_{max} are given in Table 2. The calculation label IBM-1 corresponds to the one performed in the U(5)-O(6) model[9], while the one labelled IBM-2 is performed using the standard semi-microscopic method[27]. In the U(5)-O(6) model the second and third 2^+ states in the configuration mixing calculations even after the introduction of the mixing hamiltonian have very low wentropy with respect to their unperturbed basis since all states possess an O(5) dynamical symmetry. In the concrete example discussed above the wentropy for the second and third 0^+ states would be nearly maximal for two level mixing in four level mixing this is reduced to about 0.4. One notices from Table 2 that in the IBM-2 calculation the wentropy is generally higher, but that its variations, except for the third 2^+

J_i^π	0_1^+	2_1^+	0_2^+	2_2^+	4_1^+	0_3^+	2_3^+	4_2^+	0_4^+	2_4^+
IBM-1	0.01	0.03	0.28	0.08	0.08	0.27	0.14	0.14	0.13	0.08
IBM-2	0.07	0.10	0.36	0.24	0.16	0.43	0.55	0.50	0.31	0.55

Table 2: The wentropy ratio W/W_{max} for the lowest states in ^{112}Cd obtained from mixing calculations using the IBM-1 in an U(5)-O(6) approach and the IBM-2 in a semi-microscopic approach.

and second 4^+ states, are similar to the ones obtained in the U(5)-O(6) approach. The higher wentropy ratio in the IBM-2 calculation is partly due to the higher mixing strenght as apparent in the wentropy ratio of the groundstate or the third and fourth 4^+ states.
    In our third application of wentropy we come back to the results shown in paragraph 2 and 3, and try to quantify them in a much broader region of application. As these results are obtained from the projection of a wavefunction obtained in the IBM-1 onto the U(5) basis we have to calculate the U(5) wentropy of the states in the Casten triangle. First we calculate the U(5) wentropy of the wavefunctions used in the calculation represented in Table 1 to investigated whether the U(5) breaking is large. The result for the 292 states up to J=12 is an averaged wentropy ratio of 0.73 indicating a very small U(5) breaking. Next we parametrise the Casten triangle by the following simplified hamiltonian:

$$H(\alpha,\gamma) = (1-\gamma)\{\alpha\, a\, C_2[O(6)] + b\, C_2[O(5)] + (1-\alpha)\, c\, C_1[U(5)]\} + \gamma\, d\, C_2[SU(3)]$$
$$+ e\, C_2[SO(3)] \qquad (14)$$

In the hamiltonian (14) the parameters a to e are fixed parameters. They are taken to be a = -100 , b= 60 , c = 300 , d =-20 and e= 6 (all in keV) for a 10 boson calculation. The parameters α and γ which describe the actual position on the Casten triangle are allowed to vary between 0 to 1. The U(5), O(6) and SU(3) limits correspond to the choices $\gamma = 0$, $\alpha = 0$; $\gamma = 0$, $\alpha = 1$ and $\gamma = 1$, respectively, while the three legs of the triangle are obtained for $\gamma = 0$, α free (U(5) to O(6) transition); $\alpha = 0$, γ free (U(5) to SU(3) transition) and $\alpha = 1$, γ free (O(6) to SU(3) transition). To describe nuclei in the triangle α was allowed to range from 0 to 1 for any given value of γ. The U(5) wentropy is obtained from the projection:

$$|\Psi(i,J)\rangle = \sum_{j=1}^{n} a_{ij}|(n_d,v,v_\Delta)j,J\rangle \, . \qquad (15)$$

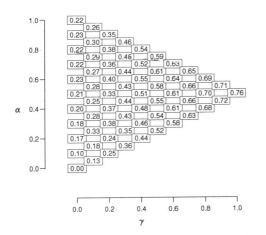

Figure 7: The averaged wentropy ratio W/W$_{max}$, as a function of the position on the Casten triangle for the 195 states with spins up to 15+.

Using formula (10) we have calculated the U(5) wentropy for each individual state with spin < 16 as a function of α and γ. Then we have calculated the ratio of this wentropy to the maximal wentropy since due to the varying number of states having a particular value of J the wentropy in itself is not a good measure. Having obtained the wentropy ratio for all 195 possible states with angular momentum up to 15, we have calculated the average of this quantity per state. This is represented in fig 7. Inspection of this figure reveals some interesting features:
a) The highest wentropy is obtained in the SU(3) limit, where the averaged wentropy ratio reaches a tremenduous 75%. This means that the concept of s and d bosons for rotational nuclei has lost its sense, as the U(5) wentropy measures in fact the amount to which it has sense to talk about s and d bosons.
b) Due to the O(5) symmetry the U(5) to O(6) transition has the lowest U(5) wentropy ratio going up to about 22%. The picture given in Fig. 6 shows some less expected effects for this transition. One notices that the wentropy ratio reaches its maximal value long before the O(6) limit is reached. In fact it saturates already half-way between U(5) and O(6). This saturation effect even increases over the whole triangle when approaching the SU(3) limit. At present an easy explanation of this effect is lacking, but it is clear that such an explanation has to go beyond the O(5) symmetry.
c) For a given value of γ the lowest wentropy ratio is found at or near to the U(5) to SU(3) leg. Deviations from the leg do, however exist. They most intriguingly indicate a departure from this leg in the sense of the new 'regular region within the Casten triangle' found in ref 30. Detailed inspection shows that the departure is much more pronounced at low spin[28].
d) In general low spin states have a larger averaged wentropy ratio than the high spin states. For the spin range J=11-15, for instance, the O(6) limit has an averaged U(5) wentropy ratio of only 11% and the saturation effect is very pronounced. As the wentropy ratio corrects for the possible states that can be formed, this distochomy is quite surprising. This observation is also similar to the reduction of the degree of chaoticity with increasing spin[30]. Closer inspection shows also that the wentropy of states with J = 0, 3 and 5 is much higher than the one for states with J=2 and 4[28].
e) Shallow wentropy minima as a function of γ occur in the central parts of the Casten triangle. The most important is situated around $\alpha = 0.5$ and $\gamma = 0.15$. Whether they are accidental or not is still an open question.

In the last two paragraph an attempt was made to measure the effect of perturbations on quantum mechanical systems by using the concept of wentropy. Among other things wentropy measures the dynamical symmetry content or the amount of mixing of wavefunctions. The analysis of the wentropy ratio of the Casten triangle showed a strong correlation with the results on the degree of chaoticity determined by Whelan and Alhassid[30].

Acknowledgements

The results presented in paragraph 2 were obtained from many collaborations with experimentators and theoreticians. I would like to thank them all for this. In particular, I want to express my gratitude to K. Heyde, P. Van Isacker, H.G. Börner, J. Kern and A. Van Bogaert. For the subjects treated in paragraph 3 I'm indebted to H. Lehmann for many discussions. This work was supported by the Swiss National Science Foundation.

References

1) W.M.Zhang, C.C.Martens, D.H.Feng, and J.M.Yuan, Phys.Rev.Lett. **61** (1988)2167.
2) F.Iachello and A.Arima, *The Interacting Boson Model*, (The Cambridge University Press, 1987).
3) H.G.Börner and J.Jolie, J. of Phys. **G19** (1993) 217.
4) A.Leviatan, A.Novoselsky, I.Talmi, Phys. Lett. **B172** (1986) 144.
5) H.G.Börner, J.Jolie, S.J.Robinson, R.F.Casten, J.A.Cizewski, Phys.Rev. **C42** (1990) R2271.
6) J. Klora et al. Nucl. Phys. **A561** (1993) 1.
7) H.G.Börner, J.Jolie, S.J.Robinson, B.Krusche, R.Piepenbring, R.F.Casten, A. Aprahamian, J.P.Draayer, Phys. Rev. Lett. **66** (1991) 691, 2837.
8) R.F.Casten, J.Jolie, H.G.Börner, D.S.Brenner, N.V.Zamfir, W.-T.Chou, and A. Aprahamian. Phys. Lett. **B297** (1992) 19.
9) M.Délèze, S.Drissi, J.Jolie, J.Kern, J.P.Vorlet, Nucl.Phys.**A554** (1993) 1.
10) M. Bertschy, S.Drissi, P.E. Garrett, J.Jolie, J.Kern, S.J. Mannanal, J.P.Vorlet, N. Warr, J. Suhonen, subm. to Nucl. Phys. A.
11) J.Jolie, in Proc. of the 8th. Int. Symp. on Capture Gamma-Ray Spectroscopy and related topics, Ed. J. Kern (World Scientific 1994) p43.
12) K.Heyde, C.De Coster, J.Jolie, J.L.Wood, Phys. Rev. **C46** (1992) 541.
13) K.Heyde, contr. to this conf.
14) I.Bauske et al. Phys. Rev. Lett. **71** (1993) 975.
15) J .Cizewski et al. Phys. Rev. Lett. **58** (1987) 10.
16) K.Heyde, C.De Coster, D.Ooms, A.Richter, Phys. Lett. **B312** (1993) 267.
17) U.Mayerhofer et al. Nucl Phys. **A492** (1989) 1.
18) F.Hoyler et al. Nucl. Phy. **A512** (1990) 189.
19) J.Jolie, U.Mayerhofer, T. von Egidy, H.Hiller, J.Klora, H.Lindner, H.Trieb, Phys. Rev. **C43** (1991) R16.
20) G.Rotbard et al. Phys. Rev. **C47** (1993) 1921.
21) P.Van Isacker, K.Heyde, J.Jolie, A.Sevrin, Ann. of Phys. **171** (1986) 253.
22) F.Iachello, Phys. Rev. Lett. **44** (1980) 772.
23) P.Van Isacker, J.Jolie, K.Heyde, A.Frank, Phys.Rev.Lett.**54** (1985) 653.
24) J.Jolie and H.Lehmann, subm. to Phys.Rev.Lett
25) H.Lehmann, Diplomarbeit, Uni Fribourg 1994, unpublished.
26) H.Lehmann, J.Jolie, to be publ.
27) M.Délèze et al. Nucl. Phys. **A551** (1993) 269.
28) J.Jolie, to be publ.
29) F.Iachello, D.Levine, Europhys. Lett. **4** (1987) 389.
30) N.Whelan, Y.Alhassid, Nucl. Phys. **A556** (1993) 42

THE SCISSORS MODE AND RELATED MODES
REVISITED IN THE IBM AND OTHER MODELS*

A. RICHTER

Institut für Kernphysik, Technische Hochschule Darmstadt
D-64289 Darmstadt, Germany

ABSTRACT

Recent developments in the field of magnetic dipole excitations in heavy deformed nuclei are discussed by using electromagnetic and hadronic probes of different selectivity. Particular emphasis is given on the physics of the so called scissors mode in terms of the Interacting Boson Model. Further topics are magnetic dipole sum rules, the dependence of orbital M1 strength on deformation and its relation to nuclear monopole properties, the M1 transitions as a probe of F–spin symmetry, the direct detection of spin M1 strength and the possible existence of a high lying scissors mode, i.e. the $K^\pi = 1^+$ component of the isovector giant quadrupole resonance.

1. Introduction

This talk is one of three consecutive talks at international meetings I have been asked to give on the same subject, i.e. the magnetic dipole response of nuclei. There is no point in reiterating what I tried to formulate as best as I could once and I sincerely hope the reader will understand that each of the three manuscripts is only a slightly modified version of the other. Furthermore, considering the wealth of experimental data presented in the actual talk the rather limited space allowed for its written version in these proceedings forces me to restrict myself essentially only to a summary of those points which I did discuss orally. Also, the list of references given at the end will necessarily not be complete, and I focus the attention of the reader to recent articles[1,2] which deal with the same topic.

To set the proper tone at a meeting on perspectives for the Interacting Boson Model (IBM) on the occasion of its 20th anniversary I remind the reader that almost exactly ten years ago an article appeared in the literature[3] with the title "New magnetic dipole excitation mode studied in the heavy deformed nucleus ^{156}Gd by inelastic electron scattering". This article has led truly to a renaissance of high resolution, low energy spectroscopy with electrons, photons and protons, to the development of novel theoretical ideas and to the improvement of already existing nuclear models. The large amount of scientific articles published since the discovery of the mode and the sizable number of articles still appearing (about 20 per year) is remarkable and signals that experimentalists and theoreticians are both fascinated by this elementary nuclear excitation and are driven to understand its very nature.

---

*Work supported by the German Federal Minister for Research and Technology (BMFT) under contract number 06DA641I.

The experimental search for it about a decade ago was driven by a theoretical predictiction by Lo Iudice and Palumbo in terms of the so called Two Rotor Model[4] (TRM) and by a fairly cute estimate by Iachello for its expected transition strength in the IBM, and the latter termed it[5] at first "Nuclear Wobble". Soon after, for the obvious out of phase movement of protons against neutrons, the mode has been called "Scissors Mode". A most natural framework for studying proton and neutron degrees of freedom in collective states of nuclei is the IBM–2 (ref. [6]). When including proton and neutron bosons explicitly, besides the symmetric combinations that turn out to be equivalent to the IBM–1 description of nuclear structure, nonsymmetric couplings give rise to a totally new family of states of mixed symmetry[5]. In even-even nuclei the scissors mode leads through small angle vibrations of protons against neutrons to the excitation of $J^\pi = 1^+$ states that are the best examples for mixed symmetry states known so far. The discovery of those states made thus the F–spin concept[6] really meaningful and allowed a unique determination of the strength of the Majorana force which is responsible for the splitting between states of F_{max} and $F_{max} - 1$.

2. Qualitative Nature of the Magnetic Dipole Response

Both the TRM and the IBM are clearly the simplest macroscopic and microscopic approaches, respectively, towards an understanding of the basic features of the orbital magnetic dipole mode, and a large number of experiments since its discovery have revealed ample information on its excitation energy, its fragmented transition strength, its form factor and on the relative importance of its orbital vs. spin content. For a proper description of all those features more refined theoretical descriptions in terms of the shell model, RPA and QRPA had to be developed (for recent references, see [1,2,7–11] and references therein).

What is the simplest approach towards the nature of the magnetic dipole response in nuclei? Let us briefly recall the structure of the M1 operator

$$T(M1) = \sum_i \left\{ g_l(i)\vec{l}_i + g_s(i)\vec{s}_i \right\} \mu_N \tag{1}$$

with the g's being the usual g–factors for neutrons and protons. After rewriting (1) as a sum of isoscalar and isovector pieces using $t_z(i) = \pm 1/2$ for protons and neutrons, respectively, and neglecting further the small isoscalar piece because the g's of proton and neutron are of about equal magnitude but opposite sign we end up with the following structure of the isovector M1 operator:

$$\begin{aligned} T(M1)_{IV} &= \left\{ \sum_i t_z(i)\vec{l}_i + (g_p - g_n)\sum_i t_z(i)\vec{s}_i \right\} \mu_N \\ &= \left\{ \frac{1}{2}(\vec{L}_p - \vec{L}_n) + 4.71\, T(M1)_{\Delta T_z = 0} \right\} \mu_N \end{aligned} \tag{2}$$

This equation yields already some insight[1]. The isovector strength splits into orbital and spin parts, the first involving $\vec{L}_p - \vec{L}_n$, which, viewed as rotation generator, immediately suggests the **scissors notion** in a qualitative way. The spin–flip piece is the $\Delta T_z = 0$ component of the Gamow–Teller operator which is enhanced because its coefficient is $(g_p - g_n)$.

Let us look next, very schematically, what happens in an RPA calculation of the excitation strength in a medium heavy or heavy even–even nucleus. The unperturbed particle–hole strengths are scattered from the ground state up to say 10 MeV in excitation energy and the orbital and spin–flip contributions are thoroughly mixed. By turning on the well known particle–hole interaction in the spin-isospin channel, the spin–flip piece of the excitation, carrying the majority of the total strength, is swept up to excitation energies of 10 MeV or higher. The **orbital strength**, however, **hardly moves at all**. It remains low-lying, is scissors-like and weakly collective, but its observability is a strong collective effect as a consequence of the fact that the $p - h$ force has moved the competing stronger spin–flip strength up to higher excitation energy.

The weakly collective M1 excitation now becomes an ideal test of microscopic models of nuclear vibrations. Shell models are usually calibrated to reproduce properties of strong collective excitations ($2^+, 3^-$, electric giant resonances). Weakly collective phenomena, however, force the models to make real predictions and the fact that the transitions in question are strong on the single–particle scale makes it impossible to dismiss failures as a mere detail. This should be kept in mind in an assessment of the wide variety of models which this new excitation mode has already inspired. The above discussion nowhere mentions deformation which is introduced alongside the discussion of the experimental data.

Returning to the IBM, the M1 operator of Eq. (1) in fermion space has its image in boson space

$$T^B(M1) = \{g_\pi L_\pi + g_\nu L_\nu\} \mu_N \tag{3}$$

with g_π and g_ν being the respective proton and neutron boson pair g–factors and L_π and L_ν the corresponding orbital angular momenta. These pair g–factors can be estimated from an analysis of g–factors of first excited 2^+–states[12]. The expected M1 strength in the SU(3) limit of IBA–2 (most nuclei to be discussed below are good rotors and are sufficiently well described in this limit) is given (in μ_N^2) by

$$B(M1)\uparrow = \frac{3}{4\pi} \frac{4N_\pi N_\nu}{N_\pi + N_\nu} (g_\pi - g_\nu)^2. \tag{4}$$

We note here for completeness that the transition strength of electric quadrupole and magnetic octupole transitions is of similar structure[13] as the expression for the M1 strength, i.e.

$$B(E2)\uparrow \sim \cdots (e_\pi \pm e_\nu)^2 \tag{5}$$
$$B(M3)\uparrow \sim \cdots (f_\pi \pm f_\nu)^2 \tag{6}$$

where the role of the g–factors in M1 transitions is now played by effective electric quadrupole and magnetic octupole boson charges, e and f, respectively, and the minus (plus) sign denotes transitions from the symmetric ground state into mixed symmetric (symmetric) boson model states. In Eq. (4) we have fortunately[12] $g_\pi \approx 1$ and $g_\nu \approx 0$ and M1 transitions into mixed symmetry states are therefore (with a strength depending upon the boson numbers N_π and N_ν) observable. For E2 transitions, however, e_π seems to be about equal to e_ν in the rotational limit[14], and the r.h.s. of Eq. (5) thus indicates immediately the inhibition of transitions into mixed symmetric 2^+ states, i.e. the 2^+ member of the $K = 1$ band. The situation is presently much less clear in the case of M3 transitions, and experimental searches for both E2 and M3 transitions in heavy nuclei into states of mixed symmetry should be undertaken.

That the simple picture of the nuclear magnetic dipole response is at least approximately correct is shown in Fig. 1 by using the three nuclei ^{56}Fe, ^{156}Gd and ^{238}U as examples. The mean excitation energy of the orbital mode is approximately

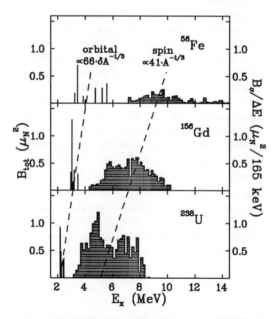

Fig. 1. The nuclear orbital and spin magnetic dipole response in a medium heavy, a heavy and a very heavy nucleus derived from experiments with electromagnetic and hadronic probes, respectively.

given by $E_x \simeq 66\delta A^{-1/3}$ MeV with δ being the ground state quadrupole deformation. The spin dipole strength recently found in inelastic proton scattering [15] lies at $E_x \simeq 41 A^{-1/3}$ MeV and thus exhibits an excitation energy dependence reminiscent of the shell model. As is seen in Fig. 1 the ratio of orbital to spin strength is indeed small, indicating that the spin strength is the really collective part of the total M1 strength.

3. Magnetic Dipole Response in Heavy Deformed Nuclei

3.1. Overview

I will now discuss recent advances in the study of the magnetic dipole response in heavy deformed nuclei. The salient features of the scissors mode unraveled in high-resolution electron and photon scattering experiments are the following:

- The center of gravity of the orbital M1 strength distribution lies in rare earth nuclei at $E_x \simeq 3$ MeV.
- The total strength is $\sum B(M1) \simeq 3\mu_N^2$ and the maximum strength that is carried in the transition to an individual state is roughly $1.5\,\mu_N^2$.
- In the nuclear transition current the orbital part dominates over the spin part and one has typically $B_l(M1)/B_\sigma(M1) \simeq 10/1$.
- The summed transition strength up to $E_x \simeq 4$ MeV is proportional to the quadrupole ground state deformation.

Extreme forward angle inelastic scattering experiments of protons at medium energy indicate the following with respect to the M1 spin-flip resonance:

- It is located at excitation energies $E_x \simeq 5 - 10$ MeV and is characterized mostly by a double-humped structure.
- The total transition strength is $\sum B(M1) \simeq 11\mu_N^2$.

3.2. Search for the Scissors Mode in Even-Odd Nuclei

As the situation with even-A nuclei is fairly well understood, the next question concerns odd-A nuclei for which only scarce data exist. The search for the scissors mode has so far been negative[16] in ^{165}Ho but clearly positive in ^{163}Dy as has been demonstrated beautifully by a Giessen/Cologne/Stuttgart collaboration[17] in a nuclear resonance fluorescence experiment. The measured ground state transition strengths with $\Delta K = 1$ in a chain of Dy isotopes is shown in Fig. 2. For ^{163}Dy it has not been possible to determine the spin of the excited state due to nearly identical angular correlation functions for three possible decay chains. Therefore the ground state decay widths have been multiplied with the appropriate spin weighing factors g. From the figure it is obvious, however, that the detected strength fits both, energetically and in its magnitude, into the systematics of the neighbouring even-even isotopes. Also, IBFM predictions[17] by Arias, Frank and Van Isacker are in essential agreement with the experimental observations in ^{163}Dy.

We have recently started a search for the scissors mode in ^{167}Er (see Fig. 3) in order to study the influence of the nucleon-core interaction on the transition strength and the fragmentation. Strong orbital M1 transitions are known in the neighboring nuclei ^{166}Er and ^{168}Er.

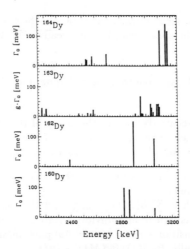

Fig. 2. Dipole strength distributions in four Dy isotopes. The strength is given in units of the g.s. decay width Γ_0 and for ^{163}Dy in form of the spin dependent decay width $g\Gamma_0$ (from [17]).

The combined results from nuclear resonance fluorescence spectra taken at the Superconducting **D**armstadt **E**lectron **Lin**ear **Ac**celerator (S–DALINAC) at bremsstrahlung endpoint energies $E_0 = 3.5, 4.6$ and 5.8 MeV are presented in Fig. 4. The mostly weak, but clearly identifiable transitions indicate a strong fragmentation of the measured strength. Under the (most likely very reasonable) assumption that the strength observed is indeed M1 strength the indicated three groups of transitions at $E_x \simeq 2.9, 3.5$ and 4.0 MeV yield a total strength of $\sum B(M1) \simeq 2.5\mu_N^2$. This value

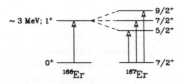

Fig. 3. Expected fragmentation of the scissors mode in ^{167}Er.

is comparable with the summed M1 strength observed in the neighboring nucleus ^{168}Er (upper part of Fig. 4) and also in ^{166}Er (not shown). However, the running sum of the strengths taken in the odd–A nucleus ^{167}Er saturates later than in the adjacent even–A nuclei indicating again a much larger fragmentation in the former nucleus.

In passing I note two possible improvements of the measurements of such weak transitions. Firstly, a Darmstadt/Cologne/Rossendorf collaboration will in the future at the low energy ($E_0 = 2.5 - 10$ MeV) photon scattering setup of the

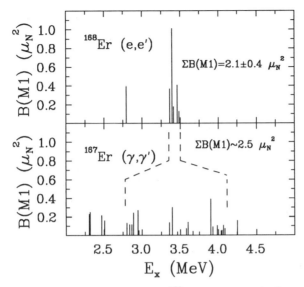

Fig. 4. Magnetic dipole strength distribution in ^{168}Er derived from (e,e') measurements at the DALINAC (upper part) and measured dipole strength in ^{167}Er (converted into M1 strength) from (γ,γ') measurements at the S-DALINAC (lower part).

S-DALINAC use an EUROBALL CLUSTER–Detector that has a high detection efficiency and the necessary background suppression needed to detect weak transitions. Secondly, a 180°–(e,e') high–resolution scattering facility which has just been installed at the S-DALINAC by a Catholic University of Washington/Darmstadt collaboration is essentially a "spin–filter" whereby the multipolarity of transverse excitations can be determined rather quickly. How well the spectrometer works on a light nucleus like ^{28}Si already is shown without any explanation in Fig. 5.

3.3. Phenomenology of Summed Orbital Magnetic Dipole Strengths

I now describe some advances that have recently been made in understanding the physical origin of the measured orbital M1 strengths. These advances are not great and they have to be seen, of course, against the background of a large body of work on the subject, reflecting ten years of developement in both experiment and theory. In the restricted space available to me, it is not possible to review this background, and to put what is to follow into its proper perspective.

The most important discovery made recently in the field of the scissors mode has come out from experiments[18] at the S-DALINAC done on a chain of the even–even ^{148,150,152,154}Sm isotopes. The orbital M1 strength varies quadratically with the deformation parameter δ. This result — also verified in corresponding experiments on a series of even–even Nd isotopes with varying deformation by a

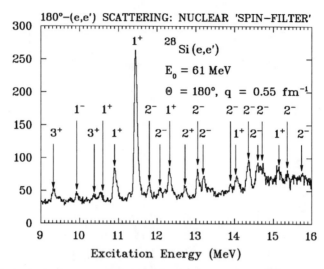

Fig. 5. First inelastic electron scattering spectrum taken at $\theta = 180°$ in ^{28}Si at the S-DALINAC. The spins and parities of all states shown can be determined straightforwardly by form factor measurements at two or three momentum transfers q.

Cologne/Giessen/Stuttgart collaboration[19] — has been anticipated in a systematic study[20] of M1 strength in the rare earth region within the Nilsson model where quantitatively a direct correlation between the quadrupole g.s. deformation and the orbital magnetic dipole strength was shown. Since the neutron–proton interaction is mainly responsible for the quadrupole deformation of the nuclear ground state the experimental observation is of great interest for the development of nuclear models of deformation.

Rather than comparing the original experimental data[18] with results from recent model calculations - as noted in[2] practically all models, some of them after appropriate modifications of earlier versions, yield a strength more or less quadratic in the deformation parameter — I present in Fig. 6 a correlation plot[7] of the summed orbital M1 strengths in the nuclei indicated vs. the corresponding E2 strengths between the ground state and the first 2^+ state of the ground state band. This striking manifestation of quadrupole collectivity in the magnetic dipole strength has also been looked at within the IBM-2 using a sum rule approach[7].

The strong M1/E2 correlation has first been discovered[21] when the respective M1 and E2 strengths were investigated in the frame of Casten's $N_p \cdot N_n$ scheme[22] in which a factor $P = N_p N_n/(N_p + N_n)$ with N_p and N_n being the number of valence protons and neutrons outside closed shells is considered. The factor P, a normalized form of $N_p \cdot N_n$, can be viewed as counting the average number of $p - n$ interactions compared to like nucleon interactions. A correlation persists between the $B(M1)$ and $B(E2)$ values for the entire region of $0 \leq P \leq 8$.

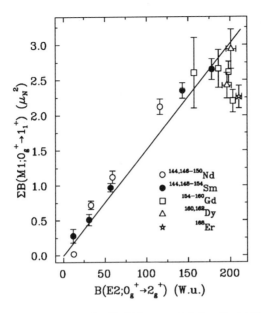

Fig. 6. Correlation plot of the summed orbital M1 strength below $E_x \leq 4$ MeV vs. the E2 strength from the g.s. to the first excited 2^+ state for the nuclei indicated in the figure. The solid line is a fit of the data points.

An updated version[23] of the original plot[21] of the summed M1 strength vs. the P-factor is shown in Fig. 7. The rapid increase of transition strength for $4 < P < 5$ is correlated with the onset of deformation. It can be viewed in terms of increasingly dominant quadrupole interaction strength over the pairing strength. Furthermore, there is a saturation of the strength for $P \geq 6$ and the physical origin of this phenomenon is not immediately obvious. A number of suggestions have already been made in the literature (see e.g. [24,25] and for a rather extensive treatment of quadrupole collectivity in M1 transitions[7] and other references cited therein).

In the context of the observed M1/E2 correlation it is interesting to note that the IBM-2 fails to describe it[21] since for large boson numbers N the respective transition strengths behave like $B(M1) \simeq g^2 P$ and $B(E2) \simeq e_B^2 N^2$.

The physics content of Fig. 7 can be displayed slightly different[26] if the summed orbital M1 strength is plotted vs. the mass number of the investigated nuclei (Fig. 8). The figure shows the increase of strength towards midshell, then a saturation of the strength at about $3\,\mu_N^2$ and a drop for the heaviest rare earth nuclei, i.e. the three W isotopes. It is, of course, conceivable that this drop is a real nuclear structure effect but another possibility might just be that the strength in those nuclei is so fragmented that part of it might have just escaped detection.

The discussed experimental observations of the correlation of M1 and E2 strength

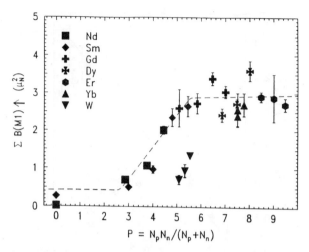

Fig. 7. Summed M1 strength vs. Casten's P–factor (this is an updated[23] version of a figure in [21]).

and the saturation of those strengths at midshell are very interesting. Although theoretical interpretations of these facts are still uncertain one could still use a simple phenomenological approach in form of a sum rule that is essentially model independent and parameter free[27]. This sum rule is based on an expression of the M1 strength which, though first derived within the TRM is valid in a general context of the following sum rule of Lipparini and Stringari[28]:

$$B(M1)\uparrow \simeq \frac{3}{16\pi} \Theta_{sc} \omega_{sc} (g_p - g_n)^2 \mu_N^2 \qquad (7)$$

Here ω_{sc} is the excitation energy of the scissors mode (called E_x before) and Θ_{sc} the mass parameter, which is very close to the moment of inertia. This latter quantity important in the isovector rotation can be estimated from the "classical" sum rule for E2 strength derived by Bohr and Mottelson long ago[29]. One arrives at an expression[27] for the summed strength

$$B(M1)\uparrow \simeq \left\{ 0.0042 \frac{4NZ}{A^2} \omega_{sc} A^{5/3} \delta^2 (g_p - g_n)^2 \right\} \mu_N^2 \qquad (8)$$

which is not only extremly simple and transparent but also contains the experimentally observed dependence of the M1 strength on the square of the deformation parameter.

As Fig. 9 from [26] shows convincingly that expression (8) works equally well for transitional and strongly deformed nuclei, and the ratio of the experimental over the calculated strength being unity for the majority of nuclei indicates that **all** the orbital magnetic dipole strength has been detected. This is certainly the second major discovery made recently in the field.

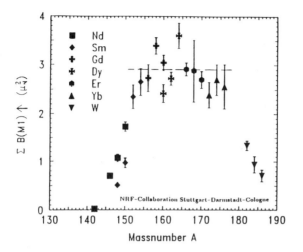

Fig. 8. Summed orbital M1 strengths in the nuclei indicated on the figure as a function of their mass number.

3.4. Some Implications of those Results

3.4.1. M1 strength and isotope shift

What can we learn with respect to nuclear monopole properties from the fact that the experimentally determined orbital M1 strength is related to the square of the quadrupole deformation of the ground state? Phenomenologically the nuclear isotope shift within a liquid drop model approach including a quadrupole shape deformation characterized by the usual deformation variable β is given by

$$\Delta \langle r^2 \rangle = \frac{4}{5} r_0^2 A^{-1/3} + \frac{3}{4\pi} r_0^2 A^{2/3} \Delta \langle \beta^2 \rangle, \qquad (9)$$

and since (as shown **experimentally**) $\sum B(M1) \sim \langle \beta^2 \rangle$, Iachello[30] and Otsuka[31] argue that nuclear monopole properties are also related to the summed orbital M1 strength. This is purely empirical though, and from such a relation variations of the nuclear radius can follow (or the other way around).

On the contrary, in a true IBM–2 approach[32] we can start from the non–energy weighted M1 sum rule of Ginocchio[25]

$$\sum B(M1) = \frac{9}{4\pi} (g_\pi - g_\nu)^2 \frac{P}{N-1} \langle 0_1^+ | \hat{n}_d | 0_1^+ \rangle \qquad (10)$$

which relates the orbital M1 strength precisely to the d–boson expectation value in the nuclear g.s. using the concept of F–spin symmetry and in particular pure F–spin for the ground state. Since the E0 operator in the IBM is[30,33]

$$T(E0) = \gamma_0 \hat{n}_s + \beta_0 \hat{n}_d = \gamma_0 \hat{N} + \beta_0' \hat{n}_d \qquad (11)$$

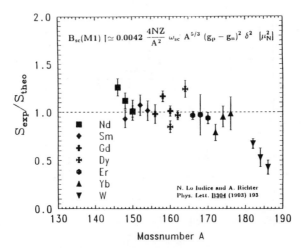

Fig. 9. Ratio of the summed experimental M1 strength and the one calculated from the expression of Eq. (8), listed also on top of the figure. For the excitation energy of the scissors mode a value of 3 MeV was taken and for the gyromagnetic factors $g_p = 2\,Z/A$ and $g_n = 0$.

and since
$$\Delta\langle r^2 \rangle = \gamma_0 + \beta'_0 \Delta\langle \hat{n}_d \rangle \tag{12}$$
we can rewrite the d–boson expectation value in Eq. (10) and obtain
$$\sum B(M1) = \frac{9}{4\pi}(g_\pi - g_\nu)^2 \frac{1}{\beta'_0} \frac{P}{N-1} \left[\langle r^2 \rangle - \gamma_0 N\right]. \tag{13}$$

So, while the phenomenological approach[30,31] argues that the summed M1 strength is related to deformation and thus to the nuclear radius do we[32] start from a rather precise relation between the summed M1 strength and a particular measure of deformation, i.e. the expectation value of d–bosons in the nuclear ground state. Since in actual cases the $\gamma_0 N$ term is rather small we expect a linear relation between $\sum B(M1)$ and $\langle r^2 \rangle$ with the factor $P/(N-1)$ determining the slope.

In Fig. 10 the linear relation between the summed magnetic dipole strength and the isotope shift is indeed observed for the particular nuclei indicated in the figure. It appears that an almost constant β'_0 "deformation" strength results for the particular region of rare earth nuclei considered ($86 < N < 96$). The decrease in slope when approaching the $N = 82$ closed shell spherical nuclei signals probably the breakdown of the simple expression in Eqs. (10) and (13).

The above arguments can be turned around. Large isotope shifts (as e.g. observed in the region of neutron deficient Au nuclei) should imply large M1 strengths. Furthermore, since $\sum B(M1) \sim \beta^2$ large M1 strengths might be observed in superdeformed nuclei.

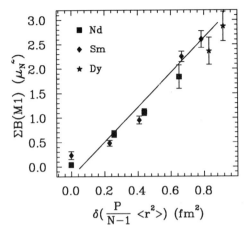

Fig. 10. Relation between the experimental orbital M1 strength and the quantity $\delta(P/(N-1)\langle r^2 \rangle)$ which is related to the isotope shift $\Delta\langle r^2\rangle$. Values for the radii are taken from Otten[34].

3.4.2. M1 transitions as a probe of F-spin symmetry

The investigation of the F-spin purity of the nuclear wave functions, i.e. the existence of F-spin symmetry in nuclei, is strongly connected to the investigation of M1 transitions. Due to the fact that in sd–boson space the M1 transition operator in Eq. (4) has necessarily F-vectorial character, M1 transitions from a presumable symmetric $F = F_{max}$ g.s. lead to a $F = F_{max} - 1$ mixed symmetry excited state (Fig. 11). How big is the amount of F-spin in nuclear wave functions? Various

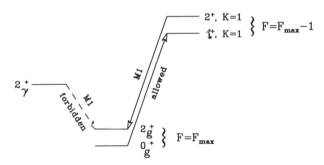

Fig. 11. Schematic illustration of allowed and forbidden M1 transitions within the F-spin scheme of the IBM.

numerical investigations in the past have yielded F-spin admixture probabilities ranging from 2% – 28% (see [35] for a summary).

With the very extensive knowledge on the scissors mode that is now available we have decided to look again at the relation between F-spin breaking components in the Hamilton operator, the strength of F-spin forbidden M1 transitions within rotational bands and the excitation energy and transition strength of the scissors mode[35].
Briefly, starting from an IBM-2 Hamiltonian

$$H = H^{(0)} + H_1 + M \qquad (14)$$

containing F-tensor terms and the Majorana term, those M1 transition matrix elements, which would vanish if H_1 in Eq. (14) is zero, are calculated in 1st order perturbation theory and compared to **experimental values**. In contrast to other approaches to the problem[36] we are here not interested in the origin of the F-spin symmetry breaking but primarily in the magnitude of the perturbing matrix element.
Since the F-spin forbidden and allowed matrix elements are related as follows

$$\langle 2_g^+ \| T(M1) \| 2_\gamma^+ \rangle = \Delta\alpha \langle 1^+, K = 1 \| T(M1) \| 0_g^+ \rangle \qquad (15)$$

the perturbing F-spin matrix element $\langle |H_1^2| \rangle^{1/2}$ is determined by the impurity coefficient

$$\Delta\alpha = \frac{\langle |H_1^2| \rangle^{1/2}}{\Delta E}. \qquad (16)$$

We see from Eqs. (15) and (16) that because the problem is treated in 1st order perturbation theory, **the transition strength and the excitation energy of the scissors mode play the role of a major scale** and replace any parameterization of the hamiltonian and the M1 transition operator, especially the use of effective g-factors.

The results for the four good rotational nuclei ^{158}Gd, ^{162}Dy, ^{164}Dy and ^{168}Er are that the F-spin impurity coefficients $\Delta\alpha$ for members of the ground and γ bands are of order $10^{-3} - 10^{-4}$. The parameter free analysis performed thus indicates an extremly high F-spin purity of the low lying states involved. The derived values for $\Delta\alpha$ are also much smaller than other estimates (for the most recent one, see [37]).

Possible reasons for this difference might be our restriction to the SU(3) limit, the statistical ansatz for the perturbing matrix elements and the level adjusting due to the strong Majorana force in numerical treatments of Eq. (14). To be on the safe side the estimated F-spin impurities should thus be considered as lower limits.

What have we learned from this exercise for the interacting boson model? In comparing the F-spin breaking matrix elements with the strong F-spin restoring matrix element of the Majorana operator and taking into account the smallness of the derived impurities $\Delta\alpha$ to describe the F-spin forbidden M1 transitions we can conclude that the IBM with sd-bosons and good F-spin is a well based description scheme for rotational nuclei.

3.5. Some Remarks on Summed Spin Magnetic Dipole Strengths

We have also in recent years started a comprehensive study of the spin magnetic dipole strengths in medium heavy and heavy deformed nuclei using both unpolarized and polarized protons. There is again no space here for extensive remarks on that topic and I will only list some basic features relevant to the scope of this talk.

One of the key nuclei we have looked at with unpolarized[15] and polarized[38] proton scattering at extreme forwared angles and medium energies in a Darmstadt/Münster/TRIUMF collaboration is ^{154}Sm. As an example, the upper part of Fig. 12 shows the measured cross section for excitation energies $E_x \simeq 4 - 32$ MeV, the lower part the corresponding transverse spin–flip probability S_{nn}. The relative maximum in S_{nn} near 8 MeV confirms the spin–flip character of the structures located at the low energy tail of the giant dipole resonance. From the angular

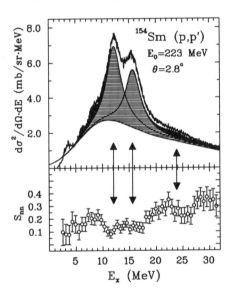

Fig. 12. Differential cross section and transverse spin–flip probability for inelastic polarized proton scattering on ^{154}Sm. The hatched areas show the double humped GDR. Visible on the low energy side of the GDR is the spin–flip M1 resonance between $E_x \simeq 5 - 12$ MeV and at $E_x = 23.4$ MeV the IVGQR. The arrows visualize the connection between the electric resonances and dips in the spin–flip probability.

distribution of the cross section the actual extracted spin–flip strength amounts to

$$B_\sigma(M1) = 10.5 \pm 2.0 \, \mu_N^2 \, . \qquad (17)$$

A comparison of the experimental strength distribution with various theoretical predictions[39,40,41] is given in Fig. 13. The theoretical distributions have been folded

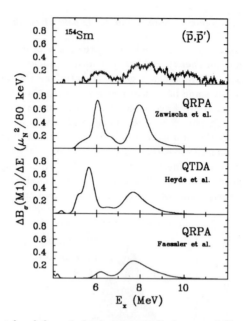

Fig. 13. Experimental and theoretical spin magnetic dipole strength distributions in ^{154}Sm.

with a Gaussian of variable width in order to facilitate the comparison. They were already calculated at a time when the experiment[15] revealed a double humped strength distribution. Now we find even more strength resting in a third bump which no calculation shows. From the comparison one can thus conclude that at present the agreement between the measured and the calculated distributions is still on a qualitative level.

3.6. Orbital Magnetic Dipole Mode and the Isovector Giant Quadrupole Resonance

As can be seen in Fig. 12 we also identified the isovector giant quadrupole resonance (IVGQR), about which our knowlegde is still rather scarce as compared to other giant resonances of low multipolarity, at $E_x = 23.4 \pm 0.6$ MeV in ^{154}Sm. It exhausts $76 \pm 11\%$ of the isovector E2 sum rule.

In the context of the still unsolved problems concerning the magnetic dipole modes in heavy deformed nuclei the deduced IVQGR properties play an important role. Firstly, it has been argued[42,43] that the $K^\pi = 1^+$ component of this resonance is the real manifestation of the classical scissors mode, for which within a RPA approach an E2 ground state transition strength of 1380 e^2fm^4 has been estimated[42]. Considering that the contribution of the $K^\pi = 1^+$ component to the IVQGR is expected to be 40% of the total strength[44] we deduce 1050 ± 180 e^2fm^4 from our experiment and hence a value in fairly good agreement with the theoretical prediction. Secondly, the deduced IVGQR strength nearly completes the determined E2 strength in terms of sum rules. We are therefore able to test the completeness of the experimentally observed orbital M1 strength I spoke about in sect. 3.3 further by applying a recently formulated new energy weighted sum rule[45] by Moya de Guerra and Zamick to the data which connects the orbital M1 strength to the difference of the total isoscalar and isovector E2 strength.

$$\sum B(M1)\uparrow = \frac{9}{16\pi}\chi \left\{ \sum_{isoscalar} B(E2)\uparrow - \sum_{isovector} B(E2)\uparrow \right\} \text{MeV}\mu_N^2 \qquad (18)$$

where χ ist the strength of the quadrupole–quadrupole interaction and $B(E2)$ values are in units of $e^2\text{fm}^4$ (for further details, see [45]). From the experimental data for ^{154}Sm we get for the l.h.s. of Eq. (18) 7.71 ± 0.44 μ_N^2MeV and for the r.h.s. 9.32 ± 0.31 μ_N^2MeV, i.e. we have detected below $E_x \leq 4$ MeV over 80% orbital M1 strength of the sum rule limit. This result is fully compatible with the one displayed in Fig. 9.

4. Summary

I have discussed an important example for an elementary excitation mode that has been and still is a major field of nuclear structure research at low energy. As I have already pointed out in [46], contrary to the well known E1 response of stable nuclei, the M1 response in a total of 12 open–shell nuclei — it is shown for six of them in Fig. 14 — could only in recent years be measured through a combination of different high–resolution probes, i.e. inelastic electron, photon and proton scattering. Remember that it took about thirty years after its discovery to understand the E1 giant resonance and it is no surprise that the data of Fig. 14 present a great challenge to nuclear structure theories.

The weakly collective **orbital** strength at around 3 MeV of excitation energy is called scissors mode strength made up from small angle vibrations of neutrons vs. protons. It is dependent on the square of the quadrupole deformation of the ground state and also strongly correlated to the E2 strength between the ground state and the first excited 2^+ state. Those facts are still not fully understood and the nature of the quadrupole–quadrupole force driving the nucleus to deformation (or alternatively the different neutron and proton deformation which might in the future be studied if pion beams of high resolution became available) needs to be clarified. Note that the $J^\pi = 1^+$ states excited from the ground state of open–shell nuclei through the scissors mode constitute still the most clear–cut examples for the mixed symmetry states of the IBM–2 and further searches for those states should be performed.

As is demonstrated beautifully in Fig. 14 the **spin** strength is removed from the orbital strength by the repulsive spin–isospin force up to a mean excitation energy of about 7.5 MeV. It is independent of deformation, strongly collective and it shows

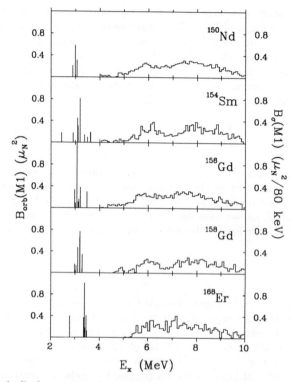

Fig. 14. Magnetic dipole response of several deformed rare earth nuclei determined by inelastic electron, photon and proton scattering.

the phenomenon of quenching, which is still with us since about two decades and only partly explained. The spin strength furthermore is double humped and shows the "isoscalar–isovector" excitation picture reminiscent of the famous M1 strength problem in ^{208}Pb many people have been looking at for years. For a long time the question "Where has all the M1 strength gone?" has been with us. As Fig. 14 shows impressively: It is there, but quenched.

5. Acknowledgements

I am grateful to my collaborators at the S–DALINAC and elsewhere for discussions and their many valuable contributions to the material of this talk. Special thanks are due to P. von Brentano, D. Frekers, K. Heyde, U. Kneissl, N. Lo Iudice, C. Lüttge, P. von Neumann-Cosel, C. Rangacharyulu, H. Wörtche and A. Zilges for sharing their insight into magnetic excitations, and last but not least to Friedrich Neumeyer for his great help and skill during the preparation of this manuscript.

6. References

1. A. Richter, *Nucl. Phys.* **A522** (1991) 139c.
2. A. Richter, in *The Building Blocks of Nuclear Structure*, ed. A. Covello (World Scientific, Singapore, 1993) p. 335.
3. D. Bohle, A. Richter, W. Steffen, A.E.L. Dieperink, N. Lo Iudice, F. Palumbo and O. Scholten, *Phys. Lett* **137B** (1984) 27.
4. N. Lo Iudice and F. Palumbo, *Phys. Rev. Lett.* **41** (1978) 1532.
5. F. Iachello, *Phys. Rev. Lett.* **53** (1984) 1427.
6. A. Arima, T. Otsuka, F. Iachello and I. Talmi, *Phys. Lett.* **66** (1977) 205.
7. K. Heyde, C. De Coster, A. Richter and H.-J. Wörtche, *Nucl. Phys.* **A549** (1992) 103.
8. R. Nojarov, A. Faessler, P. Sarriguren, E. Moya de Guerra and M. Grigorescu, *Nucl. Phys.* **A563** (1993) 349.
9. R. Nojarov, *Nucl. Phys.* **A571** (1994) 93.
10. A.A. Raduta, I.I. Ursu and N. Lo Iudice, *Nucl. Phys.* **A551** (1993) 73.
11. D. Zawischa and J. Speth, *Nucl. Phys.* **A569** (1994) 343c.
12. M. Sambataro, O. Scholten, A.E.L. Dieperink and G. Piccitto, *Nucl. Phys.* **A413** (1984) 333.
13. O. Scholten, A.E.L. Dieperink, K. Heyde and P. van Isacker, *Phys. Lett.* **149B** (1984) 279.
14. D. Bohle, A. Richter, K. Heyde, P. van Isacker, J. Moreau and A. Sevrin, *Phys. Rev. Lett.* **55** (1985) 1661.
15. D. Frekers, H.J. Wörtche, A. Richter, R. Abegg, R.E. Azuma, A. Celler, C. Chan, T.E. Drake, R. Helmer, K.P. Jackson, J.D. King, C.A. Miller, R. Schubank, M.C. Vetterli and S. Yen, *Phys. Lett.* **B244** (1990) 178.
16. N. Huxel, W. Ahner, H. Diesener, P. von Neumann-Cosel, C. Rangacharyulu, A. Richter, C. Spieler, W. Ziegler, C. de Coster and K. Heyde, *Nucl. Phys.* **A539** (1992) 478.
17. I. Bauske, J.M. Arias, P. von Brentano, A. Frank, H. Friedrichs, R.D. Heil, R.-D. Herzberg, F. Hoyler, P. Van Isacker, U. Kneissl, J. Margraf, H.H. Pitz, C. Wesselborg and A. Zilges, *Phys. Rev. Lett.* **71** (1993) 975.
18. W. Ziegler, C. Rangacharyulu, A. Richter, and C. Spieler, *Phys. Rev. Lett.* **65** (1990) 2515.
19. J. Margraf, R.D. Heil, U. Maier, U. Kneissl, H.H. Pitz, H. Friedrichs, S. Lindenstruth, B. Schlitt, C. Wesselborg, P. von Brentano, R.-D. Herzberg and A. Zilges, *Phys. Rev.* **C47** (1993) 1474.
20. C. De Coster and K. Heyde, *Phys. Rev. Lett.* **63** (1989) 2797.
21. C. Rangacharyulu, A. Richter, H.J. Wörtche, W. Ziegler and R.F. Casten, *Phys. Rev.* **C44** (1991) R949.
22. R.F. Casten, D.S. Brenner and P.E. Haustein, *Phys. Rev. Lett.* **58** (1987) 658.
23. P. von Brentano, A. Zilges, U. Kneissl and H.H. Pitz, preprint, Universität zu Köln, (Nov. 1993).

24. L. Zamick and D.C. Zheng, *Phys. Rev.* **C44** (1991) 2522.
25. J.N. Ginocchio, *Phys. Lett.* **B265** (1991) 6.
26. P. von Brentano, A. Zilges, R.-D. Herzberg, U. Kneissl, J. Margraf and H.H. Pitz, *Nucl. Phys.* **A**, in press.
27. N. Lo Iudice and A. Richter, *Phys. Lett.* **B304** (1993) 193.
28. E. Lipparini and S. Stringari, *Phys. Lett.* **B130** (1983) 139.
29. A. Bohr and B.R. Mottelson, *Nucleare Structure. Vol.II* (Benjamin, New York, 1975) ch. 6.
30. F. Iachello, *Nucl. Phys.* **A358** (1981) 89c.
31. T. Otsuka, *Hyperfine Ints.* **74** (1992) 93.
32. K. Heyde, C. De Coster, D. Ooms and A. Richter, *Phys. Lett.* **B312** (1993) 267.
33. F. Iachello and A. Arima, *The Interacting Boson Model* (Cambridge University Press, New York, 1987)
34. E. Otten, in *Treatise on Heavy-Ion Science, Vol.8*, ed. D.A. Bromley (Plenum, New York, 1989) p. 517.
35. O. Engel, A. Richter and H.J. Wörtche, *Nucl. Phys.* **A565** (1993) 596.
36. A.E.L. Dieperink, O. Scholten and D.D. Warner, *Nucl. Phys.* **A469** (1987) 173.
37. A. Wolf, O. Scholten and R.F. Casten, *Phys. Lett.* **B312** (1993) 372.
38. H.J. Wörtche, C. Rangacharyulu, A. Richter, D. Frekers, O. Häusser, R.S. Henderson, C.A. Miller, A. Trudel, M.C. Vetterli and S. Yen, to be published.
39. D. Zawischa and J. Speth, *Phys. Lett.* **B252** (1990) 4.
40. C. De Coster and K. Heyde, *Phys. Rev. Lett.* **66** (1991) 2456.
41. P. Sarriguren, E. Moya de Guerra, R. Nojarov and A. Faessler, *J. Phys.* **G19** (1993) 291.
42. N. Lo Iudice and A. Richter, *Phys. Lett.* **B228** (1989) 291.
43. D. Zawischa and J. Speth, *Z. Phys.* **A339** (1991) 97.
44. D. Zawischa, J. Speth and D. Pal, *Nucl. Phys.* **A311** (1978) 445.
45. E. Moya de Guerra and L. Zamick, *Phys. Rev.* **C47** (1993) 2604.
46. A. Richter, *Nucl. Phys.* **A553** (1993) 417c.

FROM SEMICLASSICAL TO MICROSCOPIC DESCRIPTIONS OF SCISSORS MODES

N. LO IUDICE

Dipartimento di Scienze Fisiche, Università Federico II di Napoli
and Istituto Nazionale di Fisica Nucleare, Sezione di Napoli
Mostra d'Oltremare Pad.19, Napoli, Italy

ABSTRACT

A quantum mechanical definition inspired by the semiclassical two-rotor model is proposed for the M1 strength of the scissors mode. Such a definition is consistent with the deformation properties of the mode and is effectively adopted in most phenomenological and schematic models. This is illustrated here for the interacting boson model and schematic RPA. Mention is made of semiclassical and RPA predictions of additional scissors like excitations in deformed and superdeformed nuclei and of a RPA calculation formulated in the laboratory frame.

1. Introduction

Since their first discovery in ^{154}Gd by Richter and coworkers through (e,e') experiments[1], the low lying $M1$ excitations observed in deformed nuclei[2] have been ascribed to a scissors like rotational oscillation among deformed proton and neutron fluids as suggested by the semiclassical two-rotor model (TRM)[3] which together with the algebraic proton-neutron interacting boson model (IBM2)[4] inspired the experiment. A clear maybe conclusive, evidence in favour of their scissors nature is offered by the deformation properties recently discovered in Sm isotopes[5,6] and later confirmed in Nd nuclei[7]. This is indeed the conclusion of most model descriptions[8-18], with one exception[19]. We mention here the studies performed in IBM2[11-14] and RPA[9,18] or TDA[12].

Of considerable interest for our purposes is a phenomenological analysis based on the use of the M1 strength in the form borrowed from the TRM[17]. Such a form will be shown to lead to a quantum definition of the scissors summed M1 strength of general validity which is consistent with IBM2 and schematic RPA. IBM2 proved to be quite successful in accounting for the main properties of the mode, namely the magnitude of the M1 strength and the form factor. RPA is widely used for studying the mode in its detailed properties. In its schematic form, RPA does not describe adequately the deformation properties. It has however the merit of pointing out the possible existence of an additional scissors like excitation in the region of the isovector giant resonance[20] and of strongly collective modes of the same nature in superdeformed nuclei[21]. We will see that exactly the same predictions can be made within the semiclassical approach whose complete equivalence with schematic RPA is explicitly shown. For a detailed description of the mode realistic RPA calculations are needed. These present however a series of problems[20]. A notable one is posed by the occurrence of spurious rotational admixtures[22], whose removal requires the development of apposite techniques[22] or the use of a selfconsistent basis[23]. A RPA calculation carried out in the laboratory frame as a way of solving some of the intrinsic RPA problems has been proposed recently[24] and will be illustrated briefly here.

2. The scissors M1 strength

The quantum mechanical definition of the $M1$ scissors strength given here is inspired by the TRM. In this model protons and neutrons form two axially symmetric rotors with mass parameters $\Im_p$ and $\Im_n$ and angular momenta $\vec{J}_p$ and $\vec{J}_n$ respectively, whose relative rotational oscillations are described by the angle 2ϑ between their symmetry axes. Such a motion is determined by a two dimensional harmonic oscillator Hamiltonian

$$H = \frac{1}{2\Im_{sc}}(S_1^2 + S_2^2) + \frac{1}{2}C\vartheta^2 , \qquad \Im_{sc} = \frac{4\Im_p\Im_n}{\Im_p + \Im_n} , \qquad (1)$$

where ϑ plays the role of a radial variable ($\vartheta^2 = \vartheta_1^2 + \vartheta_2^2$) and $S_i = J_i^{(p)} - J_i^{(n)} = id/(d\vartheta_i)$ are the variables conjugate to ϑ_i. The mass parameter $\Im_{sc}$ and the restoring force constant C determine the properties of the mode. This is the first $K^\pi = 1^+$ excited state with an excitation energy $\omega = \sqrt{C/\Im_{sc}}$. It is excited by the S_μ component of the magnetic dipole operator with a strength ($g_r = g_p - g_n$)

$$B(M1)\uparrow = \frac{3}{16\pi}\sum_{\mu=\pm 1}|<\mu|S_\mu|0>|^2 \, g_r^2 \, \mu_N^2 \simeq \frac{3}{16\pi}\Im_{sc}\,\omega g_r^2 \, \mu_N^2 . \qquad (2)$$

The above expression is easily obtained by exploiting the harmonic relations

$$\Im_{sc} = \frac{1}{\omega}<0|S^2|0> = \frac{1}{\omega}\sum_\mu |<\mu|S_\mu|0>|^2 , \qquad C = \Im\omega^2 = \omega\sum_\mu |<\mu|S_\mu|0>|^2 . \qquad (3)$$

These can be written in the more general form

$$\Im_{sc} = \sum_\mu <0|S_\mu^\dagger \frac{1}{H-E_0} S_\mu|0> , \qquad C = \frac{1}{2}\sum_\mu <0|\left[S_\mu^\dagger,[H,S_\mu]\right]|0> , \qquad (4)$$

The above relations, first derived in a sum rule approach[25], are valid for any Hamiltonian and an operator $S_\mu = J_\mu^{(p)} - J_\mu^{(n)}$ of general form. It is now a simple matter to derive the following scissors sum rule

$$\sum_n \omega_n B_n(M1)\uparrow = \frac{3}{16\pi}\sum_{n,\mu}\omega_n |<n\mu|S_\mu|0>|^2 \, g_r^2 \, \mu_N^2$$

$$= \frac{3}{32\pi}\sum_\mu <0|\left[S_\mu^\dagger,[H,S_\mu]\right]|0> \simeq \frac{3}{16\pi}\Im\omega^2 g_r^2 \, \mu_N^2 . \qquad (5)$$

This is generally valid as long as the M1 transition is promoted by its component in S_μ. Under the (experimentally supported) assumption of small fragmentation of the mode, it yields a summed M1 strength given by eq.(2) with $\Im_{sc}$ and C defined by eqs.(4). Eq.(2) can therefore be assumed to define the scissors M1 summed strength.

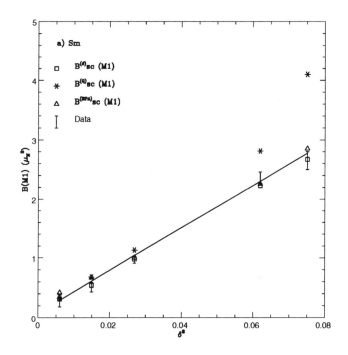

Figure 1. Summed M1 strength versus δ^2 in Sm isotopes. The line connects the experimental points. The data are taken from Ref.5.

This definition has been shown to be consistent with the deformation properties of the measured $M1$ summed strength[17]. We have extracted the mass parameter from the E2 classical energy weighted sum rule[26] and inserted into eq.(2) with the result

$$B(M1)\uparrow \simeq 0.0065 \frac{A^{1/3}}{\chi_D Z^2} \omega B(E2)\uparrow \frac{\mu_N^2}{e^2 fm^4}(g_p - g_n)^2. \qquad (6)$$

where $\chi_B = B_{irr}/B_{rot} \simeq 1/5$ is the ratio between the irrotational and rotational mass parameters. Making use of the standard expression $B(E2)\uparrow = 5/(16\pi)\ Q_0^2 = (1/(5\pi)\ Z^2 R^4 \delta^2 e^2$, with $R = 1.2 A^{1/3}\ fm$, we obtain the quadratic deformation law

$$B_{sc}(M1)\uparrow \simeq 0.004\ \omega A^{5/3}\ \delta^2\ g_r^2 \mu_N^2, \qquad (7)$$

The above equations show the strict link between the δ^2 law[5,7] and the $M1 - E2$ relation in agreement with the alike saturation properties observed for the two strengths[6].

Numerical calculations carried out with $g_n = 0$ and $g_r = g_p = 2g_R = (2Z)/A$ yield results in good agreement with experiments[5], if we use the deformation parameter adopted in Ref.5 ($B_{sc}^{(\delta)}(M1)$ in fig.1). The strength is instead increasingly overestimated as the deformation increases if the parameter is extracted directly from the experimental E2 strength ($B_{sc}^{(Q)}(M1)$ in figure). The two M1 strengths differ by terms of higher order in δ.

3. Scissors strength in IBM2

The connection with model descriptions other than the TRM can be more easily established once the variable θ is replaced with the shape variables α_μ by means of the relation[27] $\alpha_{21} = \alpha_{2-1} = -i\sqrt{3/2}\beta\vartheta$. The TRM Hamiltonian (eq.1) becomes now harmonic in α with new constant B and C_α related to the old ones by $\Im_{sc} = 3\beta^2 B$ and $C = 3\beta^2 C_\alpha$.

It has been shown already[28] that in its classical limit, IBM2 yields the TRM Hamiltonian. It is on the other hand straightforward to express the scissors strength (eq.2) in terms of valence proton and neutron pairs N_π and N_ν as in IBM2 by putting $\Im_p \simeq N_\pi/N\ \Im$ and $\Im_n \simeq N_\nu/N\ \Im$, with $N = N_\pi + N_\nu$, in the mass parameter $\Im_{sc}$ (eq.1)

$$B(M1)\uparrow \simeq \frac{3}{16\pi}\Im_{sc}\,\omega g_r^2\,\mu_N^2 \simeq \frac{3}{16\pi}\frac{4N_\pi N_\nu}{N^2}\Im\,\omega g_r^2\,\mu_N^2 \qquad (8)$$

We will show now at least in a heuristic fashion, that this is the form of the IBM2 M1 strength in the classical limit (large N). To this purpose we start with the quite general IBM expression of the total M1 strength[11,14]

$$B(M1)\uparrow = \frac{3}{16\pi}<0|S^2|0>g_r^2\mu_N^2 \simeq \frac{9}{8\pi}\frac{4N_\pi N_\nu}{N(N-1)}<N_d>g_r^2\mu_N^2\ , \qquad (9)$$

where $<N_d>$ is the ground state average of the quadrupole boson number operator $N_d = N_d^{(\pi)} + N_d^{(\nu)} = d_\pi^\dagger \cdot d_\pi + d_\nu^\dagger \cdot d_\nu$. Being this a scalar, its mean value can be computed in the intrinsic frame, where for axial symmetric systems only the $\mu = 0$ component survives. Using as intrinsic ground state a coherent state ψ_c

$$d_{\tau,0}\psi_c = d'_\tau \psi_c \qquad\qquad (\tau = \pi,\nu) \qquad (10)$$

with d'_τ pure numbers, we have

$$\beta_\tau = <\alpha_\tau> = 2\alpha_\tau^{(0)}d'_\tau\ ,\qquad \alpha_\tau^{(0)} = \sqrt{\frac{1}{2B_\tau\omega}}\ . \qquad (11)$$

Assuming equal deformation for protons and neutrons ($\beta_p = \beta_n = \beta$) we get

$$3<N_d> = \frac{3\beta^2}{4\alpha_0^2} = \frac{3}{2}B\beta^2\omega = \frac{1}{2}\Im\omega\ ,\quad \Im = 3\beta^2(B_p + B_n) = 3\beta^2 B\ . \qquad (12)$$

This inserted into eq.9 yields the scissors M1 strength given by eq.8 if we put $N - 1 \simeq N$, which is justified in the classical limit. In this new form, the IBM-2 M1 strength displays a δ^2 dependence. The strength so derived can be computed by the phenomenological procedure already outlined in sect.2, with good results if we use for the gyromagnetic factor the value $g_r \simeq 2g_R$ (and not $g_\pi = 1$ as erroneously stated in Ref.27).

4. Semiclassical approach and schematic RPA

4.1 Low and high energy scissors modes in deformed and superdeformed nuclei

A complete equivalence between the semiclassical approach and schematic RPA can be stated once the TRM Hamiltonian is expressed in terms of the shape variables, by imposing that the zero-point amplitude of α_μ is equal to that of the quadrupole field[26]. We can therefore obtain the schematic RPA results directly from the defining eqs.3 using an anisotropic harmonic oscillator basis with frequencies ω_1 and ω_2 obeying the volume preserving condition $\omega_1^2 \omega_3 = \omega_0^3$ ($\omega_0 = 41 A^{-1/3}$). We may choose for the unperturbed scissors energy $\omega_{sc}^0 \simeq 2E = \sqrt{(\delta\omega_0)^2 + (2\Delta)^2}$, obtaining for the mass parameter a superfluid moment of inertia $\Im_{sc} \simeq \Im_{sf} \simeq (\delta\omega_0/2E)^3 \Im_{rig}$, where $\Im_{rig} = 2/(\delta\omega_0) \sum_{ph \in \delta\omega_0} |(S_1)_{ph}|^2 \simeq 2/5 \, mAR^2$. The unperturbed restoring force constant follows from the relation $C_0 \simeq (2E)^2 \Im_{sc}$, while the potential component is fixed from the ratio $b = C_1/C_0 = -V_1/(4V_0) \simeq 0.5$ between the static isovector and isoscalar potentials. Energy and M1 strength result to be

$$\omega_- \simeq 1.26(2\Delta)\sqrt{1+x^2}, \quad B_-(M1) \uparrow \simeq 0.001(2\Delta) A^{5/3} \frac{x^3}{1+x^2} g_r^2 \mu_N^2, \quad (13)$$

where $x = \delta\omega_0/(2\Delta)$. The strength goes like δ^3 for small deformations ($x \ll 1$) and becomes linear for very large $\delta's$ ($x \gg 1$). In the range of deformations observed in Sm isotopes the strength increases almost linearly with δ^2 but with a slope which leads to values twice as large as the experimental data in ^{154}Sm. A quenching factor is needed. This can be effectively obtained only through realistic RPA[9,18] or TDA[12] calculations, which account for spin admixtures.

We may alternatively put in eq.3 $\omega_{sc}^0 = 2\omega_0$ obtaining an irrotational mass parameter $\Im_{sc} = \Im_{irr} = \delta^2 \Im_{rig}$ and an unperturbed restoring force constant $C_0 \simeq (2\omega_0)^2 \Im_{irr}$. The potential component of C is fixed from the symmetry energy potential which yields $b = C_1/C_0 \simeq 1.9$. The final result is

$$\omega_+ = \sqrt{\frac{C}{\Im_{irr}}} \simeq 139 A^{-1/3} \, MeV, \quad B_+(M1) \uparrow \simeq \frac{3}{16\pi} \omega_+ \Im_{irr} \simeq 0.12 \delta^2 A^{4/3} g_r^2 \mu_N^2. \quad (14)$$

This new scissors mode is the $K^\pi = 1^+$ member of the isovector giant resonance. The δ^2 square dependence of its M1 strength is to be noted.

The occurrence of strongly collective low and high energy scissors excitations in superdeformed nuclei suggested recently in a RPA calculation[21], is naturally predicted in the semiclassical context. Let us assume that K is a good quantum number and that the transition goes from K to $K+1$. The $M1$ operator couples the state $|IMK>$ to the states $|I'M'K+1>$ with $I' = I-1, I, I+1$. Using the standard expression for the transition matrix elements and the TRM intrinsic wave function[3], we obtain for $I \gg K$

$$\sum_{I'} B(M1, IK \to I'K+1) \simeq \frac{3}{16\pi} \Im_{sc} \omega \frac{1}{K+1} g_r^2 \, \mu_N^2. \quad (15)$$

For $K = 0$ we gain the standard scissors strength given by eq.2. For the low energy mode we may use eqs(13) with a vanishing pairing gap (rigid body assumption) and $g_n = 0$ and $g_p = 1$ in analogy with RPA. For the high energy mode where protons and neutron behave as irrotational fluids it is appropriate to use eqs.(14) with $g_p = 2g_R = 2Z/A$ as suggested by schematic RPA. We obtain for the superdeformed ^{152}Dy ($\delta \simeq 0.62$) in substantial agreement with the RPA results[21]

$$\omega_- \simeq 6.1 MeV, \quad B_-(M1)\uparrow \simeq 22.6\mu_N^2, \qquad \omega_+ \simeq 26 MeV, \quad B_+(M1)\uparrow \simeq 26.1\mu_N^2. \quad (16)$$

4.2 RPA in the Laboratory frame

In order to avoid some of the problems encountered in intrinsic RPA, we have carried out a RPA calculation directly in the laboratory frame[24] using as single particle basis states angular momentum projected wave functions of the form[29]

$$\Phi_{\alpha IM}(d) = \mathcal{N}_{\alpha I} P^I_{MI}[\varphi_{\alpha I}\psi_g] \quad (17)$$

where $\varphi_{\alpha I} = \varphi_{nlj I}$ are spherical single particle states, $\mathcal{N}_{\alpha I}$ is a normalization factor, P^I_{MI} a projection operator of standard form and ψ_g a quadrupole boson coherent state describing the core and inducing deformation. These states are mutually orthogonal with respect to I and M. The single particle energies are obtained by taking the mean value of a rotationally invariant particle-core Hamiltonian. They depend not only on (nlj) but also on I which plays the same role as $|\Omega|$ in the Nilsson basis. The close connection with such a scheme can be guessed since the Nilsson Hamiltonian can be obtained by taking the mean value of the particle-core hamiltonian in the coherent state ψ_g. For a given deformation it is possible to obtain the Nilsson level scheme to a good approximation by a suitable choice of the particle-core coupling strength. The correspondence with the Nilsson states is not one to one. Because of the degeneracy in M, $2I + 1$ states of the present basis will correspond to a $|\Omega| = I$ Nilsson state. We can keep however all M-degenerate projected single particle states as long as we normalize them to $2/(2I+1)$ rather than 1.

We adopted this projected basis in a QRPA calculation using a rotational invariant Hamiltonian with a two-body potential composed of proton (p) and neutron (n) monopole pairing treated in BCS and p-p, n-n and p-n quadrupole and spin interactions. The $J^\pi = 1^+$ QRPA states are not coupled to the ground state by the total angular momentum and are therefore free of spurious rotational admixtures.

The numerical applications have been made for Sm isotopes. As shown in fig.1, the summed M1 strength of the orbital excitations, all falling below 4 MeV, is linear in δ^2, in good agreement with the experimental data. As in the other microscopic approaches, pairing is crucial for enforcing such a deformation law. The detailed properties of the mode have also been studied with satisfactory results[24]. The computed distribution of the orbital strength is indeed in qualitative agreement with the experimental M1 spectrum. The $M1$ spin distribution with its characteristic double-hump structure observed recently[30] in ^{154}Sm is well reproduced.

5. Concluding remarks

We have given a quantum mechanical definition of the scissors M1 strength. This is dictated by the TRM and is consistent with the observed deformation properties of the mode if the mass parameter is chosen to be close to the empirical moment of inertia. We feel therefore entitled to conclude that the observed M1 excitations are promoted by a scissors like motion between protons and neutrons behaving roughly as superfluid rotors in agreement with the conclusions reached in many model descriptions.

The definition applies to all the approaches based on the assumption that the M1 transition is dominantly if not totally promoted by the generator of the mode S_μ. We have indeed seen that the M1 strength derived in IBM-2 and schematic RPA can be cast into the form given by the defining equation.

For a detailed study of the properties of the mode we have proposed a RPA calculation formulated in the laboratory frame which avoids some of the problems encountered in intrinsic RPA. The results obtained are quite encouraging.

According to the semiclassical approach and in agreement with a RPA analysis, there is room also for a high energy scissors mode. The two modes should exist and be strongly collective in super-deformed nuclei. These semiclassical predictions should be sufficiently reliable since these new modes are supposedly little affected by pairing and spin admixtures.

References

1. D. Bohle, A. Richter, W. Steffen, A.E.L. Dieperink, N. Lo Iudice, F. Palumbo and O. Scholten, *Phys. Lett.* **B137** (1984) 27.
2. For a review A. Richter, in "The building blocks of nuclear structure", A. Covello ed., (World Scientific, Singapore, 1992) p.135.
3. N. Lo Iudice and F. Palumbo, *Phys. Rev. Lett.* **41** (1978) 1532;
   G. De Franceschi, F. Palumbo and N. Lo Iudice, *Phys. Rev. C* **29** (1984) 1496.
4. F. Iachello, *Nucl. Phys.* **A358** (1981) 89c; see F. Iachello and A. Arima, *The interacting boson model* (Cambridge University Press, Cambridge,1987), for a list of references.

5. W. Ziegler, C. Rangacharyulu, A. Richter and C. Spieler, *Phys. Rev. Lett.* **65** (1990) 2515.
6. C. Rangacharyulu, A. Richter, H.J. Wörtche, W. Ziegler and R.F. Casten, *Phys. Rev. C* **43** (1991) R949.
7. J. Margraf et al., *Phys. Rev. C* **47**; F.R. Metzger, *ibid.* **18** (1978) 1603.
8. S.G. Rohozinski and W. Greiner, *Z. Phys.* **A322** (1985) 271.
9. I. Hamamoto and C. Magnusson, *Phys. Lett.* **B260** (1991) 6.
10. L. Zamick and D.C. Zheng, *Phys. Rev. C* **44** (1991) 2522; **C46** (1992) 2106.
11. J.N. Ginocchio, *Phys. Lett.* **B265** (1991) 6.
12. K. Heyde and C. De Coster, *Phys. Rev. C* **44** (1991) R2262 ; K. Heyde, C. De Coster, A. Richter and H.-J. Wörtche *Nucl. Phys.* **A549** (1992) 103.
13. T. Mizusaki, T. Otsuka and M. Sugita, *Phys. Rev. C* **44** (1991) R1277.
14. K. Heyde, C. De Coster, C. Ooms and A. Richter, *Phys. Lett* **B312** (1993) 267.
15. E. Garrido, E. Moya de Guerra, P. Sarriguren and J.M. Udias, *Phys. Rev. C* **44** (1991) R1250
16. N. Lo Iudice, A.A. Raduta and D.S. Delion, *Phys. Lett.* **B300** (1993) 195; *Phys. Rev. C* **50** (1994) in press.
17. N. Lo Iudice and A. Richter, *Phys. Lett.* **B304** (1993) 193.
18. P. Sarriguren, E. Moya de Guerra, R. Nojarov and A. Faessler. *J. Phys: Nucl. Phys.* **G19** (1993) 291.
19. R.R Hilton, W. Höhenberger and H. J. Mang, *Phys. Rev. C* **47** (1993) 602.
20. N. Lo Iudice and A. Richter, *Phys. Lett* **B228** (1989) 291.
21. I. Hamamoto and W. Nazarewicz, *Phys. Lett.* **B297** (1992) 25.
22. R. Nojarov and A. Faessler, *Nucl. Phys.* A484 (1988) 1.
23. K. Sugawara - Tanabe and A. Arima, *Phys. Lett.* **B206** (1988) 573.
24. A.A. Raduta, N. Lo Iudice and I.I. Ursu, *Nucl. Phys.* submitted to.
25. E. Lipparini and S. Stringari, *Phys. Lett.* **B130** (1983) 139.
26. A. Bohr and B.R. Mottelson, *Nuclear Structure* (Benjamin, N.Y. 1975), Vol. II, ch.6.
27. N. Lo Iudice, in *Capture Gamma-ray Spectroscopy*, J. Kern ed., (World Scientific, Singapore, 1994) p.154.
28. See for instance H.R. Walet, P.J. Brussard and A.E.L. Dieperink, *Phys. Lett.* **B163** (1985) 4.
29. A. A. Raduta and N. Sandulescu, *Nucl. Phys.* **A591** (1991) 299; A.A. Raduta, D.S. Delion and N. Lo Iudice, *Nucl. Phys.* **A551** (1993) 73.
30. D. Frekers et al., *Phys. Lett.* **B244** (1990) 178; see also A. Richter, *Nucl. Phys.* **A553** (1993) 417c.

MIXED-SYMMETRY STATES IN O(6) NUCLEI

G. MOLNÁR, T. BELGYA, B. FAZEKAS
Institute of Isotopes
Budapest H-1525, Hungary

D.P. DIPRETE, R.A GATENBY, S.W. YATES
Department of Chemistry, University of Kentucky
Lexington, KY 40506-0055, USA

and

T. OTSUKA
Department of Physics, University of Tokyo
Hongo, Bunkyo-ku, Tokyo 113, Japan

ABSTRACT

Mixed-symmetry states have been searched for in the two "classical" O(6) type nuclei using the (n,n'γ) reaction. The ^{134}Ba nucleus provides the first example of a mixed-symmetry 2^+ state in this class. The observed total M1 strength of about 0.2 μ_N^2 is unequally shared between two close-lying states, at 2029 keV and 2088 keV energy. The centroid and the summed B(M1) strength to the first 2^+ state are nicely reproduced by IBM-2 calculations. The situation is more complex for ^{196}Pt in that the M1 strength seems to be fragmented among a large number of 1^+ and 2^+ states, with none of the individual strengths exceeding the order of 0.01 μ_N^2. More conclusive data and a new IBM-2 calculation are needed, however.

1. Introduction

Excited states, not fully symmetric with respect to proton and neutron degrees of freedom were predicted[1] by the Interacting Boson Model[2] and have subsequently been observed[3] in deformed nuclei. These are the 1^+ states associated with the low-energy collective magnetic dipole mode, which is of isovector type and corresponds to a scissors like motion of the deformed proton and neutron fluids, hence the name "scissors mode". The isovector 1^+ states have been observed in a number of deformed nuclei from light to heavy, and their properties have been thoroughly studied[4].

Much less is known, however, about the existence of mixed-symmetry states in spherical-like nuclei, close to the SU(5) or O(6) limits of the Interacting Boson Model. In both classes of nuclei the lowest state of mixed-symmetry type is expected[5,6] to be a 2^+ level according to the IBM-2. In the SU(5) limit this is typically the third 2^+ state, characterized by M1 decay to the first 2^+ level and by weak E2 branches to the second

2^+ as well as to the 0^+ ground state[7]. Examples of such 2^+ levels have been found[7-10] in some nuclei close to the SU(5) limit. M1 rates in the range of $(0.05-0.5)\mu_N^2$ have been observed, values of $(0.1-0.2)\mu_N^2$ being typical. This is one order of magnitude weaker than the ground state M1 decay rates of 1^+ states in SU(3) nuclei, that makes the identification of mixed-symmetry 2^+ levels in spherical nuclei more difficult.

A somewhat similar but more complex situation is predicted[11,12] for O(6) nuclei where the lowest-lying mixed-symmetry configurations of [N-1,1], <N-1,1> are a 2^+ state and a 1^+ state, with the 2^+ state being the analog of the lowest mixed-symmetry state in the SU(5) limit. On the other hand, the somewhat higher lying 1^+ state is the O(6) analog of the M1 scissors mode in the SU(3) limit. It is worth mentioning that mixed-symmetry β vibrations characterized by sigma quantum numbers <N-2,0> have also been predicted[11,12] to exist. For the O(6) case, however, no experimental evidence for mixed-symmetry states has been found prior to our work.

Here we present some recent results on the search for mixed-symmetry states in γ-unstable nuclei corresponding to the O(6) limit of the IBM. The two classical examples[1], i.e. ^{134}Ba and ^{196}Pt have been investigated in the (n,n'γ) reaction using the University of Kentucky Van de Graaff facility. Mixed-symmetry states were attempted to be identified by determining E2/M1 multipole mixing ratios from angular distributions and B(M1) values whenever the DSAM measurements provided level lifetimes. Conclusions were drawn on the basis of comparison with IBM-2 calculations.

2. The A~130 region: ^{134}Ba

The A~130 mass region is located just above the Z=50 proton shell and below the N=82 neutron shell. It is so far the best example of an extended O(6) region, comprising the Xe-Ba chain[13]. In fact, the nucleus ^{134}Ba has first been suggested to obey the newly predicted O(6) symmetry in one of the classical papers by Arima and Iachello[14]. Moreover, the first theoretical predictions for mixed-symmetry states in spherical-like nuclei were also made[15] for this nucleus, along with ^{130}Ba.

The nonselective inelastic neutron scattering (INS) or (n,n'γ) reaction was chosen to study the level scheme of ^{134}Ba with the main purpose to locate possible mixed-symmetry states. Experimental procedures and results have been described in detail elsewhere[16,17], hence only the most important results will be discussed here.

Already the first γ-ray angular distribution measurements revealed[16] that there are two close-lying 2^+ states of 2029 keV and 2088 keV energy, respectively, decaying by essentially pure M1 transitions to the first 2^+ level. Their lifetimes could be obtained in subsequent DSAM experiments[17]. The reduced M1 transition probabilities gave further support to our early suggestion[16] that the newly characterized two 2^+ states share the properties of the lowest mixed-symmetry state of ^{134}Ba. Such a fragmentation has also been observed[18] in ^{56}Fe and it could be accounted for by a microscopic calculation[19].

For a quantitative analysis the early IBM-2 calculations[20] were repeated. A slight change in the Majorana parameters provided a reasonable agreement with experiment for both unequal and equal Majorana constants. In Table 1 a comparison between experiment and theory is made for the 2^+ states. It is clear that for the 2029 keV - 2088 keV pair both the centroid energy and the summed B(M1) strength of about 0.2 μ_N^2 agrees with the theoretical prediction for a single mixed-symmetry 2^+ state. More details, also for other low-lying states, can be found in the original paper[17]. The mixed-symmetry 1^+ state should occur substantially higher, at about 2.9 MeV, according to our calculation.

It is remarkable that an independent calculation[21] using an F-spin symmetric hamiltonian and a single-parameter Majorana term with a constant derived from systematics yielded an energy of 2.16±0.25 MeV for the mixed-symmetry 2^+ level of ^{134}Ba, in agreement with our result. The same calculation gave a 2.55±0.25 MeV energy and a B(M1) rate of 0.16 μ_N^2 for the mixed-symmetry 1^+ level. According to a recent resonance fluorescence experiment[22,23] the new 2940 keV state might be a candidate for such a level if positive parity can be confirmed. Its energy is closer to our prediction and to the RPA result of Ref. 21, according to which a pair of orbital M1 excitations occur at 2.88 MeV and 3.20 MeV with B(M1) strengths of 0.20 μ_N^2 and 0.29 μ_N^2, respectively.

Table 1. M1 transition rates for the lowest 2^+ states in ^{134}Ba

E_{exp} keV	J_i	J_f	$B(M1)_{exp}$ μ_N^2	E_{th} keV	$<F \cdot F>/max$	$B(M1)_{th}$ μ_N^2
605	2_1			607	98	
1168	2_2	2_1	0.0003(1)	1285	90	0.062
2029	2_3	2_1	0.062(8)	2078	49	0.233
2088	2_4	2_1	0.137(12)			
2371	2_5	2_1	0.001(1) 0.008(4)	2476	60	0.026

3. The A~190 region: ^{196}Pt

The other classical region of O(6) symmetry occurs below the Z=82 and N=126 shell closures and encompasses the Os-Pt chain. The ^{196}Pt is another classical case[1], it is probably the best studied O(6) nucleus[24-29]. Nevertheless, nothing is known about mixed-symmetries in this case, except for some angular correlation measurements[29] which have provided the first hint that large M1 admixtures may characterize the decays of first 3^+ and third 2^+ levels. Hence it seemed reasonable to perform a more detailed experimental study, using the same technique as for ^{134}Ba, in order to locate the lowest mixed-symmetry states. Some preliminary results[30] are discussed below.

The new angular distribution measurements enabled us to determine uniquely many of the relevant E2/M1 mixing ratios (Table 2). While the first 3^+ state decays to the 2_2^+ state via an almost pure E2 transition, the forbidden branch to 2_1^+ can be either E2 or M1. More interesting is the third 2^+ level. While decay to 2_2^+ proceeds via a 70% M1 transition, deexcitation to the first 2^+ level could be either E2 or M1. Of course, the latter would be a possible signature of mixed-symmetry.

The DSAM lifetime measurements had to be conducted at a sufficiently high bombarding energy to produce enough recoil. Even so, lifetimes longer than about 0.5 ps could not be determined, hence this value was used as an upper limit when calculating the transition probabilities, partly shown in Table 2. Moreover, for levels below 1.5 MeV cascade feedings have made lifetime determination impossible.

Table 2. Some multipole mixing ratios and M1 transition rates in ^{196}Pt

E_i keV	J_i	J_f	E_γ keV	$\delta(E2/M1)_{exp}$	$B(M1)_{exp}$ μ_N^2	$B(M1)_{th}$ [31] μ_N^2
689	2_2	2_1	333	-4.8±0.2	~1x10^{-3} [26]	5.4x10^{-2}
1015	3_1	2_2	326	-4.2±0.3		1.1x10^{-1}
		2_1	659	-0.06 -(4.6÷2)		1.7x10^{-5}
1293	4_2	4_1	416	-(1.8÷0.9)		1.3x10^{-1}
1362	2_3	3_1	347			1.2x10^{-2}
		2_2	673	-0.63±0.07		6.2x10^{-2}
		2_1	1006	10÷∞ -0.37±0.06		1.9x10^{-1}
1605	2_4	2_2	916	-0.31±0.05 7÷16	<3x10^{-2} <7x10^{-4}	3.1x10^{-2}
		2_1	1249	-0.15±0.10 2.8÷6.3	<4x10^{-2} <3x10^{-3}	1.2x10^{-1}
1677	2_5	2_2	989	-0.14÷0.91 1.4÷10	<6x10^{-3} <2x10^{-3}	2.9x10^{-2}
		2_1	1322	0 1.0÷2.3	<1x10^{-2} <5x10^{-3}	6.3x10^{-2}

The early IBM-2 calculations of Bijker et al.[31] have been repeated in order to compare the obtained E2/M1 mixing ratios and reduced transition probabilities with the present data. Part of the results is included in Table 2. Unfortunately, the lifetime of the 2_3^+ state, which is predicted to contain only 53 percent of the maximal F-spin component, is not known. Nevertheless, the 2_4^+ case indicates that the strongest possible B(M1) value is nearly one order of magnitude weaker than the prediction, which is about 10^{-1} μ_N^2 for this state as well. A number of higher 2^+ states, up to 2784 keV, have been identified which decay to the first 2^+. If the choice of small δ were unique the corresponding B(M1) values would be of the order of 10^{-2} μ_N^2. This may indicate that the 2^+ mixed-symmetry state is highly fragmented.

The situation is similar with respect to the 1^+ states where the lowest candidate is the 1969 keV level and more than twenty other states up to 3131 keV energy may have a ground state B(M1) rate of the order of 10^{-2} μ_N^2. More stringent experimental tests are needed to see if a strong M1 has been missed. On the other hand, the IBM-2 calculations of Ref.[31] also pose problems, as nonstandard values have been taken for the χ parameters which should also affect predictions for mixed-symmetry states.

4. Summary

Observation of the predicted mixed-symmetry states in deformed nuclei and in some spherical nuclei has been one of the great successes of the Interacting Boson Model. We have attempted to locate such states in O(6) type nuclei which are intermediate between the SU(5) and SU(3) cases in that both 2^+ and 1^+ mixed-symmetry states are expected to occur at low energy and decay predominately by M1 transitions to the first 2^+ and to the ground state, respectively.

In the A~130 region, ^{134}Ba provides the first example of mixed symmetry as far as O(6) nuclei are concerned[16,17]. Two close-lying 2^+ states just above 2 MeV share the M1 strength, in a manner similar to the case of the vibrational ^{56}Fe nucleus[18,19]. There is at least a candidate for a mixed-symmetry 1^+ level[22,23].

The A~190 region is still less clear. So far data for ^{196}Pt seem to indicate a large fragmentation of M1 strength for both 2^+ and 1^+ states[30]. More lifetime data and unique mixing ratio determinations are needed to clarify the situation with the 2^+ case, especially concerning the third 2^+ state. A resonance fluorescence study of 1^+ states would also be most useful.

5. Acknowledgements

We wish to thank U. Kneissl for communicating preliminary results to us. This work has been supported in part by the Hungarian Academy of Sciences and the U.S. National Science Foundation Grants No. INT-901770 and Nos. PHY-9001465, PHY-9300077 as well as by the Hungarian OTKA Grants Nos. 1895 and T-4162.

6. References

1. F. Iachello, *Nucl. Phys.* **A358** (1981) 89c.
2. F. Iachello and A. Arima, *The interacting boson model* (Cambridge University Press, Cambridge, 1987).
3. D. Bohle, A. Richter,W. Steffen, A. E. L. Dieperink, N. Lo Iudice, F. Palumbo and O. Scholten, *Phys. Lett.* **137B** (1984) 27.
4. A. Richter, contribution to this Conference.
5. O. Scholten, K. Heyde, P. Van IsackerJ. Jolie, J. Moreau and M. Warouquier, *Nucl. Phys.* **A438** (1985) 41.
6. P. Van Isacker, K. Heyde, J. Jolie and A. Sevrin, *Ann. Phys.* **171** (1986) 253.
7. P. O. Lipas, P. von Brentano and A. Gelberg, *Rep. Prog. Phys.* **53** (1990) 1353.
8. W. D. Hamilton, in *Capture Gamma-Ray Spectroscopy 1987*, ed. K. Abrahams and P. Van Assche (The Institute of Physics, Bristol, 1988) p. 577.
9. A. Giannatiempo, A. Nannini, A. Perego, P. Sona and G. Maino, *Phys. Rev.* **C44** (1991) 1508.
10. A. Giannatiempo, A. Nannini, A. Perego and P. Sona, *Phys. Rev.* **C44** (1991) 1844.
11. A. B. Balantekin, B. R. Barrett and P. Halse, *Phys. Rev.* **C38** (1988) 1392.
12. B. R. Barrett and T. Otsuka, *Phys. Rev.* **C42** (1990) 2438.
13. R. F. Casten and P. von Brentano, *Phys. Lett.* **152B** (1985) 22.
14. A. Arima and F. Iachello, *Ann. Phys. (NY)* **123** (1979) 468.
15. A. van Egmond, K. Allaart and G. Bonsignori, *Nucl. Phys.* **A436** (1985) 458.
16. G. Molnár, R. A. Gatenby and S. W. Yates, *Phys. Rev.* **C37** (1988) 898.
17. B. Fazekas, T. Belgya, G. Molnár, Á. Veres, R. A. Gatenby, S. W. Yates and T. Otsuka, *Nucl. Phys.* **A548** (1992) 249.
18. S. A. A. Eid, W. D. Hamilton and J. P. Elliott, *Phys. Lett.* **166B** (1986) 267.
19. H. Nakada, T. Otsuka and T. Sebe, *Phys. Rev. Lett.* **67** (1991) 1086.
20. G. Puddu, O. Scholten and T. Otsuka, *Nucl. Phys.* **A348** (1980) 109.
21. H. Harter, P. O. Lipas, R. Nojarov, Th. Taigel and A. Faessler, *Phys. Lett.* **205B** (1988) 174.
22. P. von Brentano et al., Institut für Strahlenphysik Ann. Rep. 1993 (Univ. Stuttgart, 1993), p. 16.
23. U. Kneissl, private communication.
24. J. A. Cizewski, R. F. Casten, G. J. Smith, M. L. Stelts, W. R. Kane, H. G. Börner and W. F. Davidson, *Phys. Rev. Lett.* **40** (1978) 167.
25. J. A. Cizewski, R. F. Casten, G. J. Smith, M. R. MacPhail, M. L. Stelts, W. R. Kane, H. G. Börner and W. F. Davidson, *Nucl. Phys.* **A323** (1979) 349.
26. H. H. Bolotin, A. E. Stuchbery, I. Morrison, D. L. Kennedy and C. G. Ryan, *Nucl. Phys.* **A370** (1981) 146.
27. C. Lim, R. H. Spear, M. P. Fewell and G. J. Gyapong, *Nucl. Phys.* **A548** (1992) 308.

28. R. Levy, N. Tsoupas, N. K. B. Shu, A. Lopez-Garcia, W. Andrejtscheff and N. Benczer-Koller, *Phys. Rev.* **C25** (1982) 293.
29. A. M. Bruce and D. D. Warner, in *Capture Gamma-Ray Spectroscopy and Related Topics-1984*, ed. S. Raman (AIP New York, 1985), p. 431.
30. D. P. DiPrete, Ph. D. Thesis (University of Kentucky, 1994).
31. R. Bijker, A. E. L. Dieperink, O. Scholten and R. Spanhoff, *Nucl. Phys.* **A344** (1980) 207.

Asymptotic evaluation of F-Spin content and M1 sum rule

A. F. Diallo
Instituto de Estudios Nacionales, Universidad de Panama, Republic of Panama

B. R. Barrett
Physics Department, University of Arizona, Tucson AZ 85721

D. E. Davis
Physics Department, University of Stellenbosch, Stellenbosch 76000, South Africa

Abstract

An IBM-2 version of the $1/\Lambda$-Expansion Method is used to evaluate the F-Spin content of nuclear states with bosons of arbitrary even multipolarity present in the system. A M1 sum rule is derived and the need for a Majarana term (or lack of it) in the study of magnetic properties is discussed. The 1^+ energy centroids, the summed transition strengths and gyromagnetic ratios of the Samarium isotopes are determined using a no free parameter formula with plausible angular single particle g-factors.

1 Introductory remarks

The concept of F-spin initially introduced[1] to facilitate the discussion of the proton-neutron symmetry in the IBM-2 provides a useful tool in the description of mixed symmetry states and the magnetic properties of nuclei. The discovery of strong collective M1 transitions[2] was seen as a confirmation of the model and as an experimental evidence of the **Majarana** term. However, this term does not find a shell model grounding, its introduction based on the need to push up those states of lesser symmetry with respect to proton and neutron exchange. Traditionally, the F-spin content is evaluated numerically within an NPBOS calculation; Kuyucak and Morrison[3] provided an analitical exact expression for $\mathbf{F} \cdot \mathbf{F}$ within the framework of the angular-momentum projected intrinsic states formalism and this procedure is further simplified by an aproximation (without angular-momentum projection) due to Ginocchio and Kuyucak[4].

A new expression for $\mathbf{F} \cdot \mathbf{F}$ is derived in the present work using a Laplace realization of the Intrinsic States formalism with angular-momentum projection. This is an IBM-2 generalization of the $1/\Lambda$-expansion method[5]. It allows a twofold evaluation of both the F-spin content and of the Majarana contribution to the Hamiltonian. The general procedure of the method is outlined in section 2 while section 3 deals with derivation of an M1 sum-rule. In section 4, we present an application of the model to the Samarium isotopes and some concluding remarks. The energy centroids, transition strengths and gyromagnetic ratios are determined and compared to the results obtained by Mizusaki and Otsuka[6].

2 The $1/\Lambda$ expansion and F-spin content

The $1/\Lambda$ method is an asymptotic realization of the Hartree-Bose formalism within the Variation-After-Projection scheme. The details of the procedure are outlined by Diallo, David and Barrett[5]. The intrinsic boson creation operators

$$b_m^\dagger = \sum_{\substack{l=0 \\ \text{even}}}^{2p} x_{lm} b_{lm}^\dagger \tag{1}$$

are introduced, where the structure coefficients x_{lm} represent the wave-functions. The bands can then be described by repeated action of these operators on the vaccum $|0\rangle$. In particular, the ground-state band is dominated be the $m = 0$ component and we thus use

$$|\Phi_N\rangle = \frac{1}{\sqrt{N!}}(b_0^\dagger)^N |0\rangle \qquad (2)$$

to describe the boson condensate. We will drop the index 0 from now on and write x_l and $b^\dagger$ instead of x_{l0} and $b_0^\dagger$ respectively.

All members of the band are then obtained by projection onto good angular-momentum of this state

$$|\Phi_N(L)\rangle \propto P_{00}^L |\Phi_N\rangle \,,$$

where P_{MK}^L is the standard angular-momentum projector.

The expectation value of an operator O(N,L) (in the ground state band) is now cast into the form:

$$\langle O(N,L)\rangle = \frac{\langle \Phi_N | O(N,L) P_{00}^L | \Phi_N \rangle}{\langle \Phi_N | P_{00}^L | \Phi_N \rangle} \qquad (3)$$

In practice, the Variation-After-Projection method is reduced to evaluating integral ratios of the type

$$R(J,L,n) = \frac{\int_0^\pi d\beta \sin\beta P_J(\cos\beta)\, P_L(\cos\beta)\, [Z(\cos\beta)]^{N-n}}{\int_0^\pi d\beta \sin\beta\, P_L(\cos\beta)\, [Z(\cos\beta)]^N} \,, \qquad (4)$$

where the boson condensate angular distribution is

$$Z(x) = \sum_{\substack{l=0 \\ \text{even}}}^{2p} x_l^2 P_l(x) \,, \qquad (5)$$

$p_k(x)$ being the Legendre polynomial of order k.

This ratio of integrals can be approximated by an asymptotic expansion in

$$\Lambda = N \sum_{\substack{l=0 \\ \text{even}}}^{2p} l(l+1) x_l^2 \,.$$

The final result[7] is

$$\begin{aligned}
R(J,L,n) \;\sim\; & 1 + \frac{S - \bar{J}}{\Lambda}\left(1 - \frac{\bar{L}}{\Lambda}\right) \\
& + \frac{1}{\Lambda^2}\left\{ 32\alpha_2^{(J)} + 4\alpha_1^{(J)}\left[1 + S + \frac{Nt}{\Lambda}\right]\right. \\
& \left. + \left[2S(2 + S + \frac{Nt}{2\Lambda}) - nt\right]\right\} \,.
\end{aligned} \qquad (6)$$

In (6), we have used the following definitions:

$$m_k = \sum_{l=2}^{2p} (\bar{l})^k x_l^2 \qquad \Lambda = N m_1$$

$$S = n m_1 \qquad t = m_1^2 + \frac{m_2}{2}$$

and the $\alpha_k^{(l)}$ are computed recursively from

$$\alpha_{k+1}^{(l)} = -\frac{\bar{l} - 2k(2k+1)}{4(k+1)^2}\alpha_k^{(l)}. \tag{7}$$

with $\alpha_0^{(l)} = 1$ and $\alpha_1^{(l)} = -\bar{l}/4$

This expansion is easily generalized to the IBM-2 where the integral-ratio to be aproximated are:

$$R(J, L, n, n') = \frac{\int_0^\pi d\beta \sin\beta P_J(\cos\beta) P_L(\cos\beta) \left[Z_\pi(\cos\beta)\right]^{N_\pi - n} \left[Z_\nu(\cos\beta)\right]^{N_\nu - n'}}{\int_0^\pi d\beta \sin\beta P_J(\cos\beta) P_L(\cos\beta) \left[Z_\pi(\cos\beta)\right]^{N_\pi} \left[Z_\nu(\cos\beta)\right]^{N_\nu}} \tag{8}$$

The final result is:

$$\begin{aligned}R(J, L, n, n') \sim\ & 1 + \frac{S - \bar{J}}{\Lambda}(1 - \frac{\bar{L}}{\Lambda}) \\ & + \frac{1}{\Lambda^2}\left\{32\alpha_2^{(J)} + 4\alpha_1^{(J)}\left[1 + S + \frac{N_\pi t_\pi + N_\nu t_\nu}{\Lambda}\right]\right. \\ & + \left.\alpha_0^{(J)}\left[2S(2 + S + \frac{N_\pi t_\pi + N_\nu t_\nu}{2\Lambda}) - (nt_\pi + n't_\nu)\right]\right\}, \end{aligned} \tag{9}$$

where $S = nm_{\pi 1} + n'm_{\nu 1}$ and $t_\rho = m_{\rho 1}^2 + m_{\rho 2}/2$.

Note that this ratio reduces to (6) for $m_{\pi k} = m_{\nu k}$ when there is only one type of boson.

We can now use (9) to determine different quantities. In particular, the value of $\mathbf{F} \cdot \mathbf{F}$ can be written as

$$\begin{aligned}\langle N_\pi N_\nu L | \mathbf{F} \cdot \mathbf{F} | N_\pi N_\nu L\rangle =\ & \sum_{\substack{l=0 \\ \text{even}}}^{2p} \{N_\pi N_\nu R(l, L, 1, 1) + N_\pi R(l, L, 1, 0)\} \\ & + F_0(F_0 + 1) R(l, L, 0, 0). \end{aligned} \tag{10}$$

Using eq. (9) to calculate $\delta F = F_m(F_m + 1) - \langle FF_0 | \mathbf{F} \cdot \mathbf{F} | FF_0\rangle$ we obtain:

$$\delta F \sim \mathcal{F}^{(0)} + \frac{\mathcal{F}^{(1)}}{\Lambda}(1 - \frac{\bar{L}}{\Lambda}) + \frac{\mathcal{F}^{(2)}}{\Lambda^2}, \tag{11}$$

where

$$\begin{aligned}\mathcal{F}^{(0)} &= N_\pi N_\nu \left[1 - (\mathbf{X}_\pi \cdot \mathbf{X}_\nu)^2\right] \\ \mathcal{F}^{(1)} &= N_\pi N_\nu \left[(m_{\pi 1} + m_{\nu 1})(\mathbf{X}_\pi \cdot \mathbf{X}_\nu)^2 - 2m_{\pi\nu 1}(\mathbf{X}_\pi \cdot \mathbf{X}_\nu)\right] \\ \mathcal{F}^{(2)} &= N_\pi N_\nu \left\{m_{\pi\nu 2}(\mathbf{X}_\pi \cdot \mathbf{X}_\nu)^2 + 2m_{\pi\nu 1}^2 + 2m_{\pi\nu 1}(\mathbf{X}_\pi \cdot \mathbf{X}_\nu)\left[1 - 2(m_{\pi 1} + m_{\nu 1})\right]\right. \\ &\quad + \left.(\mathbf{X}_\pi \cdot \mathbf{X}_\nu)^2\left[(m_{\pi 1} + m_{\nu 1})^2 - m_{\pi 1} - m_{\nu 1} - \frac{m_{\pi 2} + m_{\nu 2}}{2} + (m_{\pi 1}^2 + m_{\nu 1}^2)\right]\right\}. \end{aligned} \tag{12}$$

The result derived by Ginocchio[4] is a particular case (sd limit) of our leading order term $\mathcal{F}^{(0)}$ because no angular-momentum projection is done in that work. It is easy to see that if $m_{\pi\nu k} = m_{\pi k} = m_{\nu k}$, $F = F_m$ and we reproduce the IBM-1 results. The present approach also has the advantage of providing a single prescription for determining the properties of all band-members.

To verify the validity of the present approach we need separate proton and neutron distributions, and it would be desirable to obtain their deformation using techniques such as double π experiments. Here we limit ourselves to comparing the exact NPBOS diagonalization result with the corresponding $1/\Lambda$ approximation for the ^{148}Sm. For the yrast band lowest members (0^+, 2^+ and 4^+) we obtain an F-spin purity of 97.1, 96.5 and 96.2 % compared with 96.2, 96.1 and 96.9 % of the exact NPBOS calculation respectively.

3 B(M1) Sum Rule: Generalities

In recent years there has been increased interest in the study of the mixed-symmetry states, particularly the 1^+ state and the large B(M1) transition strength associated with it[6,7,8]. In the present work, we look at the summed excitation strength, derive an energy-weighted sum rule and determine the mean excitation-energy of the 1^+ *scissor states*. The derivation of the main results will be done in the generalized IBM-2, but the IBM-1 projected results will be used for numerical evaluations.

The M1 excitation-strength function is defined in the standard way as:

$$S_{M1}(E) = \sum_{j,\mu} |\langle j|T_\mu(M1)|0^+\rangle|^2 \delta(E - E_j), \tag{13}$$

which can be alternatively written in terms of the Hamiltonian as

$$S_{M1}(E) = \sum_\mu \left\langle 0^+ \left| (T_\mu(M1))^\dagger \delta(E - H) T_\mu(M1) \right| 0^+ \right\rangle. \tag{14}$$

Of more interest, however, are the summed-excitation strength and the strength density defined by

$$S_{M1} = \int_{-\infty}^{\infty} S_{M1}(E) dE = \langle 0^+|T(M1) \cdot T(M1)|0^+\rangle \tag{15}$$

and

$$\rho_{M1}(E) = \frac{S_{M1}(E)}{S_{M1}}, \tag{16}$$

respectively. They allow us to determine the cumulant moments, of which the lowest two are of particular interest, since they represent the mean-excitation energy and its fragmentation width. Written in terms of commutators, these quantities are given by:

$$E_c = \int_{-\infty}^{\infty} \rho_{M1}(E) E dE = \frac{1}{2} \frac{\left\langle 0^+ \left| \left[(T(M1))^\dagger, [H, T(M1)]\right] \right| 0^+ \right\rangle}{S_{M1}}, \tag{17}$$

and

$$(\Delta E)^2 = \int_{-\infty}^{\infty} \rho_{M1}(E - E_c)^2 dE = \frac{1}{2} \frac{\left\langle 0^+ \left| \left[(T(M1))^\dagger, [H, [H, T(M1)]]\right] \right| 0^+ \right\rangle}{S_{M1}} - E_c^2 \tag{18}$$

Our interest will be in the summed strength and the mean-excitation energy, the fragmentation being left for a later investigation. For algebraic convenience, the generalized IBM-2 is used in the derivation. The one-body $M1$ operator is defined in the usual way by:

$$T(M1) = \sqrt{\frac{3}{4\pi}} \sum_{l>0} (g_{\pi l} L_{\pi l} + g_{\nu l} L_{\nu l}), \tag{19}$$

where, as before, $g_{\rho l}$ and $L_{\rho l}$ represent the g-factor and angular-momentum operator, for a boson ρ (π, ν) of spin $l(2,4,\cdots)$. The choice of the boson-gyromagnetic ratios is quite intricate and warrants careful consideration.

A generalized Talmi Hamiltonian can be considered with the boson spins spanning the values $l = 0, \cdots, 2p$.

$$H = \sum_{\rho,l} \epsilon_{\rho l} \hat{n}_{\rho l} - \sum_{k=2}^{4p} \kappa_k T_\pi^{(k)} \cdot T_\nu^{(k)} + \xi M. \tag{20}$$

The result for a more F-spin invariant Hamiltonian is easily obtainable along the same lines. It is customary to use the same single-particle boson energies for both protons and neutrons. In any case, this term does not contribute to the commutators, because the $T(M1)$ commutes with the number operator.

There is strong evidence that the ground state is nearly F-spin pure[6], better than 97% for the Sm isotopes. This justifies the IBM-2 to IBM-1 projection which allows us to simplify the evaluation of the different quantities, and the solution of the variational problem. The Hartree-Bose intrinsic-state formalism can be used in evaluating the expectation values. This mean field approach is suitable for deformed nuclei, and it can be used in the $1/\Lambda$ approximation, provided the expansion parameter Λ that represents the mean-squared angular momentum carried by the boson condensate is reasonably large (> 10).

Let us now turn our attention to the summed strength. First, the $M1$ operator is rewritten in terms of F-spin tensor operators of rank 0 (scalar) and rank 1 (vector):

$$T(M1) = [T(M1)]_0^0 + [T(M1)]_0^1, \quad (21)$$

where

$$[T(M1)]_0^0 = \sqrt{\frac{3}{4\pi}} \sum_l (\frac{g_{\pi l} + g_{\nu l}}{2})(L_{\pi l} + L_{\nu l}), \text{ and } [T(M1)]_0^1 = \sqrt{\frac{3}{4\pi}} \sum_l (\frac{g_{\pi l} - g_{\nu l}}{2})(L_{\pi l} - L_{\nu l}). \quad (22)$$

Note that the F-spin scalar component is proportional to the angular momentum: it contributes to S_{M1}, but not to the transitions.

Through the use of the Wigner-Eckart theorem, S_{M1} is evaluated. The final result is:

$$S_{M1} = -\left\langle 0^+ \left| A \cdot A - \frac{4N_\pi N_\nu}{N(N-1)} \left[\frac{B \cdot B}{N} + C\right] \right| 0^+ \right\rangle \quad (23)$$

$$A = \sqrt{\frac{3}{4\pi}} \sum_l g_l^{eff} L_l, \quad g_l^{eff} = \frac{N_\pi g_{\pi l} + N_\nu g_{\nu l}}{N}$$

$$B = \sqrt{\frac{3}{4\pi}} \sum_l \delta g_l L_l; \quad C = \frac{3}{4\pi} \sum_l (\delta g_l)^2 l(l+1) \hat{n}_l$$

$$\delta g_l = \frac{g_{\pi l} - g_{\nu l}}{2}.$$

Note that the expression (23) does not involve any approximation and the boson g factors are completely arbitrary. No assumption has been made about the wavefunctions (other than F-spin purity). The term A corresponds to the $T(M1)$ operator in the IBM-1 projection and its evaluation gives excited-state g factors in that limit.

Using the microscopically plausible values $g_{\pi l} = 1$ and $g_{\nu l} = 0$ for all values of $l > 0$, we find that the first two terms in Eq. (23) are proportional to L^2, and their expectation values in the ground state vanish. We are left with:

$$S_{M1} = \frac{4N_\pi N_\nu}{N(N-1)} \langle 0^+ |C| 0^+ \rangle = \frac{3}{4\pi} \frac{N_\pi N_\nu}{N-1} \sum_l l(l+1) \frac{\langle 0^+ |\hat{n}_l| 0^+ \rangle}{N}. \quad (24)$$

The summed strength is proportional to the mean-squared angular momentum carried by the boson condensate m_1. The result by Ginocchio[9] is a special case of this expression, corresponding to the sd IBM.

Now, let us consider the problem of the mean-excitation energy. The result of the double commutator for a term of the form $H_k = T_\pi^k \cdot T_\nu^k$ is:

$$\left[(T(M1))^\dagger, [H_k, T(M1)]\right]$$

$$= -\frac{3}{4\pi} \sum_{j,l,j',l'} t_{\pi j l}^k t_{\nu j' l'}^k G_\pi^{(k)}(j,l) \cdot G_\nu^{(k)}(j',l')$$

$$\times [(g_{\pi j} - g_{\pi l})(\bar{j}g_{\pi j} - \bar{l}g_{\pi l}) + \bar{k}g_{\pi j}g_{\pi l} + [(\pi,j,l) \longleftrightarrow (\nu,j',l')]$$

$$-\frac{1}{2\bar{k}}\{(\bar{j} - \bar{l})(g_{\pi j} - g_{\pi l}) + \bar{k}(g_{\pi j} + g_{\pi l})\}$$

$$\{(\bar{j'} - \bar{l'})(g_{\nu j'} - g_{\nu l'}) + \bar{k}(g_{\nu j'} + g_{\nu l'})\}]. \quad (25)$$

For the special choices $g_{\pi l} = g_\pi$ and $g_{\nu l} = g_\nu$, this expression acquires the exceptionally simple form:

$$\left[(T(M1))^\dagger, [H_k, T(M1)]\right] = -\frac{3}{4\pi}(g_\pi - g_\nu)^2 \overline{k} T^k \cdot T^k, \tag{26}$$

which is a generalization of the result obtained by K. Heyde and C. De Coster[8].

The contribution from the Majarana term is determined in a similar way. First, the Majarana operator is written in terms of raising and lowering F-spin operators, $F_\pm$, by writing the scalar product as:

$$F \cdot F = F_+ F_- + F_0(F_0 - 1). \tag{27}$$

The F_0 term does not contribute and the commutators with the scalar part of Eq. (21) vanish. The Majarana contribution is then determined by repeated use of the identity

$$\left[F_\mu, T_q^{(k)}\right] = (\overline{k})^{\frac{1}{2}} \langle k\,q\,1\,\mu |\, k\,q + \mu\rangle T_{q+\mu}^{(k)}, \tag{28}$$

for an F-spin tensor of rank k. The final result is:

$$\left[T^\dagger(M1), [M, T(M1)]\right] = 8\frac{3}{4\pi}\sum_l \overline{l}(\delta g_l)^2 (F \cdot F - F_0^2) + 2(T_1^1 T_{-1}^1 + T_{-1}^1 T_1^1), \tag{29}$$

where

$$T_1^1 = -\sqrt{\frac{3}{4\pi}}\sum_l \delta g_l \sqrt{\frac{2\overline{l}(2l+1)}{3}}\left[b_{\pi l}^\dagger \tilde{b}_{\nu l}\right]^1 \text{ and } T_{-1}^1 = \sqrt{\frac{3}{4\pi}}\sum_l \delta g_l \sqrt{\frac{2\overline{l}(2l+1)}{3}}\left[b_{\nu l}^\dagger \tilde{b}_{\pi l}\right]^1. \tag{30}$$

The final expression for the mean-excitation energy or centroid energy (after F-spin projection) is given by:

$$E_c = \sum_k \overline{k}\, \kappa_k \frac{\langle 0^+ | T^k \cdot T^k | 0^+ \rangle}{2\sum_l l(l+1)\langle 0^+ | \hat{n}_l | 0^+ \rangle} + \xi N. \tag{31}$$

The Majarana contribution is, as expected, the splitting between F_m and $F_m - 1$ F-spin headbands. Note, however, the inverse dependence of the first term on
$\Lambda = \sum_l l \langle 0^+ | \hat{n}_l | 0^+ \rangle$, the squared angular momentum carried by the boson condensate. The consequence of this inverse dependence is that the model cannot be used to describe spherical nuclei ($\Lambda = 0$). We expect, on the other hand, a good agreement for deformed nuclei where the intrinsic-state formalism is a good approximation to the problem.

4 Application to the Samarium Isotopes

Equations (24) and (31) have been applied to a description of the Samarium isotopes. The parameter set is the IBM-1 projection of the one used in Ref.[6]. No hexadecupole term is included in the present calculation. The summed strength is depicted in Fig. (1.a), we note a variation related to a spherical to deformed shape transition. The experimental values are shown with error bars and the present calculation is represented by $\diamond$. The summed strength is independent of the choice of the Hamiltonian (except for the dependence of the wavefunctions upon the Hamiltonian), and the qualitative trend is well described by the present model with no evidence of subshell closure. Also shown in this figure is the $SU(3)$ prediction of Barrett and Halse[10] in the sdg limit. As can be observed, the $SU(3)$ calculation overestimates the summed strength. This is because of an overly large prediction of the g boson weight in that limit. In the case of spherical nuclei ($^{146-148}$Sm), we have a vanishing value that can be understood because of the orbital nature of the transition in which we are interested. The experimental value is mainly a spin contribution in this region. The isotope ^{150}Sm is a transitional case, and we do not expect the intrinsic state formalism to hold for transitional nuclei.

The results for the centroid energy are plotted in Fig. (1.b). They show the mean-excitation-energy contribution of the quadrupole part (⋆), the quadrupole plus Majarana (□), and the experimental value reported in[7] (•). Again, the experimental results are well described by the model in the deformed region using a Majarana strength of 50 keV, which is half that used by Mizusaki and Otsuka[11]. This small strength is consistent with Talmi's argument about a weak Majarana strength[12]. The main contribution comes from the multipole interaction of the Hamiltonian.

Figure (1.c) shows the 2_1^+ g-factors calculated in the intrinsic states formalism (⋆) and the results of the exact diagonalization by Otsuka[11] (•). Also shown are the experimental values with error bars[7]. It should be noted that, in principle, the excited-states 2^+ wavefunctions are needed here, rather than those of the ground state. There is, nevertheless, good agreement in the deformed region.

To conclude, we have derived a simple formula for the F-spin content of nuclear states that relates to the proton and neutron distributions. A simple prescription is also proposed for the study of magnetic properties in the deformed region using a set of naive but plausible values for the orbital g-factors of the bosons. The experimental results of the Samarium isotopes are well reproduced with a minor majarana contribution to the energy centroid of the scissor states.

References

[1] A. Arima, T. Otsuka, F. Iachello, and I. Talmi, *Phys. Lett.* **B66** (1977) 205.

[2] D. Bohle, A. Richter, W. Steffen, A. E. L. Dieperink, N. Loiudice, F. Palumbo and O. Scholten, *Phys. Lett.* **B137** (1984) 27.

[3] S. Kuyucak and I. Morrison, *Ann. of Phys.* **195** (1989) 126.

[4] J. Ginocchio and S. Kuyucak, LA-UR **91-2812** 1992

[5] A.F.Diallo, E.D.Davis and B. R. Barrett, *Ann. of Phys* **222**(1993) 159.

[6] T. Otsuka and M. Sugita, *Phys. Lett.* **B215** (1988) 205.

[7] W. Ziegler, C. Rangacharyulu, A. Richter, and C. Spieler, *Phys. Rev. Lett.* **65** (1990) 2515.

[8] K. Heyde and C. de Coster, *Phys. Rev.* **C44** (1991) R2262.

[9] J. Ginocchio, *Phys. Lett.* **B265** (1991) 6; J. Ginocchio, *Nucl. Phys.* **A541** (1992) 211.

[10] B. R. Barrett, and P. Halse, *Phys. Lett.* **B155** (1985) 133.

[11] T. Mizusaki, T. Otsuka and M. Sugita, *Phys. Rev.* **C44** (1991) 1277.

[12] A. Novoselski and I. Talmi, *Phys. Lett.* **B160** (1988) 13.

[13] S. Kuyucak and I. Morrison, *Phys. Rev. Lett.* **58** (1987) 315.

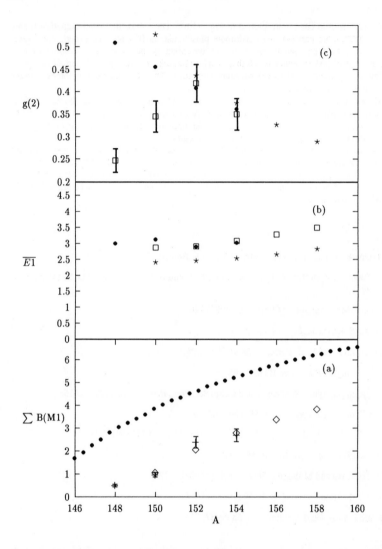

Figure 1: Summed B(M1) (a), Energy centroids (b) and g-factors of 2^+ (c) for the Sm isotopes. See legend in text.

PARTICLE-HOLE EXCITATIONS IN THE INTERACTING BOSON MODEL

K.HEYDE

Institute for Theoretical Physics
Vakgroep Subatomaire en Stralingsfysica
Proeftuinstraat 86, B-9000 Gent (Belgium)

ABSTRACT

We discuss the experimental evidence for particle-hole excitations near closed shells. The possibilities to incorporate particle-hole pair excitations within the interacting boson model are presented and ample reference is made to both numerical IBM configuration mixing studies and to a systematic study of the variation of the low-lying 0^+ intruder states throughout the whole nuclear mass region. We present a classification scheme in which both particle and hole s and d bosons are incorporated. This study leads to the extensive discussion of various symmetries (intruder spin and K-spin) when constructing many-particle many-hole (mp-nh) configurations. We finally indicate possible extensions to treat mp-nh excitations in an isospin invariant way.

1. Introduction

The Interacting Boson model (IBM) in its intial stage aimed at a description of low-lying collective quadrupole motion in nuclei where many valence protons and neutrons were playing a major role in determining the nuclear properties. Starting from a truncation of the model space into a space of interactiong s(l=0) and d(l=2) bosons, an algebraic model governed by the U(6) group and its various reductions proved to exhibit a very rich structure of possible excitations. Many extensions as well as the close connection to the underlying shell-model have later on been considered in extending the original idea of an sd-IBM and it is outside the scope of this contribution to discuss them[1].

In a rather early stage, when constructing the IBM, it became clear too that particle-hole excitations across major closed shells were playing an important role in understanding all low-lying excited states in nuclei, particularly when approaching the nuclear closed shells. These phenomena became apparent both in the odd-mass (e.g. the In, Sb, Tl, Bi, ...) and even-even (e.g. the Sn, Cd, Pb, Hg, Pt, ...) nuclei and the name 'intruder excitations' was coined to those excitations. A study of the appearance and characteristics of such intruder excitations has been carried out throughout the whole nuclear mass region[2,3].

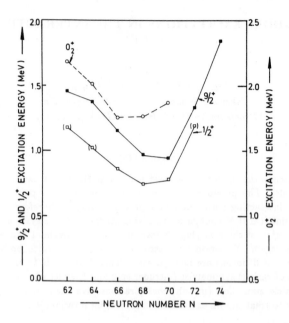

Figure 1: Excitation energy for the lowest $9/2^+$ and $1/2^+$ intruder state in the odd-mass Sb and In nuclei (left-hand axis) and for the 0^+ intruder state in the even-even Sn nuclei (right-hand axis).

2. Particle-hole pair excitations

From the observation of very low-lying $1/2^+$ and $9/2^+$ intruder excitations with a possible 2h-1p and 2p-1h origin in the In and Sb odd-mass nuclei, respectively, a step towards the suggestion for low-lying 2p-2h 0^+ excitations in the intermediate even-even Sn nuclei could be made (Fig.1). The 0^+ intruder states acting as the band head of a collective band, extending to high spin (10^+, 12^+, ...) were observed soon after and firm evidence later on appeared in the even-even very neutron deficient Pb nuclei too[3]. An estimate for the energy of such 2p-2h excitations near closed shells can be made from shell-model origin as[4]

$$2(\epsilon_p - \epsilon_h) - \Delta E_{pair} - \Delta E_{mon.} + \Delta E_{quad.} \qquad (1)$$

where the unperturbed p-h excitation energy is related to nucleon separation energy differences i.e. $\epsilon_p - \epsilon_h = S_p(Z,N) - S_p(Z+1,N)$ for proton excitations. The extra pairing energy gain can also be expressed using particle and two-particle separation energies i.e. $\Delta E_{pair}(holes) = 2S_p(Z,N) - S_{2p}(Z,N)$ and $\Delta E_{pair}(particles) = 2S_p(Z+$

$1, N) - S_{2p}(Z+2, N)$ for a proton 2p-2h excitation. Similar relations hold for neutron 2p-2h excitations.

Besides corrections from a variation in the unperturbed p-h excitation energy coming from the monopole part of the proton-neutron interaction (A-dependence of the changing average field), called the monopole correction $\Delta E_{mon.}$[5], a very important energy term is, however, originating from the strongly attractive residual proton-neutron interaction. Since the excited-core configuration contains more "valence" (particles or holes counted from the nearest closed shell) nucleons outside the closed shell, the relative binding energy gain $\Delta E_{quad.}$ originating from the neutron-proton interaction can become a dominant term. If we use a schematic quadrupole- quadrupole interaction $\kappa Q_\pi \cdot Q_\nu$ one can evaluate this energy difference in closed form, in perturbation theory if using a specific ansatz for the regular state and intruder state configuration[4].

It is precisely here that contact with the IBM can be made. If the extra pairs that are formed in creating the intruder 0^+ configuration are counted as two additional bosons and the wave function for the regular state is depicted as $|N_\pi, N_\nu; 0^+\rangle$, the intruder configuration can be denoted as the state $|N_\pi+2, N_\nu; 0^+\rangle$ if we specialize to proton 2p-2h excitations for the moment. The quadrupole energy gain $\Delta E_{quad.}$ then becomes

$$\Delta E_{quad.} = \langle N_\pi+2, N_\nu; 0^+|\kappa Q_\pi \cdot Q_\nu|N_\pi+2, N_\nu; 0^+\rangle - \langle N_\pi, N_\nu; 0^+|\kappa Q_\pi \cdot Q_\nu|N_\pi, N_\nu; 0^+\rangle \ . \tag{2}$$

For a large number of neutron pairs (bosons) and making the assumption of SU(3) wavefunction to describe the s,d pair (boson) distribution in both the regular and intruder state, an exact value for $\Delta E_{quad.}$ is obtained as

$$\Delta E_{quad.} = 4\kappa N_\nu + \frac{6\kappa N_\nu (2N_\nu - 1)}{(2N_\pi + 2N_\nu + 3)(2N_\pi + 2N_\nu - 1)} \tag{3}$$

with the dominant part given by the number of neutron pairs (bosons) present[4]. This result expresses the fact that the quadrupole binding energy maximizes at mid-shell in agreement with the present data in both the Sn and Pb mass regions (Fig.2).

The present shell-model approach presents a reasonable first-order attempt to desribe the appearance and mass variation of the lowest-lying 0^+ intruder excitations near closed shells in even-even medium-heavy and heavy nuclei with a clear-cut neutron excess. Because of the major attention given to the lowest 0^+ intruder state only, there is no easy way to obtain good information about the collective band structure built on top of the intruder states.Moreover, no mixing between the regular and intruding excitations is invoked since we only calculate the energy difference between the regular and intruder 0^+ excitations.

An important step is obtained by studying the full dynamics within the space spanned by the regular modes of motion and in the model space including the 2p-2h pair excited configurations, separately. To perform this, the boson model assumption

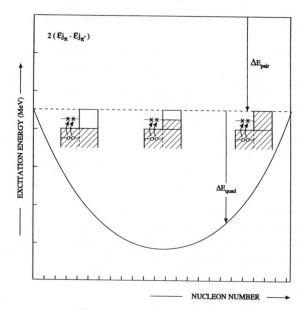

Figure 2: The variation in the 0$^+$ intruder energy as a function of the number of valence nucleons (see also eq.3). Both the unperturbed, the pairing and quadrupole energy contribution as well as a pictorial representation of the proton and neutron distributions are given.

of a fixed boson number has to be relaxed and separate IBM calculations for each boson number have to be carried out independently at first. A serious problem relates to (i) the relation between the energy eigenvalues in the separate model spaces and, (ii) the mixing between these model spaces. The term configuration mixing IBM was used and first extensive calculations for the Hg nuclei were carried out by Duval and Barrett[6]. A critical parameter Δ was used to describe the energy difference between the lowest energy eigenvalues in both model spaces and played the role mainly of a free parameter as well as the mixing parameters of a mixing Hamiltonian connecting the N and $N+2$ model spaces i.e.

$$H_{mix} = \alpha(s^+s^+ + h.c.)^{(0)} + \beta(d^+d^+ + h.c.)^{(0)} \quad . \tag{4}$$

The above shell-model description of 0$^+$ intruder excitations, however, gives a connection of the quantity Δ to be used and typical shell-model quantitities (single-particle energies, pairing and proton-neutron interactions,..) with a definition[7]

$$\Delta = 2\left(\epsilon_p - \epsilon_h\right) - \Delta E_{pair} - \Delta E_{mon.} \quad . \tag{5}$$

The quadrupole binding interaction energy is then taken care of by the dynamics of the full IBM Hamiltonian acting in the various model spaces corresponding to a different boson number N.

Many configuration mixing calculations have been carried out in various mass regions. The more recent studies of Délèze et al.[8], where a large number of the IBM parameters were determined through mapping from a shell-model description and where the quantity Δ was determined in a consistent way are quite impressive in the level of detailed agreement for the even-even Cd nuclei. Both the regular, anharmonic quadrupole vibrational excitations, the more collective intruder bands as well as the situations where strong mixing between the two subspaces occurs, are well desribed. The most serious drawback remains the large number of parameters and the fact that from the very beginning the extra particle and hole pair created in the formation of the intruding excitations are treated as bosons (s or d boson) undistinguishable from the other N bosons (symmetrical representations used only).

The above method has even been extended to treat high-spin states by enlarging the model spaces even further: one treats a chain of $N \to N+2 \to N+4 \to N+6 \to ...$ boson model systems that are governed by a Hamiltonian in each model space and taking into account both shifts in energy eigenvalues between the different subspaces as well as the mutual mixing. The energy matrix obtained has the following structure

$$\begin{bmatrix} E_i(N) & H_{mix}(N, N+2) & 0 & \cdots \\ H_{mix}(N, N+2) & E_i(N+2) + \Delta' & H_{mix}(N+2, N+4) & \cdots \\ 0 & H_{mix}(N+2, N+4) & E_i(N+4) + \Delta'' & \cdots \\ \vdots & \vdots & \vdots & \ddots \\ & & & & E_i(N+k) + \Delta^k \end{bmatrix} \quad (6)$$

Studies along these lines were carried out quite some time ago by Heyde et al.[9] and, more recently, systematic studies of high-spin states within an IBM configuration approach have been studied by Barfield et al.[10] and Vretenar et al.[11,12,13].

3. Intruder analog multiplets and intruder-spin

Guided from the fact that many of the above intruder excitations and the collective bands built on them are very similar in structure(level spacings, B(E2) values, ...) to the low-lying collective bands in adjacent nuclei with the same number of active nucleons, the idea of an underlying symmetry is tempting to be formulated[14].

The basic starting point is that particle and hole fermion pair excitations relative to a closed shell are treated as 'particle' and 'hole' s and d bosons. So, any dynamical study of an interacting boson sytem will be extended by doubling the number of degrees of freedom: we now use s_p^+, d_p^+ (p stands for 'particle' boson) and s_h^+, d_h^+ (h stands for 'hole' boson) explicitely. Our first interest is situated in classifying nuclear collective bands in nuclei where the total number of 'valence' nucleons remains constants i.e. we study all possible partitions of a given number of bosons in particle and hole bosons. Since in the medium-heavy and heavy nuclei the intruder states are mainly formed by proton pair excitations, we consider a constant neutron boson number and a distribution of proton boson number as Np, $(N-1)p-1h$, $(N-2)p-2h$, ...Nh bosons. In using a short-hand notation for the particle and hole s and d bosons

$$p_i^+ \equiv \{s_p^+, d_p^+\} \quad ; h_j^+ \equiv \{s_h^+, d_h^+\} \tag{7}$$

the algebra formed by considering all generators of the set

$$\{p_i^+ p_j\,,\, h_i^+ h_j\,,\, p_i^+ h_j\,,\, h_i^+ p_j\} \tag{8}$$

is the group U(12). A meaningful algebraic 'horizontal' reduction is expressed as

$$U(12) \supset U_{ph} \otimes U(2) \simeq U_{ph}(6) \otimes SU_I(2) \otimes U_+(1) \tag{9}$$

The algebra $U_{ph}(6)$ consists of the 36 summed generators $\{p_i^+ p_j + h_i^+ h_j\}$. The similarity between the well-known SU(2) isospin (T-spin) realization is obvious and becomes even more transparent if we associate the projection +1/2 to the elementary intruder spin 1/2 for a particle boson p_i^+ and projection -1/2 for the corresponding hole boson h_j^+. A general state $N(N_p, N_h)$ can then be constructed and characterized by its intruder spin and z-component eigenvalues I (I_z) as

$$|I, I_z\rangle = |N/2, (N_p - N_h)/2\rangle \quad . \tag{10}$$

Multiplets are characterized by a given I-spin and the I_z value ($= (N_p - N_h)/2$) defines the particular multi-particle mult-hole excitation and the corresponding nucleus (Fig.3).

It is clear that the actual realization of such multiplets will depend on (i) the variation of the IBM parameters in going from nucleus to nucleus and, (ii) on the proximity of states of different I-spin in a given nucleus. The first changes are mostly rather smooth and give rise to deviations from absolutely identical bands in the multiplet : much like in isospin an energy expression of second order in I results. For an IBM Hamiltonian containing one-and two-body s and d boson interactions, the tensor rank in I-spin space can couple up to two. Moreover, the specific variation in the χ_π parameters mainly can give rise to a smooth change from the similar bands: a typical example can be found in the Ru, Cd, Te, Ba I=3/2 multiplet (see Fig.4) where a change from O(6) character towards U(5) and SU(3) is rather clear. The proximity of levels in a given nucleus differing in I-spin by one unit can now result in locally strongly perturbed wave functions. This will in general not influence the energy spectra very much since coupling matrix elements are normally very small (order of 50 - 100 keV in the Sn, Cd region) and the various levels need to approach each other very well. The $|\Delta I| = 1$ coupling caused by the mixing Hamiltonian has been studied both numerically in the ^{112,114}Cd nuclei[8, 15, 16] and recently, an analytical study of coupling between a U(5) regular band and a O(6) intruder band has

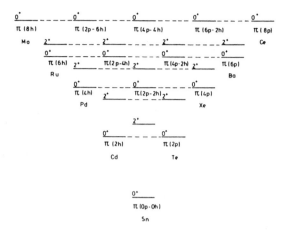

Figure 3: The various multi-particle multi-hole excitations for an intruder spin invariant Hamiltonian, ordered in a number of multiplets with constant $N_p + N_h$ value and constant I-spin quantum number.

been worked out[17]. It was shown that extra selection rules appear to regulate the local strong coupling giving rise to strongly admixed wave functions, in particular for the 0^+ and 0^+ states and thereby strongly modifying the well-known E2 and E0 vibrational intensity and selection rules[18] (see Fig.5). Quite some attention has been given to ways of analyzing the quintuplet of states in the Cd nuclei within a strongly perturbed harmonic vibrational description purely. It was in particular Casten et al.[16] who pointed out that there is a difficulty in comparing results obtained in an unmixed (or weakly coupled case) and maximally mixed (strong coupling case) description. Putting the appearance of particle-hole excitations in the Cd nuclei in a larger context of a systematic behaviour of particle-hole excitations in the whole Sn region, both in even-even and in odd-mass nuclei, a largely coherent picture shows up[8]. Recently, the possibilty of applying the above classification scheme to the description of many-particle many-hole excitations in the neutron deficient Pb region was pointed out[19, 20]. Evidence for I=2 bands in ^{186,188}Pb nuclei, related through the 'horizontal' U(12) algebraic structure to bands in ^{182,184}Pt and ^{178,180}W nuclei was shown[21] (Fig.6). Because of the rather scarce amount of data at present, a further experimental exploration of this mass region is highly interesting in order to test the appearance of rather complex mp-nh excitations and their fitting into the intruder multiplet classification scheme. Finally we stress the rather stable 'global' symmetry generated by the I-spin structure, even though strong configuration mixing of very close-lying levels can cause 'locally' strong perturbations of the intruder analog picture.

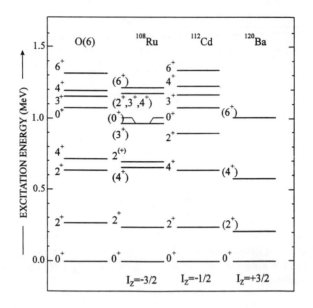

Figure 4: The various members of the I=3/2 multiplet (^{108}Ru, ^{112}Cd, ^{120}Ba). A comparison with a pure O(6) spectrum is indicated.

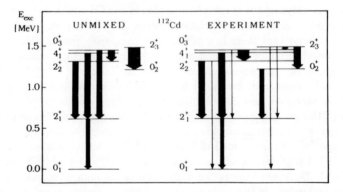

Figure 5: Comparison of the unmixed situation of an anharmonic vibrational and intruder band, including the B(E2) transitions, together with the corresponding experimental situation in ^{112}Cd. The thickness of the lines is proportional to the absolute B(E2) values.

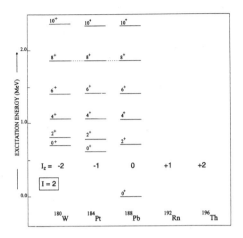

Figure 6: The various members of a I=2 multiplet in the Pb region encompassing a possible 4p-4h band in ^{188}Pb[21].

The classification of intruder multi-particle multi-hole (mp-nh) excitations can be extended to encompass also the various mp-nh excitations in a given nucleus where $m = a + r$ and $n = b + r$ with r=0,2,4,.... and a and b two even integers such that the $ap - bh$ configuration corresponds to the ground-state configuration of the nucleus under consideration. The operators connecting these various configurations are of the form

$$\{p_i^+ p_j, h_i^+ h_j, p_i^+ h_j^+, p_i h_j\} \qquad (11)$$

and taken together, form the non-compact algebra U(6,6). In this 'vertical' scheme, the possible reduction can be realized

$$U(6,6) \supset U_{ph}(6) \otimes U_K(1,1) \simeq U_{ph}(6) \otimes SU_K(1,1) \otimes U_-(1) \quad . \qquad (12)$$

The SU(1,1) is a non-compact algebra to which we associate the new quantum number, K-spin. Its generators are given by the expressions

$$\hat{K}_z \equiv \frac{1}{2} \sum_i \left(p_i^+ p_i + h_i^+ h_i \right) + 3 \quad , \quad \hat{K}_+ \equiv \sum_i p_i^+ h_i^+ \quad , \quad \hat{K}_- \equiv \sum_i p_i h_i \qquad (13)$$

These generators satisfy the commutation relations

$$\left[\hat{K}_z, \hat{K}_\pm \right] = \pm \hat{K}_\pm \qquad \left[\hat{K}_+, \hat{K}_- \right] = -2\hat{K}_z \qquad (14)$$

Remark the change of sign in the last commutator as compared to the corresponding relation for SU(2); this expresses the non-compact character of the vertical reduction.

The $U_-(1)$ algebra is generated by the difference of the number operators for the particle and hole bosons, which is a constant of motion in a given nucleus, and is given as the eigenvalue of the operator

$$\sum_i \left(p_i^+ p_i - h_i^+ h_i \right) \quad , \tag{15}$$

an operator which commutes with the generators of U(6,6).

The IBM studies by Barfield et al.[10] and Vretenar et al.[11,12,13] are calculations taking into account 2p-2h, 4p-4h, 6p-6h, ... configurations and can be used to test the validity of the above classification scheme. In these calculations, the aim was to reproduce the high-spin behaviour of collective bands and study the band-crossing regions. By studying the parameters used in the various boson number subspaces, the most particular change is related to (i) an overall reduction of the d-boson energy with increasing boson number, (ii) a general decrease in the quadrupole coupling strength with increasing boson number and,(iii) a decrease in χ_π with increasing boson number. These choices are clear indications for smooth changes in the K-structure (or of the boson number) in proceeding through the various members of the yrast band. At the band-crossing regions, sudden changes in K show up and, depending on the interaction strength between the various independent bands, a strong backbending (for weak coupling) and very smooth change in the yrast band structure (for strong coupling) can result.This feature is showing up in the yrast B(E2) values too.

A schematic model with a linear decrease of the d-boson energy as a function of the boson number i.e. has been studied quite some time ago by Heyde et al.[9] for the Dy nuclei in an IBM-1 framework using the same ideas discussed above. In Fig.7 the changing structure of the yrast members as a function of boson number with N=13 as starting value. Up to spin 12, even though the wave function is containing mixtures of $N+2$ and $N+4$ components, the relative amplitudes do not change very much. Above the crossing region, from spin 18 onwards, again rather stable mixing amplitudes appear with $N+6$ as the dominant amplitudo. Only in the spin region 14, 16 a rapid change in the structure of the wave functions (local perturbation) results.

At present, more detailed studies of this vertical classification need to be performed. Moreover, both the horizontal and vertical algebras (U(12) and U(6,6)) can be put into a larger algebraic structure by introducing the additional operators

$$\left\{ p_i^+ p_j^+, h_i^+ h_j^+, p_i p_j, h_i h_j \right\} \quad . \tag{16}$$

Together with the generators from (8) and (11), one obtains the non-compact algebra Sp(24) which has 300 generators. The various reductions are combined in Fig.8

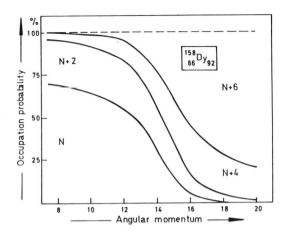

Figure 7: The various probabilities for a given boson number N in ^{158}Dy describing the wave function along the yrast band ($8 \leq J \leq 20$).

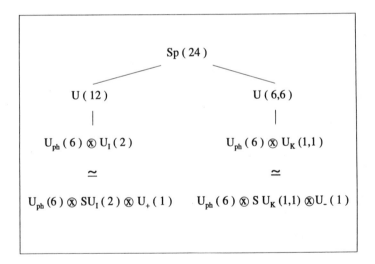

Figure 8: Group chain for the classification of intruder multi-particle multi-hole excitations indicating both the 'horizontal' U(12) and the 'vertical' non-compact U(6,6) group

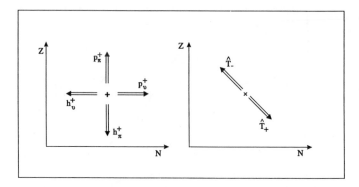

Figure 9: The various particle and hole boson creation operators acting on a given nucleus and the corresponding action in the (N,Z) plane for light nuclei. The motion in the (N,Z) plane implied by the isospin $\hat{T}_\pm$ operators are also indicated.

4. Supermultiplet structures and possible extension to light nuclei

In the discussion up to now, pair excitations all correspond to a given charge (proton, neutron). In the extension to study light nuclei where no longer a neutron excess is present to separate proton and neutron pair excitations, we should allow for both proton (π) and neutron (ν) pair excitations in describing particle-hole excitations in these light nuclei (Fig.9). Thereby, a still more general classification will be needed containing both the isospin and intruder-spin limiting symmetries in describing the multitude of many-particle many-hole (mp-nh) excitations.

It is our aim, in future, to construct a formalism where it is possible to connect the various mp-nh configurations including both isospin (T) and intruder spin (I). The most general case needs to study the s,d boson degrees of freedom (i,j=1,6) as well as the three charge pair states ($\pi\pi, \nu\nu$ and $\pi\nu$: $\rho = \pi, \nu, \delta$) in order to obtain a fully isospin invariant algebra. The algebra will be formed by the generators

$$\left\{ p_i^+ p_j, h_i^+ h_j, p_i^+ h_j, h_i^+ p_j \right\}_\rho \to U(36) \tag{17}$$

forming the U(36) algebra.

The possible reduction of this U(36) as well as a detailed study of the possible dynamical symmetries is a topic for study to be carried out in near future.In giving a close look the nuclei between ^{40}Ca and ^{56}Ni in which the $1f_{7/2}$ shell-model orbital is filled, one can observe strong similarities in band structures that give possible evidence for experimental realization of a combined isospin and intruder spin classification e.g.comparing the 4p-4h intruder band in ^{40}Ca and the ground-state band in ^{48}Cr. Here too, a much enlarged study of the experimental data is needed to look for more examples that could put a firm ground to an enlarged multiplet (supermultiplet?) structure in these light nuclei.

5. Conclusion

Particle-hole excitations in nuclei near closed shells are a major factor in determining nuclear structure at low energies. We have discussed various possibilities to incorporate the particle-hole pair excitations within the interacting boson model. Besides a discussion of the systematic variation of the lowest-lying 0^+ intruder state throughout the whole nuclear mass region, the collective bands built on these intruder states can be obtained by considering the particle-hole pairs as two additional bosons and carrying out configuration mixing studies. In contrast to these numerical studies, we propose a classification scheme in which particle and hole bosons are treated explicitly leading to the concept of intruder analog multiplets. We also point out the possibility to characterize many-particle many-hole excitations in a given nucleus using an extension towards non-compact groups. High-spin studies are a possible application of the latter scheme. Finally, we suggest a possible extension in constructing an isospin invariant way of making particle-hole pair excitations possible with applications in the light $1f_{7/2}$ nuclei.

The author is grateful to P.Van Isacker, J.Jolie, C.De Coster and J.L.Wood regarding the study of intruder analog states and the proposed new avenues in studying these topics. He likes to thank the NFWO and the IIKW for financial support.

6. References

1. Algebraic approaches to nuclear structure: Interacting Boson and Fermion models, Contemporary Concepts in Physics, vol.6 , ed.R.F.Casten (Harwood Academic Publ., 1993)

2. K.Heyde, P.Van Isacker, M.Waroquier, J.L.Wood and R.A.Meyer, *Phys.Repts.* **102** (1983), 291

3. J.L.Wood, K.Heyde, W.Nazarewicz, M.Huyse and P.Van Duppen, *Phys.Repts.* **215** (1992), 101

4. K.Heyde, J.Jolie, J.Moreau, J.Ryckebusch, M.Waroquier, P.Van Duppen, M.Huyse and J.L.Wood, *Nucl. Phys.* **A466** (1987), 189

5. K.Heyde, P.Van Isacker, R.F.Casten and J.L.Wood, *Phys. Lett.* **155B** (1985), 303

6. P.D.Duval and B.R.Barrett, *Phys. Lett.* **100B** (1981), 223; *Nucl. Phys.* **A376** (1982), 213

7. J.Jolie and K.Heyde, *Phys. Rev.* **C42** (1990), 2034

8. M.Délèze et al., *Nucl. Phys.* **A551** (1993), 269; *ibid* **A554** (1993), 1

9. K.Heyde, J.Jolie, P.Van Isacker, J.Moreau and M.Waroquier, *Phys. Rev.* **C21** (1984), 1428

10. A.F.Barfield and B.R.Barrett, *Phys. Rev.* **C44** (1991), 1454

11. F.Iachello and D.Vretenar, *Phys. Rev.* **C43** (1991), R945

12. D.Vretenar, V.Paar, G.Bonsignori and M.Savoia, *Phys. Rev.* **C44** (1991), 223

13. D.Vretenar, G.Bonsignori and M.Savoia, *Phys. Rev.* **C47** (1993), 2019

14. K.Heyde, C.De Coster, J.Jolie and J.L.Wood, *Phys. Rev.* **C46** (1992), 541

15. J.Kumpulainen et al., *Phys. Rev.* **C45** (1992), 640

16. R.F.Casten, J.Jolie, H.G.B"rner, D.S.Brenner, W.V.Zamfir,W.-T.Chou and A. Aprahamian, *Phys. Lett.* **B297** (1993), 19

17. J.Jolie, Proc. of this Conference and H.Lehman, Diplomarbeit, Univ. de Fribourg, (1994), unpubl.

18. K.Heyde, C.De Coster, J.L.Wood and J.Jolie, *Phys. Rev.* **C46** (1992), 2113

19. J.Heese et al., *Phys. Lett.* **B302** (1993), 390

20. W.Nazarewicz, *Phys. Lett.* **B305** (1993), 195

21. K.Heyde, P.Van Isacker and J.L.Wood, *Phys. Rev.* **C49** (1994), 559

FORMULATION AND APPLICATION OF MULTIPARTICLE-MULTIHOLE CONFIGURATIONS WITHIN THE PROTON-NEUTRON INTERACTING BOSON MODEL

B. R. BARRETT and A. F. BARFIELD

Department of Physics, Bldg. 81, University of Arizona, Tucson, AZ 85721

ABSTRACT

A straightforward extension of the two-configuration-mixing approach of Duval and Barrett within the proton-neutron Interacting Boson Model (IBM-2) to multiparticle-multihole configuration mixing is presented. An application of this extended model to ^{192}Hg is then given and discussed. Simply extrapolating the IBM-2 parameters with N_π does not yield a superdeformed (SD) band in ^{192}Hg. By taking ^{232}U as an N analogue for the SD band in ^{192}Hg, we can obtain IBM-2 parameters, based on ^{232}U, which produce an SD-like band in ^{192}Hg.

1. Introduction

The existence of low-lying intruder configurations[1] (i.e., nuclear shape coexistence) in nuclear structure is well established in several mass regions. In several cases, such as the mercury (Hg) isotopes, these intruder levels can be simply interpreted as configurations built from two-particle-two-hole excitations.[2,3] In other cases,[4,5] these low-lying intruders appear to be multiparticle-multihole configurations. Such low-lying multiparticle-multihole configurations may play a role in heavy-mass nuclei, particularly regarding superdeformed (SD) bands, as pointed out by Iachello.[6]

Here we describe the extension of the proton-neutron Interacting Boson Model (or IBM-2) to include several excited multiparticle-multihole (np-mh) configurations along with the ground-state (gs) band within the same formalism. The basic idea is one of expanding the two-configuration-mixing method of Duval and Barrett[2] within the IBM-2 to several configurations, where each configuration represents a different np-mh state. Part of the motivation for this extension is to see if np-mh configurations will produce SD bands and, if so, at what levels of particle-hole excitation for both the protons and the neutrons. Because of earlier IBM-2 investigations for the Hg isotopes,[2,3] we choose ^{192}Hg[7,8] as a test case. Identical band structure[9] in the gs band in ^{232}U and in the SD band in ^{192}Hg suggests that the N analogue approach [6,10] can be utilized in determining appropriate IBM-2 parameter values for describing the SD band.

In section 2 the multi-configuration-mixing formalism is developed and then applied, as an example, to ^{192}Hg in section 3. In section 4 the concept of N analogues is described and applied to the SD band in ^{192}Hg by way of ^{232}U. Conclusions are given in section 5.

2. Multi-Configuration-Mixing Formalism

The basic IBM-2 of interacting s ($J=0$) and d ($J=2$) proton and neutron bosons has been described in the literature[11,12] and elsewhere in these proceedings. We assume a standard IBM-2 Hamiltonian of the form:

$$H = \epsilon(\hat{n}_{d_\pi} + \hat{n}_{d_\nu}) + \kappa \hat{Q}_\pi \cdot \hat{Q}_\nu + V_{\pi\pi} + V_{\nu\nu} + M_{\pi\nu}, \tag{1}$$

where ϵ is the excitation energy of a d boson, $\hat{n}_{d_\pi}$ is the number operator for d bosons, κ is the strength of the quadrupole-quadrupole interaction between proton and neutron bosons, $V_{\rho\rho}$ is the residual interaction between alike bosons ($\rho=\pi$ or ν), and $M_{\pi\nu}$ is a Majorana term that separates configurations of different neutron-proton symmetry and, in particular, insures that the low-lying configurations are mainly symmetric. The quadrupole operator is given by

$$\hat{Q}_\rho = (s^\dagger \tilde{d} + d^\dagger s)^{(2)}_\rho + \chi_\rho (d^\dagger \tilde{d})^{(2)}_\rho, \tag{2}$$

where $\rho=\pi$ or ν, and $s^\dagger$, s, $d^\dagger$, and $\tilde{d}$ are spherical tensor operators that create and annihilate s and d bosons, respectively. The alike boson interaction is of the form

$$V_{\rho\rho} = \sum_{L=0,2,4} \frac{1}{2}(2L+1)^{1/2} C_{L\rho} \left[(d^\dagger d^\dagger)^{(L)} (\tilde{d}\tilde{d})^{(L)} \right]^{(0)}_\rho, \tag{3}$$

while the Majorana term is given by

$$M_{\pi\nu} = \lambda \left[(s_\nu^\dagger d_\pi^\dagger - d_\nu^\dagger s_\pi^\dagger)^{(2)} \cdot (s_\nu \tilde{d}_\pi - \tilde{d}_\nu s_\pi)^{(2)} - 2 \sum_{k=1,3} (d_\nu^\dagger d_\pi^\dagger)^{(k)} \cdot (\tilde{d}_\nu \tilde{d}_\pi)^{(k)} \right]. \tag{4}$$

Within the basic IBM-2 formalism one obtains a low-lying nuclear spectrum utilizing the Hamiltonian (1) and some given number of valence proton and neutron bosons. Nuclear shape coexistence then results within this formalism by exciting pairs of alike nucleons across the major-shell gap, so as to increase the effective number of valence nucleons and, hence, of "bosons".

The original shape coexistence calculations of Duval and Barrett[2] considered the Hg isotopes, which have only one-proton-boson hole, since $Z=80$ for Hg. Because the nuclear deformation is driven by the $\hat{Q}_\pi \cdot \hat{Q}_\nu$ term in the IBM-2 Hamiltonian (1) and its strength is proportional to $N_\pi N_\nu$, the nuclear spectrum looks vibrational or U(5)-like for $N_\pi=1$ and arbitrary values of N_ν, while it becomes more rotational or SU(3)-like for larger values of N_π as N_ν increases. (N_ρ = the number of bosons of type ρ, where $\rho=\pi$ or ν.)

The method of Duval and Barrett (hereafter referred to as DB) considers the lowest possible excitation, that of one proton pair across the major-shell gap. This leads to a 2p-4h proton configuration for the Hg isotopes (i.e., one-proton-particle boson and two proton-hole bosons), which is treated as being "effectively" three proton-hole bosons.[13] However, the schematic-model calculations of Kaup and Barrett[5] indicate that multiparticle-multihole excitations are energetically favored. In fact, in the Kaup-Barrett method it is the maximum allowed excitation, namely, the half filling of the empty shell above the gap, that leads to the deformed configuration of lowest energy.

The extension of the basic DB method is straightforward, namely, to consider excitations of more pairs of protons across the major-shell gap, leading successively to (4p-6h), (6p-8h), etc., boson configurations. The IBM-2 calculations for increasing effective boson number are performed in the usual manner with appropriate changes in the parameters based on their empirically determined variations with N_π or on semimicroscopic arguments.

One then needs to consider the mixing of these multiparticle-multihole configurations. In the DB approach, only two configurations (i.e., $N_\pi=1$ and $N_\pi=3$) are considered for a given value of N_ν. A general mixing interaction of the following form is employed for connecting the two configurations in their effective boson spaces:

$$V_{\mathrm{mix}} = \alpha(s_\pi^\dagger s_\pi^\dagger + s_\pi s_\pi)^{(0)} + \beta(d_\pi^\dagger d_\pi^\dagger + \tilde{d}_\pi \tilde{d}_\pi)^{(0)}. \quad (5)$$

The generalization of this interaction to include the mixing of several configurations, all with different boson numbers (i.e., N_π, $N_\pi+2$, $N_\pi+4$, etc.), would be quite difficult and complicated. For simplicity, we assume that only neighboring configurations couple strongly and neglect any coupling between configurations that differ by more than two proton (or neutron) bosons. That is, $N_\pi=1$ couples with $N_\pi=3$, and $N_\pi=3$ with $N_\pi=5$, etc., but $N_\pi=1$ does *not* couple with $N_\pi=5$ (all such mixing matrix elements are *assumed* to be zero), etc.

A separate IBM-2 calculation is done for each value of N_π, using the computer code NPBOS[14] and the results for the various configurations are then combined. The mixing Hamiltonian matrix has the form

$$H_{\mathrm{mix}} = \begin{bmatrix} (\lambda_1) & [V_{\mathrm{mix}}(1,2)] & (0) & (0) & \cdots \\ [V_{\mathrm{mix}}(1,2)] & (\lambda_2+\Delta_1) & [V_{\mathrm{mix}}(2,3)] & (0) & \cdots \\ (0) & [V_{\mathrm{mix}}(2,3)] & (\lambda_3+\Delta_2) & [V_{\mathrm{mix}}(3,4)] & \cdots \\ (0) & (0) & [V_{\mathrm{mix}}(3,4)] & (\lambda_4+\Delta_3) & \cdots \\ \cdots & \cdots & \cdots & \cdots & \cdots \end{bmatrix}, \quad (6)$$

where λ_1 is same subset of the IBM-2 (NPBOS) eigenvalues for the normal configuration, the λ_i contain the IBM-2 eigenvalues for the excited configurations, $[V_{\mathrm{mix}}(i,j)]$ is defined by (5), and the Δ_i are the relevant pair-excitation energies. Diagonalization of this matrix for each value of the angular momentum leads to the eigenenergies and eigenvectors of the mixed configurations.

In addition to the strength parameters α and β in (5), the mixing calculation requires Δ_1, Δ_2, etc., which represent the energies needed to excite the pair(s) across the gap. The simplest choice for the multiple-pair excitations would be to take multiples of Δ_1: Δ_1 for 2p-2h, $2\Delta_1$ for 4p-4h, $3\Delta_1$ for 6p-6h, etc., but this is not realistic because the nucleus deforms with increasing particle-hole excitations leading, in general, to a smaller shell gap and, hence, a lower excitation energy for the next pair of nucleons. Heyde et al.[15] have studied this change in Δ in some detail but have not considered the problem for multiple-pair excitations and the mixing of several configurations. In the next section, we present a simple scaling argument, based on the Nilsson model, to estimate the changes in the (np-mh) excitation energy with increasing N_π.

3. Application to ^{192}Hg

Because of our previous studies of configuration mixing in the mercury isotopes[2,3] and because of Iachello's suggestion[6] that the mixing of the multiparticle-multihole configurations in the IBM might be a useful approach to investigating SD bands in nuclei such as ^{192}Hg, we have chosen ^{192}Hg as a test for our multi-configuration-mixing approach in the IBM-2.

Extensive calculations have been performed for ^{192}Hg, so as to test this approach. The main points of interest are to study how the results change 1) when the number of (np-mh) configurations is varied and 2) when different choices and/or assumptions are made for the multipair excitation energies Δ_i. Detailed agreement with the experimental data is not the goal of this particular investigation. The complete results of these exploratory studies are presented in Ref.[16]. Typical results will be shown here.

Two cases are considered regarding the first point: (1) $N_\pi=1$, 3, 5, and 7 and $N_\nu=7$ (four configurations) and (2) $N_\pi=1$, 3, 5, 7 and 9 and $N_\nu=7$ (five configurations). For convenience the configurations are referred to as 1π (normal configurations), 3π (2p-4h), 5π (4p-6h), etc. We note that approximating the higher excited configurations by $(n+m)/2$ proton-boson holes is a great oversimplification, because the particles and holes are in different major shells and because the "structure of the bosons" will change as more and more nucleons are excited across the major-shell gap. However, this assumption is a reasonable first approximation for testing the model, before attempting a more complicated description, and can be justified by F-spin arguments[13] and by studies on intruder analogue states.[10]

The second point is illustrated by two cases: (1) $\Delta_n = n\Delta_1$, i.e., equally spaced two-particle excitation energies, and (2) $\Delta_n < n\Delta_1$, for $n > 1$. The latter is consistent with the expectation that the energy needed to excite the second, third, etc., pair of nucleons should be less than the energy needed to excite the first pair, because the nucleus has become deformed in the process.

For the second case, we estimate the energies Δ_n (the energy to excite n pairs) relative to Δ_1 from the Nilsson diagram for protons, $50 \leq Z \leq 82$.[17] For simplicity, we consider only differences in single-particle (or quasiparticle) energies and ignore pairing and other effects.

The energy to excite the first pair of protons across the $Z=82$ shell gap is roughly twice the difference in energy of the $h_{9/2}$ and $d_{3/2}$ levels, for zero deformation, so that $E_1=1\hbar\omega_0$ (from the Nilsson diagram). The nucleus, with $Z=80$, is now deformed and can be approximately described by two protons in $\frac{3}{2}^-[532]$, two holes in $\frac{3}{2}^+[651]$, and

Table 1: IBM-2 Hamiltonian parameters for ^{192}Hg employed in the calculations. For all configurations, λ=0.15 MeV (FS=0.15, FK=0). All parameters are in MeV, except for χ_ν and χ_π, which are dimensionless.

N_ν	N_π	ϵ	κ	χ_ν	χ_π	$C_{0\nu}$	$C_{2\nu}$	$C_{4\nu}$
7	1	0.68	-0.165	0.4	1.0	0.63	0.05	0.14
7	3	0.35	-0.145	0.4	-1.3	0.00	0.00	0.04
7	5	0.35	-0.12	0.4	-1.8	0.00	0.00	0.00
7	7	0.35	-0.10	0.4	-2.0	0.00	0.00	0.00
7	9	0.35	-0.08	0.4	-2.0	0.00	0.00	0.00

two holes in $\frac{1}{2}^+$[660], with lower orbits filled. Exciting another pair of protons from $\frac{1}{2}^-$[651] to $\frac{1}{2}^-$[530] requires additional energy E_2=0.4$\hbar\omega_0$ (for deformation ϵ=0.27), to give a 4p-6h configuration. Subsequently, exciting a pair from $\frac{9}{2}^-$[514] to $\frac{11}{2}^-$[505] requires E_3=0.6$\hbar\omega_0$ and yields a 6p-8h configuration, etc. The phenomenological fits for the Hg chain have employed the value Δ_1=4 MeV to describe the pair excitation energy for the 3π configuration. If this corresponds to E_1, then we estimate the energy needed to excite two pairs of protons to be $\Delta_2 = E_1 + E_2 \simeq 6$ MeV, rounding to the nearest MeV. Similarly, $\Delta_3 = E_1 + E_2 + E_3 \simeq 8$ MeV, etc.

A more difficult problem is the determination of the parameters to use in the IBM-2 Hamiltonian (1). Finding appropriate parameters with which to describe the excited configurations is difficult because little, if any, data are available. One possible description employs parameter values that give reasonable fits to known nuclei that have the same number of valence protons and neutrons, e.g., ^{192}Hg$_{112}$ (Z=80), ^{188}Os$_{112}$ (Z=76), ^{168}Hf$_{96}$ (Z=72), and ^{164}Er$_{96}$ (Z=68), for N_ν=7 and N_π=1, 3, 5, and 7, respectively. This method was rejected because the resulting energies for the 3π configuration are much too high, relative to the deformed bands in $^{182-190}$Hg, for which data are available. This poor result is not surprising, in view of the simplistic effective boson number approximation for the particle-plus-hole situation. Instead, we utilize previously determined ^{192}Hg parameter values for the 1π and 3π configurations either by keeping them fixed at the 3π values or by extending the trends for the 1π and 3π cases.

The parameter values of Barfield[18] are used in tact to describe both the 1π (normal) and the 3π (2p-4h) configurations. These parameters result from an overall fit to the even mercury isotopes $182 \leq A \leq 198$, which gives a good description both for the normal vibrational-like states and for the shape-coexisting rotational-like band seen in the light nuclides $^{180-190}$Hg.

The proton-proton interaction $V_{\pi\pi}$ in Eq.(1) is *not* included for any of the configurations, and the neutron-neutron interaction $V_{\nu\nu}$ is dropped for the 5π and higher configurations. The mixing strengths in Eq.(6) are held constant and taken from previous calculations[3], α=0.15 MeV, β=0.10 MeV. The IBM-2 parameter values for the various configurations are given in Table 1.

Although the mixing strengths α and β are held constant, the matrix elements of V_{mix} [Eqs.(5) and (6)] are state dependent, increasing in magnitude with increasing boson number. For example, for angular momentum $J=0$ and the IBM-2 parameter values utilized, $V_{\text{mix}}(1,2)=-0.163$ MeV, $V_{\text{mix}}(2,3)=0.437$ MeV, $V_{\text{mix}}(3,4)=-0.638$ MeV, and $V_{\text{mix}}(4,5)=-0.856$ MeV. The magnitude of these matrix elements decreases somewhat with increasing angular momentum. (The sign of the matrix elements depends on the phase convention adopted for the IBM-2 eigenfunctions and is unimportant.)

Figure 1 shows the results for four configurations with the successive multipair excitation energies taken to be Δ, 2Δ and 3Δ, where $\Delta=4$ MeV is the value employed in the two-configuration-mixing calculations. For simplicity, only the gs bands for each configuration are included in the calculation, with spin up to $20\hbar$. There is very little mixing among the excited bands because of the initial, large separation energies assumed between the unmixed bands. The experimental yrast levels,[19] denoted as ×'s, are shown for reference. We note that the calculated levels for $J \geq 10$ do not correspond to the experimental levels shown, because the latter states are known to be two-quasiparticle (2qp) in nature[19] and are outside of our model space. However, the yrast states for $J \leq 6$, which are clearly collective, are well reproduced by our model. (Rotation-aligned bands cannot be reproduced within the IBM unless 2qp degrees of freedom are explicitly added to the model space, as discussed elsewhere in these proceedings.)

Figure 2 shows results after mixing for the same four configurations, with the multipair excitation energies $\Delta_1=4$ MeV, $\Delta_2=6$ MeV, and $\Delta_3=8$ MeV, as estimated from the Nilsson diagram. There is now considerable mixing among the excited configurations, even for spin $J=0$. The addition of a 9π configuration has the effect of lowering the energies of the other states, as shown in Ref.[16].

From theoretical investigations,[20] it is known that the "effective moment of inertia" of the gs band in IBM-2 calculations is inversely proportional to the values of both ϵ and κ in Eq.(1). This effect is clearly seen in Figs.1 and 2, where the moment of inertia is significantly smaller for the 1π band than for the other three bands, because its ϵ value is about twice as big as the ϵ value for the other bands. The moments of inertia for the excited bands are not very different, but do increase with N_π, as κ decreases.

In principle, one should also include neutron-pair excitations. We have, however, not found a way to predict the change in IBM-2 parameter values when increasing both the proton and neutron boson numbers, and, therefore, will not attempt the addition of excited neutron pairs to the model at this time.

As a general feature of the results, we note that increasing the number of excited pairs does increase the effective moment of inertia, so that the band with the largest number of excited pairs does possess the largest moment of inertia. On the other hand, none of the bands obtained has an effective moment of inertia large enough to correspond to that of a SD band, as observed in ^{192}Hg.[7,8]

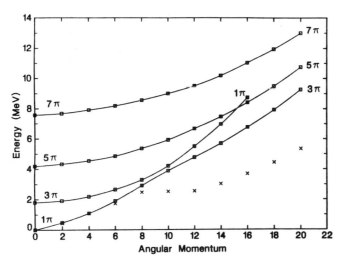

Figure 1: Calculated energy levels after mixing for the ground-state bands of the 1π, 3π, 5π, and 7π configurations of ^{192}Hg with multipair excitation energies Δ_1 through Δ_3 equal to 4.0, 8.0, and 12.0 MeV, respectively. The experimental yrast band, denoted by crosses, is shown for comparison. For $J > 8$, these states are 2qp in nature and should not be reproduced by the calculation because they are outside of the model space. The data are from Ref.[19].

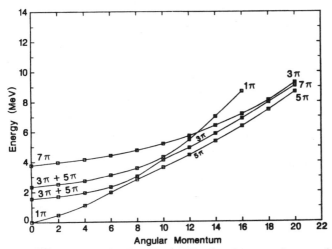

Figure 2: Calculated energy levels after mixing for the same four configurations as in Fig.1 (1π, 3π, 5π, and 7π), with Δ_1 through Δ_3 equal to 4.0, 6.0, and 8.0 MeV, respectively.

Table 2: IBM-2 parameters for ^{232}U ($N_\pi=5$, $N_\nu=7$) from Czarnowski's fit[21] (1981) and from the present fit (1994) to the low-lying experimental spectrum. All parameter values are in MeV, except χ_ν and χ_π, which are dimensionless.

Nuclide	ϵ	κ	χ_ν	χ_π	$C_{0\nu}$	$C_{2\nu}$	$C_{4\nu}$	λ
^{232}U(1981)	0.26	-0.046	-0.95	-1.55	-0.1	0.0	0.0	0.04
^{232}U(1994)	0.27	-0.049	-1.15	-1.20	0.0	0.0	0.0	0.15

4. N Analogues, the Uranium Isotopes, and SD bands

A comparison of the gamma-ray energies for the SD band in ^{192}Hg with those for the lower part of the gs band of ^{232}U indicates a striking similarity, suggesting a similar structure.[9] As mentioned earlier, Iachello[6] and Heyde and coworkers[10] have pointed out the significance of multiparticle-multihole configurations as N analogues of gs configurations. In particular, the gs band in ^{232}U and the excited $N_\pi=5$ band in ^{192}Hg (as described in section 3) are N analogues, because they have the same number of active (interactive) protons and neutrons. For example, $^{232}_{92}$U$_{140}$ has five valence proton bosons and seven valence neutron bosons ($N=N_\pi+N_\nu=12$), while $^{192}_{90}$Hg$_{112}$ has one valence (hole) proton bosons and seven valence (hole) neutron bosons. If the SD band in ^{192}Hg is considered to be a (4p-6h) excited proton state, then the "effective" valence (hole) proton boson number is 5 ($N_\pi=5$), so that $N=N_\pi+N_\nu=12$. If nuclei with the same number of active protons and neutrons have the same structure, then we should be able to describe the $N_\pi=5$ band in ^{192}Hg with the IBM-2 parameters for the gs band in ^{232}U.

An IBM-2 calculation has been performed[9] for ^{232}U, utilizing parameters by Czarnowski.[21] It was found that this yields a gs band for ^{232}U which is in good agreement with experiment but is too spread out for the high-spin states. On the other hand, these parameters yield a band for the higher-spin states which has gamma-ray energies that are similar to the SD band in ^{192}Hg. Czarnowski's parameter fits included the first-excited 0$^+$ state in ^{232}U as being the bandhead of the β-band, while it is now known that many low-lying 0$^+$ states in the actinides are not β vibrations but are of different collective structure.[22] The work of Daley[23] indicates that these first-excited 0$^+$ states are really cluster-like configurations and not the β-bandhead states.

A new determination of the IBM-2 parameters [in Eq.(1)] for the uranium isotopes has been undertaken, starting with the Czarnowski values,[21] but ignoring the excited 0$^+$ states; including other data besides energy levels, such as branching ratios; and updating some of the other parameter values. This is consistent with the two configuration mixing calculations for the Cd isotopes by Délèze, et al.,[24] who found that reliable IBM-2 parameter values were obtained only when all available experimental data were included in the fit. Table 2 shows the present (1994) set of parameter values for ^{232}U, along with those of Czarnowski (1981). The recent IBM-2 fits to the uranium isotopes yield parameter values which produce a stiffer

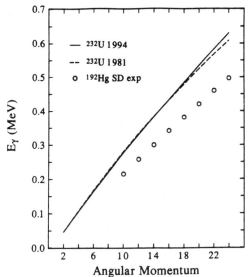

Figure 3: Gamma transition energies, $E_\gamma = E_J - E_{J-2}$, for ^{232}U calculated with 1981 parameter values (dashed line) (Ref.[21]) and with 1994 parameter values (solid line) (See Table 2).

band (i.e., a more nearly constant moment of inertia), but the trend in the energy spectrum is roughly the same as that obtained using Czarnowski's parameters, as can be seen in Figure 3, which compares the calculated γ-ray energies for both ^{232}U fits with the experimental SD band in ^{192}Hg. It should be noted that the SD band in ^{192}Hg needs to be shifted by two units of angular momentum in order to coincide with the gs band in ^{232}U. Figure 4 shows the results of a three-configuration ($N_\pi=1$, 3 and 5, $N_\nu=7$) calculation for ^{192}Hg, using Czarnowski's parameter values for the $N_\pi=5$ band. The 1π and 3π configurations are mixed, but the 3π and 5π configurations are not. This calculation yields a $N_\pi=5$ band with more-or-less the correct moment of inertia and E_γ to correspond to the SD band in ^{192}Hg.

Hence, the assumption of the using gs-band N analogues to determine the IBM-2 parameters of the multiparticle-multihole excited configurations appears to be a reasonable one. Although a $N_\pi=5$, $N_\nu=7$ configuration in ^{192}Hg is probably not collective enough to produce an SD band, one can interprete this configuration in terms of "bosons" whose structure is different than that of the other configurations, so that they are like the superbosons[25] or the supershells [26] of other approaches. Thus, the uranium isotopes serve as a means of determining the IBM-2 parameters for the "superbosons" appropriate for the SD band in ^{192}Hg. If the $N_\pi=5$ band consists of bosons of a different structure from those in the $N_\pi=1$ and 3 bands, then it would be consistent with our assumed lack of mixing between the $N_\pi=3$ and 5 bands.

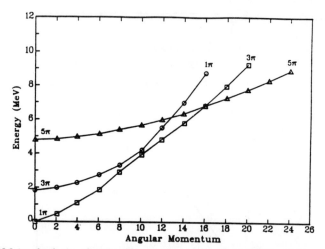

Figure 4: IBM-2 calculation for the $N_\pi=1$, 3 and 5 bands in ^{192}Hg with configuration mixing between the $N_\pi=1$ and 3 bands. The $N_\pi=1$ and 3 parameters are those of Table 1. The 1981 parameters,[9,21] given in Table 2, are utilized for the $N_\pi=5$ band.

5. Conclusions

The generalization of the IBM-2 configuration-mixing method of Duval and Barrett[2] from two to several configurations, each corresponding to a different (np-mh) excitation of the nucleus, has been presented. The main assumption is that only "nearest-neighbor" configurations, defined as N_ρ and $N_\rho+2$ ($\rho=\pi$ or ν), mix strongly. The principal problem in performing calculations is that of choosing the IBM-2 parameter values to use with each multiparticle-multihole configuration included in the calculation. It is assumed that the excitation energy of the first pair of nucleons across the shell gap is larger, in general, than the energy needed to excite subsequent pairs.

As a first test, this generalized configuration-mixing method is applied to ^{192}Hg, where parameter values for the low-lying configurations are estimated from trends for the first two configurations and from semimicroscopic considerations. The final results strongly depend on the choice of the multipair excitation energies. Assuming large, equally spaced excitation energies produces no significant mixing of the excited bands. Decreasing the multipair energies relative to the one-pair excitation energy results in significant mixing at low spins. Using parameters based on an extrapolation of the parameters for the first two configuration leads to no results resembling an SD band for any of the mixed configurations, indicating that this approach would require a larger number of excited proton-pairs and neutron-pairs in order to produce a SD band.

On the other hand, the N analogue argument provides a way of determining IBM-2 parameter values for a gs band, which can then be used for describing excited-state configurations consisting of the same number N of interacting protons and neutrons. In this case the bosons in the excited-state configurations can be thought of as "superbosons", which require a smaller number in order to have a larger collectivity. An example of this N analogue approach was given in terms of the gs band in ^{232}U and the $N_\pi=5$ excited band in ^{192}Hg.

6. Acknowledgment

This work was supported in part by the U.S. NSF Grant PHY-9103011.

7. References

1. J.L. Wood, K. Heyde, W. Nazarawicz, M. Huyse, and P. Van Duppen, Phys. Rep. **215** (1992) 1, and references therein.
2. P.D. Duval and B.R. Barrett, Nucl. Phys. **A376** (1982) 213; Phys. Lett. **100B** (1981) 223.
3. A.F. Barfield, B.R. Barrett, K.A. Sage, and P.D. Duval, Z. Phys. A **311** (1983) 205; A.F. Barfield, Ph.D. dissertation, University of Arizona, 1986.
4. H. Liu and L. Zamick, Phys. Rev. C **29** (1984) 1040; H. Liu, L. Zamick, and H. Jaqaman, *ibid.* **31** (1985) 2251; D.C. Zheng, D. Berdichevsky, and L. Zamick, *ibid.* **38** (1988) 437.
5. U. Kaup and B.R. Barrett, Phys. Rev. C **42** (1980) 981.
6. F. Iachello, Nucl. Phys. **A522** (1991) 83c.
7. J.A. Becker, *et al.*, Phys. Rev. C **41** (1990) 9.
8. D. Ye, *et al.*, Phys. Rev. C **41** (1990) 13.
9. A.F. Barfield and B.R. Barrett, Nucl. Phys. **A 557** (1993) 551c.
10. K. Heyde, C. De Coster, J. Jolie, and J.L. Wood, Phys. Rev. C **46** (1992) 541; K. Heyde, contribution to this conference.
11. A. Arima and F. Iachello, in *Advances in Nuclear Physics,* edited by J.W. Negele and E. Vogt (Plenum, New York, 1984), Vol.13.
12. F. Iachello and A. Arima, *The Interacting Boson Model* (Cambridge University, London, 1987).
13. B.R. Barrett, Mod. Phys. Lett. A **7** (1992) 1391.
14. T. Otsuka and O. Scholten, KVI Internal Report No. 253, 1979.
15. K. Heyde, J. Jolie, J. Moreau, J. Ryckebush, M. Waroquier, P. Van Duppen, M. Huyse, and J.L. Wood, Nucl. Phys. **A466** (1987) 189; K. Heyde, J. Jolie, J. Moreau, J. Ryckebush, M. Waroquier, and J.L. Wood, Phys. Lett. **176B** (1986) 255; K. Heyde, P. Van Isacker, R.F. Casten, and J.L. Wood, *ibid.* **155B** (1985)303.
16. A.F. Barfield and B.R. Barrett, Phys. Rev. C **44** (1991) 1454.
17. *Table of Isotopes,* 7th ed., edited by C.M. Lederer and V.S. Shirley (Wiley, New York, 1978).
18. A.F. Barfield, Contributed Papers to the International Conference on Contemporary Topics in Nuclear Structure Physics, Cocoyoc, 1987 (unpub-

lished).
19. H. Hübel, A.P. Byrne, S. Ogaza, A.E. Stuchbery, and G.D. Dracoulis, Nucl. Phys. **A453** (1986) 316.
20. S. Kuyucak and I. Morrison, Ann. Phys. (NY) **195** (1989) 216.
21. W.M. Czarnowski, private communication. Final results unpublished, preliminary results reported in W.M. Czarnowski, B.R. Barrett, P.D. Duval, and K.A. Sage, Bull. Am. Phys. Soc. **26** (1981) 482.
22. Y.A. Ellis-Akozali, Nucl. Data Sheets **40** (1983) 523; E.N. Shurshikov, Nucl. Data Sheets **53** (1988) 601; and references therein.
23. H.J. Daley and B.R. Barrett, Nucl. Phys. **A449** (1986) 256; H.J. Daley, Ph.D. dissertation, University of Arizona, 1984.
24. M. Délèze, et al., Nucl. Phys. **A 551** (1993) 269; ibid. **A 554** (1993) 1.
25. T. Otsuka and M. Honma, Phys. Lett. **B268** (1991) 305; M. Honma, Ph.D. dissertation, University of Tokyo, 1994.
26. C.L. Wu, D.H. Feng, and M.W. Guidry, Ann. Phys. (NY) **272** (1993) 187.

PARTIAL DYNAMICAL SYMMETRY IN THE INTERACTING BOSON MODEL

AMIRAM LEVIATAN
Racah Institute of Physics, The Hebrew University
Jerusalem 91904, Israel

ABSTRACT

The concept of partial symmetry is explained and motivated. We show the relevance of partial dynamical $SU(3)$ symmetry for the spectroscopy of axially deformed nuclei as well as to studies of mixed systems which exhibit coexistence of chaotic and regular forms of dynamics.

1. Introduction

The Interacting Boson Model[1] (IBM) describes collective states in even-even nuclei in terms of interacting monopole $(s^{\dagger})$ and quadrupole $(d_{\mu}^{\dagger})$ bosons. A conserved boson number, $\hat{N} = s^{\dagger}s + \sum_{\mu} d_{\mu}^{\dagger} d_{\mu}$, confers on the model a group structure of $U(6)$. One of the attractive features of the IBM is the occurrence of dynamical symmetries for which the Hamiltonian is written in terms of the Casimir operators of a chain of nested subgroups of $U(6)$. A dynamical symmetry provides considerable insight since it allows all properties of the system, e.g. spectrum and transition rates, to be calculated in closed analytic form. The labels of irreducible representations (irreps) of the groups in the chain serve as quantum numbers to classify the eigenvalues and eigenstates of the Hamiltonian. The corresponding wave functions are determined solely by symmetry and are therefore independent of parameters in the Hamiltonian.

The IBM exhibits three dynamical symmetries associated with the chains

$$\begin{aligned} U(6) &\supset U(5) &\supset O(5) &\supset O(3) \\ U(6) &\supset SU(3) &\supset O(3) \\ U(6) &\supset O(6) &\supset O(5) &\supset O(3) \end{aligned} \quad (1)$$

These exact symmetries impose severe constraints on the corresponding spectrum (e.g. particular band degeneracies) which are rarely observed in realistic nuclei. More often one finds that these symmetries are not obeyed uniformly, i.e., some levels fulfill the symmetry while other levels do not. At times, we see an indication for a good symmetry in some states, although there is no justification for assuming the Hamiltonian to be invariant (e.g. F-spin invariance is incompatible with a microscopic interpretation of the IBM). These observations lead naturally to the notion of partial symmetries (i.e. exact symmetries but only for a selected set of states) which is the subject of the present contribution. We present an explicit construction of IBM Hamiltonians with partial dynamical $SU(3)$ symmetry and show its relevance to the spectroscopy of axially deformed nuclei, e.g ^{168}Er. In addition to discrete spectroscopy, partial dynamical symmetries (pds) are useful for studying statistical aspects of mixed systems which are partly chaotic and partly regular. Part of the work reported here was done in collaboration with N. Whelan (Niels Bohr Institute) and Y. Alhassid (Yale University).

2. Exact Dynamical SU(3) Symmetry

The $SU(3)$ dynamical symmetry is relevant for axially deformed nuclei and corresponds to the chain,

$$\begin{array}{cccc} U(6) & \supset & SU(3) & \supset & O(3) \\ \downarrow & & \downarrow & & \downarrow \\ [N] & & (\lambda,\mu) & K & L\, M \end{array} \quad (2)$$

The basis states are are labeled by $|[N](\lambda,\mu)KLM\rangle$, where N is the total number of bosons, L the angular momentum and (λ,μ) denote the $SU(3)$ irreps. For $SU(3)$ dynamical symmetry, the Hamiltonian and spectrum have the form

$$H(SU(3)) = a_1 \hat{C}_{SU(3)} + a_2 \hat{C}_{O(3)} \quad (3a)$$

$$E = a_1 \left[\lambda^2 + \mu^2 + \lambda\mu + 3(\lambda + \mu)\right] + a_2 L(L+1) \quad (3b)$$

where $\hat{C}_G$ denotes the Casimir operator of the group G, $\hat{C}_{SU(3)} = 2Q^{(2)} \cdot Q^{(2)} + (3/4)L^{(1)} \cdot L^{(1)}$, and $\hat{C}_{O(3)} = L^{(1)} \cdot L^{(1)}$. The generators of $SU(3)$ are the quadrupole $Q^{(2)}$ and angular momentum $L^{(1)}$ operators

$$Q^{(2)}_\mu = d^\dagger_\mu s + s^\dagger \tilde{d}_\mu - \frac{1}{2}\sqrt{7}\,(d^\dagger \tilde{d})^{(2)}_\mu \;\; ; \;\; L^{(1)}_\mu = \sqrt{10}(d^\dagger \tilde{d})^{(1)}_\mu \quad (4)$$

The lowest $SU(3)$ irrep $(2N,0)$ has rotational states with $L = 0^+, 2^+, 4^+, \ldots, (2N)^+$ ($K = 0$) resembling the ground band of an axially deformed prolate nucleus. The first excited $SU(3)$ irrep $(2N-4,2)$ contains both the β ($K=0, L=0^+, 2^+, 4^+, \ldots$) and γ ($K=2, L=2^+, 3^+, 4^+, \ldots$) bands which are degenerate. This characteristic degeneracy in the $SU(3)$ dynamical symmetry, is not observed in most deformed nuclei[2] as is evident from the experimental spectrum of ^{168}Er, shown in Fig. 3. To conform with the experimental data one is therefore compelled to break the $SU(3)$ symmetry. To do so, the usual approach has been to include in the Hamiltonian terms from other chains, in particular, adding the Casimir of $O(6)$ to the $SU(3)$ Hamiltonian (3a) so as to lift the undesired β-γ degeneracy[3]. In this procedure the $SU(3)$ symmetry is completely broken, all eigenstates are mixed and no analytic solutions are retained. The question we wish to explore is whether it is possible to break the symmetry but in a very particular way so that **part** of the states (but not all!) will still be solvable with good symmetry. We will refer to this situation as partial dynamical symmetry[4], to indicate that the virtues of a dynamical symmetry (e.g. solvability) are fulfilled but by only a subset of states. The construction of Hamiltonians with partial symmetries is non-trivial since such hamiltonians cannot be scalars of the symmetry group (otherwise all states will have the symmetry). The IBM played a decisive role in developing an algorithm for such a novel symmetry construction.

3. Partial Dynamical SU(3) Symmetry

The key ingredient in constructing IBM Hamiltonians with $SU(3)$ partial dynamical symmetry is an intrinsic state which is a condensate of N bosons[5]

$$|c; N\rangle = (N!)^{-1/2}[(s^\dagger + \sqrt{2}d_0^\dagger)/\sqrt{3}\,]^N|0\rangle \quad . \tag{5}$$

It is the lowest weight state in the $SU(3)$ irrep $(\lambda, \mu) = (2N, 0)$. The rotational members of the ground band with good angular momentum can be obtained by projection. IBM Hamiltonians which have $|c; N\rangle$ as an eigenstate were encountered in the study of the intrinsic structure of the model[6]. They have the generic form,

$$H(h_0, h_2) = h_0 P_0^\dagger P_0 + h_2 P_2^\dagger \cdot \tilde{P}_2 = h_0 P_0^\dagger P_0 + h_2 \sum_\mu P_{2,\mu}^\dagger P_{2,\mu} \,, \tag{6}$$

where h_0, h_2 are arbitrary constants and $\tilde{P}_{2,\mu} = (-)^\mu P_{2,-\mu}$. The boson pair operators of angular momentum $L = 0$ and 2 transform as $(0, 2)$ under $SU(3)$ and are defined as

$$\begin{aligned} P_0^\dagger &= d^\dagger \cdot d^\dagger - 2(s^\dagger)^2 \,, \\ P_{2,\mu}^\dagger &= \sqrt{2}\, s^\dagger d_\mu^\dagger + \sqrt{\frac{7}{2}}(d^\dagger d^\dagger)_\mu^{(2)} \quad . \end{aligned} \tag{7}$$

For $h_2 = 2h_0$, the Hamiltonian in eq. (6) is related to the $SU(3)$ Casimir operator

$$H(h_2 = 2h_0) = \frac{1}{2}h_2\bigl[-\hat{C}_{SU(3)} + 2\hat{N}(2\hat{N} + 3)\bigr] \,, \tag{8}$$

and thus becomes an $SU(3)$ scalar. For $h_2 = -2h_0/5$ it transforms as a $(2,2)$ $SU(3)$ tensor component. For arbitrary h_0, h_2 coefficients, $H(h_2, h_0)$ is therefore not an $SU(3)$ scalar, nevertheless it always has $|c; N\rangle$ (which has good $SU(3)$ $(2N, 0)$ species) as an exact zero-energy eigenstate. This property is a direct outcome of the following relations observed by the bosons pair operators (7)

$$P_{L,\mu}|c; N\rangle = 0 \qquad L = 0, 2 \,. \tag{9}$$

Since $H(h_0, h_2)$ is an $O(3)$ scalar, it follows that states of good L projected from $|c; N\rangle$ are also zero energy eigenstates which span the $(2N, 0)$ representation. Thus, when the coefficients h_0, h_2 are positive, the Hamiltonian (6) becomes positive definite by construction, and even though it is not an $SU(3)$ scalar, it still has an exactly degenerate zero-energy ground state band whose rotational members possess good $SU(3)$ symmetry. Even more surprising is the presence (to be shown below) of additional eigenstates in excited bands of $H(h_0, h_2)$ with good $SU(3)$ character. Further examination shows that the boson pair operators (7) satisfy also the following relations for single commutators acting on the condensate

$$\left[P_{L,\mu}, P_{2,2}^\dagger \right] |c; N\rangle = \delta_{L,2} \delta_{\mu,2} (6N + 9) |c; N\rangle \tag{10}$$

and the following operator identities for double commutators

$$\left[\left[P_{L,\mu}, P_{2,2}^\dagger \right], P_{2,2}^\dagger \right] = \delta_{L,2} \delta_{\mu,2} 12 P_{2,2}^\dagger \tag{11}$$

By combining relations (9)–(11) we find that the sequence of states

$$|k\rangle \propto \left(P_{2,2}^\dagger \right)^k |c; N - 2k\rangle \quad , \tag{12}$$

are eigenstates of $H(h_0, h_2)$ in eq. (6) with eigenvalues

$$E_k = 3h_2 \Big(2N + 1 - 2k \Big) k \tag{13}$$

By comparing the latter eigen-energies with the $SU(3)$ eigenvalues (3b), and recalling eq. (8), it is easy to verify that these $|k\rangle$ states are in the $SU(3)$ irreps $(2N - 4k, 2k)$ with $2k \leq N$. It can be further shown that they are lowest weight states in these representations. The states $|k\rangle$ are deformed and serve as intrinsic states representing γ^k bands with angular momentum projection $(K = 2k)$ along the symmetry axis[7]. In particular, $|k = 0\rangle$ represents the ground-state band $(K = 0)$ and $|k = 1\rangle$ is the γ-band $(K = 2)$. The intrinsic states break the $O(3)$ symmetry but since the Hamiltonian in (6) is an $O(3)$ scalar, the projected states $|(2N - 4k, 2k) K = 2k; L, M\rangle$, with good $L \geq K$ are also eigenstates of $H(h_0, h_2)$ with good $SU(3)$ symmetry. It should be noted that for $k \neq 0$ the states projected from $|k\rangle$ span only part of the corresponding $SU(3)$ irreps. There are other states originally in these irreps which do not preserve the $SU(3)$ symmetry and therefore get mixed. This situation corresponds precisely to that of partial $SU(3)$ symmetry. An Hamiltonian $H(h_0, h_2)$ which is not an $SU(3)$ scalar has a subset of **solvable** eigenstates which continue to have good $SU(3)$ symmetry. All of the above discussion is applicable also to the case when we add to the Hamiltonian (6) the Casimir operator of $O(3)$, and by doing so converting the partial $SU(3)$ symmetry into partial dynamical $SU(3)$ symmetry. The additional rotational term contributes just an $L(L + 1)$ splitting but does not affect the wave functions.

An example of this $SU(3)$ partial symmetry is displayed in Fig. 1, for $N = 7$ bosons. Typical spectra of the Hamiltonian (6) are shown for two cases: (a) $h_2 = 2h_0$ and (b) $h_2 = 1.25 h_0$. As noted above, case (a) corresponds to a full $SU(3)$ symmetry and all states are arranged in degenerate $SU(3)$ multiplets. For case (b) the $SU(3)$ symmetry is partially broken, so that most of the eigenstates have a

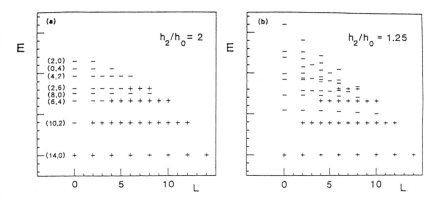

Figure 1. Spectra (energy E vs. angular momentum L) of the Hamiltonian (6). (a) For $h_2/h_0 = 2$. $SU(3)$ symmetry labels (λ, μ) are shown on the left. Some levels with $L \neq 0$ exhibit multiplicity which is not shown. (b) For $h_2/h_0 = 1.25$. Levels which continue to exhibit $SU(3)$ symmetry are marked by a $(+)$ symbol. The energy scale is arbitrary and the same value of h_2 was used in both cases. Taken from ref [4].

mixture of $SU(3)$ irreps and the above $SU(3)$ degeneracy is lifted. However, some of the eigenstates (marked by a + in Fig. 1) continue to carry good $SU(3) \supset O(3)$ representation labels and are arranged in multiplets. In particular, members of the the $K = 0$ ground band still transform as $(14, 0)$ under $SU(3)$. The $\beta(K = 0)$ and $\gamma(K = 2)$ bands, which originally were degenerate in the $SU(3)$ symmetry limit (and belonging to the $(10, 2)$ representation), split. Only the members of the γ-band continue to carry good $SU(3)$ labels $(10, 2)$, while the β-band has a mixture of several $SU(3)$ irreps. Similarly, the $\gamma^2(K = 4)$ band and the $\gamma^3(K = 6)$ band preserve their $SU(3)$ character. All other states exhibit mixing. In Fig. 2 we show the $SU(3)$ content of three eigenstates of case (b) which originally, in case (a), belonged to the $(2, 6)$ irrep of $SU(3)$. The state $L = 6_7$ is one of the solvable states in the $\gamma^3(K = 6)$ band and remains 100% in the $(2, 6)$ irrep. In contrast, the $L = 6_8$ and $L = 0_4$ states exhibit a significant spread over several $SU(3)$ irreps.

The Hamiltonian of eq. (6) which has partial $SU(3)$ symmetry, corresponds to particular relations among the parameters of the general IBM Hamiltonian. This is clearly evident from its multipole form,

$$H(h_0, h_2) = \epsilon \hat{n}_d + \kappa\, Q^{(2)}(\chi) \cdot Q^{(2)}(\chi) + \alpha\, T^{(2)} \cdot T^{(2)} + \eta\, L^{(1)} \cdot L^{(1)}$$
$$+ 2h_0\, \hat{N}(2\hat{N} + 3) \quad (14a)$$
$$\epsilon = (h_2 - 2h_0)(2\hat{N} + 3) \quad, \quad \kappa = -2h_0 \quad, \quad \chi = -\sqrt{7}h_2/4h_0 \quad,$$
$$\alpha = 7(h_2 - 2h_0)^2/8h_0 \quad, \quad \eta = -(h_2 + h_0)/4 \quad (14b)$$

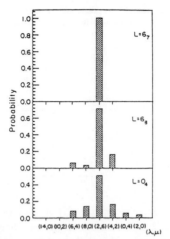

Figure 2. $SU(3)$ decomposition (probability vs. $SU(3)$ irrep (λ,μ)) for selected states in the spectrum of Fig. 1(b). Taken from ref. [4].

with $T^{(2)} = (d^\dagger \tilde{d})^{(2)}$ and $Q^{(2)}(\chi) = (d^\dagger s + s^\dagger \tilde{d}) + \chi T^{(2)}$. All the various parameters in eq. (14) are expressed in terms of h_0 and h_2.

4. Partial Dynamical Symmetry in Nuclear Spectroscopy

Partial dynamical symmetries (pds) are not just a formal notion but are actually realized in nuclei. As such they may serve as a useful tool in realistic applications of algebraic methods to nuclear spectroscopy. We consider ^{168}Er as a a typical example of an axially deformed prolate nucleus in the rare earth region, and show the relevance of $SU(3)$ pds to its description.

4.1. Energy Spectrum

The experimental spectra[3] of the ground (g), β, and γ bands in ^{168}Er is shown in Fig. 3. We attempt a description in terms of an IBM Hamiltonian with partial dynamical $SU(3)$ symmetry

$$H = h_0 P_0^\dagger P_0 + h_2 P_2^\dagger \cdot \tilde{P}_2 + \lambda\, L^{(1)} \cdot L^{(1)}, \qquad (15)$$

According to the discussion in section (3), the spectrum of the ground and γ bands is solvable and is given by

$$\begin{aligned} E_g(L) &= \lambda L(L+1) \\ E_\gamma(L) &= 3h_2(2N-1) + \lambda L(L+1) \end{aligned} \qquad (16)$$

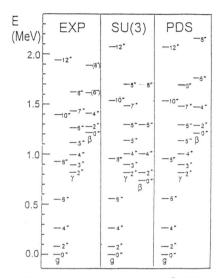

Figure 3. Spectra of ^{168}Er. Experimental energies[3] (EXP) compared with an IBM calculation in an exact $SU(3)$ dynamical symmetry ($SU(3)$) and in a partial dynamical $SU(3)$ symmetry (PDS). The latter employs the Hamiltonian in eq. (15) with $h_0 = h_2 = 0.008$, $\lambda = 0.013$ MeV.

The Hamiltonian in eq. (15) is specified by three parameters (N=16 for ^{168}Er according to the usual boson counting). We extract the values of λ and h_2 from the following experimental energy differences

$$\lambda = \frac{1}{6}(E_{2_\gamma^+} - E_{0_g^+})$$
$$h_2 = (E_{2_\gamma^+} - E_{2_g^+})/3(2N-1) \qquad (17)$$

For an exact $SU(3)$ dynamical symmetry, $h_0 = h_2/2$ implying $E_\beta(L) = E_\gamma(L)$ for $L \geq 2$, even. The corresponding spectrum in this case (shown in Fig. 3) deviates considerably from the experimental data since empirically the β and γ bands are not degenerate. On the other hand, when the dynamical $SU(3)$ symmetry is partial, one can vary h_0 so as to reproduce the β bandhead energy $E_\beta(L = 0)$ (The large-N estimate[6] $\epsilon_\beta = 2N(4h_0 + h_2)$ is a convenient initial guess). Having determined the three parameters λ, h_0, h_2, the prediction for other rotational members of the ground β and γ bands is shown in Fig. 3. Clearly, the $SU(3)$ pds spectrum is an improvement over the schematic, exact $SU(3)$ dynamical symmetry description since the β-γ degeneracy is lifted. The good $SU(3)$ character for the ground and γ bands is retained in the pds calculation, while the β band is mixed. The $SU(3)$ decomposition of selected states in these bands are shown in Table 1.

Table 1. SU(3) decomposition of lowest members of the ground (g) γ and β bands in the partial dynamical $SU(3)$ calculation of Fig. 3.

	(32,0)	(28,2)	(24,4)	(26,0)
0^+_g	1.0			
2^+_g	1.0			
2^+_γ		1.0		
3^+_γ		1.0		
0^+_β		0.87	0.03	0.1
2^+_β		0.87	0.03	0.1

4.2. Wave Functions and E2 transitions

The usual (but not unique) choice for orthonormal basis in case of $SU(3)$ symmetry is the Vergados basis[8] $\Psi_V((\lambda,\mu)\chi LM)$. It is obtained by known linear combinations from the non-orthogonal Elliott basis[9] $\phi_E((\lambda,\mu)KLM)$. For states in the ground (g), γ and β bands we have

L even : $\quad \Psi_V((2N,0)\chi = 0, LM) \quad = \quad \phi_E((2N,0)K = 0, LM)$

L even : $\quad \Psi_V((2N-4,2)\chi = 0, LM) \quad = \quad \phi_E((2N-4,2)K = 0, LM)$

L even : $\quad \Psi_V((2N-4,2)\chi = 2, LM) \quad = \quad x_{20}^{(L)} \phi_E((2N-4,2)K = 0, LM)$ $\quad$ (18)
$\quad\quad\quad\quad\quad\quad\quad\quad\quad\quad\quad\quad\quad\quad\quad\quad\quad\quad + x_{22}^{(L)} \phi_E((2N-4,2)K = 2LM)$

L odd : $\quad \Psi_V((2N-4,2)\chi = 2, LM) \quad = \quad \phi_E((2N-4,2)K = 2, LM)$

where $x_{20}^{(L)}$, $x_{22}^{(L)}$ are known coefficients[8]. For the Hamiltonian in eq. (15) with partial dynamical $SU(3)$ symmetry, the solvable states are those projected from the intrinsic states $|(\gamma)^k(2N-4k, 2k)K = 2k\rangle$. In particular, members of the ground and γ bands are the Elliott states $\phi_E((2N,0)K = 0, LM)$ and $\phi_E((2N-4,2)K = 2, LM)$ respectively. Using eq. (18), they can be expressed in terms of states in the Vergados basis. Since matrix elements in the Vergados basis of the general IBM $E2$ operator

$$T(E2) = \alpha Q^{(2)} + \theta (d^\dagger s + s^\dagger \tilde{d}) \quad (19)$$

are known in closed from[10,11], it is possible to obtain **analytic** expressions for the E2 rates between the subset of solvable states. Thus, for intraband transitions in the $(2N,0)K = 0$ ground band (L even, $L' = L \pm 2$), we get

$$B_E(E2; K = 0, L_g \to K = 0, L'_g) = B_V(E2; \chi = 0, L_g \to \chi = 0, L'_g) \quad (20)$$

Table 2. $B(E2)$ branching ratios from states in the γ band in ^{168}Er. The experimental ratios (EXP) and the broken $SU(3)$ calculation of Warner Casten and Davidson (WCD) are taken from ref [3]. (PDS) are the partial dynamical $SU(3)$ symmetry calculation.

J_i^π	J_f^π	EXP	PDS	WCD	J_i^π	J_f^π	EXP	PDS	WCD
2_γ^+	0_g^+	54.0	64.27	66.0	6_γ^+	4_g^+	0.44	0.89	0.97
	2_g^+	100.0	100.0	100.0		6_g^+	3.8	4.38	4.3
	4_g^+	6.8	6.26	6.0		8_g^+	1.4	0.79	0.73
3_γ^+	2_g^+	2.6	2.70	2.7		4_γ^+	100.0	100.0	100.0
	4_g^+	1.7	1.33	1.3		5_γ^+	69.0	58.61	59.0
	2_γ^+	100.0	100.0	100.0	7_γ^+	6_g^+	0.74	2.62	2.7
4_γ^+	2_g^+	1.6	2.39	2.5		5_γ^+	100.0	100.0	100.0
	4_g^+	8.1	8.52	8.3		6_γ^+	59.0	39.22	39.0
	6_g^+	1.1	1.07	1.0	8_γ^+	6_g^+	1.8	0.59	0.67
	2_γ^+	100.0	100.0	100.0		8_g^+	5.1	3.57	3.5
5_γ^+	4_g^+	2.91	4.15	4.3		6_γ^+	100.0	100.0	100.0
	6_g^+	3.6	3.31	3.1		7_γ^+	135.0	28.64	29.0
	3_γ^+	100.0	100.0	100.0					
	4_γ^+	122.0	98.22	98.5					

For interband transitions between the $(2N-4,2) K = 2$, γ band and the $(2N,0) K = 0$, ground band (L even, $L' = L, L \pm 2$) we get,

$$B_E(E2; K=2, L_\gamma \to K=0, L'_g) =$$

$$\left[\frac{\sqrt{B_V(E2;\chi=2, L_\gamma \to \chi=0, L'_g)} + x_{20}^{(L)} \sqrt{B_V(E2;\chi=0, L_\gamma \to \chi=0, L'_g)}}{x_{22}^{(L)}} \right]^2 \quad (21)$$

and for L odd, $L' = L \pm 1$ we have

$$B_E(E2; K=2, L_\gamma \to K=0, L'_g) = B_V(E2; \chi=2, L_\gamma \to \chi=0, L'_g) \quad (22)$$

Here $B_V(E2)$ and $B_E(E2)$ are $B(E2)$ rates calculated in the Vergados and Elliott bases respectively.

In general the parameters α and θ of the E2 operator (19) can be extracted from the experimental values of $B(E2; 0_g^+ \to 2_g^+)$ and $B(E2; 0_g^+ \to 2_\gamma^+)$. The resulting $SU(3)$ pds E2 rates for ^{168}Er are found to be in agreement with experiment and are similar to the calculation by Casten Warner and Davidson[3] (where the $SU(3)$ symmetry is broken for all states) as shown in Table 2 ($\theta/\alpha = 4.256$). The $SU(3)$

pds calculation reproduces correctly the $(\gamma \to \gamma)/(\gamma \to g)$ strengths and the dominance of $\beta \to \gamma$ over $\beta \to g$ transitions. If we recall that only the ground band has $SU(3)$ components $(\lambda, \mu) = (2N, 0)$ (see Table 1) and that $Q^{(2)}$ in eq. (19) is a generator of $SU(3)$, it follows that $\beta, \gamma \to g$ $B(E2)$ ratios are independent of both α and θ. Furthermore, since the ground and γ bands have pure $SU(3)$ character, $(2N, 0)$ and $(2N-4, 2)$ respectively, the corresponding wave-functions do not depend on parameters of the Hamiltonian and are determined solely by symmetry. Consequently, the B(E2) ratios for $\gamma \to g$ transitions quoted in Table 2 are parameter-free predictions of $SU(3)$ pds. The agreement between these predictions and the data confirms the relevance of partial dynamical $SU(3)$ symmetry to the spectroscopy of axially deformed nuclei.

5. Statistical Aspects of Partial Dynamical Symmetry

Partial dynamical symmetries may play a role not only for discrete spectroscopy but also for analyzing statistical aspects of nonintegrable systems. If a system posses a dynamical symmetry, it is integrable. Hence dynamical symmetry breaking is connected to nonintegrability and may give rise to chaotic motion[12] It follows that Hamiltonians with partial dynamical symmetry are not completely integrable and may be used as a tool to study mixed systems which are partly chaotic and partly regular. The current interest in such systems is driven by their being in some sense generic[13]. Most realistic systems are neither completely integrable nor fully chaotic. The dynamics of a generic classical Hamiltonian system is mixed. KAM islands of regular motion and chaotic regions coexist in phase space. If no separation between regular and irregular states is done in the associated quantum system, the statistical properties of the spectrum are usually intermediate between the Poisson and the Gaussian orthogonal ensemble (GOE) statistics.

The IBM is particularly useful to study the onset of chaos in dynamical systems since it provides a framework for tractable calculations both at the quantum and classical levels. The classical limit of the IBM Hamiltonian is obtained through the use of coherent states parametrized by the six complex numbers $\{\alpha_s, \alpha_\mu; \mu = -2, \ldots, 2\}$ and taking $N \to \infty$. The classical Hamiltonian is then obtained by substituting $s^\dagger, d_\mu^\dagger \to \alpha_s^*, \alpha_\mu^*$ and $s, d_\mu \to \alpha_s, \alpha_\mu$ with $1/N$ playing the role of $\hbar$.

To study the impact of partial (but exact) symmetries on the interplay between order and chaos, we consider[14] a family of Hamiltonians similar in form to eq. (6) but with the following boson pair operators

$$\begin{aligned} P_0^\dagger &= d^\dagger \cdot d^\dagger - \beta_0^2 (s^\dagger)^2 \,, \\ P_{2,\mu}^\dagger &= \beta_0 \, s^\dagger d_\mu^\dagger + \sqrt{\frac{7}{2}} (d^\dagger d^\dagger)_\mu^{(2)} \,. \end{aligned} \quad (23)$$

When $\beta_0 = \sqrt{2}$, the above Hamiltonian reduces to that of eq. (6) and has $SU(3)$ partial dynamical symmetry. In this case, the number of solvable states for a given spin L and boson number N is known[14] and for large N it is possible to estimate

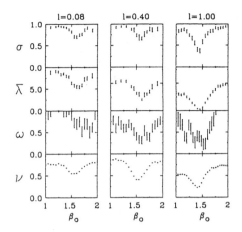

Figure 4. Classical ($\sigma, \bar{\lambda}$) and quantal (ω, ν) measures of chaos versus β_0 for the Hamiltonian mentioned in the text with $h_2/h_0 = 15$. Shown are three cases with classical spins $l = 0.08, 0.4$ and 1. The quantal calculations (ω, ν) are done for $N = 25$ bosons and spins $L = 2, 10$, and 25, respectively. Taken from ref [14].

the fraction $f(l, N)$ of solvable states as a function of the spin per boson $l \, (= L/N)$. For a given l, the fraction of solvable states is found to decrease like $1/N^2$ with boson number. However, at a given boson number N, this fraction increases with l.

To study the effect of the $SU(3)$ partial dynamical symmetry on the dynamics, we fix the ratio h_2/h_0 at a value far from the exact $SU(3)$ symmetry (for which $h_2/h_0 = 2$, $\beta_0 = \sqrt{2}$). We then change β_0 in the range $1 \leq \beta_0 \leq 2$. Classically, we determine the fraction σ of chaotic volume and the average largest Lyapunov exponent $\bar{\lambda}$. To analyze the quantum Hamiltonian, we study the spectral and transition intensity distributions. The nearest neighbors level spacing distribution is fitted by a Brody distribution $P_\omega(S) = AS^\omega exp(-\alpha S^{1+\omega})$ where $\omega = 0$ for the Poisson statistics and for the GOE $\omega = 1$, corresponding to integrable and fully chaotic motion respectively. The intensity distribution of the $SU(3)$ quadrupole (E2) operator (4) is fitted by a χ^2 distribution in ν degrees of freedom, $P_\nu(y) = \left[\left(\nu/2 \langle y \rangle \right)^{\nu/2} / \Gamma(\nu/2) \right] y^{\nu/2-1} \exp(-\nu y/2\langle y \rangle)$. For the GOE, $\nu = 1$ and ν decreases as the dynamics becomes regular.

Fig. 4 shows the two classical measures $\sigma, \bar{\lambda}$ and the two quantum measures ω, ν for the above mentioned Hamiltonian as a function of β_0. The parameters of the Hamiltonian are taken to be $h_2/h_0 = 15$ and the number of bosons is $N = 25$. Shown are three classical spins $l = 0.08, 0.4$ and 1 which correspond in the quantum case to $L = 2, 10$ and 25. All measures show a pronounced minimum which gets deeper and closer to $\beta_0 = \sqrt{2}$ (where the partial $SU(3)$ symmetry occurs) as the classical spin increases. This behavior is correlated with the fraction of solvable states (at a constant N) being larger at higher l. As expected, the occurrence of partial symmetries leads to suppression of chaos. It is interesting to note that the classical measures show a clear enhancement of regular motion near $\beta_0 = \sqrt{2}$ even though the fraction of solvable states vanishes as $1/N^2$ in the classical limit $N \to \infty$.

6. Summary

The notion of partial dynamical symmetry (pds) generalizes the familiar concept of exact (dynamical) symmetries. We have constructed explicitly IBM Hamiltonians with $SU(3)$ pds. Such Hamiltonians are not invariant under $SU(3)$ but have a subset of eigenstates with good $SU(3)$ symmetry. The special states are solvable and span part of particular $SU(3)$ irreps. Their wave-functions, eigenvalues and $E2$ rates are known analytically. A general algorithm how to construct Hamiltonians with partial dynamical symmetry for any semi-simple group is available[4].

Partial dynamical symmetries (pds) can overcome the schematic features of exact dynamical symmetries (e.g. non-degenerate β and γ bands in ^{168}Er) and simultaneously retain their virtues (i.e. solvability) for some states. Pds can address situations where a subset of levels exhibit a symmetry which is not shared by all states and therefore does not arise from invariance of the Hamiltonian (e.g an F-spin invariant Hamiltonian may be too restrictive). These attributes transform the mathematical notion of pds into a working tool for practical applications to spectroscopy and enhance the scope of algebraic methods for realistic systems. In addition to discrete spectroscopy, the notion of partial symmetries is useful for studying statistical aspects of mixed systems exhibiting coexistence of regular and chaotic motions. We have demonstrated that partial dynamical symmetry can suppress chaos even when the fraction of solvable states vanishes in the classical limit. Hamiltonians with partial symmetries (such as the IBM) can play an important role in studying the influence of a symmetry on the interplay between order an chaos in dynamical systems both at the quantum and classical levels.

On the occasion of its 20th anniversary, it is gratifying to witness the vitality of the IBM in providing a proper environment for nurturing new concepts of symmetry and spectroscopic tools. It is the combined computational simplicity and rich symmetry structure embedded in the IBM that makes it an ideal testing ground for further developments.

7. References

1. F. Iachello and A. Arima, *The Interacting Boson Model* (Cambridge Univ. Press, Cambridge, 1987).
2. R.F. Casten and D.D. Warner, *Rev. Mod. Phys.* **60** (1988) 389.

3. D.D. Warner, R.F. Casten and W.F. Davidson, *Phys. Rev.* **C24** (1981) 1713.

4. Y. Alhassid and A. Leviatan, *J. Phys.* **A25** (1992) L1265.
   A. Leviatan, in *Symmetries in Science VII* (B. Gruber and T. Otsuka Eds.), (Plenum Press, NY, 1994), p. 383.

5. J.N. Ginocchio and M.W. Kirson, *Nucl. Phys.* **A350** (1980) 31.

6. A. Leviatan, *Z. Phys.* **A321** (1987) 201.
   M.W. Kirson and A. Leviatan, *Phys. Rev. Lett.* **55** (1985) 2846.

7. H.T. Chen and A. Arima, *Phys. Rev. Lett.* **51**, 447 (1983).

8. J.D. Vergados, *Nucl. Phys.* **A111** (1968) 681.

9. J.P. Elliott, *Proc. Roy. Soc.*, **A245** (1958) 128, 562.

10. A. Arima and F. Iachello, *Ann. Phys.* **111** (1978) 201.

11. P. Van Isacker, *Phys. Rev.* **C27** (1983) 2447.

12. W.M. Zhang, C.C. Martens, D.H. Feng and J.M. Yuan, *Phys. Rev. Lett.* **61** (1988) 2167.

13. O. Bohigas, S. Tomsovic and D. Ullmo, *Phys. Rep.* **223** (1993) 43.

14. N. Whelan and Y. Alhassid and A. Leviatan, *Phys. Rev. Lett.* **71** (1993) 2208.

1/N EXPANSION IN BOSON MODELS: PRESENT STATUS AND FUTURE APPLICATIONS

SERDAR KUYUCAK
Department of Theoretical Physics, Research School of Physical Sciences
Australian National University, Canberra, ACT 0200, Australia

ABSTRACT

We review the basic structure of the $1/N$ expansion method in boson models and present some selected applications which illustrate the usefulness of the method in nuclear structure and reaction problems.

1. Introduction

The interacting boson model[1] (IBM) is inherently an algebraic model and does not have an immediate physical picture associated with it as in the Bohr-Mottelson model[2] (BMM). This shortcoming has been overcome with the introduction of intrinsic states which have provided a geometric picture for the IBM.[1,3] In addition, intrinsic sates have been used in deriving various matrix elements which are accurate to the leading order in N, the boson number.[1] Because the rotational symmetry is broken in the intrinsic frame, in order to bbtain more accurate results, one has to restore it by performing angular momentum projection. This program has been carried out in the IBM and was shown to lead to a $1/N$ expansion for all matrix elements.[4] Here, we review the basic structure of the $1/N$ expansion formalism and discuss some of its pertinent applications in nuclear and molecular spectroscopy. Recent advances in computer algebra have revolutionized the way algebra is done, expanding the boundaries of algebraic techniques to higher levels of complexity. Special attention will be paid to these new developments and how they affect the $1/N$ expansion calculations. A note on notation; throughout the paper, zero subscripts are suppressed for convenience and bar denotes angular momentum eigenvalues, e.g. $b_0 \equiv b$ and $\bar{L} \equiv L(L+1)$.

2. Review of 1/N Expansion Method

We consider a set of boson creation and annihilation operators $\{b_{lm}^\dagger, b_{lm}\}$ which describe a variety of quantum systems depending on the values of l. For example, $l = 0, 2$ (sd IBM) describes quadrupole collectivity in nuclei. By adding $l = 4$ or $l = 1, 3$ bosons, the sd IBM can be extended to include hexadecapole or octupole excitations. The system with $l = 0, 1$ bosons is known as the vibron model,[5] and describes the molecular rotation-vibration spectra. Each boson system has an algebra $U(n), n = \sum_i (2l_i + 1)$ associated with it, which, through the use of group theoretical techniques, provides analytic solutions at certain dynamical symmetry limits. The $1/N$ expansion method extrapolates between these limits, providing analytic solutions for intermediate cases which are usually more realistic.

The ground state of a boson system can be written as a condensate

$$|N, \mathbf{x}\rangle = (N!)^{-1/2}(b^\dagger)^N|0\rangle, \quad b^\dagger = \sum_{lm} x_{lm} b^\dagger_{lm}, \qquad (1)$$

where x_{lm} are the (normalized) boson mean fields which can be associated with the shape variables of the system. As a trial state, Eq. (1) is complete and would give the exact ground energy if varied after projection (VAP). However, such a calculation is prohibitively difficult to perform, and the assumption of axial symmetry is necessary to obtain analytical results. The contribution from the non-axial parts to the ground energy is found to be of order $1/N^2$ from comparisons with the exact diagonalization results,[6] hence this is a well justified approximation. (Diatomic molecules have axial symmetry, so the formalism is exact for the vibron model.) For axially symmetric systems, K, the projection to the symmetry axis is a good quantum number, and the intrinsic boson operators can be labeled by K, i.e. $b^\dagger_K = \sum_l x_{lK} b^\dagger_{lK}$. For the ground state which has $K = 0$, this amounts to suppressing the m sum in Eq. (1). Other bands are obtained from the ground band by acting with the orthogonal intrinsic boson operators, for example, the one-phonon bands are given by $|\phi_K\rangle = b^\dagger_K|N-1,\mathbf{x}\rangle$.

2.1. Normalization

The fundamental quantity in the $1/N$ expansion is the projected norm integral for the ground band, $\mathcal{N}_g(N, L)$. Since all the other matrix elements can be evaluated by manipulations of $\mathcal{N}_g(N, L)$, it is worthwhile to discuss its basic structure in some detail. The norm with angular momentum projection is given by

$$\mathcal{N}_g(N,L) = \langle N, \mathbf{x}|P^L_{00}|N,\mathbf{x}\rangle,$$

$$= \frac{2L+1}{2} \int_0^\pi d\beta \sin\beta \, d^L_{00}(\beta) \Big[\sum_l x_l^2 \, d^l_{00}(\beta)\Big]^N, \qquad (2)$$

where P^L_{00} is the projection operator and d^L_{00} denote Wigner d matrices. The integral in Eq. (2) has been evaluated exactly in some special cases. For example, in the vibron model, one obtains the following closed form expression

$$\mathcal{N}_g(N,L) = \frac{(2L+1)}{a(N+1)} \Big[{}_2F_1(-L, L+1; N+2; 1/a)$$

$$-(-1)^L (x_0^2 - x_1^2)^{N+1} {}_2F_1(-L, L+1; N+2; (x_1^2 - x_0^2)/a)\Big], \qquad (3)$$

where ${}_2F_1$ is a hypergeometric function, and $a = \sum_l \bar{l} x_l^2$ is ubiquitous to the angular momentum projection and represents the "average angular momentum squared" carried by a single boson. Note that the second term in Eq. (3) vanishes in the $O(4)$ limit of the vibron model ($x_1 = x_0$), and is completely negligible in general.

(For positive parity bosons, it is identical to the first term which leads to a factor of 2.) Eq. (2) was also evaluated in the SU(3) limit and in the sd IBM without approximations.[4] In all cases, the norm integral leads to a unique expansion in N, L given by

$$\mathcal{N}_g(N, L) = \frac{2(2L+1)}{aN} \sum_{n=0}^{n} \frac{(-1)^n}{n!(aN)^n} \sum_{m=0}^{n} \alpha_{nm} \bar{L}^m,$$

$$= \frac{2(2L+1)}{aN} \Big[1$$
$$- \frac{1}{aN} \left(\bar{L} + \alpha_{10} \right)$$
$$+ \frac{1}{2(aN)^2} \left(\bar{L}^2 + \alpha_{21} \bar{L} + \alpha_{20} \right)$$
$$- \frac{1}{6(aN)^3} \left(\bar{L}^3 + \alpha_{32} \bar{L}^2 + \alpha_{31} \bar{L} + \alpha_{30} \right) + \ldots \Big]. \quad (4)$$

The explicit form of the norm will be useful in illustrating the concept of layers in the $1/N$ expansion. As a general terminology, the coefficients $\alpha_{nn} \equiv 1$ in the first column are called "first layer", α_{nn-1} in the second column "second layer", etc. The most elegant way to obtain these coefficients is not a direct integration of Eq. (2), which is by no means easy in general, but through algebra by exploiting the symmetries of the system. Namely, boson number conservation and rotational invariance demands that the expectation values of the number and angular momentum operators must satisfy $\langle \sum_l \hat{n}_l \rangle = N$ and $\langle \mathbf{L}.\mathbf{L} \rangle = \bar{L}$, respectively. Each of these conditions leads to a set of $k-1$ linear equations for k coefficients, and hence together (over) completely determine them. The use of Mathematica[7] has made this task rather easy and α_{nm} have been determined up to the fourth layer. Here we quote the second and third layer coefficients which are used below

$$\alpha_{10} = 1 + a - a_1/2a,$$
$$\alpha_{21} = 4 + 6a - 3a_1/a,$$
$$\alpha_{32} = 10 + 18a - 9a_1/a,$$
$$\alpha_{43} = 20 + 40a - 20a_1/a,$$
$$\alpha_{54} = 35 + 75a - 75a_1/2a,$$
$$\alpha_{65} = 56 + 126a - 63a_1/a, \quad (5)$$

$$\alpha_{20} = 2 + 6a + 2a^2 - 3a_1 - 10a_1/3a + 3a_1^2/2a^2 - a_2/3a,$$
$$\alpha_{31} = 18 + 72a + 42a^2 - 54a_1 - 40a_1/a + 45a_1^2/2a^2 - 4a_2/a,$$
$$\alpha_{42} = 88 + 400a + 300a^2 - 360a_1 - 220a_1/a + 135a_1^2/a^2 - 20a_2/a,$$
$$\alpha_{53} = 308 + 1500a + 1300a^2 - 1500a_1 - 2450a_1/3a + 525a_1^2/a^2 - 200a_2/3a,$$
$$\alpha_{64} = 868 + 4410a + 4200a^2 - 4725a_1 - 2380a_1/a + 1575a_1^2/a^2 - 175a_2/a, \quad (6)$$

where $a_n = \sum_l \bar{l}^{n+1} x_l^2$ denotes higher moments of a. Normalizations for all other bands are obtained from $\mathcal{N}_g(N, L)$, and have similar expansions.

2.2. Matrix Elements

Once the norm of an intrinsic state $|\phi_K\rangle$ is known, matrix elements (m.e.) of a k-body operator $\hat{O}$ can be evaluated in a straightforward manner using boson calculus and angular momentum algebra techniques.[4] The calculations are, however, rather lengthy, and we refer to the original references for details. Here we concentrate on the general form of the m.e. which is given by

$$\langle \hat{O} \rangle_L = N^k \sum_{n,m} \frac{O_{nm}}{(aN)^m} \left(\frac{\bar{L}}{a^2 N^2}\right)^n$$

$$= N^k \Big\{ O_{00} + \frac{O_{01}}{aN} + \frac{O_{02}}{(aN)^2} + \frac{O_{03}}{(aN)^3} + \cdots$$

$$+ \frac{\bar{L}}{a^2 N^2} \left(O_{10} + \frac{O_{11}}{aN} + \frac{O_{12}}{(aN)^2} + \cdots \right)$$

$$+ \left(\frac{\bar{L}}{a^2 N^2}\right)^2 \left(O_{20} + \frac{O_{21}}{aN} + \cdots \right)$$

$$+ \left(\frac{\bar{L}}{a^2 N^2}\right)^3 \left(O_{30} + \cdots \right) + \cdots \Big\}. \tag{7}$$

The expansion coefficients O_{nm} in Eq. (7) involve various quadratic forms of the mean fields x_{lm} corresponding to the single-boson m.e. of $\hat{O}$ and its moments. Again, the explicit form is given to facilitate the illustration of layers in the $1/N$ expansion. Notice that the i coefficients O_{nm} in the i'th column have $n + m = i - 1$ constant, and are referred as the layer "$i - 1$". The leading term in Eq. (7) thus forms the zeroth layer. This name is appropriate since calculations in the intrinsic frame give the same result independent of projection. The connection between the layers in the normalization (4) and the m.e. (7) is that in order to calculate the m.e. up to the i'th layer, one needs to know the coefficients α_{nm} up to that layer. This is very useful in higher order calculations as it restricts the number of terms in the expansion, cutting down the amount of algebra. To make this point clear, we note that Eq. (7) is complete to the third layer whereas a complete calculation to order $1/N^6$ would require 6 more terms belonging to the fourth, fifth and sixth layers. As will be clear from Eq. (9) below, the complexity of the coefficients O_{nm} increases exponentially with layers, and each of the extra terms would lead to expressions pages long. From a practical point of view, such accuracy is never required, and hence use of layers is a more sensible approach than a complete calculation to a given order in $1/N$.

As an expansion in $\bar{L}$, the form of Eq. (7) follows from rotational invariance. Similar expansions in $\bar{L}$ are also used in the BMM in fitting experimental data.[2] In the IBM, they follow from a basic Hamiltonian which is intellectually more satisfying. To address the question of convergence with $\bar{L}$, we note that for deformed systems, $\bar{L}/(aN)^2 < 1$ and O_{n0} gets progressively smaller with increasing n.

In the original papers on the $1/N$ expansion,[4] a gaussian approximation was used in the evaluation of the norm integral which gave only the first layer coefficients correctly, and hence the m.e. to that layer. With a few exceptions (e.g. g-factors), the first layer results were mostly sufficient for the purposes of nuclear spectroscopy at that time. Since then, we have applied the $1/N$ expansion method to the vibron model where accuracy of spectroscopic data is such that one needs expressions up to the third layer for meaningful comparisons. A similar level of accuracy is needed for the description of high spin states in superdeformed nuclei. The amount of algebra required in such a calculation is rather horrendous, enough to put off even the most enthusiastic algebraist. Fortunately, the Mathematica software has come in very handy in the $1/N$ calculations, taking the drudgery out of the "algebra crunching". As alluded to in the last subsection, it has been instrumental in the evaluation of the norm integral to higher orders, and the third layer calculations that we perform routinely now, would have never been attempted without computer algebra. As an example, we give here the ground energy for a general quadrupole Hamiltonian

$$H = \kappa Q.Q, \qquad Q = \sum_{j,l} q_{jl}[b_j^\dagger \tilde{b}_l]^{(2)}, \tag{8}$$

to the third layer, which will be used in a study of superdeformed nuclei in the next section

$$\begin{aligned}
E = \kappa N^2 \Big\{ & U + \frac{1}{aN}\Big(aU - U_1 + aC\Big) \\
& + \frac{1}{a^2N^2}\Big((-2a + a_1)U + (1 - a - a_1/a)U_1 + U_2/2 + a^2C - aC_1\Big) \\
& + \frac{1}{a^3N^3}\Big((2a + 2a^2 - 14a_1/3 - aa_1 + 5a_1^2/2a - 2a_2/3)U \\
& \quad + (-1 + a - a_1/2 + 7a_1/2a - 5a_1^2/2a^2 + a_2/2a)U_1 \\
& \quad + (-7/6 + 5a_1/4a)U_2 - U_3/6 \\
& \quad + (-a^2 + aa_1/2)C + (a - a_1)C_1 + aC_2/2\Big) \\
& + \frac{\bar{L}}{a^2N^2}\Big[-2aU + U_1 \\
& \quad + \frac{1}{aN}\Big((4a + 2a^2 - 4a_1)U + (-2 + a + 3a_1/a)U_1 - U_2 - a^2C + aC_1\Big) \\
& \quad + \frac{1}{a^2N^2}\Big((-6a - 16a^2 - 4a^3 + 21a_1 + 15aa_1 - 15a_1^2/a + 3a_2)U \\
& \quad\quad + (3 + 2a - 2a^2 - 4a_1 - 14a_1/a + 25a_1^2/2a^2 - 2a_2/a)U_1 \\
& \quad\quad + (7/2 + 2a - 5a_1/a)U_2 + U_3/2 \\
& \quad\quad + (2a^2 + 2a^3 - 2aa_1)C + (-2a - 2a^2 + 3a_1)C_1 - aC_2\Big)\Big] \\
& + \frac{\bar{L}^2}{2a^4N^4}\Big[(-2a - 2a^2 + 3a_1)U + (1 - 2a_1/a)U_1 + U_2/2
\end{aligned}$$

$$+\frac{1}{aN}\Big((8a + 30a^2 + 14a^3 - 32a_1 - 37aa_1 + 26a_1^2/a - 4a_2)U$$
$$+(-4 - 8a + 2a^2 + 29a_1/2 + 20a_1/a - 39a_1^2/2a^2 + 5a_2/2a)U_1$$
$$+(-4 - 9a/2 + 13a_1/2a)U_2 - U_3/2$$
$$+(-a^2 - 2a^3 + 3aa_1/2)C + (a + 2a^2 - 2a_1)C_1 + aC_2/2)\Big]$$
$$+\frac{\bar{L}^3}{3a^6 N^6}\Big[(-8a - 36a^2 - 24a^3 + 100a_1/3 + 54aa_1 - 30a_1^2/a + 10a_2/3)U$$
$$+(4 + 12a - 24a_1 - 20a_1/a + 21a_1^2/a^2 - 2a_2/a)U_1$$
$$+(10/3 + 6a - 6a_1/a)U_2 + U_3/3\Big]\Big\}. \tag{9}$$

Here U_n are given by

$$U_n = \sum_{jlj''l'I} \bar{I}^n \langle j0j'0|I0\rangle\langle l0l'0|I0\rangle \begin{Bmatrix} j & j' & I \\ l' & l & 2 \end{Bmatrix} q_{jl}q_{j''l'}x_j x_l x_{j'} x_{l'}. \tag{10}$$

Using the angular momentum algebra techniques described in Ref.[4], these sums can be evaluated in closed form. The first four terms needed in Eq. (9) are given by

$$U = A^2,$$
$$U_1 = (2A_1 - 3A)A,$$
$$U_2 = (2A_2 - 24A_1 + 18A)A + (A_{11} - A_2 + 7A_1)A_1 + (A_{11} - A_2)^2/12,$$
$$U_3 = (2A_3 - 36A_2 - 18A_{11} + 192A_1 - 144A)A$$
$$+(3A_{21} - 3A_3 + 48A_2 + 24A_{11} - 146A_1)A_1/2$$
$$-(3A_{21} - 3A_3 + 25A_2 - 14A_{11})A_2/12$$
$$+(3A_{21} - 3A_3 + 11A_{11})A_{11}/12, \tag{11}$$

where the various quadratic forms in Eqs. (9,11) are defined as

$$A_{mn} = \sum_{jl} \bar{j}^m \bar{l}^n \langle j0l0|20\rangle q_{jl}x_j x_l, \quad C_n = \sum_{jl} 5\bar{l}^n (q_{jl}x_l)^2/(2l+1), \tag{12}$$

and correspond to various moments of the single-boson quadrupole m.e.. Note that the zero subscripts are suppressed for convenience as usual.

Similar expressions are obtained for the excited band energies. One consequence of the dominance of the ground intrinsic boson operator in excited bands is that all bands have the same leading order coefficients O_{n0}, but differ in the higher order ones, O_{n1}, O_{n2}, etc..

The in-band transition m.e. also have the same structure as Eq. (7) except $\bar{L}$ is replaced by $\bar{L}_i, \bar{L}_f$, the initial and final angular momenta. The interband transitions have a somewhat different structure. Compared to the in-band ones, they are suppressed by $N^{-n/2}$ where n is the number of different intrinsic boson operators in the two bands, and the spin dependence occurs at the $1/N$ level instead of $1/N^2$.

There are also important band mixing effects which will be further discussed in the next section.

3. Selected Applications of $1/N$ Expansion

The $1/N$ expansion offers a controlled, step by step alternative to full diagonalization of the boson Hamiltonian. Therefore, it has been especially useful in cases where an exact diagonalization is difficult because of large basis space, e.g. sdg IBM where most of the applications have been concentrated so far. In this section, we give some selected applications in nuclear structure (energy systematics, $E2$ transitions) and reaction problems (proton scattering) which highlight the power of analytic methods in global systematic studies. Some applications, e.g. g-factors,[8] and subbarrier fusion[9] are discussed by others in these proceedings.

3.1. Energy Systematics

The analytic expressions obtained for the ground and excited band energies have been very useful in systematic studies of experimental data.[10,11] As a first example, we consider "the β band moment of inertia anomaly",[10] that is the moment of inertia of the β band in deformed nuclei is found to be larger than that of the ground band, in some cases up to 50%. The IBM calculations predicted the opposite which has been a source of concern. Numerical calculations are not very illuminating in such situations, and one needs guidance from analytical results to get to the core of the problem. As discussed in the last section, $1/N$ expansion allows variations of order $1/N$ at most between the ground and excited band quantities, and hence the observed differences which are much larger than $1/N$ can not be explained in the framework of the IBM, regardless of the parameters used in the Hamiltonian. This raises the possibility that such bands may not be the β band but some other $K = 0$ band. Analytic results have also shed light on the direction of the moment of inertia ratio $R = \mathcal{I}_\beta/\mathcal{I}_g$ which should be greater than one. Inclusion of one-body energy with $\epsilon_d > 0$ in a quadrupole Hamiltonian reduces this ratio further. (A negative ϵ_d would go in the right direction but unfortunately is not allowed!) A similar situation is encountered in diatomic molecules where $R > 1$ but the variation is of the order $1/N$. Cross-fertilization from the vibron model calculations suggests that the increase in the moment of inertia is due to stretching effects, and inclusion of an attractive monopole interaction would lead to a ratio larger than one.

A global study of shapes in the IBM at high spins is another area where the $1/N$ expansion has contributed significantly. The leading terms O_{00} and O_{10} in the ground energy expression (9) were found to be accurate enough for this purpose, which enabled a rather simple analysis of shapes at high spins. The major findings of this study were that i) in the sd model, no dynamical shape transitions were possible, ii) in the sdg model, well deformed nuclei showed a similar behaviour to the sd model, but transitional nuclei could exhibit dynamical shape transitions (prolate to oblate or viceversa) under certain conditions. Later, we performed fits to all the available data (energies, $E2$ and $E4$ transitions) in Os and Pt isotopes using the sdg model,[13] and found that these conditions are satisfied in actual cases, confirming the prediction of dynamical shape transitions in transitional nuclei. The origin of the dynamic shape transitions is the competition between the quadrupole and hexadecapole degrees of freedom. In deformed nuclei, the quadrupole collectivity dominates and the hexadecapole collectivity is not strong enough to compete even at high spins. In contrast, in transitional nuclei, the former gets relatively weaker while the latter gets stronger so that a shape transition is likely to happen at high spins. The empirical situation with regard to occurence of dynamical shape transitions is not very clear at present. While the quadrupole moments drop significantly with increasing spin in some cases (e.g. ^{196}Pt[14]), the error bars are too large to reach a definite conclusion. Accurate measurements of quadrupole moments are now possible with the new detector systems, and hopefully tests of this prediction will be carried out in the near future.

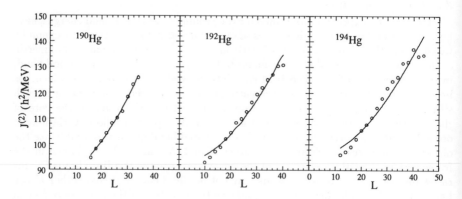

Fig. 1. Comparison of the dynamic moment of inertia in $^{190-194}$Hg with the super IBM calculations. The parameters used in the fits are N_{super} = 29, 30, 31, κ = -35, -34, -33 keV, q = 0.68, 0.72, 0.72 for $^{190-194}$Hg respectively.

Recently, a super IBM has been proposed for the study of superdeformation.[15] In super IBM, normal bosons are supplemented with superdeformed bosons which correspond to the Cooper-pairs in the superdeformed potential. The number of superdeformed bosons, N_{super} is typically around 30-40, and because of large deformation, g boson effects are important. Thus, the super IBM offers a fertile ground for the application of the $1/N$ expansion. Here we use the energy formula (9) to describe the superdeformed bands in the Hg isotopes. At high spins ($L \sim 40$), the second layer result is found to be not accurate enough, and one needs the full third layer expression. We have compared Eq. (9) with exact numerical projection results and found the deviation to be less than 0.1%. In Fig. 1, we show the dynamic moment of inertia, $\mathcal{J}^{(2)}$ results obtained with a quadrupole Hamiltonian. The three quadrupole parameters q_{22}, q_{24}, q_{44} are scaled from their SU(3) values with a single factor q. N_{super} is determined from microscopic calculations,[15] and κ and q are fitted to the experimental data. A reasonably good description of $\mathcal{J}^{(2)}$ is obtained. It is worthwhile emphasizing that $\mathcal{J}^{(2)}$ involves a double difference of the level energies, and therefore it is a much more sensitive quantity to compare than the level energies themselves. We note that the SU(3) limit corresponds to a rigid rotor and would give a flat line for $\mathcal{J}^{(2)}$. The deviation from a constant $\mathcal{J}^{(2)}$ is due to stretching effects and could be explained by the breaking of the SU(3) symmetry. The q values obtained in the above fits indicate that this breaking is around 30%.

Finally, we mention that the expressions derived in the $1/N$ expansion are quite general and can be used in any spectrum-generating algebra. For example, the ground energy in the vibron model has exactly the same form as Eq. (9) and has been used in a study of rotational parameters in diatomic molecules. The O(4) limit of the vibron model, like the SU(3) limit of IBM, corresponds to a rigid rotor, and one needs to break the O(4) symmetry to describe the spectroscopic data. Our initial results indicate that this breaking is not given by a simple O(4) →U(3) transition, and more general Hamiltonians are currently being considered.

3.2. E2 Transitions

The $1/N$ expansion formalism was instrumental in uncovering the nature of interband E2 transitions in the IBM.[6] Empirically, the $\beta \to g$ and $\gamma \to g$ E2 transitions in deformed nuclei show systematic spin dependent deviations from the Alaga rules which are ascribed to band mixing effects in the BMM.[2] A calculation of the interband E2 m.e. to order $1/N$, including the band mixing effects, have led to expressions[6] that are similar in form to those used in the BMM

$$\langle L'_g \parallel T(E2) \parallel L_K \rangle = e[N(2-\delta_K)(2L+1)]^{1/2} \langle LK\,2-K|L'0\rangle$$
$$\times \left(M_{00} + \frac{M_{01}}{N}\right)\left[1 + \frac{z}{N}(\bar{L}' - \bar{L})\right], \qquad (13)$$

where we have factored out the spin independent part to conform with the Mikhailov plot representation, but left the N dependence explicit otherwise. The coefficients M and z in Eq. (13) involve quadratic forms of the mean fields as in (12) and are given in Ref.[6]. An analysis of Eq. (13) employing a quadrupole Hamiltonian indicated that i) for $\gamma \to g$ E2 transitions, contribution from band mixing accounts for only half of the z coefficient, the rest being direct, ii) for $\beta \to g$ E2 transitions, there are no band mixing contributions to leading order in $1/N$, and z is entirely direct. Thus the pure band mixing assumption for the z coefficient in the BMM is not supported by a consistent Hamiltonian formalism.

From the early days, the boson cut off in the in-band E2 m.e. was thought to provide a signature for the finite boson number, but this was never verified experimentally. Systematic study of interband E2 transitions in rare-earth nuclei provided the first evidence for finite N effects.[16] Eq. (13) furnishes an analytic proof for the $1/N$ dependence of spin dependent terms conjectured, in Ref.[16] from numerical studies.

Table 1. Comparison of Mikhailov plot parameters M and z in Eq. (13) with those obtained from an exact diagonalization of the Hamiltonian for various values of N. The last row shows the leading order $1/N$ results to which the "exact" results converge with increasing N. The "$1/N$" column gives the values for $M_{00} + M_{01}/N$.

N	$M_{\beta g}$ Exact	$1/N$	$z_{\beta g}$ Exact	$M_{\gamma g}$ Exact	$1/N$	$z_{\gamma g}$ Exact
8	0.200	0.174	0.154	0.520	0.481	0.218
10	0.179	0.160	0.149	0.485	0.459	0.232
12	0.163	0.151	0.150	0.464	0.444	0.236
14	0.153	0.144	0.152	0.451	0.434	0.236
16	0.146	0.139	0.153	0.441	0.426	0.237
$1/N$		0.105		0.155	0.372	0.238

The analytic expressions obtained in the $1/N$ expansion are not only useful for qualitative interpretation of experimental results but can also be used with confidence in the actual analysis of data. To demonstrate this point, we present in Table 1 a comparison of the exact diagonalization results with Eq. (13) in the case of a quadrupole Hamiltonian with a consistent $E2$ operator. In sd IBM, the only relevant parameter is q_{22} which has been chosen half the SU(3) value ($-\sqrt{7}/4$) as a realistic value in the rare-earth region. An inspection of Table 1 shows that the $1/N$ expansion results are accurate to within a few percent. Considering the effort required in extracting the exact Mikhailov parameters, namely, first diagonalizing the Hamiltonian, then calculating a number of $E2$ transitions between the two bands, and finally doing a least square fit to those m.e., this would appear to be a small price to pay. A curious observation is that the z coefficient, which is given only to the leading order, is more accurate than the M coefficient which includes the $1/N$ corrections. This happens because z involves a ratio, and the higher order terms in the numerator and the denominator, were they included, would almost cancel each other!

3.3. Medium Energy Proton Scattering

The $1/N$ expansion technique is ideally suited to reaction problems since one is primarily interested in the low-lying excitations of a system. Due to space limitations, we refer to Refs.[17,18] for details of the algebraic-eikonal approach to reactions and the implementation of the $1/N$ expansion method. In Fig. 2, we present an application of the formalism to 800 MeV (p,p') scattering from ^{176}Yb. Consistent with the small β_4 deformation in ^{176}Yb, a small hexadecapole interaction is needed to explain all three cross sections.

4. Future Developments

Some of the applications described in the last section are still under active development (e.g., superdeformation, vibron model, and scattering problems), and will

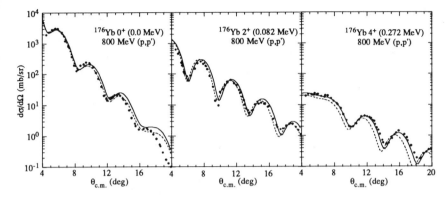

Fig. 2. Comparison of (p,p') scattering cross sections in ^{176}Yb with the sdg model calculations. The dashed line is for a pure quadrupole interaction with a strength $e_2 = 0.14$ eb. The solid line includes a hexadecapole interaction and the strengths are $e_2 = 0.16$ eb, $e_4 = 0.01$ eb^2.

be pursued in the near future. There are some other areas which have been partly or entirely neglected due to various reasons. As stressed in the introduction, the availability of computer algebra has opened up new possibilities, and in concluding, we would like to sketch some of these research areas that we hope to undertake in the future.

Although the 1/N formalism for the proton-neutron IBM (IBM-2) was established in the original papers,[4] it has not been applied to any specific problems until very recently.[8] The IBM-2 is essential in description of $M1$ properties, which are complimentary to those of $E2$, and directly probe the proton and neutron distributions in nuclei. Thus, new insights into nuclear structure can be obtained from a study of $M1$ properties. Another interesting question is the dynamical evolution of shapes in the IBM-2. If the proton and neutron shapes have opposing tendencies, the competition between the two could drive a dynamical shape transition at high spins. A rudimentary study has shown that this is possible in transitional nuclei (again!). Finally, the 1/N formulas can be used in constructing an accurate mass formula for nuclei far from the stability line, a problem of current interest in both nuclear and astrophysics.

So far, the 1/N expansion formalism has not been developed for the IBFM, the primary reason being the complexity of calculations. With the availability of computer algebra, however, this becomes possible, and we are planning such a research program. Study of odd nuclei is important, not only because IBFM compliments IBM, but also it has important applications in nuclear astrophysics (e.g., astrophysical s- and r-processes, dark matter detection), diversions that may be necessary for the survival our kind.

Boson approximations have become very popular in condensed matter physics

(CMP). One advantage of boson systems on a lattice is that one doesn't need to worry about rotational invariance (and hence all the complications arising from the angular momentum projection). The lattice interactions used in CMP have an uncanny resemblance to the IBM ones, and it would be interesting to apply some of the techniques developed for the IBM in condensed matter problems.

5. Acknowledgements

This research is supported by the Australian Research Council. It is a pleasure to thank Iain Morrison and Vi-Sieu Lac, my past collaborators who helped to develop the $1/N$ expansion, and Baha Balantekin, Jonathan Bennett, Taka Otsuka, Michio Honma, Siu-Cheung Li and Matthew Roberts who are expanding its horizons. Finally, I am ever grateful to Feza Gürsey and Franco Iachello for setting me on the path of algebraic methods, and for support and encouragement over the years.

6. References

1. F. Iachello and A. Arima, *The Interacting Boson Model* (Cambridge University Press, Cambridge, 1987).
2. A. Bohr and B.M. Mottelson, *Nuclear Structure*, Vol. 2 (Benjamin, Reading, 1975).
3. J.N. Ginocchio and M.W. Kirson, *Nucl. Phys.* **A350** (1980) 31.
4. S. Kuyucak and I. Morrison, *Ann. Phys.* (N.Y.) **181** (1988) 79; **195** (1989) 126.
5. F. Iachello, *Chem. Phys. Lett.* **78** (1981) 581.
6. S. Kuyucak and I. Morrison, *Phys. Rev.* **C41** (1990) 1803.
7. S. Wolfram, *Mathematica* (Addison-Wesley, Redwood City, 1991).
8. A.E. Stuchbery, these proceedings.
9. A.B. Balantekin, these proceedings.
10. S. Kuyucak and I. Morrison, *Phys. Rev.* **C38** (1988) 2482.
11. S. Kuyucak, I. Morrison, and T. Sebe, *Phys. Rev.* **C43** (1991) 1187.
12. S. Kuyucak, V-S. Lac, and I. Morrison, *Phys. Lett.* **B263** (1991) 146; S. Kuyucak, *Prog. Part. Nucl. Phys.* **28** (1992) 391.
13. V-S. Lac and S. Kuyucak, *Nucl. Phys.* **A539** (1992) 418.
14. A. Mauthofer *et al.*, *Z. Phys.* **A336** (1990) 263.
15. T. Otsuka and M. Honma, *Phys. Lett.* **B268** (1992),305; M. Honma, *Superdeformation in the Hg-Pb region and an extension of IBM* (Ph.D. thesis, University of Tokyo, 1993), and these proceedings.
16. R.F. Casten, D.D. Warner, and A. Aprahamian, *Phys. Rev.* **C28** (1983) 894.
17. J.N. Ginocchio *et al.*,*Phys. Rev.* **C33** (1986) 247.
18. S. Kuyucak and I. Morrison, *Phys. Rev.* **C48** (1993) 774.

FROM SENIORITY TO COLLECTIVITY

Akito Arima
RIKEN, Japan

ABSTRACT

A prehistory of the Interaction Boson Model is reviewed and the condition which must be satisfied by effective interactions in semi-magic nuclei is discussed

1. A Prehistory of the Interacting Boson Model in Japan

Sakai collected experimental data of excitation energies of even–even nuclei. From his systematics he concluded that those low–lying excited levels can be classified as ground–state bands, β and γ bands[1]. He then asked me if harmonic oscillator states can be classified in such a way in 1967. My answer was "yes". The reason is as follows.

One can classify, in general, d^N configurations in terms of SU_5. For bosons, only fully symmetric states $[N]$ are allowed. There is a supergroup O_5 for the group SU_5. For bosons again, the irreducible representation is determined as $(\lambda, 0)$ by using the seniority. The seniority can then be related to n_β as $N = \lambda + 2n_\beta$, where n_β can denote the number of the nodes of the β–vibration or that of 0^+ pairs of d bosons. Here, we include O_4 into the canonical chain, so as to classify O_5 completely. Since O_4 is isomorphic to $SU_2 \times SU_2$, the classification of O_4 can be made in terms of four parameters. In a fully symmetric case, as is the present one, one of these four parameters can be dropped, and we need only three quantum numbers. Hence, the states of the present O_5 system can be classified completely by λ, μ, L and M. One can then prove that $L = \mu, \mu + 1, \ldots, 2\mu - 2, 2\mu$. Note that $L = 2\mu - 1$ is missing. Here λ can be written as $\lambda = \mu + 3n_\Delta$ where n_Δ is intuitively interpreted as the number of three d bosons coupled to angular momentum $0(d^3(0))$. Thus, one obtains a classification as shown in Fig. 1. When one views this figure in the horizontal direction keeping N constant, the $n_\beta = 0$ part gives rise to the classification of O_5.

```
6⁺ ———     3⁺ ———     4⁺ ———     2⁺ ———     0⁺ ———
4⁺ ———     2⁺ ———                 0⁺ ———
2⁺ ———        γ                      β             δ
0⁺ ———
```

Fig. 1. The classification of states with d^N configuration

We shall look at Fig. 1 differently, sweeping from the bottom upwards. As N increases one by one, we find more states, and the values of L change. One finds a clear regularity in the pattern of these variations, which actually appear to be similar to the pattern of rotational levels.

I then went a step further and introduced a heuristic s–boson so that the total number of bosons is conserved. Then, the above arguments can be applied to the classification of the states, $s_0^{N_0-N} d^N$. I introduced further a boson–boson interaction of the $Q \cdot Q$ type between bosons, and then diagonalized it. It turns out that the classification in terms of the SU_3 group is more convenient, producing the rotational level scheme. I published a paper about this result in a Japanese Journal in 1967 but never translated it into English[2]. Very recently Otsuka kindly translated this article into English.

I then became very much interested in this boson model and read a paper written by Brink, Toledo Piza and Kerman[3]. I then found that a general two body interaction of two bosons can be expressed as

$$V_{ij} = a\mathbf{1}_{ij} + bP_{ij} + c(\vec{\ell}_i \cdot \vec{\ell}_j)$$

where $\mathbf{1}$ is a constant interaction, P is a pairing interaction and $\vec{\ell}$ is the orbital angular momentum operator of a d boson. This idea came from Racah who treated the p–shell nuclei in the same way[4]. The eigenvalues of this interaction can be immediately found. I, however, did not write any paper.

Kishimoto was one of my graduate students at that time. He brought these ideas including the classification of d bosons and the diagonalization of the general two body interaction into Texas. There he and Tamura used these facts in their paper on the boson expansion technique[5].

These days, I did not know what are bosons, especially s bosons. Everybody believed that the d boson, namely quadrupole boson is made of a particle and a hole. Then no one imagined the necessity of s bosons and knew what this is.

When I listened to Talmi in 1971 in Padua where he told us the generalized seniority scheme[6], I was convinced that the monopole pair must correspond to the s boson and then as a natural consequence the d boson, too, corresponds to a quadrupole pair which is excited by the quadrupole operator. However, I did not pursue this model further until the summer of 1974, when KVI in Groningen invited me.

Iachello showed me an exciting result of his calculation in which he diagonalized a general d–boson Hamiltonian. He fitted parameters to reproduce the level structure of certain nuclei such as the Er isotopes. Agreement between his predictions and observed data were surprisingly good. Iachello's results convinced me that the d–boson model works well. I then told Franco that his Hamiltonian can be diagonalized analytically. This was the starting of our collaboration. Without this discovery of Franco's, I would never have come back to this model. Talmi was a concluding remarker of the Amsterdam Conference in the fall of 1974 where Franco reported his results. Talmi then got interested in this model. Otsuka also joined us in 1976.

Franco and I introduced the s boson in addition to the d boson in 1975.
This is a short prehistory of the Interacting Boson Model in Japan.

2. Effective Interactions Which Conserve Seniority

Talmi has emphasized that the excitation energies of the Sn isotopes are almost constant (Fig. 2). This fact indicates that either seniority or generalized seniority is a good approximation. It is then a very important question what interaction conserves the seniority (or generalized seniority).

Here I would like to show the most general form of the seniority conserving interactions. Before that, let me show the Fig. 3 from which one can learn that the pairing plus $Q \cdot Q$ interaction cannot keep the 2^+ energy constant.

If the pairing interaction is strong, the seniority is a good quantum number. The most convenient way to handle such the case is the use of the quasi–spin formalism. Using the creation operator a_{jm}^+ and the annihilation operator a_{jm}, one can construct the following three operators:

$$S_+ = \frac{1}{2}\sum_{j,m}(-1)^{j-m}a_{jm}^+ a_{j-m}^+$$

$$S_0 = \frac{1}{2}(\hat{N} - \Omega)$$

$$S_- = \frac{1}{2}\sum_{j,m}(-1)^{j-m}\tilde{a}_{jm}\tilde{a}_{j-m}$$

where

$$\hat{N} = \sum_{j,m} a_{jm}^+ a_{jm}$$

$$\Omega = \frac{1}{2}\sum_{j}(2j+1)$$

and

$$\tilde{a}_{jm} = (-1)^{j-m} a_{j-m}.$$

These three operators satisfy the commutations rule of the SU_2 generators. Under the quasi–spin formalism, the creation operator a_{jm}^+ and the annihilation operator $\tilde{a}_{jm}$ behave as spinors:

$$[S_0, a_{jm}^+] = \frac{1}{2}a_{jm}^+ \quad , \quad [S_0, \tilde{a}_{jm}] = -\frac{1}{2}\tilde{a}_{jm}$$
$$[S_+, a_{jm}^+] = 0 \quad , \quad [S_+, \tilde{a}_{jm}] = a_{jm}^+$$
$$[S_-, a_{jm}^+] = \tilde{a}_{jm} \quad , \quad [S_-, \tilde{a}_{jm}] = 0$$

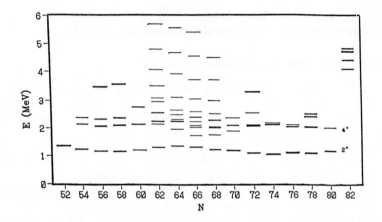

Fig. 2. Observed energies of J=2 and J=4 levels of even Sn isotopes

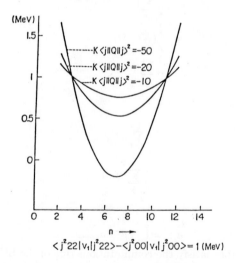

$\langle j^2 22|v_1|j^2 22\rangle - \langle j^2 00|v_1|j^2 00\rangle = 1$ (MeV)

Fig. 3. Calculated energies of J=2 levels of j=13/2 configuration. As an effective interaction, a linear combination of monopole pairing and $Q \cdot Q$ interactions is assumed

The creation and annihilation operators at the same time satisfy the normal commutation relations which define spherical tensor operators of half–integer rank. Namely the creation and annihilation operators behave like double tensors,

$$S^{(1/2,j)}_{1/2,m} = a^+_{jm} \quad \text{and} \quad S^{(1/2,j)}_{-1/2,m} = \tilde{a}_{jm}.$$

Operators bilinear in a^+ and $\tilde{a}$ are again either vectors or scalars in quasi–spin space. These are

$$S^{(1,\lambda)}_{1,\mu}(j_1, j_2) = -[a^+_{j_1} \times a^+_{j_2}]^{(\lambda)}_\mu / \sqrt{1 + \delta(j_1, j_2)}$$

$$S^{(1,\lambda)}_{0,\mu}(j_1, j_2) = -\{[a^+_{j_1} \times \tilde{a}_{j_2}]^{(\lambda)}_\mu - (-1)^{j_1+j_2-\lambda}[a^+_{j_2} \times \tilde{a}_{j_1}]^{(\lambda)}_\mu\} / \sqrt{2(1 + \delta(j_1, j_2))}$$
$$+ \delta(j_1, j_2)\delta(\lambda, 0)\sqrt{\Omega_{j_1}/2}$$

$$S^{(1,\lambda)}_{-1,\mu}(j_1, j_2) = -[\tilde{a}_{j_1} \times \tilde{a}_{j_2}]^{(\lambda)}_\mu / \sqrt{1 + \delta(j_1, j_2)}$$

$$S^{0,\lambda}_{0,\mu} = \{[a^+_{j_1} \times \tilde{a}_{j_2}]^{(\lambda)} + (-1)^{j_1+j_2-\lambda}[a^+_{j_2} \times \tilde{a}_{j_1}]^{(\lambda)}_\mu\} / \sqrt{2(1 + \delta(j_1, j_2))}$$

It is now clear how to construct general quasi–spin scalar two–body interactions which conserve seniority;

$$V_1 = \sum \chi_\lambda(j_1 j_2 j'_1 j'_2)[S^{(0,\lambda)}_0 \times S^{(0,\lambda)}_0]^{(0)}_0$$

and

$$V_2 = \sum G_\lambda(j_1 j_2 j'_1 j'_2)\{[S^{(1,\lambda)}_1 \times S^{(1,\lambda)}_{-1}]^{(0)}_0$$

$$-[S^{(1,\lambda)}_0 \times S^{(1,\lambda)}_0]^{(0)}_0 + [S^{(1,\lambda)}_{-1} \times S^{(1,\lambda)}_1]^{(0)}_0\},$$

$j_1 j_2 j'_1$ and j'_2 being omitted in S's.

As an example, let me take an interaction containing the $Q \cdot Q$ interaction which is schematically written as

$$V = G_2\{(D^+ \cdot \tilde{D}) + (Q \cdot Q) + (\tilde{D} \cdot D^+)\}$$

where

$$D^+ = S^{(1,2)}_1$$

$$Q = S^{(1,2)}_0.$$

Since

$$\tilde{D} \cdot D^+ = D^+ \tilde{D} + \sum_j \alpha_j \hat{N}_j,$$

where α is a constant,

$$V = G_2\{(D^+ \cdot \tilde{D}) + (Q \cdot Q)\} + \sum_j \alpha_j \hat{N}_j.$$

This means that we must add the quadrupole pairing interaction $(D^+ \cdot \tilde{D})$ to the $(Q \cdot Q)$ interaction to conserve seniority. At the same time, one must be careful that an appropriate single particle energy term must be added. This last statement should be interesting, because one could absorb an observed single particle energy differences in this term by choosing $G_\lambda(j_1 j_2, j_1' j_2')$ appropriately.

One simple example is as follows. Let me take a system consisting two single particle levels j_1 and j_2. The total quasi-spin S is the sum of $S(j_1)$ and $S(j_2)$:

$$S_+ = S_+(j_1) + S_+(j_2)$$

where

$$S_+(j_1) = \frac{1}{2}\sum_m (-1)^{j_1-m} a^+_{j_1 m} a^+_{j_1 -m}.$$

Addition of the following operator

$$\epsilon(S(j_1) \cdot S(j_1)) = \epsilon\{S_+(j_1)S_-(j_1) - S_0(j_1)S_0(j_1) + S_-(j_1)S_+(j_1)\}$$

does not break seniority, because this operator is a quasi-spin scalar. This addition, however, solves the degeneracy of two seniority-one states with $J = j_1$ and j_2. However this kind of addition produces only constant shifts of energies of seniority-one states.

If we are lucky, the observed single particle energies can be absorbed into a quasi-spin scalar two-body operator. Then all levels of even-even nuclei have to have the same excitation energies independently of n, and the same for odd-A nuclei. However, nature is not so simple.

Let me have a single particle energy term which cannot be absorbed into a quasi-spin scalar interaction

$$\sum \epsilon_j N_j.$$

This term is a quasi-spin vector. Therefore in first order perturbation, this produces the following energy

$$<j^n vJ|\sum \epsilon_j N_j|j^n vJ> = \frac{\Omega - n}{\Omega - v} <j^v vJ|\sum \epsilon_j N_j|j^v vJ>.$$

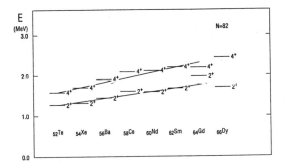

Fig. 4. Excitation energies of J=2 and 4 levels of N=82 isotones.

Figs. 4 and 5 show that the 2^+ and 4^+ excitation energies of the isotones with N = 82 and 126 change their values almost linearly. Namely the perturbation seems to work well.

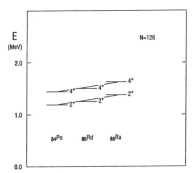

Fig. 5. Excitation energies of J=2 and 4 levels of N=126 isotones.

3. Conclusions

In this talk, I reviewed a prehistory of the Interacting Boson Model concerning mainly Japan.

Then the easiest way to find an effective interaction which conserves seniority was explained. Two things must be pointed out here. One concerns the $Q \cdot Q$ interaction. One needs the quadrupole pairing interaction in addition to the $Q \cdot Q$ one in order to keep seniority as a good quantum number. The second point is that there seems to absorb single particle energy differences into a seniority conserving interaction. Therefore the seniority scheme may work better than commonly thought.

Finally I would like to stress the importance of finding a good effective interaction which describes single closed shell nuclei as well as possible.

4. References

1. M. Sakai, Nucl. Data Tables, **15** (1975) 513.
2. A. Arima, Soryushiron Kenkyu (in Japanese) **35** (1967) E47.
3. D.M. Brink, A. F. R. de Toledo Piza, and A. K. Kerman, Phys. Lett. **19** (1965)413.
4. G. Racah, L. Farkas Memorial Volume, Research Council of Israel Jerusalem (1952).
5. T. Kishimoto and T. Tamura, Nucl. Phys. **A192** (1972) 246.
6. I. Talmi, Nucl. Phys. **A172** (1971) 1.

QUADRUPOLE "PHONON" EXCITATIONS IN EVEN XENON NUCLEI

P. VON BRENTANO, O. VOGEL, N. PIETRALLA, A. GELBERG,
AND I. WIEDENHÖVER
Institut für Kernphysik, Universität zu Köln
D-50937 Köln, Germany

ABSTRACT

A systematic study of low lying positive parity states in even-even Xe nuclei with $A = 116-132$ is presented. New experimental results for low spin states in ^{128}Xe are shown. These are compared with IBA calculations in the framework of the Consistent Q Formalism (CQF). A short description of the multiple Q-phonon scheme near the O(6) symmetry and its extension to nuclei outside the symmetries are given. The purity of several yrast states is examined in the framework of the CQF Hamiltonian. The empirical phonon purity of the first 2^+ states is deduced from the branched electromagnetic decay of the first 3^+ state in the frame of the phonon model.

1. Introduction

In recent years many investigations on the nuclear structure in the mass region around $A = 130$ were done see e.g. refs.[1-5] and the references given therein. We will discuss in particular the physics of the low lying positive parity collective states in the even-even Xenon nuclei with mass number $A = 116-132$. Here it is of interest to identify the band heads of many different excitations. For the positive parity states in many even-even nuclei up to 6 band heads were identified. These are the g-band,then the even and odd spin γ-band and the second "K=0" bands as well as the even and odd spin "K=4" band. In fig. 1 a survey of low lying positive parity states, retrieved from the NNDC online data bank,[13] is shown. These data provide an ideal testing ground for for various collective models as in particular the Interacting Boson Model (IBA-1 and IBA-2),[7] the General Collective Model (GCM),[10] the Rotation Vibration Model (RVM),[11] and the Fermion Dynamical Symmetry model (FDSM).[12] These experiments have confirmed in particular the suggestion[14] that the collective levels in even Xe-Ba nuclei are good examples of the dynamical O(6) symmetry[7] of the Interacting Boson Model (IBA) as suggested in refs. .[1,2,14-19] The low lying collective bands in the odd mass Xe-Ba and Cs nuclei have been recently investigated in detail in Köln, Stony Brook, and other labs. They were interpreted with great success in the frame of the Triaxial Rotor plus Particle Model (TRPM)[20,21] and more recently also in the framework of the Interacting Boson Fermion Model (IBFM) (see ref.[31] and references therein). The successful interpretation of these nuclei both in the frame of a model in which the nuclear shape is γ unstable as in the O(6) symmetry of the IBA (and the IBFM respectively) and in models where the nucleus has rigid triaxial shape, like in the particle rotor model, is surprising. Thus one has to ask whether these nuclei are gamma soft or gamma rigid. Of course there remains the possibility that the even nuclei are gamma soft and the odd nuclei are gamma rigid due to core polarization. Furthermore it is of interest to find observables which can decide this question. The answer to this second question is that most observables are sensitive only to the average γ deformation. A survey of the average γ deformation in this region has been given in refs.[3,22,21]. This survey was based on the E2 branching ratio from the decay of 2^+_γ state[6,22] and on the energy staggering of the yrast band of the odd nuclei.[23,21] As an observable of the gamma softness the energy staggering in the quasi-gamma band can be used.[6,23,18,16] A much stronger proof for the gamma softness can be found in the observation of a low lying 0^+ state, which decays by a strong B(E2) to the 2^+_2 state and only weakly to the 2^+_1 state.[1–3,5,16] The nature of this 0^+ state in the vicinity of the O(6) dynamical symmetry has been clarified in the multiple quadrupole phonon scheme suggested by Otsuka et al.,[4,24] which will be discussed below. The investigation of this special 0^+ state and its electromagnetic decay properties is very important for the discussion of the shape in the Xe-Ba nuclei. It turns out that this state is usually the second 0^+ state in the Xe isotopes as shown in fig. 1

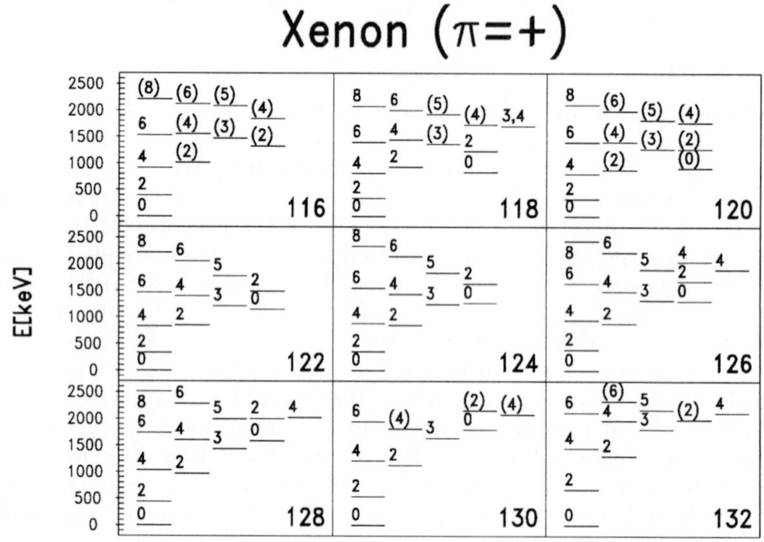

Xenon (π=+)

Fig. 1. Low lying positive parity states in even-even Xe nuclei. Data taken from NNDC online data bank.[13]

2. Low Spin States in ^{128}Xe

Although in beam gamma spectroscopy experiments using fusion reactions with heavy ion projectiles give a very rich spectrum with many rotational bands, they are not very suitable for the spectroscopy of the low lying non yrast collective states.

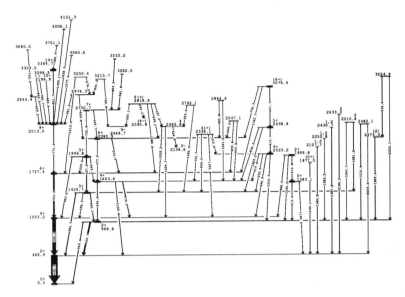

Fig. 2. Partial experimental level scheme of ^{128}Xe from the ^{125}Te$(\alpha,n)^{128}$Xe reaction.[32]

Normally only the ground band and the quasi-gamma-band are observed in these experiments, but none of the other states predicted by collective models. Useful tools for low spin spectroscopy are fusion reactions with light projectiles as (p,n) or (α,n) reactions[15] which populate low spin levels far above the yrast line. Also the investigation of the beta decay from medium spin isomers using multi detector coincidence arrays e.g. the OSIRIS 6 Cube spectrometer has proved to be a very useful tool for these investigations see e.g. ref..[24] In a recent experiment at the Cologne FN Tandem accelerator, we reexamined low spin states in ^{128}Xe by the reaction ^{125}Te$(\alpha,n)^{128}$Xe with the OSIRIS 6 Cube gamma spectrometer. Thus it was possible to observe 154 levels in a spin interval from 0 to $10\hbar$. From these 70 were previously unknown. A total of 258 transitions were observed. Due to this experiment the first $K = 4$ band is established in ^{128}Xe, and multipolarity mixing ratios for many transitions were determined. In figure 2 part of the experimental level scheme is shown[32].

In figure 3 experimental level energies are compared with those calculated from the IBA. For our calculation of the effective γ deformation of nuclei in the $A \approx 130$ mass region we have used the special CQF (Consistent Q Formalism) Hamiltonian[25,6]

$$H = \kappa Q(\chi).Q(\chi) + \kappa' L.L \qquad (1)$$

where

$$Q_\mu = Q_\mu(\chi) = s^\dagger \tilde{d}_\mu + d_\mu^\dagger s + \chi (d^\dagger d)_\mu^{(2)}. \qquad (2)$$

The variation of the parameter χ allowed us to go from the O(6) symmetry ($\chi = 0$) to SU(3) symmetry ($\chi = -\sqrt{7}/2$). Due to the choice of the special CQF Hamiltonian

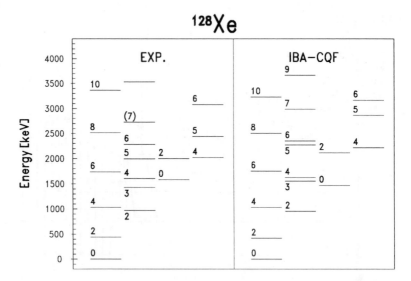

Fig. 3. Experimental and IBA calculated excitation energies for ^{128}Xe with the CQF Hamiltonian of eq. 1 and 2. The data is from refs..[32,33] The used parameters are $\kappa = 0.097$MeV, $\kappa' = 0.011$MeV, $\chi = -0.10$ and $\alpha = 0.111$MeV^{-1}.

the wave functions and thus the transition probabilities depend only on the parameter χ (for fixed N), which was determined from the B(E2) ratio R_γ:

$$R_\gamma = \frac{B(E2, 2^+_\gamma \to 2^+_1)}{B(E2, 2^+_\gamma \to 0^+_1)}, \qquad (3)$$

This allowed a unique determination of the parameter χ for many nuclei, since R_B is accurately available from the experiments and depends rather sensitively on the structural parameter χ. For the energy fits the ratio κ/κ' was fitted to the relative positions of the 4^+_2 state to the 4^+_1 and 6^+_1 state. In order to reproduce the excitation energies of states with higher spins up to $10\hbar$, we renormalized the energies E^0_x from the IBA calculation according to eq. 4. This renormalization has been proposed by Lipas et al. for rotational bands.[27]

$$E_x = \frac{E^0_x}{1 + \alpha E^0_x} \qquad (4)$$

As shown in fig. 3 the overall agreement between experiment and theory is quite good. The reproduction of the staggering in the quasi gamma band is not perfect, however. It is well known that a description of the staggering can be given by the use of an additional cubic term $(d^\dagger d^\dagger d^\dagger)^{(3)}(ddd)^{(3)}$ in the Hamiltonian (e.g. ref.[28])

3. Q–Phonon Scheme for Nuclei near to the $O(6)$ symmetry

The Interacting Boson Model has been solved analytically for the three symmetries. Thus for energies and for branching ratios of electromagnetic quadrupole transitions of the three symmetry limits analytical formulas have been given.[7,8,29] Outside the symmetries such calculations can be done numerically. With the Quadrupole phonon scheme recently proposed by Otsuka et al.[24] a more intuitive picture of the $O(6)$ wave functions is given. In table 1 the multi Q-phonon structure of the lowest excitations in the $O(6)$ symmetry is given.[24] We note that the maximum number of Q-phonons is equal to the τ quantum number. One should note that these wave functions are not normalized. In this case the Q operator of eq. 2 with $\chi = 0$ is used. We note that the Q-phonon scheme applies also to the nuclei with a $U(5)$ symmetry and to the g-band of the nuclei with a $SU(3)$ symmetry if a value of the parameter $\chi = -1.32$ is used. One of the suggestive advantages of the Q-phonon scheme is that the selection rules for quadrupole transitions become very transparent. Theses

Table 1. Q-phonon configuration of low lying $O(6)$ states.

$\|I^\pi, \tau, \nu_\Delta\rangle$		
$\|0_1^+, 0, 0\rangle$		$\|0\rangle$
$\|2_1^+, 1, 0\rangle$	Q	$\|0\rangle$
$\|4_1^+, 2, 0\rangle$	$[QQ]^{(4)}$	$\|0\rangle$
$\|6_1^+, 3, 0\rangle$	$[QQQ]^{(6)}$	$\|0\rangle$
$\|2_2^+, 2, 0\rangle$	$[QQ]^{(2)}$	$\|0\rangle$
$\|4_2^+, 3, 0\rangle$	$[QQQ]^{(4)}$	$\|0\rangle$
$\|3_1^+, 3, 0\rangle$	$[QQQ]^{(3)}$	$\|0\rangle$
$\|5_1^+, 4, 0\rangle$	$[QQQQ]^{(5)}$	$\|0\rangle$
$\|0_2^+, 3, 1\rangle$	$[QQQ]^{(0)}$	$\|0\rangle$

selection rules allow only quadrupole transitions for which the number of Q-phonons changes by one. In table 3 the relative $B(E2)$ values of several low lying states in ^{126}Xe ref.[16] and ^{128}Xe ref.[32] taken from recent experiments in Köln using the (α,n) reaction are compared with B(E2) values, calculated with CQF IBA Hamiltonian. Furthermore the phonon configuration of the states is given. One notes that the resemblance between experiment and the IBA calculated ones is very good, and the selection rules of the Q-phonon scheme are only slightly broken.

4. Extension of the Q-phonon scheme as an approximate scheme to nuclei outside the symmetries

It is natural to attempt to extend the Q-phonon scheme as an approximate scheme to nuclei outside the $O(6)$ and $U(5)$ symmetries. In this the generalized CQF Hamiltonian was used

Table 2. Experimental and IBA calculated relative B(E2) ratios for ^{126,128}Xe. Also shown are the Q-phonon configurations to stress the selection rules. The data are from refs.[16,32,33].

					^{126}Xe $\chi=-0.09$		^{128}Xe $\chi=-0.10$	
			I_i	$\rightarrow I_f$	exp.	IBA	exp.	IBA
QQ	$\rightarrow$	Q	2_2^+	$\rightarrow 2_1^+$	100	100	100	100
$\not\rightarrow$				$\rightarrow 0_1^+$	1.5(4)	1.5	1.3(6)	1.3
QQQ	$\rightarrow$	QQ	3_1^+	$\rightarrow 2_2^+$	100	100	100	100
$\not\rightarrow$		Q		$\rightarrow 2_1^+$	2.4(7)	2.2	1.48(8)	1.9
	$\rightarrow$	QQ		$\rightarrow 4_1^+$	34(10)	37	34.0(22)	37
QQQ	$\rightarrow$	QQ	4_2^+	$\rightarrow 2_2^+$	100	100	100	100
$\not\rightarrow$		Q		$\rightarrow 2_1^+$	0.4(1)	0.02	1.7(7)	0.01
	$\rightarrow$	QQ		$\rightarrow 4_1^+$	83(23)	80	98(15)	81
$QQQQ$	$\rightarrow$	QQQ	5_1^+	$\rightarrow 3_1^+$	100	100	100	100
$\not\rightarrow$		QQ		$\rightarrow 4_1^+$	2.9(8)	0.8	2.6(3)	0.7
	$\rightarrow$	QQQ		$\rightarrow 4_2^+$	76(21)	45	76(6)	45
	$\rightarrow$	QQQ		$\rightarrow 6_1^+$	75(23)	43	64(10)	43
QQQ	$\rightarrow$	QQ	0_2^+	$\rightarrow 2_2^+$	100	100	100	100
$\not\rightarrow$		Q		$\rightarrow 2_1^+$	7.7(2)	1.1	8(3)	1.0

$$H = \kappa(\frac{\epsilon}{\kappa}\hat{n}_d + Q(\chi).Q(\chi) + \frac{\kappa'}{\kappa}L.L) \qquad (5)$$

We note that the $L.L$ term is diagonal and thus does not affect the wavefunctions. Thus one does not consider this term in the following discussion. By varying the two parameters $(\epsilon/\kappa,\chi)$ one can interpolate between the three symmetries. The variation of the parameters can be displayed in the Casten triangle[6] (see fig. 4), where the unshaded area corresponds to parameters appropriate to existing nuclei. Thus one can investigate the purity of the Q-phonon assignments of table 1 in the whole Casten triangle and one can study the validity of the Q-phonon selection rules outside the symmetries

In order to investigate the phonon purity of a state with wavefunction $|I^+\rangle$ one considers the rest wave functions $|r_I\rangle$, which is the difference between the wave function of the eigenstate and the pure Q-phonon configuration. The configurations used are the following

$$|2_1^+\rangle = \alpha_2[Q]^{(2)}|0_1^+\rangle + |r_2\rangle \qquad (6)$$
$$|3_1^+\rangle = \alpha_3[QQQ]^{(3)}|0_1^+\rangle + |r_3\rangle \qquad (7)$$
$$|4_1^+\rangle = \alpha_4[QQ]^{(4)}|0_1^+\rangle + |r_4\rangle \qquad (8)$$

where the state wavefunctions $|I_i^\pi\rangle$ are normalized and where the rest wave function

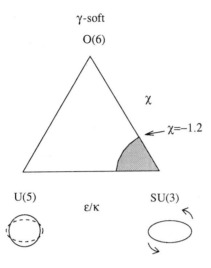

Fig. 4. Casten triangle of the three dynamical symmetries of the IBA.[6] The area which does not correspond to existing nuclei is shaded

$|r_I\rangle$ is orthogonal to the Q-phonon configuration, i.e.

$$\langle r_2|[Q]^{(2)}|0_1^+\rangle = 0 \qquad (9)$$
$$\langle r_3|[QQQ]^{(3)}|0_1^+\rangle = 0 \qquad (10)$$
$$\langle r_4|[QQ]^{(4)}|0_1^+\rangle = 0. \qquad (11)$$

Our interest lies in determining the square modules of $|r_I\rangle$. which measures the impurity of the phonon configuration of the state $|I\rangle$. A detailed investigation on the phonon purity of the 2_1^+ state was done in a paper by Pietralla, von Brentano, Casten and Zamfir.[30] There the rest term $|r_2\rangle$ as defined by eq. 7 was investigated in the parameter space of the Casten triangle. This paper showed that the 2_1^+ remains rather pure for the parameters in the Casten triangle corresponding to actual nuclei. One finds that $\langle r_2|r_2\rangle$ is less than 8% for all nuclei, and for most nuclei less than 4% . The result of the calculations of the variation of the square modulus of the rest wave function $|r_4\rangle$: $R^{(4,2)} = \langle r_4|r_4\rangle$ with the parameters ϵ/κ and χ is shown in fig. 5. Again a rather high phonon purity $\geq 93\%$ is found. As the 2_1^+ and the 4_1^+ states have a surprisingly good phonon purity it is of great interest to study the phonon purity of the odd spin yrast states. As an example we show here in fig. 4 the results for the 3_1^+ state, where $R(3,3) = \langle r_3|r_3\rangle$. In this case the impurity is always less than 4% for parameters corresponding to actual nuclei. One finds in general that all examined yrast states have a very good phonon purity for parameters corresponding to existing nuclei.

It was shown, that the multiple Q-phonon configuration is a very good descrip-

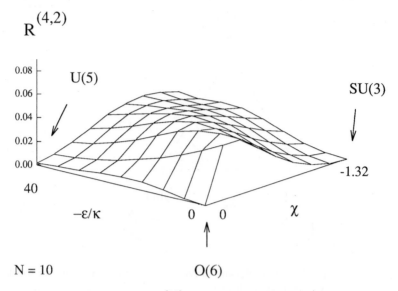

Fig. 5. Q-phonon purity $R^{(4,2)} = \langle r_4 | r_4 \rangle$ of the 4_1^+ state of the IBA.

tion of the low spin yrast states in collective nuclei, i.e. nuclei which are well described by the IBA.

After having discussed the phonon purity of the yrast states, we have to check the validity of the selection rules for the quadrupole transitions in the Q-phonon scheme. As an example we examine the B(E2) ratio of the quadrupole transition in the $(\epsilon/\kappa, \chi)$ space for the following configuration

$$|3^+, 3\rangle = \mathcal{N}^{(3,3)}[QQQ]^{(3)}|0_1^+\rangle, \tag{12}$$

which describes the 3_1^+ state in the O(6) limit. In figure 7 the B(E2) ratio of the quadrupole decay

$$\frac{B(E2; [QQQ]^{(3)}|0_1^+\rangle \to [Q]^{(2)}|0_1^+\rangle)}{B(E2; [QQQ]^{(3)}|0_1^+\rangle \to [QQ]^{(2)}|0_1^+\rangle)} \tag{13}$$

to the one phonon 2^+ configuration and the two phonon 2^+ configuration is plotted in the $(\epsilon/\kappa, \chi)$ space. As above these configurations are exact eigenstates of the Hamiltonian only in the U(5) and O(6) limits, where they describe the 2_2^+ and 2_1^+ states. The calculations show that the selection rules for the decay of the Q-phonon configurations do not lose their validity outside the O(6) symmetry, since the Q forbidden transition has only a strength of less 0.3% in the non shaded area of the triangle.

We further assume that near the O(6) limit the two lowest 2^+ states are essentially composed of one and two phonon configurations, i.e. :

$$|2_1^+\rangle = \alpha \mathcal{N}_\infty Q|0\rangle + \beta \mathcal{N}_\epsilon [QQ]^2|0\rangle \tag{14}$$

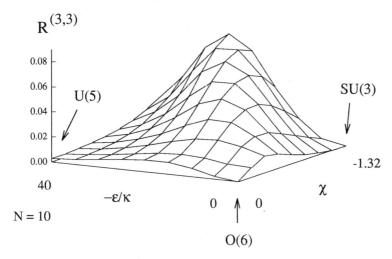

Fig. 6. Q-phonon purity of the 3^+_1 state of the IBA. Calculated for a boson number $N = 10$.

$$|2^+_2\rangle = -\beta \mathcal{N}_\infty Q|0\rangle + \alpha \mathcal{N}_\epsilon [QQ]^2|0\rangle \qquad (15)$$

Where the configurations $\mathcal{N}_\infty Q|0\rangle$ and $\mathcal{N}_\epsilon [QQ]^2|0\rangle$ are normalized. Form this and the

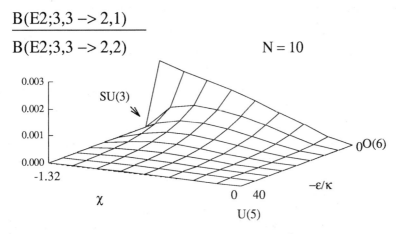

Fig. 7. B(E2) ratio for the decay of an three Q-phonon 3^+ configuration to one and two phonon configurations. (see eq. 13). Calculated for a boson number $N = 10$.

Table 3. Impurity of the first 2_1^+ state in even Xe nuclides. The values of R_{th} were derived from IBA calculations using eq. 1 and the parameters χ determined as discussed in the text. The values of R_{exp} were calculated from experimental branching and multipolarity mixing ratios of the 3_1^+ state.

Nucleus	N	χ	R_{th}	R_{exp}
^{122}Xe	9	-0.116	4.3%	3.7(9)%
^{124}Xe	8	-0.115	3.4%	2.8(4)%
^{126}Xe	7	-0.086	1.7%	2.4(7)%
^{128}Xe	6	-0.101	1.4%	1.5(1)%
^{130}Xe	5	-0.090	0.7%	1.4(1)%
^{132}Xe	4	-0.061	0.2%	1.1(1)%

above assumptions one obtains for the E2 branching ratio R_{exp} of the 3^+ state:

$$R_{\text{exp}} = \frac{B(E2; 3_1^+ \to 2_1^+)}{B(E2; 3_1^+ \to 2_2^+)} = \frac{\beta^2}{\alpha^2} \qquad (16)$$

In table 3 we compare the ratio R^3 from the experiments with values calculated with the IBA. One notes that the agreement between experimental and theoretical impurities is quite good. An alternative way to obtain the value of β^2 is to use the measured $B(E2; 0^+ \to 2_k^+)$ values. This has been done in the work by Pietralla et al.[30] for many nuclei and it will be reported here in the talk by R.F. Casten. In many Xe nuclei these B(E2) values are however unknown. This shows that the above method is quite useful and it can be used in future work with radioactive beams.

5. Conclusions

1. The Q-phonon scheme is a very useful scheme in the IBA for nuclei near the O(6) symmetry.

2. As far as the low spin positive parity yrast states are considered, the Q-phonon scheme is for a very useful approximate scheme in the IBA also outside the O(6) and U(5) symmetries.

3. The low spin positive parity odd spin yrast states in particular the 3_1^+ and 5_1^+ states have a very pure Q-phonon configuration in particular near the O(6) symmetry.

4. The low spin positive parity even spin yrast states in particular the 2_1^+ and 4_1^+ states have a reasonably pure Q-phonon configuration. For parameters corresponding to actual nuclei the squared modulus of the purity is above 96%.

5. The non yrast states quickly loose the Q–phonon purity outside the O(6) symmetry.

6. The selection rules of the Q-phonon scheme for electromagnetic quadrupole transitions work also outside the symmetries to a very good accuracy. The Deviation are less than 0.3 % for the 3 Q-phonon 3^+ configuration.

7. The branching ratio from the 3^+_1 state to the 2^+ states is correlated with the impurity of the 2^+_1 state.

8. The Q-phonon scheme makes evident nuclear properties that are common to all collective nuclei, independent of their actual shapes.

We still have to understand why the Q-phonon scheme works also for the yrast states outside the symmetries to rather good accuracy. Furthermore a general approximative description of the non yrast states similar to the Q-phonon scheme should be searched.

For discussions on the Q-phonon scheme we thank in particular T. Otsuka. We would like to acknowledge further R.F. Casten, A. Dewald, R.V. Jolos, A. Mertens, P. Petkov N.V. Zamfir and K.O. Zell for stimulating discussions. This work was partly supported by the BMFT under contract no. O60 K 602 I and by the cooperation agreement between the DFG and the JSPS.

References

1. P. von Brentano, A. Gelberg, O. Vogel, A. Dewald, and I. Wiedenhöver, *Int. Conf. High Spin Physics and Gamma soft Nuclei,* September 17-21, 1990, University of Pittsburgh, Carnegie Mellon, eds. J.X. Saladin, R.A. Sorensen, C.M. Vincent, World Scientific, (Singapore) p. 344.
2. P. von Brentano, A. Gelberg, O. Vogel, A. Dewald, and O. Stuch, *Proc. Summer School on Recent Advances in Nuclear Structure,* Predeal, Romania 1990, eds. D. Bucurescu, G. Cata-Danil, N.V. Zamfir, World Scientific, (Singapore 1991) p. 129.
3. P. von Brentano, A. Dewald, A. Gelberg, P. Sala, G. Siems, I. Wiedenhöver, R. Wirowski, and J. Yan, *Int. Symp. on the Frontier of Nuclear Spectroscopy*, Kyoto, Japan, October 23-24 (1992), ed. H. Eijri, World Scientific, (Singapore 1994).
4. P. von Brentano, O. Vogel, A. Dewald, A. Gelberg, and N. Pietralla, *Int. Symp. on the Frontiers of nuclear structure Physics,* Tokyo, Japan, 1994. eds. T. Otsuka and K. Yasakai, World Scientific, (Singapore 1994).
5. P. von Brentano, K. Kirch, U. Neuneyer, G. Siems, and I. Wiedenhöver, *Proc. Summer School on Frontier Topics in Nucl. Phys.,* Predeal, Romania, 1993, eds. A. Sandulescu, W. Scheid, Plenum Publ.
6. R.F. Casten, *Nuclear Structure from a Simple Perspective,* Oxford Univ. Press (1990).
7. F. Iachello and A. Arima, *The Interacting Boson Model,* Oxford Univ. Press, Cambridge, 1987, and refs. given therein.
8. R.F. Casten, J.P. Draayer, K. Heyde, P.O. Lipas, T. Otsuka, and D.D. Warner, *Algebraic Approaches to Nuclear Structure,* Harwood Publ., Chur, Switzerland, 1993, and refs. given therein.

9. I. Talmi, *Simple Models of Complex Nuclei*, Harwood Publ., Chur, Switzerland, 1993.
10. J.M. Eisenberg and W. Greiner, *Nuclear Theory Vol. I*, North Holland Publ., (Amsterdam 1987).
11. A. Faessler, W. Greiner, and R.K. Sheline, *Nucl. Phys.* **80** (1965) 417.
12. C.-L. Wu, D.-H. Feng, and M. Guidry, *The Fermion Dynamical Symmetry Model*, Plenum Press, (New York 1994).
13. NNDC online data bank, National Nuclear Data Center, Brookhaven, New York.
14. R. F. Casten and P. von Brentano, *Phys. Lett.* **B 152**(1985) 22.
15. W. Lieberz, A. Dewald, W. Frank, A. Gelberg, W. Krips, D. Lieberz, R. Wirowski and P. von Brentano, *Phys. Lett.* **B 240** (1990) 38.
16. F. Seiffert, W. Lieberz, A. Dewald, S. Freund, A. Gelberg, A. Granderath, D. Lieberz, R. Wirowski, and P. von Brentano, *Nucl. Phys.* **A 554**(1993) 287.
17. R. Wyss, A. Granderath, R. Bengtsson, P. von Brentano, A. Dewald, A. Gelberg, A. Gizon, J. Gizon, S. Harissopulos, A. Johnson, W. Lieberz, W. Nazarewicz, J. Nyberg and K. Schiffer, *Nucl. Phys.* **A 505** (1989) 337.
18. N.V. Zamfir and R. Casten, *Phys. Lett.* **B260** (1991) 265.
19. A. Arima and F. Iachello, *Annals of Phys.* **123** (1979) 468.
20. A. Gelberg, D. Lieberz, P. Brentano, I. Ragnarsson, P.B. Semmes, and I. Wiedenhöver, *Nucl. Phys.* **A557** (1993) 439c.
21. O. Vogel, A. Gelberg, R.V. Jolos, and P. von Brentano, *Nucl. Phys.* **A** in press.
22. J. Yan, O. Vogel, P. von Brentano, and A. Gelberg, *Phys. Rev.* **C48** (1993) 1046.
23. D. Lieberz, A. Gelberg, P. von Brentano, I. Raganarssona and P.B. Semmes, *Phys. Lett.* **B282** (1992) 7.
24. G. Siems, U. Neuneyer, I. Wiedenhöver, S. Albers, M. Eschenauer, R. Wirowski, A. Gelberg, P. von Brentano and T. Otsuka, *Phys. Lett.* **B320** (1994) 1.
25. R.F. Casten and D.D. Warner, *Phys. Rev.* **C28** 1798 (1983).
26. L. Wilets and M. Jean, *Phys. Rev.* **102** (1956) 3.
27. P. Holmberger and P.O. Lipas, *Nucl. Phys.* **A117** (1968) 552.
28. R.F. Casten, P. von Brentano, K. Heyde, P. Van Isacker and J. Jolie, *Nucl. Phys.* **A439** (1985) 289.
29. A. Arima and F. Iachello, *Ann. Phys.* **123** (1979) 468.
30. N. Pietralla, P. von Brentano, R.F. Casten, T. Otsuka, and N.V. Zamfir, submitted to *Phys. Rev. Lett.* .
31. N. Yoshida, A. Gelberg, T, Otsuka, and H. Sagawa, Int. Conf. on Perspectives for the Interacting Boson Model, Padova, Italy, 13-17 June 1994.
32. I. Wiedenhöver, A. Mertens, R. Kühn, K. Kirch, W. Lieberz, E. Radermacher, M. Wilhelm, and P. von Brentano, to be published.
33. R. Reinhardt, A. Dewald, A. Gelberg, W. Lieberz, K. Schiffer, K.P. Schmittgen, K.O. Zell, and P. von Brentano, *Z. Phys.* **A329** (1988).

THE PAIR-COUPLING MODEL

DAVID J. ROWE
Department of Physics, University of Toronto,
Toronto, Ont. M5S 1A7, Canada

ABSTRACT

The foundations of the IBM are discussed briefly followed by a review of the pair-coupling model in a coherent-state representation. Pair-coupled states correspond to number-projected BCS states. Their coherent-state wave functions are well known in character theory as Schur functions. The coherent-state representation also gives an expansion for nucleon pairs in which the leading terms are bosons of the IBM type. We conclude with a discussion of the pairing-plus-quadrupole problem and the apparent persistence of dynamical symmetries in the face of strong symmetry mixing interactions.

1. Introduction

I was asked to talk about pairing and its relationship to the IBM[1]. The subject is of interest to this conference because it is thought to be central to the microscopic foundations of the IBM. My interest in the IBM derives from a desire to learn how to include pairing degrees of freedom in the microscopic collective model (the symplectic model[2]). In particular, I want to describe the rotations of nuclei that are both deformed and superconducting without independent-particle or independent-quasiparticle constraints. The problem is that there is no dynamical symmetry to make this possible. But nuclei can do it and they do it in an intuitively simple manner. I conclude that we should be able to describe the phenomenon in correspondingly simple terms. Moreover, if our intuition is compatible with the many-nucleon structure of the nucleus, then so should be our description. However, it may be necessary to take limits; e.g., let the number of nucleons go to infinity or approximate particle-hole pairs or pairs of nucleons as bosons. These are the kinds of approximations on which the IBM is based.

So why does the IBM work so well? A simple answer is that its u(6) algebraic structure, customarily expressed in terms of six (s and d bosons)

$$u(6) = \langle s^\dagger s, s^\dagger d_\nu, d_\nu^\dagger s, d_\mu^\dagger d_\nu \rangle, \qquad (1)$$

is sufficiently versatile that it includes subalgebras which contract to dynamical algebras for all the essential types of collective motions with which the model is concerned. Moreover, u(6) appears to be the simplest Lie algebra with this property.

The relationship of the IBM to the BMF (Bohr-Mottelson-Frankfurt[3,4]) model, known from the formulation of Janssen, Jolos and Dönau[5], can be expressed algebraically as follows. The BMF model has a dynamical algebra

$$[hw(5)]u(5) = \langle d_\nu^\dagger, d_\nu, I, d_\mu^\dagger d_\nu \rangle, \qquad (2)$$

which is a semidirect sum of a Heisenberg-Weyl (d-boson) algebra and u(5). Now, an $(N, \dot{0})$ irrep of the IBM u(6) algebra contracts as $N \to \infty$ according to the equations (easily derived by coherent-state methods)

$$
\begin{aligned}
\text{u}(6) \to [\text{hw}(5)]\text{u}(5) \; ; \quad s^\dagger s &\mapsto NI \\
s^\dagger d_m &\mapsto \sqrt{N}\, d_m \\
d_m^\dagger s &\mapsto \sqrt{N}\, d_m^\dagger \\
d_m^\dagger d_n &\mapsto d_m^\dagger d_n \, .
\end{aligned}
\tag{3}
$$

Thus, the IBM contracts to the BMF model. Conversely, the IBM is a compactification of the BMF model. This means that the two models have the potential to describe the same kinds of physics. A remarkable fact is that the contraction/compactification relationships between the IBM and BMF algebras extend to their dynamical subalgebra chains, as indicated in Fig. 1.

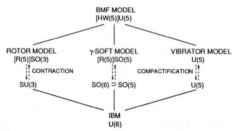

Fig. 1. Subalgebras (submodels) of the dynamical algebra u(6) of the IBM and corresponding subalgebras (submodels) of the dynamical algebra [hw(5)]u(5) of the BMF model.

The s-boson substructure of the IBM also relates to the pair-coupling models and Kerman's quasispin algebra[6]. Indeed, it is the relationship of the IBM bosons to fermion pairs that has been emphasized in considering the microscopic origins of the model; cf. in particular, the derivation of Otsuka, Arima, and Iachello[7]. Kerman's quasispin algebra is spanned by the su(2) operators

$$
\begin{aligned}
S_+ &= \sum_{j_i, m_i > 0} (-1)^{j_i + m_i} a^\dagger_{j_i, m_i} a^\dagger_{j_i, -m_i} \, , \\
S_- &= \sum_{j_i, m_i > 0} (-1)^{j_i + m_i} a_{j_i, -m_i} a_{j_i, m_i} \, , \\
S_0 &= \tfrac{1}{2} \sum_{j_i, m_i > 0} (a^\dagger_{j_i, m_i} a_{j_i, m_i} - a_{j_i, -m_i} a^\dagger_{j_i, -m_i}) \, .
\end{aligned}
\tag{4}
$$

Su(2) also has a Holstein-Primakof realisation in terms of s-boson operators

$$
S_+ = s^\dagger \sqrt{2S - s^\dagger s}, \quad S_- = \sqrt{2S - s^\dagger s}\, s, \quad S_0 = s^\dagger s - S \, ,
\tag{5}
$$

where S is any positive half integer. Thus, for states of s-boson number small compared to $2S$, we have the contraction as $S \to \infty$

$$\mathrm{su}(2) \to [\mathrm{hw}(1)]\mathrm{u}(1) \,; \quad S_+ \mapsto \sqrt{2S}\, s^\dagger,$$
$$S_- \mapsto \sqrt{2S}\, s,$$
$$S_0 \mapsto s^\dagger s - SI, \qquad (6)$$

where I is the identity operator.

One can do better than this[7] following the proposal of Arima and Iachello[1] to map $J = 0$ and $J = 2$ coupled nucleon pairs to bosons;

$$\sum_j C_j^0 (a_j^\dagger a_j^\dagger)_0 \to s^\dagger, \quad \sum_{ij} C_{ij}^2 (a_i^\dagger a_j^\dagger)_{2M} \to d_M^\dagger. \qquad (7)$$

This mapping can also be expressed as a group contraction which is valid for a large number of single-particle levels and a small number of nucleons. More generally, the boson approximation is the leading term in a coherent-state (boson) expansion.

The above relationships suggest two completely different derivations of the IBM. One derivation, as indicated above, follows from a microscopic derivation of the BMF model (which we have). Another derives the boson algebra as a subalgebra of a contracted fermion-pair algebra. The exciting possibility is that, although inequivalent, the two derivations yield similar IBM algebraic structures and raise the possibility of combining them. This is something we have not been able to do at a microscopic level. Indeed, an outstanding problem in the microscopic theory of nuclear collective motion is to include simultaneously the pairing and geometrical collective degrees of freedom without making independent-particle or independent-quasiparticle approximations (as in the unified model and models based on Hartree-Bogolyubov theory). In the following, I shall discuss the pair-coupling model[8] as an interpolation between two incompatible algebraic structures which become compatible in a contraction limit. A more detailed treatment of the subject will appear in a journal article by Chen, Song and myself[9].

2. The pair-coupling model

2.1. The Hamiltonian

We consider the Hamiltonian

$$H = \sum_{jm} \varepsilon_j a_{jm}^\dagger a_{jm} - GS_+ S_-. \qquad (8)$$

In terms of the su(2) quasispin operators

$$S_+^i = \sum_{m_i > 0} (-1)^{j_i + m_i} a_{j_i m_i}^\dagger a_{j_i, -m_i}^\dagger$$
$$S_-^i = \sum_{m_i > 0} (-1)^{j_i + m_i} a_{j_i, -m_i} a_{j_i m_i}$$
$$S_0^i = \tfrac{1}{2} \sum_{m_i > 0} (a_{j_i m_i}^\dagger a_{j_i m_i} - a_{j_i, -m_i} a_{j_i, -m_i}^\dagger), \qquad (9)$$

the Hamiltonian (8) is expressed

$$H = 2\sum_i \varepsilon_i S_0^i - G \sum_{ij} S_+^i S_-^j + \text{const.} \qquad (10)$$

Thus, H is quadratic in the elements of the (multiple quasispin) dynamical algebra

$$\mathcal{A} = \text{su}(2)_1 + \ldots + \text{su}(2)_d. \qquad (11)$$

2.2. Dynamical subalgebra chains

The Hamiltonian (10) has two dynamical subalgebra chains. When G is zero, the eigenstates of H belong to irreps of the chain, with quantum numbers $s_i = \frac{1}{4}(2j_i+1)$,

$$\begin{array}{cccc} \text{su}(2)_1 + \ldots + \text{su}(2)_d & \supset & \text{u}(1)_1 + \ldots + \text{u}(1)_d. \\ s_1 \qquad\qquad s_d & & m_1 \qquad\qquad m_d \end{array} \qquad (12)$$

And, when the single-particle energies are all equal, so that (to within a constant)

$$H = 2\varepsilon S_0 - G S_+ S_-, \qquad (13)$$

the eigenstates have good coupled quasispin and belong to irreps of the chain

$$\begin{array}{cccc} \text{su}(2)_1 + \ldots + \text{su}(2)_d & \supset & \text{su}(2) & \supset & \text{u}(1), \\ s_1 \qquad\qquad s_d & & s & & m \end{array} \qquad (14)$$

where su(2) is the summed quasispin algebra with

$$S_0 = \sum_i S_0^i, \quad S_\pm = \sum_i S_\pm^i. \qquad (15)$$

2.3. Ground states

When $G \neq 0$ and the single-particle energies are not all equal, there is no dynamical subalgebra other than the full algebra $\mathcal{A}$. Nevertheless, the generalized seniority[10] (pair-coupling[8]) approximation provides an extremely good approximation for ground-state wave functions and broken-pair excited states. In the generalized seniority approximation the ground state of the $2n$-particle nucleus is approximated by the pair-coupled state

$$|n(\alpha)\rangle = [S_+(\alpha)]^n |0\rangle = [\sum_j \alpha_j S_+^j]^n |0\rangle, \qquad (16)$$

where $|0\rangle$ is the zero-particle vacuum state and the α_j coefficients are adjusted to minimize the energy of the state. To illustrate the accuracy of the pair-coupling approximation, we show in Fig. 2 the ground-state energy in comparison with exact numerically computed results; the corresponding results of the BCS approximation are also shown.

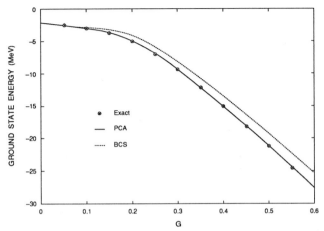

Fig. 2. The ground-state energy, calculated in the pair-coupling model, shown as a function of the coupling constant G in comparison with the corresponding results of exact numerical calculations and the BCS approximation (taken from ref. 9).

2.4. Independent pairs and broken-pair states

Independent-pair states are defined relative to the $(2n-2)$-particle ground state $|n-1\rangle$ by diagonalizing the Hamiltonian in the space of states of the form

$$X^\dagger_{JM}(\gamma)|n-1\rangle = \sum_{j_1 j_2}\gamma_{j_1 j_2 J}(a^\dagger_{j_1}\times a^\dagger_{j_2})_{JM}|n-1\rangle. \qquad (17)$$

Such an approximation allows all interactions between the two extra-core particles plus interactions with the static field generated by the $(2n-2)$ core particles. It is a natural generalization of the independent-particle approximation. The pair states of interest in the present context are the $J=0$ states for which

$$X^\dagger_0(\gamma) = S_+(\gamma) = \sum_i \gamma_i S^i_+. \qquad (18)$$

These states are given by the solutions of the equation

$$\sum_j \langle n-1|S^i_-(H-E_\gamma)S^j_+|n-1\rangle\gamma_j = 0. \qquad (19)$$

Let α denote the lowest energy solution for γ. Then, a self-consistent solution of the independent-pair equations is given by approximating the state $|n-1\rangle$ in eq. (19) by the pair-coupled state $|(n-1)(\alpha)\rangle = [S_+(\alpha)]^{n-1}|0\rangle$. We then observe the remarkable fact that the state

$$|n(\alpha)\rangle = S_+(\alpha)|(n-1)(\alpha)\rangle, \qquad (20)$$

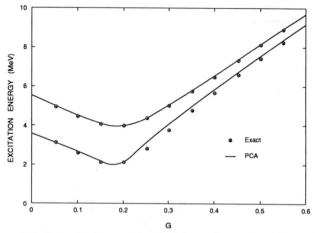

Fig. 3. 0^+ states calculated in the independent-pair approximation shown in comparison with results of exact numerical calculations (taken from ref. 9).

obtained by a self-consistent solution of the independent-pair equations,

$$\sum_j \langle (n-1)(\alpha)|S_-^i(H-E_0)S_+^j|(n-1)(\alpha)\rangle \alpha_j = 0 \tag{21}$$

is simultaneously a solution of the variational equation

$$\frac{\partial}{\partial \alpha_i}\langle n(\alpha)|(H-E_0)|n(\alpha)\rangle = 0. \tag{22}$$

Recall a parallel correspondence for the Hartree-Fock equations; the HF variational equations are simultaneously self-consistent mean-field equations for single-particle states. This is one reason the pair-coupling model works so well.

2.5. Number-projected BCS states

An (unnormalized) BCS state can be expressed in the form

$$|\Phi(\alpha)\rangle = \exp[S_+(\alpha)]|0\rangle. \tag{23}$$

This state is the vacuum of the quasiparticle operators

$$\eta_{jm}^\dagger = u_j a_{jm}^\dagger - v_j a_{j\bar{m}}, \quad \eta_{j\bar{m}}^\dagger = u_j a_{j\bar{m}}^\dagger + v_j a_{jm}, \tag{24}$$

where

$$u_j^2 = \frac{1}{1+\alpha_j^2}, \quad v_j^2 = \frac{\alpha_j^2}{1+\alpha_j^2}. \tag{25}$$

It follows that the $2n$-particle component of $|\Phi(\alpha)\rangle$ is the pair-coupled state

$$P_{2n}|\Phi(\alpha)\rangle = \frac{1}{n!}[S_+(\alpha)]^n |0\rangle = \frac{1}{n!}|n(\alpha)\rangle. \tag{26}$$

Thus, the pair-coupling model is equivalent to a number-projected BCS approximation albeit with the α_j coefficients optimized to minimize the energy of the state after number projection. As we now indicate, number projection is easy to carry out in a coherent-state representation.

3. Coherent-state representation

3.1. The multiple quasispin algebra $\mathcal{A}$

The zero-particle vacuum state $|0\rangle$ is the lowest-weight state for an irrep of the dynamical algebra $\mathcal{A}$ of eq. (11). With $S_-(\beta) = \sum_i \beta_i S_-^i$, an arbitrary state $|\Psi\rangle$ in the irrep, has coherent-state wave function[11]

$$\psi(\beta) = \langle 0|e^{S_-(\beta)}|\Psi\rangle. \tag{27}$$

Moreover, an element $W \in \mathcal{A}$ is represented by the operator $\Gamma(W)$ where

$$[\Gamma(W)\psi](\beta) = \langle 0|e^{S_-(\beta)}W|\Psi\rangle. \tag{28}$$

Hence, we determine that

$$\begin{aligned}
\Gamma(S_-^i) &= \partial/\partial\beta_i \\
\Gamma(S_0^i) &= -s_i + \beta_i\partial/\partial\beta_i \\
\Gamma(S_+^i) &= 2s_i\beta_i - \beta_i^2\partial/\partial\beta_i
\end{aligned} \tag{29}$$

and

$$\Gamma(H) = \sum_i 2\varepsilon_i\left(-s_i + \beta_i\frac{\partial}{\partial\beta_i}\right) - G\sum_{ij}\beta_i\left(2s_i - \beta_i\frac{\partial}{\partial\beta_i}\right)\frac{\partial}{\partial\beta_j}. \tag{30}$$

3.2. The BCS wave function

The BCS state $|\Phi(\alpha)\rangle$ has coherent-state wave function

$$\chi^\alpha(\beta) = \langle 0|e^{S_-(\beta)}|\Phi(\alpha)\rangle = \langle \Phi(\beta)|\Phi(\alpha)\rangle = \prod_i (1 + \beta_i\alpha_i)^{2s_i}, \tag{31}$$

which is easily derived by group theoretical methods (cf. ref. 9).

3.3. Pair-coupled (Schur) functions

By definition, the pair-coupled states $\{|n(\alpha)\rangle\}$ satisfy the equation

$$|n(\alpha)\rangle = S_+(\alpha)|(n-1)(\alpha)\rangle. \tag{32}$$

Thus, if $|n(\alpha)\rangle$ has coherent-state wave function

$$\varphi_n^\alpha(\beta) = \langle 0|e^{S_-(\beta)}|n(\alpha)\rangle, \tag{33}$$

it follows that $\varphi_n^\alpha(\beta)$ satisfies the recursion relation

$$\varphi_n^\alpha(\beta) = \sum_i \alpha_i\beta_i\left(2s_i - \beta_i\frac{\partial}{\partial\beta_i}\right)\varphi_{n-1}^\alpha(\beta). \tag{34}$$

Alternatively, if we define

$$\varphi_n(z) = \varphi_n^\alpha(\beta), \quad \partial_i = \partial/\partial z_i, \tag{35}$$

where $z_i = \alpha_i \beta_i$, we obtain the recursion relation

$$\varphi_n(z) = \sum_i z_i(2s_i - z_i \partial_i)\varphi_{n-1}(z). \tag{36}$$

This equation can be expressed in the form[11]

$$\varphi_n = \sum_{m=1}^{n} (-1)^{m+1} \frac{(n-1)!}{(n-m)!} \varphi_{n-m} \phi_m, \tag{37}$$

where ϕ_m are the so-called *power functions*

$$\phi_m(z) = \sum_i 2s_i z_i^m. \tag{38}$$

The solutions are then easily determined to be given by the determinantal functions

$$\varphi_n = \begin{vmatrix} \phi_1 & 1 & 0 & 0 & \cdots & 0 \\ \phi_2 & \phi_1 & 2 & 0 & \cdots & 0 \\ \phi_3 & \phi_2 & \phi_1 & 3 & \cdots & 0 \\ & & \cdots & & & \\ \phi_n & \phi_{n-1} & \phi_{n-2} & \phi_{n-3} & \cdots & \phi_1 \end{vmatrix}. \tag{39}$$

These functions are well known in the theory of group characters; they are the so-called *antisymmetric Schur functions*. Thus, in the coherent-state representation, both the BCS state $|\Phi(\alpha)\rangle$ and its number-projected components, the pair-coupled states, have simple well-known coherent-state wave functions.

3.4. The harmonic (boson) limit

Let $b_i^\dagger$ and b_i denote the boson operators

$$b_i^\dagger = \sqrt{2s_i}\, \beta_i, \quad b_i = \frac{1}{\sqrt{2s_i}} \frac{\partial}{\partial \beta_i}. \tag{40}$$

The coherent-state representation of the quasispin operators then becomes

$$\Gamma(S_-^i) = \sqrt{2s_i}\, b_i,$$
$$\Gamma(S_0^i) = b_i^\dagger b_i - s_i,$$
$$\Gamma(S_+^i) = \sqrt{2s_i}\, b_i^\dagger - \frac{1}{\sqrt{2s_i}} b_i^\dagger b_i^\dagger b_i, \tag{41}$$

corresponding to a Dyson boson expansion. Likewise, the coherent-state representation of the Hamiltonian become

$$\Gamma(H) = \sum_i 2\varepsilon_i b_i^\dagger b_i - 2G \sum_{ij} \sqrt{s_i s_j}\, b_i^\dagger b_j - G \sum_{ij} \sqrt{\frac{s_j}{s_i}}\, b_i^\dagger b_i^\dagger b_i b_j. \tag{42}$$

Now for states for which the number of b_i bosons is small compared to s_i (assuming s_i is large), we can neglect the second term of $\Gamma(S_+^i)$. Then

$$\Gamma(H) \approx \Gamma(H_{\text{HO}}) = \sum_i 2\varepsilon_i b_i^\dagger b_i - 2G \sum_{ij} \sqrt{s_i s_j}\, b_i^\dagger b_j, \qquad (43)$$

becomes the Hamiltonian of a coupled harmonic oscillator. This Hamiltonian is easily diagonalized and expressed in the form

$$\Gamma(H_{\text{HO}}) = \sum_\nu E_\nu s_\nu^\dagger s_\nu, \quad s_\nu^\dagger = \sum_i C_{\nu i} b_i^\dagger. \qquad (44)$$

Moreover, in the coherent-state representation, the independent-pair operators $X_0^\dagger(\gamma)$ reduce to

$$\Gamma(X_0^\dagger(\gamma)) = \sum_i \gamma_i \Gamma(S_+^i) \approx \sum_i \gamma_i \sqrt{2s_i}\, b_i^\dagger \qquad (45)$$

and become s-boson operators. Thus, we have the remarkable result that the pair-coupling approximation is exact in the harmonic (boson) limit.

We now understand why the pair-coupling (generalized seniority) approximation is so good. It is because it is valid in three different limits: (i) when the coupling constant G is zero and independent-pair states reduce to independent-particle states; (ii) when G is large compared to single-particle energy differences, and (iii) when the boson limit is valid. In the first two cases, the dynamical subalgebra chains of eqs. (12) and (14) apply. The third case, is interesting because it shows that the existence of a limiting dynamical symmetry, which may only be realized very approximately in a practical finite system (i.e., for finite values of $2s_i$), can cause the results of a related dynamical symmetry (in this case the su(2) quasispin symmetry) to persist much longer than expected. This is especially significant in view of the fact that many dynamical algebras have boson limits.

4. The pairing plus quadrupole problem

We now speculate on possible extensions of the above arguments to a system governed by a pairing-plus-quadrupole Hamiltonian

$$H = \sum_{jm} \varepsilon_j a_{jm}^\dagger a_{jm} - G S_+ S_- - \tfrac{1}{2} \chi Q \cdot Q, \qquad (46)$$

where Q is Elliott's su(3) quadrupole operator[12].

The full dynamical symmetry algebra for this Hamiltonian is large. It is the full orthogonal (fermion-pair) algebra

$$o(2\Omega) = \langle a_{j_i m_i}^\dagger a_{j_j m_j}^\dagger, a_{j_i m_i} a_{j_j m_j}, a_{j_i m_i}^\dagger a_{j_j m_j} - a_{j_j m_j} a_{j_i m_i}^\dagger \rangle, \qquad (47)$$

where $\Omega = \sum_i (2j_i + 1)$. There are now three dynamical subalgebras which correspond to the three components of H. The first two, applicable when χ is zero, have already been mentioned in section 2.2 in connection with the pairing Hamiltonian (10). The third is Elliott's su(3) algebra which applies when the single-particle energies are all equal and the pairing coupling constant G is zero. The incompatibility of these dynamical algebras is reflected in the fact that the only dynamical algebra for the general mixed-symmetry Hamiltonian (46) is, in general, the full fermion-pair algebra, o(2Ω). It is therefore of interest to consider the algebraic structure of the approximate pair-coupled solutions to the mixed symmetry problem.

When χ is small, the spectrum of a $2n$-particle nucleus can be obtained from the $J = 0$ pair-coupling model with excited states given by the RPA (random-phase approximation). Specifically, if $X^-(\alpha)$ denotes the pair-annihilation operator defined such that

$$X_0^-(\alpha)|n(\alpha)\rangle \propto |(n-1)(\alpha)\rangle,\tag{48}$$

then excitation operators can be constructed of the RPA form

$$O_{JM}^\dagger(\gamma) = x X_{JM}^\dagger(\gamma) X_0^-(\alpha) - y(-1)^{J+M} X^\dagger(\alpha) X_{J,-M}^-(\gamma).\tag{49}$$

Such RPA excitation operators imply a so-called *correlated* ground state for the $2n$-particle nucleus of the form

$$|\tilde{n}\rangle = N\left(|n(\alpha)\rangle + \sum_{\gamma J} C_{\gamma J}\left[O_J^\dagger(\gamma) \times O_J^\dagger(\gamma)\right]_0 |(n-2)(\alpha)\rangle\right).\tag{50}$$

Now it is known that the correlated RPA ground and excited states can all be generated by angular-momentum projection from an *intrinsic* state of the form

$$|\Phi_n\rangle = \left[X_0^\dagger(\alpha) + \sum_{\gamma J} c_{\gamma J} X_{J0}^\dagger(\gamma)\right]^n |0\rangle.\tag{51}$$

In fact, it is known that one can improve considerably on the RPA by angular-momentum projecting states from such a generator state[13]. The significance of this observation is that generator states of the type (51) are generalized pair-coupled states of the type used, for example by Sheikh[14], and can be expressed in the form

$$|n(\alpha)\rangle = \left[X^\dagger(\alpha)\right]^n |0\rangle,\tag{52}$$

where now $X^\dagger(\alpha)$ is a pair-creation operator of mixed angular momentum.

With only pairing-plus-quadrupole interactions, it is tempting to include only $J = 0$ and $J = 2$ pairs in the expansion of $X^\dagger(\alpha)$. However, such a restriction may not be justified if χ is not small.

When the single-particle energies are all degenerate and the pair coupling constant G is zero, one knows that the eigenstates of H belong to Elliott su(3) irreps. In this case, states can again be angular-momentum projected from an intrinsic generator state. One also knows that, for the leading su(3) irrep, i.e., the one that gives the lowest-energy states, the intrinsic state is a Slater determinant of single-particle states. Thus, it is a special case of a deformed pair-coupled state.

Now, one expects that the spectrum of H will continue to be of a rotational type when χ is large, even when the single-particle energies are nondegenerate and G is not zero. Moreover, from the success of Hartree-Bogolyubov theory, one expects that states can be generated rather well by number projecting and angular-momentum projecting states from a deformed BCS vacuum state, particularly if the projected states are optimized following projection. Thus, the evidence supports the case for using pair-coupled wave functions as generator states for the low-energy spectrum of a Hamiltonian of the type (46).

Within the near future, we hope to examine the validity of the general pair-coupling model, for a Hamiltonian of the type (46), by comparison with exact numerically computed results where the latter are possible. Numerical calculations of the type we have in mind have already been carried out by Bahri and Draayer[15] and our hope is to make use of their results and computer codes.

5. Concluding remarks

One may well ask why we feel the need to go through such complicated algebraic arguments to arrive at an approximation that has been known and used for many years. Our objective, in fact, has not been to justify or derive the generalized-seniority/pair-coupling approximation so much as to discover what make the approximation so successful. We believe that its success is not accidental but arises from an underlying symmetry which may be more generally applicable. Moreover, such symmetries are needed in the construction of a realistic microscopic theory of nuclear collective motion which includes all the important independent-particle, pair-coupling and phonon degrees of freedom.

We have also been motivated to understand why dynamical symmetries seem to persist in situations where, a priori, one expects them to be badly violated. For example, one observes rotational bands in nuclei even when the spin-orbit and pairing interactions strongly mix representations of the rigid-rotor algebra. The physical reason for this persistance can be attributed to the adiabaticity of nuclear rotations. But, physical understanding can usually be given a mathematical expression. And indeed we find that the simple physical concept of *decoupled adiabatic rotations* leads to an elegant mathematical concept, the concept of *embedded representations*, and a new kind of coherent (adiabatic) mixing of irreps[16]. We find a similar persistence of a dynamical (quasispin) symmetry in superconducting nuclei. Thus, we seek a parallel symmetry principle to understand why it happens.

Finally, one may ask what all this has to do with the IBM. The answer is perhaps not a great deal at the present time, except perhaps for the fact that the IBM seems to correspond to a contraction limit in which the approximations of the pair-coupling model become precise.

6. Acknowledgements

This work is supported in part by the Natural Sciences and Engineering Research

Council of Canada. I am pleased to acknowledge important contributions to this work from Hong Chen and permission to show her results prior to publication. I also acknowledge helpful discussions with Lev Shekhter.

7. References

1. A. Arima and F. Iachello, *Phys. Rev. Lett.* **35** (1975) 1069; *Ann. of Phys.* **99** (1976) 253; **111** (1978) 201; **123** (1979) 468; F. Iachello and A. Arima, *The Interacting Boson Model* (Cambridge Univ. Press, 1987).
2. G. Rosensteel and D. J. Rowe, *Phys. Rev. Lett.* **38** (1977) 10; *Ann. Phys.*, *NY* **126** (1980) 343; D. J. Rowe, *Rep. Prog. in Phys.* **48** (1985) 1419.
3. A. Bohr, Mat. Fys. Medd. Dan. Vid. Selsk. **26**, (1952) No. 14; *Nuclear Structure*, Vol. 2 (Benjamin, New York, 1975).
4. G. Gneuss and W. Greiner, *Nucl. Phys.* **A171** (1971) 449.
5. D. Janssen, R.V. Jolos and F. Dönau, *Nucl. Phys.* **A224** (1974) 93.
6. A.K. Kerman, *Phys. Rev.* **120** (1961) 300; A.K. Kerman, R.D. Lawson and M.W. Macfarlane, *Phys. Rev.* **124** (1961) 162.
7. T. Otsuka, A. Arima and F. Iachello, *Nucl. Phys.* **A309** (1978) 1.
8. B. Lorazo, *Nucl. Phys.* **A153** (1970) 255; B. Lorazo and C. Quesne, *Nucl. Phys.* **A440** (1985) 397; Y.K. Gambhir, A. Rimini and T. Weber,*Phys. Rev.* **188** (1969) 1573; Y.K. Gambhir, S. Haq and J.K. Suri, *Ann. Phys.* (N.Y.) **133** (1981)
9. H. Chen, T. Song and D.J. Rowe, *The pair-coupling model*, to be published.
10. I. Talmi, *Nucl. Phys.* **A172** (1971) 1.
11. D.J. Rowe, T. Song and H. Chen, *Phys. Rev.* **C44** (1991) R598.
12. J. P. Elliott, *Proc. Roy. Soc.* **A245** (1958) 128, 562.
13. D.J. Rowe, *Phys. Rev.* **175** (1968) 1283. 154.
14. J.A. Sheikh, *Phys. Rev.* **C43** (1991) 1733.
15. C. Bahri and J.P. Draayer, private communication and to be published.
16. D.J. Rowe, P. Rochford and J. Repka, *J. Math. Phys.* **29** (1988) 572; P. Rochford and D.J. Rowe, *Phys. Lett.* **B210** (1988) 5.

BOSON MAPPINGS AND PHENOMENOLOGICAL BOSON MODELS

HENDRIK B GEYER and PETR NAVRÁTIL*
Institute of Theoretical Physics, University of Stellenbosch
Stellenbosch 7600, South Africa

and

JACEK DOBACZEWSKI
Institute of Theoretical Physics, Warsaw University
Hoża 69 PL-00-681 Warsaw, Poland

ABSTRACT

The boson mapping formalism has not only been shown to constitute a viable calculational tool for the nuclear many-body problem, but has also furthered insight into the structure of phenomenological models such as the interacting boson model (IBM). Two aspects which have been highlighted through boson mappings are the role of possible spurious states in boson analyses of fermion systems and the uncovering of additional dynamical symmetries. These developments and recent applications are discussed with some emphasis on the role of similarity transformations, also in the recent construction of boson-fermion mappings appropriate to the interacting boson-fermion model (IBFM) and supersymmetric versions of the model.

1. Introduction

Although boson mappings, or boson expansions as they were mostly referred to previously, have been investigated and applied for quite some time (see the recent review by Klein and Marshalek[1] for some earlier references), most recent developments in the formalism and its applications have been stimulated by the advent of and progress linked to the phenomenological interacting boson model (IBM)[2].

Here we review aspects where boson phenomenology and boson mappings have mutually stimulated important developments or insight in the respective fields and also illustrate this by some recent examples. Firstly we recall how boson mappings, following phenomenology, have been developed into a many-body formalism which can be applied in its own right to solve fermion many-body problems. We illustrate this by applying the formalism to fermion systems with SO(12) and Sp(10) symmetries[3]. These models are generalizations of the Ginocchio SO(8) and Sp(6) models[4] and now include G-pairs of fermions in addition to S- and D-pairs as building blocks of a decoupled collective subspace.

In such phenomenologically inspired applications of boson mappings the matter of spurious states inevitably crops up. These states, which may arise because of an effective overcompleteness of the boson basis (chosen for its convenience), their role, and identification, have been extensively discussed [5,6,7,8,9,10] and we only briefly touch upon this aspect in the context of the SO(12) and Sp(10) models.

Dynamical symmetry, one of the attractive properties associated with the IBM, refers to those hamiltonians where one can capitalise on the factorisation of the hamiltonian and subsequent rewriting in terms of Casimir operators associated with a single chain of subalgebras. That this notion of dynamical symmetry allows the identification of IBM dynamical symmetries other than the traditional U(5), SU(3) and O(6) limits (all rather easily identifiable by simple operator re-ordering in the IBM hamiltonian), was already pointed out some time ago[11]. We briefly recall the arguments that lead to this identification and emphasise the role of a class of similarity transformations which also become important in the next category of the phenomenology-microscopy interplay which we address, namely supersymmetry.

The notion of supersymmetry is firmly linked to a system of both fermions and bosons, exhibiting an invariance with respect to exchange between these two classes of particles. On a phenomenological level supersymmetry[12] has in fact proved to be fruitful in nuclear structure physics within the framework of the interacting boson-fermion model (IBFM)[13]. Since the bosons involved are intimately linked to collective *fermion* pairs, one might conclude that although supersymmetry may be a valid and useful property on the phenomenological level, it could not persist as an exact property of a fermon system only. It may therefore be somewhat surprising to discover that a fermion system on its own can also exhibit *exact supersymmetry*. This is discussed in general and demonstrated for a specific SO(8)⊗SO(5) model in section 4.

2. Boson mappings as a many-body tool

When one compares applications of boson mappings as a many-body formalism with a phenomenological boson analysis, significant similarities and differences can be identified which we briefly address.

Mapping of a given (chosen) fermion hamiltonian typically leads to higher order boson interactions than the assumed 1-plus-2-body nature of an IBM hamiltonian. In the case of Dyson mappings, which we exclusively address here, the general structure is indeed of 1-plus-2-body nature, but with respect to a simple boson basis which spans the boson Fock space, generally non-hermitian. Since the strength coefficients are not arbitrary, but fixed by the fermion interaction and mapping, the structure can be identified as quasi-hermitian[14], which, if no truncation of either the number of degrees of freedom or the basis is involved, guarantees the usual properties required from quantum mechanical systems, such as real eigenvalues and orthogonality of eigenstates[14]. Since a non-hermitian hamiltonian on the phenomenological level will generally not have these properties, the mapped Dyson hamiltonians seem to be excluded when dynamical symmetries of the IBM are considered. That this is too hasty a conclusion, is discussed further in the next section.

Here we concentrate further on how a Dyson mapped hamiltonian is typically diagonalised and how this compares with the phenomenological program. In the boson mapping formalism in the so-called Beliaev-Zelevinsky-Marshalek (BZM) approach (see Ref.[1]), which includes the Dyson mapping, the mapping of operators has priority above the mapping of states. The equivalence between the original fermion and the mapped boson problems then emerges in a transparent way when the boson basis is constructed by associating with each fermion basis state a boson state which is obtained by replacing each fermion operator by its mapped image and switching from the fermion vacuum $|0\rangle$ to the boson vacuum $|0\rangle^{5,1}$. This construction of the so-called physical basis can become rather cumbersome and impractical and preference is given to a boson basis which is dictated, as in the phenomenological case, by the degrees of freedom appearing in H, and particle number. Generally this reflects a change of basis by similarity transformation on the physical basis, except for the complication that the transformation can be singular[5]. The simplicity of using the boson basis may therefore be at the expense of obtaining spurious states from the diagonalisation of the mapped hamiltonian.

It is known, however, that diagonalisation in the boson basis yields all the exact eigenstates and eigenvalues, together with some spurious solutions[5], and that at least two practical methods exist to identify spurious solutions. Firstly there is an analysis of matrix elements introduced by Geyer, Engelbreht and Hahne[5], based on the invariance of the physical subspace under mapped operators and the orthogonality of this subspace to its complement. This is summarised in the properties[5,6]

$$(\tilde{\varphi}_{\text{spur}}|\Theta_{\text{D}}|\psi_{\text{phys}}) = 0 \quad ; \quad (\tilde{\psi}_{\text{phys}}|\Theta_{\text{D}}|\varphi_{\text{spur}}) \neq 0 \tag{1}$$

Secondly there is the direct $\mathcal{R}$-projection method of Dobaczewski[15]. Application of these methods can be found in [5,6,7,8,9,10].

The conclusion to be drawn from this is that the relative simplicity of the phenomenological boson analysis may be retained when applying boson mappings. This is especially true when spurious states can be identified exactly. It also applies to large scale realistic calculations[16,17,18], provided that one still refines insight into either the role of possible spurious states or when (and why) they may be safely ignored, following truncation of either or both of the number of degrees of freedom or the basis.

We conclude this section by briefly referring to some recent results[9,3] which illustrate application of the above program to a non-trivial problem. The application is to generalised Ginocchio models with SO(12) and Sp(10) symmetry, with building blocks the fermion pair and multipole operators

$$F_{JM}^{\dagger}(KI;ki) = \frac{1}{\sqrt{2}}\hat{K}\hat{I}\sum_{j_1 j_2}\hat{j}_1\hat{j}_2\begin{Bmatrix} k & i & j_1 \\ k & i & j_2 \\ K & I & J \end{Bmatrix}(a_{j_1}^{\dagger}a_{j_2}^{\dagger})_M^{(J)}, \tag{2}$$

$$P_M^J(KI;ki) = -\hat{k}\hat{i}\hat{K}\hat{I}\sum_{j_1 j_2}\hat{j}_1\hat{j}_2\begin{Bmatrix} k & i & j_1 \\ k & i & j_2 \\ K & I & J \end{Bmatrix}(a_{j_1}^{\dagger}\tilde{a}_{j_2})_M^{(J)}. \tag{3}$$

Taking $K = 0$ and $i = 5/2$ defines the SO(12) model, whereas for $I = 0$ and $k = 2$ the corresponding model is Sp(10). Clearly these models have S-, D- and

G-pairs of collective fermions and hamiltonians constructed from these generators will have decoupled collective subspaces of states constructed from these pairs. We refer to Ref.[9] where it is shown that diagonalisation of the hamiltonian

$$H = \kappa_2 P^2 \cdot P^2 + \kappa_3 P^3 \cdot P^3 + \kappa_4 P^4 \cdot P^4 + \kappa_1 P^1 \cdot P^1 \longrightarrow H_D \quad , \quad (4)$$

in an sdg-boson basis appropriate to ^{146}Sm gives a spectrum which describes the low-lying states reasonably well, while it can also be shown that all these states are spurious. The conclusion to be drawn is that a successful phenomenological boson application does not necessarily imply that spurious states have been avoided.

In Ref.[9] diagonalisation of the hamiltonian

$$H = (1-x)(G_0 S^\dagger S + G_2 D^\dagger \cdot D) + x(\kappa_2 P^2 \cdot P^2 + \kappa_4 P^4 \cdot P^4 + \kappa_1 P^1 \cdot P^1) , \quad (5)$$

with x varying between 0 and 1, shows that when the pairing part dominates (small x), spurious states are generally high up in the spectrum, while spurious states are the lowest states when the multipole part dominates.

As a final example from Ref.[3] the hamiltonians

$$H = (1-x)\left(G_0 S^+ S + G_2 D^+ \cdot D\right) + x\kappa_2 Q \cdot Q, \quad (6)$$

and

$$H = (1-x)\left(G_0 S^+ S + G_2 D^+ \cdot D\right) + x\kappa_2 Q \cdot Q + G_4 G^+ \cdot G \quad (7)$$

have been diagonalised for 8 fermions (4 sdg-bosons) for which it is known that no spurious states appear. The role of the $G^+ \cdot G$ term is to push up low-lying 4^+ vibrational states. The results for the interaction strengths $G_0=-80$ keV, $G_2=-30$ keV, and $\kappa_2=-25$ keV are summarised in Table 1.

We stress that these are exact results for the fermion models considered and have been obtained by Dyson mapping and rather simple adaption of existing codes for diagonalisation in a boson basis.

3. Dynamical symmetries of the IBM

As the starting point to address the question of which dynamical symmetries the (scalar) boson IBM hamiltonian (assumed to be hermitian) may contain, it seems sufficient to notice that when at most two-body interactions between s-bosons and d-bosons are involved, the most general form can be analysed in terms of U(6) generators. This observation requires of course only a simple rewriting of the hamiltonian utilising the commutation relations between boson creation and annihilation operators. The general hamiltonian can then be rewritten in terms of the first and second order Casimir invariants of those groups which appear in group chains of U(6) containing SO(3). This means that H_{IBM} can be written either as

$$H_{\mathrm{IBM}} = \sum_i \lambda_i g_i(U(6)) + \sum_{ij} \lambda_{ij} g_i(U(6)) g_j(U(6)) \quad (8)$$

Table 1: Ratios of excitations $R_I=(E_{I^+}-E_{0^+})/(E_{2^+}-E_{0^+})$ in yrast bands of algebraic SD and SDG models in the vibrational (x=0) and rotational (x=1) limit. Results for the vibrational, rotational, and γ-unstable quadrupole model are also given for comparison.

Model		$I=4$	$I=6$	$I=8$	$I=10$
vibrational		2	3	4	5
Sp(6)	$x=0$	1.91	2.74	3.69	
SO(8)	$x=0$	1.89	2.68	3.26	
Sp(10)	$x=0$	1.86	2.61	3.09	5.02
SO(12)	$x=0$	1.87	2.62	3.25	5.58
rotational		3.33	7	12	18.33
Sp(6)	$x=1$	3.33	7	12	
SO(8)	$x=1$	2.5	4.5	7	
Sp(10)	$x=1$	2.56	4.7	7.47	11.16
SO(12)	$x=1$	3.16	6.47	11.19	22.40
γ-unstable		2.5	4.5	7	10

or

$$H_{\text{IBM}} = \epsilon\mathcal{C}_1(U(5))+\alpha\mathcal{C}_2(U(5))+\beta\mathcal{C}_2(O(5))+\gamma\mathcal{C}_2(O(3))+\delta\mathcal{C}_2(SU(3))+\eta\mathcal{C}_2(O(6)) \quad (9)$$

where $g_i(U(6))$ is a generator of SU(6), with the $\mathcal{C}$'s Casimir operators of the order and associated with the algebras as indicated. As already reflected in expression (9) above, it is well known that only three of the group chains referred to above exist, namely those which have either U(5), SU(3) or O(6) as second link in the chain. Associated with each chain one has a boson basis in which a general IBM hamiltonian can be diagonalised. Dynamical symmetry then refers to those cases where the interaction parameters are such that, when the hamiltonian is written in terms of Casimir invariants, it is only the Casimir invariants of one particular chain which appear. In such instances the hamiltonian is trivially diagonal in the basis corresponding to the particular chain involved, leading to analytical expressions in terms of eigenvalues of the Casimir invariants.

From this description it seems that the three group chains mentioned exhaust the possibilities for the occurrence of a dynamical symmetry within the IBM framework. If one reconsiders expression (8), however, it is clear that there is no reason why for some choices of λ_i and λ_{ij} it should not be possible write H_{IBM} as

$$H_{\text{IBM}} = \sum_i \mu_i g_i(\text{G}) + \sum_{ij} \mu_{ij} g_i(\text{G})g_j(\text{G}) \quad (10)$$

with G$\neq$U(6) and the μ's functions of the λ's. Whereas the original form (8) might for a set of λ's not lead to any of the three already mentioned dynamical symmetries, the equivalent form in terms of the μ's might be such that a dynamical symmetry associated with a chain starting with G can in fact be identified. If this is to be a distinct dynamical symmetry, then of course $G \not\subset U(6)$.

That this situation is indeed possible for an SO(7) dynamical symmetry, and indeed all dynamical symmetries associated with the Ginocchio model[4] or fermion dynamical symmetry model (FDSM)[19], is discussed further below. Although SO(8) is clearly not a subalgebra of SU(6), these further dynamical symmetries of IBM are possible because for particular interaction strengths the IBM hamiltonian can indeed be factorised in terms of the generators of SO(8) (or Sp(6), which will not be considered here) as in expression (10)

A prerequisite for this statement to be valid, is of course that it must be possible to construct realisations of SO(8) generators in terms of s- and d-bosons. For the particular SO(8) representation considered in Ref.[4] this has indeed been done by a number of authors[20,21,22]. Since those (Dyson) mappings that lead to a finite one-plus two-body sd-boson hamiltonian, turn out to be non-hermitian in general, one's initial reaction is to consider such hamiltonians as outside the scope of the IBM framework. Indeed, for arbitrary strength coefficients, not fixed by the mapping, one could generally face the unacceptable situation that the spectrum will include complex values.

On the other hand the work of Refs.[11,23] shows that it is often possible to find similarity transformations which hermitise these Dyson-type boson hamiltonians while retaining their one- plus two-body nature. Correspondingly it therefore becomes possible to analyse some IBM hamiltonians in terms of SO(8) generators.

Consider as an example the IBM hamiltonian

$$H = -G_0 n_d(\Lambda - 2N + n_d + 5) + G_0 N(\Lambda - N + 6) + 2 \sum_{r=1,3} b_r [d^\dagger \tilde{d}]^r \cdot [d^\dagger \tilde{d}]^r \quad (11)$$

with N the total boson number and Λ a parameter. Alternatively H can be written as

$$\begin{aligned}H &= -G_0 \left[(d^\dagger(\Lambda - 2N) + d^\dagger \cdot d^\dagger \tilde{d}) \cdot (\tilde{d}) + (N - \Lambda/2)(N - \Lambda/2 - 5)\right] \\ &+ b_3 \sum_{r=1,3} \left[(2\sqrt{2}[d^\dagger \tilde{d}]^r) \cdot (2\sqrt{2}[d^\dagger \tilde{d}]^r)\right] \\ &+ \tfrac{1}{4}(b_1 - b_3) \left[(2\sqrt{2}[d^\dagger \tilde{d}]^1) \cdot (2\sqrt{2}[d^\dagger \tilde{d}]^1)\right] \ .\end{aligned} \quad (12)$$

The set of bracketed operators (...) above, namely

$$(D^\mu)_Z \equiv d^\mu(\Lambda - 2n_s - 2n_d) + d^\dagger \cdot d^\dagger \tilde{d}_\mu \quad (13)$$
$$(D_\mu)_Z \equiv d_\mu \quad (14)$$
$$(P^1_\mu)_Z \equiv 2\sqrt{2}[d^\dagger \times \tilde{d}]^1_\mu \quad (15)$$
$$(P^3_\mu)_Z \equiv -2\sqrt{2}[d^\dagger \times \tilde{d}]^3_\mu \quad (16)$$
$$(S_0)_Z \equiv N - \Lambda/2 \quad (17)$$

(with $d^\mu \equiv d^\dagger_\mu$) can be identified as the 21 generators of SO(7). In fact, H can also be written as

$$H = -G_0 \mathcal{C}_2(SO(7)) + \tfrac{1}{5}(b_1 - b_3)\mathcal{C}_2(O(3)) + b_3 \mathcal{C}_2(O(5)) \qquad (18)$$

which implies that H is diagonal in a basis defined by $G \supset SO(7) \supset O(5) \supset O(3)$ with G still to be identified. H is obviously also diagonal in the usual U(5) basis (which is equivalent to the basis defined above) and the eigenspectrum is simply given in terms of the usual quantum numbers of the U(5) basis as

$$E(N, n_d, v, L) = G_0(\Lambda - N + 6) - G_0 n_d(\Lambda - 2N + n_d + 5) + b_3 v(v+3) + \tfrac{1}{5}(b_1 - b_3)L(L+1) \quad . \qquad (19)$$

The *factorisation* of an IBM hamiltonian such as in expression (12) and subsequent identification of a *dynamical symmetry* as in (18) indeed comply with the standard notion of these concepts, although it may in general not be so simple to do this by inspection, as the factorisation is asymmetrical as compared to the standard SU(6) factorisation.

The set of operators (13) – (17) have in fact been obtained from a Dyson mapping of SO(8) generators derived from the standard Dyson mapping followed by similarity transformation[11,23]. This particular similarity transformation is the one which transforms the standard Dyson mapped[11] SO(7) hamiltonian[4,11]

$$(H_{SO(7)})_D = G_0(n_s(\Omega - n_s + 5) + n) + G_0 s^\dagger s^\dagger \tilde{d} \cdot \tilde{d} + \tfrac{1}{4} \sum_{r=1,3} b_r (P^r)_D \cdot (P^r)_D \quad . \qquad (20)$$

into the hermitian IBM type hamiltonian (11), where the association $\Lambda \equiv \Omega$ has been made. We emphasise that on the microscopic fermion level Ω has a specific interpretation (2Ω being the size of the fermion space as usual), whereas on the phenomenological level Λ can be treated as a parameter. The similarity transformation, required to transform away the non-hermitian part $G_0 s^\dagger s^\dagger \tilde{d} \cdot \tilde{d}$ in the Dyson mapped hamiltonian (20), is given by

$$Z = \frac{(\Omega + 5 - 2n_s)!!}{(\Omega + 5 - n_s - \hat{n}_s)!!} \exp(\tfrac{1}{2} s^\dagger s^\dagger \tilde{d} \cdot \tilde{d})_\Lambda \; , \qquad (21)$$

$$Z^{-1} = {_\vee} \exp(-\tfrac{1}{2} s^\dagger s^\dagger \tilde{d} \cdot \tilde{d}) \frac{(\Omega + 3 - n_s - \check{n}_s)!!}{(\Omega + 3 - 2n_s)!!} \; . \qquad (22)$$

(See Refs.[23,11], also for a discussion of the notation linked to the positional operators indicated with a hat or inverted hat.)

Much of the original indication that SO(7) symmetry is realised in the Pd-Ru region is linked to the behaviour of B(E2) values[24], the calculaton of which of course requires an appropriate transition operator. It is clear that the set of SO(7) generators contains *no* candidate for a quadrupole operator T^{E2}_μ, which must therefore be chosen either from G or constructed in an *ad hoc* fashion. This is of course also true in the fermion analysis[24] where G is (by default) defined to be SO(8) from the outset, as the analysis is done for an SO(8) model.

It should be stressed, however, that the choice and/or structure of transition or transfer operators is *not* uniquely determined by the structure of the phenomenological hamiltonian, whether it posesses dynamical symmetry or not. Of course one endeavours to make the simplest possible choice for such operators. For electromagnetic transition operators this usually implies that one chooses a linear combination of generators which have the correct SO(3) transformation property, but higher order combinations of generators are sometimes required. To realize that for a given dynamical symmetry of the hamiltonian the restriction of transition operators to generators of a *sub-algebra* is too restrictive, one only has to look at the U(5) boson symmetry. A restriction of T_μ^{E2} to U(5) generators would imply for vibrational nuclei the non-physical result $B(E2; 2_1^+ \to 0_1^+) = 0$, a clear indication that one should relax the requirement that transition operators be generators of that sub-algebra that defines a dynamical symmetry.

For the SO(7) symmetry it was found that the single SO(8) generator P_μ^2 is already a good choice[24]. The exact fermion model results will therefore be reproduced by the boson operator $(P_\mu^2)_Z \equiv Z(P_\mu^2)_D Z^{-1}$, with

$$(P_\mu^2)_Z = 2\left[d^\mu s + \frac{\Lambda + 6 - n_s}{\Lambda + 7 - n_s} s^\dagger \tilde{d} - \frac{\Lambda + 6 - n_s}{(\Lambda + 5 - n_s)(\Lambda + 7 - n_s)} d^\mu s^\dagger \tilde{d} \cdot \tilde{d}\right] \quad (23)$$

obtained from the Dyson image $(P_\mu^2)_D$ by similarity transformation, although it is found that the simplest hermitised form also gives results which compare favourably with those of Ref.[24]. For further discussion about the structure and implementation of $(P_\mu^2)_Z$ we refer to the original literature[11,25].

We conclude this section by pointing out that the question as to which possible dynamical symmetries may be accommodated in an IBM framework, is not completely settled, although the consideration that higher rank algebras will probably involve bosons with $L > 2$, seems to indicate that all symmetries following from SU(6), SO(8) and Sp(6) indeed exhaust the list. Nevertheless, the present analysis may also play a role when dynamical symmetries associated with higher order boson interactions are considered.

4. Exact supersymmetry in a fermion system

The clue to the identification of exact supersymmetry in a fermion system lies in the construction of appropriate boson-fermion mappings of collective states. As shown in Refs.[26,27] such mappings can be *systematically derived* from supercoherent states

$$|C, \phi\rangle = \exp\left(C_i A^i + \phi_\mu a^\mu\right)|0\rangle \quad (24)$$

where $C_i = (C^i)^*$ are complex numbers, and $\phi_\mu = (\phi^\mu)^*$ are complex Grassmann variables *anti-commuting* with fermion operators a^μ and a_μ.

Direct application leads to the mapping

$$A^j \longleftrightarrow R^j \equiv \mathcal{A}^j + B^i[\mathcal{A}_i, \mathcal{A}^j] - \tfrac{1}{2} c_{ik}^{jl} B^i B^k B_l, \quad (25)$$

$$[A_i, A^j] \longleftrightarrow [\mathcal{A}_i, \mathcal{A}^j] - c_{ik}^{jl} B^k B_l, \qquad (26)$$
$$A_j \longleftrightarrow B_j, \qquad (27)$$
$$a^\nu \longleftrightarrow \alpha^\nu + B^i [\mathcal{A}_i, \alpha^\nu], \qquad (28)$$
$$a_\nu \longleftrightarrow \alpha_\nu, \qquad (29)$$

where $\mathcal{A}^j = \frac{1}{2}\chi^j_{\mu\nu}\alpha^\mu \alpha^\nu$ are collective pairs of *ideal fermions* which commute with all the boson operators[26,27,1].

The resulting physical subspace is, however, not satisfactory as it has a large redundancy between bosons and pairs of ideal fermions. This is the result of both a boson and an ideal fermion pair creation operator appearing in the image of a fermion pair creation operator.

Similar to the SO(7) construction, this problem may be circumvented by applying to the above set of boson-fermion images an appropriate similarity transformation, X, which here takes the form

$$X = \sum_{n=0}^{\infty} (\frac{1}{C_F - \hat{C}_F} \mathcal{A}^i B_i)^n_{\sim}, \qquad (30)$$

while similar constructions can be given for collective algebras such as SO(8), to which the following application refers. (See Refs.[27,28] for further detail.) We note that X is of similar structure as the SO(7) similarity transformation Z, which could be summed to a closed form because of the simple number operator dependence of C_F there.

In the ideal space one now has both boson and non-redundant fermion degrees of freedom and may therefore expect boson-fermion symmetry and, under favourable circumstances, even dynamical supersymmetry as we now sketch for the boson-fermion mapping of an SO(8) model[4] (or fermion dynamical symmetry model[19]), defined by collective monopole and quadrupole fermion pairs S^+ and D^+. In expressions (2) and (3) $i = \frac{3}{2}$, and k takes on integer values, so that in (2) only S $(J = 0)$ and D $(J = 2)$ pairs occur, while in (3) J can be 0, 1, 2, or 3. By the standard Dyson prescription[1,20,22], the even collective SO(8) states can then be mapped onto sd-boson states.

In order to generalize the model to odd systems we include $2\Omega = (2i+1)(2k+1)$ creation and annihilation operators $a^\dagger_{jm}$ and a_{jm}, where $j = |k-i|, \ldots, k+i$ for an integer value of the inactive angular momentum k and the active angular momentum $i = 3/2$[4]. A boson-fermion mapping of this algebra can be derived, similar to the images obtained from expressions (24) – (28) by similarity transformation. In this case the series defining the similarity transformation (30) cannot be explicitly summed up because the denominator $C_F - \hat{C}_F$, with $C_F = \frac{1}{2}n(\Omega + 6 - \frac{1}{2}n) - C_{2\text{Spin}_F(6)}$, contains the quadratic Casimir operator of the Spin(6) group which is not expressible in terms of number operators.

A system of collective fermion pairs is now mapped onto a system of $s^\dagger$ and $d^\dagger$ bosons, and a system of collective fermion pairs with an odd fermion onto $s^\dagger$ and $d^\dagger$

bosons and one ideal fermion. This situation is reminiscent of the phenomenological IBFM and we now indicate that in the SO(8) model with additional fermions, supersymmetry is in fact manifest in the mapped systems.

The six bosons above fix the boson sector of the supersymmetric structure in terms of U(6). The size of the fermion sector depends on k, and for $k=0$ ($j=3/2$) the four fermion states lead to an overall IBFM U(6/4) supersymmetry[30,29]. However, in SO(8) all particles occupy the same $j=3/2$ level while in the IBFM it is assumed that the bosons occupy the whole valence shell with only the fermion restricted to $j=3/2$.

As a more realistic situation, consider $k=2$ corresponding to $j=1/2$, 3/2, 5/2, and 7/2. In the IBFM, a related supersymmetry with the same single-particle content is U(6/20), realized in the Au-Pt isotopes[31]. The group reduction chain is $U_B(6) \otimes U_F(20) \supset SO_B(6) \otimes SU_F(4) \supset \text{Spin}_{B+F}(6) \supset \text{Spin}_{B+F}(5) \supset \text{Spin}_{B+F}(3)$. A hamiltonian chosen as a linear combination of quadratic Casimir operators in the chain yields the analytic U(6/20) IBFM energy formula[31]

$$E = A\sigma(\sigma + 4) + \tilde{A}[\sigma_1(\sigma_1 + 4) + \sigma_2(\sigma_2 + 2) + \sigma_3^2]$$
$$+ B[\tau_1(\tau_1 + 3) + \tau_2(\tau_2 + 1)] + CJ(J + 1). \quad (31)$$

In the SO(8) Ginocchio model all $j=1/2, 3/2, 5/2$, and $7/2$ states are degenerate which leads to unrealistic odd spectra. One can lift this degeneracy by adding to the SO(8) algebra multipole operators corresponding to interchanged active and inactive angular momenta k and i. Suppose we add two such operators $\bar{P}_J$ for $J=1$ and 3, which form the SO(5) algebra and commute with all SO(8) generators. We may then consider the fermion group reduction chain $SO(8) \otimes SO(5) \supset \text{Spin}(6) \otimes SO(5) \supset \text{Spin}(5) \otimes SO(5) \supset \widetilde{\text{Spin}}(5) \supset \widetilde{\text{Spin}}(3)$. Here $\widetilde{\text{Spin}}(5)$ is generated by $G^{(3)} = \sqrt{5}P_3 - 2\sqrt{2}\bar{P}_3$ and $G^{(1)} = \sqrt{5}P_1 + 2\sqrt{2}\bar{P}_1$, and $\widetilde{\text{Spin}}(3)$ by $G^{(1)}$, where P_J are the original SO(8) multipole operators.

To arrive at an equivalent boson-fermion description we perform the SO(8) boson-fermion mapping discussed above, whereas the new SO(5) algebra is simply mapped from the original fermion space to the ideal fermion space. It can be shown that the boson-fermion images of generators of $\widetilde{\text{Spin}}(5)$ are (up to a normalization factor) just the generators of the subgroup Spin(5) of U(6/20) in the IBFM [32]. Consequently, on the boson-fermion level we have now the following group chain $U_B(6) \otimes U_F(20) \supset U_B(6) \otimes U_F(4) \otimes U_F(5) \supset SO_B(6) \otimes SU_F(4) \otimes SO_F(5) \supset \text{Spin}_{B+F}(6) \otimes SO_F(5) \supset \text{Spin}_{B+F}(5) \otimes SO_F(5) \supset \widetilde{\text{Spin}}_{B+F}(5) \supset \widetilde{\text{Spin}}_{B+F}(3)$. The associated energy expression is then

$$E = A[\sigma_1(\sigma_1 + 4) + \sigma_2(\sigma_2 + 2) + \sigma_3^2] + B[\tau_1(\tau_1 + 3) + \tau_2(\tau_2 + 1)]$$
$$+ \tilde{B}[\tilde{\tau}_1(\tilde{\tau}_1 + 3) + \tilde{\tau}_2(\tilde{\tau}_2 + 1)] + CJ(J + 1), \quad (32)$$

where (τ_1, τ_2) are the $\text{Spin}_{B+F}(5)$ irreps with $\tau_2 = 1/2$ and $(\tilde{\tau}_1, \tilde{\tau}_2)$ are the irreps of $\widetilde{\text{Spin}}_{B+F}(5)$.

As a simple application of the supersymmetric SO(8) ⊗ SO(5) energy formula (32), we have compared even and odd spectra for a hamiltonian with parameters

$B = 35$ keV, $\tilde{B} = 12.63$ keV, and $C = 18$ keV chosen so that the even part coincides with the one of Ref. [31]. In the odd spectrum we obtain more low-lying $J=3/2^+$ states than one gets in the IBFM (31). (See also Refs.[27,28]

This then sketches how one may derive a new generalised Dyson boson-fermion mapping for a collective algebra extended by single fermion operators. The mapping gives finite non-hermitian boson-fermion images of collective pairs and single fermion operators expressed in terms of ideal boson and fermion annihilation and creation operators. As an example we discussed a fermionic model, namely $SO(8) \otimes SO(5)$, which extends the $SO(8)$ model by a non-trivial interaction between the collective pairs and decoupled particles. In this model an exact supersymmetric structure is revealed, analogous to the interacting boson-fermion model, but with full recognition of the Pauli principle.

We thus find that exact supersymmetry, which in principle mixes bosonic and fermionic degrees of freedom, may equally well appear in purely fermionic models on the dynamical level.

5. Conclusions

We have illustrated how boson phenomenology and boson mappings have mutually stimulated significant developments in the respectve fields by focusing in turn on boson mappings as a many-body tool, the role of spurious states in microscopy and phenomenology, additional dynamical symmetries of the IBM and the emergence of exact supersymmetry in fermion systems analogous to the existence of phenomenological supersymmetry in an IBFM framework.

From these discussions it is clear that there is still much scope for further developments of the boson mapping formalism and its applications. As was also concluded in that review paper, we confirm that for boson mappings there is life after Klein and Marshalek's comprehensive review[1].

6. References

[*] On leave of absence from The Institute of Nuclear Physics, Czech Academy of Sciences, Řež near Prague, Czech Republic

1. A. Klein and E.R. Marshalek, Rev. Mod. Phys. **63** (1991) 375.
2. A. Arima and F. Iachello, The Interacting Boson Model, (Cambridge University Press, Cambridge, 1987).
3. P. Navrátil, H.B. Geyer, J. Dobeš and J. Dobaczewski, Ann. Phys. (N.Y.) (submitted).
4. J.N. Ginocchio, Ann. Phys. (N.Y) **126** (1980) 234.
5. H.B. Geyer, C.A. Engelbrecht, and F.J.W. Hahne, Phys. Rev. C **33** (1986) 1041.
6. J. Dobaczewski, H.B. Geyer, and F.J.W. Hahne, Phys. Rev. C **44** (1991) 1030.

7. H.B. Geyer, P. Navrátil, F.J.W. Hahne and J. Dobaczewski, *Proceedings of the 4th International Spring Seminar on Nuclear Physics*, Amalfi, 1992, ed. A. Covello (World Scientific, Singapore, 1993) p. 281.
8. P. Navrátil and H.B. Geyer, Nucl. Phys. **A556** (1993) 165.
9. P. Navrátil, H.B. Geyer, J. Dobeš and J. Dobaczewski, Nucl. Phys. **A556** (1994) 225c.
10. J. Joubert, F.J.W. Hahne, P. Navrátil and H.B. Geyer, Phys. Rev. **C 50** (1994) (in press).
11. H.B. Geyer, F.J.W. Hahne and F.G. Scholtz, Phys. Rev. Lett. **58** (1987) 459.
12. F. Iachello, Phys. Rev. Lett. **44** (1980) 772.
13. F. Iachello and P. van Isacker, *The Interacting Boson-Fermion Model*, (Cambridge University Press, Cambridge, 1991).
14. F.G. Scholtz, H.B. Geyer and F.J.W. Hahne, Ann. Phys. (N.Y.) **213** (1992) 74.
15. J. Dobaczewski, Nucl. Phys. **A369** (1981) 237.
16. K. Takada and S. Tazaki, Nucl. Phys. **A448** (1986) 56.
17. K. Takada and K. Yamada, Nucl. Phys. **A496** (1989) 224.
18. K. Takada and R. Shimuzu, Nucl. Phys. **A523** (1991) 354.
19. C.-L. Wu, D.H. Feng, and M. Guidry, Adv. Nucl. Phys. (to be published).
20. H.B. Geyer and F.J.W. Hahne, Nucl. Phys. **A363** (1981) 45.
21. A. Arima, N. Yoshida and J.N. Ginocchio, Phys. Lett. **101B** (1981) 209.
22. J. Dobaczewski, Nucl. Phys. **A380** (1982) 1.
23. G.K. Kim and C.M. Vincent, Phys. Rev. **C 37** (1987) 1517.
24. R.F. Casten, C.L. Wu, D.H. Feng, J.N. Ginocchio and X.L. Han, Phys. Rev. Lett. **56** (1986) 2578.
25. H B Geyer, F J W Hahne and F G Scholtz, in *Second International Spring Seminar on Nuclear Physics*, Capri, Ed. A Covello (World Scientific, Singapore, 1988) 493.
26. J. Dobaczewski, F.G. Scholtz, and H.B. Geyer, Phys. Rev. **C48** (1993) 2313.
27. P. Navrátil, H.B. Geyer and J. Dobaczewski, Phys. Rev. Lett. (1994) (submitted).
28. P. Navrátil, H.B. Geyer and J. Dobaczewski, (in preparation).
29. A.B. Balantekin, I.Bars, and F.Iachello, Nucl. Phys. **A370** (1981) 284.
30. F. Iachello and S. Kuyucak, Ann. Phys. (N.Y.) **136** (1981) 19.
31. Y-S. Ling *et al.*, Phys. Lett. **B148** (1984) 13.
32. P. Navrátil and J. Dobeš, Phys. Rev. **C37** (1988) 2126.

FINITE BOSON MAPPINGS OF FERMION SYSTEMS

CALVIN W. JOHNSON AND JOSEPH N. GINOCCHIO
Theoretical Division, Los Alamos National Laboratory
Los Alamos, NM 87545

ABSTRACT

We discuss a general mapping of fermion pairs to bosons that preserves Hermitian conjugation, with an eye towards deriving finite and usable boson Hamiltonians—that is, deriving the Interacting Boson Model from the fermion shell model.

1. Introduction

The dynamics of strongly interacting many-fermion systems is at the heart of nuclear physics, and gives rise to both the riches and the difficulties therein. A basic goal is simple to state: given an interaction (which itself is a difficult and separate problem) and the corresponding fermion Hamiltonian $\hat{H}_F$, solve the stationary Schrödinger equation and find the (low-lying) eigenstates,

$$\hat{H}_F |\Psi_\lambda\rangle = E_\lambda |\Psi_\lambda\rangle, \qquad (1)$$

and transition matrix elements between the eigenstates $T_{\lambda'\lambda} = \langle \Psi_{\lambda'} | \hat{T} | \Psi_\lambda \rangle$. For more than three or four particles integration of the Schrödinger or equivalent equation is numerically intractable. Instead, what one often does is devise a many-body basis which can be truncated to a computationally managable size, and then solve (1) as a *matrix diagonalization problem*. In the standard fermion shell model the many-body basis, and its truncation, is built from single-particle configurations which in turn arise from, or so at least we pretend, a mean-field or Hartree-Fock calculation. This picture encompasses many basic properties of nuclei, such as deformation, but for detailed correlations the number of configurations needed becomes simply enormous. For example for full-shell calculations of rare earth nuclides, the size of the fermion Fock space, i.e. the dimension of the matrix to be diagonalized, is of the order[1,2] 10^{15-21}! While some very clever techniques[2,3] are being employed to attack this problem, it is sensible to ask about truncations other than in the single-particle picture.

One alternative approach is to build/truncate the Fock space based on two-particle degrees of freedom. We know pairwise correlations are important from the BCS theory of superconductivity[4], where the wavefunction is a condensate of boson-like Cooper pairs of electrons coupled to zero linear momentum, from the binding-energy systematics of even and odd nuclei, and of course from the success of the phenomenological Interacting Boson Model[5] (IBM). In the latter many states and transition amplitudes are successfully described using only s- and d- (angular momentum $J = 0, 2$) bosons, which are widely thought to represent coherent nucleon pairs[1,6]. In both cases an enormous number of fermion degrees of freedom are well modeled by only a handful of boson degrees of freedom. Note that although the algebraic limits of the IBM—SU(3), SU(5), O(6), and so on—illuminate collective behavior, in the most general case it, like the fermion shell model, also reduces to a matrix diagonalization problem, with the critical advantage of a computationally tractable model space.

Our goal then is to find a much smaller, and thus managable, boson Fock space, and a set of boson images of the Hamiltonian $(\hat{h}_B)$ and transition operators $(\hat{t}_B)$ in that boson space that reproduce or at least approximate well the low-energy dynamics of the original fermion system,

$$\hat{h}_B \left| \Phi_\lambda \right\rangle = E_\lambda \left| \Phi_\lambda \right\rangle, \quad \left\langle \Phi_{\lambda'} \right| \hat{t}_B \left| \Phi_\lambda \right\rangle = T_{\lambda'\lambda}. \quad (2)$$

The mapping of fermion systems to bosons has a long history[7], starting at least with Holstein and Primakoff[8] in 1940 and continuing through Dyson[9] and Belyaev and Zelevinskii[10] to name just a few. After the introduction of the IBM Otsuka, Arima, Iachello, and Talmi[1,6] (OAIT) investigated its microscopic foundations by mapping fermion shell model states to bosons.

Because of the importance and controversy of this topic we are revisiting boson mappings. Most boson expansion techniques concentrate on mapping operators and algebras[9,10]. Instead we follow Marumori[11] and OAIT[6] by mapping matrix elements.

2. Fermion pairs, matrix elements, and truncation

The first step is to define many-body states built from pairs of fermions and then calculate matrix elements. We work in a fermion space with 2Ω single-particle states. For the fermion shell-model basis states one often uses Slater determinants, antisymmeterized products of single-fermion wavefunctions which we can write using Fock creation operators: $a_j^\dagger, j = 1, \cdots, 2\Omega$ on the vacuum $a_{i_1}^\dagger \cdots a_{i_n}^\dagger \left|0\right\rangle$ for n fermions. For an even number of fermions we instead construct basis states from $N = n/2$ fermion pair creation operators,

$$\left|\psi_\beta\right\rangle = \prod_{m=1}^N \hat{A}_{\beta_m}^\dagger \left|0\right\rangle, \quad \hat{A}_\beta^\dagger \equiv \frac{1}{\sqrt{2}} \sum_{ij} \left(\mathbf{A}_\beta^\dagger\right)_{ij} a_i^\dagger a_j^\dagger. \quad (3)$$

We always choose the $\Omega(2\Omega - 1)$ matrices $\mathbf{A}_\beta$ to be antisymmetric to preserve the underlying fermion statistics, thus eliminating the need later on to distinguish between 'ideal' and 'physical' bosons. Generic one- and two-body fermion operators we represent by $\hat{T} \equiv \sum_{ij} T_{ij} a_i^\dagger a_j$, $\hat{V} \equiv \sum_{\mu\nu} \langle \mu | V | \nu \rangle \hat{A}_\mu^\dagger \hat{A}_\nu$, where $T_{ij} = \langle i | \hat{T} | j \rangle$; from such operators one can construct a fermion Hamiltonian H_F.

Now one needs matrix elements of these states, including the overlap: $\langle \Psi_\alpha | \Psi_\beta \rangle$, $\langle \Psi_\alpha | \hat{H}_F | \Psi_\beta \rangle$, $\langle \Psi_\alpha | \hat{T} | \Psi_\beta \rangle$, and so on. These matrix elements are much more difficult to compute than the corresponding matrix elements between Slater determinants. Silvestre-Brac and Piepenbring[12], laboriously using commutation relations, derived a Wick theorem for fermion pairs. Rowe, Song and Chen[13] using 'vector coherent states' (we would say fermion-pair coherent states) found matrix elements between pair-condensate wavefunctions, states of the form $\left(\hat{A}^\dagger \right)^N |0\rangle$. Using a theorem by Lang et al.[3], we have generalized[14] the method of Rowe, Song and Chen and recovered (actually discovered independently) the expressions of Silvestre-Brac and Piepenbring. One could now solve the Schrödinger equation (1) numerically, after truncating the fermion Fock space by restricting the set of pairs, denoted by $\{\bar{\alpha}\}$, used to construct the many-body states.

Before moving on we make two comments. The first regards the choice of truncation. Rowe, Song and Chen[13] give a variational principle that seems obviously useful in this regards. Otsuka and Yoshinaga[15] start from HFB states; the two approaches can probably be related in some approximation. The second comment is that naive truncations motivated by the IBM may not be successful: while phenomenology does not require $J = 4$ pairs microscopy does[16]. Recently Halse, Jaqua and Barrett[17] found that $J = 0, 2$ pairs do not describe well the low-lying spectra of a quadrupole $Q \cdot Q$ Hamiltonian in a single $j = 17/2$ shell, and that while inclusion of $J = 4$ pairs improves the situation considerably even that model space is lacking. One could address the situtation by renormalization of g-boson effects into d-bosons[18] or through effective interaction theory[19,20], neither of which we address here.

3. Boson representations of matrix elements

We now want to translate the fermion matrix elements into boson space. We take the simple mapping of fermion states into boson states

$$|\psi_\beta\rangle \to |\phi_\beta\rangle = \prod_{m=1}^{N} b_{\beta_m}^\dagger |0\rangle, \qquad (4)$$

where the $b^\dagger$ are boson creation operators. We construct boson operators that preserve matrix elements, introducing boson operators $\hat{T}_B$, $\hat{V}_B$, and most important importantly the *norm operator* $\hat{\mathcal{N}}_B$ such that $(\phi_\alpha | \hat{T}_B | \phi_\beta) = \langle \psi_\alpha | \hat{T} | \psi_\beta \rangle$, $(\phi_\alpha | \hat{V}_B | \phi_\beta) = \langle \psi_\alpha | \hat{V} | \psi_\beta \rangle$. and $(\phi_\alpha | \hat{\mathcal{N}}_B | \phi_\beta) = \langle \psi_\alpha | \psi_\beta \rangle$. We term $\hat{T}_B, \hat{V}_B$ the boson *representations* of the fermion operators $\hat{T}, \hat{V}$. The details of the construction is given in Reference 14, and one finds the 'linked-cluster' (a la Kishimoto and Tamura[19,21] although with differences) expansion of the representations to be of the form

$$\hat{\mathcal{N}}_B = 1 + \sum_{\ell=2}^\infty \sum_{\{\sigma,\tau\}} w_\ell^0(\sigma_1,\ldots,\sigma_l; \tau_1,\ldots,\tau_l) \prod_{i=1}^\ell b_{\sigma_i}^\dagger \prod_{j=1}^\ell b_{\tau_j}. \quad (5)$$

and similarly for $\hat{V}_B, \hat{T}_B$. In the norm operator the ℓ-body terms embody the fact that the fermion-pair operators do not have exactly bosonic commutation relations, and act to enforce the Pauli principle.

The norm operator can be conveniently and compactly expressed[22] in terms of the kth order Casimir operators of the unitary group SU(2Ω), $\hat{C}_k = 2^k \operatorname{tr}(\mathbf{P})^k$, $\mathbf{P} = \sum_{\sigma\tau} b_\sigma^\dagger b_\tau \mathbf{A}_\sigma \mathbf{A}_\tau^\dagger$ (and so is both a matrix and a boson operator; the trace is over the matrix indices and not the boson Fock space)

$$\hat{\mathcal{N}}_B = :\exp\left(-\frac{1}{2} \sum_{k=2}^\infty \frac{(-1)^k}{k} \hat{C}_k \right): \quad (6)$$

where the colons ':' refer to normal-ordering of the boson operators. Similarly — and this is a new result we have not seen elsewhere in the literature — the representations $\hat{T}_B, \hat{V}_B$ can also be written in compact form[14,23]:

$$\hat{T}_B = 2\sum_{\sigma,\tau} :\operatorname{tr}\left[\mathbf{A}_\sigma \mathbf{T} \mathbf{A}_\tau^\dagger \mathbf{G} \right] b_\sigma^\dagger b_\tau \hat{\mathcal{N}}_B:, \quad (7)$$

$$\hat{V}_B = \sum_{\mu,\nu} \langle \mu | V | \nu \rangle \sum_{\sigma,\tau} :\left\{ \operatorname{tr}\left[\mathbf{A}_\sigma \mathbf{A}_\mu^\dagger \mathbf{G} \right] \operatorname{tr}\left[\mathbf{A}_\nu \mathbf{A}_\tau^\dagger \mathbf{G} \right] \right.$$
$$\left. + 4 \operatorname{tr}\left[\mathbf{A}_\sigma \mathbf{A}_\mu^\dagger \mathbf{P} \mathbf{G} \mathbf{A}_\nu \mathbf{A}_\tau^\dagger \mathbf{G} \right] \right\} b_\sigma^\dagger b_\tau \hat{\mathcal{N}}_B:, \quad (8)$$

where $\mathbf{G} = (1 + 2\mathbf{P})^{-1}$. These compact forms are useful for formal manipulation[14,23]. Furthermore they have the powerful property of exactly expressing the fermion matrix elements under *any* truncation, a fact not previously appreciated in the literature even for the norm operator[22]. By this we mean the following: suppose we truncate our fermion Fock space to states constructed from a restricted set of pairs $\{\bar{\sigma}\}$. Such a truncation need *not* correspond to any subalgebra. Then the representations in the corresponding truncated boson space, which still exactly reproduce the fermion matrix elements and which we denote by $[\mathcal{N}_B]_T$ etc., are the same as those given above, retaining only the 'allowed' bosons with unrenormalized coefficients. This invariance of the coefficients under truncation will not hold true for the boson *images* introduced below.

With the boson representations of fermion operators in hand one can express the fermion Schrödinger equation as a generalized boson eigenvalue equation,

$$\hat{\mathcal{H}}_B |\Phi_\lambda\rangle = E_\lambda \hat{\mathcal{N}}_B |\Phi_\lambda\rangle. \tag{9}$$

Here $\hat{\mathcal{H}}_B$ is the boson representation of the fermion Hamiltonian Every physical fermion eigenstate in (1) has a corresponding eigenstate, with the same eigenvalue, in (9). Because the space of states constructed from pairs of fermions is overcomplete, there also exist spurious boson states that do not correspond to unique physical fermion states. The overcompleteness also means that (9) is harder to solve exactly than (1). So one truncates the model space. As mentioned in the previous section such truncations can seriously distort the spectrum, and so one in principle one should employ some renormalization or effective interaction scheme.

4. Boson images

In general the boson representations given in (6), (7) and (8) do not have good convergence properties, so that simple termination of the series such as (5) in ℓ-body terms is impossible and use of the generalized eigenvalue equation (9), as written, is problematic. Instead we "divide out" the norm operator to obtain the *boson image*:

$$h_B \sim \text{``}\mathcal{H}_B/\mathcal{N}_B\text{.''} \tag{10}$$

That this is reasonable is suggested by the explicit forms of (7) and (8). The hope of course is that h_B is finite or nearly so, so that a 1+2-body fermion Hamiltonian is mapped to an image

$$h_B \sim \theta_1 b^\dagger b + \theta_2 b^\dagger b^\dagger bb + \theta_3 b^\dagger b^\dagger b^\dagger bbb + \theta_4 b^\dagger b^\dagger b^\dagger b^\dagger bbbb + \ldots \tag{11}$$

with the ℓ-body terms, $\ell > 2$, zero or greatly suppressed.

4a. Exact results

It turns out that for a number of cases the image of the Hamiltonian is exactly finite. In particular, for the full boson Fock space the representations factor in a simple way: $\hat{T}_B = \hat{\mathcal{N}}_B \hat{T}_B = \hat{T}_B \hat{\mathcal{N}}_B$ and $\hat{V}_B = \hat{\mathcal{N}}_B \hat{V}_B = \hat{V}_B \hat{\mathcal{N}}_B$, where the factored operators $\hat{T}_B$, $\hat{V}_B$, which we term the boson images of $\hat{T}$, $\hat{V}$, have simple forms:

$$\hat{T}_B = 2 \sum_{\sigma\tau} \text{tr} \left(\mathbf{A}_\sigma \mathbf{T} \mathbf{A}_\tau^\dagger \right) b_\sigma^\dagger b_\tau, \tag{12}$$

$$\hat{V}_B = \sum_{\mu\nu} \langle \mu | V | \nu \rangle \left[b_\mu^\dagger b_\nu + 2 \sum_{\sigma\sigma'} \sum_{\tau\tau'} \text{tr} \left(\mathbf{A}_\sigma \mathbf{A}_\mu^\dagger \mathbf{A}_{\sigma'} \mathbf{A}_\tau^\dagger \mathbf{A}_\nu \mathbf{A}_{\tau'}^\dagger \right) b_\sigma^\dagger b_{\sigma'}^\dagger b_\tau b_{\tau'} \right] \tag{13}$$

In general one can find an image Hamiltonian $\hat{H}_B = \hat{T}_B + \hat{V}_B$. This result, and its relation to other mappings, was noted by Marshalek[24].

Thus any boson representation of a Hamiltonian factorizes: $\hat{\mathcal{H}}_B = \hat{\mathcal{N}}_B \hat{H}_B$. Since the norm operator is a function of the SU(2Ω) Casimir operators it commutes with the boson images of fermion operators[14], and one can simultaneously diagonalize both $\hat{\mathcal{H}}_B$ and $\hat{\mathcal{N}}_B$. Then Eqn. (9) becomes

$$\hat{H}_B |\Phi_\lambda\rangle = E'_\lambda |\Phi_\lambda\rangle. \tag{14}$$

where $E'_\lambda = E_\lambda$ for the physical states, but E'_λ for the spurious states is no longer necessarily zero. The boson Hamiltonian $\hat{H}_B$ is by construction Hermitian and, if one starts with at most only two-body interactions between fermions, has at most two-body boson interactions. All physical eigenstates of the original fermion Hamiltonian will have counterparts in (14). It should be clear that transition amplitudes between physical eigenstates will be preserved. Spurious states will also exist but, since the norm operator $\hat{\mathcal{N}}_B$ commutes with the boson image Hamiltonian $\hat{H}_B$, the physical eigenstates and the spurious states will not admix.

The boson Schrödinger equation (14), though finite, is not much use as the boson Fock space is still much larger than the original fermion Fock space, and we still must truncate the boson Fock space. Although the representations remain exact under truncation, the factorization into the image does not persist in general: $\left[\hat{\mathcal{H}}_B\right]_T \neq \left[\hat{\mathcal{N}}_B\right]_T \left[\hat{H}_B\right]_T$. This was recognized by Marshalek[24]. (An alternate formulation[24] does not require the complete Fock space, but mixes physical and spurious states and so always requires a projection operator.)

One can however find sufficient conditions such that a factorization

$$\left[\hat{\mathcal{H}}_B\right]_T = \left[\hat{\mathcal{N}}_B\right]_T H_D \tag{15}$$

does exist, in particular, if the truncated subset $\{\bar{\alpha}\}$ of fermion-pair creation and annihilation operators form a closed subalgebra, that is,

$$\left[\hat{A}_{\bar{\alpha}}, \left[\hat{A}_{\bar{\beta}}, \hat{A}_{\bar{\gamma}}^{\dagger}\right]\right] = \sum_{\bar{\delta}} C_{\bar{\alpha}\bar{\beta}\bar{\gamma}}^{\bar{\delta}} \hat{A}_{\bar{\delta}}. \tag{16}$$

Examples include the SO(8) and Sp(6) models[25] which specify how to choose the pair operators. In general H_D, though finite, is not Hermitian; we call it a *Dyson* image[9].

That one can find a Dyson image for a closed subalgebra is no surprise. A more intriguing result is that under certain conditions which we give elsewhere [14,23] even in the truncated space the Dyson image of some interactions is in fact Hermitian; furthermore this Hermitian image is related to the truncation of the image in the full boson Fock space by a simple renormalization of the two-body piece:

$$H_D = \left[H_B^{1\,\text{body}}\right]_T + f \left[H_B^{2\,\text{body}}\right]_T, \tag{17}$$

where $1 \leq f \leq 2$ and f is a function of the number of pairs in the excluded and allowed spaces[23]. The conditions are complicated but hold true for important cases, for example, the quadrupole-quadrupole and other multipole-multipole interactions in the SO(8) and Sp(6) models (that is, interactions of the generic form $P^r \cdot P^r$ in the notation of Ginocchio[25]) have Hermitian Dyson images. Not all interactions in these models have Hermitian Dyson images, such as pairing in any model and, in the SO(8) model, the particular combination $g_0(S^{\dagger}S + \frac{1}{4}P^2 \cdot P^2)$ which is the SO(7) limit. It so happens that these particular cases nonetheless can be brought into finite, Hermitian form as discussed in the next section.

4b. Approximate or numerical images

The most general image Hamiltonian one can define is

$$h_B \equiv U \left[\tilde{\mathcal{N}}_B\right]_T^{-1/2} [\mathcal{H}_B]_T \left[\tilde{\mathcal{N}}_B\right]_T^{-1/2} U^{\dagger}, \tag{18}$$

which is manifestly Hermitian for any truncation scheme and any interaction, with U a unitary operator. (Because the norm is a singular operator it cannot be inverted. Instead $\left[\tilde{\mathcal{N}}_B\right]_T^{-1/2}$ is calculated from the norm only in the physical subspace, with the zero eigenvalues which annihilate the spurious states retained. Then h_B does not mix physical and spurious states.)

This prescription is, we argue, useful for practical derivation of boson image Hamiltonians. Consider the expansion (11) of h_B. The operators $[\mathcal{H}_B]_T$ and $\left[\tilde{\mathcal{N}}_B\right]_T^{-1/2}$ have similar expansions, and by multiplying out (18) one sees immediately that the coefficient θ_ℓ depends only on up to ℓ-body terms in $[\mathcal{H}_B]_T$ and $\left[\tilde{\mathcal{N}}_B\right]_T^{-1/2}$, derived from 2ℓ-fermion matrix elements which are tractable for ℓ small. Ideally h_B would have at most two-body terms, and our success in finding finite images in the previous section gives us hope that the high-order many-body terms may be at most small; at any rate the convergence can be calculated and checked term-by-term.

We contrast this procedure to that of OAIT[1,6]. They construct, for $2N$ fermions, low-seniority basis states of S and D fermion pairs, $\left|S^{N-n_d}D^{n_d}\right\rangle$, and then orthonormalize the states such that the zero-seniority state is mapped to itself, and states of higher seniority are orthogonalized against states of lower seniority,

$$|v\rangle \to |\text{``}v\text{''}\rangle = |v\rangle + |v-2\rangle + |v-4\rangle + \ldots \tag{19}$$

Then OAIT calculate the matrix elements $\left\langle \text{``}S^{N-n'_d}D^{n'_d}\text{''}\left|H_F\right|\text{``}S^{N-n_d}D^{n_d}\text{''}\right\rangle$ and obtain the coefficients for their boson Hamiltonian. These coefficients have an explicit N-dependence (and for large N and arbitrary systems such matrix elements are not trivial to calculate, especially in analytic form!) and thus implicitly a many-body dependence. This may appear advantageous but has two drawbacks. The first is that it's not clear how to systematically calculate many-body contributions beyond that contained in the OAIT prescription, whereas our method is fully and rigorously systematic. The second, and more important, is that the OAIT prescription can induce many-body effects where none are needed. We'll illustate this latter point shortly.

The Hermitian image h_B is related to the Dyson image H_D by a similarity transformation $\mathcal{S} = U\left[\mathcal{N}_B\right]_T^{1/2}$,

$$h_B = \mathcal{S} H_D \mathcal{S}^{-1}. \tag{20}$$

The similarity transformation $\mathcal{S}$ orthogonalizes the fermion states $|\Psi_{\tilde{\alpha}}\rangle$ inasmuch $\mathcal{S}\mathcal{N}_B\mathcal{S}^{-1} = 1$ in the physical space (and $= 0$ in the spurious space). This is akin to Gram-Schmidt orthogonalization and the freedom to choose U, and $\mathcal{S}$, corresponds to the freedom one has in ordering the vectors to be Gram-Schmidt orthgonalized. The OAIT procedure is just one particular choice out of many. We can use the freedom in the choice of U to our advantage. Consider the SO(8) model[25] and its three algebraic limits: the pure pairing interaction, the quadrupole $P^2 \cdot P^2$ interaction, which can be written in terms of SO(6) Casimir operators, and the linear combination of pairing and quadrupole $S^\dagger S + \frac{1}{4} P^2 \cdot P^2$ which can be written in terms of SO(7) Casimirs (see Ginocchio[25] for details and notation). The Dyson image of the quadrupole interaction is Hermitian and $H_D = h_B$ with $U = 1$. The Dyson images of the pairing and SO(7) interactions are finite but non-Hermitian. We have found U's $\neq 1$ for both these cases (but not the same U) such that their respective Hermitian images h_B are finite; the one for pairing is exactly the OAIT prescription, while that for SO(7) is exactly opposite, orthogonalizing states of low seniority against states of higher seniority. These general Hermitian images do not have a simple relation to the truncated full image, as do the Hermitian Dyson image (17). The one-body piece remains unchanged but there can be signficant renormalization, and even change of sign, of the two-body piece. For example, for the pairing interaction $[H_B]_T = s^\dagger s + \frac{1}{2\Omega^2} s^\dagger s^\dagger s s +$ additional terms, including off-diagonal terms such as $d^\dagger d^\dagger s s$, whereas $h_B = s^\dagger s - \frac{1}{\Omega} s^\dagger s^\dagger s s +$ (depending on the truncation) terms such as $d^\dagger d^\dagger \tilde{d}\tilde{d}$ but *no* off-diagonal terms. Hence we see here the renormalization is not just a simple overall factor f as it was for the Hermitian Dyson image: it is -2Ω for the $s^\dagger s^\dagger s s$ term but 0 for terms such as $d^\dagger d^\dagger s s$.

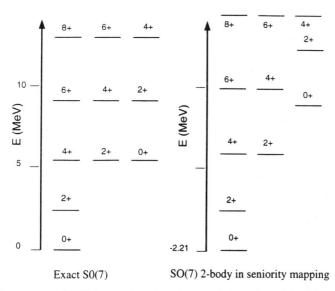

Exact S0(7)   SO(7) 2-body in seniority mapping

Fig. 1: Spectrum of SO(7) interaction, for 7 bosons, in SO(8) model with exact (left) and approximate (right) two-body boson Hamiltonians.

If one uses an "inappropriate" transform S it can induce an unneeded and unwanted many-body dependence. This principle we illustrate in the SO(8) model with the SO(7) interaction, whose spectrum is exactly known and for which we can derive a finite Hermitian image with no many-body dependence; this is the left-hand spectrum in Figure 1. For the right-hand side we took the U appropriate for pairing, that is the OAIT prescription, and calculated the spectrum keeping only the strict two-body terms. The distortion in the spectrum from the exact result, such as the overall energy shift and the large perturbation in the third band, indicates missing many-body terms. That is, if one mapped the SO(7) interaction using the canonical OAIT procedure one would find of necessity a many-body dependence in the interaction coefficients. By orthogonalizing the basis in the a different way, however, as expressed by a different choice of U, the many-body dependent vanishes. It is at least possible that some of the N-dependence of OAIT is induced by their choice of orthogonalization and could be minimized with a different choice. We are currently exploring how to exploit this freedom to best effect.

5. Summary

In order to investigate rigorous foundations for the phenomenological Interacting Boson Model, we have presented a rigorous microscopic mapping of fermion pairs to bosons, paying special attention to exact mapping of matrix elements, Hermiticity, truncation of the model space, and many-body term. In particular: • We presented new, general and compact forms for boson *representations* that preserve fermion matrix elements. • We discussed several analytic cases, both in the full boson Fock space and in truncated Fock spaces, in which the boson *image* Hamiltonian which results from "dividing out" the norm from the representation, is finite. • In the most general case for truncated spaces the image Hamiltonian may not be finite but we have demonstrated there is some freedom in the mapping that one could possibly exploit to minimize the many-body terms. This freedom, which manifests itself in a similarity transformation that orders the orthogonalization of the underlying fermion basis, depends on the Hamiltonian.

Several open questions remain: • What in general is the best way to find U? • If the Dyson image is finite though non-Hermitian, does there always exist a U such that the Hermitian image h_B is also finite? • Finally, questions of spuriosity and effective interactions have not been addressed here.

This research was supported by the U. S. Department of Energy. The boson calculations for Figure 1 were performed using the PHINT package of Scholten[26].

References

1. T. Otsuka, A. Arima, F. Iachello and I. Talmi, *Phys. Lett.***76B** (1978) 139.

2. D. J. Dean et al, *Phys. Lett. B* **317** (1993) 275.

3. G. H. Lang, C. W. Johnson, S. E. Koonin, and W. E. Ormand, *Phys. Rev. C* **48** (1993) 1518.

4. J. Bardeen, L. N. Cooper and J. R. Schrieffer *Phys. Rev.* **108** (1957) 1175.

5. F. Iachello and A. Arima, *The Interacting Boson Model* (Cambridge University Press, 1987).

6. T. Otsuka, A. Arima, and F. Iachello, *Nucl. Phys.* **A309** (1978) 1.

7. A. Klein and E. R. Marshalek, *Rev. Mod. Phys.* **63** (1991) 375; P. Ring and P. Schuck, *The Nuclear Many-Body Problem* (Springer-Verlag, 1980).

8. T. Holstein and H. Primakoff *Phys. Rev.* **58** (1940) 1098.

9. F. J. Dyson *Phys. Rev.* **102** (1956) 1217.

10. S. T. Belyaev and V. G. Zelevinskii, *Nucl. Phys.* **39** (1962) 582.

11. T. Marumori, M. Yamamura, and A. Tokunaga, *Prog. Theor. Phys.* **31** (1964) 1009; T. Marumori, M. Yamamura, A. Tokunaga, and A. Takeda, *Prog. Theor. Phys.* **32** (1964) 726.

12. B. Silvestre-Brace and R. Piepenbring *Phys. Rev. C* **26** (1982) 2640.

13. D. J. Rowe, T. Song and H. Chen, *Phys. Rev. C* **44**(1991) R598.

14. J. N. Ginocchio and C. W. Johnson, to be published.

15. T. Otsuka and N. Yoshinaga, *Phys. Lett.* **168B** (1986) 1.

16. T. Otsuka,, *Phys. Rev. Lett.* **46** (1981) 710.

17. P. Halse, L. Jaqua and B. R. Barrett *Phys. Rev. C* **40** (1989) 968.

18. T. Otsuka and J. N. Ginocchio, *Phys. Rev. Lett.* **55** (1985) 276.

19. H. Sakamoto and T. Kishimoto, *Nucl. Phys.* **A486** (1988) 1.

20. P. Navratil and H. B. Geyer, *Nucl. Phys.* **A556** (1993) 165.

21. T. Kishimoto and T. Tamura, *Phys. Rev.* **C 27** (1983) 341.

22. J. Dobaczewski, H. B. Geyer, and F. J. W. Hahne, *Phys. Rev.* **C 44** (1991) 1030.

23. C. W. Johnson and J. N. Ginocchio, Los Alamos preprint LA-UR-94-1469, submitted to *Physical Review C*, 1994.

24. E. R. Marshalek, *Phys. Rev.* **C 38** (1988) 2961.

25. J. N. Ginocchio, *Ann. Phys.* **126** (1980) 234.

26. O. Scholten in *Computational Nuclear Physics 1*, K. Langanke, J. A. Maruhn and S. E. Koonin, eds., (Springer-Verlag 1991) Chapter 5.

FAVORED PAIRS AND RENORMALIZATION EFFECTS FROM NON-COLLECTIVE PAIRS

Naotaka Yoshinaga
Department of Physics, College of LiberalArts, SaitamaUniversity, Urawa 338, Japan

ABSTRACT

The favored pairs with angular momentum 0 and 2 are enough to reproduce the low-lying states in the light Sn isotopes. A two-j shell model is used to investigate side bands of a deformed nucleus, which shows that the SDG-pair subspace correctly reproduces the bandhead energies. The renormalized SDG-pair subspace is sufficient to reproduce moments of inertia which were previously thought as difficult physical quantities to be obtained in the pair alignment model.

1. Introduction

When I first joined the history of the interacting boson model, it was 1981 and I was a second year graduate student. It was just after Warner, Casten and Davidson made a famous n-γ experiment [1] on ^{168}Er. Complete energy bands were constructed below 2 MeV with spin up to 6. They also analyzed their results in terms of the IBM [2]. The point is that they reproduced all the positive-parity bands without any missing band below 2 MeV except the K=3$^+$ band. Also E2 transition rates were quite well reproduced. Especially they observed the dominance of γ→g over β→g B(E2) values and the dominance of β→γ over β→g transitions. This is an inherent feature of the IBM near the SU(3) limit.

The success of the IBM in deformed region provoked the argument concerning the microscopic foundation of the IBM. Bohr, Mottelson and other Copenhagen people raised questions on the validity of the IBM from a microscopic point of view. Let me briefly explain what was the problem. They analyzed the intrinsic wave function of a deformed nucleus [3]. If the pairing interaction is off, particles occupy from the lowest intrinsic single particle level. It is called particle alignment. If one assumes the wave function is made of S and D pairs, they claim its wave function (it is called pair alignment) cannot reproduce the original wave function. They pointed out that the SD pairs account for only a tiny fraction of the wave function of a deformed nucleus, and some physical quantities (for instance, moments of inertia) are very sensitive to the behavior of the Fermi surface which cannot be described by the SD pairs. Many works had started after this. And I myself was inspired by their paper. In this talk I would like to show how far we can go using the pair alignment model.

2. Favored pairs

First let me explain what the favored pairs or the dominant collective pairs are. The two-nucleon creation operators with each orbit coming from j_1 and j_2 are defined in the following

$$A_M^{\dagger(J)}(j_1 j_2) = [a_{j_1}^\dagger a_{j_2}^\dagger]_M^{(J)} \tag{1}$$

Using these creation operators one can define the collective pairs of angular momenta 0 and 2

$$S^\dagger = \sum_j \alpha_j A_0^{\dagger(0)}(jj) \quad , D_M^\dagger = \sum_{j_1 j_2} \beta_{j_1 j_2} A_M^{\dagger(2)}(j_1 j_2) \quad . \tag{2}$$

Here the structure constants α and β should be determined by an optimum method in order to maximize the collectivity of S- and D-pairs. If necessary, one can include a collective pair of angular momentum four (G-pair). After determining these favored pairs, we assume that any low-lying collective states of even-even nuclei are well approximated by the following states constructed by the S- and D-pairs,

$$| S^{n_s} D^{n_d} \gamma J > , \tag{3}$$

where γ is an additional quantum number. This is called the pair approximation.

3. Sn isotopes

Sn isotopes give a good testing ground for a model or an approximation. This morning there was a beautiful talk by Prof. Arima about how one can use the concept of seniority in nuclear physics. It was shown that by modifying the quadrupole and quadrupole pairing interaction, one can effectively get any desired single particle energies. As far as a phenomenological model is concerned, we can have an arbitrary interaction. Here I would like to show that the generalized seniority of Talmi [4] becomes a good concept even in the case that the seniority of Racah [5] is largely broken.

Here the effective interaction is assumed to take the following form:

$$H = \sum_j \varepsilon_j N_j - G_0 P^{\dagger(0)} P^{(0)} - G_2 P^{\dagger(2)} \cdot \tilde{P}^{(2)} - \kappa : Q \cdot Q : \tag{4}$$

These terms are the single particle energies, pairing interaction and quadrupole-pairing interaction and quadrupole interaction, respectively. The precise definition of the interaction is written in reference[6] . The S-pair is determined by solving the number conserved BCS equation

$$\delta < S^N | H | S^N > = 0 \qquad . \qquad (5)$$

The structure of the D-pair is determined by number conserved Tamm-Dancoff method

$$\delta < S^{N-1} D | H | S^{N-1} D > = 0 \qquad . \qquad (6)$$

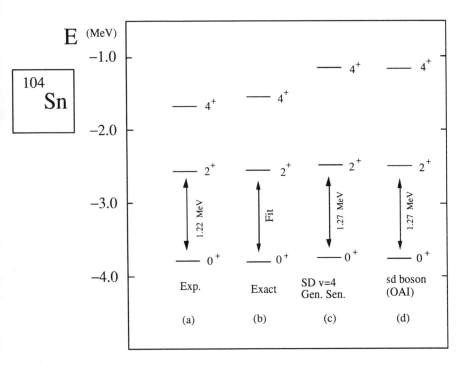

Fig.1 Comparison of energy levels in ^{104}Sn; (a) Experiment, (b) Full Shell model, (c) Generalized seniority, (d) OAI boson calculation. Single particle energies are taken as $\varepsilon_{1d_{5/2}} = 0.0$, $\varepsilon_{0g_{7/2}} = 0.5$, $\varepsilon_{2s_{1/2}} = 2.0$, $\varepsilon_{1d_{3/2}} = 2.5$, $\varepsilon_{0h_{11/2}} = 2.9$ (MeV) . The strengths of the two-body interactions are G_0=0.232, G_2=κ=0.036 (MeV). Experimental binding energies are adjusted to the corresponding theoretical ones in fig.1 and 2.

After determining the structure constants, we construct the SD basis states. The construction of the states in the SD space was discussed in detail in ref.[7]. Then we diagonalize the Hamiltonian in the SD-pair subspace.

The pairing strength is determined to reproduce the even-odd mass difference and single particle energies are taken from Sagawa et al [8]. In this example I set two strengths of the quadrupole pairing and quadrupole interactions equal for simplicity. Therefore if the spreading of the single particle energies is small, this interaction preserves seniority [6]. They are determined to fit the experimental excited energy of the first 2^+ state in ^{104}Sn.

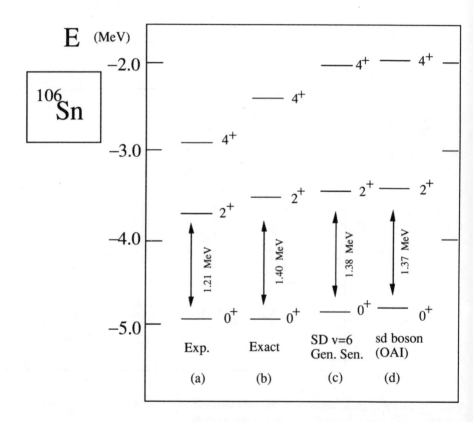

Fig.2 The same comparison as in fig. 1 for ^{106}Sn

In fig. 1 energy levels are shown in experiment, the full shell model space, SD-pair subspace and sd-boson space by the OAI mapping. A good agreement is seen between the exact and the SD calculation. It also shows that the OAI mapping works extremely well in this case. In this example the dimension in the full shell model space is 1504 for ^{104}Sn and 31124 for ^{106}Sn in the m-scheme, while the dimension in the SD -pair subspace for 2^+ states is 2 and 3 for ^{104}Sn and ^{106}Sn, respectively. One can reduce the size of the gigantic shell model space to a very tiny one [9]. The first 4^+ state is reproduced at a reasonable position.

Using the same parameters the same comparison is made in fig. 2 for ^{106}Sn. The result in the SD subspace gives a good agreement, which means that generalized seniority is a good concept in this example. The OAI simulates the results of the generalized seniority well.

4. Single-j shell case and renormalization effect

As one sees in the Sn case, we need only collective S and D pairs in the spherical region. To investigate what about in other regions, I take up a single-j shell. Here I use a renormalization method invented by Nakada. Let me briefly explain his renormalization method[10]. Original orthonormal basis states (L-dimension) are defined in terms of the SD pairs for each angular momentum. If one takes L=0 states, they are denoted as

$|\varphi_1\rangle = |S^N \ L=0\rangle$, $|\varphi_2\rangle = |S^{N-2}D^2 \ L=0\rangle$, ..., $|\varphi_L\rangle = |D^N \ L=0\rangle$. Now for each $|\varphi_i\rangle$, another state is made by multiplying H

Block 1	Block 2	...	Block L
$|\varphi_1\rangle$,	$|\varphi_2\rangle$,	...	$|\varphi_L\rangle$
$H|\varphi_1\rangle$,	$H|\varphi_2\rangle$,	...	$H|\varphi_L\rangle$

Diagonalizing the Hamiltonian in i-th block and picking up the lowest state in energy, we define the renormalized SD-pair subspace, that is, the i-th basis state of which is a linear combination of $|\varphi_i\rangle$ and $H|\varphi_i\rangle$ with certain amplitudes. This is the renormalization procedure with order n=1. If one considers up to $H^2|\varphi_i\rangle$ in addition, it has order n=2. Schmidt's orthonormalization procedure is taken finally. In fact a more elaborate way of renormalization can be considered. After diagonalizing the Hamiltonian in this SD-pair subspace, the eigenstate $|\Psi_i\rangle$ is a linear combination of $|\varphi_i\rangle$'s. The $|\Psi_i\rangle$'s are still the basis states in the SD-pair subspace. Then for each $|\Psi_i\rangle$, we apply the procedure stated before. In this paper this latter method is referred as the "deformed" renormalization (denoted as DR), which means a renormalization in deformed basis states, while I call the former one the "spherical" renormalization (denoted as SR) hereafter.

The pairing plus QQ Hamiltonian $H = -x\, P^{(0)} P^{(0)} + (1-x)\, QQ$ is diagonalized in the full-, SD-, SDG-, renormalized SD- and renormalized SDG-subspaces. Here x is the variable parameter where x is changed from 1 to 0 to simulate various nuclear regions.

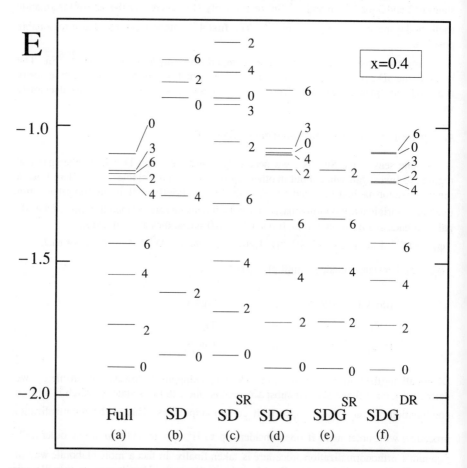

Fig.3 Comparison of energy levels in various spaces in a single-j shell model; (a) Exact, (b) SD (c) renormalized SD (d) SDG (e) renormalized SDG (d) SDG with deformed renormalization. SR means "spherical" renormalization and DR means "deformed" renormalization.

The precise definition of the interaction is written in ref. [11]. Here I take x=0.4, j=23/2 and 6 particles. This gives an example in the transitional region. The 4^+ state has a nature of one hexadecapole phonon state and it is well described by one G-pair. It is seen that the renormalized SD-pair subspace can incorporate the effect of one G-pair even if their original space does not contain any G pair. The low-lying energy levels in the SDG subspace with deformed renormalization are almost identical to the exact ones.

Now we take $x=0.0$. This is a complete deformed case. The spherical renormalization does not work well in this case. Especially the order of levels in the gamma band is incorrect. The original SDG subspace can reproduce bandhead energies, but moments of inertia are difficult to reproduce. The SDG subspace with deformed renormalization almost perfectly reproduces the exact results. Particularly the moment of inertia in the ground band is almost identical.

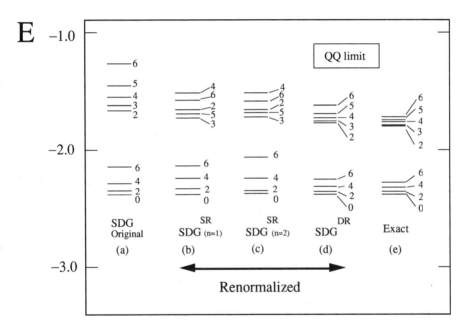

Fig.4 Similar comparison of energy levels as in fig. 3, but $x=0.0$ (QQ limit).

5. Two-j model

The single-j shell model has some deficiencies. In the first place there is only one kind of pair for each angular momentum and it is impossible to check the effect from non-collective pairs with the same spin. Secondly the value of single-j is usually taken as large to represent a large major shell so that there is a possibility that some unrealistic effects appear because of such high angular momentum pairs. Moreover, the β band appears high in energy. Here I take up a two-j model to investigate side bands in deformed nuclei.

In order to get the structure constants of S and D pairs, we consider the following

intrinsic state $|(\Lambda)^N>$ where the pair creation operator $\Lambda^\dagger$ is defined as
$\Lambda^\dagger = x_0 S^\dagger + x_2 D^\dagger_0$. Then the structure constants are determined by minimizing energy

$$\delta < \Lambda^N | H | \Lambda^N > = 0 \qquad (7)$$

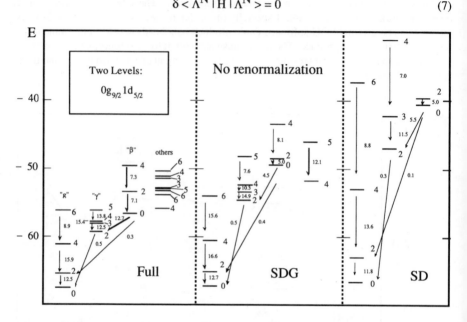

Fig.5 Energy levels in exact SDG and SD subspaces with 6 particles. The numbers beside arrows indicate B(E2) transition ratios.

Here I use H= $-$ Q.Q. Namely, this is the deformed limit. Two levels are taken as $0g_{9/2}$ and $1d_{5/2}$. This example may not be a good example because $Ex(4^+)/Ex(2^+)$ (= R)=3.0 and irregularity can be seen in the "gamma" band, but rotational bands can be seen for the g, γ and β. Bands are constructed so that each member of bands is connected through strong E2 transitions. In both SDG and SD cases the ground band energies are well reproduced. Concerning intraband transitions, they are pretty well described by the SDG-pair approximation, for instance, B(E2) from 2^+ to 0^+ in the ground band is 12.5 in full space and 12.7 in the SDG subspace. This number reduces to 11.8 in the SD case. Concerning interband transitions, they are well described by the SDG approximation except for the 0^+ of β → 2^+ of γ transition. This is an example which shows strong transitions of β → γ band compared to the β → g and γ → g, which is a feature near the SU(3) limit in the IBM. (Note: Some discrepancies can be seen in comparison with the fig. 7 in ref. 6.

This is mainly because in this present work we identify the second 4^+ as a member of the γ band while in ref. 6 we identified it as a member of the $K=4_1^+$ band).

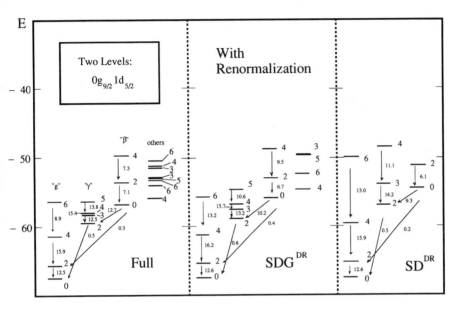

Fig.6 Same comparison of energy levels as in fig.5 with renormalization.

Next we see effects from other non-collective pairs using the deformed renormalization. The bandhead energies of the gamma band and beta bands are now well reproduced. The transition from the beta to gamma is strong in the SDG subspace. I show probabilities of finding the original SDG pairs in the renormalized SDG subspace in Table 1. For members of the ground bands, the probabilities of SDG pairs in the renormalized eigenstates are more than 95 %, but the moment of inertia is almost identical to the exact one. For instance, the ground state has only 0.5 % components outside the original SDG subspace. Therefore only a tiny change in the wave function refines the moment of inertia substantially. In the SD space intraband transitions from the beta to gamma bands are stronger than gamma to ground and beta to ground, which is an inherited feature from the exact calculation.

Table 1
Comparison is made for various physical quantities in each space. $E_0(g)$, $\Delta E_2(g)$ means the binding energy and excited energy of the 2^+ state, respectively. Here $I_J = J(J+1)/2(E_J-E_0)$.

	Exact	SD	SDG	SD^{DR}	SDG^{DR}
$E_0(g)$	−67.45	−66.92	−67.15	−67.38	−67.43
$\Delta E_2(g)$	2.00	3.76	2.16	2.16	2.01
R	3.08	3.52	3.03	3.46	3.09
I_2	1.50	0.80	1.40	1.39	1.49
I_4	1.62	0.76	1.53	1.34	1.61
$E_2(\gamma)$	−59.3	−47.0	−54.7	−57.1	−58.9
$E_0(\beta)$	−56.7	−40.7	−49.4	−54.6	−55.8
Prob. $0_1^+(g)$				98.6%	99.5%
Prob. $2_1^+(g)$				94.4%	99.0%
Prob. $4_1^+(g)$				80.4%	97.9%
Prob. $2_2^+(\gamma)$				57.5%	72.4%
Prob. $0_2^+(\beta)$				57.4%	53.5%

The OAI mapping is carried out in the sdg-boson space in fig. 7. Except for the moments of inertia bandhead energies of the ground band and the gamma bands are well reproduced.

6. Summary

In summary the favored S and D pairs are sufficient to reproduce the low-lying states in the light Sn isotopes. Even in the non-degenerate case the SD pair approximation and the OAI works well. There seems less effect from the non-collective pairs in the spherical case.

A two-j shell model has been used to investigate side bands of a deformed nucleus, which shows that the SDG-pair subspace correctly reproduces the bandhead energies. The role of G pairs is relevant in well deformed case and Nakada's renormalization method is useful to get moments of inertia in the renormalized SDG-pair subspace.

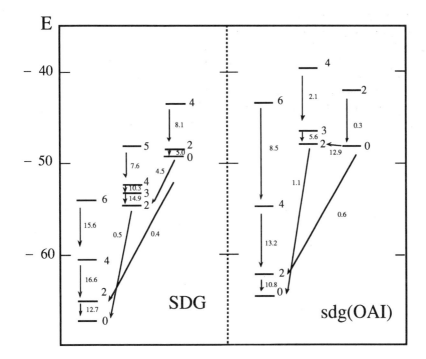

Fig.7 The same comparison as in fig. 6 in the SDG subspace and in the sdg boson space

7. Acknowledgements

I would like to thank Prof. T. Sebe for his shell model code and express my gratitude to Dr. H. Nakada for discussion.

8. References

1. D. D. Warner, R. F. Casten and W. F. Davidson, *Phys. Rev. C* **24** (1981)1713
2. D. D. Warner, R. F. Casten and W. F. Davidson, *Phys. Rev. Lett.* **45** (1980)1761
3. A. Bohr and B.R. Mottelson, *Phys. Scrip.* **22** (1980)468
4. I. Talmi, *Nucl. Phys.* **A172** (1971)1
5. G. Racah, *Phys. Rev.* **62** (1942)438, **63** (1943)367
6. N. Yoshinaga and D. M. Brink, *Nucl. Phys.* **A515** (1990)1
7. N. Yoshinaga, *Nucl. Phys.* **A503** (1989)65
8. H. Sagawa, O. Scholten, B.A. Brown and B.H. Wildenthal,

Nucl. Phys. **A462** (1987)1
9. N. Yoshinaga, Nucl. Phys. **A570** (1994)421c
10. H. Nakada, Ph. D. thesis (University of Tokyo, 1991)
    H. Nakada, in Proc. 4th Int. Spring Seminar on nuclear physics: The building blocks of nuclear structure, ed. by A. Covello (World Scientific, Singapore, 1993) p.271.
11. N. Yoshinaga, Nucl. Phys. **A493** (1989)323

BROKEN PAIRS IN THE INTERACTING BOSON MODEL DESCRIPTION OF HIGH SPIN STATES

D.Vretenar
Physics Department, University of Zagreb, 41000 Zagreb, Croatia

G.Bonsignori and M.Savoia
I.N.F.N. Sez. di Bologna and Department of Physics,
University of Bologna, Bologna, Italy

Abstract

The structure of high-spin states in nuclei is described in the framework of the interacting boson model. The IBM is extended to the physics of high-spin states by including selective noncollective fermion states through the successive breaking of the correlated S and D pairs. High angular momentum states are generated in this way, and their structure is described by the coupling between fermions in broken pairs and the bosonic core. We present a model that is based on the IBM-1/IBFM-1 and includes one-, two-, three- and four-fermion states (one and two broken pairs). The model is applied to two transitional regions of nuclei where the structure of high-spin states close to the yrast line is characterized by the presence of a unique-parity high-j orbital: $h\frac{11}{2}$ for Sm and Nd nuclei, and $i\frac{13}{2}$ for even and odd Hg isotopes. In both cases the model reproduces in detail the experimental data on high-spin bands.

1 The Model

The interacting boson model (IBM)[1], and the interacting boson fermion model (IBFM)[2] provide an excellent description of the structure of states of relatively low angular momentum ($J \leq 10\hbar$) in medium heavy and heavy nuclei. To extend the models to the physics of high-spin states one has to include, in addition to bosons and the unpaired fermion, selective noncollective fermion pairs that occupy part of the original shell model space. This is done by breaking the correlated S and D pairs (s and d bosons) to form fermion pairs. High-spin states are described in terms of broken pairs.

Several extensions of the IBM have been investigated that include two-fermion states (one broken pair) in addition to bosons. These initiated with the work of Gelberg and Zemel,[3] who used an empirical model to incorporate two-particle states in an SU(3) boson basis and investigated backbending phenomena. Faessler et al.[4] have proposed a semimicroscopic model, based on the IBM-1, for the inclusion of two-quasiparticle states in a boson basis. The model has been successfully applied to the description of high-spin states in Hg, Ba and Ce isotopes. This approach was also used to study yrast high-spin states in odd-mass Hg isotopes by extending the IBFM to include three-quasiparticle states.[5] Yoshida, Arima and Otsuka[6] extended the proton-neutron IBM (IBM-2) to include states with two-fermions. The model has been used to analyze high-spin states in Ba and Ce[6], Ge[7], and Dy[8] isotopes. Zemel and Dobes[9] have used the IBM-2 plus two-quasiparticle model to describe properties of low-spin states in Po and Rn isotopes. More recently, using an interacting boson-plus-fermion pair model, Chuu, Hsieh and Chiang have investigated in a series of papers the structure of high-spin states in Pt[10], Dy[11], Er[12], Ge[13] and U[14] isotopes. In Refs.[15-23] we have further extended the IBM to include two- and four-fermion noncollective states (one and two broken pairs), and applied the model in the description of high-spin states in the Hg[16,20], Sr-Zr[18,21,22], and Nd-Sm[23] regions.

The model is based on the simplest version of the interacting boson (fermion) model: IBM-1/IBFM-1. The boson space consists of s and d bosons, with no distinction between protons and neutrons. The bosons can be regarded as collective fermion pair states (correlated S and D pairs) that approximate the valence nucleon pairs. To generate high-spin states, the model allows one or two bosons to be destroyed and form noncollective pairs, represented by two- and four-quasiparticle states that couple to the boson core. The model space for an even-even nucleus with $2N$ valence nucleons is

$$| N\ bosons > \oplus\ |\ (N-1)bosons \otimes 1\ broken\ pair >$$
$$\oplus\ |\ (N-2)bosons \otimes 2\ broken\ pairs >$$

Although generally the fermions in broken pairs occupy all the valence single-particle orbitals from which the bosons have been mapped, for the description of high-spin states close to the yrast line the most important are the unique parity orbitals $(g\frac{9}{2}, h\frac{11}{2}, i\frac{13}{2})$. By allowing the fermions in broken pairs to couple to angular momenta $J_F = 0$ and 2, spurious states are introduced in the model, i.e. the basis does not strictly obey the Pauli principle. Projection of the spurious components from the model space requires that all valence fermion orbitals are included in the basis, making it thus prohibitively large. The procedure consists in constructing microscopically the s and d bosons in the basis of valence fermion orbitals, and removing these particular linear combinations from the basis of broken pairs. Therefore the projection of spurious states is normally not performed. In most cases the percentage

of spurious components for states close to the yrast line is negligible.
For an odd-even nucleus with $2N + 1$ valence nucleons, we consider the model space

$$| \; N \; bosons \otimes 1 \; fermion \; > \; \oplus$$
$$| \; (N - 1) bosons \otimes 1 \; broken \; pair \otimes 1 \; fermion \; >$$

Although in the IBFM-1 there is no distinction between proton and neutron bosons, we have two possibilities for the fermion subspace. The two fermions in the broken pair can be of the same type as the unpaired fermion, resulting in a space with three identical fermions. If the fermions in the broken pair are different from the unpaired fermion, the fermion basis contains two protons and one neutron or vice versa. In many odd-even nuclei the blocking effect of the unpaired nucleon makes energetically more favorable the alignment of a pair of nucleons of the opposite type.

The model Hamiltonian is

$$H = H_B + H_F + V_{BF} + V_{MIX} \tag{1}$$

H_B is the boson Hamiltonian of IBM-1[1]

$$H_B = \epsilon \hat{n}_d + \sum_{L=0,2,4} \tfrac{1}{2} \sqrt{2L+1} \, C_L \, [(d^+ \times d^+)^{(L)} \times (\tilde{d} \times \tilde{d})^{(L)}]^{(0)}$$
$$+ \tfrac{1}{\sqrt{2}} V_2 \{[(d^+ \times d^+)^{(2)} \times (\tilde{d} \times s)^{(2)}]^{(0)} + h.c.\}$$
$$+ \tfrac{1}{2} V_0 \{[(d^+ \times d^+)^{(0)} \times (s \times s)^{(0)}]^{(0)} + h.c.\} \tag{2}$$

The fermion Hamiltonian H_F contains the single-fermion energies and fermion-fermion interactions

$$H_F = \sum_\alpha E_\alpha a_\alpha^+ \tilde{a}_\alpha + \frac{1}{4} \sum_{abcd} \sum_{JM} V_{abcd}^J A_{JM}^+(ab) A_{JM}(cd) \tag{3}$$

The interaction between the unpaired fermions and the boson core contains three terms,

$$V_{BF} = \Gamma_0 \sum_{j_1 j_2} (u_{j_1} u_{j_2} - v_{j_1} v_{j_2}) \langle j_1 \| Y_2 \| j_2 \rangle (a_{j_1}^+ \times \tilde{a}_{j_2})^{(2)} \cdot \hat{Q}_B^{(2)}$$
$$-\Lambda_0 \, 2\sqrt{5} \sum_{j_1 j_2 j_3} (2j_3+1)^{-\tfrac{1}{2}} (u_{j_1} v_{j_3} + v_{j_1} u_{j_3})(u_{j_2} v_{j_3} + v_{j_2} u_{j_3}) \times$$
$$\langle j_3 \| Y_2 \| j_1 \rangle \langle j_3 \| Y_2 \| j_2 \rangle \; : \; [(a_{j_1}^+ \times \tilde{d})^{(j_3)} \times (\tilde{a}_{j_2} \times d^+)^{(j_3)}]^{(0)} \; :$$
$$+ A_0 \sqrt{5} \sum_j (2j+1) (a_j^+ \times \tilde{a}_j)^{(0)} \cdot (d^+ \times \tilde{d})^{(0)} \tag{4}$$

representing the dynamical, exchange, and monopole interaction of the interacting boson-fermion model, respectively. The only new ingredient needed in the calculation of high-spin states is the pair-breaking interaction V_{MIX} that mixes states with different number of fermions, conserving only the total nucleon number

$$V_{MIX} = -U_0 \sum_{j_1 j_2} u_{j_1} u_{j_2} (u_{j_1} v_{j_2} + u_{j_2} v_{j_1}) \langle j_1 \| Y_2 \| j_2 \rangle^2 (2j_2+1)^{-\tfrac{1}{2}} [(a_{j_2}^+ \times a_{j_2}^+)^{(0)} \cdot s]$$
$$-U_2 \sum_{j_1 j_2} (u_{j_1} v_{j_2} + u_{j_2} v_{j_1}) \langle j_1 \| Y_2 \| j_2 \rangle [(a_{j_1}^+ \times a_{j_2}^+)^{(2)} \cdot \tilde{d}] + h.c. \tag{5}$$

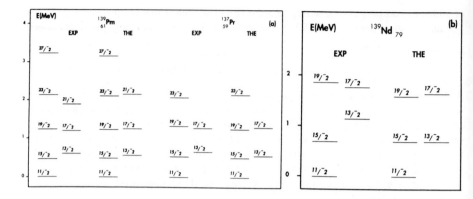

Figure 1: Negative-parity states in ^{139}Pm, ^{137}Pr (a), and ^{139}Nd (b), compared with results of IBFM calculations. Excitation energies are relative to $E(\frac{11}{2}\overset{-}{_1})$.

One obvious application of the model is the description of high-spin phenomena in well-deformed nuclei (crossing between rotational bands, backbending of the moment of inertia, reduction of B(E2) values in the region of crossing, etc.).[15,17] As compared to traditional models based on the cranking scheme, the present approach has the advantage that all calculations are performed in the laboratory system and provide results directly comparable with experimental quantities. Even more important is the possibility to apply the IBM -plus- broken pairs to transitional nuclei in which fermion degrees of freedom couple to a complex mixture of vibrational and rotational collective modes of excitations.

2 The Nucleus ^{138}Nd

Rare-earth nuclei with mass number A≈140 are expected to be soft with respect to the γ degree of freedom. In all the even-even nuclei of this mass region the alignments of both neutron and proton pairs in the $h_{11/2}$ orbital generate low-lying 10$^+$ states, in many cases isomeric. These excitations have been localized in the N=78 ^{140}Sm and ^{142}Gd nuclei and identified as the proton or neutron $h^2_{11/2}$ configurations through g-factor measurements. The structure of high-spin states in these nuclei is characterized by ΔJ=2 decoupled bands based on 10$^+$ isomers. In N=78 nuclei Sm, Gd, Dy the two low-lying 10$^+$ states are found at about the same excitation energy. These states are also observed in the ^{138}Nd nucleus but, because of the lowering of the proton number and consequently of the lowering of the Fermi energy surface respect to the $h_{11/2}$ proton orbital, we expect the $(\pi h^2_{11/2})10^+$ to lie at higher energy compared to the 10⁺ state of neutron character which should have constant energy in all the N=78 isotones.

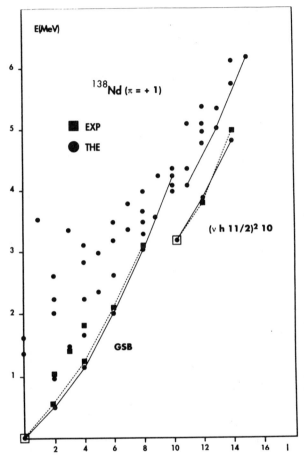

Figure 2: Energy *vs* angular momentum diagram for calculated (circles) and experimental (squares) yrast positive-parity states in ^{138}Nd.

In the calculation for ^{138}Nd we have used the set of parameters of the boson hamiltonian: $\epsilon = 0.37$, $C_0 = 0.2$, $C_2 = -0.036$, $C_4 = 0.187$, $V_2 = 0$, and $V_0 = -0.2$ (all values in MeV).[23] The number of bosons is $N = 7$. The boson parameters are very close to the values that are derived from an IBM-2 calculation for ^{136}Ce. The resulting spectrum is close to the O(6) dynamical symmetry limit. The systematics of experimental data from neighboring nuclei indicates that the yrast 2qp band is a two-neutron band $(\nu\, h\frac{11}{2})^2$, and the band based on the 10⁺ state at 3700 keV is a two-proton configuration $(\pi\, h\frac{11}{2})^2$. Our model space does not include simultaneously protons and neutrons in broken pairs. Instead we have to perform two separate calculations: one for the neutron bands, and one with protons in the broken pairs. In both calculations we use the same boson core. In order to reduce the large size

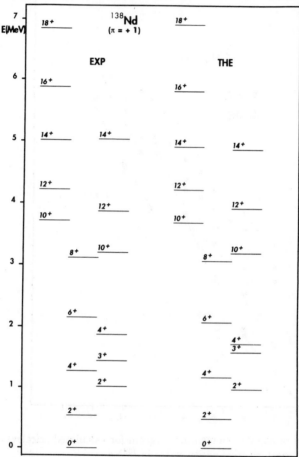

Figure 3: Comparison between experimental and calculated positive-parity levels in ^{138}Nd.

of the fermion space, only the $h\,\frac{11}{2}$ proton and neutron orbitals are included in the calculation of positive-parity bands. The single-quasiparticle energies and occupation probabilities are obtained by a simple BCS calculation using Kisslinger-Sorensen single-particle energies. For the proton $h\,\frac{11}{2}$ orbital: $v^2 = 0.03$ and $E = 1.75\ MeV$. The occupation probability of the neutron $h\,\frac{11}{2}$ orbital is $v^2 = 0.82$ and $E = 1.6$ MeV.

The parameters of the boson-fermion interactions are determined from interacting boson-fermion model (IBFM) calculations of low-lying negative-parity states in neighboring odd-A nuclei. For the proton states we consider $^{137}_{59}\text{Pr}_{78}$ and $^{139}_{61}\text{Pm}_{78}$. The corresponding odd-neutron nucleus is $^{139}_{60}\text{Nd}_{79}$. In Fig. 1 we compare the negative-parity levels of these nuclei with results of model calculation. The decoupled structures are based on the $h\frac{11}{2}$ proton and neutron orbitals. For ^{137}Pr we take as core nucleus ^{138}Nd, the occupation probability of the proton $h\frac{11}{2}$ orbital is $v^2 = 0.05$ and the parameters of the boson-fermion interactions are: $\Gamma_0 = 0.05$ MeV and $\chi = +1$ for the dynamical interaction, and $\Lambda_0 = 2.0$ MeV for the exchange interaction. For ^{139}Pm the core nucleus is ^{140}Sm, and the occupation probability and parameters of the boson-fermion interaction are the same as in ^{137}Pr. The core for ^{139}Nd is taken to be ^{138}Nd, $v^2(\nu h \frac{11}{2}) = 0.82$, $\Gamma_0 = 0.4$ MeV, $\chi = -1$, $\Lambda_0 = 3.0$ MeV. The calculation reproduces the experimental spectra, especially for the two odd-proton nuclei (Fig. 1a). For ^{139}Nd (Fig. 1b), we obtain the inversion $E(\frac{13}{2}^-_1) > E(\frac{15}{2}^-_1)$, but the large splitting between these levels is not reproduced by the calculation.

In Fig. 2 we display the results of model calculation for ^{138}Nd with neutrons in broken pairs. For simplicity, the model space contains only one broken pair $(h\frac{11}{2})^2$. Only few lowest levels of each spin are shown in the energy vs angular momentum diagram. The calculated levels are compared with the experimental yrast states. The collective ground-state band is the yrast band up to angular momentum $I = 8^+$. The calculation reproduces the experimental positions of states of the ground-state band, as well as the excitation energies of the states 2^+_2, 3^+_1, and 4^+_2. The two neutron band $(h\frac{11}{2})^2$ becomes the yrast band at the state 10^+_1. Only the first three states of this band are seen in the experiment. The calculated levels reproduce the excitation energies. The main components in the wave functions of the states of this band are: $|(\nu h\frac{11}{2})^2 J_F = 10, J_B; J = J_F + J_B >$, where $|J_B>$ denotes a collective state of the boson system belonging to the ground-state band with angular momentum J_B. The two $h\frac{11}{2}$ neutrons are completely decoupled from the core and align their angular momenta along the axis of rotation. The fermion angular momentum J_F is a good quantum number for 2qp states close to the yrast line. Above the yrast we follow the odd-spin band $|(\nu h\frac{11}{2})^2 J_F = 10, J_B; J = J_F + J_B - 1 >$. This band is not seen in the experiment. For states higher above the yrast line the Coriolis mixing is much stronger and classification into bands is no longer possible. In Fig. 3 we compare in a more usual form the lowest calculated levels with experimental data. Here the two-proton band $|(\pi h\frac{11}{2})^2 J_F = 10, J_B; J = J_F + J_B >$ is also included. The proton states are obtained in a separate calculation, that is, there is no mixing between two-proton and two-neutron states. The calculated two-proton band reproduces the corresponding experimental levels up to the highest observed member of this band: 18^+. The agreement between calculated and experimental levels is better than for the two-neutron band $(\nu h\frac{11}{2})^2 10$. The reason is that the transition energies in the two-proton band are very similar to the energy spacings in the ground-state band. The transition energies in the two-neutron band follow the ground-state band in ^{140}Nd, rather than in ^{138}Nd.

3 High-Spin States in $^{190-194}$Hg

In the last couple of years there has been a renewed interest in the structure of the neutron-deficient Hg isotopes with $A \approx 190$. In most isotopes superdeformed bands have been found at high angular momenta / rotation frequency. At the same time the level structure of states with normal deformation has been considerably extended and new level sequences have been established. The $A \approx 190$ Hg isotopes are located in a transitional region which, in the IBM representation, corresponds to a change from the $O(6)$ to the $SU(3)$ symmetry limit. The structure of high-spin states with normal deformation in these weakly oblate and γ-soft nuclei has been described in the cranked shell model, by coupling one- and two-quasiparticle states to a Bohr-Mottelson core, and in the interacting boson model with one and two broken pairs.

The yrast bands of the even Hg isotopes reveal an interesting anomaly around angular momenta 8^+, 10^+ and 12^+. The energies of these states are distributed in an interval of ≈ 150 keV. If one plots the moment of inertia as a function of angular velocity for the yrast states a very strong backbending is observed. The moment of inertia increases by a factor of 20 for the triplet of states 8^+, 10^+, 12^+ as compared to the ground state band. This anomaly is explained by the alignment of a $i13/2$ neutron pair. The experimental values of g-factors of the 12_1^+ states confirm the $(\nu\ i13/2)^2$ structure. A second backbending is observed at $J = 20\hbar$.

The energy spectra of the even isotopes ^{190,192,194}Hg are very similar. We have used the same set of parameters of the boson Hamiltonian (2) for all three nuclei. Their values are $\epsilon = 0.2775$, $C_0 = 0.6082$, $C_2 = 0.2065$, $C_4 = 0.2013$, $V_2 = -0.0219$, $V_0 = -0.2080$ (all values in MeV). The number of bosons is: $N = 7$ for ^{194}Hg, $N = 8$ for ^{192}Hg, and $N = 9$ for ^{190}Hg. The yrast two-quasiparticle structures are based on the states 8_1^+. For the description of 2qp and 4qp positive parity states close to the yrast line the most important fermion orbital is $\nu\ i13/2$. The occupation probabilities are obtained by a BCS calculation with Kisslinger and Sorensen single-particle energies: $v_{i13/2}^2(^{190}Hg) = 0.66$, $v_{i13/2}^2(^{192}Hg) = 0.74$, and $v_{i13/2}^2(^{194}Hg) = 0.81$. $E_{i13/2} \simeq 1.2$ MeV for all three isotopes. In Fig.4 we compare the result of model calculation of positive-parity states with experimental data for ^{190}Hg.[20] The parameters of the boson- fermion interaction (Eq.4) are: $\Gamma_0 = 0.45$ MeV, $\chi = 0.4$, $A_0 = 0$, $\Lambda_0 = 1.0$ MeV. For the residual interaction between fermions (Eq.3) we take the surface δ-interaction with strength parameter $v_0 = -0.2$ MeV. The parameters of the mixing interaction are: $u_2 = 0.2$ MeV, $u_0 = 1.3$ MeV. The yrast states 0^+, 2^+, 4^+, and 6^+ belong to the collective ground-state band (GSB). The two-quasiparticle structure $(\nu\ i13/2)^2$ is based on the triplet of isomeric states 8^+, 10^+, and 12^+. These states are band-heads of three different $(\nu\ i13/2)^2$ bands. The leading components in the wave functions of states belonging to these bands are $|(\nu\ i13/2)^2 J_F, J_B; J = J_F + J_B >$, where $J_F = 12$ for the yrast 2qp band, and we also identify states that belong to the band $(\nu\ i13/2)^2 J_F = 10$. The state 8_1^+ is the band-head of the strongly mixed band $(\nu\ i13/2)^2 J_F = 8$. Just above the yrast we find the odd-spin band $|(\nu\ i13/2)^2 J_F = 12, J_B; J = J_F + J_B - 1 >$. The second backbending of the moment of inertia at $J = 20_1^+$ is due to the alignment of a second pair of neutrons in the $i13/2$ orbital. The state 20_1^+ is the band-head of the band $|(\nu\ i13/2)^4 J_F = 20, J_B; J = 20 + J_B >$. Similar results are obtained for ^{192}Hg and ^{194}Hg.

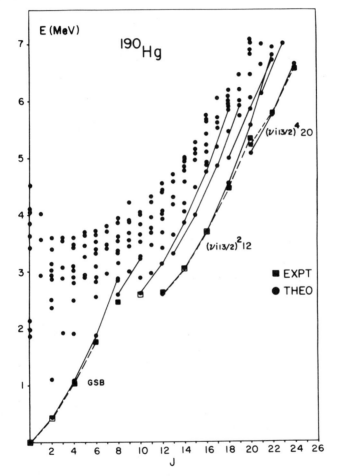

Figure 4: Yrast levels in ^{190}Hg compared with model calculation.

A very sensitive test for the interpretation of the triplet of isomeric states is provided by the B(E2) values for the transitions $12_1^+ \to 10_1^+$ and $10_1^+ \to 8_1^+$. In Table I we compare the experimental B(E2) values with results of model calculation. For the neutron effective charge we take $e = 0.5$ and $\chi = 0.4$ is the same value that is

Table 1: Reduced E2 transition rates in ^{190,192,194}Hg.

Transition	Isotope	Transition energy (keV)	$B(E2)(e^2fm^4)$ experiment	$B(E2)(e^2fm^4)$ theory
$12_1^+ \to 10_1^+$	190	23.9	640(70)	123
	192	28.4	1190(100)	523
	194	52.0	1601(133)	264
$10_1^+ \to 8_1^+$	190	131.9		521
	192	60.1	2750(470)	1561
	194	59.5	2068(400)	953

used in the boson quadrupole operator of the dynamical boson-fermion interaction (Eq.4). The vibrational charges ($e^{VIB} = 1.0$ for ^{190}Hg, $e^{VIB} = 1.5$ for ^{192}Hg, and $e^{VIB} = 1.2$ for ^{194}Hg) are fixed to reproduce $B(E2; 2_1 \to 0_1) \approx 2000 \ e^2fm^4$, a value extrapolated from experimental data on ^{196,198}Hg. The experimental values are much larger than those usually found in the region of backbending, and especially if one thinks of them as describing interband transitions. The calculated B(E2)'s are of the correct order of magnitude, but are smaller than the experimental values. This indicates that the calculated wave functions are not enough mixed, although part of the discrepancy might be accounted for by the uncertainty in $B(E2; 2_1 \to 0_1)$. The best result is obtained for ^{192}Hg. The calculated B(E2)'s are just a factor two smaller than the experimental values. In all three isotopes we obtain $B(E2; 10_1 \to 8_1) > B(E2; 12_1 \to 10_1)$, in agreement with experimental data.

The moments of inertia of yrast states in odd-Hg isotopes display a strong backbending between $J = \frac{29}{2}^+$ and $J = \frac{33}{2}^+$, at approximately the same excitation energy as the triplet of isomeric states in the even isotopes. In Fig. 5 we compare the calculated[24] and experimental positive parity states for ^{191}Hg. The core nucleus is ^{192}Hg with $N = 8$ bosons. The one- and three-fermion space contains only the $i13/2$ neutron orbital. The occupation probability is $v_{i13/2}^2(^{191}Hg) = 0.70$ and $E_{i13/2} = 1.28 \ MeV$. The parameters of the boson- fermion interaction (Eq.4) are: $\Gamma_0 = 0.25 \ MeV$, $\chi = 0.4$, $A_0 = 0.01 \ MeV$, $\Lambda_0 = 1.0 \ MeV$. The residual surface δ-interaction is included with strength parameter $v_0 = -0.2 \ MeV$. The parameters of the mixing interaction are $u_2 = 0.2 \ MeV$ and $u_0 = 1.0 \ MeV$. The only difference with respect to the parameters used for the even isotopes is the reduction of the strength of the dynamical interaction. The inclusion of the monopole boson-fermion interaction which slightly compresses the spectrum of high-spin states in the odd-A isotopes simulates the effect of two-broken pairs.

The lowest structure is the decoupled band of favored states based on the neutron $i13/2$ orbital. Above the yrast band two levels that belong to the unfavored band are seen: $\frac{15}{2}^+$ and $\frac{19}{2}^+$. The calculation reproduces both bands, and in particular the inversions: $\frac{15}{2}$ is above $\frac{17}{2}$, and $\frac{19}{2}$ is above $\frac{21}{2}$. The three-quasiparticle structure $(\nu\, i13/2)^3$ is based on the doublet of isomeric states $\frac{29}{2}^+$ and $\frac{33}{2}^+$. This doublet corresponds to the triplet of states 8^+, 10^+, and 12^+ in the even Hg isotopes (the state $(i13/2)^3 \frac{31}{2}$ is not allowed by angular momentum coupling). The states $\frac{29}{2}^+$ and $\frac{33}{2}^+$ are bandheads of two three-neutron bands. The yrast band is $|(\nu\, i13/2)^3 \frac{33}{2}, J_B; J = \frac{33}{2} + J_B >$. The state $\frac{29}{2}^+_1$ is the bandhead of the band $|(\nu\, i13/2)^3 \frac{29}{2}, J_B; J = \frac{29}{2} + J_B >$. Only the bandhead is seen in the experimental data. Above the yrast one can also follow the band of unfavored states $|(\nu\, i13/2)^3 \frac{33}{2}, J_B; J = \frac{33}{2} + J_B - 1 >$. Higher above the yrast states are much more mixed and it is not possible to group them into bands.

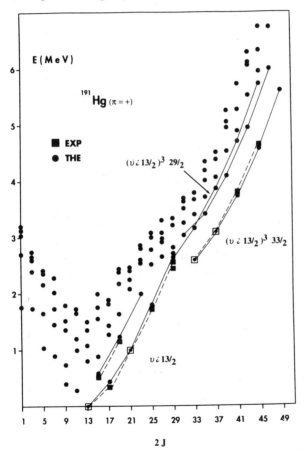

Figure 5: Comparison of experimental yrast levels in ^{191}Hg with results of model calculation.

The lowest three-neutron negative-parity structures in ^{191}Hg and ^{193}Hg are two bands based on $\frac{21}{2}^-_1$ and $\frac{23}{2}^-_1$. The moments of inertia of these bands are higher than those of the positive-parity bands and of the ground-state band of the core nucleus. This indicates that there is more mixing in the negative-parity states and the structure of the bands is more complex. The ground state in both ^{191}Hg and ^{193}Hg is a $\frac{3}{2}^-$, with a low-lying $\frac{5}{2}^-$ at $40 keV$ in ^{193}Hg (this state is not observed in ^{191}Hg). In Fig. 6 we compare the experimental $\pi = -1$ bands in ^{193}Hg with the lowest three neutron negative parity states. The core nucleus is ^{194}Hg with $N = 7$ bosons. The fermion space contains the neutron orbitals: $\nu\ p3/2$ ($E = 1.23 MeV$, $v^2 = 0.38$), $\nu\ f5/2$ ($E = 1.38 MeV$, $v^2 = 0.22$), $\nu\ i13/2$ ($E = 1.38 MeV$, $v^2 = 0.78$). For the boson-fermion interaction only the dynamical interaction is included with $\Gamma_0 = 0.42\ MeV$. The parameters of the mixing interactions are $u_2 = 0.2\ MeV$ and $u_0 = 0$. The surface

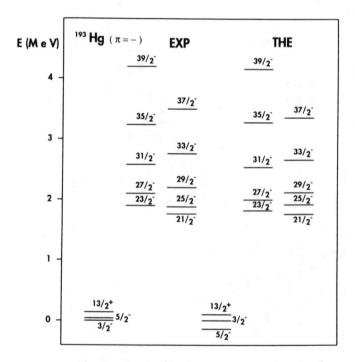

Figure 6: Experimental and calculated negative-parity levels in ^{193}Hg.

δ-interaction has the strength parameter $v_0 = -0.2 MeV$. The calculation reproduces the moments of inertia of the two bands, as well as their relative position. The wave function of states of both bands are dominated by components with two neutrons in the i13/2 orbital coupled to the maximal angular momentum $J = 12$, and the third neutron occupies p3/2 or f5/2. Except for the bandhead $\frac{21}{2}^-_1$, the wave functions of all states have as the main fermion component $[(\nu\ i13/2)^2 12, \nu\ f5/2]$. The leading fermion component of $\frac{21}{2}^-_1$ is $[(\nu\ i13/2)^2 12, \nu\ p3/2]$, and for the lowest states (below $\frac{27}{2}^-$) the mixing between components based on p3/2 and f5/2 is still relatively strong. It is more difficult to reproduce at the same time the structure of the high-spin bands and the correct ordering of low-lying one neutron-states. The calculated $\frac{5}{2}^-_1$ is slightly below $\frac{3}{2}^-_1$ ($\approx 140 keV$), while in the experiment $E(\frac{5}{2}^-_1) - E(\frac{3}{2}^-_1) \approx 40 keV$, with $\frac{3}{2}^-_1$ being the ground state.

References:

1. F.Iachello and A.Arima, *The Interacting Boson Model* (Cambridge University Press, Cambridge, 1987).

2. F.Iachello and O.Scholten, *Phys. Rev. Lett.* **43** (1979) 679; F.Iachello and P.Van Isacker, *The Interacting Boson-Fermion Model* (Cambridge University Press, Cambridge, 1991).

3. A.Gelberg and A.Zemel, *Phys. Rev.* **C22** (1980) 937.

4. I.Morrison, A.Faessler, and C.Lima, *Nucl. Phys.* **A372** (1981) 13; A.Faessler, S.Kuyucak, A.Petrovici, and L.Petersen, *ibid.* **A438** (1985) 78.

5. S.Kuyucak, A.Faessler, and M.Wakai, *Nucl. Phys.* **A420** (1984) 83.

6. N.Yoshida, A.Arima, and T.Otsuka, *Phys. Lett.* **114B** (1982) 86.

7. N.Yoshida and A.Arima, *Phys. Lett.* **164B** (1985) 231.

8. C.E.Alonso, J.M.Arias, and M.Lozano, *Phys. Lett.* **177B** (1986) 130.

9. A.Zemel and J.Dobes, *Phys. Rev.* **C27** (1983) 2311.

10. D.S.Chuu and S.T.Hsieh, *Phys. Rev.* **C38** (1988) 960.

11. D.S.Chuu, S.T.Hsieh, and H.C.Chiang, *Phys. Rev.* **C40** (1989) 382.

12. S.T.Hsieh and D.S.Chuu, *Phys. Rev.* **C43** (1991) 2658.

13. S.T.Hsieh, H.C.Chiang, and D.S.Chuu, *Phys. Rev.* **46** (1992) 195.

14. H.C.Chiang, S.T.Hsieh, and H.Z.Sun, *Phys. Rev.* **C49** (1994) 1917.

15. D.Vretenar, V.Paar, G.Bonsignori, and M.Savoia, *Phys. Rev.* **C42** (1990) 993.

16. F.Iachello and D.Vretenar, *Phys. Rev.* **C43** (1991) 945.

17. D.Vretenar, V.Paar, G.Bonsignori, and M.Savoia, *Phys. Rev.* **C44** (1991) 223.

18. P.Chowdhury, *et al.*, *Phys. Rev. Lett.* **67** (1991) 2950.

19. Y.Alhassid and D.Vretenar, *Phys. Rev.* **C46** (1992) 1334.

20. D.Vretenar, G.Bonsignori, and M.Savoia, *Phys. Rev.* **C47** (1993) 2019.

21. C.J.Lister, P.Chowdhury, and D.Vretenar, *Nucl. Phys.* **A557** (1993) 361c.

22. A.A.Chisthi, *et al.*, *Phys. Rev.* **C48** (1993) 2607.

23. G.de Angelis, *et al.*, *Phys. Rev.* **C49** (1994) .

24. D.Vretenar, G.Bonsignori, and M.Savoia, *Phys. Rev.* **C** *(to be published)*.

TESTING THE I.B.M. PLUS BROKEN PAIRS MODEL AT HIGH SPIN: AN EXPERIMENTAL VIEWPOINT

C. J. LISTER
Physics Division, Argonne National Laboratory, Argonne, IL 60439 U.S.A.

and

P. CHOWDHURY
*Wellesley College, Wellesley, MA 02181 U.S.A.
and University of Massachusetts, Lowell, MA 01854 U.S.A.*

and

D. VRETENAR
Physics Department, University of Zagreb, 4100 Zagreb, Croatia

ABSTRACT

Extensive measurements and calculations have been made on neutron deficient A = 80-90 nuclei. The measurements focused on extracting electromagnetic matrix elements between states of angular momentum $J > 12\,\hbar$, which provided stringent tests of the IBM plus broken pairs model in the four quasi-particle regime. Starting from a detailed analysis of ^{86}Zr, the model was extended to both isotones and isotopes in the region. The model is especially powerful for interpreting transitional nuclei where the competition between vibrational and single particle degrees of freedom determine the structure of states and their decays. It appears entirely appropriate for analysis of the new features revealed by the "third generation" gamma arrays.

1. Introduction

We all agree that the formally correct starting point for describing low-lying excitations in nuclei is through the nuclear shell model. However, except for the lightest nuclei and a few exceptional heavier cases, this approach is generally numerically intractable and we are forced to adopt stratagems which are to a greater or lesser degree phenomenological in nature. Several approaches have been fruitful: sweeping the majority of residual interactions into a deformed mean field and considering the rotation of the field and a few nucleons in that potential, developing nuclear Hartree-Fock and Hartree-Fock-Bogoliubov calculations, truncating the shell model space etc., etc.

In the last decade the cranked Nilsson-Strutinsky type calculations have been particularly powerful in interpreting the behavior of highly deformed nuclei, where the shape of the potential to a large degree determines the wave functions of states and thus the matrix elements between them. However, the approach has shortcomings, particularly when the nuclear potential is soft to shape oscillations and the states are not dominated by rotational components, and when configurations are mixed strongly. In these more complicated situations other, more microscopic, approaches might be more powerful and elegant in interpreting the important degrees of freedom. It may well be that half of all nuclei fall into this latter category.

In the realm of approaches using truncations of the shell model the IBM has proved unusually robust, both in its ability to track the changes of nuclear behavior from vibrational through triaxial to deformed, and in allowing the essential symmetries of states to be identified. As this is the topic of this conference, I do not need to say too much about it. The essential truncation of pairs of nucleons to form J=0 and J=2 bosons, does however impact the ability to predict the structure of states of high angular momentum. In the full shell model space -- and in nature -- high spin states on the yrast line (that is the states of each spin with lowest energy) are usually formed by unpaired nucleons occupying high angular momentum single particle levels, often rather few of them, but these are excluded from the IBM space. Thus, if we wish to restore these components to our model we must start un-truncating the boson space.

This truncation followed by un-truncation is not totally elegant. However, as we will see, it may provide the best description of high spin states that we have for transitional nuclei at the moment. Further, it allows the investigation of the influence of unpaired particles on the boson core. We thus have the possibility of investigating the development of the core IBM space under the influence of different numbers of unpaired particles in the same sense that traditionally we have investigated the IBM space with different numbers of bosons. Further, the modification of the residual interactions between the core and configurations with different numbers of nucleons can be investigated in analogy with the modification of pairing with spin in the deformed mean field approach.

The proceeding paper in this conference by Dario Vretenar describes the model in detail. It has been tested most extensively on transitional nuclei in the mercury (A=190) neodymium (A=140) and zirconium (A=80-90) regions. In this paper we will discuss our work in the mass 80-90 region where we have focused our attention. We have worked almost exclusively on even-even nuclei, but a model with one and three quasi-particles has been written and awaits detailed testing. At present the model is an extension of the IBM-1 and can only consider one type of quasi-particles. This is a major shortcoming as we will see and a IBM-2 version, allowing proton and neutron excitations is an obvious and interesting future extension.

2.   **The Starting Point:** ^{86}Zr

^{86}Zr lies towards the middle of the fp-g shell model space, as is illustrated in fig. 1a. Counting from the major shell closures we have seven bosons, five proton and two neutron. We will return later to the question of subshell closure -- particularly at nucleon number 40. This is a particularly easy case to study, as all the positive parity yrast states have wave functions dominated by fermion occupation of the $g_{9/2}$ single particle state and the fp states have little influence. We can schematically illustrate what we may expect, in fig. 1b. There should be three sets of states, with zero, two, and four quasi-particles. As we are considering like particles in the $g_{9/2}$ shell, the lowest fully aligned two quasi-particle state has J=8 and the lowest four quasi-particle state J=12. Naively the states should be separated by the pair-breaking energy.

Now the questions we can address begin to become more clearly defined. What are the relevant parameters for the IBM core? Do they change in the higher quasi-particle regime? Are the four particle states at twice the two quasi-particle energy? What parameterization do we need for the core-fermion coupling? How much do the different configurations mix? As we do not have the space to discuss these features in detail we

241

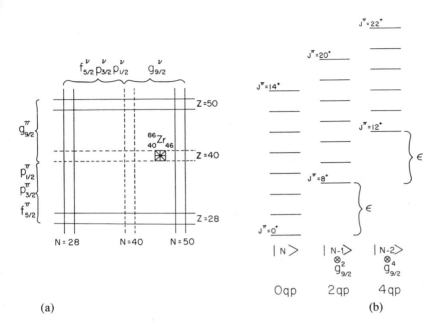

Fig. 1 Orientation to the structure of transitional A=80-90 nuclei. (a) illustrates the main relevant shell closures we will discuss while (b) illustrates the positive parity yrast states to expect.

should mention our longer presentations in refs. 1-3. Some salient points are worthy of discussion.

First, the parameterization. We have tried not to "fine tune" our calculations, and have attempted to stick to the general trends of parameterization which have been investigated for the region at lower spin[4]. The core-fermion coupling is that of the IBFM for odd nuclei which have been investigated in the region[5]. Thus, we are left with the shell model calculation and the mixing parameterization. The shell model part follows the Kisslinger-Sorensen model with a surface delta interaction. The configuration mixing is not constrained by previous studies and we must resort to experimental data to gain insight into its strength.

At the start of this work there was almost no information on mass 80-90 nuclei in the higher spin (J > 12) regime. We have conducted experiments at Chalk River, using the 8-π spectrometer and at GAMMASPHERE. Fortunately, nuclei in this region have also been studied by other groups, especially the Gottingen group[6], and the database on these nuclei has grown rapidly. The center panel of fig. 2 shows a partial level scheme for ^{86}Zr. Most of the features we anticipated are present. The core is that of an anharmonic vibrator with modest collectivity; the B(E2:2 → 0) is only 14 W.u. and decreases with spin. It does not seem to change dramatically in the two or four quasi-particle regime -- and in our calculations we have not varied this.

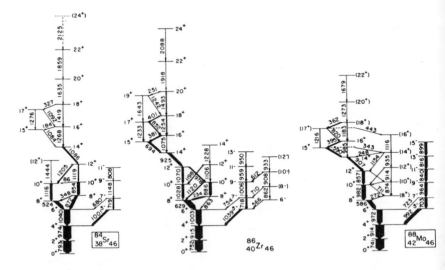

Fig. 2 Partial level schemes of N=46 isotones ^{84}Sr, ^{86}Zr and ^{88}Mo showing the similarity of the positive parity states in zero, two, and four quasi-particle regimes. Data are taken from refs. 3, 1, and 6, respectively.

The complexity of having TWO sets of states is instantly apparent -- one due to neutron configurations and one due to protons. In the two quasi-particle regime these do not, fortunately, mix too strongly, so our IBM-1 based approach remains useful even if we have to perform two separate calculations for the different states and cannot yet incorporate their mixing. The g-factors of the J=8, two quasi-particle band heads have been recently measured[7,8] and show the lower state to correspond to that one may expect for pure aligned $g_{9/2}$ neutrons and the upper to be pure aligned $g_{9/2}$ protons.

It is in the four quasi-particle regime that this nucleus is most interesting and illustrative. The level spacing appears to indicate a significant change in structure; the vibrational-like spacing of levels is replaced by a sequence which appears more rotational. From level spacing alone one would be tempted to assume the quasi-particles have deformed the core and it has undergone a U(5) to SU(3) change in symmetry. However, such a change would be accompanied by a change in electromagnetic matrix elements; the B(E2) transitions should become enhanced. Only through measuring the transition rates can we infer the true structure of the states. This we have done, using the Doppler shift attenuation method, as is illustrated in fig. 3. The E2 matrix elements show no increase at high spin over the lower spin states. If interpreted in the framework of geometric models they correspond to rather modest deformation, with a quadrupole deformation $\beta_2 \approx 0.1$. In fact, the vibrational core persists and the change in level spacing arises from the increased core-fermion coupling.

The magnetic transitions are quite different. Instead of showing a smooth trend, a sequence with alternate very large and rather small matrix elements is found. This is shown in fig. 3(b). The staggered pattern is described quite naturally in our model, with

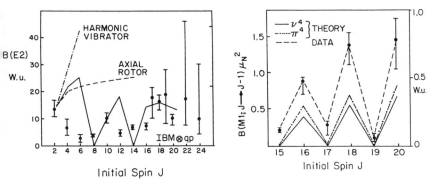

Fig. 3 (a) The electric quadrupole transition matrix elements along the yrast line compared to our calculation and simple geometric models. (b) The magnetic dipole transitions in the four quasi-particle regime.

big B(M1) matrix elements between states with the same number of aligned core bosons and vanishing matrix elements between states with different numbers of core bosons. The predictions for four quasi-protons and for four quasi-neutrons in the $g_{9/2}$ shell are also shown. The fact that the predicted strengths for protons and neutrons are similar is surprising, as proton states usually have larger magnetic matrix elements. In this case there is a cancellation between terms, which leads to matrix elements which have different signs but similar magnitudes and which result in very similar transitional strengths. It is interesting to note that the predictions for pure neutron and pure proton strengths lie well below the experimental value, but a constructive superposition may not.

In conclusion, for our first test case, we find remarkably good predictive power for our model if we fix the mixing of zero, two, and four quasi-particle configurations to be rather small. The strengths and weaknesses of our model are apparent: the model is simple and predicts many states of each spin, and the projected wave functions reveal the mixing of aligned configurations. The pattern of electromagnetic transitions is well reproduced, especially at high spin and can be traced to large vibrational amplitudes in the yrast and near-yrast wave functions. Evidence for the need for the inclusion of proton-neutron configurations is clear. The four quasi-particle states are bound by 1 MeV more than the simple expectation of double the two quasi-particle energy. Initially, we investigated both deformation and reduction of pairing as the cause of this lowering. The g-factor measurements indicate the states are of mixed proton-neutron nature (which are not described in our model), and not quasi-particles of the same type (which is what we calculate). Thus, neutron-proton residual interaction would appear to be the cause of the enhanced binding.

Returning to our original theme, which is that all nuclear models are approximations to the exact state of nature, we shouldn't be surprised or shocked that some of these features can be reproduced in other frameworks. In ^{88}Zr and ^{88}Mo large spherical shell model calculations have been performed[9,10] and many of the features, like the large M1 transitions, can be described quite well, but only for mixed proton-neutron configurations. Similarly, in a geometric picture rotationally aligned protons and deformation aligned neutrons can lead to large, staggered M1 transitions. However, one may argue that the most appropriate model in any situation is the one which most elegantly reveals the structure and symmetries of the nucleus -- and for these transitional nuclei the IBM plus broken pairs model seems to do rather well.

3.   The Isotones ^{84}Sr and ^{88}Mo

Of course, a model is only useful if it has some predictive power. In our IBM framework it is clear that the neighboring isotones of ^{86}Zr should look remarkably similar. This is because the number of bosons remains almost constant and always has N=46, i.e., 2 neutron bosons, while the fermi level for protons changes rather little as we are at the midshell. In fig. 2 we can see that the features we see in ^{86}Zr are reproduced in both the lighter isotope ^{84}Sr and the heavier ^{88}Mo. Not only are the level patterns the same, but the E2 and M1 electromagnetic matrix elements and the bandhead magnetic moments are all very similar. Thus, in this respect the model seems to do very well. Some subtle differences, like the change in staggering of the odd and even spin positive parity states may be attributed to the shift in the proton fermi level and are being studied in detail.

4.   The Isotopes of Zirconium ^{86,84,82}Zr

The isotopes of zirconium present a more stringent test of the model, as both the core and the neutron fermi level are changing. The change between ^{86}Zr and ^{84}Zr is most striking. The very complicated decay pattern of ^{86}Zr is replaced by a simple yrast cascade which is more collective. The cause of the difference lies in a change in the relative energies of the aligned proton and neutron J=8 states as one moves further from the N=50 shell closure. In N=46 isotopes the aligned states are almost degenerate and both well populated in heavy ion reactions. In contrast, for N=44 and below, the neutron aligned states lie well above the yrast line and are not populated. Thus, a comparison of proton states in these isotopes is what is most informative. We are again fortunate that the experimental situation has recently shown healthy progress.[2,11] We find that the core anharmonicity increases, but the nuclei never become good SU(3) rotors. As the nuclei become more collective the wave functions have increasingly complicated structure with many different fermion boson aligned configurations in each state. Thus, the striking simplicity of the N=46 isotopes is lost, even though our calculations are able to reproduce the experimental features rather well.[2]

Our discussion of states to this point has centered on the yrast cascades. However, the calculations we have performed have many states of each spin, corresponding to the diagonalization of many configurations with different combinations of fermion and boson alignment. To date, the vast majority of data has been on yrast states and the non-yrast states in this region are not really known. Fig. 4 shows the state of calculation for ^{84}Zr, indicating the position of the first ten positive parity states of each spin. Exploring this new non-yrast dimension is now possible with the new third generation spectrometers, as is discussed below.

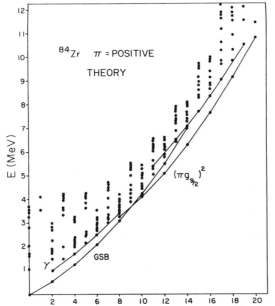

Fig. 4 The first ten states of each spin calculated for ^{84}Zr. (Ref. 2)

A key question which remains is whether or not the core is modified in the two and four quasi-particle regimes in N < 46 nuclei. A high precision lifetime measurement has been performed on ^{84}Zr using the GAMMASPHERE facility. This data is undergoing analysis, but is of sufficient statistical quality to allow measurements to spins near J=30 where fully aligned terminations of structure corresponding to exhaustion of the shell model space are expected.

5. Saturation and Termination Effects

At the highest spins we might hope to see the other end of the saturation problem. The highest states we can make in our model space consist of four aligned fermions which are aligned with the remaining bosons. For ^{86}Zr this state would have spin J=22 for identical fermions or, as the g-factors indicate, J=26 for mixed proton-neutron configurations. Fig. 2 shows we only see up to spin J=24, so the data from the "second generation" gamma ray spectrometers like the 8-π and Tessa have not been able to provide definitive data on terminations in this region. In the case of ^{84}Zr, we have eight core bosons at spin zero and the fully aligned mixed proton-neutron state is expected to have spin J=28. Our experiment using the "third generation" spectrometer GAMMASPHERE allows us to look at this termination problem in detail. Fig. 5 shows a spectrum generated from thin target triple gamma-gamma-gamma coincidences and reveals the ground state band does indeed seem to change above spin J=28. The regularly spaced sequence of levels ends and a gap of about 2.4 MeV is found before other configurations are found. In nature, we can always form higher states, either by entering the six quasi-particle regime, or by exciting particles from our shell model "core" at Z=28 and N=50. We hope to show

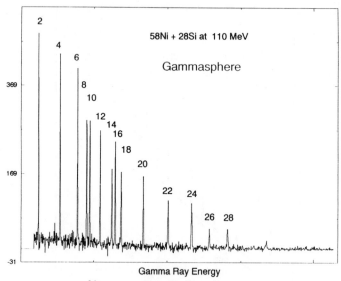

Fig. 5 The spectrum of the ^{84}Zr groundstate band obtained by selecting pairs of coincidences between states above spin J=16 and projecting the spectrum of third coincident gamma rays.

from our lifetime measurements that the collectivity falls away as the four quasi-particle shell model space becomes fully aligned and true terminations are being seen.

The data from GAMMASPHERE reveal another feature which may be appropriately studied in our broken pairs model. We find that many of the states in the yrast sequence are populated from discrete, observable non-yrast levels. As we have seen in fig. 5, our calculation predicts much more than the yrast sequence, and the levels above the yrast line may provide us with further insight into the structure of these transitional nuclei. The position and mixing of these levels, relative to the yrast states depends critically on the residual interactions which cause mixing. These features can thus be investigated in our model. This extension of our knowledge of the high spin properties "sideways" (as opposed to "upwards") in spin seems to be a ubiquitous feature of data from the new third generation arrays. To understand it we will need both careful data analysis and microscopic models which properly incorporate the mixing of configurations. This area is one where our approach may be uniquely useful.

In order to interpret the data at the highest spins, we think that at least the full fp-g shell model space is needed. However, at lower spin this is not the case. A ubiquitous feature in these A=80-90 transitional nuclei is that the B(E2: $2 \rightarrow 0$), B(E2:$4 \rightarrow 2$), and B(E2: $6 \rightarrow 4$) in the yrast sequence have a trend which is decreasing in strength. This cannot be accounted for by any trivial application of collective models. On the other hand, if we invoke a sub-shell gap at N=40 and have a very limited number of core bosons (only 2 for ^{86}Zr!) the trends are all well reproduced.[3] This "opening-up" of the relevant shell model space with increasing excitation energy and spin in a further interesting area which we are studying.

6. Where We Stand

Many questions are left unresolved and are a challenge for the future. The GAMMASPHERE data will provide a rigorous test of the model in the areas of band terminations and of the collectivity of non-yrast states. We hope to report on this soon. There is a wealth of data on neighboring odd-A nuclei which need detailed investigation with our one and three quasi-particle model in order to get a better understanding of core-multi-fermion coupling. More ambitiously, we may aim for the development of an IBM-2 core plus mixed proton and neutron configurations, for although this would be considerably more complex, it is the direction nature is telling us we must eventually go.

Nuclei are complicated and even our best models of structure are approximate. Thus, in order to clearly see the strengths and weaknesses of each it is important that we investigate as wide a variety of approximations as we can. The IBM plus broken pairs model discussed here should be seen in this light. For transitional nuclei, it seems to offer a uniquely powerful tool for categorizing states and understanding their underlying symmetry and structure.

7. Acknowledgements

First, we would like to thank Franco Iachello for all his strategic wisdom, advice and encouragement in this and other projects. We are also very thankful for all the help from Dan Blumenthal, Ben Crowell, Amir Chishti, Pat Ennis, Sean Freeman and Chris Winter who formed a wonderful group to work with. We would like to thank the operations staff at the ESTU at Yale and the accelerators at Chalk River and Lawrence Berkeley Lab. Finally, thanks to the scientists at the gamma ray spectrometers at Chalk River and LBL who taught us how to get the job done. This work was supported in part by the U.S. Department of Energy grants DE-FG02-91ER-40609, and DE-FG02-94ER-40848, and by the U.S. Department of Energy, Nuclear Physics Division, under Contract W-31-109-Eng-38.

References

1. P. Chowdhury et al., *Phys. Rev. Lett.* **67** (1991) 2950.
2. A. A. Chishti et al., *Phys. Rev.* **C48** (1993) 2607.
3. C. J. Lister, P. Chowdhury, and D. Vretenar, *Nucl. Phys.* **A557** (1993) 361.
4. F. Iachello and A. Arima, *The Interacting Boson Model*, (Cambridge University Press 1987) and *Ann. Phys.* (NY) **99** (1976) 253, **111** (1978) 201, and **123** (1979) 468.
5. D. Bucuresch et al., *Nucl. Phys.* **A401** (1983) 22 and references therein.
6. K. P. Lieb et al., *Z. Phys.* **A342** (1992) 257, **A338** (1991) 139, **A340** (1991) 125, and private communication (1994).
7. J. Billowes et al., private communication (1993).
8. N. Benczer-Koller et al., to be published (1994).
9. E. K. Warburton et al., *Phys. Rev.* **C31** (1985) 1211.
10. M. Weiszflog et al., *Z. Phys.* **A342** (1992) 257.
11. S. Mitarai et al., *Z. Phys.* **A344** (1993) 405.

249

BOSON MAPPINGS AND MICROSCOPY
OF THE INTERACTING BOSON MODEL *

J. DOBEŠ and P. NAVRÁTIL
*Institute of Nuclear Physics, Czech Academy of Sciences
CS 250 68 Řež, Czech Republic*

ABSTRACT

The seniority boson mapping as given by the similarity transformation of the Dyson mapping with a successive application of boson mean-field techniques is discussed. The problem of the unphysical states is noted. Results of the microscopically derived IBM are confronted with experimental data for a few sample nuclei. In calculations, g- and s'-boson degrees of freedom outside the sd-IBM space are explicitly taken into account.

1. Introduction

Low-lying collective states in medium-heavy and heavy nuclei provide an example of bosonlike behaviour in a many-fermion system. Such a situation is described phenomenologically with models in which boson degrees of freedom appear. Of course, it is an important ambition of the theory to link a phenomenological description to the underlying many-fermion physics and give thus a microscopic justification of boson models.

Boson degrees of freedom are included very explicitly in the Interacting Boson Model (IBM).[1] Even at the phenomenological stage, there is an aspect of the IBM that makes it different from alternative model approaches. Namely, the total number of the IBM bosons is a conserved quantity and equals the number of valence-shell fermion pairs. Relation to the valence-shell fermions invokes an association of the IBM bosons with fermion pairs and suggests thus a direction towards the IBM microscopy.[†]

2. Truncation and bosonization

In the passage from the shell model to the IBM, truncation and bosonization should be encountered. The truncation means an identification of and restriction to the subspace of the relevant or most important degrees of freedom. The bosonization transforms the original description of the fermion problem in the fermion space into an equivalent one that uses the boson space and boson operators. The succession of the truncation and bosonization might depend on the particular approach chosen.

---
*Supported by the Grant Agency of Czech Republic under grant No. 202/93/2472.
†Here, one should perhaps note that the shell model within one major valence shell is also only a model and may not be particularly microscopic.

In most studies, first the fermion space is truncated. One employs a notion that the coherent fermion S and D pairs are preponderant in the structure of the low-lying collective states. The collective space composed namely of these pairs is selected. Then the bosonization is performed using a technique that is usually called the Marumori mapping. A correspondence between the fermion states and mapped boson states is imposed and the boson images of fermion operators are looked for to reproduce the fermion matrix elements in the boson space. In practice, only states with up to one (or two) D pairs are considered in the mapping procedure. This method has first been devised for the single-j shell case[2] and is referred to as the OAI scheme. It has frequently been used by many authors to study the IBM microscopy.[3]

Generally, the bifermion operators do not close under commutations in the truncated space. From this point of view, the use of the Marumori mapping seems to be necessary in the bosonization step. There is a different bosonization technique, called usually the Belyaev-Zelevinsky (BZ) mapping, that is based on a reproduction of the bifermion commutation relations by the corresponding boson images. The BZ mapping can be applied first to bosonize the full fermion space. Afterward, the truncation is performed in the boson space.

Among mappings of the BZ type, the Dyson boson mapping (DBM) seems to be particularly attractive due to its finiteness whereas its nonunitarity can easily be treated. It is not, however, possible to truncate the boson space when the DBM is used. Reason can be seen for example from the DBM image of the S-pair creation operator, which in the single-j shell reads as

$$\left(S^\dagger\right)_D = s^\dagger \left(1 - \frac{2}{\Omega}N + \frac{1}{\Omega}n_s\right) - \frac{1}{\Omega}\sum_{\lambda \neq 0} \hat{\lambda} \left(B_\lambda^\dagger B_\lambda^\dagger\right)^{(0)} s$$

$$+ \sqrt{\frac{2}{\Omega}} \sum_{\lambda_i \neq 0} \hat{\lambda}_1 \hat{\lambda}_2 \hat{\lambda}_3 \begin{Bmatrix} \lambda_1 & \lambda_2 & \lambda_3 \\ j & j & j \end{Bmatrix} \left[\left(B_{\lambda_1}^\dagger B_{\lambda_2}^\dagger\right)^{(\lambda_3)} \tilde{B}_{\lambda_3}\right]^{(0)} \quad (1)$$

Here, s relates to the boson with $L=0$, B_λ denotes bosons with $\lambda \neq 0$, and Ω is pair degeneracy of the shell. Bosons of all angular momenta enter the DBM image of the seniority $v = 0$ state $(S^\dagger)^N|0\rangle$. The truncation to the s and d bosons modifies results considerably. In contrast, one aims to establish simple relations between fermion states with good seniority and boson states with a fixed number of d bosons, such as

$$|n_F = 2N, v = 0\rangle \quad \rightarrow \quad |n_s = N\rangle, \quad (2)$$
$$|n_F = 2N, v = 2\rangle \quad \rightarrow \quad |n_s = N - 1, n_d = 1\rangle. \quad (3)$$

Such a clear relation already holds for the DBM images of the fermion bra-states. To achieve this also for the ket-states, one could impose the condition that the image of the $S^\dagger$ operator is given by the DBM of the SU(2) subalgebra of the full shell model algebra

$$S^\dagger \to \sqrt{\Omega}\Big(s^\dagger - \frac{1}{\Omega}s^\dagger s^\dagger s - \frac{2}{\Omega}s^\dagger d^\dagger \cdot \tilde{d}\Big) \tag{4}$$

The mapping thus obtained we call the seniority mapping (SBM). The task of its construction has many solutions. One of them is found by noticing that the images of the monopole pairing hamiltonian resulting from the DBM and the SBM should be related by a similarity transformation.[4] This transformation then enables one to construct the seniority images Θ_{sen} of fermion operators from their Dyson forms Θ_D

$$\Theta_{\text{sen}} = Z\Theta_D Z^{-1} \tag{5}$$

For the transformation matrix Z, the perturbational series expansion is found

$$Z^{-1} = \sum_{k=0}^{\infty} \left(\frac{1}{\widehat{H_0} - H_0}W\right)^k \wedge \tag{6}$$

where H_0 and W are diagonal and off-diagonal parts of the DBM image of the monopole pairing hamiltonian, respectively, and the wedge mark $\wedge$ determines at which position the operator $\widehat{H_0}$ is to be evaluated. In the single-j shell case, the OAI results are recovered if the approximation is made to retain only the lowest-order $k=0$ and 1 terms in the above sum. In the multi-j shell case, further approximations and truncation are done by employing the boson mean-field techniques.[5] In the truncation procedure, the collective s boson is given by the Hartree-Bose method and the collective d boson as well as the other non-zero-L bosons are obtained within the Tamm-Dancoff Approximation. The whole scheme thus represents calculationally very simple approximation to the shell model in the $v=0$ and 2 space.

In Fig.1, the exact shell-model $v=2$ energies are compared with those given by the SBM procedure. The energies of the negative-parity two-nucleon states are shown in the inset graph. One deduces that the many-fermion spectrum manifests the complicated Pauli correlations and these are well described in the SBM scheme. A more detailed comparison of methods approximating the shell model in the $v=0$ and 2 space is given in Ref.6.

3. Unphysical states

The boson image of the fermion hamiltonian has to be diagonalized in a boson basis. When this basis is constructed from images of the fermion basis, the problem becomes as complicated as the original fermion one. In practice, the ideal boson basis (full boson Fock space) is used. Such a basis may, however, be overcomplete including the unphysical (spurious) states in which the Pauli principle is violated. In this section, we mention some results of the model study of a simple system of four neutrons and four protons, each kind of nucleons occupying the single-j shell.[7]

The DBM and the SBM given by a similarity transformation of the DBM, of course, provide identical spectra when used in the full ideal boson space. When the

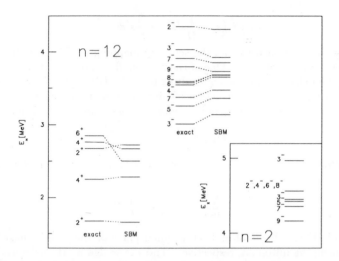

Fig. 1. Exact shell-model energies calculated in the seniority $v=2$ space are compared with results of the SBM procedure for 12 protons in the $Z=50$-82 shell. The BSSP single-particle energies and SDI with $G=0.235$ MeV and $\varphi=1.4$ are used (see Ref.5 for notation). The same hamiltonian is used in the inset graph for system of 2 protons.

fermion interaction in the pairing form is mapped, the unphysical states are pushed up and the lowest states in the boson spectra are physical. When, however, the boson image of the fermion interaction in the multipole-multipole form (as the $Q \cdot Q$ proton-neutron interaction) is used, the spurious states become numerous at low energies. In Fig.2, the lowest 0^+ and 2^+ states as obtained from the exact fermion and SBM calculations are compared. With the increasing $Q \cdot Q$ proton-neutron strength, unphysical states come to be the lowest one in the SBM approach. Such a behaviour stresses the importance of an identification of unphysical states and the necessity of the careful treatment of bosonized fermion tasks.

The SBM results are close to those obtained when the OAIT expression[8] for the quadrupole operator is adopted. The OAIT procedure extends the one-body form of the OAI operator by including the dependence on the number of s bosons and improves thus the boson mapping for the physical sector of the $v=4$ states. On the other hand, it gives strong quadrupole matrix elements between unphysical states. As a result, the unphysical states become the lowest-lying ones when the $Q \cdot Q$ strength is increased. If one would modify the mapping and weaken matrix elements in unphysical sector, one should obtain an improved description. Indeed, one approximation to such a modified mapping appears to follow from considering only the lowest-order one-body terms in the SBM quadrupole operator (see dotted lines in Fig.2), which essentially is the OAI procedure.

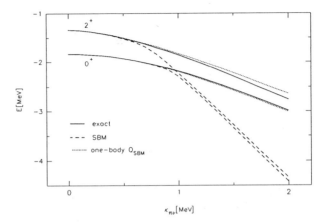

Fig. 2. Dependence of the lowest 0^+ and 2^+ energies of the strength of the $Q \cdot Q$ proton-neutron interaction for the system four protons and four neutrons in the orbits with $j_\pi = j_\nu = 9/2$. The strengths of the monopole pairing interaction between like nucleons are $G_\pi = 0.13$ MeV and $G_\nu = 0.10$ MeV. Exact fermion results (full lines) are compared with the SBM calculations in the complete form (dashed lines) and in the simplified form with the one-body terms retained only in the SBM quadrupole operator (dotted lines).

In the Marumori-type OAI scheme, one should properly orthonormalize the fermion states composed of S and D pairs and relate them to the boson states.* The physical boson sector would thus be defined clearly. In practice, however, the mapping is stopped at some point and usually this is done at the $v=2$ level. Then questions of the convergency of the whole procedure and of the spurious admixtures should be answered. Unfortunately, no studies exist on these points.

In this respect, the SBM might to be more amenable. The expansion (6) can be restricted up to terms with $k=4$ without modifying the mapping in the physical sector. Several methods can be used for the detection of the unphysical states. In combimation with the effective operator techniques, these methods prove to be of some applicability also in the truncated boson space.[7,9]

4. Calculations with microscopic IBM hamiltonian

In applications to realistic nuclei, we start from the shell-model hamiltonian containing the single-particle terms, the like-nucleon interaction in the form of the surface-delta interaction with the increased quadrupole pairing part, and the quadrupole-quadrupole and hexadecapole-hexadecapole proton-neutron interactions. First, the problem of like nucleons is solved within the seniority $v=0$ and 2 space either directly or by the SBM method. The one-body boson operators,

---

*The othonormalization of the fermion basis may not be unique when additional pairs are included or when states with $v>4$ are considered.

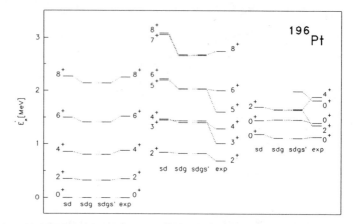

Fig. 3. Experimental and microscopically calculated IBM energies for the ^{196}Pt.

that are the single-boson energies and the multipole boson operators, are then obtained. From these, the proton-neutron IBM hamiltonian is constructed. We consider also the g boson with $L=4$ and the second $L=0$ s' boson outside the original sd-IBM space. Finally, the projection onto the IBM-1 space is performed in calculations presented below.

Since the mapping is constructed for the lowest-seniority states, the whole procedure is mainly tailored to the spherical region. Indeed, a good agreement of the microscopic IBM calculations with experimental data can be obtained for the vibrational nuclei. Example for the ^{148}Sm nucleus is given in another contribution to this Conference.[10]

Also for the gamma-soft nuclei, the microscopically derived IBM hamiltonian predicts data satisfactorily as is shown in calculations of Otsuka et al.[11] for the Xe-Ba region. In Fig.3, our results for the ^{196}Pt spectra are displayed.* The ground-state band is well reproduced. In the quasi-gamma band, however, a pronounced odd-even staggering emerges which is not observed experimentally. To a certain degree, similar problem occurs also in the phenomenological IBM analysis. For the $B(\text{E2})$ probabilities, good agreement with measurements was obtained using reasonable values of the fermion effective E2 charges.[5]

In Fig.3, spectra with the g and s' bosons included in the model space are shown. The effect of these bosons is not too pronounced so that it might be included into the renormalization procedure. There are, however, states in calculated spectra that are mainly based on the g- or s'-boson configurations. The present analysis suggests that $4^+(1.887\ \text{MeV})$ and $0^+(1.823\ \text{MeV})$ states could be of such a kind.

---

*The same shell-model hamiltonian is used as in calculations for ^{196}Pt in Ref.5. The present results differ slightly from those of Ref.5 because the "exact" method of Ref.6 is adopted to get the IBM parameters.

Table 1. $B(E4)$ reduced matrix elements for the ^{196}Pt in e^2b^4. Experimental[12] and microscopically calculated IBM values are shown. IBM results from sd- and sdg-calculations are presented. The proton and neutron effective E4 fermion charges are equal to $1.5e$ and $0.7e$, respectively.

	exp	sdg	sd
$0_1^+ \to 4_1^+ (0.88$ MeV$)$	0.024(5)	0.030	0.006
$0_1^+ \to 4_2^+ (1.29$ MeV$)$	0.020(4)	0.008	$1 \cdot 10^{-5}$
$0_1^+ \to 4_3^+ (1.89$ MeV$)$	0.044(13)	0.051	$5 \cdot 10^{-9}$

In Table 1, the $B(E4)$ values are shown. With reasonable E4 fermion charges, we explain the E4 strengths rather well. Of course, here the inclusion of the g boson is essential.

In regions of deformed nuclei, microscopic IBM calculations encounter problems. The main difficulty is caused by too large microscopically calculated single-boson energies. For large single-boson energies, the quadrupole-quadrupole interaction is not able to admix configurations with many d (or g) bosons into the low-lying bands substantially. On the other hand, the IBM phenomenology requires these admixtures to be strong in the rotational region. Resulting microscopically calculated spectra are too decompressed in comparison to experiment. As the case might be, one could increase the quadrupole-quadrupole strength with an effect of increasing the calculated moment of inertia. Then, however, positions of the beta and gamma bands move up excessively. One can also modify the input shell-model parameters to get a better agreement with experiment.[5] The justification of such a procedure is not, however, too clear.

In Table 2, results of the microscopically determined IBM are compared with experiment and with mean-field fermionic calculations[13] for the deformed ^{156}Gd nucleus. The same shell-model hamiltonian comprising the major valence proton and neutron shells is used both calculations. Disagreement between two calculations and disagreement between calculations and experimental value are considerable. Those microscopic approaches to the IBM that are based on the seniority concept and are appropriate for vibrational region seem to fail and not to reproduce the (expected) shell-model results in the deformed case. Moreover, results of Ref.13 suggest that the shell model within the single major valence shells, or at least such a model with parameters empirically determined from the near-to-magic nuclei, may not be a sufficient tool to study deformed nuclei.

Table 2. Moment of inertia $\mathcal{I}$ (in $\hbar^2 \text{MeV}^{-1}$) of the g.s. band and excitation energy (in MeV) of the 0_2^+ state for the ^{156}Gd. Experimental values are compared with the results of the fermionic mean field (FMF) calculations[13] and of the microscopic IBM approach.

	exp	FMF	IBM		
			sd	sdg	$sdgs'$
$\mathcal{I}$	33.7	15.7	5.41	5.95	6.24
$E(0_2^+)$	1.05	1.63	2.63	2.53	2.43

5. Conclusions

The general approach to the IBM microscopy, which would be applicable in the whole range of nuclei where the IBM phenomenology is successful, is not yet available. The boson mapping techniques provide a powerful tool for transcription of the fermion problem into the boson language. They prove to work well in model calculations. In realistic many-nucleon cases, however, the questions of convergency and of the unphysical admixtures remain to be answered.

The procedures based on the seniority concept have been discussed in the present contribution. As a result, the microscopically calculated IBM parameters appear to be reasonable in cases of spherical and gamma-soft nuclei but fail for the deformed region. For the latter, alternative bosonization methods have been suggested.[14] These, however, have not up to now been used in detailed analysis of nuclear properties. Moreover, the suitability of the starting shell-model assumptions should also be examined.

The g and s' bosons have been considered in our analysis. Their effect might mostly be included into the sd-IBM space by renormalization. There are, however, levels and properties for which the explicit account of these extra degrees of freedom is essential. The corresponding extension of the phenomenological IBM contains a large number of parameters. Therefore, microscopically based or motivated estimates for these parameters are certainly helpful.

References

1. A.Arima and F.Iachello, *The Interacting Boson Model* (Cambridge University Press, Cambridge, 1987).
2. T. Otsuka, A. Arima, and F. Iachello, *Nucl. Phys.* **A309** (1978) 1.
3. F. Iachello and I. Talmi, *Rev. Mod. Phys* **59** (1987) 339 and references therein.
4. H.B. Geyer, *Phys. Rev.* **C34** (1986) 2373.
5. P. Navrátil and J. Dobeš, *Nucl. Phys.* **A507** (1990) 340; **A533** (1991) 223.
6. J. Dobeš, P. Navrátil, and O. Scholten, *Phys. Rev.* **C45** (1992) 2795.
7. P. Navrátil and J. Dobeš, *Phys. Rev.* **C40** (1989) 2371; *Phys. Rev.* **C46** (1992) 220.
8. T. Otsuka, A. Arima, F. Iachello, and I. Talmi, *Phys. Lett.* **B76** (1978) 139.
9. P. Navrátil and H.B. Geyer, *Nucl. Phys.* **A556**, (1993) 165.
10. E.D. Davis and P. Navrátil, contribution to these Proceedings
11. T. Otsuka, *Nucl. Phys.* **A557**, (1993) 531c.
12. W.T.A. Borghols et al.,*Phys. Lett.* **B152** (1985) 330.
13. L.C. de Winter et al., *Phys. Lett.* **B179** (1986) 322.
14. T. Otsuka and N. Yoshinaga, *Phys. Lett.* **B168** (1986) 1; J. Dukelsky and S. Pittel, *Nucl. Phys.* **A456** (1986) 75

MICROSCOPIC GENERALIZATION OF STANDARD INTERACTING BOSON MODEL IN THE RESTRICTED DYNAMICS APPROACH

J.A.CASTILHO ALCARAS
Instituto de Física Teórica, Universidade Estadual Paulista
São-Paulo, 01405, Brasil

J.TAMBERGS, J.RUŽA, T.KRASTA
Nuclear Research Center, Latvian Academy of Sciences
Salaspils, LV-2169, Latvia

O.KATKEVIČIUS
Institute of Theoretical Physics and Astronomy, Lithuanian Academy of Sciences
Vilnius, 2600, Lithuania

ABSTRACT

The evaluation of microscopic Generalized Interacting Boson Model (GIBM), deduced from the general microscopic nuclear Hamiltonian via collective O_{A-1} invariant microscopic Hamiltonian of the Restricted Dynamics Model in the case of central multipole-Gauss type effective NN-potential is briefly discussed. The GIBM version, which includes all 6th order terms in the expansion of the collective part of NN-potential, has been obtained. This GIBM version contains additional terms in comparison with the standard (sd-boson) IBM. The microscopic expressions for the standard IBM Hamiltonian parameters in terms of multipole-Gauss type NN-potential parameters have been obtained as well.

1. Introduction

The relationships between microscopic and phenomenological collective nuclear models have been studied in several papers[1,2,3]. Our studies are directed to the evaluation of specific connections allowing to calculate the characteristics of phenomenological collective models starting from the microscopical ones. This work is devoted to the deduction of such relationships and microscopical generalizations of Interacting Boson Model (IBM) in the framework of restricted dynamics approach, continuing the investigations, described in papers[2,3] and conference contribution[4].

2. Microscopic Evaluation of Generalized Interacting Boson Model

Starting from the general microscopic nuclear Hamiltonian

$$H^{micr} = -\frac{\hbar^2}{2m}\frac{1}{A}\sum_{i<j=2}^{A}\left(\vec{\nabla}_i - \vec{\nabla}_j\right)^2 + \sum_{i<j=2}^{A} V_W(|\vec{r}_i - \vec{r}_j|), \quad (1)$$

where for the sake of simplicity we have assumed that all nucleons are equivalent and that the nucleon-nucleon (*NN*) potential includes only the Wigner interaction term, one can project the collective O_{A-1} invariant microscopic Hamiltonian of the Restricted Dynamics Model (RDM)[5]

$$H_{coll}^{RDM} = T_{coll}\left(\rho^{(i)}, \upsilon^{(i)}\right) + V_{coll}\left(\rho^{(i)}\right), \qquad (2)$$

where $\rho^{(i)}$ - three radial type variables, and $\upsilon^{(i)}$ - three Euler angles ($i = 1, 2, 3$), forming the set of Dzublik-Zickendraht collective microscopic variables. If one considers only central Wigner interaction with multipole-Gauss type *NN*-potential and introduces another collective orbital variables[2,3,4,5] - ρ^2, β, γ, which are the linear combinations of $\left(\rho^{(i)}\right)^2$, one can obtain from Eq.(2) the microscopic phenomenological Hamiltonian of Generalized Bohr-Mottelson Model (GBMM):

$$H^{GBMM} = H_{kin}^{GBMM}\left(\rho^2, \beta, \gamma, \upsilon^{(i)}\right) + H_{pot}^{GBMM}\left(\rho^2, \beta, \gamma\right). \qquad (3)$$

If one expresses the collective coordinates $\rho^2, \beta, \gamma, \upsilon^{(i)}$ via six creation operators $\eta^{\kappa\mu}$ and six annihilation operators $\eta^{+\kappa\mu}$, one can introduce in Eq.(3) the U_6 group structure. The U_6 scalar part of H^{GBMM} ($\eta^{\kappa\mu}, \eta^{+\kappa\mu}$) gives the generalized version of Interacting Boson Model (GIBM) (see Eq.(25) in[4]):

$$H^{GBMM}\left(\eta^{\kappa\mu}, \eta^{+\kappa\mu}\right) = H^{GIBM}\left(\eta^{\kappa\mu}, \eta^{+\kappa\mu}\right) + H^{GBMM'}\left(\eta^{\kappa\mu}, \eta^{+\kappa\mu}\right), \qquad (4)$$

where $H^{GBMM'}$ denotes the remaining part of H^{GBMM}.

The Hamiltonian of GIBM, entering in Eq.(4), can be presented as follows (see Eqs.(28),(30) in[4]):

$$H^{GIBM}\left(\eta^{\kappa\mu}, \eta^{+\kappa\mu}\right) = H_{kin}^{GIBM} + H_{pot}^{GIBM}, \qquad (5)$$

where the kinetic energy term is

$$H_{kin}^{GIBM} = \frac{C_k}{4}\left(6 + 2\sum_{\kappa\mu}\eta^{\kappa\mu}\eta^{+\kappa\mu}\right) = \frac{C_k}{4}\left(6 + 2\hat{N}\right). \qquad (6)$$

Here $\hat{N}$ is the operator of the total number of U_6 bosons $\left(\hat{N} = \hat{n}_s + \hat{n}_d\right)$ and C_k is the kinetic energy constant ($C_k = h/m_{eff}$).

The potential energy term of Eq.(5) can be deduced analytically in the case of multipole-Gauss type *NN*-potential expanded in Taylor series:

$$V_{NN}(r) = V_0 r^{2a} e^{-\alpha r^2} = V_0 \sum_t b_t r^{2t} = V_0 \sum_t \frac{(-\alpha)^t}{(t-a)!} r^{2t}, \qquad (7)$$

where $r = |\vec{r}_{A-1} - \vec{r}_A|$, $a = 0, 1, 2, \ldots$ and $t = a, a+1, a+2, \ldots$ Operator H_{pot}^{GIBM} then can be obtained, if one takes the U_6-scalar part of H_{pot}^{GBMM}:

$$H_{pot}^{GIBM} = \left(H_{pot}^{GBMM}\right)_{|U_6}. \qquad (8)$$

Then (for details see[3,4,5])

$$H_{pot}^{GBMM} = \frac{A(A-1)}{2}\left[V_0\sum_t b_t r^{2t}\right]_{\substack{\text{collective}\\\text{part}}} = C_p \sum_{t=0,2,4,...} b_t R_t \sum_{i,l,m} C_t(i,l,m)F_t(i,l,m), \tag{9}$$

where

$$b_t R_t = \frac{(-\alpha)^t}{(t-a)!}\frac{(A-3)!!}{(2t+A-3)!!}. \tag{10}$$

In Eq.(9) one can write

$$C_t(i,l,m)F_t(i,l,m) = C_t(i,l,m)\left[\left(\rho^2\right)^i \times \left(\beta^2\right)^l \times \left(\beta^3\cos 3\gamma\right)^m\right] =$$

$$= \sum_t (2t+1)!!\left(\frac{2}{3}\right)^t (\rho)^{2t}\sum_{k=0}^{t}\left(\frac{t!}{k!(t-k)!}\right)(-\sqrt{6})^k\left(\frac{\beta}{\rho^2}\right)^k (B_k(\cos 3\gamma)), \tag{11}$$

where (see[6])

$$B_k(\cos 3\gamma) = \frac{k!}{(2k+1)!!}\sum_{j=\Delta,\Delta+2,...}^{j_{max}}\frac{(k-j-1)!!}{[(k-3j)/2]!j!}\left(\frac{3}{2}\right)^{\frac{k-3j}{2}}(\cos 3\gamma)^j \tag{12}$$

with $\Delta = 0(1)$, if k is even(odd), and $j_{max} = \Delta + 2\{[k/6] + [(k+4)/6] - [(k+5)/6]\}$. (Symbol $[x]$ denotes here the integer part of x).

The potential energy constant C_p, entering in Eq.(9), for the Wigner type central NN-interaction assumes simple form

$$C_p = N_0 V_0 = \frac{A(A-1)}{2}V_0. \tag{13}$$

In more general case, when central NN-interaction contains Wigner, Majorana, Bartlett and Heizenberg forces of equal potential depth V_0, C_p is (see[3])

$$C_p = [N_0 c_W + \Lambda(f)c_M + \Lambda(S)c_B + \Lambda(T)c_H]V_0, \tag{14}$$

where $\Lambda(f), \Lambda(S), \Lambda(T)$ are the eigenvalues of the symmetric group class operators, consisting of orbital, spin and isospin exchange operators[7,8] and c_W, c_M, c_B, c_H are the exchange constants of the central interaction, normalized with the condition[8] $c_W + c_M + c_B - c_H = -1$.

Then GIBM Hamiltonian Eq.(5) for multipole-Gauss type NN-potential, taking into account the kinetic energy operator Eq.(6) and $t = 0, 2, 4, 6$ terms of the collective potential energy operator expansion Eqs.(8) and (9), can be written as

$$H^{GIBM} = \frac{1}{2}C_k(\hat{N}+3) + C_p\left[b_0 R_0 + b_2 R_2\left(\frac{20}{3}F_2(2,0,0)_{|U_6} + 8F_2(0,1,0)_{|U_6}\right) + \right.$$

$$+ b_4 R_4 \left(\frac{560}{3} F_4(4,0,0)_{|U_6} + 576 F_4(0,2,0)_{|U_6} + 1344 F_4(2,1,0)_{|U_6} - \right.$$

$$- 256\sqrt{6} F_4(1,0,1)_{|U_6}\right) + b_6 R_6 \left(\frac{320320}{27} F_6(6,0,0)_{|U_6} + \frac{640640}{3} F_6(4,1,0)_{|U_6} - \right.$$

$$- \frac{732160\sqrt{6}}{9} F_6(3,0,1)_{|U_6} + +549120 F_6(2,2,0)_{|U_6} - 133120\sqrt{6} F_6(1,1,1)_{|U_6} +$$

$$+ 115200 F_6(0,3,0)_{|U_6} + 20480 F_6(0,0,2)_{|U_6}\Big)\Big]. \tag{15}$$

The U_6-scalar terms $F_t(i,l,m)_{|U_6}$ in Eq.(15) one can obtain (see Sec.4) expressing the SO_3-scalar invariants ρ^2, β^2, $\beta^3 \cos 3\gamma$, entering in terms $F_t(i,l,m)$ (see Eq.(11)), via creation-annihilation operators (see[2,5]):

$$\rho^2 = \sqrt{\frac{3}{2}} \left(\eta^{00} + \eta^{+00}\right), \tag{16}$$

$$\beta^2 = \frac{1}{2} \sum_\mu \left(\eta^{2\mu} + \eta^{+2\mu}\right)\left(\eta^{2\mu} + \eta^{+2\mu}\right), \tag{17}$$

$$\beta^3 \cos 3\gamma = \frac{\sqrt{7}}{4} \sum_{\mu\mu'\mu''} \left(\eta^{2\mu} + \eta^{+2\mu}\right)\left(\eta^{2\mu'} + \eta^{+2\mu'}\right)\left(\eta^{2\mu''} + \eta^{+2\mu''}\right) C \begin{smallmatrix} 2 & 2 & 2 \\ \mu & \mu' & \mu'' \end{smallmatrix}. \tag{18}$$

Therefore, the GIBM Hamiltonian Eq.(15) will be presented in the form suitable for the comparison with the standard IBM Hamiltonian.

3. Standard IBM Hamiltonian

The standard IBM Hamiltonian, taking into account s and d bosons, one can write using two one-boson operators $\hat{N}$, $\hat{n}_d$ (where $\hat{N} = \hat{n}_0 + \hat{n}_d$) and seven two-boson interaction operators $(\hat{A}_0, \hat{A}_2, \hat{A}_4, \hat{D}, \hat{E}, \hat{B}, \hat{C})$ (see[9,10]):

$$H^{IBM} = e_0 + \varepsilon_s(\hat{N} - \hat{n}_d) + \varepsilon_d \hat{n}_d + \frac{1}{2} c_0 \hat{A}_0 + \frac{\sqrt{5}}{2} c_2 \hat{A}_2 + \frac{3}{2} c_4 \hat{A}_4 +$$

$$+ \frac{1}{2} v_0 \hat{D} + \frac{1}{\sqrt{2}} v_2 \hat{E} + u_2 \hat{B} + \frac{1}{2} u_0 \hat{C}, \tag{19}$$

where $e_0, \varepsilon_s, \varepsilon_d, c_0, c_2, c_4, v_0, v_2, u_2, u_0$ are the model parameters. The two-boson interaction operators can be presented via first and second order Casimir operators of three subgroup chains (SU_5, SU_3, O_6) of U_6 group:

$$\hat{A}_0 = \frac{1}{5} \left\{\hat{n}_d(\hat{n}_d + 3) - \frac{1}{2} C_2(O_5)\right\}; \tag{20}$$

$$\hat{A}_2 = \frac{1}{7\sqrt{5}} \left\{2\hat{n}_d(\hat{n}_d - 2) + C_2(O_5) - \hat{L}^2\right\}; \tag{21}$$

$$\hat{A}_4 = \frac{1}{210} \left\{36\hat{n}_d(\hat{n}_d - 2) - 3 C_2(O_5) + 10 \hat{L}^2\right\}; \tag{22}$$

$$\hat{B} = \frac{1}{\sqrt{5}}\hat{n}_0 \cdot \hat{n}_d; \tag{23}$$

$$\hat{C} = \hat{n}_0(\hat{n}_0 - 1); \tag{24}$$

$$\hat{D} = \frac{1}{\sqrt{5}}\left\{2\hat{n}_0 \cdot \hat{n}_d + 5\hat{n}_0 + \hat{n}_d + \frac{1}{2}\left(C_2(O_5) - C_2(O_6)\right)\right\}; \tag{25}$$

$$\hat{E} = \frac{3}{4\sqrt{35}}\left\{\hat{Q}^2 - \frac{1}{6}\hat{L}^2 - \frac{1}{3}C_2(O_5) + \frac{2}{3}C_2(O_6) - \frac{1}{3}\left(2\hat{n}_d^2 + 16\hat{n}_0 \cdot \hat{n}_d + 40\hat{n}_0 + 18\hat{n}_d\right)\right\}, \tag{26}$$

where (see[10]) $\hat{Q}^2 = C_2(SU_3) - \hat{L}^2/2$.

In future we shall use the equivalent form of the standard IBM Hamiltonian, written using directly Casimir operators (see Eq.(2.4) in[3]):

$$H^{IBM} = e_0 + e_1\hat{N} + e_2\hat{N}(\hat{N} + 5) + \varepsilon\hat{n}_d + \alpha\hat{n}_d(\hat{n}_d + 4) + \varepsilon_1\hat{N} \cdot \hat{n}_d +$$
$$+ \beta C_2(O_5) + 2\gamma\hat{L}^2 + \delta C_2(SU_3) + \eta C_2(O_6), \tag{27}$$

where

$$C_1(U_6) = \hat{N} = \sum_{\kappa\mu}\eta^{\kappa\mu}\eta^{+\kappa\mu} = \eta^{00}\eta^{+00} + \sum_{\mu}\eta^{2\mu}\eta^{+2\mu} = \hat{n}_0 + \hat{n}_d;$$
$$C_2(U_6) = \hat{N}(\hat{N} + 5); \quad C_2(U_5) = \hat{n}_d(\hat{n}_d + 4); \tag{28}$$
$$C_1(U_5) = \hat{n}_d = \sum_{\mu}\eta^{2\mu}\eta^{+2\mu}; \quad C_2(O_3) = \hat{L}^2.$$

Casimir operators $C_2(O_3)$, $C_2(O_5)$, $C_2(SU_3)$, $C_2(O_6)$ also can be expressed using $\eta^{\kappa\mu}$ and $\eta^{+\kappa\mu}$ combinations. The parameters of this Hamiltonian are related with parameters of Eq.(19) via Eq.(2.6),(2.7) given in[3].

4. The U_6 Restriction of Invariants $F_t(i, l, m)$

The most complicated stage in the studies of the relationships between the GIBM and standard IBM Hamiltonians, given by the formulae Eq.(15) and (27), respectively, is to find the U_6 restricted (scalar) parts $F_t(i, l, m)_{|U_6}$ of terms $F_t(i, l, m)$. For every function of operators $\eta^{\kappa\mu}$ and $\eta^{+\kappa\mu}$, denoted as $f(\eta^{\kappa\mu}, \eta^{+\kappa\mu})$, its restriction to U_6 means the extraction of the part, expressed entirely as a function of U_6 group generators. Therefore each of its monomials must have the same number of operators $\eta^{\kappa\mu}$ and $\eta^{+\kappa\mu}$.

Let us denote the U_6 restricted part of f as "b" and the rest as "a". So we have $f(\ldots) =$ "a" $+$ "b" with "b" $= f(\ldots)_{|U_6}$. It is relevant also, when we have $f(\ldots) = f_1(\ldots) \cdot f_2(\ldots) = (a_1 + b_1) \cdot (a_2 + b_2) = a_1 \cdot a_2 + a_1 \cdot b_2 + b_1 \cdot a_2 + b_1 \cdot b_2$. It is obvious, that $b_1 \cdot b_2$ is a part of "b", $(a_1 \cdot b_2 + b_1 \cdot a_2)$ is a part of "a" and $a_1 \cdot a_2$ may have a part of "a" and a part of "b".

Such an analysis was used in the evaluation of $F_t(i, l, m)_{|U_6}$ terms, entering in Eq.(15), and the results are as follows:

1) For $(\rho^2)^i_{|U_6}$ type terms:

$$\rho^4_{|U_6} = F_2(2,0,0)_{|U_6} = \frac{3}{2}(2\hat{n}_0+1) = 3\left(\hat{N}-\hat{n}_d\right) + \frac{3}{2}, \qquad (29)$$

$$\rho^8_{|U_6} = F_4(4,0,0)_{|U_6} = \frac{27}{4}\left(2\hat{n}_0^2 + 2\hat{n}_0 + 1\right), \qquad (30)$$

$$\rho^{12}_{|U_6} = F_6(6,0,0)_{|U_6} = \frac{135}{8}\left(4\hat{n}_0^3 + 6\hat{n}_0^2 + 8\hat{n}_0 + 3\right). \qquad (31)$$

2) For $(\beta^2)^l_{|U_6}$ type terms:

$$\beta^2_{|U_6} = F_2(0,1,0)_{|U_6} = \hat{n}_d + \frac{5}{2}, \qquad (32)$$

$$\beta^4_{|U_6} = F_4(0,2,0)_{|U_6} = \frac{1}{4}\left\{6\hat{n}_d^2 + 30\hat{n}_d + 35 - C_2(O_5)\right\}, \qquad (33)$$

$$\beta^6_{|U_6} = F_6(0,3,0)_{|U_6} = \frac{1}{8}\left\{\left(20\hat{n}_d^3 + 150\hat{n}_d^2 + 376\hat{n}_d + 315\right) - C_2(O_5)\left(6\hat{n}_d + 15\right)\right\}. \qquad (34)$$

3) For $\left[(\rho^2)^i \times (\beta^2)^l\right]_{|U_6}$ type terms:

$$(\rho^4\beta^2)_{|U_6} = F_4(2,1,0)_{|U_6} = \frac{3}{4}\left\{6\hat{n}_0\cdot\hat{n}_d + 15\hat{n}_0 + 3\hat{n}_d + 5 + \frac{1}{2}\left(C_2(O_5) - C_2(O_6)\right)\right\}, \qquad (35)$$

$$(\rho^8\beta^2)_{|U_6} = F_6(4,1,0)_{|U_6} = \frac{9\sqrt{5}}{4}\left\{(2\hat{n}_0+1)\cdot\hat{D} - [\hat{n}_0, \hat{D}]\right\} +$$
$$+ \frac{27}{8}\left(2\hat{n}_0^2 + 2\hat{n}_0 + 1\right)(2\hat{n}_d + 5), \qquad (36)$$

$$(\rho^4\beta^4)_{|U_6} = F_6(2,2,0)_{|U_6} = \frac{3}{8}\left\{\sqrt{5}\left(\hat{D}(4\hat{n}_d+10) - 2[\hat{n}_0,\hat{D}]\right) +\right.$$
$$\left. + (2\hat{n}_0+1)\left(6\hat{n}_d^2 + 30\hat{n}_d + 35 - C_2(O_5)\right)\right\}, \qquad (37)$$

where $[\hat{A}, \hat{B}]$ denotes the commutator.

4) For the group of most complicated terms, containing SO_3-invariant $\beta^3\cos 3\gamma$:

$$\left(\rho^2\beta^3\cos 3\gamma\right)_{|U_6} = F_4(1,0,1)_{|U_6} = -\frac{3\sqrt{105}}{4\sqrt{2}}\hat{E}, \qquad (38)$$

$$\left(\rho^6\beta^3\cos 3\gamma\right)_{|U_6} = F_6(3,0,1)_{|U_6} = \frac{3\sqrt{105}}{32\sqrt{2}}\left\{-\frac{\sqrt{5}}{2}\left[\hat{E},[\hat{n}_0,\hat{D}]\right] -\right.$$
$$\left. - \sqrt{5}\left[[\hat{n}_0,\hat{E}],\hat{D}\right] + 18(2\hat{n}_0+1)\hat{E} - 18[\hat{n}_0,\hat{E}]\right\}, \qquad (39)$$

$$\left(\rho^2\beta^5\cos 3\gamma\right)_{|U_6} = F_6(1,1,1)_{|U_6} = -\frac{\sqrt{105}}{8\sqrt{2}}\left\{\frac{5\sqrt{5}}{8}\left(\left[\hat{E},[\hat{n}_0,\hat{D}]\right] - 2\left[[\hat{n}_0,\hat{E}],\hat{D}\right]\right) -\right.$$

$$-\frac{5}{2}\left[[\hat{n}_0, \hat{E}], \hat{A}_0\right] + (10\hat{n}_0 + 6\hat{n}_d + 19)\hat{E} - 2[\hat{n}_0, \hat{E}]\right\}, \qquad (40)$$

$$\left(\beta^6(\cos 3\gamma)^2\right)_{|U_6} = F_6(0,0,2)_{|U_6} = \frac{35}{16}\left\{\frac{18}{5}(\hat{n}_0 + 1)\left(-\sqrt{5}\hat{A}_2 + 4\sqrt{5}\hat{B}_2 - 2\hat{n}_d\right) - \right.$$

$$\left. - 9\left[[\hat{n}_0, \hat{E}], \hat{E}\right] + \frac{3}{5}\left(6\sqrt{5}\hat{A}_2 - 12\sqrt{5}\hat{B}_2 + 12\hat{n}_d + 10\right) + \hat{T}\right\}, \qquad (41)$$

where

$$\hat{T} = \frac{2}{3(3\hat{n}_0^2 + 3\hat{n}_0 + 2)}\left\{\frac{3}{5}\left(3\sqrt{5}\hat{A}_2 + 6\hat{n}_d + 10\right) \cdot \hat{n}_0(\hat{n}_0 - 1)(\hat{n}_0 - 2) - \right.$$

$$\left. - \frac{5}{32}\left[\left([\hat{E}, \hat{D}] + \frac{1}{2}[[\hat{n}_0, \hat{E}], [\hat{n}_0, \hat{D}]]\right), \left(\frac{1}{2}\left[\hat{E}, [\hat{n}_0, \hat{D}]\right] + [[\hat{n}_0, \hat{E}], \hat{D}]\right)\right]\right\} \qquad (42)$$

and

$$\hat{B}_2 = \frac{1}{14\sqrt{5}}\left\{4\hat{n}_d^2 + 20\hat{n}_d - 2C_2(O_5) + \hat{L}^2\right\}. \qquad (43)$$

The two-boson interaction operators $\hat{A}_0, \hat{A}_2, \hat{D}, \hat{E}$, used in Eqs.(36)–(42), are the same ones, which enter the standard IBM Hamiltonian Eq.(19).

Let us note that U_6 restricted terms (29)–(33),(35) and (38) for the first time have been evaluated in [3] and our more general approach has confirmed these results.

5. Microscopic Evaluation of Standard IBM Hamiltonian Parameters

In order to obtain the microscopic expressions for the standard IBM Hamiltonian parameters one needs to transform the H^{GIBM} expression Eq.(15) (inserting the explicit expressions Eqs.(29)–(41) for terms $F_t(i, l, m)_{|U_6}$) to the form of Eq.(27), which gives

$$H^{GIBM} = H^{IBM} + H^{GIBM'}, \qquad (44)$$

where the first non-vanishing GIBM term $H^{GIBM'}$ appears only, when one takes into account also the $t = 6$ terms of Eq.(15).

Then, if one compares in Eqs.(44) and (27) the coefficients at corresponding operator expressions ($\hat{N}, \hat{N}(\hat{N} + 5), \ldots$), one obtains:

$$e_0 = \frac{3}{2}C_k + C_p(b_0 R_0 + 30 b_2 R_2 + 11340 b_4 R_4 + 162162000 b_6 R_6);$$

$$e_1 = \frac{1}{2}C_k + C_p(20 b_2 R_2 + 10800 b_4 R_4 - 517348000 b_6 R_6);$$

$$e_2 = 120\, C_p(21 b_4 R_4 + 137150 b_6 R_6);$$

$$\varepsilon = -4\, C_p(3 b_2 R_2 - 1314 b_4 R_4 + 980840 b_6 R_6); \qquad (45)$$

$$\alpha = -120\, C_p(39 b_4 R_4 + 27110 b_6 R_6); \qquad \varepsilon_1 = 48\, C_p(69 b_4 R_4 - 218950 b_6 R_6);$$

$$\beta = 24\, C_p(21 b_4 R_4 + 27820 b_6 R_6); \qquad \gamma = 48\, C_p(3 b_4 R_4 - 2555 b_6 R_6);$$

$$\delta = -144\, C_p(3 b_4 R_4 - 2795 b_6 R_6); \qquad \eta = -72\, C_p(11 b_4 R_4 + 13910 b_6 R_6).$$

In our analysis we have omitted the term $\hat{T}$, given by Eq.(42), since in first approximation it refers only to $H^{GIBM'}$ term of Eq.(44).

6. Conclusions

Our study of microscopic generalization of Interacting Boson Model and evaluation of standard IBM parameters continues the investigations started in[3,4]. In comparison with earlier studies[2,3], we take into account the kinetic energy term of GIBM and include the $t = 6$ terms of the potential energy expansion for multipole-Gauss type collective NN potential. It allows us to obtain first terms, which do not enter in the standard IBM Hamiltonian.

The extraction of collective degrees of freedom in Restricted Dynamics Approach allows to introduce in GIBM the spin-isospin variables. Therefore, it is possible to take into account explicitly all exchange components of the central effective NN potential, as it was done in our microscopic evaluation of the standard IBM Hamiltonian parameters.

The important problem of such an analysis, which appears starting with 6th order terms, is an unambiguousity of standard IBM parameter evaluation procedure. It can be solved only, when one chooses the basis of higher (third) order invariants (e.g. $\hat{N}^3$, $\hat{n}_d^3$) of GIBM. Therefore, our results can be regarded only as the first step in the investigation of GIBM, microscopically derived from the effective NN-potential.

7. Acknowledgements

The research described in this publication was made possible in part by Grant No.LBD000 from the International Science Foundation.

8. References

1. M. Moshinsky, *Nucl.Phys.* **A338** (1980) 156.
2. V.Vanagas,*Bulg.J. of Phys.* **9** (1982) 231.
3. V.Vanagas,*Lietuvos fizikos rinkinys* **31** (1991) 545.
4. J.Ruza et al., in *Proceedings of the 8th International Symposium on Capture Gamma-Ray Spectroscopy and Related Topics. Fribourg, Switzerland, 20–24 September 1993*, ed. J.Kern (World Scientific, 1994) p. 289.
5. V.Vanagas, *Algebraic Foundations of the Microscopic Nuclear Theory* (Nauka, Moscow, 1988) (in Russian).
6. O.Katkevicius, *Lietuvos fizikos rinkinys* **32** (1992) 220 (in Russian).
7. V.Vanagas, *Algebraic Methods in Nuclear Theory* (Mintis, Vilnius, 1971) (in Russian).
8. L.Sabaliauskas et al., *Izvestiya AN SSSR, ser.fiz.* **52** (1988) 838 (in Russian).
9. A.Arima, F.Iachello, *Ann. of. Phys.* **99** (1976) 253.
10. O.Castanos et al., *J. Math. Phys.* **20** (1979) 35.

265

SYMMETRIES IN ODD-MASS NUCLEI AND THEIR APPLICATIONS

P. VAN ISACKER
GANIL, BP 5027, F-14021 Caen Cedex, France

ABSTRACT

Some aspects of the application of symmetries in odd-mass nuclei are reviewed in the context of the interacting boson–fermion model. Results concerning scissors states in odd-mass nuclei are presented with reference to a recent nuclear resonance fluorescence experiment on ^{163}Dy. The observed dipole strength in this nucleus around 3 MeV is consistent with a scissors-like interpretation.

1. The Role of Symmetry

The word 'symmetry' is derived from the greek expression $\sigma \upsilon \nu \ \mu \epsilon \tau \rho o \nu$, meaning 'with proportion' or 'with order'. In modern theories of physics it has acquired a definition much more precise than that in antiquity but the general idea of seeking to order physical phenomena still remains.

Symmetries and the formal theory of them—group theory—have been applied to nuclear structure problems at numerous occasions, starting with the pioneering work of Wigner on the spin–isospin invariance in light nuclei.[1] Subsequent high points were the studies of Racah on the classification of complex spectra[2] and of Elliott on rotational properties of nuclei[3] and their connection with SU(3). In each of these applications the existence of a symmetry implies certain regularities in predicted nuclear properties, most notably the energies. In other words, a symmetry provides an ordering criterion of our observations and, as such, its meaning still covers the one it had in ancient Greece. However, in the context of quantum mechanics a symmetry means much more than that, and most importantly, it implies the existence of 'good' quantum numbers, which are the quantum extension of the classical classical concept of conserved quantities.

During the last two decades the use of symmetry-based considerations in nuclear physics has received a renewed impetus with the advent of the interacting boson model of Arima and Iachello.[4] It is certainly not the only group-theoretical model in nuclear physics that has been developed over this period but, given its simplicity and elegance, it has become the most popular. One of the appealing aspects of the interacting boson model is that it can be readily extended to deal with more complex phenomena that lie outside the scope of the model in its simplest version. One particularly important extension has been towards odd-mass nuclei, achieved by considering, in addition to the boson core, a fermion coupled to the core through an appropriate boson–fermion interaction[5,6]. The resulting interacting boson–fermion model lends itself very well to a study based on symmetry considerations whereby certain classes of model hamiltonians can be solved analytically. Again this type of study has two aspects. The first is that these solvable hamiltonians provide simple energy formulas through which one may attempt to classify ('to order') observed spectra of odd-mass nuclei. Secondly, and arguably of more importance, the symmetry hamiltonians provide quantum numbers, the goodness of which can be tested—albeit indirectly—through the existence of associated selection rules in transition and reaction rates. If there is one criticism to be uttered of the methodology employed in this field of research, it is that all too often classifications have been proposed solely on the basis of energy fits, without due consideration being given to the quantum numbers involved.

It is not intented here to give a comprehensive review of the symmetries of the interacting boson–fermion model. In fact, such an overview makes up the bulk of a recent monograph treating the interacting boson–fermion model[7] and as such, it is clearly outside the scope of this contribution. Rather, the general procedure is illustrated with one example, ^{195}Pt. Next, some recent results are discussed on mixed-symmetry states in odd-mass nuclei and their interpretation in terms of the interacting boson–fermion model.

2. An Example: ^{195}Pt

The nucleus ^{195}Pt is situated in a spherical-to-deformed transitional region. Its core, ^{196}Pt, can be reasonably well described as an $O(6)$ nucleus; the natural-parity single particle orbits for the neutrons, dominant in the low-energy spectrum, are $p_{1/2}$, $p_{3/2}$ and $f_{5/2}$, which can be decomposed in the pseudo-orbital angular momenta 0 and 2, and pseudo-spin $1/2$. (For a review of pseudo-spin, including relevant references, see for example Ref. [8].)

Because of the similarity in the angular momentum structure of the bosons and of the pseudo-orbital part of the fermion, an analytical decomposition is possible.[9] The corresponding hamiltonian contains five parameters which must be adjusted to reproduce the experimental[10,11] energies. The resulting fit is shown in Fig. 1, which also gives the various quantum numbers that are associated with the different levels. (Sect. 3.2.3 of Ref. [7] includes a thorough discussion of this classification and its quantum numbers.)

One should not underestimate the difficulty of obtaining such level of agreement, even with five parameters: all states are calculated at about the correct energy and for each spin the correct number of levels is found. Nevertheless, with five parameters there is considerable freedom in changing the level energies and hence the following question arises: How do we know that, in this fit, the correct theoretical quantum numbers are associated with the various experimental levels? The answer is that we cannot know as long as we do not involve properties other than level energies.

The question of assigning quantum numbers to observed levels is closely related to the goodness of these quantum numbers, which can be probed by means of selection rules in electromagnetic transition and transfer reaction rates. An example is given in Table 1, where 25 measured[10,11] $B(E2)$ values in ^{195}Pt are compared with the predictions of the symmetry classification for this nucleus. The level assignment is identical to the one in Fig. 1 but, for simplicity, levels are characterised by their energy and spin, and not by the full set of quantum numbers.

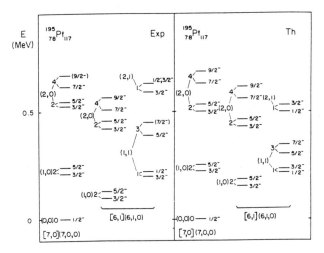

Fig. 1. The experimental spectrum of negative-parity states in ^{195}Pt compared to an analytical calculation with a symmetry hamiltonian of the interacting boson–fermion model.

The results shown in the table do not depend on the parameters entering the hamiltonian. This is so because, whatever the values of these parameters, the wave function of a given state invariably is characterized by the same set of quantum numbers. As a result, the calculated transition rates only depend on the parameters in the transition operator and, in the case of Table 1, on a single parameter in the E2 transition operator. The selection rules in the E2 transitions are clear from the table: several of the transitions are predicted to be zero. One generally finds that the corresponding measured values are small indeed (typically, a factor 10 smaller than the large allowed ones) confirming the approximate validity of the quantum numbers in this symmetry scheme. There is a single exception to this agreement: the 420-to-99 keV transition is predicted to be large but observed to be small, casting doubt on the classification of the 420 keV $3/2^-$ level.

Results of similar quality are obtained[12] for intensities of one-neutron transfer reactions starting from and leading to ^{195}Pt and confirm the proposed assignment of quantum numbers. More recently this symmetry scheme was used to calculate magnetic dipole moments in ^{195}Pt and again agreement with the data was found.[13]

In the next section a recent application of the interacting boson–fermion model is discussed, which concerns mixed-symmetry states in odd-mass deformed nuclei.

3. Mixed-Symmetry States in Odd-Mass Nuclei

Given that the occurrence of mixed-symmetry or 'scissors' states in doubly-even nuclei is by now an established experimental fact,[14] it is natural to ask whether these excitations have a counterpart in odd-mass spectra. This problem was analysed from

Table 1. E2 transition rates in ^{195}Pt

E_i^a	J_i	E_f^a	J_f	$B(E2; J_i \to J_f)^b$		E_i^a	J_i	E_f^a	J_f	$B(E2; J_i \to J_f)^b$	
				Expt	Calc					Expt	Calc
211	3/2	0	1/2	0.190(10)	0.179	667	9/2	239	5/2	0.200(40)	0.239
239	5/2	0	1/2	0.170(10)	0.179	563	9/2	239	5/2	0.091(22)	0.022
525	3/2	0	1/2	0.017(1)	0	239	5/2	99	3/2	0.060(20)	0
544	5/2	0	1/2	0.008(4)	0	525	3/2	99	3/2	≤ 0.033	0.007
99	3/2	0	1/2	0.038(6)	0.035	613	7/2	99	3/2	0.005(3)	0.009
130	5/2	0	1/2	0.066(4)	0.035	420	3/2	99	3/2	0.005(4)	0.177
420	3/2	0	1/2	0.015(1)	0	508	7/2	99	3/2	0.240(50)	0.228
455	5/2	0	1/2	≤ 0.00004	0	389	5/2	99	3/2	0.200(70)	0.219
199	3/2	0	1/2	0.025(2)	0	525	3/2	130	5/2	0.009(5)	0.003
389	5/2	0	1/2	0.007(1)	0	667	9/2	130	5/2	0.012(3)	0.010
613	7/2	211	3/2	0.170(70)	0.215	563	9/2	130	5/2	0.240(40)	0.253
508	7/2	211	3/2	0.055(17)	0.020	389	5/2	130	5/2	≤ 0.014	0.055
525	3/2	239	5/2	≤ 0.019	0.072						

^{a} In units of keV.
^{b} In units of e^2b^2.

a theoretical point of view, first by considering a specific combination of single-particle orbits available to the odd fermion that allows for an analytic solution,[15] and next by coupling a single orbit to a deformed core.[16] The latter situation, in a slightly modified form, is the one relevant for a comparison with recently obtained data[17] in ^{163}Dy and will be discussed here.

The core of a deformed odd-mass nucleus can be described in terms of bosons by taking the SU(3) limit of the interacting boson model. Since it is intended to study scissors states of the odd-mass nucleus, it is essential to distinguish between neutrons and protons and hence to adopt a neutron–proton interacting boson model description of the core. To a good approximation it can be assumed that the core states are characterized by definite values of F spin,[18] which is equivalent to labelling them by the U(6) representation $[N - f, f]$, where $N = N_\nu + N_\pi$ is the total number of (neutron and proton) bosons and $f = \frac{1}{2}N - F$. A classification of the odd-mass nucleus, which assumes a weak coupling only between the particle and the core, can now be written down readily,

$$\begin{pmatrix} U^B(6) & \supset & SU^B(3) & \supset & O^B(3) \end{pmatrix}$$
$$\downarrow \qquad \quad \downarrow \qquad \qquad \downarrow$$
$$[N-f,f] \qquad (\lambda,\mu) \qquad \quad \chi L$$

$$\otimes \begin{pmatrix} Sp^F(2j+1) & \supset & SU^F(2) \end{pmatrix} \supset SU^{BF}(2) \qquad , \quad (1)$$
$$\downarrow \qquad \qquad \qquad \downarrow \qquad \qquad \quad \downarrow$$
$$[1] \qquad \qquad \qquad j \qquad \qquad \quad J$$

where χ is a missing label in the SU(3) $\supset$ O(3) reduction. It is (approximately) equivalent to K_L, the projection of the core angular momentum L on the intrinsic axis of symmetry,[3] and that is how it henceforth will be denoted. Weak-coupling states are thus written as

$$|[N-f,f](\lambda,\mu)K_L L,j;JM_J\rangle. \qquad (2)$$

The basis (2) is not very suitable for deformed nuclei since they require a strong particle–core coupling. In that case the odd-mass eigenstates acquire the form

$$|[N-f,f](\lambda,\mu)K_L,j,K=K_j\pm K_L JM_J\rangle$$
$$= \sum_L \sqrt{2L+1}[1+(-1)^L\delta_{K_L 0}]^{1/2} \begin{pmatrix} L & j & J \\ -K_L & \mp K+K_L & \pm K \end{pmatrix}$$
$$\times |[N-f,f](\lambda,\mu)K_L L,j;JM_J\rangle, \qquad (3)$$

where K_j and K are the projections of j and J, respectively, on the axis of symmetry and are conserved quantities in the strong-coupling basis. The expression (3) is familiar from the strong-coupling adiabatic model of Bohr and Mottelson[19] and arises from the assumption that the particle moves adiabatically in a deformed potential defined with respect to an intrinsic coordinate system.

It is not possible to give an algebraic classification (in the laboratory frame) yielding the states (3) nor can an exact analytic expression be obtained for the energies (although the latter can be done in the large-N limit[20]). This, however, is only of secondary importance; more crucial is the occurrence in (3) of quantum numbers (such as K) which again can be tested with selection rules. On the other hand, it is possible to construct a hamiltonian which has the states (2) and (3) as eigenstates for two limiting choices of its parameters. Apart from the standard SU(3) hamiltonian to describe the boson core, it contains the boson–fermion interaction

$$\hat{V}^{\mathrm{BF}} = \alpha \hat{Q}^{\mathrm{B}} \cdot \hat{Q}^{\mathrm{F}} + \beta \hat{L}^2, \qquad (4)$$

where $\hat{Q}^{\mathrm{B}}$ ($\hat{Q}^{\mathrm{F}}$) is the boson (fermion) quadrupole operator and $\hat{L}$ is the boson angular momentum operator. For $\alpha \ll \beta$ the weak-coupling basis (1) is obtained, while the strong-coupling limit (3) arises for $\alpha \gg \beta$.[21]

It is now straightforward to compute electromagnetic transition probabilities using the standard transition operators of the neutron–proton interacting boson model.[4] Of particular interest for the excitation of scissors states are the expressions for the M1 transition matrix elements.[16] The results are summarised in Fig. 2, where the $B(\mathrm{M1};K_i J_i \to K_f J_f)$ values are shown, from a ground state with $K_i = J_i$ to several mixed-symmetry states with different combinations of K_f and J_f. The $B(\mathrm{M1})$ values depend only weakly on the specific j value used for the single-particle orbit, which in the figure is taken as $j = 13/2$. Furthermore, the boson numbers are $N_\nu = 7$ and $N_\pi = 8$ and the boson g factors $g_\nu = 0.05~\mu_{\mathrm{N}}$ and $g_\pi = 0.65~\mu_{\mathrm{N}}$, obtained from a fit to magnetic dipole moments of 2_1^+ states in rare-earth nuclei[22] and appropriate for the

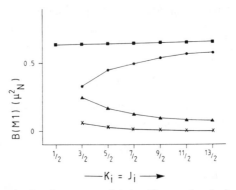

Fig. 2. B(M1;$K_i J_i \to K_f J_f$) values from a ground state with $K_i = J_i$ to mixed-symmetry states with $K_f = K_i-1, J_f = J_i-1$ (circles), $K_f = K_i-1, J_f = J_i$ (triangles), $K_f = K_i-1, J_f = J_i+1$ (crosses) and $K_f = K_i+1, J_f = J_i+1$ (squares). The B(M1) values are calculated for $j = 13/2$ with boson numbers and boson g-factors as discussed in the text.

nucleus ^{163}Dy, discussed below. Other cases can be easily obtained, since the B(M1) values scale as

$$(g_\nu - g_\pi)^2 \frac{N+1}{N(2N-1)} N_\nu N_\pi. \qquad (5)$$

A first experiment searching for M1 scissors strength in ^{165}Ho was described by Huxel et al.[23]; no transition with $B(M1)\uparrow \geq 0.1\mu_N^2$ could be detected in the energy range around 3 MeV. In a subsequent nuclear resonance fluorescence experiment[17] the nucleus ^{163}Dy was chosen as a first candidate since the neighbouring even–even nuclei ^{162}Dy and ^{164}Dy are well investigated.[24] In both isotopes the orbital M1 strength is concentrated in two or three strong transitions and in ^{164}Dy the M1 strength is the largest of all rare-earth nuclei.

Figure 3 shows a comparison of the transition strengths observed in ^{163}Dy with previous data[24] for ^{160,162,164}Dy. Due to the unknown J_f in the case of ^{163}Dy the quantity $g \cdot \Gamma_0$ is plotted, where Γ_0 is the transition width and g the factor $(2J_f + 1)/(2J_i+1)$ which amounts to $2/3$, 1 and $4/3$ for spins $J_f = 3/2$, $5/2$ and $7/2$, respectively. There is a clear concentration of dipole strength in ^{163}Dy near 3 MeV which fits nicely into the systematics of the even Dy isotopes, where the corresponding peaks are claimed to have a scissors-like character.[24,25,26]

The single-j analysis cannot be applied as it stands to ^{163}Dy, but some simple assumptions allow its generalization. The two dominant neutron orbits $f_{7/2}$ and $h_{9/2}$ in ^{163}Dy can be considered as pseudo-spin partner orbits, that is, with $j = \tilde{l} \pm 1/2$ where $\tilde{l}$ is the pseudo-orbital angular momentum of the odd particle ($\tilde{l} = 4$ in ^{163}Dy). The resulting expressions for wave functions and M1 matrix elements are then a straightforward generalisation of the ones obtained in the single-j case.

The results for ^{163}Dy are presented in Table 2. In the left half of the table the three states are listed which have largest $B(M1)\uparrow$ values; all other states are excited with significantly smaller strengths. The three states that are appreciably excited

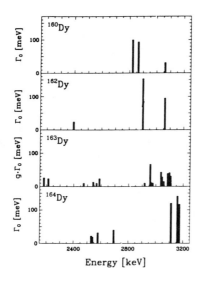

Fig. 3. Dipole strength distribution in ^{163}Dy and in neighbouring even–even Dy isotopes.

Table 2. Calculated M1 excitation and decay of mixed-symmetry (ms) states in ^{163}Dy

J_i	J_f	$B(M1; J_i \to J_f)$ (μ_N^2)			J_i	J_f	$B(M1; J_i \to J_f)$ (μ_N^2)		
		$l=4$	$j=7/2$	$j=9/2$			$l=4$	$j=7/2$	$j=9/2$
$5/2_1$	$3/2_{ms}$	0.41	0.45	0.45	$3/2_{ms}$	$5/2_1$	0.61	0.68	0.68
$5/2_1$	$5/2_{ms}$	0.20	0.18	0.18	$3/2_{ms}$	$7/2_1$	0.00	0.00	0.00
$5/2_1$	$7/2_{ms}$	0.62	0.66	0.65	$5/2_{ms}$	$5/2_1$	0.20	0.18	0.18
					$5/2_{ms}$	$7/2_1$	0.44	0.29	0.50
					$5/2_{ms}$	$9/2_1$	0.00	0.00	0.00
					$7/2_{ms}$	$5/2_1$	0.47	0.49	0.49
					$7/2_{ms}$	$7/2_1$	0.17	0.09	0.16
					$7/2_{ms}$	$9/2_1$	0.02	0.02	0.02

have spins $J = 7/2$, $3/2$ and $5/2$ (in order of decreasing strength). Their decay into the ground-state band is given in the right half of Table 2. Also shown, in the columns $j = 7/2$ and $j = 9/2$, are the results of calculations in which only one single-particle orbit (either $f_{7/2}$ or $h_{9/2}$) is strongly coupled to the core. The M1 excitation results do not differ significantly from each other or from those for $\tilde{l} = 4$; the decay, however, is more sensitively dependent on the single-particle j and/or the coupling scheme. On the basis of these results one may attempt an interpretation of some of the observed states. For example, the 2958 keV level is strongly M1 excited (relative to other levels) and has an M1 branching ratio of 0.23; both features are in qualitative agreement with the calculated $J = 7/2$ scissors mode state.

It should be emphasised that the E1 character of the transitions observed in ^{163}Dy cannot be ruled out on experimental grounds. In the neighbouring even–even isotopes, however, the positive parities of the levels around 3 MeV are deduced from electron scattering experiments[25] in the case of ^{164}Dy and for ^{162,164}Dy from photon linear polarization measurements.[26] Given the smooth variation of the energy of these levels as a function of neutron number, this strongly suggests an M1 character of the transitions to the 3 MeV levels in ^{163}Dy. The situation is less clear for the other levels in ^{163}Dy observed around 2.2 and 2.5 MeV. For example, the latter might be related to the 2.5 MeV levels in ^{162}Dy (not shown in Fig. 3, see Ref. [26]), in which case the associated transitions would have E1 character.

It is not clear as yet whether these strong M1 excitations are as common a phenomenon in odd-mass nuclei as they are in even–even isotopes. Evidence from two very recent experiments[27] is mixed: while dipole states are clearly seen in ^{161}Dy and fit into the systematics of neighbouring Dy isotopes, no strong M1's are found in ^{157}Gd. The explanation of the isotope dependence of the fragmentation is clearly outside the scope of the schematic model presented here and calls for a more microscopic treatment.

4. Concluding Remark

It might seem that there is little connection between the approach sketched in Sect. 2 and the one presented in Sect. 3. There are certainly differences. The former uses the by now standard method of defining a classification starting from a chain of algebras, and subsequently deriving an energy formula, the wave functions, expressions for matrix elements of various observables and corresponding selection rules. In the approach of Sect. 3, there is no chain of algebras and no exact energy formula. However, the features of defining a set of good quantum numbers and deriving the associated selection rule are still present. These features represent the essential characteristics of a (dynamical) symmetry and as a consequence, the strong-coupling scheme discussed in Sect. 3 can be regarded as a genuine symmetry classification.

5. Acknowledgements

I wish on this occasion to thank Franco Iachello for the encouragement and support he has given me over many years; it has been a privilege for me to work with him. Many thanks are also due to Alejandro Frank and Pepe Arias with whom the work on odd-mass scissors states was done. The latter work has benefited greatly from a collaboration with several experimental groups; I wish to thank all those involved and especially Ulrich Kneissl.

6. References

1. E. P. Wigner, Phys. Rev. **51** (1937) 106.
2. G. Racah, Phys. Rev. **61** (1942) 186; **62** (1942) 438; **63** (1943) 367; **76** (1949) 1352.
3. J. P. Elliott, Proc. Roy. Soc. A **245** (1958) 128; **245** (1958) 562.
4. F. Iachello and A. Arima, *The Interacting Boson Model* (Cambridge University Press, Cambridge, 1987).
5. A. Arima and F. Iachello, Phys. Rev. Lett. **35** (1975) 1069.
6. F. Iachello and O. Scholten, Phys. Rev. Lett. **43** (1979) 679.
7. F. Iachello and P. Van Isacker, *The Interacting Boson-Fermion Model* (Cambridge University Press, Cambridge, 1991).
8. P. Van Isacker and D. D. Warner, J. Phys. G, to be published.
9. A. B. Balantekin, I. Bars, R. Bijker and F. Iachello, Phys. Rev. C **27** (1983) 1761.
10. A. M. Bruce, W. Gelletly, J. Lukasiak, W. R. Phillips and D. D. Warner, Phys. Lett. B **165** (1985) 43.
11. A. Mauthofer, K. Stelzer, J. Gerl, Th. W. Elze, Th. Happ, G. Eckert, T. Faestermann, A. Frank and P. Van Isacker, Phys. Rev. C **34** (1986) 1958.
12. M. Vergnes, G. Berrier-Ronsin and G. Rotbard, Phys. Rev. C **36** (1987) 1218.
13. S. Kuyucak and A. E. Stuchbery, Phys. Rev. C **48** (1993) R13.
14. A. Richter, in *Proceedings of the International Conference on Contemporary Topics in Nuclear Structure Physics*, ed. by R. F. Casten, A. Frank, M. Moshinsky and S. Pittel (World Scientific, Singapore, 1988), p. 127.
15. P. Van Isacker and A. Frank, Phys. Lett. B **225** (1989) 1.
16. A. Frank, J. M. Arias, and P. Van Isacker, Nucl. Phys. A **531** (1991) 125.
17. I. Bauske, J. M. Arias, P. von Brentano, H. Friedrichs, R. D. Heil, R.-D. Herzberg, F. Hoyler, P. Van Isacker, U. Kneissl, J. Margraf, H. H. Pitz, C. Wesselborg and A. Zilges, Phys. Rev. Lett. **71** (1993) 975.
18. T. Otsuka, A. Arima, F. Iachello and I. Talmi, Phys. Lett. B **76** (1978) 139.
19. A. Bohr and B. R. Mottelson, *Nuclear Structure. II Nuclear Deformations* (Benjamin, New York, 1975).
20. R. Bijker and V. K. B. Kota, Ann. Phys. (NY) **187** (1988) 148.
21. D. M. Brink, B. Buck, R. Huby, M. A. Nagarajan and N. Rowley, J. Phys. G

13 (1987) 629.
22. A. Wolf, D. D. Warner and N. Benczer-Koller, Phys. Lett. B **158** (1985) 7.
23. N. Huxel, W. Ahner, H. Diesener, P. von Neumann-Cosel, C. Rangacharyulu, A. Richter, C. Spieler, W. Ziegler, C. De Coster and K. Heyde, Nucl. Phys. A **539** (1992) 478.
24. C. Wesselborg, P. von Brentano, K. O. Zell, R. D. Heil, H. H. Pitz, U. E. P. Berg, U. Kneissl, S. Lindenstruth, U. Seemann and R. Stock, Phys. Lett. B **207** (1988) 22.
25. D. Bohle, G. Küchler, A. Richter and W. Steffen, Phys. Lett. B **148** (1984) 260.
26. H. Friedrichs, B. Schlitt, J. Margraf, S. Lindenstruth, C. Wesselborg, R. D. Heil, H. H. Pitz, U. Kneissl, P. von Brentano, R. D. Herzberg, A. Zilges, D. Häger, G. Müller and M. Schumacher, Phys. Rev. C **45** (1992) R892.
27. U. Kneissl, private communication.

LOW-LYING STATES IN ODD-A Ce ISOTOPES :
EXPERIMENTAL RESULTS AND DESCRIPTION BY THE IBFM-1 MODEL

Jean GIZON
Institut des Sciences Nucléaires, IN2P3-CNRS
Université Joseph Fourier, F-38026 Grenoble, France

ABSTRACT

The present state of our knowledge of low-lying states in odd-A Ce nuclei is reviewed. The experimental methods employed in decay studies are described and recent results shown. Levels in $^{125-135}$Ce are discussed in terms of collective bands and interpreted within the multishell IBFM-1 model.

1. Introduction

An inspection of a Nilsson diagram shows that many orbitals are available for transitional nuclei in the A ~ 130 region. The structure of odd-A, even-Z nuclei from Xe to Nd is dominated at moderate deformation and low spin by the $h_{11/2}$ and several positive parity orbitals. These latter orbitals, namely $s_{1/2}$, $d_{3/2}$, $d_{5/2}$, $g_{7/2}$, mix strongly with each other producing complex level structures.

The first evidence of the $h_{11/2}$ neutron-hole band[1] comes from in-beam measurements on ^{133,135}Ce and ^{135,137}Nd. Then a $\Delta I = 1$ positive parity band generated from the $g_{7/2}$ shell was observed first in ^{129}Ba (ref.[2]). The levels of these two band structures are strongly populated in heavy-ion induced fusion-evaporation reactions and are easily observed. On the contrary, 1/2+ and 3/2+ low-spin bands are weakly fed in this type of reactions. They are unknown or poorly known in most of these nuclei. To observe them, one must employ light-ion beams in in-beam γ-ray spectroscopy or study radioactive decays.

A research programme has been especially developed at ISN, Grenoble to study low-lying states of very neutron-deficient nuclei in the above mentioned transitional region, these states being populated by radioactive decay.

In the first part of the lecture, I describe experimental techniques used. Then I present some experimental results for odd-A Ce nuclei and discuss the main collective bands. The last part of the lecture is devoted to calculations made with the IBFM-1 model insisting on the mixed character of the positive parity states.

2. Techniques employed in decay studies

Experiments were performed to investigate the structure of neutron deficient Pr, Ce and La isotopes from the β/EC decays of the Nd → Pr → Ce → La radioactive chains[3]. Neodymium to cerium precursors were produced bombarding thin (1-3 mg/cm^2)

self supporting isotopically enriched targets of ^{94}Mo and ^{96}Mo with 5-6 MeV/nucleon ^{40}Ca or ^{37}Cl beams from the SARA accelerator in Grenoble.

When needed, the mass identification of the nuclei is made by means of the on-line isotope separator coupled to a He-jet ion-source system or an ion-guide. A full description of the experimental set-up is given in ref.[4]. The mass separated ions are transported from the collecting point of the magnet to a low background site at 6 meters, through seven double Einzel lenses. The activity is transferred in front of the detectors by means of a flexible automatic tape driver system.

The level schemes were obtained from several types of measurements using the He-jet system alone : γ-ray singles multianalysis, internal conversion electron spectra and γ-X-t, γ-γ-t and γ-e⁻ coincidences. The conversion electrons were detected by means of the ISOCELE electron selector from Orsay and an electron guide designed recently.

3. The level structure of odd-A neutron-deficient Ce nuclei

In the last years many experiments were devoted to the identification of low-lying states in the Ce (Z = 58) isotopic chain. In this section, I show only two examples which are selected to illustrate the type of results we obtain : i) ^{131}Ce for which a complex level structure was established and ii) ^{127}Ce which is the most neutron-deficient Ce isotope for which new results are becoming available.

3.1. The nucleus ^{131}Ce

Two high spin bands based on 7/2$^+$ and 9/2$^-$ states, respectively were known from in-beam γ-ray studies[5]. From experiments performed to study the low-energy states, two separated groups of levels (fig. 1) without any connections are found. They are based on a 1/2$^+$ level and on the 7/2$^+$ ground state (g.s.), respectively. Varying collecting - and counting-times of the mother nuclei activity, these two groups have been associated with the decay of the 3/2$^+$ g.s. ($T_{1/2}$ = 95 s) and 11/2$^-$ isomer ($T_{1/2}$ = 5.7 s [ref.[6]]) of ^{131}Pr, respectively.

In addition to the 9/2$^-$, 11/2$^-$, 13/2$^-$ yrast levels in the negative parity system generated by the coupling of a $h_{11/2}$ neutron-hole to a prolate core, other levels are observed, for example (7/2$^-$), (9/2$^-$), (11/2$^-$), (13/2$^-$) states which are very weakly populated in fusion-evaporation heavy-ion induced reactions.

The ΔI=1 bands based on 1/2$^+$ and 3/2$^+$ states result from the $s_{1/2}$ and $d_{3/2}$ orbitals. Their connections to the $g_{7/2}$ g.s. band were not seen in the decay studies[4] and their level energies were unknown. Recently it has been possible in an in-beam γ-ray study made with the multidetector EurogamI to find links between this 1/2$^+$ band and a (19/2$^+$) three quasiparticle band lying at high excitation energy[7]. This fixes the energy of the lowest 1/2$^+$ level 61 keV above the ground state. The 11/2$^+$→ 7/2$^+$ and 15/2$^+$→ 11/2$^+$ transitions are found to have energies of 508 and 623 keV, respectively.

In ^{131}Ce, many low-lying states are established including four well defined bands. It is interesting to note that the whole information is obtained combining results from both β-decay and in beam γ-ray studies.

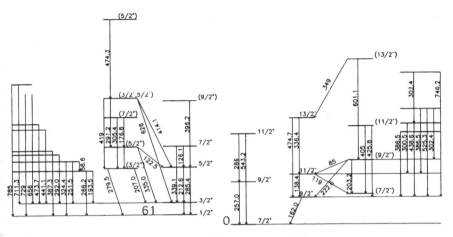

Fig. 1 : Level scheme of ^{131}Ce deduced from ^{131}Pr decay.

3.2. The nucleus ^{127}Ce.

Two high-spin bands were known from previous in-beam studies[8]. They were assigned to ^{127}Ce on the basis of identification of γ-rays made with the recoil separator at Daresbury[9] and also with the He-jet plus on-line isotope separator at Grenoble. In the latter experiment, the internal conversion coefficients of the 126 and 160 keV lines were measured. These transitions which have a M1 multipolarity were placed at the bottom of the $h_{11/2}$ and $g_{7/2}$ neutron-hole structures, respectively (fig. 2).

Only the three lowest levels of the $\nu h_{11/2}$ band are populated from the ^{127}Pr decay. The rest of the feeding is divided between two positive parity bands : 11 % of the decay goes to the four lowest levels of the 5/2$^+$ band and 57 % to a new band whose level spacings and the staggering are typical of the 1/2$^+$ band structure observed in heavier isotopes.

The absolute excitation energies of the three bands are undetermined. However the energy difference between the 7/2$^-$ and 5/2$^+$ base states is small and estimated to be less than 20 keV. Indeed there is no E1 transition with energy greater than 20 keV showing up in γ and electron spectra.

3.3. Collective band structures

Experimental features of collective bands are given in this subsection and the main trends described along the isotopic chain.

The $h_{11/2}$ band structure is produced by the coupling of a $h_{11/2}$ quasi-neutron hole to an even-even triaxial core. It is known for all isotopes from ^{137}Ce to ^{125}Ce. The base states changes from $I^\pi = 11/2^-$ to 9/2$^-$ and 7/2$^-$ with decreasing mass i.e. with the penetration of the Fermi level into the $h_{11/2}$ shell. The Ce isotopes and their neighbours are characterized by triaxiality which decreases when going away from the Z = 82 shell

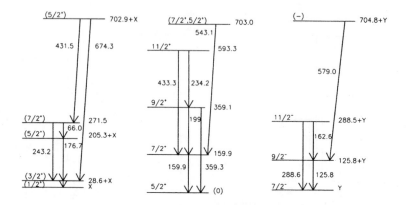

Fig. 2 : Levels in ^{127}Ce observed after β-decay of ^{127}Pr.

closure. This is clearly proved by the reduction of the signature splitting between the favoured ($\alpha = -1/2$) and unfavoured ($\alpha = +1/2$) levels, e.g. from ~ 250 keV for ^{135}Ce to ~ 25 keV for ^{125}Ce at 0.25 MeV rotational frequency. In fact the total $h_{11/2}$ band structure is much more complex than the main $\Delta I = 1$ spin sequence generally observed. Indeed, the triaxiality manifests itself by the existence of several level subsystems. This has been shown in the A ~ 130 region for the first time[2] in ^{129}Ba.

The bands based on 7/2+ and 5/2+ states in $^{125-131}$Ce have no signature splitting. They are apparently generated from the $g_{7/2}$ subshell but a detailed analysis shows that their level configurations are mixed $g_{7/2}$ and $d_{5/2}$.

The remaining positive parity levels found in ^{129,131}Ce constitute a rather complicated structure which can be arranged in several groups. For example two apparent bands based on 1/2+ and 3/2+ states show up in ^{131}Ce with many connections (fig. 1). This so-called 1/2 band exists also in $^{127, 129, 133}$Ce (fig. 3) with signature splitting increasing drastically from ^{133}Ce to ^{127}Ce. If useful for understanding or organizing the level structure itself, this simplifying classification into apparent bands does not hold when considering the intrinsic nature of these states which are extremely mixed (see section 4.2).

4. IBFM-1 model applied to light Ce nuclei

Theoretical works within geometric and algebraic models are rather numerous in the A ~ 130 transitional region. Successful descriptions of the La, Ba, Cs, Xe isotopes were obtained within the IBFM-1 (ref.[11]) and IBFM-2 (ref.[12]) models. Except for the calculations especially made to describe the 3 qp band above the backbending[13,14] within the broken pair model[15], the odd-A Ce have never been analyzed in detail so far. In this section, I present results concerning the collective bands and configuration mixings of the Ce isotopes.

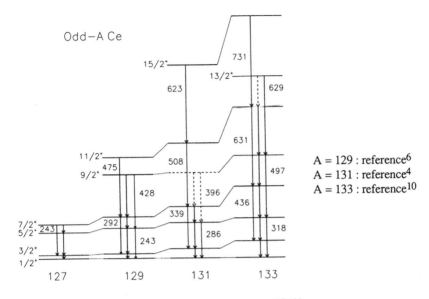

Fig. 3 : Band based on $1/2^+$ state in $^{127-133}$Ce.

4.1. Model and method

The IBFM-1 model and procedure used previously for odd-A Ba (ref.[16]) and Xe (ref.[17]) isotopes were applied to describe both the positive - and negative - parity states in odd-A Ce nuclei.

The Hamiltonian[18] of the odd-A system is $H = H_B + H_F + V_{BF}$ where H_B describes the even-even core in the IBA-1 model, H_F the single-particle energies and V_{BF} the boson-fermion interaction[18]. The full expansion of these terms can be found e.g. in ref.[19].

The aim of the calculations is to obtain, using a minimum number of parameters, a unitary description of the properties of the odd-A Ce nuclei considered as a neutron-hole coupled to even-even cores. The cores (H_B) are reproduced by comparisons to the known even-even Ce nuclei. The IBA-1 parameters for the Ce cores are found to vary smoothly from ^{126}Ce to ^{136}Ce. The V_{BF} parameters A_0, Γ_0 and Λ_0 are determined by an iterative process fitting experimental data relative to the odd-A nuclei. They are allowed for slight variations along the isotopic chain.

The quasiparticle energies and occupation probabilities were determined from a BCS calculation. We used the same single particle level diagram as for Ba isotopes. Multischell calculations were performed considering the $2d_{5/2}$, $1g_{7/2}$, $3s_{1/2}$ and $2d_{3/2}$ shells for positive-parity states and $1h_{11/2}$, $2f_{7/2}$ and $1h_{9/2}$ for negative-parity ones. The calculations provide a good description of the experimental level energies and branching ratios. Due to a lack of experimental data, it is impossible to test the values of magnetic and quadrupole moments in Ce nuclei.

4.2. Positive parity states

The agreement between calculated and experimental level structures is illustrated for ^{131}Ce (fig. 4). The 7/2$^+$ g.s. band(C) is well reproduced up to the first band crossing. The excitation energies and groupings of other levels are also well described. The $\Delta I = 1$ band (B) based on the 1/2$^+$ state identifies with the $s_{1/2} + d_{3/2}$ band. Other levels (A) including those on top of the 3/2$^+$ level at 341 keV could correspond to states denoted "strongly mixed".

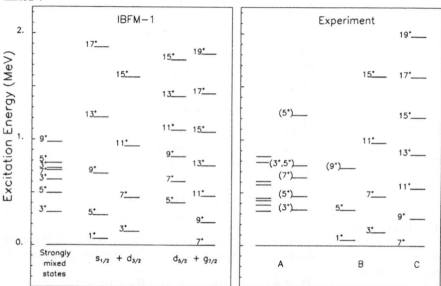

Fig. 4 : Comparisons of calculated (left) and experimental (right) positive-parity levels in ^{131}Ce. The spin values are multiplied by 2.

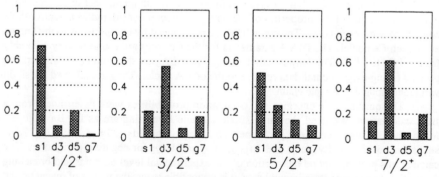

Fig. 5 : Components of the wave functions of 1/2$^+$, 3/2$^+$, 5/2$^+$, 7/2$^+$ states in the $s_{1/2} + d_{3/2}$ band of ^{131}Ce.

The levels of the g.s. band (C) are essentially generated from the $g_{7/2}$ shell (more than 93% up to I = 17/2) with a contribution from $d_{5/2}$ shell increasing with spin. On the contrary the wavefunctions of states B are rather mixed. Their main components are $s_{1/2}$ and $d_{3/2}$ (~ 75% of the total), the rest being distributed over the $d_{5/2}$ and $g_{7/2}$ shells (fig. 5). The states with I + 1/2 = odd (even) are principally $s_{1/2}$ ($d_{3/2}$). The states labelled A are also strongly mixed with ~ 65% from the $s_{1/2}$ and $d_{3/2}$ orbitals. This large amplitude agrees with the preferential deexcitation towards states B.

The main gamma decays are well reproduced by the model but, due to the underestimation of the M1 operator[20], quantitative differences remain between experimental and calculated branching ratios. The same conclusions apply to the other isotopes.

4.3. Negative parity states

The nature of the negative parity states is less complex than that of the positive parity ones. Indeed the major component $h_{11/2}$ represents more than 90% of the wavefunctions, the amplitudes of the $f_{7/2}$ and $h_{9/2}$ components increasing above I ~ 21/2 ℏ.

The complex experimental $h_{11/2}$ level structure of ^{131}Ce (fig. 6) is well described by the model, in particular excitation energies and gamma decays of the yrast band (A) and the parallel cascade (B). It is important to underline that the counterpart of the weakly fed sequence (C) exists in the calculations.

The level spectra of the other isotopes are also interpreted in a similar acceptable way. The $h_{11/2}$ level pattern observed in this mass region can be also explained by the triaxial-rotor-plus-particle model[21] and the IBFM model in the Spin(6) symmetry limit[22].

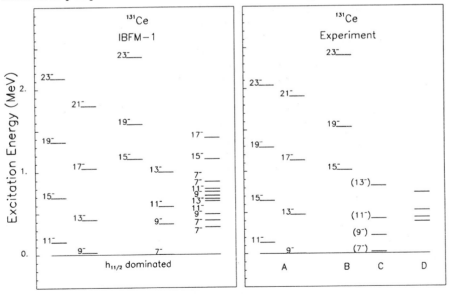

Fig. 6 : Comparisons of calculated (left) and experimental (right) negative-parity levels in ^{131}Ce. The spin values are multiplied by 2.

5. Conclusion

Experimental results have been obtained for odd-A Ce nuclei. They constitute a part of our investigation of the A ~ 130 region by β-decay studies. Data relative to intrinsic states and collective bands are available down to mass 125. Experimental features like level excitation energies, main gamma decays and band structures of positive and negative parities are reasonably well described by the IBFM-1 model.

Acknowledgments

The results presented here have been obtained in the frame of a collaboration between physicists of ISN Grenoble, CSNSM Orsay, AFI Bucharest, INRNE Sofia, IEP Warsaw and IPN Lyon. I would like to thank all the members of this collaboration, especially J. Genevey and A. Gizon. I am also greatly indebted to G. Cata-Danil for the IBFM-1 calculations he performed and for many discussions we had during the preparation of this lecture.

References

1. J. Gizon et al., *Nucl. Phys.* **A222** (1974) 557.
2. J. Gizon, A. Gizon and J. Meyer-ter-Vehn, *Nucl. Phys.* **A277** (1977) 464.
3. J. Genevey et al., *Contribution to this Conference*.
4. A. Gizon, in *Proc. of Internat. School on Nucl. Phys.*, Varna, Bulgaria, Oct. 1993.
5. J. Gizon et al., *Nucl. Phys.* **A290** (1977) 272.
6. J. Genevey et al., *Inst. Phys. Conf. Ser.* **No132** : Section 5, p.671.
7. M. Palacz et al., *Z. Phys.* **A338** (1991) 467.
8. B.M. Nyako et al., *Z. Phys.* **A334** (1989) 513.
9. A.N. James et al., *Daresbury Ann. Rep., Nucl. Struct. Appendix* (1985/86) p.103.
10. R. Ma et al., *Phys. Rev.* **C36** (1987) 2322.
11. A. Arima and F. Iachello, *Phys. Rev. Lett.* **35** (1975) 1069.
12. A. Arima et al., *Phys. Lett.* **66B** (1977) 205.
13. A. Faessler et al., *Nucl. Phys.* **A456** (1986) 381.
14. H.C. Chiang and S.T. Hsieh, *Phys. Rev.* **C43** (1991) 2445.
15. N. Yoshida, A. Arima and T. Otsuka, *Phys. Lett.* **114B** (1982) 86.
16. D. Bucurescu et al., *Phys. Rev.* **C43** (1991) 2610.
17. G. Cata-Danil et al., *J. Phys. G.* in press and *this Conference*.
18. F. Iachello and O. Scholten, *Phys. Rev. Lett.* **43** (1979) 679.
19. F. Iachello and P. Van Isacker, in *The Interacting Boson-Fermion Model* (Cambridge University Press, 1991).
20. L.D. Wood and I. Morrison, *J. Phys.* **G15** (1989) 997.
21. J. Meyer-ter-Vehn, *Nucl. Phys.* **A249** (1975) 111, 141.
22. F. Iachello, *Phys. Rev. Lett.* **44** (1980) 772.

INTRINSIC FRAME DESCRIPTION OF COMPOSED

BOSON-FERMION SYSTEMS

C.E. ALONSO and J.M. ARIAS

Departamento de Física Atómica, Molecular y Nuclear, Sevilla, Spain

and

A. VITTURI

Dipartimento di Fisica and INFN, Padova, Italy

ABSTRACT

The study of the Interacting Boson-Fermion Model by using the Intrinsic State Formalism is considered. We analyse the coupling of the odd particle (fermion) to the quadrupole mean field generated by the bosons. Connection with the traditional geometrical models for odd-even nuclei is stablished. The extension of the formalism to include the study of excited bands is also presented. Finally, we investigate the cranking of the boson-fermion hamiltonian to produce a single particle scheme in a rotating frame.

1. Introduction

There are many problems in Physics in which a description in terms of bosons and fermions is suitable. Special interest is devoted to systems in which the bosons represent Cooper pairs and the fermions individual particles. In this case, whenever the number of bosons is large it is meaningful to introduce the concept of mean field and describe the problem in terms of the more pictorical and intuitive geometrical variables. This can be achieved, among other methods, by studying the problem in the intrinsic frame of reference (where symmetries may break spontaneously even though they are symmetries of the full hamiltonian and give rise to good quantum numbers in the laboratory frame) and by introducing the concept of intrinsic state[1]. In this talk we will discuss the use of this formalism to study some aspects of the Interacting Boson-Fermion Model[2] (IBFM) of nuclear structure for odd-even nuclei. For the case of even-even nuclei, the connection of the usual Interacting Boson Model[3] (IBM) with the traditional geometrical models has been widely investigated by Leviatan and collaborators[4]. Here we will concentrate in the IBFM, for which some studies in the intrinsic frame have already been done[5,6], and will study the case of the odd fermion coupled to a system with only quadrupole bosonic degrees of freedom (s and d bosons)[7].

The intrinsic state formalism has several advantages: i) it is illustrative since it gives a geometrical image of the rather abstract picture given by the second quantized form of the IBFM; ii) it provides a considerable reduction in the complexity of the usual calculations in the laboratory frame; and iii) it leaves room for introducing new relevant degrees of freedom in the problem. We will try to show these advantages along this paper.

In section 2 we present briefly the formalism and make connections with the traditional geometrical models (Nilsson model, Meyer–ter–Vehn model, ...). In section 3 we discuss the coupling of the odd–particle to intrinsic bosonic excitations (β- and γ-vibrations). In section 4, we present the rotating intrinsic frame formalism. Finally, section 5 is devoted to present some conclusions.

2. Coupling the Fermion to Different Quadrupole Deformed Fields

The IBFM, in its traditional formulation, is a model of a single particle (fermion) coupled to N bosons (s and d bosons, L=0 and 2 respectively). Its study as a fermion moving in an external field generated by the bosons will allow us to stablish a link with traditional models of deformed odd–even nuclei written in terms of shape variables as the well known Nilsson model and Meyer-ter-Vehn model.

We start with a simple hamiltonian,

$$H = H_B + H_F + V_{BF} , \tag{1}$$

where H_B is the boson Hamiltonian for which we will take the simple form,

$$H_B = -\kappa_B \, \hat{Q}_B \cdot \hat{Q}_B . \tag{2}$$

With the boson quadrupole operator of the form

$$\hat{Q}_{B\,\mu} = (s^\dagger \tilde{d} + d^\dagger \tilde{s})_{2\mu} - \chi \, (d^\dagger \tilde{d})_{2\mu} , \tag{3}$$

with $\tilde{d}_\mu = (-1)^\mu d_{-\mu}$. H_F is the usual fermion hamiltonian, $H_F = \sum_j \epsilon_j \, a_j^\dagger \tilde{a}_j$ with $\tilde{a}_{j\mu} = (-1)^{j-\mu} a_{j-\mu}$, and V_{BF} is the boson–fermion interaction taken as a simple quadrupole-quadrupole one,

$$V_{BF} = -\kappa \, \hat{Q}_B \cdot \hat{Q}_F \tag{4}$$

with

$$\hat{Q}_F = \sum_{jj'} \langle j||Y^{(2)}||j'\rangle (a_j^\dagger \tilde{a}_{j'})^{(2)} , \tag{5}$$

where j,j' denote the (spherical) single particle orbitals.

The basic idea is to construct a boson intrinsic state for the ground state and to take the expectation value of H in this state. The intrinsic state expression for the ground state can be written as,

$$|\Psi_g(\beta,\gamma)\rangle = \frac{1}{\sqrt{N!}} \, (\Gamma_g^\dagger(\beta,\gamma))^N |0\rangle \tag{6}$$

with

$$\Gamma_g^\dagger(\beta,\gamma) = \frac{1}{(1+\beta^2)^{1/2}}[s^\dagger + \beta\cos\gamma\, d_0^\dagger + \sqrt{\frac{1}{2}}\beta\sin\gamma(d_2^\dagger + d_{-2}^\dagger)]. \tag{7}$$

The mean-field Hamiltonian for the additional fermion is therefore

$$H = E_B + \sum_j \epsilon_j a_j^\dagger \bar{a}_j - \kappa\, \langle \Psi_g | \hat{Q}_B | \Psi_g \rangle \cdot \hat{Q}_F \tag{8}$$

where $E_B = \langle \Psi_g | H_B | \Psi_g \rangle$ is the expectation value of the boson hamiltonian. We note that the boson energy surfaces E_B depend quadratically on the boson number N, while the boson-fermion interaction depends only linearly on N. For the discussion in this paper we shall assume that (when N is large) the structures of the basic bosons are determined by minimizing the boson part E_B. We shall therefore assume as equilibrium forms the same we had obtained for the pure boson part, neglecting the changes in the equilibrium due to the extra particle.

With this mean field hamiltonian we can determine the single particle levels in the presence of quadrupole deformation. For doing that, one diagonalizes H for fixed values of β and γ. The explicit expressions for the expectation values of the boson quadrupole operator are needed, (we use the notation $\hat{Q}_{ij}(\beta,\gamma) = \langle \Psi_g(\beta,\gamma) | \hat{Q}_{ij} | \Psi_g(\beta,\gamma) \rangle$)

$$\hat{Q}_{20}(\beta,\gamma) = \frac{N}{1+\beta^2}(2\beta\cos\gamma - \sqrt{\frac{2}{7}}\beta^2\chi\cos 2\gamma) \tag{9}$$

$$\hat{Q}_{22}(\beta,\gamma) = \hat{Q}_{2-2}(\beta,\gamma) = \frac{N}{1+\beta^2}(\sqrt{2}\beta\sin\gamma + \sqrt{\frac{1}{7}}\beta^2\chi\sin 2\gamma). \tag{10}$$

It is worth noting that the external field contains, in general, quadratic terms in β whose importance is governed by χ. Different elections of χ, β and γ will lead to the different cases of the odd particle coupled to an axial deformed, triaxial or γ-instable fields.

a) Axially deformed nuclei

For axially deformed field $\gamma_e = 0°$ and only the term $\hat{Q}_{20}(\beta,\gamma)$ remains for the external field, with the value

$$\hat{Q}_{20}(\beta,\gamma = 0°) = \frac{N}{1+\beta^2}(2\beta - \sqrt{\frac{2}{7}}\beta^2\chi). \tag{11}$$

In this case, is clear that the hamiltonian of Eq. (8) does not connect states with different values of K, projection of the angular momentum on the intrinsic axis. K is a good quantum number.

In Fig. 1a it is shown a calculation of the single particle energies in the intrinsic frame with χ fixed, as it is suggested by microscopic calculations and used in ref.[5] In Fig. 1b the same is shown but with $\chi = \sqrt{\frac{7}{2}}(1-\beta^2)/\beta$. This last election bring us to a kind of Nilsson model. The detailed study of wheter the experimental data show a preference for one scheme or the other is still missing.

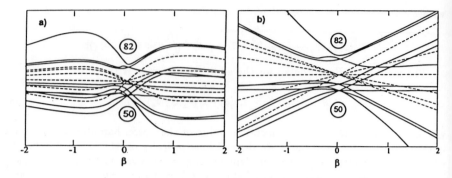

Fig. 1. Single–particle energies in the intrinsic frame for a particle coupled to an axially symmetric core as a function of β in the 50–82 major shell. In a) the parameter χ is kept fixed to the value 0.5, while in b) χ varies according to $\chi = \sqrt{\frac{7}{2}}(1-\beta^2)/\beta$.

b) Triaxial nuclei

In this case $\gamma \neq 0°$ and $\beta \neq 0$. In this case the hamiltonian of Eq. (8) includes $\hat{Q}_{20}(\beta,\gamma)$ as well as $\hat{Q}_{22}(\beta,\gamma)$ and $\hat{Q}_{2-2}(\beta,\gamma)$. Thus, the final states do not have good value of K. As in the previous item one can take several elections of χ and obtain different schemes. One can take χ fixed, as suggested by microscopic calculations, or leave χ varying in an appropriate way. In particular, it can be shown that if one chooses χ varying with γ as $\chi(\gamma) = \chi(\gamma = 0)\cos 3\gamma$ one obtains a scheme like that provided by the Meyer-ter-Vehn model. It is worth noting that the quadratic boson hamiltonian of Eq. (2) cannot produce minima with $\gamma \neq 0°$. For having triaxial deformations one has to introduce cubic terms. However, with the present formalism one can just simulate a triaxial external field by choosing $\gamma \neq 0°$ and $\beta \neq 0$.

c) γ–instable nuclei

This situation corresponds to $\chi = 0$. Again in this case the hamiltanian of Eq. (8) includes $\hat{Q}_{20}(\beta,\gamma)$, $\hat{Q}_{22}(\beta,\gamma)$ and $\hat{Q}_{2-2}(\beta,\gamma)$ and, consequently, the final states do not have good value of K. Although the external field in which the odd particle moves is γ–instable the coupling of the odd particle destroys the γ-instability. In Fig. 2 we illustrate this fact for two cases.

3. Intrinsic Excitations

As stated before, the simplification in the calculation provided by the intrinsic frame formalism presented is that it can incorporate easily other properties of the coupled system. For instance, intrinsic $\beta - -$ and $\gamma - -$ excitations. These can be associated to the intrinsic states,

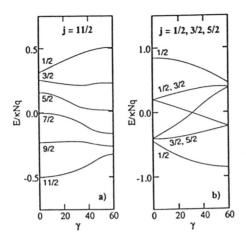

Fig. 2. Single-particle energies in the intrinsic frame for a particle coupled to an γ-instable core ($\chi = 0$ and $\beta_e = \sqrt{2}$) as a function of γ. In a) the odd particle can occupy a single j=11/2 shell, while in b) the space for the odd particle is a degenerate j=1/2, 3/2, 5/2 shell.

$$|\Psi_\beta(\beta,\gamma)\rangle = \frac{1}{\sqrt{(N-1)!}} (\Gamma_g^\dagger(\beta,\gamma))^{N-1}(\Gamma_\beta^\dagger(\beta,\gamma))|0\rangle \qquad (12)$$

$$|\Psi_\gamma(\beta,\gamma)\rangle = \frac{1}{\sqrt{(N-1)!}} (\Gamma_g^\dagger(\beta,\gamma))^{N-1}(\Gamma_\gamma^\dagger(\beta,\gamma))|0\rangle \qquad (13)$$

with

$$\Gamma_\beta^\dagger(\beta,\gamma) = \frac{1}{(1+\beta^2)^{1/2}}[-\beta s^\dagger + \cos\gamma\, d_0^\dagger + \sqrt{\frac{1}{2}}\sin\gamma(d_2^\dagger + d_{-2}^\dagger)] , \qquad (14)$$

and

$$\Gamma_\gamma^\dagger(\beta,\gamma) = [-\sin\gamma\, d_0^\dagger + \sqrt{\frac{1}{2}}\cos\gamma(d_2^\dagger + d_{-2}^\dagger)] . \qquad (15)$$

In this case, the calculation is more involved since apart of the matrix elements of the boson quadrupole operator in the ground state, the matrix elements in the excited bands as well as the crossing terms are needed. All of them are given explicitly in ref.[7]. In Fig. 3 one example of such a calculation including the one β and γ intrinsic excitations is presented. We have tested the goodness of this mean field treatment by comparing with the exact results obtained in the laboratory frame of reference. This calculation is also presented in Fig. 3. For the case presented in Fig. 3 the difference of the intrinsic and laboratory calculations is of the order of

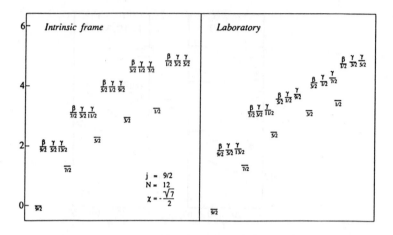

Fig. 3. Single-particle energies obtained in the intrinsic frame compared to the band heads obtained in the laboratory frame for the case of single j-shell ($j = \frac{9}{2}$). The parameters used are: $\beta = \sqrt{2}$ and $\gamma = 0$.

8%. Better accuracies can be obtained by including next order terms in the $1/N$ expansion.

4. Rotating the Intrinsic Frame

The intrinsic frame offers a simple formalism for studying phenomena related to high-spin states. This can be done easily by cranking the hamiltonian of Eq. (1) by including a Coriolis term

$$H = H_B + H_F + V_{BF} - \omega(L_x + j_x) ,\qquad(16)$$

where ω is the cranking frecuency and L_x (j_x) is the x-component of the boson (fermion) angular momentum. For the case of $\omega \neq 0$ terms related to the components d_1 and d_{-1} should be included in the boson condensate,

$$|\Psi_g(\beta,\gamma)\rangle = \frac{1}{\sqrt{N!}}(\Gamma_g^\dagger)^N|0\rangle ,\qquad(17)$$

with

$$\Gamma_g^\dagger(\beta,\gamma) = \frac{1}{\sqrt{(1+\beta^2+a_1^2)}}[s^\dagger + \beta\cos\gamma d_0^\dagger + \sqrt{\frac{1}{2}}\beta\sin\gamma(d_2^\dagger + d_{-2}^\dagger) + a_1(d_1^\dagger + d_{-1}^\dagger)].\qquad(18)$$

The cranking problem for even–even systems has already been studied[8,9]. It has been found that the a_1 parameter depends linearly in ω. For the odd–even systems the hamiltonian of Eq. (16) is diagonalized in the space obtained by coupling the boson condensate of Eqs. (17-18) to the single particle states characterized by (j, m). It can be shown that if we restrict to leading order in ω the dependence of the hamiltonian on the cranking frecuency is

$$H = E_B(\beta,\gamma) + H_F - \kappa \, \hat{Q}(\beta,\gamma) \cdot \hat{Q}_F - \omega j_x \; . \qquad (19)$$

The states obtained by diagonalizing this hamiltonian are labelled by parity and signature. In Fig. 4 one such a calculation is presented for a symmetric prolate boson condensate core $(\gamma = 0°, \beta = \sqrt{2}, \chi = -0.5)$ and the odd fermion in the 50–82 major shell. All the characteristics of the cranked Nilsson model are in this figure. For $\omega = 0$ the usual Nilsson levels are obtained, they are twofold degenarate. For $\omega \neq 0$ this degeneracy is broken by the Coriolis term and a split is observed. This split is particularly rapid for the orbits with large value of j and small projection of the odd-fermion angular momentum along the intrinsic symmetry axis. For large values of ω some levels move rapidly down so as new magic numbers can appear for high angular momenta.

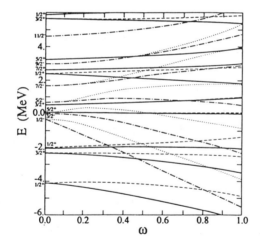

Fig. 4. Single–particle energies in the 50–82 shell as a function of ω and for the case of an axially symmetric prolate boson condensate. Different lines correspond to different signature and parities.

The inclusion of second order corrections in ω (ω^2 terms) is straightforward in the intrinsic frame formalism. These terms come from several sources: i) ω^2 dependence appears in the expectation value of the boson part of the hamiltonian in the intrinsic state of Eqs. (17-18), ii) The expectation value of the boson operators $\hat{Q}_{22}$, $\hat{Q}_{2-2}$ and $\hat{Q}_{20}$ in the boson intrinsic state of Eqs. (17- 18) include a term with a_1^2 which is proportional to ω^2, iii) The expectation value of L_x is proportional to a_1 and consequently the expectation value of ωL_x in the boson condensate of Eqs. (17-18) goes as ω^2. The inclusion of these terms is very easy and is presented in ref[10].

5. Conclusions

The study of odd-even nuclear systems by using the intrinsic frame formalism provides a simple and comprehensive way of describing them. We have shown as in this way the formalism contains as particular cases the more familiar geometrical models, i.e. the Nilsson model for particle coupled to an axially symmetric field, the Meyer-ter-Vehn model for the case of coupling the particle to a triaxial field, the cranked Nilsson model, etc. The scope and limitations of the method presented remains to be done by applying it to actual nuclei.

6. Acknowledgements

This work was performed in part with grants from CE contract CHRX-CT92-0075, DGICYT project PB92-0663 and INFN-DGICYT exchange collaborations.

7. References

1. J.N.Ginocchio and M.W.Kirson, *Nucl. Phys.* **350** (1980) 31.
2. F. Iachello and P. Van Isacker, *The Interacting Boson-Fermion Model* (Cambridge University Press, Cambridge, 1991).
3. F.Iachello and A.Arima, *The Interacting Boson Model* (Cambridge University Press, Cambridge, 1987).
4. A.Leviatan,*Ann. Phys. (N.Y.)* **179** (1987) 201.
5. A.Leviatan,*Phys. Lett.* **209B** (1988) 415.
6. J. Dukelsky and C. Lima,*Phys. Lett.* **182B** (1986) 116.
7. C.E. Alonso, J.M. Arias, F. Iachello and A. Vitturi, *Nucl. Phys.* **A539** (1992) 59.
8. H. Schaaser and D.M. Brink, *Nucl. Phys.* **A452** (1986) 1.
9. J.Dukelsky, G.G.Dussel, R.P.J.Perazzo, S.L.Reich and H.M.Sofia, *Nucl. Phys.* **A425** (1984) 93.
10. C.E. Alonso, J.M. Arias and A. Vitturi, *to be published*.

THE BOSON-FERMION INTERACTION IN THE GENERALIZED HOLSTEIN-PRIMAKOFF BOSON EXPANSION

J. DUKELSKY
Grupo de Física Nuclear, Universidad de Salamanca,
37008 Salamanca, Spain

and

S. PITTEL
Bartol Research Institute, University of Delaware,
Newark, DE 19716, USA

ABSTRACT

The generalized Holstein-Primakoff (GHP) boson expansion is used to derive the images of the one particle creation and annihilation operators in terms of collective bosons and ideal fermions. The collective transformation is obtained from a one-to-one mapping between a Hartree-Bogoliubov fermion coherent state and a Hartree-Bose coherent state, warranting the validity of the procedure for well-deformed nuclei. The pair-fermion operator related to the like-particle interaction, is then mapped. Particular attention is paid to the exchange and mixing terms.

1. Introduction

The interacting boson-fermion model[1] (IBFM) has been highly successful in correlating and explaining the low-energy spectra of a large variety of odd-mass nuclei[2]. A major step in the development of the model was the introduction of an exchange interaction between the odd fermion and the collective bosons, originating in the Pauli exclusion principle. This interaction, first introduced phenomenologically[1], was later derived microscopically in the framework of generalized seniority[3] (GS).

In the GS treatment, also referred as the Otsuka-Arima-Iachello (OAI) method, fermion states with generalized seniority $w \leq 3$ are related to corresponding states in the boson space. The boson-fermion image of the fermion operators is then defined by equating matrix elements in the two spaces. In this sense, the OAI method should be considered as a Marumori-type of boson expansion, truncated to low-generalized-seniority states.

Recently a new method[4], based on the Dyson boson expansion, was proposed as an alternative way to derive the boson-fermion images of fermion operators. This method, called the similarity-transformed Dyson mapping (STDM), first performs a mapping of the full fermion hamiltonian onto a finite but non-hermitian boson hamiltonian. The boson hamiltonian is then transformed by means of a seniority-dictated similarity transformation and finally hermitized.

Comparison of the two procedures[5] has shown an overall agreement for $w \leq 2$. It was speculated, however, that differences may appear for extensions to w=4, though they have not yet been pursued.

Truncation to low-seniority states, a common feature of the two above mentioned procedures, is known to be good in spherical or nearly spherical nuclei but not in strongly deformed nuclei where generalized seniority is badly broken. Thus, there is reason to doubt the validity of the boson-fermion images of fermion hamiltonians and other observables like the one-particle transfer operator that were derived in the OAI or the STDM when applied to deformed nuclei.

From a different point of view, the OAI method has been used to obtain a simple form of the boson-fermion interaction in terms of a reduced set of adjustable parameters, which has proven very successful for the description of the low-energy spectra of odd-mass nuclei and more recently for the high-spin spectra of even-even nuclei.

Several attempts have been reported to derive microscopically the IBFM in the deformed region with different approaches[6]. We here address this problem by using the generalized Holstein-Primakoff[7] (GHP) boson expansion and mean field techniques. We begin by reviewing earlier work done on the derivation of the boson-fermion images for the pseudo-particle and quadrupole operators[8]. We then obtain the images of the pair-fermion operator, relevant for the mapping of the like-particle interaction. Assuming a surface delta interaction (SDI) between like nucleons we discuss the physical origin of the exchange term responsible for the Pauli exclusion principle between the fermion and the like bosons, and the mixing term that couples bosons with quasi-fermion pairs. The latter term is important in efforts to extend the scope of the model to high spins.

2. Review of the GHP boson expansion procedure

The basic idea of the GHP method is to map a system of $2N+1$ fermions onto a system of N bosons and one ideal fermion, through the requirement that the original commutation algebra be preserved order-by-order by the mapping[7].

The GHP boson expansion involves the introduction of boson creation and annihilation operators B^+_{ik} and B_{ik} which take the place of the fermion pair-creation and pair-annihilation operators $c^+_i c^+_k$ and $c_k c_i$, respectively. These boson operators satisfy the following antisymmetry and commutation relations:

$$B^+_{ki} = -B^+_{ik}, \quad \left[B^+_{ik}, B^+_{jl}\right] = \delta_{ij}\delta_{kl} - \delta_{il}\delta_{kj} \quad \left[B^+_{ik}, B^+_{jl}\right] = 0 \ . \tag{1}$$

In addition, the degree of freedom of the odd fermion is taken into account by an ideal fermion creation (annihilation) operator a^+_i (a_i) that commutes with the bosons.

We concentrate here on the GHP boson-fermion image of the single-fermion creation operator, which is given by

$$\hat{c}^+_{j_1 m_1} = \sum_{j_2 m_2} (\sqrt{I-A})_{j_1 m_1 j_2 m_2} a^+_{j_2 m_2} + \sum_{j_2 m_2} B^+_{j_1 m_1 j_2 m_2} a_{j_2 m_2} \ , \tag{2}$$

where $A = B^+B$ and $\hat{c}^+$ means the GHP image of c^+. The operator $\sqrt{I-A}$, which is characteristic of the GHP method, is to be interpreted by its Taylor series expansion.

Focusing on the image of the creation and annihilation operators has the feature that it explicitly incorporates effects of the Pauli principle. Moreover it has the remarkable property of relating the classical limit in the ideal boson space with the Hartree-Fock-Bogoliubov (HFB) approximation in the fermion space[10]. In the classical limit, boson operators are replaced by c numbers:

$$B_{ij} \to v_{ij}, \quad B_{ij}^+ \to v_{ij}^* . \tag{3}$$

It is readily seen that upon this replacement eq. (2) reduces to a Bogoliubov transformation and the ideal fermion can be considered as a quasi-particle.

In ref..11, we established a relation between a HFB intrinsic state (the quasi-particle vacuum) and a boson coherent state, by equating the density matrix in the two spaces. It is important to note that the boson coherent state is not the precise GHP image of the HFB quasi-particle vacuum, which in the Bloch-Messiah basis is given by

$$\left(|\phi_0\rangle_F\right)_{GHP} \propto exp\left[\frac{1}{2}\sum_\alpha \frac{v_\alpha}{u_\alpha}\left(B^+\sqrt{I-A}\right)_{\alpha\bar\alpha}\right]|0\rangle_B . \tag{4}$$

The boson coherent state is obtained by considering the square root operator, reflecting the Pauli exclusion principle, in the classical limit:

$$|\phi_0\rangle_B \propto exp\left[\frac{1}{2}\sum_\alpha v_\alpha B_{\alpha\bar\alpha}^+\right]|0\rangle_B . \tag{5}$$

The intrinsic structure of the collective boson, determined by the v_α amplitudes, is different from the structure of the fermion condensed pairs, which have amplitudes v_α/u_α. The consequences of these differences were discussed in ref. 12, where it was shown that the boson description of the deformed fermion intrinsic state provides stronger s-d dominance.

The intrinsic boson appearing in the coherent state (5) can be decomposed into collective bosons $\gamma^+_{\lambda\mu}$ of well-defined angular momentum:

$$\gamma^+_{\lambda\mu} = \frac{1}{2}\sum_{j_1 j_2} X^\lambda_{j_1 j_2} B^+_{j_1 j_2 \lambda\mu} \quad, \quad B^+_{j_1 j_2 \lambda\mu} = \sum_{m_1 m_2} \langle j_1 m_1 j_2 m_2 | \lambda\mu \rangle B^+_{j_1 m_1 j_2 m_2} . \tag{6}$$

Further details can be found in ref. 11.

In ref. 13, it was shown how to speed the convergence of the series for the square root operator in well-deformed nuclei by performing the Taylor series expansion around the mean value of the A operator in an s-boson condensate. Keeping terms up to first order in the expansion leads to the following expression for the square root operator:

$$\left(\sqrt{I-A}\right)_{j_1 j_2} \approx \delta_{j_1 j_2}\delta_{m_1 m_2} u_{j_1} - \frac{1}{2u_{j_1}}\tilde{A}_{j_1 m_1 j_2 m_2} , \tag{7}$$

where v^2_i is the occupation probability of the single-particle state i ($v^2_i+u^2_i=1$) obtained from the spherical part of a HFB calculation. The expression for $\tilde{A}$, in terms of collective bosons is

$$\tilde{A}_{j_1 m_1 j_2 m_2} = \sum_{j_3 L_1 L_2 \lambda \mu}{}' \hat{\lambda} \hat{L}_1 \hat{L}_2 (-)^{j_3+m_1+L_2+\lambda+\mu} X^{L_1}_{j_2 j_3} X^{L_2}_{j_1 j_3} \begin{pmatrix} j_1 & j_2 & \lambda \\ m_1 & -m_2 & -\mu \end{pmatrix}$$

$$\times \begin{Bmatrix} L_1 & L_2 & \lambda \\ j_1 & j_2 & j_3 \end{Bmatrix} \left(\gamma^+_{L_1} \gamma_{L_2}\right)^\lambda_\mu , \qquad (8)$$

where the prime in the summation means that the term with $L_1=L_2=0$ is not included. It was already taken into account in the classical limit though the u_j factor in the first term of eq. (7).

3. Boson-fermion images of the fermion operators

Substituting the approximate expression (7) for the square root into eq. (2), using expression (8) for the $\tilde{A}$ operator, and introducing the collective transformation (6) for the boson operators, yields the following result for the pseudo-fermion creation operator:

$$\hat{c}^+_{j_1 m_1} = u_{j_1} a^+_{j_1 m_1} + \sum_{j_2 \lambda} X^\lambda_{j_1 j_2} \frac{\hat{\lambda}}{\hat{j}_1} \left(\gamma^+_\lambda a_{j_2}\right)^{j_1}_{m_1} - \frac{1}{2 u_{j_2}} \sum_{j_2 j_3 L_1 L_2 \lambda}{}' (-)^{j_2+j_3} X^{L_1}_{j_1 j_3} X^{L_2}_{j_2 j_3}$$

$$\times \frac{\hat{\lambda} \hat{L}_1 \hat{L}_2}{\hat{j}_1} \begin{Bmatrix} j_1 & j_2 & \lambda \\ L_2 & L_1 & j_3 \end{Bmatrix} \left[\left(\gamma^+_{L_1} \gamma_{L_2}\right)^\lambda a^+_{j_2}\right]^{j_1}_{m_1} . \qquad (9)$$

From this expression we see that the terms $\left[(s^+ d) a^+_{j_2}\right]^{j_1}_{m_1}$ and $\left[(d^+ s) a^+_{j_2}\right]^{j_1}_{m_1}$ appear at the same order in the expansion, as also found in ref. 6, but in contrast with the OAI or the STDM procedures where only the first term appears at this order in the GS classification. The second term, which arises in the GS scheme from configurations with $w \geq 3$, is expected to be important for deformed systems.

The boson-fermion image of the one-particle creation operator (9) is precisely the operator that arises in treating single-particle transfer from an even-even IBM system to its odd-mass IBFM neighbor.

We are now in position to evaluate the boson-fermion images of the pair-fermion creation operator. [The image of the particle hole operator can be found in Ref. 8.] We obtain this image by mapping each individual fermion operator using eq. (9). Up to first order in the boson operators, the pair-fermion image is

$$\left(\hat{c}^+_{j_1} \hat{c}^+_{j_2}\right)^\lambda_\mu = u_{j_1} u_{j_2} \left(a^+_{j_1} a^+_{j_2}\right)^\lambda_\mu + u_{j_2} X^\lambda_{j_1 j_2} \gamma^+_{\lambda \mu}$$

$$-\frac{1}{2u_{j_2}}\sum_{j_3j_4L_1L_2L_3L_4}{}' X^{L_1}_{j_1j_3}X^{L_2}_{j_2j_4}X^{L_3}_{j_3j_4}\hat{L}_1\hat{L}_2\hat{L}_3\hat{L}_4\begin{Bmatrix}j_1 & j_2 & \lambda \\ j_3 & j_4 & L_3 \\ L_1 & L_2 & L_4\end{Bmatrix}\left[\left(\gamma^+_{L_1}\gamma^+_{L_2}\right)^{L_4}\gamma_{L_3}\right]^\lambda_\mu$$

$$+u_{j_1}\sum_{j_3L}X^L_{j_3j_2}\frac{\hat{L}}{\hat{j}_2}\left\{a^+_{j_1}\left[\gamma^+_L a_{j_3}\right]^{j_2}\right\}^\lambda_\mu - (-)^{j_1+j_2-\lambda}u_{j_2}\sum_{j_3L}X^L_{j_3j_1}\frac{\hat{L}}{\hat{j}_1}\left\{a^+_{j_2}\left[\gamma^+_L a_{j_3}\right]^{j_1}\right\}^\lambda_\mu . \quad (10)$$

The image of the pair-annihilation operator is obtained by hermitian conjugation of (10).

The first term corresponds to a two-ideal-fermion creation operator reduced by the two u factors, and the second to a pure one-boson creation operator. The following two terms are of pure exchange character. There is no a priori restriction on the angular momentum of the pair operator; in particular, a monopole pair operator can lead to strong exchange effects, as will be discussed below.

4. Exchange and mixing terms from the like-particle interaction

The exchange term of the IBFM hamiltonian plays a crucial role in the structure of odd-mass nuclei. It was originally derived from the quadrupole neutron-proton interaction within the OAI procedure, and subsequently simplified in terms of a single parameter Λ_0 for semi-realistic IBFM1 applications[2]:

$$V_{ex} = \sum_{j_1j_2j_3}\frac{\Lambda^{j_3}_{j_1j_2}}{\hat{j}_3}:\left[\left(d^+a_{j_1}\right)^{j_3}\left(da^+_{j_2}\right)^{j_3}\right]^0_0 \quad , \quad \Lambda^{j_3}_{j_1j_2} = -2\sqrt{5}\,\Lambda_0\,\beta_{j_1j_3}\beta_{j_2j_3} . \quad (11)$$

Here $\beta_{j_1j_2}$ defines the collective structure of the D fermion pair in the OAI.

Phenomenological fits give values of Λ_0 an order of magnitude greater than those obtained from a microscopic GS treatment. It was later shown that the origin of the exchange force can be traced instead to the quadrupole pairing interaction, at least for vibrational nuclei[14].

The contribution of the quadrupole neutron-proton interaction to the exchange term of deformed nuclei has already been derived in the GHP formalism[8]. Based on the conclusions from the GS treatment of vibrational nuclei, we expect that the like-particle interaction should here too make an important (and perhaps the dominant) contribution to the exchange term. Thus, we now extend the GHP formalism to the like-particle interaction, assuming for concreteness a surface delta interaction:

$$H_{SD} = -G\sum_{\lambda\mu}\frac{1}{\hat{\lambda}^2}P^+_{\lambda\mu}P_{\lambda\mu} , \quad (12)$$

where

$$P^+_{\lambda\mu} = \sum_{j_1 j_2} \langle j_1 \| Y_\lambda \| j_2 \rangle \left(c^+_{j_1} c^+_{j_2} \right)^\lambda_\mu . \tag{13}$$

Mapping (13) by means of (10), we get to first order in the boson operators

$$\hat{P}^+_{\lambda\mu} = \sum_{j_1 j_2} \langle j_1 \| Y_\lambda \| j_2 \rangle \left\{ u_{j_1} u_{j_2} \left(a^+_{j_1} a^+_{j_2} \right)^\lambda_\mu + \frac{1}{2} \left(u_{j_1} + u_{j_2} \right) X^\lambda_{j_1 j_2} \gamma^+_{\lambda\mu} \right.$$

$$\left. + 2 u_{j_1} \sum_{j_3 L} X^L_{j_3 j_2} \frac{\hat{L}}{\hat{j}_2} \left[a^+_{j_1} \left(\gamma^+_L a_{j_3} \right)^{j_2} \right]^\lambda_\mu \right\} . \tag{14}$$

The structure coefficients of the collective bosons [see eq. (6)] can be obtained from a deformed HFB calculation. Alternatively, it is possible to fix the structure of the s boson from a BCS calculation and that of the d boson by requiring that it exhausts the E2 sum rule[2]. Here we follow the latter prescription, whereby the s and d structure coefficients are given by

$$X^0_{jj} = \frac{\hat{j} v_j}{2\sqrt{N}} \quad ; \quad X^2_{j_1 j_2} = \frac{1}{\eta} (v_{j_1} + v_{j_2}) \langle j_1 \| Y_2 \| j_2 \rangle . \tag{15}$$

Here N is the number of bosons and η is chosen to satisfy the normalization condition given in Ref. 11. This parametrization of the structure coefficients gives an overlap with those determined directly from a deformed HFB calculation greater than 90%.

With the use of (14), we can now isolate the contributions to the exchange interaction that arise in lowest order from an SDI hamiltonian:

$$H_{ex} = -G \sum_\lambda K_\lambda \left\{ \sum_{j_1 j_2 j_3 L} \frac{\langle j_2 \| Y_\lambda \| j_1 \rangle}{\hat{j}_2} u_{j_1} \hat{L} X^L_{j_3 j_2} \left[\left(a^+_{j_1} \gamma_\lambda \right)^{j_2} \left(\gamma^+_L a_{j_3} \right)^{j_2} \right]^0_0 + h.c. \right\} , \tag{16}$$

where

$$K_\lambda = \sum_{jj'} \frac{\langle j \| Y_\lambda \| j' \rangle}{\hat{\lambda}} (u_j + u_{j'}) X^\lambda_{jj'} . \tag{17}$$

Truncating to s and d bosons only leads to

$$H_{ex} = \sum_{j_1 j_2} \Gamma_{j_1 j_2} \left[\left(a^+_{j_1} s \right)^{j_1} \left(d^+ a_{j_2} \right)^{j_1} \right]^0_0 + h.c.$$

$$+ \sum_{j_1 j_2 j_3} \Delta^{j_3}_{j_1 j_2} \left[\left(a^+_{j_1} d \right)^{j_3} \left(d^+ a_{j_2} \right)^{j_3} \right]^0_0 , \tag{18}$$

where

$$\Gamma_{j_1 j_2} = -G\sqrt{5}\, K_0 \frac{u_{j_1}}{\hat{j}_1} \langle j_1 \|Y_0\| j_1 \rangle X^2_{j_2 j_1} - G K_2 \frac{u_{j_2}}{\hat{j}_1} \langle j_2 \|Y_2\| j_1 \rangle X^0_{j_1 j_1} ,$$

$$\Delta^{j_3}_{j_1 j_2} = -G K_2 \sqrt{5}\, \frac{u_{j_1}}{\hat{j}_3} \langle j_3 \|Y_2\| j_1 \rangle X^2_{j_2 j_3} . \tag{19}$$

It is important to note here that the first term in (18) breaks generalized seniority. Though it does not appear in the low order expansion of the OAI or the STDM mappings, it contributes to first order in the GHP procedure.

The mixing terms arising from the SDI hamiltonian (12) can likewise be obtained by analyzing eq. (14); the resulting expression is

$$H_{mix} = -\frac{G}{2} \sum_\lambda \left(\sum_{j_3 j_4} (u_{j_3} + u_{j_4}) X^\lambda_{j_3 j_4} \frac{\langle j_3 \|Y_\lambda\| j_4 \rangle}{\hat{\lambda}} \right)$$

$$\times \sum_{j_1 j_2} \langle j_1 \|Y_\lambda\| j_2 \rangle u_{j_1} u_{j_2} \left[(a^+_{j_1} a^+_{j_2})^\lambda \gamma_\lambda \right]^0_0 + h.c. . \tag{20}$$

Truncation of (20) to s and d bosons gives

$$H_{mix} = -U_0 \sum_j u_j^2 \hat{j} \left[(a^+_j a^+_j)^0_0 s + h.c. \right] - U_2 \sum_{j_1 j_2} \langle j_1 \|Y_2\| j_2 \rangle u_{j_1} u_{j_2} \left\{ \left[(a^+_{j_1} a^+_{j_2})^2 d \right]^0_0 + h.c. \right\}, \tag{21}$$

where

$$U_0 = -\frac{G}{8\pi\sqrt{N}} \sum_j u_j v_j \hat{j}^2 \quad ; \quad U_2 = -\frac{G}{2\pi} \sum_{j_1 j_2} \frac{\langle j_1 \|Y_2\| j_2 \rangle}{\sqrt{5}} (u_{j_1} + u_{j_2})(v_{j_1} + v_{j_2}) . \tag{22}$$

Both mixing terms are similar to the ones derived by Morrison et. al.[15] and recently used to describe high-spin phenomena[16]. It should be noted, however, that the monopole term vanishes in the spherical limit, since it corresponds to the $H_{02}+H_{20}$ part of the BCS hamiltonian. It may play an important role in the description of high-spin states in deformed nuclei.

5. Summary

We have developed in this paper the generalized Holstein-Primakoff boson expansion method to treat deformed odd-mass nuclei. We have focused on the coupling between boson and fermion degrees of freedom, as reflected in the boson-fermion exchange interaction of the IBFM and the mixing interaction that couples bosons to quasi-fermion pairs.

In earlier work, we showed how the quadrupole-quadrupole interaction between neutrons and protons contributes to these boson-fermion-coupling interactions. Here we extend the analysis to like-nucleon interactions, motivated by the observation that in the vibrational regime they appear to play the dominant role.

Our derivations lead to boson-fermion couplings that in lowest order contain generalized-seniority-breaking terms. We expect that such terms, though not present in OAI derivations, should be important in the deformed regime.

The interactions that result from our analysis are amenable to convenient parametrization for phenomenological fits. However, in such calculations, it is important to use a boson hamiltonian that is compatible with the GHP procedure.

6. Acknowledgments

This work was supported by the National Science Foundation under grant # PHY-9303041 and by NATO under grant # CRG.900466. We wish to acknowledge C. E. Alonso and J. M. Arias for their important contributions to our earlier work on the GHP procedure for odd-mass systems.

7. References

1. O. Scholten and F. Iachello, *Phys. Rev. Lett.* **43** (1979) 679.
2. O. Scholten, *Prog. Part. Nucl. Phys.* **14** (1985)189.
   F. Iachello and P. Van Isacker, *The interacting boson-fermion model* (Cambridge Univ. Press, Cambridge, 1991).
3. O. Scholten and A. E. L. Dieperink, *Interacting Bose- Fermi systems in nuclei* (Plenum Press, New York, 1981), pag. 343.
4. H. B. Geyer and I. Morrison, *Phys. Rev.* **C40** (1989) 2383.
5. J. Dobes, P. Navrátil and O. Scholten, *Phys. Rev.* **C45** (1992) 2795.
6. H. Sofia and A. Vitturi, *Phys.Lett* **B222** (1989) 317.
   V. Paar and S. Brandt, *Phys. Lett* **B143** (1984) 1.
7. E. R. Marshalek, *Nucl. Phys.* **A347** (1980) 253.
8. C. E. Alonso, J. M. Arias, J. Dukelsky and S. Pittel, *Nucl. Phys.* **A539** (1992) 391.
9. E. R. Marshalek, *Nucl. Phys.* **A224** (1974) 221.
10. E.R Marshalek and G. Holzwarth, *Nucl. Phys.* **A191** (1972) 438.
11. J. Dukelsky, S. Pittel, H. M. Sofia and C. Lima, *Nucl. Phys.* **A456** (1986) 75.
12. J. Dukelsky, G. G. Dussel and H. M. Sofia, *Nucl. Phys.* **A373** (1982) 267.
13. J. Dukelsky and S. Pittel, *Phys. Lett.* **B177** (1986) 125.
14. T. Otsuka, N. Yoshida, P. Van Isacker, A. Arima and O. Scholten, *Phys. Rev.* **C35** (1987) 328.
15. I. Morrison, A. Faessler and C. Lima, *Nucl. Phys.* **A372** (1981) 13.
16. D. Vretenar, G. Bonsignori and M. Savoia, *Phys. Rev.* **C47** (1993) 2019.

TERACTING BOSON-FERMION MODEL FOR STRONGLY DEFORMED NUCLEI

R.V.JOLOS
Bogoliubov Theoretical Laboratory, Joint Institute for Nuclear Research
141980 Dubna, Russia

and

A.GELBERG
Institut für Kernphysik, Universität zu Köln
D-50937, Köln, Germany

ABSTRACT

A method is developed to calculate the interaction matrix elements of the boson–fermion Hamiltonian, basing on the Nilsson single particle hamiltonian

1. Introduction

The IBFM Hamiltonian[1,2] has the same structure as any core–particle type model Hamiltonian. There is a boson (even–even core) part, a fermion part and a term, describing an interaction of a fermion with bosons (core vibrations). Every term in the total Hamiltonian is rotational invariant. A fermion Hamiltonian includes the pairing effects. In spite of the fact that the boson–fermion interaction can be very strong as in the case of deformed nuclei and therefore the mean field has a nonzero deformation, the pairing effects are treated and the boson –fermion interaction matrix elements are determined for spherical single particle states, using the $u - v$ Bogoliubov transformation for spherical orbits or the seniority scheme[3]. This approximation is not satisfactory for deformed nuclei because of a competition between pairing and a particle–vibration coupling, which is important and not taken into account in this case. The last effect is included into consideration by the equation of motion method[4,5]. However, in this case the dimension of the Hamiltonian matrix increases in two times and one should take care of excluding spurious states.

Below we develop a method which allows to consider pairing in the intrinsic frame, i.e. as the forces acting between particles occupying Nilsson single particle states and nevertheless to get a total Hamiltonian written in the laboratory frame, i.e. in a rotationally invariant form. This Hamiltonian has the same structure as a usual IBFM Hamiltonian, however, the boson–fermion interaction matrix elements are calculated in a different way.

2. Method

Let c_{jm}^+ and $\tilde{c}_{jm}$ be the particle creation and annihilation operators, where m is a momentum projection on the intrinsic axis. The Nilsson particle operators $\alpha_{\nu m}^+$ and $\tilde{\alpha}_{\nu m}$ are determined by the transformation

$$\alpha_{\nu m}^+ = \sum_j N_{j\nu}^{(m)} c_{jm}^+ \tag{1}$$

where $N_{j\nu}^{(m)}$ are Nilsson transformation coefficients. Let us introduce also the quasi-particle operators $a_{\nu m}^+$ and $\tilde{a}_{\nu m}$

$$a_{\nu m}^+ = u_\nu^{(m)} \alpha_{\nu m}^+ - v_\nu^{(m)} \tilde{\alpha}_{\nu m} \tag{2}$$

which we will consider as operators creating ideal fermions appearing in the IBFM Hamiltonian. Here $u_\nu^{(m)}$ and $v_\nu^{(m)}$ are the coefficients of the u–v Bogoliubov transformation

$$(u_\nu^{(m)})^2 + (v_\nu^{(m)})^2 = 1 \tag{3}$$

Using a unitarity of Nilsson transformation we get from the inverse of (2) and (1)

$$c_{jm}^+ = \sum_\nu N_{j\nu}^{(m)} (u_\nu^{(m)} a_{\nu m}^+ + v_\nu^{(m)} \tilde{a}_{\nu m}). \tag{4}$$

We can rewrite the last relation in the following way

$$c_{jm}^+ = \sum_\nu (u_{\nu j}^{(m)} a_{\nu m}^+ + v_{\nu j}^{(m)} \tilde{a}_{\nu m}) \tag{5}$$

where

$$u_{\nu j}^{(m)} \equiv N_{j\nu}^{(m)} u_\nu^{(m)}, \tag{6}$$

$$v_{\nu j}^{(m)} \equiv N_{j\nu}^{(m)} v_\nu^{(m)}. \tag{7}$$

When the deformation β vanishes

$$N_{j'\nu}^{(m)} \to \delta_{j'j(\nu)} \tag{8}$$

Thus, there is one-to-one correspondence between ν and $j(\nu)$ and we can write $j(\nu)$ instead of ν

$$c_{jm}^+ = \sum_{j'(\nu)} (u_{j'(\nu)j}^{(m)} a_{j'(\nu)m}^+ + v_{j'(\nu)j}^{(m)} \tilde{a}_{j'(\nu)m}) \tag{9}$$

For the sake of simplicity in the following we will use j instead $j(\nu)$. It is convenient to put $u_{j'j}^{(m)}$ and $v_{j'j}^{(m)}$ in the form

$$u_{j'j}^{(m)} = \sum_l u_{j'j}^{(l)}(jml0|j'm), \tag{10}$$

$$v_{j'j}^{(m)} = \sum_l v_{j'j}^{(l)}(jml0|j'm) \tag{11}$$

where $u_{j'j}^{(l)}$ and $v_{j'j}^{(l)}$ are determined by the relations

$$u_{j'j}^{(l)} = \frac{(2l+1)}{(2j'+1)} \sum_m u_{j'j}^{(m)}(jml0|j'm) \tag{12}$$

$$v_{j'j}^{(l)} = \frac{(2l+1)}{(2j'+1)} \sum_m v_{j'j}^{(m)}(jml0|j'm) \tag{13}$$

Substituting (10) and (11) into (9) we get

$$c_{jm}^+ = \sum_{j',l} \sqrt{\frac{2j'+1}{2j+1}} (l0j'm|jm)(u_{j'j}^{(l)} a_{j'm}^+ + v_{j'j}^{(l)} \tilde{a}_{j'm}), \tag{14}$$

where a_{jm}^+ creates a quasiparticle in a Nilsson orbital. Because of the symmetry properties of the Nilsson coefficients

$$N_{jj'}^{(-m)} = (-1)^{j-j'} N_{jj'}^{(m)} \tag{15}$$

l in (14) takes only even values.

The relation (14) is written in the intrinsic frame. Let us assume that this relation is a consequence of a more general one written in the laboratory frame

$$c_{j'm}^+ = \sum_{j,l} \sqrt{\frac{2j+1}{2l+1}} \sum_{\mu,m'} (l\mu jm'|j'm)(\hat{u}_{jj'}^{(l\mu)} a_{jm'}^+ + \hat{v}_{jj'}^{(l\mu)} \tilde{a}_{jm'}) \tag{16}$$

where $\hat{u}_{jj'}^{(l\mu)}$ and $\hat{v}_{jj'}^{(l\mu)}$ are operators acting on collective variables only and a_{jm}^+ creates spherical quasiparticles. We suppose also that these operators are expressed in terms of s^+, s, d_m^+, $\tilde{d}_m$ and higher multipolarity boson operators. In order to derive (14) from (16) we should averaged (16) over the boson coherent state, corresponding to the intrinsic wave function. The ideal fermion operators a_{jm}^+ and $\tilde{a}_{jm}$ commute with $s-$, $d-$ and other boson operators.

The simplest form of the operator equation (16) with the minimal degree of d-boson operators, which gives (14) after averaging over boson coherent state

$$|N;\beta_I> = \frac{1}{\sqrt{N!(1+\beta_I^2)^N}} (s^+ + \beta_I d_0^+)^N |0> \tag{17}$$

is

$$c_{jm}^+ = u_{jj}^{(0)} a_{jm}^+ + \sum_{j'} \sqrt{\frac{2j'+1}{2j+1}} \frac{u_{j'j}^{(2)}}{\beta_I \sqrt{\frac{N}{1+\beta_I^2}}} s^+ (\tilde{d} a_{j'}^+)_{jm} + \frac{v_{jj}^{(0)}}{\sqrt{\frac{N}{1+\beta_I^2}}} s^+ \tilde{a}_{jm}$$

$$+ \sum_{j'} \sqrt{\frac{2j'+1}{2j+1}} \frac{v_{j'j}^{(2)}}{\beta_I \sqrt{\frac{N}{1+\beta_I^2}}} (d^+ \tilde{a}_{j'})_{jm} + ... \quad (18)$$

Terms of higher multipolarity in bosons are omitted in (18).

3. Approximation

Let us analyse relations (10) and (11). Using asymptotic expressions for Clebsch–Gordan coefficients we get

$$u_{j'j}^{(l)} = \frac{2l+1}{2j'+1} \sqrt{\frac{(l-j+j')!}{(l+j-j')!}} \sum_m u_{j'}^{(m)} N_{jj'}^{(m)} P_l^{j-j'}(\cos\theta = \frac{m}{j'+1/2}) \quad (19)$$

and the analogous relation for $v_{j'j}^{(l)}$. For large j we can also put (19) into integral form

$$u_{j'j}^{(l)} \simeq (l+\frac{1}{2}) \sqrt{\frac{(l-j+j')!}{l+j-j')!}} \int d\cos\theta P_l^{j'-j}(\cos\theta) u_j^{((j'+1/2)\cos\theta)} N_{jj'}^{((j'+1/2)\cos\theta)} \quad (20)$$

It can be shown that the Nilsson transformation coefficients for the fixed values of j and j' are monotonous functions of m without oscillations. The same is true for the occupation numbers of the single particle states considered as functions of m when j is fixed. Thus, the products $u_{j'}^{(m)} N_{jj'}^{(m)}$ and $v_{j'}^{(m)} N_{jj'}^{(m)}$ are monotonous functions of m. Therefore as is seen from (20), the most important terms in the expansion (10) and (11) are the terms with low multipolarities $l=0$ and 2.

4. Hamiltonian

Let us put (18) into the expression for the fermion quadrupole operator

$$\hat{Q}_{2\mu} = \sum_{jj'} Q_{jj'} (c_j^+ \tilde{c}_{j'})_{2\mu} \quad (21)$$

Separating the boson part we get the following expression for the fermion part of the quadrupole operator

$$Q_{2\mu}^F = \sum_{jj'} (u_{jj}^{(0)} u_{j'j'}^{(0)} - v_{jj}^{(0)} v_{j'j'}^{(0)}) (a_j^+ \tilde{a}_{j'})_{2\mu}$$

$$- \sum_{jj'j_1} Q_{jj'} (v_{j'j'}^{(0)} v_{j_1j}^{(2)} - u_{j'j'}^{(0)} u_{j_1j}^{(2)}) \sqrt{\frac{2j_1+1}{2j+1}} \frac{1}{\beta_I \sqrt{\frac{N}{1+\beta_I^2}}}$$

$$\times \{((d^+ \tilde{a}_{j_1})_j (s a_{j'}^+)_{j'})_{2\mu} + ((\tilde{d} a_{j_1}^+)_j (s^+ \tilde{a}_{j'})_{j'})_{2\mu}\} \quad (22)$$

Using the same expression for the boson–fermion interaction term as in[3] and substituting (21) we get

$$V_{BF} = \sum_{\mu,j,j'} \Gamma_{jj'}(-1)^\mu (s^+ \tilde{d}_\mu + s d_\mu^+ + \chi(d^+\tilde{d})_{2\mu})(a_j^+ \tilde{a}_{j'})_{2-\mu}$$

$$+ \sum_{j,j',j''} \Lambda_{jj'}^{j''}((d^+\tilde{a}_j)_{j''}(\tilde{d}a_{j'}^+)_{j''})_{00} \frac{1}{\sqrt{2j''+1}} \quad (23)$$

where

$$\Gamma_{jj'} = \bar{\kappa}(u_{jj}^{(0)} u_{j'j'}^{(0)} - v_{jj}^{(0)} v_{j'j'}^{(0)}) Q_{jj'} \quad (24)$$

and

$$\Lambda_{jj'}^{j''} = -\bar{\kappa}\frac{\sqrt{5}}{\beta_I}(Q_{j'j''}\sqrt{2j+1}(v_{j'j'}^{(0)} v_{jj}^{(2)} - u_{j'j'}^{(0)} u_{jj}^{(2)})$$
$$+ Q_{jj''}\sqrt{2j'+1}(v_{jj}^{(0)} v_{j'j'}^{(2)} - u_{jj}^{(0)} u_{j'j'}^{(2)})). \quad (25)$$

Her $\bar{\kappa}$ is the interaction parameter[3]. The structure of the coefficients $\Gamma_{jj'}$ and $\Lambda_{jj'}^{j''}$ shown above is different from that found in[3]. The values of these coefficients depend on the deformation of the mean field. Thus, it can influence the results of calculations, especially if the deformation is large.

5. Single j case

Let us compare numerically the matrix elements of the interaction part of the IBFM hamiltonian found in the preceeding section with those determined in[3]. If we consider for simplicity the single j case, then we get $(u_j^{(0)^2} - v_j^{(0)^2})$ instead of $(u_j^2 - v_j^2)$ and $(u_{jj}^{(0)} u_{jj}^{(2)} - v_{jj}^{(0)} v_{jj}^{(2)})$ instead of $(u_j v_j + v_j u_j)$. Here u_j and v_j are Bogoliubov transformation coefficients. The results for $j=21/2$ and $\beta=0.2$–0.3 are presented in Figs.1 and 2.

It is interesting that the coefficients found by different methods depend in a similar manner on the occupation probability of the j level. The coefficient $(u_{jj}^{(0)^2} - v_{jj}^{(0)^2})$ is a linear function of $(u_j^2 - v_j^2)$. The coefficients $(u_{jj}^{(0)} u_{jj}^{(2)} - v_{jj}^{(0)} v_{jj}^{(2)})$ and $(u_j v_j + v_j u_j)$ demonstrate a parabolic type dependence on $(u_j^2 - v_j^2)$ but with different curvatures. It is important that the renormalization of the curvature coefficient depends on β. When deformation goes to zero both coefficients are coincide. Preliminary calculations show that situation becomes more complicated in a multi-j case. Because of the levels quasicrossing with deformation increase matrix elements are changed in a more complicated way. Of course the purely phenomenological approach is possible in the single j case, i.e. the Γ and Λ parameters are fitted, and nothing is left of the j or β dependence in this special case.

The l-dependence of the $v_{jj}^{(l)}$ is shown in Fig.3, where the results for $j=13/2$ and 21/2 are presented. Chemical potential is fixed so as to get a half filled j-level. It is seen that the largest ones are the coefficients with $l=0$ and 2. The coefficient $u_{jj}^{(l)}$ depends on l in the same wave.

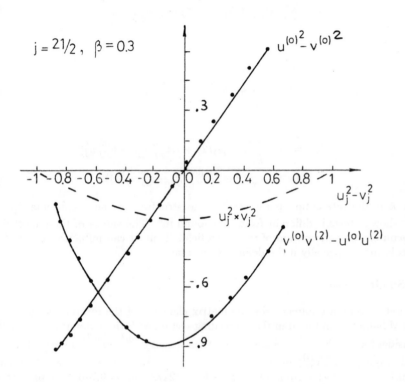

Fig. 1. Interaction matrix elements of the IBFM hamiltonian found in the present approach and in[3] as functions of the occupation number of the j-level. Here $j=21/2$ and $\beta=0.3$

6. Conclusion

In the present study we have developed a method to calculate the interaction matrix elements of the IBFM hamiltonian for strongly deformed nuclei. Considering a Nilsson hamiltonian plus pairing as obtained by averaging of the total IBFM hamiltonian over the boson coherent state we reconstruct approximately a particle-boson coupling term of the IBFM hamiltonian. The results are different from those obtained starting from spherical quasiparticles.

7. Acknowledgements

The authors would like to thank Prof.P.von Brentano for interesting discussions. One of the authors (R.V.J.) is gratefull to the Institute of Nuclear Physics of Köln University where this work has been done, for its hospitality and financial support.

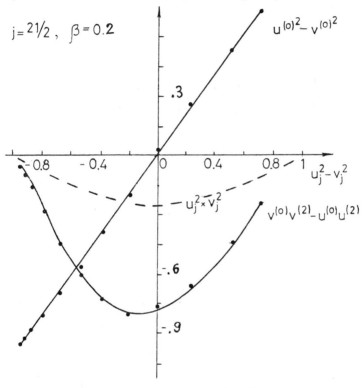

Fig. 2. The same as in Fig.1 but for $\beta=0.2$

8. References

1. A.Arima and F.Iachello, *Phys. Rev.* **C14** (1976) 761.
2. F.Iachello and O.Scholten, *Phys.Rev.Lett.* **43** (1979) 679.
3. O.Scholten, in *Progress in Particle and Nuclear Physics*, ed. A. Faessler (Pergamon, Oxford, 1985),v. 14, p. 189.

4. F.Dönau and U.Hageman, *Nucl.Phys.* bf A256 (1976) 27.
5. R.V.Jolos, *Preprint JINR, P4-7967*, (Dubna,1974).

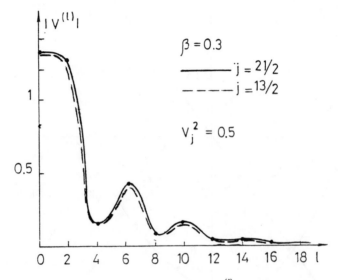

Fig. 3. Dependence of the expansion coefficient $v_{jj}^{(l)}$ on the multipole momentum l

SOFT MODE DYNAMICS IN TRANSITIONAL NUCLEI

V. G. ZELEVINSKY

*National Superconducting Cyclotron Laboratory, Michigan State University,
East Lansing, Michigan 48824-1321, USA*
and
Budker Institute of Nuclear Physics, Novosibirsk 630090, Russia

Abstract

Main features of low-lying spectra in soft spherical nuclei can be understood in the simple model of an $\mathcal{O}(5)$-symmetric nonlinear oscillator with strong quartic anharmonicity. The microscopic approach for calculating the anharmonic parameters is discussed.

1. The problem

In spite of accumulation of data and immense theoretical efforts[1-4], we still do not understand the low-lying structure of of soft (transitional) nuclei, i.e. non-magic nuclei which do not manifest well developed rotational bands (below only even-even nuclei will be discussed). The difficulties of the microscopic theory go with the fact that we deal with a self-sustained drop of the superfluid Fermi-liquid in the vicinity of static deformation. On the other hand, here one can expect universal behavior which, similar to phase transitions in macroscopic systems, can be described by a small number of parameters directly related to observables.

A successful phenomenological description should incorporate the features of the experimental pattern common for dozens of soft nuclei: (i) *quadrupole symmetry*, (ii) *collectivity* and (iii) *adiabaticity*. Almost all levels up to energy of 3-4 MeV (and frequently higher) which participate in strongly enhanced ("collective") $E2$ transitions have quantum numbers in close correspondence to the quadrupole harmonic vibrator scheme. Rare intruders are usually associated with the particular single-particle properties of a given nuclide. At the same time, the quantitative predictions for the harmonic vibrator are violated and the degeneracy is lifted.

The quadrupole collective motion is adiabatic with respect to the quasiparticle excitations: the ratio $\tau = \omega/2\bar{E}$ of collective frequency ω to the average energy $2\bar{E}$ of Cooper pair breaking is small. The enhancement factor $\Omega \gg 1$ of $E2$ transitions shows a number of quasiparticle modes coherently contributing to collective excitation. At small ω the vibrational amplitude grows $\propto \omega^{-1/2}$ (transition probability $\propto \omega^{-1}$). This agrees with the overall fit[5] of transition probabilities $B(E2; 0_1^+ \to 2_1^+)$. In a macroscopic system, a soft collective mode with $\omega \to 0$ signals onset of a second order transition to a new phase with a nonzero mean value of the vibrational coordinate which becomes an order parameter. In nuclei this would correspond to static quadrupole deformation.

However a different scenario is also possible. As $\omega \to 0$ and the zero point amplitude grows, fast noncollective degrees of freedom try to adjust themselves to slow changes of the mean field. Virtual excitation related to this readjustment costs energy which gives rise to anharmonic effects described usually by the effective hamiltonian. But in microscopic systems we might be interested in the individual features of low-lying states rather than in bulk thermodynamic properties. Even if the original mean field is already unstable, $\omega^2 < 0$, the anharmonic terms restore stability. As far as the new potential wells, which appear beyond the transition point due to symmetry breaking, are shallow enough, quantum tunneling allows the system to be still in the symmetric configuration. The fluctuations of the mean field are large so that the observable properties are governed mainly by anharmonicity.

In what follows I explain main features of the simple phenomenological model[6-9] based on strong quartic anharmonicity, give the microscopic arguments[10] justifying this model, and show how the microscopic theory can be built for a soft collective mode in a finite Fermi system.

2. Phenomenological model of collective quadrupole motion

We aim at the construction of the effective hamiltonian $\mathcal{H}(\alpha, \pi)$ expressed solely in terms of collective quadrupole coordinates α_μ and conjugate momenta π_μ. Our belief in the existence of such a hamiltonian capable of describing the broad variety of data is based on the clear correspondence between actual levels and ideal states of a harmonic vibrator although the specific degeneracies are destroyed.

The harmonic hamiltonian can be written as

$$\mathcal{H}_{harm} = \frac{1}{2}[C(\alpha^2)_{00} + \frac{1}{B}(\pi^2)_{00}] = \omega(N+5/2). \quad (1)$$

Here the force constant C, mass parameter B, boson number operator N and frequency $\omega = \sqrt{C/B}$ are introduced. The stationary states for the harmonic hamiltonian can be labeled[2-4] by total boson number N, boson seniority v (5-dimensional angular momentum or number of "active" bosons complementing the number n of boson pairs with the angular momentum zero, $N = 2n+v$), 3-dimensional angular momentum J and its projection M, corresponding to the group reduction $\mathcal{SU}(5) \supset \mathcal{O}(5) \supset \mathcal{O}(3) \supset \mathcal{O}(2)$.

For soft nuclei, the main corrections to (1) are those to potential energy since the coordinates α_μ have a large amplitude. The next term is cubic, $\propto (\alpha)^3)_{00}$. In intrinsic frame variables[1], it is proportional to $\beta^3 \cos 3\gamma$ being responsible for the sign of deformation. If it is small, the quartic term $\propto (\alpha)^4_{00} \propto \beta^4$ plays the decisive role. The quartic potential is isotropic in the 5-dimensional space and does not determine a preferential sign of deformation. This $\mathcal{O}(5)$-symmetry was assumed in the model of γ-unstable potential[11] and has its counterpart in the $\mathcal{O}(6)$-limit of the IBM[3,4,12] as well in the FDSM[13]. Here we neither introduce s-bosons nor postulate one-to-one correspondence between bosons and valence nucleon pairs; the boson number N, in contrast to v, is not conserved.

The actual transition to a permanently deformed shape may or may not happen later and it needs not to be associated with soft mode instability. It will be rather the first order transition caused by the steep downfall of energies of some excited configurations which occurs when the vibrational amplitude exceeds a critical value. Those configurations define the sign of deformation. Nuclei in a region between spherical near-magic ones and typical rotors are expected to reveal the intermediate type of structure determined by nonlinear large amplitude vibrations. This "limiting anharmonicity" domain can be characterized[10] by a relation $\Omega \tau^3 \simeq 1$ between the parameters of collectivity and adiabaticity.

Various models of anharmonic effects were suggested long ago on the phenomenological[15,16] as well as semimicroscopic level[17-19]. The experimental data imply strong anharmonicity which should be treated in a nonperturbative manner. But the direct diagonalization taking into account all possible nonlinear terms up to a certain order[16] involves too many free parameters. By those reasons, recent progress in nuclear spectroscopy turned out to be related exclusively to the IBM.

The leading role of strong quartic anharmonicity was stressed in ref.[6]. The analytical method of ref.[18] was used to find an approximate solution and to predict regularities of spectra and transition probabilities. The remarkable examples[7] of those regularities were found in actual data. In some cases the vibrational bands follow the predictions of the $\mathcal{O}(5)$-symmetry up to high spins. In the zeroth approximation[6] only quartic anharmonicity is retained (parameters $\Lambda^{(lm)}$ refer to the terms containing l coordinate operators α and m momentum operators π),

$$\mathcal{H} = \mathcal{H}_{harm} + \frac{1}{4}\Lambda^{(40)}(\alpha^4)_{00}. \qquad (2)$$

The main new effect can be visualized as creation of the boson pair condensate. Therefore one can apply the canonical transformation of Bose-operators[18,6] (in each sector with v fixed separately since the seniority remains to be conserved). The larger the value of v the more accurate the results are. Even for the lowest states (ground state with $v = 0$ and one-phonon state with $v = 1$) energies obtained deviate from exact results by $\sim 2\%$ only. One can easily develop corrections in a regular manner.

In each v-sector the excited states can be labeled by a number $\tilde{n}$ of real excitations (pairs with $J = 0$, or radial excitations in the 5-dimensional space) on top of the lowest state with given v. Energies $E(v, \tilde{n})$ can be expressed by simple analytical formulae. In the limit of extremely strong anharmonicity one can neglect the quadratic potential term in (2) and derive the energies of the lowest states

$$E(v, \tilde{0}) = const[\Lambda^{(40)}]^{1/3}(v + 5/2)(v + 7/2)^{1/3}. \qquad (3)$$

In many nuclei the yrast-line ($J = 2v$) spectrum is close to the limit (3), especially in the case of 4 valence particles or holes. Thus, for ^{100}Pd, the calculated (experimental) energies, in units of the energy $E(2_1^+)$, are 2.09 (2.13) for $J = 4$, 3.26 (3.29) for $J = 6$, 4.50 (4.49) for $J = 8$, 5.81 (5.82) for $J = 10$, 7.17 (7.16) for $J = 12$ and 8.58 (8.58) for $J = 14$. This can be considered as a clear manifestation of the new dynamical symmetry, namely that of the 5-dimensional oscillator with strong quartic anharmonicity.

As a rule, the zeroth approximation (3) is not sufficient, in particular for the side bands. The so called virtual rotation[7] was introduced to explain the deviations. The harmonic vibrator already contains orientation angles along with the shape variables β and γ. A virtual quadrupole deformation creates a (time-dependent) anisotropy and allows one to define an intrinsic frame and rotational degrees of freedom. In soft nuclei rotations and vibrations have similar excitation energies and cannot be disentangled. The virtual rotation can be described by a parameter σ responsible for $J(J+1)$ energy deviations from the pure anharmonic model. This parameter displays[7] a regular behavior along the chain of isotopes and frequently becomes very small for high spin states. It means that the upper levels manifest almost universal regularities whereas in the lower part the individual features of a given nucleus might be more important.

Since the restoring force is weak and the stability is ensured by the quartic anharmonicity, the exact landscape near the bottom of the well is of minor importance. The general character of motion is essentially the same regardless of the sign of the parameter C or ω^2. In many cases the description is improved by taking negative ω^2, see for example Xe isotopes[9]. Pushing levels with $v = 0$ up, this explains inversion of 0_2^+ and 0_3^+ levels known also in the $\mathcal{O}(6)$ limit of the IBM. It means that the random phase approximation (RPA) can serve only as a zero approximation indicating the region of its own breakdown.

Effects of lower significance like small cubic anharmonicity (parameter $\Lambda^{(30)}$) can be added. Final agreement is good[7-9] for almost all soft nuclei at small number of fitted parameters. Practically all relative energies in a given nucleus are predicted with the use of not more than 3 parameters, ω^2, σ and $\Lambda^{(30)}$; as a rule, two first of them suffice to give a good description.

The transition probabilities are to be calculated with effective operators expressed as functions of α_μ and π_μ. The quadrupole operator in the main order, $Q_\mu \to Q_\mu^{(10)}$, is simply proportional to α_μ. Strong quartic anharmonicity renormalizes the operators adding the term related to the condensate strength, $Q_\mu^{(30)} \propto (\alpha^2)_{00}\alpha_\mu$ (in some respects, the boson pairs with $J = 0$ are similar to s-bosons of IBM; here they are created by large amplitude quadrupole motion). This term suppresses $E2$ transition probabilities at high J without any artificial cut-off. In the cases where the rich information on $B(E2)$ is available, many relative transition strengths (more than two dozens in ^{104}Ru) are fit by this single parameter[8,9]. Again, small scale improvements can be made by an additional term $\sim (\alpha^2)_{2\mu}$.

3. Microscopic justification.

The model described in the previous section although extremely simple can be justified by microscopic estimates[10]. The main arguments go back to the first derivation of the quasiparticle-phonon hamiltonian[20].

Start with the microscopic hamiltonian including the mean field, pairing and particle-hole interaction with a strong quadrupole component. The RPA for 2^+ excitations gives the low-frequency, $\tau \ll 1$, mode with strong collectivity, $\Omega \gg 1$. The instability point, $C \to 0$, determines the effective quadrupole constant and its self-consistent value[2] shows that the limiting collectivity is $\Omega \simeq A(2\bar{E})/\epsilon_F$. Using empirical parameterization of the energy gap, one gets $\Omega \simeq \sqrt{A}$. The RPA near the soft pole gives, at the above collective strength, the vibrational amplitude $Q \simeq q\sqrt{\Omega/\tau}$ as it would be for the collective hamiltonian (1) where $Q \propto \alpha$ (q is a typical single-particle quadrupole matrix element). The vibrational amplitude is limited by onset of deformation due to the crossing of split single-particle orbitals. It gives $Q \simeq qA^{1/3}$. Together with the estimate for Ω, it allows one to find the limiting adiabaticity factor $\tau \simeq A^{-1/6}$ and establish the key relationship

$$\Omega\tau^3 \simeq 1. \tag{4}$$

Now one can use the phonon language and show, by comparison with the RPA, that the phonon-quasiparticle coupling is strong on the phonon scale, $\gamma^2 \simeq \omega^2$. With this estimate, the leading anharmonic interactions (fermion loops with n phonon tails) are $\Lambda^{(n0)}\alpha^n \simeq \omega\tau^{n-4}$. The odd-$n$ terms are suppressed by the particle-hole cancellation (analog of the Furry theorem). Therefore the quartic anaharmonicity indeed dominates the dynamics. In similar way, the effective operators are given by loops with one of the phonon tails substituted by an external field. For the quadrupole moment it gives $Q_\mu \sim \delta\mathcal{H}/\delta\alpha_\mu^\dagger$ and leads to the 3-phonon correction mentioned earlier.

Such a semiquantitative derivation justifies the phenomenological model and explains its success. However, the real microscopic calculation of anharmonic terms is very complicated: such a theory should account for the response of quasiparticle degrees of freedom to large amplitude collective motion, the perturbation theory is invalid and, in contrast to theory of deformed nuclei, the angular momentum has to be conserved exactly at all stages.

4. Towards the microscopic theory

The development of the microscopic theory is under way. Here I outline its main features. One starts with the exact operator equations of motion for the generalized density matrix (GDM[21]) $R_{12} = a_2^\dagger a_1$ derived with the use of the actual many-body hamiltonian H,

$$i\dot{R} \equiv [R, H] = F\{R\}, \qquad (5)$$

where the functional F contains terms linear and bilinear in R. The collective hamiltonian $\mathcal{H}(\alpha, \pi)$ is to be expressed in terms of coordinates α and conjugate momenta π. In the collective subspace the image $\mathcal{R}(\alpha, \pi)$ of the density matrix R is a matrix (with respect to single-particle orbitals) which depends on α and π. As a function of collective variables it satisfies the equations of motion governed by the supposed form of $\mathcal{H}$,

$$i\dot{\mathcal{R}} \equiv [\mathcal{R}, \mathcal{H}] = \mathcal{F}(\alpha, \pi). \qquad (6)$$

The mapping onto the collective subspace is performed by the requirement of the saturation of dynamics within this subspace. It means that eqs. (5) and (6) should be consistent,

$$\mathcal{F}(\alpha, \pi) = F\{\mathcal{R}(\alpha, \pi)\}. \qquad (7)$$

The results of mapping can be formally written as closed expressions for the coefficients of $\mathcal{H}(\alpha, \pi)$. The main quartic term is

$$\Lambda^{(40)} = (i/2)Tr\{([w^{(10)}, r^{(20)}] + [w^{(20)}, r^{(10)}])p\}. \qquad (8)$$

Here the trace is taken of matrix products in the single-particle space, $r^{(lm)}$ stands for the part of the density matrix $\mathcal{R}$ proportional to the anticommutator $[\alpha^l, \pi^m]_+$ of powers of collective operators, and $w^{(lm)}$ is the corresponding part of the mean field operator, $w_{12}^{(lm)} = \sum_{34} V_{13;42} r_{43}^{(lm)}$, where V is antisymmetrized residual interaction. The single-particle operator p is the soft mode solution of the conjugate RPA equations which, together with the similar operator x entering the (analogous to (8)) expression for the constant $\Lambda^{(04)}$ of the collective kinetic term $\propto \pi^4$, give single-particle images of collective variables $\pi = Tr(p\mathcal{R})$ and $\alpha = Tr(x\mathcal{R})$ along with the normalization

$[\alpha, \pi] = i = Tr([x,p]\rho)$. All parts $r^{(lm)}$ of the GDM, including the static density matrix $\rho \equiv r^{(00)}$, are to be found, together with $w^{(lm)}$, as solutions of corresponding integral equations. The mapping is done exactly with truncation at some stage as the only approximation which is based on the arguments of Sec. 3 and verified by the results. The proper vector coupling of angular momenta is implied.

The equations can be solved explicitly with factorizable interaction V. The quadrupole attractive forces, which create the soft mode if the two-quasiparticle excitation spectrum $e_1 - e_2$ has a gap as it is the case due to pairing, lead to the main contribution to quartic anharmonicity (κ is the quadrupole coupling constant)

$$\Lambda^{(40)} = (\kappa^2/2) \sum_{12} |q_{12}|^2 \frac{(e_1 - e_2)(\nu_1 - \nu_2)}{(e_1 - e_2)^2 - \omega^2}. \tag{9}$$

The operator parts of occupation numbers $\nu_1 \equiv r_{11}^{(20)}$ are given by the similar expression in terms of static (average over collective motion) occupation numbers $n_1 \equiv \rho_{11}$,

$$\nu_1 = -2\kappa^2 \sum_2 |q_{12}|^2 \frac{n_1 - n_2}{(e_1 - e_2)^2 - \omega^2}. \tag{10}$$

The results confirm the initial assumptions and estimates. The redistribution of particle as well as modulation of the superfluid gap by collective motion is the main source of anharmonicity. The semiclassical limit of the same equations leads to the population dynamics where energy exchange between quasiparticles and the collective mode is similar to the collision integral; this interaction makes quasiparticle motion chaotic[22] still maintaining coherence of the collective mode.

References.

1. A.Bohr and B.Mottelson, *Kgl. Dansk. Vid. Selsk. Mat.-fys. Medd.* **27** (1953) No. 16.
2. A.Bohr and B.Mottelson, *Nuclear Structure*, vol.II (Benjamin, Reading, 1975).
3. A.Arima and F.Iachello, *Phys. Rev. Lett.* **35** (1975) 1069; *Ann. Phys.* **99** (1976) 253, **111** (1978) 201, **123** (1979) 468.

4. F.Iachello and A.Arima, *The Interacting Boson Model* (Cambridge University Press, Cambridge, 1987).
5. S.Raman, C.W.Nestor,Jr., S.Kahane and K.H.Bhatt, *Phys. Rev.* **C43** (1991) 556.
6. O.K.Vorov and V.G.Zelevinsky, *Sov. J. Nucl. Phys.* **37** (1983) 830.
7. O.K.Vorov and V.G.Zelevinsky, *Nucl. Phys.* **A439** (1987) 207.
8. V.G.Zelevinsky, in *Nuclear Structure, Reactions and Symmetries*, vol.2 (Dubrovnik, 1986) p.1125; O.K.Vorov and V.G.Zelevinsky, in *Modern Developments in Nuclear Physics*, ed. O.P.Sushkov (World Scientific, Singapore, 1988) p.281.
9. V.G.Zelevinsky, *Soryushiron Kenkyu (Kyoto)* **83** (1991) D176.
10. V.G.Zelevinsky, *Int. J. Mod. Phys.* **E2** (1993) 273.
11. L.Wilets and M.Jean, *Phys. Rev.* **102** (1956) 788.
12. R.F.Casten and P. von Brentano, *Phys. Lett.* **B152** (1985) 22; P. von Brentano, A.Gelberg, S.Harissopulos and R.F.Casten, *Phys. Rev.* **C38** (1988) 2386.
13. C.-L.Wu, D.H.Feng and M.Guidry, *Adv. Nucl. Phys.* **21** (1994) 227.
14. X.-W.Pan, D.H.Feng, J.-Q.Chen and M.W.Guidry, *Preprint 001-NTG* (Drexel University, 1994).
15. A.K.Kerman and C.M.Shakin, *Phys. Lett.* **1** (1962) 151.
16. G.Gneuss and W.Greiner, *Nucl. Phys.* **A171** (1971) 449.
17. S.T.Belyaev and V.G.Zelevinsky, *Nucl. Phys.* **39** (1962) 582; *Sov. Phys. Izvestia, ser. fiz.* **28** (1964) 127.
18. V.G.Zelevinsky, *Sov. Phys. JETP* **19** (1964) 1247.
19. B.E.Stepanov, *Sov. J. Nucl. Phys.* **18** (1973) 999.
20. S.T.Belyaev, *Sov. J. Nucl. Phys.* **1** (1965) 3.
21. S.T.Belyaev and V.G.Zelevinsky, *Sov. J. Nucl. Phys.* **16** (1973) 525; V.G.Zelevinsky, *Prog. Theor. Phys. Suppl.* **74-75** (1983) 251, and in *Nuclear Structure Models*, eds. R.Bengtsson, J.Draayer and W.Nazarewicz (World Scientific, Singapore, 1992).
22. W.Bauer, D.McGrew, V.Zelevinsky and P.Schuck, to be published.

REFLECTION ASYMMETRIC SHAPES IN INTERACTING BOSON-BOSON AND BOSON-FERMION SYSTEMS

C.E. ALONSO, J.M. ARIAS and A. FRANK [‡]
Departamento de Física Atómica, Molecular y Nuclear, Sevilla, Spain

H.M. SOFIA
Comisión Nacional de Energía Atómica, Buenos Aires, Argentina

and

S.M. LENZI and A. VITTURI
Dipartimento di Fisica and INFN, Padova, Italy

ABSTRACT

An intrinsic frame analysis of octupole deformed nuclei in the SU(3) limit of the extended s-p-d-f Interacting Boson Model is presented. The results are applied to ^{226}Ra. The coupling of the odd particle to a quadrupole plus octupole deformed field is also investigated.

1. Introduction

During the last years a renewed experimental and theoretical interest on the existence of a variety of nuclear shapes has been observed. In particular, wide evidence for octupole features in the odd-mass actinides and other nuclear regions has been collected during the last few years. An octupole deformed shape has reflection asymmetry about a plane perpendicular to its intrinsic symmetry axis, which is manifested in the properties of the rotational energy spectra. Signatures of this reflection asymmetry are the presence of alternating parity levels in the rotational bands of even-even nuclei connected with strong in-band E1 transitions and parity doublets in the neighbouring odd-even nuclei.

Octupole deformation is expected to appear in addition to the dominant quadrupole deformation. Consequently, the models for describing these systems arise as extensions of those used for the description of quadrupole deformed shapes. The Interacting Boson Model[1] (IBM) has been widely used for studying these cases.

Whenever the number of bosons is large it is meaningful to introduce the concept of mean field and stablish a connection of the IBM with the traditional geometrical models written in terms of the more pictorial and intuitive geometrical variables. A proven method to do that is to study the problem in the intrinsic frame of reference[2,3], where symmetries may break spontaneously even though they are symmetries of the full hamiltonian and give rise to good quantum numbers in

[‡] On leave of absence from Instituto de Ciencias Nucleares, UNAM, México

the laboratory frame. The extension of these algebraic boson approaches to the description of reflection asymmetric shapes necessarily involves the introduction of negative parity bosons[4]. We consider in this paper the inclusion in the IBM of p- and f-bosons. Our approach is based on a general quadrupole–quadrupole plus octupole–octupole hamiltonian with no restriction on the number of negative parity bosons, but studying the problem in the intrinsic frame.

2. Positive- and Negative-Parity SU(3) Quadrupole Model

We start by considering the case of pure quadrupole interaction, in the particular case of the SU(3) hamiltonian. In the case of the SU(3) algebra based on the positive-parity s and d bosons, the hamiltonian can be written as,

$$H_Q^{sd} = -\kappa\, \hat{Q}^{sd} \cdot \hat{Q}^{sd} \quad \text{with} \quad \hat{Q}_\mu^{sd} = (s^\dagger \tilde{d} + d^\dagger \tilde{s})_{2\mu} - \frac{\sqrt{7}}{2}(d^\dagger \tilde{d})_{2\mu}\ . \tag{1}$$

The ground-state rotational band ($K^\pi = 0^+$, SU(3) representation $[2N, 0]$) of a system with N bosons is associated with the intrinsic state

$$|\Psi_g^{sd}\rangle = \frac{1}{\sqrt{N!}} (\Gamma_g^{sd\dagger})^N |0\rangle \quad \text{with} \quad \Gamma_g^{sd\dagger} = \frac{1}{\sqrt{3}}(s_0^\dagger + \sqrt{2}\, d_0^\dagger)\ . \tag{2}$$

In a similar way the lowest excited bands (β- and γ- vibrations, SU(3) representation $[2N-4, 2]$) are associated with the intrinsic states

$$|\Psi_i^{sd}\rangle = \frac{1}{\sqrt{(N-1)!}} (\Gamma_g^{sd\dagger})^{N-1} \Gamma_i^{sd\dagger} |0\rangle \quad \text{with} \quad i = \beta, \gamma\ . \tag{3}$$

These states correspond to the promotion of one of the boson of the condensate to the orthogonal $(K^\pi = 0^+)$ β-boson and $(K^\pi = 2^+)$ γ-bosons

$$\Gamma_\beta^{sd\dagger} = \frac{1}{\sqrt{3}}(-\sqrt{2}\, s_0^\dagger + d_0^\dagger) \quad \text{and} \quad \Gamma_{\gamma\pm 2}^{sd\dagger} = d_{\pm 2}^\dagger\ . \tag{4}$$

The intrinsic excitations with $K=\pm1$ generated by $\Gamma_{\pm 1}^{sd\dagger} = d_{\pm 1}^\dagger$ are instead the spurious states associated, via their combination $\Gamma_x^{sd\dagger} = \frac{1}{\sqrt{2}}(d_1^\dagger + d_{-1}^\dagger)$ and $\Gamma_y^{sd\dagger} = \frac{1}{\sqrt{2}}(d_1^\dagger - d_{-1}^\dagger)$, with rotations around the x and y axis.

A parallel treatment can be carried out for the SU(3) representation based on the negative parity p and f bosons. In this case one introduces the hamiltonian

$$H_Q^{pf} = -\kappa\, \hat{Q}^{pf} \cdot \hat{Q}^{pf}\ , \tag{5}$$

where the quadrupole operator now assumes the form

$$\hat{Q}_\mu^{pf} = \frac{9\sqrt{3}}{10}(p^\dagger \tilde{p})_{2\mu} + \frac{3\sqrt{7}}{5}(p^\dagger \tilde{f} + f^\dagger \tilde{p})_{2\mu} + \frac{3\sqrt{42}}{10}(f^\dagger \tilde{f})_{2\mu}\ . \tag{6}$$

The intrinsic state associated with the ground-state rotational band (SU(3) representation $[3N, 0]$) is

$$|\Psi_g^{pf}\rangle = \frac{1}{\sqrt{N!}} (\Gamma_g^{pf\dagger})^N |0\rangle \quad \text{with} \quad \Gamma_g^{pf\dagger} = \sqrt{\frac{3}{5}}(p_0^\dagger - \sqrt{\frac{2}{3}}\, f_0^\dagger)\ . \tag{7}$$

Intrinsic excitations can be obtained by promoting to an excited state one of the bosons of the condensate Eq. (7). Denoting with σ, π, δ and φ the excitations with $K=0,1,2$ and 3, the associated wavefunctions are

$$|\Psi_i^{pf}\rangle = \frac{(\Gamma_g^{pf\dagger})^{N-1}\Gamma_i^{pf\dagger}}{\sqrt{(N-1)!}}|0\rangle \quad \text{with} \quad i = \sigma, \pi, \delta \text{ and } \varphi, \tag{8}$$

where the excited bosons have the structure

$$\Gamma_\sigma^{pf\dagger} = \sqrt{\frac{3}{5}}(\sqrt{\frac{2}{3}}\,p_0^\dagger + f_0^\dagger) \quad , \quad \Gamma_{\delta\pm}^{pf\dagger} = f_{\pm 2}^\dagger \tag{9}$$

and

$$\Gamma_{\pi\pm}^{pf\dagger} = \sqrt{\frac{4}{5}}p_{\pm 1}^\dagger + \sqrt{\frac{1}{5}}f_{\pm 1}^\dagger \quad , \quad \Gamma_{\varphi\pm}^{pf\dagger} = f_{\pm 3}^\dagger . \tag{10}$$

Note in this case, at variance with the s-d situation, the presence of the $K = \pm 1$ operator $\Gamma_{\pi\pm}^{pf\dagger}$, which gives rise to a physical state, in addition to the orthogonal $K = \pm 1$ boson

$$\Gamma_{s\pm}^{pf\dagger} = \sqrt{\frac{1}{5}}p_{\pm 1}^\dagger - \sqrt{\frac{4}{5}}f_{\pm 1}^\dagger . \tag{11}$$

which generates the spurious states. The sets of states $|\Psi_\sigma^{pf}\rangle$ and $|\Psi_\delta^{pf}\rangle$, on the one hand, and $|\Psi_\pi^{pf}\rangle$ and $|\Psi_\varphi^{pf}\rangle$ on the other, correspond to the $[3N-4, 2]$ and $[3N-6, 3]$ representations of the SU(3) algebra, respectively.

3. The Quadrupole Plus Octupole Deformed Case

A necessary requirement of an intrinsic state wave function associated with a quadrupole plus octupole deformed system is to be characterized by mixed parity. By this we mean that our intrinsic state should simultaneously contain s-d and p-f components. From the point of view of the intrinsic ground state this can be written,

$$|\Psi_g\rangle = \frac{(\Gamma_g^\dagger)^N}{\sqrt{N!}}|0\rangle , \tag{12}$$

with a variational boson of the more general form

$$\Gamma_g^\dagger = \frac{1}{(1+\mu^2)^{1/2}}(\Gamma_g^{sd\dagger} + \mu\,\Gamma_g^{pf\dagger}) . \tag{13}$$

On the other hand, a hamiltonian with quadrupole–quadrupole interaction, although the quadrupole operator is defined as a combination of the "positive" and "negative" operators $\hat{Q}^{sd}$ and $\hat{Q}^{pf}$, i.e.

$$\hat{Q} = \hat{Q}^{sd} + \alpha\,\hat{Q}^{pf} , \tag{14}$$

does not connect states built on bosons with different parities. Our proposal for obtaining the desired mixing of parities is the use of a octupole-octupole term in the hamiltonian. We thus introduce a hamiltonian of the form

$$H = -\kappa_2\hat{Q}\cdot\hat{Q} - \kappa_3\hat{O}\cdot\hat{O} , \tag{15}$$

where the octupole boson operator takes the form

$$\hat{O}_\mu = 2(p^\dagger \tilde{d} - d^\dagger \tilde{p})_{3\mu} + \sqrt{5}(f^\dagger \tilde{s} - s^\dagger \tilde{f})_{3\mu} + \sqrt{6}(f^\dagger \tilde{d} - d^\dagger \tilde{f})_{3\mu} . \tag{16}$$

This octupole operator has been derived from the $r^3 Y_{3\mu}$ octupole matrix elements between the N=2 (s-d shell) and the N=3 (p-f shell) harmonic oscillator major shells.

By minimizing the expectation value of the total hamiltonian of Eq. (15) within the variational state Eq. (12-13) we will, in general, obtain a mixed-parity state. The equilibrium mixing parameter μ will be a function of the ratio $R = \kappa_3/\kappa_2$. From the form of the energy surface in the limit of large number of bosons N

$$E(\mu) = -\frac{N^2}{(1+\mu^2)^2} [\frac{\kappa_2}{2}(2 + 3\alpha\mu^2)^2 + \frac{8}{3}\kappa_3\mu^2] , \tag{17}$$

one finds that for small values of the octupole interaction, up to the phase transition occurring at $R^* = \frac{3(2-3\alpha)}{4}$, it is the pure sd solution which remains lowest (μ=0). For values of R larger than R^*, one obtains instead a mixing parameter

$$\mu = \left\{ \frac{4R + 3(3\alpha - 2)}{4R + 9\alpha(1 - 3\alpha/2)} \right\}^{1/2} , \tag{18}$$

reaching the asymptotic value $|\mu| = 1$, corresponding to the maximum possible mixing, in the limit of large octupole coupling. Examples of energy surfaces as a function of μ are shown in Fig. 1 for different values of the octupole strength.

The transition from the pure quadrupole to the mixed quadrupole-octupole configuration is reflected in the values of the intrinsic quadrupole and octupole moments at equilibrium, given by

$$Q_0 = N\frac{\sqrt{2} (1 + 3\alpha\mu^2/2)}{1+\mu^2} \quad \text{and} \quad O_0 = \frac{2\sqrt{2}}{\sqrt{3}} \frac{N\mu}{1+\mu^2} . \tag{19}$$

The mixing of parities, as expected, has little influence on the quadrupole moment but strongly affects the octupole moment.

The specific features of the ground state are reflected in the properties of the excited states of the system, namely in the properties of the excited bands. As in the pure quadrupole case the lowest bands will be associated to one-boson excitations, i.e. to intrinsic states where one of the boson of the condensate is promoted to an excited state. Some of the excitations will be associated with the s-d part of the basis, namely the β and γ bands and some with the p-f part of the basis. These states in a geometrical picture can be clearly interpreted as separate vibrations of the quadrupole and octupole components of the deformed surface. There are, however, two additional modes which come specifically from the interplay of the positive-parity and negative-parity degrees of freedom. The first one is a K=0 band, associated to a boson which is the antisymmetric combination with respect to the ground-state boson (13), i.e.

$$\Gamma^\dagger_\Sigma = \frac{1}{(1+\mu^2)^{1/2}}(-\mu\,\Gamma^{sd\dagger}_g + \Gamma^{pf\dagger}_g) . \tag{20}$$

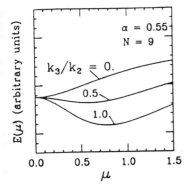

Fig. 1. Energy surfaces (Eq. (17)) as a function of μ for different values of the octupole strength.

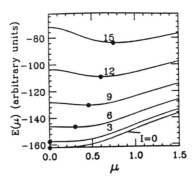

Fig. 2. Energy surfaces ($\alpha = 0.55$, $N = 9$ and $\kappa_3/\kappa_2 = 0.25$) in units of κ_2 as a function of μ for different values of the spin I. The dots mark the equilibrium configurations.

The other one is a $K = \pm 1$ band associated with the proper antisymmetric combination of the two spurious bosons associated with the s-d and p-f representation, i.e.

$$\Gamma^\dagger_{\Pi\pm} = \frac{1}{(1+3/2\mu^2)^{1/2}} \left(\sqrt{\frac{3}{2}}\mu\, d^\dagger_{\pm 1} - \frac{1}{\sqrt{5}} p^\dagger_{\pm 1} + \frac{2}{\sqrt{5}} f^\dagger_{\pm 1} \right) . \qquad (21)$$

The symmetric combination gives rise instead to the two spurious bosons $\Gamma^\dagger_{s\pm}$ of the combined space.

The predicted sequence of low-lying bands can be determined by taking the expectation values of the hamiltonian with the corresponding wave functions and are given explicitly in ref[5].

The transition to static octupole deformation and its magnitude do not only affect the sequence and the energies of the excited bands, but obviously also characterizes static multipole moments of these bands (and consequently the behaviour of electromagnetic transitions within each band) as well as the transitions connecting members of different bands. Although we have restricted our hamiltonian to include only quadrupole and octupole terms, the predominant role in the electromagnetic transitions will be played by dipole terms. In addition to the boson quadrupole and octupole operators we have therefore to introduce a dipole operator of the form

$$\hat{D}_\mu = \sqrt{8}(d^\dagger \tilde{p} - p^\dagger \tilde{d})_{1\mu} + \sqrt{10}(s^\dagger \tilde{p} - p^\dagger \tilde{s})_{1\mu} + \sqrt{42}(d^\dagger \tilde{f} - f^\dagger \tilde{d})_{1\mu} , \qquad (22)$$

which is again computed using the harmonic oscillator shells N=2 and 3, as in the case of the other multipole operators.

Expressions for all diagonal moments of all the bands and for the transition matrix elements between members of different bands can be easily calculated and are given explicitly in ref[5].

4. Spin Dependence

The considerations we have made so far are relevant in the low-spin regime. From the available experimental evidence it seems that nuclei that show octupole deformation evolve their equilibrium configuration with rising spin, passing from a situation characterized by pure quadrupole shape to a situation of coexistence of both quadrupole and octupole deformations. Within our scheme, this means that in the physical case we expect an evolution of the internal structure of the ground state band-head, with a condensate of quadrupole deformed bosons with positive parity for the lower angular momenta ($\mu = 0$, Eq. (13)) to a condensate of bosons with a mixed parity deformation ($\mu \neq 0$). This is produced by the increasing value of the moment of inertia $\mathcal{J}$ with μ, which in turn makes the excited states of higher angular momenta prefer a ground state with mixed deformation in order to decrease their energies. A simple way of showing this transition is obtained by calculating the energy surfaces $E(\mu)$ for different values of the spin I in the form

$$E_I(\mu) = E_{I=0}(\mu) + \frac{1}{2} \frac{\hbar^2 I(I+1)}{\mathcal{J}(\mu)} . \tag{23}$$

To evaluate the inertial parameter associated to the rotations, one can separate the canonical pair of coordinate and momentum that is associated to the Goldstone (zero energy) boson. This is the Thouless-Valatin[6,7] prescription, which yields the following expression for the moment of inertia:

$$\mathcal{J} = \frac{4(1 + \frac{3}{2}\mu^2)}{\kappa_2 \left[\frac{2+3\alpha\mu^2}{1+\mu^2} - \frac{2+9\alpha\mu^2}{2(2+3\mu^2)} \right] (2 + 3\alpha\mu^2) + \frac{16\kappa_3\mu^2}{3(1+\mu^2)}} . \tag{24}$$

For fixed values of the quadrupole and octupole strengths this gives rise to an increasing function of the mixing parameter μ. The resulting energy surfaces are displayed in Fig. 2, in units of κ_2, for the case of $N = 9$ bosons, with a quadrupole operator parameter $\alpha = 0.55$ and relative octupole interaction strength $R = 0.25$. The spot on each curve indicates the equilibrium configuration. As apparent from the figure, with this choice of parameters the octupole interaction is not strong enough to induce a stable octupole deformation at low spins, but this deformation becomes favoured at higher spins, due to the increasing moment of inertia.

We have applied our model hamiltonian with the parameters given above to the case of the low-lying positive-parity and negative-parity bands in ^{226}Ra. For each even spin the energy of the level has been obtained as the equilibrium minimum value of the corresponding energy surface. The calculated spectrum compared with the experimental one is shown in Fig. 3. We would like to point out that, aside from the choice of κ_2, no special search has been done for the other parameters. It was not in any case our intention to provide a precise fit to the data, but rather to show that our model hamiltonian does in fact give rise to a number of features which are present in the experimental data.

We can also compare the behaviour of the dipole interband decay $E(1)$ using the dipole operator $\hat{D}$ defined in Eq. (22). In the high spin regime, where both positive- and negative-parity states belong to the same ground-band, we have used the diagonal expression with the corresponding value of μ. In the low spin part of the spectrum the negative-parity states belong to the octupole-band "Σ", and we have used the inter-band expression (with $\mu=0$),

The calculated dipole transition matrix elements, with a common normalization factor, are compared in the lower part of Fig. 3 with the experimental values. Again the main features of the dipole transition, in particular the existence of a low-spin and a high-spin regime, are reproduced.

5. Coupling of an Odd Fermion to the Octupole Deformed Even Core

We now consider the case of the neighbouring odd system, by adding an unpaired fermion to the even boson core. Our boson-fermion system will be described by

$$H = H_B + H_F + V_{BF}, \qquad (25)$$

the different terms being associated with the purely bosonic Hamiltonian (H_B), the fermionic Hamiltonian (H_F) and the interaction between the bosons and the fermion (V_{BF}). As boson hamiltonian we consider the hamiltonian Eq. (15), which includes both quadrupole and octupole terms. For simplicity, we choose a boson-fermion interaction of the same structure, with both quadrupole and octupole components,

$$V_{BF} = -\kappa'_2 \hat{Q}_B \cdot \hat{Q}_F - \kappa'_3 \hat{O}_B \cdot \hat{O}_F, \qquad (26)$$

with

$$\hat{Q}_F = \sum_{jj'} <j||Y^{(2)}||j'\rangle (a_j^\dagger \tilde{a}_{j'})^{(2)} \quad \text{and} \quad \hat{O}_F = \sum_{jj'} <j||Y^{(3)}||j'\rangle (a_j^\dagger \tilde{a}_{j'})^{(3)}, \qquad (27)$$

where j,j' denote the (spherical) single particle orbitals.

The basic idea, as stated before, is to consider a boson coherent state and to take the expectation value of the hamiltonian in that state. The mean field hamiltonian is then diagonalized and provides a way to determine the single particle levels in presence of quadrupole–octupole deformation. In that way, it can be studied the model sensitivity in predictions about octupole deformation in nuclear ground and excited states. The mean-field hamiltonian for the additional fermion is therefore

$$H = E_B + H_F - \kappa'_2 \langle \Psi_g | \hat{Q}_B | \Psi_g \rangle \cdot \hat{Q}_F - \kappa'_3 \langle \Psi_g | \hat{O}_B | \Psi_g \rangle \cdot \hat{O}_F \qquad (28)$$

where $E_B = \langle \Psi_g | H_B | \Psi_g \rangle$ is the expectation value of the boson hamiltonian and $H_F = \sum_j \epsilon_j a_j^\dagger \tilde{a}_j$ the fermion single-particle hamiltonian. We note that the boson energy surfaces E_B depend quadratically on the boson number N, while the boson-fermion interaction depends only linearly on N. We shall assume that (when N is large) the structures of the basic bosons are determined by minimizing the boson part E_B. We shall therefore take as equilibrium forms the same ones we had obtained for the pure boson part, neglecting the changes due to the extra particle.

In the limit of large N the mean field hamiltonian Eq. (28) provides a way to determine the single-particle levels in the presence of quadrupole and octupole deformations. To obtain these levels we diagonalize H, taking the known expectation values of the quadrupole and octupole boson operators $\hat{Q}_B$ and $\hat{O}_B$ given by Eq. (19). We show in Fig. 4 the behaviour of the single particle levels in the region around neutron number 138.

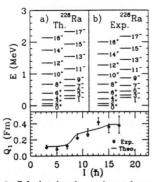
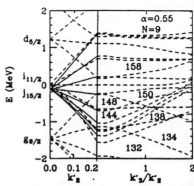

Fig. 3. Calculated and experimental spectrum of ^{226}Ra. In the lower part, experimental and calculated dipole transition matrix elements.

Fig. 4. Single particle levels as a function of the quadrupole strength (left part with κ'_2 in MeV) and then for fixed value of κ'_2 as a function of κ'_3/κ'_2 in the region around neutron number 138.

6. Acknowledgements

This work was performed in part with grants from CE contract CHRX-CT92-0075, DGICYT project PB92-0663, INFN-DGICYT and INFN-CNEA exchange collaborations, the Antorchas Foundation, and CONACyT, Mexico, under project 400340-5-3401E.

7. References

1. F.Iachello and A.Arima, *The interacting boson model* (Cambridge University Press, Cambridge, 1987).
2. J.N.Ginocchio and M.W.Kirson, *Nucl. Phys.* **350** (1980) 31.
3. A.Leviatan, *Ann. Phys. (N.Y.)* **179** (1987) 201.
4. J.Engel and F.Iachello, *Nucl. Phys.* **472** (1987) 61.
5. C.E. Alonso et al., *to be published*.
6. D.J.Thouless, and J.G.Valatin, *Nucl. Phys.* **31** (1962) 211.
7. J.Dukelsky, G.G.Dussel, R.P.J.Perazzo, S.L.Reich and H.M.Sofia, *Nucl. Phys.* **A425** (1984) 93.

E1 and E3 Transition Rates in the sdf-IBA

N.V.Zamfir
Brookhaven National Laboratory, Upton, New York 11973, USA
Clark University, Worcester, Massachusetts 01610, USA
Institute of Atomic Physics, Bucharest-Magurele, Romania

ABSTRACT

The E1 and E3 transition operators in the sdf-IBA model are discussed. An effective E1 operator containing one- and two-body terms explains very well the E1 data in transitional and rotational nuclei. The E1 parameters for rotational nuclei are obtained by imposing a constraint on the IBA model namely, the validity of Alaga rules for "pure K" states. The systematic behavior of the octupole strength is well reproduced using a one-body octupole operator. The anomalously large fragmentation of the low-energy octupole strength in non-rotational nuclei is shown to be a signature of the O(6) dynamical symmetry.

1. Introduction

Within the Interacting Boson Model (IBA), negative parity states are obtained by incorporating negative parity bosons in addition to the usual s and d bosons. Extensive studies by Scholten, Iachello and Arima [1] and by Barfield, Barrett, Wood and Scholten [2] have used only one $J^\pi = 3^-$ f boson. The purpose of this paper is to summarize several recent studies which we have done related to the dipole and octupole electromagnetic transitions in the framework of the IBA-1 sdf model. The Hamiltonian for N_B bosons in an $(sd)^{N_B-x}f^x$ (x=0,1) system is given by [2]:

$$H = H_{sd} + H_f + V_{sdf} \qquad (1)$$

Here the first term describes the even-even positive parity core upon which the negative parity excitations are built, incorporated in the last terms via an f boson:

$$H_{sd} = \epsilon_d \hat{n}_d + a_1 L_d \cdot L_d - a_2 Q_{sd} \cdot Q_{sd} \qquad (2)$$

where $\hat{n}_d = d^\dagger \tilde{d}$ is the d boson number operator and

$$L_d = \sqrt{10}(d^\dagger \tilde{d})^{(1)}, Q_{sd} = (s^\dagger \tilde{d} + d^\dagger s) + \chi_2 (d^\dagger \tilde{d})^{(2)} \qquad (3)$$

The two principal observed classes of shape transitions, spherical vibrator → symmetric rotor and γ soft → symmetric rotor, or U(5) → SU(3) and O(6) → SU(3), are obtained by varying ϵ_d/a_2 from $\infty \to 0$ along with an increase in $|\chi_2|$ from 0 to -1.32, and by setting ϵ_d =0 and varying χ_2, respectively. Of course, the boson number N_B increases as the rotational limit SU(3) is approached.

The f-boson Hamiltonian is $H_f = \epsilon_f \hat{n}_f$ where the number operator $\hat{n}_f$ is 0(1) for positive (negative) parity states. With no $f - sd$ interaction the structure and energy of the negative-parity states follow those of the positive parity states.

The $f - sd$ interaction can be considered in multipole form [2]:

$$V_{sdf} = A_1 L_d \cdot L_f + A_2 Q_d \cdot Q_f + A_3 : E_{df}^+ E_{df} : \qquad (4)$$

where:

$$L_f = 2\sqrt{7}(f^\dagger \tilde{f})^{(1)}, Q_f = -2\sqrt{7}(f^\dagger \tilde{f})^{(2)}, : E_{df}^+ E_{df} := 5 : (d^\dagger \tilde{f})^{(3)} \cdot (f^\dagger \tilde{d})^{(3)} : \qquad (5)$$

2. E1 transitions

2.1. E1 one body operator

First we consider the simplest case of an f boson weakly coupled to a U(5) core. The 3^- level is constructed by coupling an f ($l=3$) boson to the positive ground state whose boson structure consists of N_B s-bosons and no d bosons. The 1^- state belongs to the 2-phonon $2_1^+ \otimes 3_1^-$ quintuplet (fig.1). The one-body E1 operator is:

$$T_{sdf}^{(E1)} = e_1 [d^\dagger \tilde{f} + f^\dagger \tilde{d}]^{(1)} \qquad (6)$$

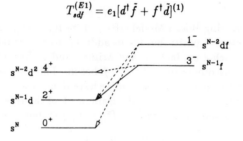

Fig. 1. Boson structure of the yrast states in the U(5) limit of the sdf-IBA. The solid (dashed) arrow corresponds to allowed (forbidden) E1 transitions for the one-body operator of eq.(6).

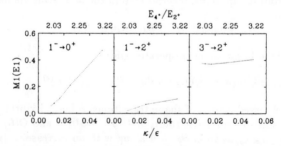

Fig. 2. Calculated matrix elements of different E1 transitions as functions of ϵ/κ (the corresponding ratio $E_{4_1^+}/E_{2_1^+}$ is given on the top scale) ($N_B=9$, $\kappa=0.05$, and $\chi_2 = -\sqrt{7}/2$)

The structure of the U(5) states leads to specific selection rules for E1 transitions (see fig.1). For example, the $1^- \to 0^+$ transition is forbidden in U(5) since it must involve the distruction of *both* an f boson *and* a d boson. This is impossible with the E1 operator in eq.(6) which creates or destroys a single f boson (and destroys or creates a d-boson). In contrast, the $3^- \to 2^+$ transition *is* allowed.

This features are changed when U(5) is broken. Any interaction that achieves such a structural change, such as the $Q \cdot Q$ interaction, will mix the U(5) basis states and, as shown in fig.2, the mixing in the 1^- and 0^+ states leads to an increasing $1^- \to 0^+$ transition strength across the spherical-deformed transition region.

2.2. E1 two-body operator

Although these schematic calculations reproduce the experimental trends of the $1^- \to 0^+$ and $3^- \to 2^+$ transition rates, in the sdf-space, a simple one-body E1 operator is unable to describe all the observed E1 transitions. This can be seen in fig. 2 : the relatively small $1^- \to 2^+$ transition strength cannot reproduce the experimentally large (~ 2) $B(E1; 1^- \to 2^+)/B(E1; 1^- \to 0^+)$ branching ratio in transitional nuclei. It thus seems necessary either to include higher-order terms [1] or to consider the full $spdf$-space [3,4,5]. In the former case, one has the problem of choosing only a few appropriate two-body terms out of the many possible to keep the number of parameters low. In the latter case, a phenomenological analysis with all terms of the $sd - pf$ interaction in the Hamiltonian is difficult. On the other hand, as pointed out in ref.[4], even small p boson admixtures lead to well behaved E1 transitions in the actinide region. In this case the p boson can be eliminated from the $spdf$-space by perturbation theory. The new [6] effective sdf-E1 operator includes two additional two-body terms relative to eq.(6):

$$T_{sdf}^{(E1)} = e_1[(d^\dagger \tilde{f} + f^\dagger \tilde{d})^{(1)} + \chi_1 O_1 + \chi_1' O_1'] \tag{7}$$

$$O_1 = [Q_{sd}^{(2)} \times (s^\dagger \tilde{f} + f^\dagger s)^{(3)}]^{(1)} \tag{8}$$

$$O_1' = \sum \sqrt{2l+1}(-1)^{l+1} \begin{Bmatrix} 2 & 1 & 1 \\ 2 & 3 & l \end{Bmatrix} [Q_{sd}^{(2)} \times (d^\dagger \tilde{f} + f^\dagger \tilde{d})^{(l)}]^{(1)} \tag{9}$$

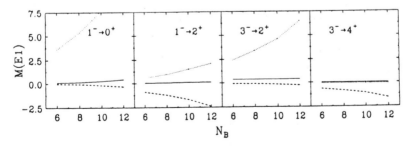

Fig. 3. Calculated E1 matrix elements indicated in different panels of the different terms in the T(E1) operator given by eq.(6) (continuous lines), eq.(8) (dotted lines), and eq.(9) (dashed lines).

In fig.3 are shown the matrix elements of the three terms in T(E1) for different transitions deexciting 1^- and 3^- states. Considering both one- and two-body terms it is possible to obtain appropriate dipole transition rates from 1- and 3- states. For example, only the two-body operator of eq.(9) gives a non-zero $3^- \to 4^+$ matrix element, but only a combination of all three contributions reproduces the experimental E1 branching ratio. The parameters χ_1 and χ'_1 in T(E1) can be determined from the data. This procedure has been applied to the transitional Sm isotopes [7,9]. Figure 4 shows the E1 transitions rates in these isotopes keeping the effective E1 charge constant ($e_1 = 0.30$ efm) in order to see if the internal structure of the E1 operator (eq.(7-9)) is able to reproduce the sharp increase of the transition strength. The other parameters used were $\chi_1 = -0.014$ and $\chi'_1 = 0.38, 0.12$ and -0.06 for $^{148,150,152}Sm$, respectively. For the rotational nucleus ^{154}Sm the values $\chi_1 = -0.030$ and $\chi'_1 = -0.060$ were deduced from an Alaga rule constraint (see below). The calculations reproduce the experimental branching ratios and give the sharp increase in both $B(E1; 3^- \to 2^+)$ and $B(E1; 1^- \to 0^+)$ values between ^{150}Sm and ^{152}Sm that reflects the changing structure of the wave functions in the shape change region at N = 90. The saturation of both B(E1) values in $^{152,154}Sm$ is well reproduced. A weak dependence of the effective charge on N ($e_1 = 0.2 - 0.3 efm$) could further improve the agreement for the B(E1) values, especially in ^{148}Sm.

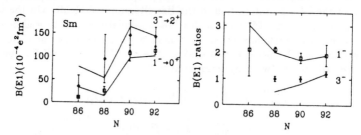

Fig. 4. Comparison of experimental (symbols) and calculated (lines) reduced E1 transition probabilities for decay of the 1_1^- and 3_1^- states in Sm: a) absolute values b) branching ratios.

2.3. Alaga rules constraints

For rotational nuclei we used a method in which the two parameters are predicted by imposing the Alaga rules on model states of rather pure K quantum number. For a large quadrupole force (A_2), the negative parity wave functions have pure K. Thus, the branching ratios of the $J_K = 1^-$ states equal the Alaga values:

$$\frac{B(E1; 1^-_{K=0-} \to 2^+_1)}{B(E1; 1^-_{K=0-} \to 0^+_1)} = R_0(\chi_1, \chi'_1) = 2.00 \tag{10}$$

$$\frac{B(E1; 1^-_{K=1-} \to 2^+_1)}{B(E1; 1^-_{K=1-} \to 0^+_1)} = R_1(\chi_1, \chi'_1) = 0.50 \tag{11}$$

The Alaga rule constraint determines the parameters χ_1 and χ'_1. The extracted χ

values are nearly independent on the Hamiltonian's parameters. They have a slight N_B dependence: χ_1 varies from -0.030 to -0.023 and $\chi'_1 = 2\chi_1$, for N_B=11-16.

In real nuclei, due to the K-mixing the experimental branching ratios deviate, sometimes significantly, from the Alaga rules. The sdf-IBA Hamiltonian, used to reproduce the experimental energies, includes this effect. The E1 operator, determined by the above procedure and containing N_B-dependent parameters, can be used to predict the B(E1) branching ratios in real nuclei. This is shown for two nuclei in fig.5 for branching ratios called $R_1(J)$, and defined analogously to eq.(11) for a sequence of J values. Figure 5 also shows the ratios $R_2(J)$, defined in a similar way, except for $K = 2^-$ initial states, as well as the B(E2)/B(E1) ratios defined in the figure. In all cases the agreement with the data is excellent [8].

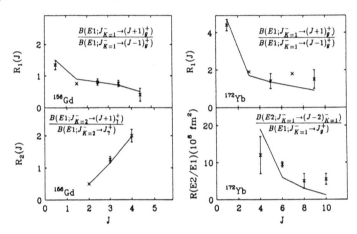

Fig. 5. Experimental and predicted (continuous lines) branching ratios in ^{156}Gd and ^{172}Yb.

3. E3 transitions

A similar approach allows us to study E3 transitions in same shape transitional regions. The E3 transition rates are calculated with the operator $T(E3) = e_3 O^{(3)}$, where e_3 is the boson octupole effective charge, and the octupole operator is[2]:

$$O^{(3)} = s^\dagger \tilde{f} + \chi_3 (d^\dagger \tilde{f})^{(3)} + h.c. \qquad (12)$$

In calculating the octupole strength some parameters are of no interest. For instance, the parameter ϵ_f merely sets the overall excitation energy of the negative-parity states and the terms $a_1 L_d \cdot L_d$ and $A_1 L_d \cdot L_f$ modify the relative energy without changing their wave functions. We therefore set ϵ_f=1, $a_1 = 0$ and $A_1 = 0$ in the following calculations. The so-called "exchange term", proportional to A_3, is important only in deformed nuclei where the correct energy ordering of the octupole bands ($K^\pi = 0^-, 1^-, 2^-$, or 3^-) cannot be obtained without such a term [2]. In fact, as was shown in ref. [10], in deformed nuclei the only critical parameter is the ratio $F_2 = A_3/A_2$ of the strength of the exchange term to that of the $Q_d \cdot Q_f$ interaction.

In order to simplify the calculations as much as possible we adopted the following schematic procedure. To reproduce the shape transition (as seen in the (sd) core states), the parameters were varied linearly: $a_2 = 0.05$ MeV, $\chi_2 = 0.22 - 0.055 N_B$ and, $\epsilon_d = 2.667 - 0.167 N_B$ for for U(5) $\to$ SU(3) or $\epsilon_d = 0$ for O(6)$\to$ SU(3). Similarly, for V_{sdf}: $A_2 = -0.04$ MeV and $A_3 = -0.02(N_B$-8) MeV for $N_B > 8$ and $A_3 = 0$ otherwise. This parameter values reproduce rather well the octupole band structure in vibrational [11], deformed [1,2,10] and in γ-soft [12] nuclei.

Although the calculated wavefunctions may be rather complex, it is possible to gain a simple qualitative understanding of the E3 strengths by exploiting the selection rules for the operator of eq.(12) whose the $3_i^- \to 0_1^+$ matrix elements are:

$$< 3_i^- |O^{(3)}|0_1^+ > \; = \; < \alpha_i(sd)^{N_B-1}f|s^\dagger \tilde{f} + \chi_3(d^\dagger \tilde{f})^{(3)} + h.c.|\alpha_0(sd)^{N_B} >$$
$$= \; < \alpha_i s^{N_B-n_i-1}d^{n_i}|k_1 s^\dagger + k_2\chi_3 d^\dagger|\alpha_0 s^{N_B-n_0}d^{n_0} >$$
$$= \; k_3 < \alpha_i s^{N_B-n_i}d^{n_i}|\alpha_0 s^{N_B-n_0}d^{n_0} > + \chi_3 k_4 < \alpha_i s^{N_B-n_i-1}d^{n_i+1}|\alpha_0 s^{N_B-n_0}d^{n_0} >$$
$$= \; k_3 \delta_{n_0,n_i}\delta_{\alpha_0,\alpha_i} + \chi_3 k_4 \delta_{n_0,n_i+1}\delta_{\alpha_0,\alpha_i} \tag{13}$$

where α_0 and α_i are additional quantum numbers and k_j are geometrical coefficients whose detailed values are not important in determining the selection rules. n_0 and n_i are the numbers of d bosons in the 0_1^+ and 3_i^- states, respectively.

Octupole transitions to the ground state can only occur between components of a 3_i^- state that have the same number of d bosons as a component of the ground state $(n_i = n_0)$, or one d boson less $(n_i = n_0 - 1)$. Thus, the distribution of octupole strength depends on the d-boson structure of the 3^- levels and the ground state, which is expected to differ in different regions. In fig. 6, the evolution of the expectation number of d bosons in the ground state and in different 3_i^- states for the U(5)$\to$ SU(3) transition (a) and for O(6) nuclei (b) is shown[13].

The essential point arises from the d-boson structure of the ground and 3^- states. In the U(5) symmetry, $< n_d > = 0$ for 0_1^+ and $3_1^- = f \otimes 0_1^+$ states and they satisfie the selection rule for $\chi_3 = 0$. In any other case, O(6), SU(3) or intermediate, the d-boson structure of the 0_1^+ and the 3^- levels, becomes more complex.

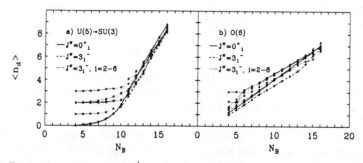

Fig. 6. Expectation values of n_d in 0_1^+ and 3_i^- states for a) U(5) $\to$ SU(3) and b) O(6) type nuclei.

The differences in octupole strength distribution between nuclei of different types are illustrated in fig. 7. In the U(5) type nuclei, the majority of the octupole strength is concentrated in the 3_1^- state. It is clear from fig. 7 that the behavior in ^{194}Pt is quite different from this, in that there is a large concentration of strength nearly 1 MeV above the 3_1^- state. However, this is readily understood. ^{194}Pt is an O(6)-like nucleus and the octupole strength in O(6) is calculated to be much more fragmented than in the U(5) → SU(3) region. This is illustrated in the middle and right panels of fig.7. The O(6) prediction is similar to the data for ^{194}Pt. This fragmentation of the octupole strength may be a signature of O(6) of nuclei.

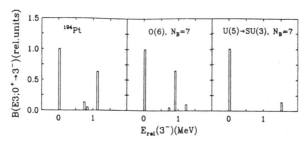

Fig. 7. Comparison of experimental octupole strength distribution in ^{194}Pt (left panel) and calculated for O(6) (middle panel) and for U(5) → SU(3) (right panel).

In rotational nuclei, a large positive value for χ_3 (ref.2) leads to a prediction of a significant octupole strength for 3^- states whose energies are 2-4 times the energy of the f boson. This is illustrated in fig.8 which shows the difference in B(E3) strength as a function of $E(3_i^-)$ for $\chi_3=0$ and 4 (a large value).

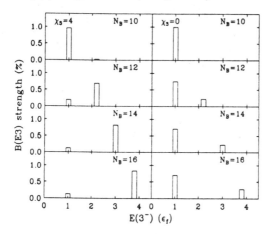

Fig. 8. The octupole strength distributions for different N_B and $\chi_3=0$ (dashed histograms) and $\chi_3=4$ (continous histograms) for non-zero (left panels) and zero (right panels) exchange interaction.

The strength corresponding to $E(3^-) = \epsilon_f$ incorporates the $B(E3)$ values of the first three octupole vibrational states ($K = 0^-, 1^-$, and 2^-). The strength at higher energy corresponds to the $K = 3^-$ state based on the lowest SU(3) irreducible representation of the positive-parity core. For all $N_B \geq 12$, a significant fraction of the strength occurs above $2\epsilon_f$. In nuclei, the Low Energy Octupole Resonance (LEOR)[14], which cannot be probably reproduced by using only one type of f boson, is located at similar energies. Thus, the high energy octupole states predicted by the model are either experimentally masked by the LEOR or they constitute a spurious prediction of the model. Assuming the latter as more likely provides a constraint on allowable χ_3 values: they cannot be so large to give significant E3 strength above the 3_1^- state. Using a smaller χ_3, the experimental octupole distribution among the low-lying states can still be reproduced by also changing the parameters in the Hamiltonian (V_{sdf}) (keeping the same A_3/A_2 ratio in order to have the same order of the octupole bands [10]).

4. Acknowledgements

Many of the aspects discussed here evolved in collaborations with P. von Brentano, R.F.Casten, O.Scholten, A.Jungclaus and P.Cottle for which I am thankful.

5. References

1. O. Scholten, F. Iachello and A. Arima, Ann. Phys. **115** (1978) 325
2. A.F.Barfield et al., Ann.Phys. **182** (1988) 344
3. J. Engel and F. Iachello, Nucl. Phys. **A472** (1987) 61
4. D. F. Kuznezov and F. Iachello, Phys. Lett. **209B** (1988) 420
5. T. Otsuka and M. Sugita, Phys. Lett. **209B** (1988) 140
6. N. V. Zamfir, O.Scholten and P.von Brentano, Z. Phys. **A337** (1990) 293
7. A. Jungclaus et al., Phys.Rev. **C47** (1993) 1020, **C48** (1993) 1005
8. P.von Brentano, N.V.Zamfir and A.Zilges, Phys. Lett. **B278** (1992) 221
9. N. V. Zamfir and P.von Brentano, Phys. Lett. **B289** (1992) 245
10. R.F.Casten, W.-T.Chou and N.V.Zamfir, Nucl.Phys. **A555** (1993) 563
11. M.Pignanelli at al, Nucl.Phys. **A519** (1990) 567
12. J.Engel, Phys.Lett. **B171** (1986) 148
13. N.V.Zamfir et al., Phys.Rev.**C48** (1993) 1745
14. M.Kirson, Phys.Lett. **108B** (1982) 237

335

sdg INTERACTING BOSON MODEL: SOME ANALYTICAL AND NUMERICAL ASPECTS

Y.D. Devi[1] and V.K.B. Kota
Physical Research Laboratory, Ahmedabad 380 009, India

Abstract

An overview of the analytical and numerical aspects of the sdg interacting boson model, the former using the dynamical symmetries and related coherent states and the latter using the SDGIBM1 code, are given with several empirical examples. The most recent developments in sdgIBM include: (i) analytical results for M1 strengths of scissors states and M1 distributions in general for odd-A nuclei; (ii) expressions for group representation matrix elements relevant for sub-barrier fusion studies; (iii) coupling schemes for describing the structure of excited rotational bands.

1. Introduction

In the past few years considerable amount of experimental data on E4 observables that involve hexadecupole degree of freedom in nuclei has started accumulating and their theoretical understanding is a challenging problem. This together with the success, both in it's analytical and numerical formulation of sd interacting boson model[1] (sdIBM) in the description of quadrupole collective properties, the importance of G ($L^\pi = 4^+$) pairs as brought out by the microscopic theories of IBM and several other indirect signatures indicating that g ($\ell^\pi = 4^+$) bosons should be included in IBM, led to the development in considerable detail, of the extended sdg interacting boson model (sdgIBM). The sdgIBM is the only pluasible model that allows one to systematically analyze hexadecupole data and understand the role of hexadecupole degree of freedom in nuclei[2].

Sections 2 and 3 summarize the various analytical and numerical aspects respectively, of sdgIBM. A schematic overview of these developments till 1992 is given in Fig. 1 and the references therein provide complete bibliography. Section 4 gives details of three most recent developments (which are analytical in nature) in sdgIBM.

2. Analytical Results in sdgIBM

The interacting boson model[1] represents a major breakthrough in nuclear structure studies as it gave rise to large number of analytical formulas for various spectroscopic observables - this implortant feature dervies from the classification of dynamical symmetries (vibrational $U_d(5)$, rotational $SU_{sd}(3)$ and γ - unstable $O_{sd}(6)$). The

---
[1]Present address: Saha Institute of Nuclear Physics, Calcutta 700 064, India

progress in this direction in sdgIBM though slow to start with, due to the complexity of the model, now substantial number of analytical formulas are available in literature which facilitate rapid analysis of data[2].

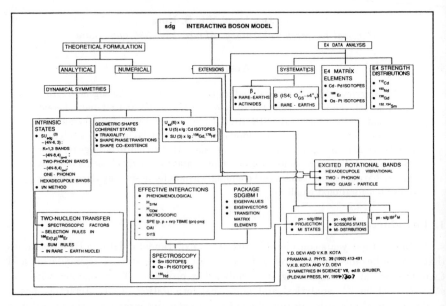

Fig.1 Schematic overview of various developments in sdgIBM

In sdgIBM the states of N-boson system belong to the totally irreducible representation (irrep) $\{N\}$ of U(15) group and it admits seven dynamical symmetries and they correspond to four strong coupling limits $SU_{sdg}(3)$, $SU_{sdg}(5)$, $SU_{sdg}(6)$ and $O_{sdg}(15)$ and three weak coupling limits $U_{sd}(6) \oplus U_g(9)$, $U_{dg}(14)$ and $U_d(5) \oplus U_{sg}(10)$ respectively. The classification of band structures and analytical formulas for energies, E2, E4 transition matrix elements and two nucleon transfer (TNT) strengths are derived in the $SU_{sdg}(3)$ and $[U_{sd}(6) \supset G] \oplus U_g(9)$ limits, in the latter case with the number of g bosons n_g=0,1 and the group G being one of the three sdIBM groups. The benchmark example for $SU_{sdg}(3)$ limit is from ^{166}Er(t,p)^{168}Er TNT data, for bands built on one phonon hexadecupole vibrations (described by $SU_{sd}(3) \times 1g$ case of $U_{sd}(6) \oplus U_g(9)$) come from ^{156}Gd, ^{178}Hf E2 data and for the $U_d(5) \times 1g$ limit from 4$^+$ states in ^{110,114,116}Cd isotopes. Another class of analytical expressions are derived by constructing sdgIBM coherent states (CS) in terms of the quadrupole and hexadecupole deformation parameters (β_2 and β_4) and the asymmetry angle γ. The equilibrium shape parameters (β_2^0, β_4^0, γ^0) in the symmetry limits and for the general sdgIBM hamiltonian show that sdgIBM admits triaxial shapes, shape coexistence and shape phase transitions. Using CS analytical formulas for a particular ratio R involving TNT strengths, for β_4's and B(IS4; $0^+_{GS} \to 4^+_\gamma$) are obtained. The corresponding data is well described by sdgIBM. A further extension of CS with angular momentum projection is easy to deal with in axial symmetry case and it has important applications among other things in sub-barrier fusion studies; see Sect. 4.2 ahead. Analytical formulation is also available in proton-neutron sdgIBM (pn-sdgIBM) and proton-neutron sdg interacting boson-fermion model (pn-sdgIBFM); the latter for odd- mass nuclei. A recent application of this is given in Sect. 4.1.

3. Numerical Aspects of sdgIBM

The interacting boson model has made systematic study of transitional nuclei tractable as the matrix dimensions in sd boson space are small and the hamiltonian contains few parameters (often being generated by interpolating the dynamical symmetry limits). Once again here also the progress in sdgIBM was rather slow till 1988 as the matrix dimensions here are large (~ 1000) and the general (1+2) body hamiltonian has far too many (3+32=35) free parameters. Both these problems are now cured[2]. In order to perform g boson mixing calculations with truncations based on g-boson number ($n_g \leq 6$), a computer code SDGIBM1 for constructing and diagonalizing hamiltonian matrices for the most general sdg IBM hamiltonian in the spherical basis ($|n_s; n_d, v_d, \alpha_d, L_d; n_g, v_g, \alpha_g, L_g; L\rangle$, where n_d, v_d, α_d and L_d are boson number, seniority quantum number, multiplicity lable and angular momentum quantum number respectively for d bosons and similarly n_g, v_g, α_g and L_g for g bosons; n_s is s boson number and $\mathbf{L} = \mathbf{L}_d + \mathbf{L}_g$) and calculating matrix elements of various transition operators using the resulting wavefunctions is developed, tested and documented. Spectroscopic calculations (spectra, E2 and E4 properties) for spherical - deformed Sm isotopes, ^{150}Nd and rotor - γ unstable transitional Pt - Os isotopes clearly established that $n_g \leq 4$ calculations are sufficient. A simple approach for deriving a few parameter sdg hamiltonian, that can be successfully used in practice, is to employ the model spe(pp+nn)-TBME(pn)-proj[5]. Starting from a quadrupole-quadrupole and hexadecupole-hexadecupole force in fermion space, this model yields a consistent-(Q^2, Q^4) hamiltonian with the relative effective charges derived from microscopic considerations. Finally it should be mentioned that SDGIBM1 code is useful for many other purposes; see Fig.2 for an example which shows the results of the calculations that are performed using the package SDGIBM1 by switching off g bosons and employing the hamiltonian $H = Q^2(\chi) \cdot Q^2(\chi);\ \ Q^2(\chi) = (s^\dagger \tilde{d} + d^\dagger s)^2 + \chi(d^\dagger \tilde{d})^2$ that interpolates $SU_{sd}(3)$ and $O_{sd}(6)$ limits for $-\sqrt{7}/2 \leq \chi \leq 0$.

4. Recent Applications of sdgIBM

4.1. Scissors States and M1 Distributions in ^{163}Dy

The discovery of M1 excited scissors 1^+ states in ^{156}Gd by Bhole et. al[7] has

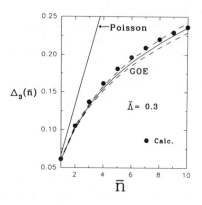

Fig. 2. Dyson-Mehta spectral rigidity statistic $\Delta_3(\bar{n})$ vs $\bar{n}$ for energy level fluctuations for 20 boson sdIBM system which exhibits GOE (Gaussian Orthogonal Ensemble) fluctuations[6]. The results are for the $SU_{sd}(3)$ to $O_{sd}(6)$ transition where the transition parameter $\bar{\Lambda} = d_L\chi^2/d_{L_{max}}\chi_1^2 = 0.3; \chi_1 = -\sqrt{7}/2$ and $d_{L_{max}} = 121$ for N = 20. In the present calculation eigenvalues from all L values are combined for fixed $\bar{\Lambda}$ which gives rise to an ensemble of 1356 eigenvalues and the ensemble averaged $\Delta_3(\bar{n})$ is what is shown in the figure. The dashed curves correspond to sample size errors on the GOE values.

led to large number of studies both theoretical and experimental on collective M1 states in even-even nuclei throughout the periodic table. The simplest description of the structure of scissors states is in terms of the pn-IBM. The growing data on several 1^+ states and the resulting M1 strength distributions led to the study of M1 states in pn-sdgIBM, to account for the observed fragmentation, as pn-sdgIBM accomodates a larger variety of 1^+ states. The question of scissors states in odd-A nuclei, employing the $SU(3) \otimes U(2)$ limit of pn-sdgIBFM was addressed for the first time in Ref.3 and detailed analytical formulation for predicting the location and M1 excitation strength of the scissors states and M1 distributions in general, is given. A European collaboration[4] has recently announced the first observation of scissors states in odd-mass nuclei - they studied ^{163}Dy nucleus with nuclear resonance flourescence technique. The data is interpreted by them using a highly simplified version of the formalism we developed earlier[3]. It is shown below that $SU(3) \otimes U(2)$ limit of pn-sdgIBFM not only keeps intact all the agreements they obtained but also accounts for the strength observed between 1.942 MeV to 2.587 MeV.

In even-odd nuclei, as is the case with ^{163}Dy, with the odd-neutron occupying the natural parity orbits in the 82-126 shell and the core nucleus described by the SU(3) limit of pn-sdgIBM, the group chain and the corresponding basis states (with F-spin) in the $SU(3) \otimes U(2)$ limit of pn-sdgIBFM are

$$\begin{aligned}
&\left| \begin{array}{cccccc} U^B_{sdg}(30) & \otimes & U^F(30) & \supset & U^B_{sdg}(15) & \otimes & SU^B(2) & \otimes & U^F(15) & \otimes & U^F(2) \\ \{N\} & & \{1\} & & \{f\} = \{f_1, f_2\} & & F = \frac{1}{2}(f_1 - f_2) & & \{1\} & & \end{array} \right. \\
&\supset \begin{array}{cccccc} SU^B_{sdg}(3) & \otimes & SU^B(2) & \otimes & SU^F(3) & \otimes & U^F(2) & \supset & SU^{BF}(3) & \otimes & SU^B(2) & \otimes & U^F(2) \\ (\lambda_B, \mu_B) & & F & & (40) & & & & (\lambda_{BF}, \mu_{BF}) & & F & & \end{array} \\
&\supset \left. \left[\begin{array}{ccc} O^{BF}(3) & \otimes & SU^F(2) & \supset & Spin^{BF}(3) \\ L & & \tilde{s} = 1/2 & & J \end{array} \right] \otimes \left[\begin{array}{cc} SU^B(2) & \supset & U^B(1) \\ F & & F_z = \frac{1}{2}(N_\pi - N_\nu) \end{array} \right] \right\rangle
\end{aligned}$$
(1)

In Eq.(1) N_π and N_ν denote proton (π) and neutron (ν) boson numbers respectively. The Bose-Fermi $SU^{BF}(3)$ irreps (λ_{BF}, μ_{BF}) for the ground state (GS) are denoted by (λ_{GS}, μ_{GS}) and they are determined by the pseudo-Nilsson versus IBFM correspondence. With this the GS in pn-gIBFM is denoted by $\left|\{N\}(4N,0)\otimes(40);(\lambda_{GS},\mu_{GS})KL;J;F=\tfrac{1}{2}N,F_z=\tfrac{1}{2}(N_\pi-N_\nu)\right\rangle$. Ignoring the single particle part, the general M1-operator in pn-sdgIBFM is

$$T^{M1} = \sum_{\rho=\pi,\nu} \sqrt{10}\left\{g_{d,\rho}\left(d^\dagger_\rho \tilde{d}_\rho\right)^1 + \sqrt{6}g_{g,\rho}\left(g^\dagger_\rho \tilde{g}_\rho\right)^1\right\} \qquad (2)$$

which has four g-factors; with respect to SU(3) the M1 operator has a piece behaving as (11) SU(3) irrep and another as (33) irrep. The M1 excited states from the GS are labelled by $\left|\{f'\}(\lambda',\mu')\otimes(40);(\lambda_f,\mu_f)K_fL_f;J_f;F'=\tfrac{1}{2}N,F'_z=\tfrac{1}{2}(N_\pi-N_\nu)\right\rangle$ where the core irreps (λ',μ') are (4N,0), (4N-4,2),(4N-6,3) with F'=N/2 (symmetric(s)) and (4N-2,1),(4N-4,2),(4N-6,3)² with F'=N/2-1 (mixed symmetric (ms)); note that the (4N-6,3) irrep appears twice for F'=N/2-1. It is easy to work out the M1 states (λ_f,μ_f) corresponding to the core irreps $\{f'\}(\lambda'\mu')$. Using F-spin SU(2) algebra and SU(3) Wigner-Racah algebra, compact expressions for the $B(M1;J_i\to J_f)$ are derived[3] and they involve the four physically meaningful g-factors $g_{R+}=N^{-1}\sum_\rho N_\rho g_{R\rho}$, $h_{R+}=N^{-1}\sum_\rho N_\rho h_\rho$, $g_{R-}=(g_{R\pi}-g_{R\nu})$, $h_{R-}=(h_\pi-h_\nu)$ where $g_{R\rho}=7^{-1}[(1+2x_0)g_{d\rho}+(6-2x_0)g_{g\rho}]$, $h_\rho=(g_{d\rho}-g_{g\rho})$, $x_0=(4N+4)/(4N-1)$ and $\rho=\pi,\nu$. The location of M1 states is calculated using the energy formula

$$E(F;(\lambda_B\mu_B)\otimes(40)(\lambda_{BF}\mu_{BF})KLJM) =$$
$$2N[\tfrac{1}{4}\Gamma(8-3\mu_{BF})+\tfrac{1}{120}\Lambda(8-3\mu_{BF})^2]+\alpha(\tfrac{1}{2}N-F)(\tfrac{1}{2}N+F+1)+\gamma_1 L(L+1)$$
$$+\gamma_2 J(J+1)+\beta\{[\lambda_B^2+\mu_B^2+\lambda_B\mu_B+3(\lambda_B+\mu_B)]-(16N^2+12N)\} \qquad (3)$$

where $\Gamma,\Lambda,\gamma_1,\gamma_2,\alpha$ and β are free parameters. The GS of ^{163}Dy nucleus belongs

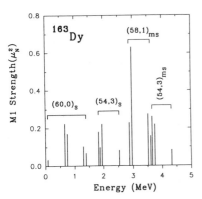

Fig. 3. M1 strength distribution in ^{163}Dy. The numbers used in the energy formula (3) are Γ = -45 keV, Λ = 450 keV, γ_1 = 0.5 keV, γ_2 = 10keV, α = 120 keV and β = -2.3 keV. Similarly the g-factors used are g_{R+} = 0.35 μ_N, h_{R+} = 0.424 μ_N, g_{R-} = 0.4 μ_N, h_{R-} = -0.8 μ_N. The final state core $(\lambda'\mu')$ $SU^B(3)$ irreps are shown in the figure; N = 15, N_π = 8 and N_ν = 7. Note that strengths less than 0.03 μ_N^2 are not shown.

to the pseudo-Nilsson configuration $[\widetilde{422}]$ with the corresponding $SU^{BF}(3)$ state being $|(4N,2)K = 2\ L = 2\ J = 5/2^-\rangle$. The prediction for M1 distribution in ^{163}Dy is shown in Fig.3. The g-factors $g_{R\pm}$ and $h_{R\pm}$ are estimated as in Ref.3.

Identifying the scissors state at 2.958 MeV as the state with $SU(3) \otimes U(2)$ configuration $|(58,1)\ F=N/2-1;\ (58,3)\ K=2\ L=3\ J=7/2^-\rangle$, the ratio R = B(M1; $J_{scis} \to 7/2^-_{GS-band}$)/B(M1; $J_{scis} \to 5/2^-_{GS}$) is 0.354. However, if the state is $|(58,1)\ F=N/2-1;\ (58,3)\ K=1\ L=2\ J=5/2^-\rangle$ the predicted value of R is 2.13. Bauske et. al.[4] values for the same in their simplified model are 0.36 and 2.2 respectively. The experimental value[4] of R is 0.23 ± 0.04 which confirms the $J=7/2^-$ state interpretation for the scissors state. Between 2.918 MeV and 3.107 MeV the observed B(M1↑) strength fragments into ten states with total strength 0.92 μ_N^2. Interpreting all these states to be scissors states, from Fig.3 it follows that the predicted summed strength is 0.86 μ_N^2 and it is distributed among nine states (concentrated at four energies). Similarly the total summed strength for all states (eight states) between 1.942 MeV and 2.587 MeV is 0.857 μ_N^2. Interpreting that all these states belong to the core $(54,3)_s$ irrep the predicted strength as can be seen from Fig.3 is 0.58 μ_N^2 (with $h_{R\pm} = 0.5\ \mu_N$, the summed strength will be 0.91 μ_N^2). Without the inclusion of g-bosons one cannot excite the (54,3) irrep and hence the B(M1↑) strength in this energy domain cannot be accounted for. For the same reason the truncated model adopted by Bauske et. al.[4]. also could not account for this strength. Thus the first observed data on M1 excited states in ^{163}Dy provides the confirmation of the $SU(3) \otimes U(2)$ limit of pn-sdgIBFM description and establishes that g-bosons are essential in describing M1 data in odd-mass nuclei.

4.2 Sub-barrier fusion in sdgIBM

In order to calculate sub-barrier fusion cross sections and other related quantities in IBM framework, the path integral approach to tunelling as developed by Balantekin and collaborators[8] provides the appropriate formalism. Here one starts with the hamiltonian $H = -\hbar^2/2\mu\Delta_R^2 + V(R) + H_{IBM} + H_{int}$ where R is the relative coordinate for the fusing nuclei, μ is the reduced mass, V(R) is the barrier potential (produced by nuclear and Coulomb forces) seen by the projectile, H_{IBM} is the target hamiltonian and H_{int} is the interaction between the target and the projectile. Under some suitable approximations, $H_{int} \to H_{int}^0$ where $H_{int}^{(0)} = g(R)\sum_{k=2,4}\sqrt{(2k+1)/4\pi} t_k Q_0^k$; g(R) takes into account the dependence on R. The H_{int} changes V(R) into a set of effective potentials which in turn, in the path integral formalism follow from the matrix elements (in target states), $< GS|e^{\frac{i}{\hbar}H_{int}^0}|GS > \to \sum_j \omega_j e^{\frac{i}{\hbar}g(R)X_j}$. With V(R) going into a set of effective potentials $V(R)+g(R)X_j$ with weight factor ω_j, the fusion cross section, for each partial wave, can be evaluated using WKB approximation. Thus the calculation of sub-barrier fusion cross sections reduces to the problem of evaluating the matrix elements $< GS|--|GS >$ above. With GS described by an IBM symmetry limit, it becomes the well defined group theoretical object called group representation matrix element (GRME). For a deformed target nucleus, (as in $^{16}O+^{154}Sm$ example) the approximate symmetry is the SU(3) limit of IBM. As there is good evidence for the importance of the hexadecupole deformation in sub-barrier fusion, it is essential that one should deal with the $SU_{sdg}(3)$ limit of sdgIBM with hexadecupole couplings (k=4 term in H_{int}^0). In practice it is also essential to deal with a general deformed intrinsic state (to define the GS rotational band) than just a $SU_{sdg}(3)$ state. Expression for the extended GRME for general $K^\pi = 0_{GS}^+$ intrinsic state, in sdgIBM is[9]

$$< N; \mathbf{x}; K = 0, L, M = 0|e^{ig(b)\bar{T}_0}|N; \mathbf{x}; K = 0, L = M = 0 > =$$
$$\frac{(2L+1)N!}{16\pi^2[\mathcal{N}(N;\mathbf{x};L)\mathcal{N}(N;\mathbf{x};L=0)]^{1/2}} \int\int\int\int d\theta_1 sin\theta_1 d\theta_2 sin\theta_2 d\phi_1 d\phi_2 P_L(cos\theta_2) \times$$
$$[\sum_{m\geq 0}(2-\delta_{m0})cos(m(\phi_1-\phi_2))R^{(m)}(\theta_1,\theta_2)]^N ;$$

$$R^{(m)}(\theta_1,\theta_2) = [P_{\mathbf{x}}^{(m)}(cos\theta_1)]^T[A^{(m)}]^T[E^{(m)}][A^{(m)}][P_{\mathbf{x}}^{(m)}(cos\theta_2)] \quad (4)$$
$$[P_{\mathbf{x}}^{(0)}(cos\theta_i)]^T = [P_0^{(0)}(cos\theta_i), x_{20}P_2^{(0)}(cos\theta_i), x_{40}P_4^{(0)}(cos\theta_i)]$$
$$[P_{\mathbf{x}}^{(m)}(cos\theta_i)]^T = [x_{20}\frac{2-m!}{2+m!}P_2^{(m)}(cos\theta_i), x_{40}\frac{4-m!}{4+m!}P_4^{(m)}(cos\theta_i)]; \quad m=1,2$$
$$[P_{\mathbf{x}}^{(m)}(cos\theta_i)]^T = [x_{40}\frac{4-m!}{4+m!}P_4^{(m)}(cos\theta_i)]; \quad m=3,4$$

In (4) i=1,2, $\bar{T}_0 = \sum_{\lambda=2,4;\, l,l'=0,2,4}\beta_{ll'}^\lambda(b_l^\dagger \tilde{b}_{l'})_0^\lambda$, $\mathcal{N}$ is the normalization factor for the angular momentum projected state $|N; \mathbf{x}; K = 0, LM\rangle$ from the intrinsic state $\phi_0 = x_{00}s^\dagger + x_{20}d_0^\dagger + x_{40}g_0^\dagger$; $x_{00} = 1$ and $\mathbf{x} = (x_{00}, x_{20}, x_{40})$. Similarly $[A^{(m)}]$ is the matrix for $\bar{T}_0$ in one boson space and with $\epsilon^{(m)}$ the corresponding eigenvalue matrix, $[E^m]_{ij} = \delta_{ij}e^{ig(b)[\epsilon^{(m)}]_{ij}}$. With g-bosons switched off, eq (4) reduces to the expression given by Balantekin et. al.[8] for sdIBM case and in the $SU_{sdg}(3)$ limit ($\mathbf{x} = (1,\sqrt{20/7},\sqrt{8/7})$ with $\bar{T}_0 = Q_0^2(SU_{sdg}(3))$) to the formula speculated earlier[2].

With $H_{int}^{(0)} = \tilde{g}(R)\, Q_0^2(SU_{sdg}(3))$, $\tilde{g}(R) = f(R)/ < 2^+||Q^2(SU_{sdg}(3))||0^+ > = (\sqrt{3}/2)f(R)/\sqrt{4N(4N+3)}$ and Eq.(4), the fusion cross section in $SU_{sdg}(3)$ limit is

$$\sigma_{fus} = \sum_m \omega_m \sigma\left(V(R) - (2N-3m)/(2\sqrt{(2N)(2N+3/2)})f(R)\right) ; \quad (5)$$
$$\omega_m = \sum_{k\geq m,\, k=0,1,2,\ldots,2N} \frac{(-1)^N(2N)!}{(2N-k)!(2k+1)} \frac{(-1)^m}{m!(k-m)!} .$$

With (4), one is in a position to carry out analysis of data including hexadecupole couplings and g-bosons.

4.3. Quasi-Particle Extension of sdgIBM

The excited rotational bands which are built on one and/or two - phonon excitations can be understood in terms of the $SU_{sdg}(3)$ limit or the $SU_{sd}(3) \times 1g$ limit of sdgIBM; in these bands there can be admixtures arising due to two-quasiparticle (2 q.p.) excitations. In addition, some of the excited rotational bands can be bands built on pure 2 q.p. excitations. For a complete understanding of the structure of these bands, one has to include 2 q.p. excitations in sdgIBM, (i.e.) to extend sdg-

IBM to sdgIBF^{2}M where two fermions or q.p.'s are coupled to sdg core. Just as in Sect.4.1 one can realize the SU(3) $\otimes$ U(2) limit of IBF^{2}M when the q.p.'s occupy the natural parity orbits. Here with the core described by the SU$_{sd}$(3) or SU$_{sdg}$(3) limits respectively, two coupling schemes are possible and they are,

$$\left|(sdg)^N\right\rangle_{SU^B(3)} \otimes \left|(sdg)^{N-1}_{SU^B(3)} \otimes (q.p.)^2_{SU^F(3)}\right\rangle_{SU^{BF}(3)}$$
$$\left|(sd)^N\right\rangle_{SU^B(3)} \otimes \left|(sd)^{N-1}_{SU^B(3)} \times 1g\right\rangle \otimes \left|(sd)^{N-1}_{SU^B(3)} \otimes (q.p.)^2_{SU^F(3)}\right\rangle_{SU^{BF}(3)} \qquad (6)$$

The correspondence between the two q.p. SU^{BF}(3) irreps vs the Nilsson configurations and the exchange force that alters the position of the SU^{BF}(3) irreps as a function of the average occupation probabilities are worked out[10]. Applications of the above two schemes in describing the observed two q.p. bands is being explored.

5. Conclusions

In conclusion, the studies made so far should mark the end of exploration of sdgIBM and establish clearly that sdgIBM is a viable and powerful tool in analyzing hexadecupole data. In addition, the three topics discussed in Sect. 4 show that sdgIBM is useful in studying many other nuclear phenomena of current interest.

6. Acknowledgements

The authors are indebted to F. Iachello for many useful and inspiring discussions all through the eight years of our investigations of sdgIBM.

7. References

1. F. Iachello and A. Arima, *The Interacting Boson Model* (Cambridge University Press, Cambridge, 1987).
2. Y.D. Devi and V.K.B. Kota, *Pramana-J. Phys.* **39** (1992) 413; V.K.B. Kota and Y.D. Devi, *Symmetries in Science VII*, ed. B. Gruber (Plenum, N.Y., 1994) p.307.
3. Y.D. Devi and V.K.B. Kota, *Nucl. Phys.* **A541** (1992) 173; *Phys. Lett.* **B287** (1992) 9.
4. I. Bauske et. al., *Phys. Rev. Lett.* **71** (1993) 975.
5. Y.D. Devi and V.K.B. Kota, *Phys.Rev.* **C48** (1993) 461.
6. Y.D. Devi, R.U. Haq and V.K.B. Kota, to be published.
7. D. Bohle et. al., *Phys. Lett.* **137B** (1984) 27.
8. A.B. Balantekin, J.R. Bennett, A.J. Deweerd and S. Kuyucak, *Phys. Rev.* **C46** (1992) 5.
9. V.K.B. Kota, *Mod. Phys. Lett.* **A8** (1993) 987.
10. Y.D. Devi and V.K.B. Kota, in *Capture Gamma-Ray Spectroscopy and Related Topics*, ed. J. Kern (World Scientific, Singapore, 1994), p.337; Physical Research Laboratory Report **PRL-TH/94-1**.

343

SUPERDEFORMATION AND IBM

MICHIO HONMA
*Center for Mathematical Science, University of Aizu, Tsuruga, Ikki-machi
Aizu-Wakamatsu, Fukushima 965, Japan*

and

TAKAHARU OTSUKA
*Department of Physics, University of Tokyo, Hongo 7-3-1
Bunkyo-ku, Tokyo 113, Japan*

ABSTRACT

The structure of the superdeformed states is studied from the viewpoint of collective nucleon pairs. The interacting boson model is extended to describe the superdeformation. A phenomenological hamiltonian is presented and applied to clarify the stabilization mechanisms and the spin dependent properties of the superdeformed states. It is shown that the unified description of two different configurations, normal low-lying states and the superdeformed states can be made within this framework.

1. Introduction

The superdeformation is characterized by extraordinary elongated prolate shape. The ratio of the longer axis to the shorter one in the usual quadrupole deformation is typically 1.3:1, while that in the superdeformation is 1.7~2.0:1. The first clear evidence for the existence of the high-spin superdeformed states was given in 1986 by the observation of beautiful rotational γ-ray sequence in ^{152}Dy.[1] Since then the superdeformation has been one of the most intriguing problems in the field of nuclear structure physics. For the normal quadrupole deformation, the Interacting Boson Model (IBM) has been successfully established as one of the most powerful frameworks. The IBM was first introduced from a phenomenological viewpoint by Arima and Iachello.[2] The microscopic foundation of the IBM has been presented later[3] on the basis of the correspondence between bosons and collective nucleon pairs. This interpretation has revealed a new aspect of the IBM, the interacting boson approximation of the nuclear shell model. Since then the IBM has been widely extended and applied to a variety of nuclear structure problems such as high-spin states and the shape coexistence.

Thus it is quite interesting to study the superdeformation from the viewpoint of the IBM. In the present study we try to extend the IBM for the superdeformation and to describe both normal low-lying states and the superdeformed states simultaneously. It should be noted that there exist no models which can treat such problems directly in the laboratory frame especially at high spin. One of the advantages of the

Table 1. Probability of each angular momentum component in the Λ-pair. Two cases of $\delta = -0.13$ and 0.40 are shown for ^{194}Hg. The probability of each boson included in the intrinsic boson of the IBM in the SU(3)-limit is also listed for comparison.

pair/boson	$\delta = -0.13$		$\delta = 0.40$		IBM-SU(3)	
	neutron	proton	neutron	proton	sd	sdg
S/s	82	93	25	32	33	20
D/d	18	7	52	51	67	57
G/g	0	0	18	14	–	23

IBM is its simplicity. This model provides us a significantly simplified approximation to the nuclear shell model, while it also keeps sufficient degrees of freedom for the phenomenological approach. We can treat the angular momentum of the many-body problem exactly, and both energies and transitions can be studied consistently within a single framework.

2. Collective pairs in the superdeformed states

We consider the relation between the superdeformation and the IBM by analyzing valence wave functions from the viewpoint of collective nucleon pairs on the basis of the Nilsson + particle-number-conserving BCS model.[4] We take ^{194}Hg as an example. In this case, in order to describe strong deformation, we take a small inert core with (Z, N)=(50, 82) and extended valence space which includes orbits up to $N = 2n + l = 12$.[5] The deformation parameter $\delta=0.40$ is adopted for the superdeformed state, which corresponds to the axis ratio 1.7 : 1. We consider also the case of $\delta = -0.13$ (slightly oblate normal deformation) for comparison. We do not take the usual "hole" picture. The "hole" picture is meaningful only when the configuration of the interested state is well described within one major shell. In the present study, we intend to discuss directly the relation of two completely different configurations, one is of normal states and the other is of the superdeformed states. In the latter case, the "hole" picture completely breaks down. From the particle picture we can take orbits $1f_{5/2}$, $2p_{3/2}$, $2p_{1/2}$ and $0i_{13/2}$ ($1d_{3/2}$, $2s_{1/2}$, and $0h_{11/2}$) as normal valence orbits for neutron (proton).[5]

The wave function can be expressed as the condensed state of coherent Cooper-pairs in the deformed potential[6]

$$P^N \mid \Psi \rangle \propto (\Lambda_\pi^\dagger)^{N_\pi}(\Lambda_\nu^\dagger)^{N_\nu} \mid \text{spherical inert core} \rangle. \tag{1}$$

In this expression $\Lambda_\pi^\dagger$ ($\Lambda_\nu^\dagger$) denotes the creation operator of a Cooper-pair in proton (neutron) orbits and N_π (N_ν) means half of the valence proton (neutron) number. These Λ-pairs can be decomposed into a linear combination of collective nucleon pairs $S^\dagger$, $D^\dagger$, $G^\dagger$, $\cdots$ with good angular momenta $J^\pi = 0^+, 2^+, 4^+, \cdots$, respectively:

$$\Lambda^\dagger = x_0 S^\dagger + x_2 D_0^\dagger + x_4 G_0^\dagger + \cdots. \tag{2}$$

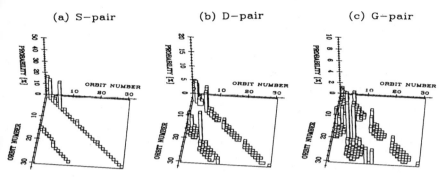

Fig. 1. Probability of each pair-component in the (a) S-pair, (b) D-pair, and (c) G-pair defined from the Λ-pair of ^{194}Hg neutron states. The orbits are ordered from $0h_{9/2}$ according to the single particle energies in a harmonic oscillator potential with weak spin-orbit force. The usual valence orbits ($1f_{5/2}$, $2p_{3/2}$, $2p_{1/2}$, and $0i_{13/2}$) correspond to the orbits No.3∼6.

The probability of each pair in the Λ-pair is given by the square of the amplitude x_J, and is listed in Table 1. It is well known that in the case of normal deformation dominant components are the S-pair and the D-pair.[7] In fact SD-dominance can be seen in the case of $\delta = -0.13$. In the case of $\delta = 0.40$, although the probability of the G-pair increases, the total probability of the S-pair and the D-pair is about 80% and we can conclude that these pairs are still dominant in the Λ-pair. This result implies that we can take the sd-boson model as a starting point, except for the description of extremely high spin states. It should be noted that the ratio of the S-pair to the other pairs is quite similar to that of s-boson to the other bosons in the SU(3)-limit of the IBM which are shown in the same table. This fact suggests the validity of taking the SU(3)-limit for the description of the superdeformed states.

We investigate further the structure of collective nucleon pairs S, D, G, $\cdots$ in detail. Each pair is written as a linear combination of many non-collective pairs $A^\dagger_{ijJM}$, which are constructed by coupling two spherical orbits i and j with the total angular momentum J and the magnetic quantum number M. Fig.1-(a), (b) and (c) show the probabilities of these non-collective pairs in the S-, D- and G-pair, respectively, in the neutron states of ^{194}Hg. In this figure we can find two important facts. The first one is that as the spin of the collective pair increases the probability distribution shifts toward higer energy orbits. In fact, the probability within or below the usual valence space is 92%, 40%, and 10% for S, D, and G-pair, respectively. The second one can be seen in the peaks which appear around the off-diagonal line apart from the components around the diagonal line. These off-diagonal components correspond to the nucleon pairs composed of two orbits which are $2\hbar\omega$ apart. These $2\hbar\omega$-components would not appear in usual treatment which includes only one valence major shell. As the spin of the collective pair increases these $2\hbar\omega$-components becomes significant in comparison with the usual $0\hbar\omega$-components. The total probability of $2\hbar\omega$-components is 14%, 36%, and 51% for S, D, and G-pair, respectively. These results suggest that in order to describe the superdeformation by bosons it is insufficient to include usual s- and d-bosons defined within the usual valence shell, and that the boson-image of the high-j, $2\hbar\omega$-components should be included explicitly.

3. An extension of IBM

We introduce superdeformed bosons in order to describe the superdeformed states. These bosons can be regarded as images of collective pairs obtained in the extended valence space and carry the collectivity of many major spherical shells. The number of bosons in the IBM is determined by half of the number of valence nucleons. Because of the small inert core, the number of superdeformed bosons increases significantly in comparison with that in the usual IBM. In fact in the case of ^{194}Hg, the boson number N_{normal} in the usual IBM becomes $(114 - 100)/2 + (80 - 64)/2 = 15$ in the particle picture mentioned in section 2. On the other hand the number of superdeformed bosons N_{super} turns to be $(114-82)/2+(80-50)/2 = 31$. In general, N_{super} is about three times larger than N_{normal}. According to the results obtained in section 2, the superdeformed collective pairs contain quite many highly excited pair components such as $2\hbar\omega$ pairs and $4\hbar\omega$ pairs. The superdeformed bosons can be written as a linear combinations of normal bosons (denoted s, d, g, $\cdots$) and their orthogonal components (denoted **s**, **d**, **g**, $\cdots$). These orthogonal components are mainly the boson images of the highly excited nucleon pairs. In order to treat the relation between normal states and the superdeformed states, we compensate the difference in the boson number between these two cases by introducing a new type of boson with $J^\pi = 0^+$, denoted σ. The σ bosons describe $J^\pi = 0^+$ nucleon pairs which consist in the lower extended valence shell. This space constitutes the upper part of the normal inert core, and becomes active in the case of the superdeformation. Since the contribution of higher spin ($J = 2, 4, \cdots$) components in this region to the coherent Λ-pair is small, we neglect these components and consider only monopole σ bosons. The number of σ bosons, denoted N_0, is determined by $N_0 = N_{\text{super}} - N_{\text{normal}}$. The boson picture for the superdeformation introduced above is schematically shown in Fig.2.

On the basis of this boson picture, we describe the superdeformed states taking into account the relation to the normal low-lying states. As a first step of our super-IBM approach we introduce a hamiltonian which is as simple as possible and yet can be understood as a natural extension of the usual IBM hamiltonian. We then consider the following hamiltonian which includes boson single particle energies and the quadrupole-quadrupole interaction:

$$H = \varepsilon_\sigma n_\sigma + \varepsilon_s n_s + \varepsilon_d n_d + \varepsilon_g n_g + \varepsilon_\mathbf{s} n_\mathbf{s} + \varepsilon_\mathbf{d} n_\mathbf{d} + \varepsilon_\mathbf{g} n_\mathbf{g} + \kappa Q \cdot Q, \quad (3)$$

where $n_b = b^\dagger \cdot \tilde{b}$ ($b = \sigma, s, d, g, \mathbf{s}, \mathbf{d}, \mathbf{g}$) denotes the boson number operator of each kind, and Q denotes a boson quadrupole operator. This quadrupole operator contains four terms:

$$Q_\mu = \sum_{i=1}^{4} e_i T_\mu^{(i)}, \quad (4)$$

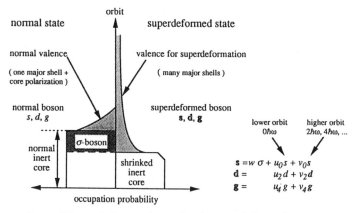

Fig. 2. Schematic boson picture for the superdeformation.

where e_i ($i=1 \sim 4$) are scaling parameters and the operators $T^{(i)}$ are given by

$$T^{(1)}_\mu = \sigma^\dagger \tilde{d}_\mu + d^\dagger_\mu \tilde{\sigma}, \tag{5}$$

$$T^{(2)}_\mu = s^\dagger \tilde{d}_\mu + d^\dagger_\mu \tilde{s} + \chi_2 [d^\dagger \tilde{d}]^{(2)}_\mu + \lambda_2 [d^\dagger \tilde{g} + g^\dagger \tilde{d}]^{(2)}_\mu + \omega_2 [g^\dagger \tilde{g}]^{(2)}_\mu, \tag{6}$$

$$T^{(3)}_\mu = s^\dagger \tilde{d}_\mu + d^\dagger_\mu \tilde{s} + \chi_3 [d^\dagger \tilde{d}]^{(2)}_\mu + \lambda_3 [d^\dagger \tilde{g} + g^\dagger \tilde{d}]^{(2)}_\mu + \omega_3 [g^\dagger \tilde{g}]^{(2)}_\mu, \tag{7}$$

$$T^{(4)}_\mu = s^\dagger \tilde{d}_\mu + d^\dagger_\mu \tilde{s} + e'_4 (s^\dagger \tilde{d}_\mu + d^\dagger_\mu \tilde{s}) + \chi_4 [d^\dagger \tilde{d} + d^\dagger \tilde{d}]^{(2)}_\mu$$
$$+ \lambda_4 [d^\dagger \tilde{g} + g^\dagger \tilde{d}]^{(2)}_\mu + \lambda'_4 [d^\dagger \tilde{g} + g^\dagger \tilde{d}]^{(2)}_\mu + \omega_4 [g^\dagger \tilde{g} + g^\dagger \tilde{g}]^{(2)}_\mu. \tag{8}$$

The term $T^{(1)}$ includes σ bosons and describes a part of the core polarization effect. In this term operators $\sigma^\dagger \tilde{d}_\mu$ and $d^\dagger_\mu \tilde{\sigma}$ are not included because σ and d are the boson images of nucleon pairs which belong to different valence shells and the matrix elements of one body operators should be vanished. The term $T^{(2)}$ is composed of normal bosons only (s, d, g) and corresponds to the normal quadrupole operator in the usual sdg-IBM. The term $T^{(3)}$ is its counterpart for orthogonal components (s, d, g). The term $T^{(4)}$ contains both components.

In order to estimate the values of parameters in the quadrupole operator, a simple mapping procedure is carried out on the basis of the "independent-pair" property of condensed coherent pairs.[8] According to this procedure, the matrix element of a quadrupole operator Q_μ with respect to the condensed state of N Λ-pairs can be simulated by the corresponding one Λ-pair matrix element with a reduction factor $(1-\epsilon)$:

$$\frac{\langle \Lambda^N | Q_\mu | \Lambda^N \rangle}{\langle \Lambda^N | \Lambda^N \rangle} = (1-\epsilon) N \frac{\langle \Lambda | Q_\mu | \Lambda \rangle}{\langle \Lambda | \Lambda \rangle}. \tag{9}$$

The factor ϵ comes from the Pauli principle and takes a value 0.3~0.4 for normal deformed states. One pair matrix elements are calculated by using the Nilsson + BCS

Table 2. Parameters in boson quadrupole operators for ^{194}Hg. Parameters e_i ($i=1 \sim 4$) are shown in the unit of fm^2 and other parameters are dimension-less.

Parm.	Value	Parm.	Value	Parm.	Value	Parm.	Value
e_1	5.854						
e_2	6.835	χ_2	0.557	λ_2	0.559	ω_2	-0.383
e_3	23.837	χ_3	-1.255	λ_3	0.959	ω_3	-1.914
e_4	5.026	χ_4	-0.850	λ_4	0.515	ω_4	-0.463
e'_4	0.183			λ'_4	0.244		

wave functions. Then we can estimate parameters in the boson operators by equating typical matrix elements of bosons and corresponding fermion pairs. Different values ϵ_i ($i = 1 \sim 4$) are taken for each term $T^{(i)}$, since they depend on the "density" of fermion pairs. Since we do not take the hole picture, the normal valence shell is almost fully occupied. On the other hand the orthogonal pairs are expected to become boson-like because the valence space has been extended significantly. Therefore, in the following estimation, we set $\epsilon_3 = 0$ and $\epsilon_1 = \epsilon_2 = \epsilon_4 = 0.4$.

The values of parameters are determined for neutron states and proton states separately. In the present case, the F-spin projection method cannot be applied to obtain IBM-1 parameters because the number of active bosons should be changed depending on the states. Thus the parameters for protons and neutrons are simply averaged. The estimated values of parameters in the boson quadrupole operators for ^{194}Hg are shown in Table 2. It can be seen that the value of e_3 is extremely large in comparison with the other matrix elements. In addition, the ratios of structure parameters χ_3, λ_3 and ω_3 are close to those of the sdg-IBM in the SU(3) limit (χ, λ, ω) = $(-11\sqrt{10}/28, 9/7, -3\sqrt{55}/14)$. This similarity to the SU(3) is expected to favor the generation of beautiful superdeformed rotational bands.

4. Results and discussions

The hamiltonian (3) is solved by the variation after projection (VAP) method. As a VAP-trial-function, we take a linear combination of many axially symmetric coherent states with different active boson numbers k:

$$| \Psi; \beta_L, u_L, C_k \rangle = \sum_{k=N_{\text{normal}}}^{N_{\text{super}}} C_k (\sigma^\dagger)^{N_{\text{super}}-k} (s'^\dagger + \beta_2 d_0'^\dagger + \beta_4 g_0'^\dagger)^k | 0 \rangle, \quad (10)$$

$$s' = u_0 s + v_0 \mathbf{s}, \qquad d' = u_2 d + v_2 \mathbf{d}, \qquad g' = u_4 g + v_4 \mathbf{g}, \quad (11)$$

where $| 0 \rangle$ denotes the vacuum for bosons and $(u_L)^2 + (v_L)^2 = 1, (L = 0, 2, 4)$. The summation with respect to k runs from the normal boson number N_{normal} up to the superdeformed boson number N_{super}. The primed bosons s', d' and g' can become both normal bosons ($u = 1$) and superdeformed bosons ($u < 1$) by varying the values

Table 3. Parameters used in the calculation. The values of ε_s, ε_d, and ε_g are fixed to be 12.0, 4.4, and 2.8MeV, respectively. The values of ε_b ($b=\sigma$, d, g) are shown in the unit of MeV, while κ is in keV/fm^4.

Param.	^{190}Hg	^{192}Hg	^{194}Hg	^{192}Pb	^{194}Pb	^{196}Pb	^{198}Pb
N_{super}	29	30	31	30	31	32	33
N_{normal}	13	14	15	5	6	7	8
ε_σ	−2.3	−2.7	−3.1	−1.6	−1.9	−2.2	−2.5
ε_d	0.6	0.6	0.6	0.9	1.0	1.1	1.1
ε_g	2.5	2.5	2.5	1.3	1.5	1.7	1.6
κ	0.1263	0.1240	0.1220	0.1224	0.1212	0.1199	0.1183

of parameters u_L (L=0, 2, 4). Thus this wave function can describe the continuous change of the structure from normal states to the superdeformed states. The variational parameters are β_2, β_4, u_0, u_2, u_4 and C_k ($k = N_{normal} \sim N_{super}$). The angular momentum projection is carried out before each variational step.

We consider superdeformed even-even nuclei in the Hg-Pb region which have been found so far. The remaining parameters, the quadrupole-quadrupole interaction strength and boson single particle energies, are determined by adjusting energies to the experimental data. At the present, the only data available for the superdeformed states is the energy of the in-band E2 transition γ-ray. We also take into account several theoretical estimations, for instance, the band-head energies of the superdeformed bands and the barrier height separating normal and the superdeformed energy minima. The adopted values of parameters are listed in Table 3. The calculated energy levels are shown in Fig.3. The spin assignment of the superdeformed states is taken from theoretical suggestions.[9] We can see that both normal states and the superdeformed states are well reproduced simultaneously.

From the viewpoint of the IBM, two different mechanisms contribute constructively to the generation of the second (superdeformed) energy minimum. The first one is the increase of the active boson number, namely, the core excitation effect. The second one is that the quadrupole matrix elements are roughly three times larger for highly excited nucleon pairs ($2\hbar\omega$-pair, $4\hbar\omega$-pair, $\cdots$) than those in the usual valence shell ($0\hbar\omega$-pair). This tendency is emphasized by the reduction factor ϵ. These two effects give rise to larger quadrupole-quadrupole energy gain, which exceed the energy loss from single particle energies under a certain condition of the active boson number and the mixing of highly excited pairs.

The large moments of inertia of the superdeformed states favor the stabilization of these states especially at high spin. In the normal energy minimum the core is quite inert. The angular momentum is gained only by transforming s bosons into higher spin (d, g) bosons with large single particle energies. Thus the energy loss increases drastically as a function of the angular momentum. While in the second

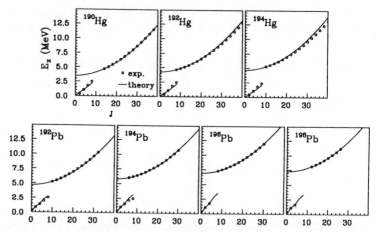

Fig. 3. Excitation energies of both normal low-lying states and the superdeformed states are shown as a function of the angular momentum. The experimental values are also shown by open circles. Since the band head energies of the superdeformed bands are not known experimentally, the lowest member of the band is located at the same energy of the corresponding calculated state.

(superdeformed) energy minimum, the angular momentum is gained mainly by the coherent contribution of many bosons without changing the gross intrinsic structure. As a result the superdeformed states have large moments of inertia.

References

1. P.J.Twin et al., Phys. Rev. Lett. **57** (1986) 811
2. F.Iachello and A.Arima, *The interacting boson model*, Cambridge U.P., Cambridge, 1987
3. A.Arima, T.Otsuka, F.Iachello and I.Talmi, Phys. Lett. **B66** (1977) 205; T.Otsuka, A.Arima, F.Iachello and I.Talmi, Phys. Lett. **B76** (1978) 139; T.Otsuka, A.Arima and F.Iachello, Nucl. Phys. **A309** (1978) 1
4. S.G.Nilsson and O.Prior, Mat.Fys.Medd.Dan.Vid.Selsk.**32** (1961) No.16
5. T.Otsuka and M.Honma, Phys. Lett. **B268** (1991) 305; M.Honma and T.Otsuka, *Proceedings of a Symposium on "Symmetries in Science VII"*, ed. B.Gruber and T.Otsuka (Plenum Press, New York, 1994); M.Honma, Doctor thesis (1994)
6. T.Otsuka and N.Yoshinaga, Phys. Lett. **B168** (1986) 1
7. T.Otsuka, A.Arima and N.Yoshinaga, Phys. Rev. Lett. **48** (1982) 387; D.R.Bes, R.A.Broglia, E.Maglione and A.Vitturi, Phys. Rev. Lett. **48** (1982) 1001
8. T.Otsuka Phys. Lett. **138B** (1984) 1
9. J.A.Becker et al., Phys. Rev. **C46** (1992) 889

MULTI-FERMION DYNAMICAL SUPERSYMMETRIES IN SUPERDEFORMED NUCLEI

J.A.Cizewski
Department of Physics and Astronomy, Rutgers University
New Brunswick, New Jersey 08903 USA

ABSTRACT

Evidence for dynamical supersymmetries which involve many fermions is observed in A≈190 superdeformed excitations.

1. Introduction

The Interacting Boson Model (IBM) was proposed 20 years ago to explain low-lying collective structure and has always predicted unexpected dynamical symmetries and supersymmetries in nuclear excitations. The O(6) limiting symmetry required spectra which were not recognized at that time to be part of the "simple" structure of nuclei. Subsequently, ^{196}Pt was recognized to be an excellent example of this symmetry, and nuclei in the Pt and Xe-Ba regions were also shown to have spectra with O(6) symmetry. Dynamical supersymmetries were derived and the best examples of supersymmetry in nature were found in excitations in odd-A nuclei near candidates for O(6) boson symmetry. However, it was assumed that these symmetries could only be realized at low angular momentum and when there was insufficient excitation energy to break a pair of nucleons coupled to L=0,2, or possibly, 4. This paper will present evidence for dynamical supersymmetries at high excitation energies and at the limits of angular momentum that the nucleus can sustain — in the spectra of superdeformed nuclei at high spin.

2. Identical bands in A≈190 nuclei

The present work will focus on the superdeformed rotational bands[1-7] in ^{192,193,194}Hg, ^{193,194}Tl, and ^{194}Pb. What was totally unexpected was the observation of SD band transitions in these nuclei with energies which were simply related to each other. As an example, one of the excited SD ^{194}Hg* bands has γ–ray energies at the arithmetic mean or "midpoint" values compared to the values for the strongly populated SD band in ^{192}Hg; the other has γ–ray energies which are identical to those of SD ^{192}Hg. In this paper "identical" will refer to γ–ray energies which are simply related to those of a reference nucleus, ^{192}Hg or ^{193}Tl. These will include energies which are identical, as well as those which occur at the midpoints and the quarter points.

Although the angular momenta of the states in these SD bands cannot be determined in a model-independent manner, fortunately, these $\Delta L=\Delta J=2$ cascades in $A\approx 190$ nuclei extend to low γ–ray energy, as low as 169 keV in ^{194}Pb.[5] Therefore, spins of individual band members can be determined[8] using simple expectations of quantum rotors, for example by a fit to the expansion in $J(J+1)$:

$$E_x(J) = \hbar^2/2\vartheta\, J(J+1) + B\,[J(J+1)]^2 + C\,[J(J+1)]^3 + \ldots \quad (1)$$

The measured γ–ray transition energies are the difference in excitation energies, $E_x(J+1) - E_x(J-1)$. The spin, J_f, of the level at the bottom of the cascade, and the moment of inertia parameters, $A=\hbar^2/2\vartheta$, B, and possibly C, can be determined from a least-squares fit of the 9 lowest transition energies to the theoretical expectations determined from eq. 1. For the $A\approx 190$ SD bands, J_f is determined[8] to be integer with high confidence for all even-even nuclei; for most odd-A nuclei J_f is determined to be half-integer.

In Fig. 1a the angular momenta as a function of frequency are plotted for the excited SD bands in ^{194}Hg and the ^{192}Hg SD band. Especially at the higher γ–ray energies, these curves are parallel, offset by an angular momentum or alignment of exactly $1.0\hbar$. The alignment has been found to be quantized for a large number of SD bands with respect to ^{192}Hg, or ^{193}Tl, as was summarized in refs. 7 and 9. The plots of the angular momenta as a function of γ–ray energy for the SD bands in ^{192,193,194}Hg and ^{193,194}Tl are displayed in Fig. 1.

There are then two aspects of the experimental results which are unexpected and need to be understood. First are the identical or related γ–ray energies in a large number of nuclei when compared to ^{192}Hg or ^{193}Tl. Second is the observed integer alignment, which sets in at moderate rotational frequencies.

3. Dynamical supersymmetries in deformed nuclei

It has been suggested[10,11] that the identical structures in superdeformed nuclei are examples of the dynamical supersymmetries that occur in the framework of the Interacting Boson Fermion Model (IBFM). The beauty of a supersymmetry is that it predicts the γ–ray energies in all members of a supermultiplet to be related. Supermultiplets that could be appropriate for the $A\approx 192$ SD nuclei are shown in Fig. 2, in which the nuclei are labeled by N_F, the number of fermions. For all members of the supermultiplet $N_B + N_F =$ constant, where N_B is the number of bosons. For a supersymmetry the yrast SD band in ^{192}Hg would be related to $N_F=1$ one-particle bands in ^{193}Hg, and $N_F=2$ two-particle bands in ^{194}Hg. The yrast SD band in ^{194}Hg would be the $N_F=0$ member of another supermultiplet and, in general, will have no relation to ^{192}Hg. It is also possible for odd-proton nuclei to be part of a supermultiplet. Then ^{193}Tl would be simply related to ^{192}Hg, and excited bands in ^{194}Pb would be simply related to ^{193}Tl. If both odd-neutron and odd-proton supermultiplets exist, then odd-odd ^{194}Tl could be expected to be simply related to ^{193}Tl.

353

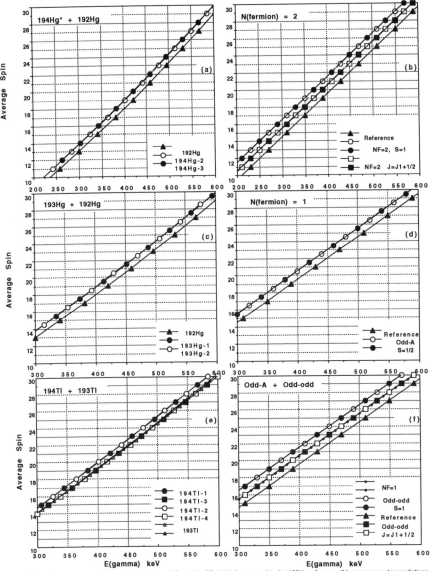

Figure 1 Average angular momenta as a function of Eg for (a) SD 194Hg* compared to the 192Hg reference; (b) supersymmetry predictions for NF=2 from eq. 2 and 4 compared to the NF=0 reference; (c) SD 193Hg compared to 192Hg; (d) NF=1 predictions from eq. 2; (e) SD 194Tl compared to 193Tl reference; (f) NF=2 predictions from eq. 2 and 4 compared to the NF=0 and 1 references. Data taken from refs. 1-7.

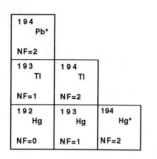

Figure 2 Supermultiplets in A≈192 nuclei.

For a supersymmetry to occur three criteria must be met. First, the core structure must be an example of a boson symmetry; SU(3) symmetry is appropriate for these highly deformed nuclei. While most tests of boson-fermion symmetries have assumed that the core structure is dominated by s and d bosons, and that the valence particle occupies a limited number of orbitals, the present work requires no such restrictions. The boson SU(3) symmetry is realized for many boson systems. Second, the valence nucleons must be in orbitals which are part of a complete major shell, which can be met if one exploits the pseudo-harmonic oscillator.[12] For all major shells with principal quantum number N>3, the highest $j=\ell+1/2$, $\ell=N$, orbital has been lowered into the N-1 shell. However, the normal parity orbitals which are left in that shell have exactly the j values of the N-1 shell. Since j is a good quantum number, rather than ℓ and s, the structure of the normal parity orbitals in a realistic N shell can be approximated by a complete pseudo-harmonic oscillator Ñ=N-1 shell. This pseudo-harmonic oscillator approximation is also valid for finite nuclear deformations, as long as the $j=\ell+1/2$ spherical state does not contribute significantly to the deformed wave function. The specific set of orbitals for the odd particle need not be specified, since the fermion excitations take on one of a number of generic types of spectra. The observed dynamic moments of inertia of the nuclei with structures related to ^{192}Hg suggest that the valence particle is not in a unique-parity configuration.

In a deformed supersymmetric system the excitation energies can be given by:[11]

$$E = E_0 + A_1 S(S+1) + B_1 L(L+1) + C_1 J(J+1) \qquad (2)$$

where the parameter E_0 gives the band-head energies, and the excitation energies depend upon the intrinsic angular momentum S, an integer orbital angular momentum L, and total angular momentum **J**=**L**+**S**. The third requirement for a supersymmetry is satisfied if the same parameters apply to all nuclei in the supermultiplet.

In the experiments only γ–ray energies, $E_\gamma=E_x(J+1)-E_x(J-1)$, are measured. The parameters A_1 and E_0 only affect the band-head energies, not the transition energies, so only 2 parameters are needed to fit γ–ray energies for all excitations in a supermultiplet.

Three generic types of spectra characterize the S=1/2, N_F=1 nuclei. Two of these involve decoupled structures, which would not give rise to the signature pairs that characterize the A≈190 "identical" SD cascades. The third generic spectrum with N_F=1, S=1/2 is shown in Fig. 3a, and will be compared to the data for ^{193}Hg. Although this figure was generated with C_1=0, the γ–ray energies can depend on C_1 without breaking the symmetry. The observed alignment $i=1\hbar$ can be reproduced when B_1=-2C_1, so that the transition energies become

$$E_\gamma(N_F=0) = B_1(2J+1) \quad J=L$$
$$E_\gamma(N_F=1) = B_1(2J-1) \quad J=L+1/2 \qquad (3)$$
$$E_\gamma(N_F=2) = B_1(2J-3) \quad J=L+1$$

As displayed in Fig. 1c the ^{193}Hg data, with ^{192}Hg as reference, are well reproduced by Fig.1d, the expectations of a $N_F=1$ supersymmetry, with $B_1=-2C_1$.

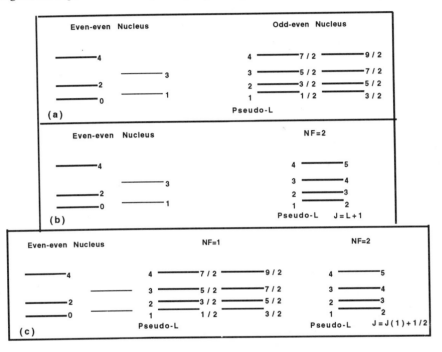

Figure 3 Generic supersymmetry spectrum from (a) eq. 2 for $N_F=1$, $S=1/2$; (b) eq. 2 for $N_F=2$, $S=1$; (c) eq. 4 for $N_F=2$, $J=J_1+1/2$.

The spectra associated with the $N_F=2$ systems of eq. 2 take on a large number of forms, since both $S=0$ and 1 are allowed, as well as the variety of ways that $\mathbf{J=L+S}$ can add vectorially. In the case of ^{194}Hg*, two bands are observed and the alignment with respect to ^{192}Hg is $1\hbar$. This suggests a supersymmetry in which the spins of the fermions are aligned, $S=1$, and both even and odd spins occur; such a spectrum is given in Fig. 3b. In Fig. 1 we compare (a) the ^{194}Hg* data, with ^{192}Hg as reference, and (b) the expectations from eq. 2 of the $N_F=2$, $S=1$ symmetry. Unfortunately, the spectrum of Fig. 3b gives $i=2\hbar$ with the same choice of parameters that fit the ^{192}Hg-^{193}Hg behavior.

However, there is no guarantee that both fermions in ^{194}Hg* will come from the same oscillator shell as the single fermion in ^{193}Hg. Therefore, the comparison between theory and experiment is not as straightforward for the $N_F>1$ systems. The level diagram for single-particle configurations at large deformations is a complicated mixture of contributions from many shells, with orbitals from the N=5 shell coexisting with those from the N=6 shell and "intruder" $j_{15/2}$ configurations. Only the isolated high-j N=7 orbitals need be considered as outside of the framework of a dynamical symmetry. Therefore, the two-fermion system can either have two particles in the same $\tilde{N}$ shell, or each fermion can come from an orbital from different $\tilde{N}$ shells.

For two fermions, from different oscillator (or pseudo-oscillator shells) one appropriate form for the excitation spectrum is:[13]

$$E = E_0 + B_2 L(L+1) + C_2 J_1(J_1+1) + D_2 J(J+1) \quad (4)$$

where J_1 is the total angular momentum of one of the fermions and J is the total angular momentum. Again, there will be a plethora of bands arising from the different ways that $J_1=L+S_1$ and $J=J_1+S_2$ can couple; one case is shown in Fig. 3c. Alignment $i=1\hbar$ can be obtained in the $N_F=1$ and $N_F=2$ systems when $B_2 = -C_2 = 2D_2$. With this requirement the γ-ray energies become

$$E_\gamma(N_F=0) = B_1 (2J+1) \qquad J=L$$
$$E_\gamma(N_F=1) = B_1 (2J-1) \qquad J=L+1/2 \quad (5)$$
$$E_\gamma(N_F=2) = B_1 (2J-1) \qquad J=L+1$$

A comparison between experiment and these predictions for ^{192}Hg and ^{194}Hg* transitions are shown in Fig. 1b and the alignment $i=1\hbar$ in the data can be reproduced.

Another system with 2 fermions that exhibits "identical" bands is ^{194}Tl, especially when related to ^{193}Tl. Displayed in Fig. 1 are the comparisons between (e) experiment and (f) the predictions for the coupling schemes of eq. 2 and 4. The data for ^{194}Tl actually require two different symmetries for this odd-odd nucleus: (1) the orbital angular momenta of both fermions are strongly coupled, and their spins couple to S=1 (eq. 2); and (2) the spin of the second fermion is weakly coupled to the total angular momentum of the first fermion (eq. 4). This should not be unexpected since the odd proton is most likely in an $i_{13/2}$ orbital,[6] while the odd neutron could be in either an N=6 or N=5 orbital, which have different radial overlaps with respect to the proton configuration. Different predictions come from these two coupling schemes. When both fermions are in the same shell eq. 2 is probably appropriate and a positive-parity band will result; when the fermions are in orbitals from different shells eq. 4 is probably valid and it is quite likely that a negative-parity band will result. A measure of the parities of these SD excitations could further test these predictions.

These candidates for supersymmetric structures were identified with the previous generation of large arrays of γ-ray detectors. In the past year there has been an explosion of new data with the first results from the larger arrays, Eurogam and Gammasphere.

One of these results was the identification[14] of SD excited bands in ^{194}Pb, a natural prediction of the supersymmetric multiplets related to ^{192}Hg. The γ-ray energies and spins of these bands indicate zero alignment with respect to ^{193}Tl. This is another example of the coupling scheme of eq. 4, which indicates that the two excited protons are probably in orbitals from different major shells.

4. Summary and perspectives

A large number of superdeformed bands in A≈190 nuclei show related structures. The observed pattern of γ-ray energies and the extracted spins of these excitations indicate that the alignment with respect to ^{192}Hg or ^{193}Tl is quantized, frequently with $i=1.0\hbar$. The simplicity of these data is a signature of a symmetry. We have presented evidence for candidates for all members of the supermultiplet shown in Fig. 2 — a very rich example of a dynamical supersymmetry with many fermions.

One consequence of the newer arrays is higher statistics for these weakly populated SD bands, so that the values of the energies of the γ rays have been determined to much higher accuracy. A surprising result has been observed for several cases:[15,16] there is a staggering in the energies of the SD band transitions in which half of the ΔJ=2 transitions have energies above the mean, and the other half have energies below the mean. In particular this has been observed for the excited SD bands in ^{194}Hg, as displayed in Fig. 4. This ΔJ=4 bifurcation could come about from a residual C_4 symmetry in these SD nuclei.[17] To date only a few cases have been identified (mostly in the A≈150 region), and the interpretation of the observed energy staggering is still in its infancy. However, it shows further that the study of nuclei at the limits of angular momenta will continue to be a fertile ground to search for the dynamical symmetries which naturally evolve from the Interacting Boson Model.

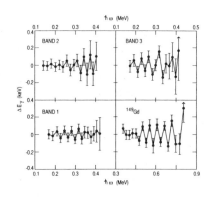

Figure 4 Staggering of the transition energies with respect to the mean for SD bands in ^{194}Hg and ^{149}Gd. This figure was taken from ref. 16.

5. Acknowledgements

The author would like to thank her colleagues at Lawrence Livermore and Lawrence Berkeley National Laboratories for their extensive efforts in the acquisition and analysis of the data on superdeformed excitations in A≈190 nuclei. In particular, I would like to thank J. A. Becker, for his work on the spin determinations, F. S. Stephens, for recognizing the importance of alignment, and J. R. Hughes and B. Cederwall for providing results prior to publication. I am also indebted to R. Bijker and F. Iachello for critical contributions to the understanding of supersymmetries in highly deformed nuclei.

Finally, I am very happy to have been given the opportunity to participate in the celebration of the Perspectives for the IBM. I look forward to many more years in which this elegant model will be used to help us understand the symmetries in nuclear excitations.

This work was supported in part by the National Science Foundation.

6. References

1. J. A. Becker, et al., *Phys. Rev.* **C 41** (1990) R9; D. Ye, et al., *Phys. Rev.* **C 41** (1990) R13.
2. C. W. Beausang, et al., *Z. Phys.* **A335** (1990) 325; M. A. Riley, et al., *Nucl. Phys.* **A512** (1990) 178.
3. E. A. Henry, et al., *Z. Phys.* **A335** (1990) 361; C. M. Cullen, et al., *Phys. Rev. Lett.* **65** (1990) 1547; M. J. Joyce, et al., *Phys. Rev. Lett.* **71** (1993) 2176.
4. M. P. Carpenter, et al., *Phys. Lett.* **240B** (1990) 44; M. W. Drigert, et al., *Nucl. Phys.* **A530** (1991) 452.
5. M. J. Brinkman, et al., *Z. Phys.* **A336** (1990) 115; K. Theine, et al., *Z. Phys.* **A336** (1990) 113.
6. P. B. Fernandez, et al., *Nucl. Phys.* **A517** (1990) 386.
7. F. Azaiez, et al., *Z. Phys.* **A336** (1990) 243; *Phys. Rev. Lett.* **66** (1991) 1030, and private communication.
8. J. A. Becker, et al., *Phys. Rev.* **C 46** (1992) 889.
9. F. S. Stephens, et al., *Phys. Rev. Lett.* **64** (1990) 2623; *Phys. Rev. Lett.* **65** (1990) 301.
10. A. Gelberg, P.von Brentano, and R. F. Casten, *J. Phys.* **G16** (1990) L143.
11. F. Iachello, *Nucl. Phys.* **A522** (1991) 83c and private communication.
12. K. T. Hecht and A. Adler, *Nucl.Phys.* **A137** (1969) 129; R. D. Ratna Raju, J. P. Draayer, and K. T. Hecht, *Nucl.Phys.* **A202** (1973) 433; A. Arima, M. Harvey, and K. Shimizu, *Phys. Lett.* **30B** (1969) 517.
13. R. Bijker, private communication.
14. J. R. Hughes, private communication and to be published.
15. S. Flibotte, et al., *Phys. Rev. Lett.* **71** (1993) 4299.
16. B. Cederwall, et al., *Phys. Rev. Lett.* **72** (1994) 3150.
17. I. M. Pavlichenkov, *Phys. Rep.* **226** (1993) 173.

CONTRIBUTION OF THE TWO–PHONON CONFIGURATIONS TO THE WAVE FUNCTION OF LOW–LYING STATES IN DEFORMED NUCLEI

V.G. SOLOVIEV

Joint Institute for Nuclear Research, Dubna, Moscow region, Russia

Abstract

The energies and wave functions of non–rotational states in ^{150}Nd, ^{156,158,160}Gd, ^{162,164}Dy, ^{166,168}Er and other nuclei have been calculated within the Quasiparticle–Phonon Nuclear Model. The wave function of excited states below 2.3 MeV has the dominating one–phonon term. The fragmentation of one–phonon states increases with excitation energy. The contribution of the double–gamma vibrational term to the wave function of the first $K^\pi = 4^+$ state in ^{168}Er equals 30%. The existence of the double–gamma vibrational term with $K^\pi = 4^+$ at (2.1-2.3) MeV in ^{166}Er has been predicted. As a rule, two–phonon states consisting of two collective phonons are strongly fragmented. The enhanced $E1$ transition rates between excited states indicate that the wave function of the initial state has a large two–phonon component consisting of the octupole phonon with $K^\pi = 0^-$ or 1^- and another phonon corresponding to the final one–phonon state. The states of deformed nuclei at energy (2-4) MeV, the wave functions of which have rather large one– or two–phonon components cannot be treated as chaotic.

1. Introduction

The vibrational states in deformed nuclei have been described within phenomenologic and microscopic models. The energies and wave functions of two–quasiparticle and one–phonon states in doubly even deformed nuclei were calculated in the period 1960-1975. A good enough description was obtained of the experimental data available at that time. Predictions were made which were later confirmed experimentally in many cases.

A new series of calculations was performed within the Quasiparticle–Phonon Nuclear Model (QPNM) [3-5]. The QPNM is used for a microscopic description of the low–spin, small–amplitude vibrational states in spherical nuclei not far from closed shells and well–deformed nuclei. The QPNM calculations were performed in nuclei with small ground state correlations. The ground state correlations increase with the collectivity of the first one-phonon states. A particle–particle interaction reduces the ground state correlation. Therefore, the energies and wave functions of many well–deformed nuclei have been calculated in the QPNM.

In recent years the Interacting Boson Model (IBM) [6,7] has been widely used for describing low–lying states of deformed nuclei [8-11]. Advances are obvious in the description of the rotational bands based on the ground and beta– and gamma–vibrational states. The important role is played by the sdg IBM and sdf IBM. It is interesting to compare the description of non–rotational states in doubly even well–deformed nuclei within the QPNM and IBM.

2. Description of the Excited States in Doubly Even Well–Deformed Nuclei in the QPNM

A specific feature of deformed nuclei is that a one–phonon state with the fixed K^π can be formed as a result of several multipole, spin–multipole and tensor interactions. For example, the one–phonon states with $K^\pi = 2^+$ can be described by using electric quadrupole and hexadecapole with $\lambda\mu = 22$ and 42 and magnetic octupole with $\lambda' L\mu = 232$ and 432 interactions. We restrict our investigation to an internal wave function with a good quantum number K. We take the Coriolis interaction into account only in the cases when its influence is very strong.

The QPNM Hamiltonian contains the average field of neutron and proton systems in a form of the axial-symmetric Woods-Saxon potential, monopole pairing, isoscalar and isovector particle-hole (ph) and particle-particle (pp) multipole interaction between quasiparticles. The procedure of calculation is the following. A canonical Bogolubov transformation is used in order to replace the particle operators by the quasiparticle ones. Then, the phonon operators $Q_{\lambda\mu i\sigma}$ are introduced and the RPA equations are solved. The phonon space is used as a QPNM basis. The RPA phonons for the $K^\pi = 0^-$ and 1^- states have been calculated in [12] with ph and pp isoscalar and isovector octupole and ph isovector dipole interactions. The RPA equation for the $K^\pi = 0^+$ states is given in [13] and for $K^\pi \neq 0^+$, 0^- and 1^- states in [4].

The QPNM wave functions consist of one- and two- phonon terms, namely,

$$\Psi_\nu(K_0^{\pi_0}\sigma_0) = \left\{ \sum_{i_0} R_{i_0}^\nu Q_{\lambda_0\mu_0 i_0\sigma_0}^+ + \sum_{\substack{\lambda_1\mu_1 i_1\sigma_1 \\ \lambda_2\mu_2 i_2\sigma_2}} \frac{(1+\delta_{\lambda_1\mu_1 i_1,\lambda_2\mu_2 i_2})^{1/2}\delta_{\sigma_1\mu_1+\sigma_2\mu_2,\sigma_0 K_0}}{2[1+\delta_{K_0,0}(1-\delta_{\mu_1,0})]^{1/2}} \right.$$

$$\left. P_{\lambda_1\mu_1 i_1,\lambda_2\mu_2 i_2}^\nu Q_{\lambda_1\mu_1 i_1\sigma_1}^+ Q_{\lambda_2\mu_2 i_2\sigma_2}^+ \right\}\Psi_0, \quad (1)$$

where $\mu_0 = K_0, \tau = \pm 1$. The secular equation for energies E_ν has the form

$$det \| (\omega_{\lambda_0\mu_0 i_0} - E_\nu)\delta_{i_0,i_0'} - \sum_{(\lambda_1\mu_1 i_1)\geq(\lambda_2\mu_2 i_2)} \frac{1+\mathcal{K}^{K_0}(\lambda_1\mu_1 i_1,\lambda_2\mu_2 i_2)}{(1+\delta_{\lambda_1\mu_1 i_1,\lambda_2\mu_2 i_2})(1+\delta_{K_0,0}(1-\delta_{\mu_1,0}))}$$

$$\frac{U_{\lambda_1\mu_1 i_1\lambda_2\mu_2 i_2}^{\lambda_0\mu_0 i_0} U_{\lambda_1\mu_1 i_1\lambda_2\mu_2 i_2}^{\lambda_0\mu_0 i_0'}}{\omega_{\lambda_1\mu_1 i_1} + \omega_{\lambda_2\mu_2 i_2} + \Delta\omega(\lambda_1\mu_1 i_1\lambda_2\mu_2 i_2) + \Delta(\lambda_1\mu_1 i_1\lambda_2\mu_2 i_2) - E_\nu} \| = 0 . \quad (2)$$

Here $\omega_{\lambda\mu i}$ is the RPA energy, the function $\mathcal{K}^{K_0}(\lambda_1\mu_1 i_1, \lambda_2, \mu_2 i_2)$ is responsible for the effect of the Pauli principle in two-phonon terms in (1), the function $U^{\lambda_0\mu_0 i_0}_{\lambda_1\mu_1 i_1, \lambda_2\mu_2 i_2}$ describes the coupling of one- and two-phonon terms in (1); $\Delta\omega(\lambda_1\mu_1 i_1, \lambda_2\mu_2 i_2)$ is the shift of the two-phonon pole due to the Pauli principle, $\Delta(\lambda_1\mu_1 i_1, \lambda_2\mu_2 i_2)$ represents the effect of three-phonon terms added to the wave function (1) and approximately equals $-0.2\Delta\omega(\lambda_1\mu_1 i_1, \lambda_2\mu_2 i_2)$. Due to this approximation, the contribution of the two–phonon configuration to the wave functions of low–lying states was calculated instead of fragmentation of the two–phonon states.

The doubly even deformed nuclei are calculated with the parameters of the Woods-Saxon potential fixed earlier. The isoscalar constants $\kappa_0^{\lambda\mu}$ of ph interactions are fixed so as to reproduce experimental energies of the first $K^\pi_{\nu=1}$ nonrotational states described by (1). The calculations were performed with the isovector constant $\kappa_1^{\lambda\mu} = -1.5\kappa_0^{\lambda\mu}$ for ph interactions and the constant $G^{\lambda\mu} = \kappa_0^{\lambda\mu}$ for pp interactions. The monopole pairing constants were fixed by pairing energies at $G^{20} = \kappa_0^{20}$. The radial dependence of the multipole interactions has the form $dV(r)/dr$, where V(r) is the central part of the Woods-Saxon potential. The phonon basis consists of ten ($i_0 = 1, 2, ...10$) phonons of each multipolarity: quadrupole $\lambda\mu = 20, 21, 22$, octupole $\lambda\mu = 30, 31, 32, 33$, hexadecapole $\lambda\mu = 43, 44$ and $\lambda\mu = 54, 55$.

The energies and wave functions of the non–rotational states in [150]Nd, [156,158,160]Gd, [162,164]Dy, [166,168]Er and other nuclei have been calculated within the QPNM and have been presented in refs. [14-20]. The gamma–ray transition probabilities between excited states in several nuclei have been calculated. Information on the two–phonon components of the wave functions of the excited states can be obtained in experimental investigations of the gamma–ray transition rates between excited states. It is a new information on the nuclear structure in addition to that from the inelastic scattering, Coulomb excitation, one- and two-nucleon transfer reactions and β–decays.

3. General Properties of the Non–Rotational States in Doubly Even Well-Deformed Nuclei

3.1. 0^+ states

There are numerous experimental data on 0^+ states in deformed nuclei. Excited 0^+ states occupy a prominent place in the microscopic nuclear theory since most of the mathematical difficulties are inherent in them. Description of several first 0^+ states in deformed nuclei in the QPNM with ph interactions cannot be thought of as satisfactory. The role of the pp interaction in the description of the excited 0^+ states is essential since with the change of G^{20} the energies of several low–lying poles of the RPA secular equations also change. The $B(E2)$ value for excitation of the $I^\pi K_\nu = 2^+ 0_1$ state and the energies of the 0_2^+, 0_3^+ and 0_4^+ states decrease at $G^{20} = \kappa_0^{20}$ in comparison with $G^{20} = 0$ and also the structure of the 0^+ states changes.

Experimental and calculated in the QPNM energies and calculated structure of the 0^+ states and branching ratio $B(E2;20_\tau \to 2^+2_1)/B(E2;2^+0_\tau \to 2^+0_{g.s.})$ are presented in Table 1.

Table 1
Energies and structure of the $K^\pi = 0^+$ states

Nuclei	K^π	exp E_ν MeV	E_ν MeV	Structure	$\frac{B(E2;2^+0_\nu \to 2^+2_1)}{B(E2;2^+0_\nu \to 2^+0_{g.s.})}$
^{168}Er	0^+_1	1.217	1.3	201 64% 202 23% {221,221} 4% {321,321} 2%	5.6
	0^+_2	1.422	1.4	201 27% 202 64% {221,221} 3%	9.0
	0^+_3	1.833	1.8	203 94% {321,321} 3%	1.6
^{166}Er	0^+_1	1.460	1.4	201 94% {321,321} 4%	0.016
	0^+_2	1.713	1.8	202 91% 203 2% {221,221} 5%	1.75
	0^+_3	1.935	2.0	203 86% 204 4% {221,221} 4% 202 3% 205 2%	74.3
^{164}Dy	0^+_1	1.655	1.6	201 85% {221,221} 6% {321,321} 4% {201,201} 3%	-
	0^+_2	1.774	1.8	202 92% 203 3%	-
^{162}Dy	0^+_1	1.400	1.4	201 97% {221,221} 2%	1.6
	0^+_2	1.670	1.7	202 89% 203 9%	1.4
^{160}Gd	0^+_1	1.380	1.3	201 84% 202 3% 203 6% {331,331} 2% {331,332} 2%	1.5
^{158}Gd	0^+_1	1.196	1.0	201 94% {201,201} 1.4%	0.08
	0^+_2	1.452	1.6	201 1% 202 93% {331,331} 2%	0.42
	0^+_3	1.743	1.8	203 97%	-
^{156}Gd	0^+_1	1.049	1.2	201 88% {221,221} 5% {201,201} 1%	0.24
	0^+_2	1.168	1.77	201 1% 202 93% 205 2%	-
	0^+_3	1.715	1.8	201 3% 203 90% {221,221} 3%	2.0
	0^+_4	1.851	2.2	203 5% 204 91% {221,221} 1%	-

The calculated structure is given as a contribution of the one–phonon $\lambda\mu i$ and two–phonon $\{\lambda_1\mu_1 i_1, \lambda_2\mu_2 i_2\}$ components to the normalization of the wave function

(1). According to our calculations, the structure of the low–lying 0^+ states is very complex. The RPA wave functions of the 0^+ states are a superposition of a great number of two–quasiparticle configurations. The first excited 0_1^+ state in several rare–earth nuclei cannot be interpreted as a beta–vibrational state due to a small $B(E2)$ value for transition to the ground state band.

The energies, $B(E2)$ values for transitions to the ground state bands and the structure of the 0^+ states which have been obtained from the one–nucleon transfer reaction were described in the QPNM correctly. The description of the two–neutron transfer reaction, ρ^2 and $X(E0/E2)$ does not contradict, as a rule, experimental data. The population of 0_2^+ and 0_3^+ states of ^{168}Er in the $^{167}Er(t,p)^{168}Er$ reaction has been rather well described in [21] within the sdg IBM.

The dominance of the $E2$ reduced transition probability from the first 0_1^+ state to the first $K_{\tau\to\gamma}^{\pi} = 2_1^+$ state over that to the ground band state $2^+0_{g.s.}$ in ^{168}Er and ^{162}Dy has been observed experimentally. The IBM calculation [8] correctly reproduced such a decay property in ^{168}Er. The dominance of the decay from the 0_1^+ state to the gamma band over that to the ground band in ^{168}Er and ^{162}Dy, calculated in the QPNM, is in agreement with experimental data. There is no dominance like that in ^{156,158}Gd and ^{166}Er. The dominance of the decay from the 0_3^+ to the 2_1^+ state over that to the ground state band in ^{166,168}Er and ^{156}Gd is due to very small $B(E2; 0^+0_3 \to 2^+0_{g.s.})$ values.

According to our calculations in the rare–earth nuclei, the contribution of the two–phonon configurations to the normalization of the wave functions of the 0^+ states with the energies below 2.3 MeV is smaller than 10%. The strength of the two phonon $0^+\{221, 221\}$ and $0^+\{201, 201\}$ states is concentrated at energies above 2.5 MeV. This result is in agreement with the calculation [22] within the multiphonon method.

3.2. $K^{\pi} = 1^+$ states

The $K^{\pi} = 1^+$ states have been calculated in the QPNM without elimination of a spurious state taking electric quadrupole and magnetic spin–octupole interactions into account. The energies and structure of the vibrational 1^+ states are mainly determined by a quadrupole interaction. In most nuclei the first $K_{\nu}^{\pi} = 1_1^+$ state is practically two-quasiparticle and the second 1_2^+ is collective. If the constant κ_0^{21} increases, then the 1_1^+ state becomes collective and the 1_2^+ state becomes a two-quasiparticle one.

According to the calculation of the $M3$ strength distribution [18], a spin–octupole interaction strongly affects the $M3$ transition rates from the ground state to the $I^{\pi}K_{\nu} = 3^-1_{\nu}$ states. The $B(M3)$ values are mostly determined by the spin part of the $M3$ operator. There is an essential difference between $B(M3)$ and $B(M1)$. According to the RPA calculation of $B(M1)$ values in refs. [23,24] and the QPNM calculation [18] the orbital part dominantes over spin ones. In the calculation [18] of the fragmentation of one–phonon states it was shown that one–phonon states with large $B(M1)$ values at energies 2.5, 3.1 and 3.3 MeV in ^{164}Dy are weakly fragmented and therefore have been observed experimentally. The $K^\pi = 1^+$ states in ^{168}Er are more strongly fragmented than in ^{164}Dy.

3.3. $K^\pi = 2^+$ states

The first $K^\pi_\nu = 2^+_1$ is a collective gamma–vibrational state in all deformed nuclei. The energies and structure of the 2^+_1 states are well described in the QPNM. The $K^\pi = 2^+$ states below 2.5 MeV are almost one–phonon ones. For example, the first five $K^\pi = 2^+$ states in ^{168}Er have been observed experimentally in ref. [25]. These one–phonon states have a different structure. Their energies and the spectroscopic factors of the (t, α) reaction are well described in ref. [26] within the QPNM.

A two–phonon state consisting of both collective phonons is strongly fragmented. For example, the {201, 221} state in ^{166}Er is strongly fragmented in the energy range (2.5–3.9) MeV. The two–phonon states {221,441}, {321,541}, {221,442} are not so strongly fragmented. These two–phonon states consist of one collective and another weakly collective phonons.

3.4. On Two–Phonon Collective States with $K^\pi = 4^+$

Let us examine two–phonon states in doubly even well–deformed nuclei. A state is determined to be a two–phonon state if the contribution of two–phonon configuration to the wave function normalization exceeds 50%. Energy centroids of two–phonon collective states in deformed nuclei are calculated in [27]. It has been shown that due to a shift of two–phonon poles, the density of levels in the energy region of the first two–phonon poles is large. Therefore, the two–phonon states should be strongly fragmented. Based on the QPNM calculation of the energy centroids of two–phonon states, it has been concluded in ref. [27] that two–phonon states consisting of two collective phonons cannot exist in well–deformed nuclei. This prediction is true in most cases. In our previous calculations [26,27] the shift $\Delta\omega(\lambda_1\mu_1 i_1, \lambda_2\mu_2 i_2)$ at $\lambda_1 = \lambda_2, \mu_1 = \mu_2$ and $i_1 = i_2$ for the $K^\pi = 4^+$ states was twice as large as it should have been. Therefore, the contribution of the two–phonon {221,221} components to the normalization of the wave functions of $K^\pi = 4^+$ was very small. The contribution of the hexadecapole 441 and 442 phonons and double–gamma vibrational {221,221} components to the normalization of the wave function of 4^+ states and $B(E2; 4^+4_\nu \to 2^+2_1)$ values are given in Table 2.

The nuclei ^{166,168}Er and ^{164}Dy are the most favourable for the observation of the $K^\pi = 4^+$ double gamma vibrational states in the energy range (2.0–2.3) MeV. Experimental investigations [28–30] have established a large double–gamma vibrational component in the first $K^\pi_\nu = 4^+_1$ state in ^{168}Er. According to the multiphonon method [22] and sdg IBM [10], the first $K^\pi_\nu = 4^+_1$ state in ^{168}Er should be a two–phonon state. According to the QPNM calculation [20], the contribution of hexadecapole {441}, one–phonon and double gamma vibrational {221,221} components to the normalization of the wave function of the 4^+_1 state in ^{168}Er equals 60% and 30%, respectively. The calculated energies of the $K^\pi_\nu = 2^+_1, 4^+_1$ and 4^-_1 states

Table 2
Energies and structure of the $K^\pi = 4^+$ states and $B(E2; 4^+4_\nu \to 2^+2_1)$ values.

Nuclei	K^π_ν	exp E_ν MeV	Calculation in QPNM E_ν MeV	Structure	$B(E2; 4^+4_\nu \to 2^+2_1)$ e^2fm^4 exp[ref.]	calc.
^{168}Er	4^+_1	2.055	2.0	441 60% {221,221} 30%	280 ± 140 [33] 315 [34] 389 ± 89 [35]	175
^{166}Er	4^+_1	1.978	2.0	441 94% {221,221} 3%	-	9
	4^+_2	-	2.14	442 12% {221,221} 82%	-	515
^{162}Dy	4^+_1	1.536	1.5	441 97% {221,221} 2,3%	17 [38]	23
^{160}Gd	4^+_1	1.070	1.2	441 98% {221,221} 1%	-	14
	4^+_2	(1.531)	1.5	442 99%	-	5
^{158}Gd	4^+_1	1.380	1.4	441 96% {221,221} 2%	-	50
	4^+_2	1.920	1.9	442 95% {221,221} 2%	-	20
^{156}Gd	4^+_1	1.511	1.5	441 94% {221,221} 5%	-	64
	4^+_2	1.861	1.9	442 90% {221,221} 4%	-	24

and the $B(E2; 2^+_1 \to 0^+_{g.s.})$, $B(E4; 4^+_1 \to 0^+_{g.s.})$ and $B(E1; 4^+_1 \to 4^-_1)$ values closely agree with experimental data analyses in ref. [37].

According to the calculation [19], the double-gamma vibrational strength with $K^\pi = 4^+$ in ^{166}Er is concentrated on one or two levels at the energy 2.1–2.3 MeV. When the energy and $B(E2; 0^+0_{g.s.} \to 2^+2_1)$ value of the $K^\pi_\nu = 2^+_1$ state are correctly described, the second $K^\pi_\nu = 4^+_2$ state at energy 2.14 MeV should be the double-gamma vibrational state. The existence of the $K^\pi = 4^+$ double-gamma vibrational state in ^{166}Er is due to a very small density of $K^\pi = 4^+$ states near the {221,221} pole and a very small value for the function $U^{441}_{221,221}$ which is responsible for a coupling between one- and two-phonon terms in the wave function (1).

A situation with the $K^\pi = 4^+$ double-gamma vibrational state in ^{164}Dy is not yet clear. According to the calculation [15,16], the $K^\pi_\nu = 4^+_1$ and 4^+_2 states in ^{156,158,160}Gd are hexadecapole ones. The admixture of the double-gamma vibrational components in their wave functions is less than 5%. The energy of the largest part of the $4^+\{221,221\}$ strength is higher than 2.4 MeV.

3.5. Octupole States and E1 Transition Rates

An octupole interaction between quasiparticles leads to the formation of collective octupole states with $K^\pi = 0^-, 1^-, 2^-$ and 3^-. The energies and wave functions of the low–lying octupole states in well–deformed nuclei in the rare–earth region are reasonably well described in the QPNM.

The origin of the $E1$ strength in the low–energy region of deformed nuclei has been investigated in ref. [12]. There are no one–phonon 1^- states below the particle threshold in spherical nuclei. The quadrupole deformation is responsible for the spliting of subshells on a spherical basis and, therefore, a part of the $E1$ strength is shifted to low–lying states. Due to an octupole interaction, the sum of the $B(E1)$ values for the transition from the ground state to the $K^\pi = 0^-$ and 1^- states up to energy 4 MeV increases by two orders of magnitudes. The isovector dipole ph interaction shifts the largest part of the $E1$ strength from the low–lying states to the region of the GDR. Nevertheless, the $B(E1)$ values, calculated with the effective charge $e^{(1)}_{eff}(p) = N/A$ and $e^{(1)}_{eff}(n) = -Z/A$, are (3–10) times as large as experimental ones.

The fragmentation of one–phonon states with $K^\pi = 0^-$ and 1^- is not as strong in the energy range (2.5–4.0) MeV. The $B(E1)$ values for the excitation of several 1^-K states are relatively large and they can be observed experimentally. The concentration of $E1$ strength in $K^\pi = 0^-$ states at energies (2.6–3.5) MeV in ^{168}Er, (3.6–3.9) MeV in ^{164}Dy and in $K^\pi = 1^-$ states at energies (3.2–3.4) MeV in ^{160}Gd have been predicted in ref. [12,16].

The $E1$ transition rates between excited states with energy below 2 MeV are mostly determined by the one–phonon terms of the wave functions of the initial and final states. A relatively large contribution of $\{221,311\}$ and $\{201,301\}$ configurations to the wave function of the initial states enhances the $E1$ transition probabilities to the $K^\pi_\nu = 2^+_1$ and 0^+_1 states compared with those between one–phonon states. The $E1$ transition rates from several 1^-1_ν states to the gamma–vibrational 221 state in ^{160}Gd, calculated in ref. [16], are demonstrated in Table 3.

Table 3

Energy, structure and E1 transition rates from the 1^-_ν states to the gamma–vibrational 2^+_1 state in ^{160}Gd, calculated in the QPNM

ν	E_ν MeV	$B(E1; 1^-1_\nu \to 2^+2_1)$ $e^2 fm^2 \cdot 10^{-3}$	Structure %
1	1.52	0.02	311 99
2	2.00	0.03	312 98
5	2.57	0.05	315 95
7	2.95	0.91	316 33 317 20 $\{221,311\}$ 30
11	3.23	0.80	317 27 318 21 $\{221,311\}$ 22
14	3.33	1.49	318 22 311 10 $\{221,311\}$ 13 $\{221,312\}$ 20 $\{211,313\}$ 8

The $B(E1)$ values for transition from $K^\pi_\gamma = 1^-_1, 1^-_2$ and 1^-_5 states to the 2^+_1 state are very small because these are transitions between one–phonon components of the wave functions. The $B(E1)$ values for transitions from $K_\nu^\pi = 1_7^-$, 1_{11}^- and 1_{14}^- to the 2_1^+ are much larger than those from $K_\nu^\pi = 1_1^-, 1_2^-$ and 1_5^- states due to the large contribution of the {221,311} configuraion to their wave functions. The experimental data on enhanced $E1$ transition between excited states gives information on the large two–phonon term of the wave function of the initial state consisting of the octupole phonon with $K^\pi = 0^-$ or 1^- and another phonon corresponding to the final state.

3.6. High–Multipolarity States

The influence of the interaction between quasiparticles with the multipolarity $\lambda = 5, 6, 7, 8$ and 9 on the mixing of two–quasineutron and two–quasiproton states has been studied in ref. [33]. High–multipolarity interactions with $\lambda = 5 - 9$ play an important role in a mixing of the two–quasiparticle states, especially when closely spaced two–quasineutron and two–quasiproton states have the same K^π and a large relevant matrix element.

4. Comparison of the QPNM with the IBM in Describing Deformed Nuclei

The collectivity of the first quadrupole and octupole states, and its absence in higher–lying states up to giant resonances, underlies all phenomenological models including the IBM. There are no essential differences in the description of the first quadrupole and octupole vibrational states in the QPNM, sdg IBM and sdf IBM. There are nuclei in which the most collective is not the first state with fixed K^π but higher lying states. Let us consider $B(E3)$ values for the excitation of $K^\pi = 3^-$ states in ^{168}Er. According to the experimental data [34], the first three 3_1^-, 3_2^- and 3_3^- states have B(E3) = 0.25, 0.60 and 0.42 s.p.u.respectively, while the fourth 3_4^- state has B(E3)=4.68 s.p.u.. The calculation [25] reproduced this distribution of $E3$ strength. The reason why the most collective state with $K^\pi = 3$ is not the first state has been described on p.305 in ref. [4]. In the sdf IBM calculation [11] of $K^\pi = 3^-$ states in ^{168}Er, the first three $K^\pi = 3^-$ states are omitted and the 3_4^- state was regarded as a first collective $K^\pi = 3^-$ state. The first three $K^\pi = 3^-$ states are collective states whose $B(E3)$ values are greater than the two–quasiparticle values by a factor between 30–60.

The wave functions of states with $K_\nu^\pi = 0_2^+, 2_2^+, 4_1^+$ in the QPNM each have one dominating one–phonon component, and in the sdg IBM each has one dominating two–boson component. Here, QPNM and IBM manifest a principal difference in the structure of these states.

The IBM singles out a subspace of collective states which constitutes a small fraction of the total space. When the subspace of collective states is singled out, all coupling to other collective and weakly collective states is cut off. One should bear in mind that this separation of collective states is not unambiguous, owing to the absence of a clear distinction between strongly collective and somewhat less collective states.

Microscopic models make use of a large, almost entire space of ph and pp two-quasiparticle states and a part of the $2p2h$ states. Coupling between collective and non–collective degrees of freedom is described as a quasiparticle–phonon interaction. The role of the quasiparticle–phonon interaction increases with an excitation energy.

The $E\lambda$ strength distribution is well described in the IBM and in the QPNM. There is a difference between the IBM and the QPNM in describing the states which lie above the first quadrupole and octupole states in well–deformed nuclei. This difference lies in the fact that the IBM quadrupole or octupole strength is fragmented but belongs to the unique $d-$ or $f-$ boson with fixed K^π. The QPNM quadrupole or octupole strength is due to a large number of one–phonon states with different quasiparticle structure. This different quasiparticle structure has been demonstrated by one–nucleon transfer reactions and β–decay measurements.

5. Fragmentation of One– and Two–Phonon States and Order–to–Chaos Transition in Deformed Nuclei

The first three–five states with fixed K^π below 2.5 MeV is well–deformed rare-earth nuclei are predominantly one–phonon states. Two closely–spaced one–phonon states having the same K^π are mixed. The fragmentation and mixing of the one-phonon states increases with an excitation energy. Fragmentation of the strongly collective one–phonon states is larger than fragmentation of the weakly collective states. The one–phonon states with a large quantum number K are weakly fragmented compared with the ones with a small K. There are levels in the energy range (3-4) MeV, the wave function of which has a large one–phonon component. These levels can be excited in inelastic scattering, (γ, γ'), one–nucleon transfer and other reactions. These states cannot be considered as chaotic states. A complete damping of one–phonon states cannot be considered as a transition to chaos due to large two–phonon or six–quasiparticle components in their wave functions.

Two–phonon states consisting of two collective phonons are fragmented strongly. As a rule, the contribution of the two–phonon configuration to the wave function normalization is less than 50%. Two–phonon states consisting of collective and weakly collective phonons are not so strongly fragmented. According to the calculation [19], the states in the energy range (3.5-3.8) MeV in ^{166}Er have the following very large two–phonon component: $2^+\{221, 443\}, 2^-\{201, 322\}$ and $4^-\{221, 322\}$. The existence of large two–phonon component in the wave function strongly enhances the $E1$ or $E2$ transition rates between relevant excited states. For example, the $E1$ transition rate from the $K^\pi = 1^-$ state, whose wave function has the two-phonon $\{221,311\}$ component equal to 20%, to the gamma–vibrational 221 state is two orders of magnitude larger than the $E1$ transition rates between one–phonon components. Two–phonon states consisting of two weakly collective phonons are fragmented weakly.

According to the present analysis of the wave functions of the excited states in the energy range (2.5–4.0) MeV in well–deformed rare–earth nuclei, the wave functions of several states have large one– or two–phonon components while other states do not. The number of states with large one– or two–phonon components in the same energy range increases with K quantum number. The mixing of closely–spaced states having the same K^π and a similar structure is large. The mixing of closely–spaced states having different structure is small. It is difficult to expect that the mixing of nuclear states can play a decisive role in an order–to–chaos transition.

It is possible to state, that the distinctly different wave functions of the states having large one– or two–phonon term at excitation energies (2-4) MeV in well–deformed doubly even nuclei contain information on specific nuclear structure, and hence, they cannot be termed as chaotic.

Fluctuation properties, generic to all systems that show chaos, are independent of the specific properties of the system. In this case, one does not need to study chaotic excited states. Therefore, it is reasonable to introduce a new approach in a study of the nuclear structure regularities against chaos at intermediate excitation energies. Instead of measuring and describing the energy and wave function of each individual state, one should investigate a strength distribution of the few– and many–quasiparticle or phonon states. The strength distribution of the many–quasiparticle, many–phonon and quasiparticle–phonon states reflects the regularity of the nuclear structure at intermediate excitation energies. The structure of nuclear states at intermediate excitation energies is very complex but not chaotic.

6. Conclusion

The above considerations allowed us to make the following conclusions:

1) It is difficult to separate collective vibrational states, except a gamma–vibrational state, and weakly collective and quasiparticles states. A unique basis is needed to treat all non–rotational states in well–deformed nuclei.

2) The wave function of the excited states of doubly even well–deformed nuclei in a rare–earth region below 2 MeV has a dominanting one–phonon term. The contribution of the two–phonon components to the normalization of their wave function is less than 10%.

3) The fragmentation and mixing of one–phonon states increase with an excitation energy. The fragmentation of strongly collective states is larger than the fragmentation of weakly collective states. The one–phonon states with a large K quantum number are weakly fragmented compared with the ones with a small K. A complete damping of one–phonon states cannot be considered as a transition to chaos.

4) The distribution of the $E\lambda$ strength among low–lying states in deformed nuclei is very interesting. Many one–phonon states, having different quasiparticle structure, contribute to the $E\lambda$ strength distribution. According to the QPNM cal-

culation, there are levels in the energy range (2–4) MeV which are strongly excited by the $E\lambda$ transition from the ground state.

5) Two–phonon states consisting of two collective phonons are, as a rule, strongly fragmented. ^{164}Dy and ^{166,168}Er are the most favourable nuclei in the rare earth region for the observation of the double gamma vibrational states with $K^\pi = 4^+$.

6) Two–phonon states consisting of collective and weakly collective phonons are not strongly fragmented. There are levels in the energy range (2-4) MeV in well–deformed nuclei, the wave functions of which have large two–phonon components. The existence of the large two–phonon component of the wave functions strongly enhances the $E1$ or $E2$ transition rates between relevant excited states.

7) Several wave function in the energy range (2.5–4.0) MeV in doubly even deformed nuclei have very large one– or two–phonon components (while others do not). These distinctly different wave functions contain specific information on nuclear structure and they cannot be treated as chaotic.

8) If the many–quasiparticle and quasiparticle–phonon components of the wave function are taken into account, a new region of regularity of nuclear states at higher excitation energies against chaos appears.

7. Acknowledgements

In conclusion, I would like to thank A.V. Shushkov and N.Yu. Shirikova for their joint investigations; some results of which were presented in this paper.

8. References

1. A. Bohr and B.R. Mottelson, Nuclear Structure (W.A. Benjamin, INC., London, 1975) v.2.
2. V.G. Soloviev, Theory of Complex Nuclei (Pergamon Press, Oxford, 1976).
3. V.G. Soloviev, Sov.J.Part.Nucl. **9** (1978) 343.
4. V.G. Soloviev, Theory of atomic nuclei: Quasiparticles and phonons (Institute of Physics, Bristol and Philadelphia, 1992).
5. V.G. Soloviev, A.V.Sushkov and N.Yu.Shirikova, Part.Nucl. **25** (1994) 377.
6. A. Arima and F. Iachello, Phys.Rev.Lett. **35** (1975) 1069.
7. A. Arime and F. Iachello, Ann.Phys. **99** (1976) 153.
8. D.D. Warner, R.F.Casten and W.F.Davidson, Phys.Rev. **C24** (1981) 1713.
9. R.F. Casten and D.D.Warner, Rev.Mod.Phys. **60** (1988) 389.
10. N. Yoshinaga, Y. Akiyama and A. Arima, Phys.Rev. **C38** (1988) 419.
11. A.F. Barfield, B.R. Barrett e.a., Ann.Phys. **182** (1988) 344.

12. V.G. Soloviev and A.V. Sushkov, Yad.Fiz. **57** (1994) N. 8.
13. V.G. Soloviev, Z.Phys. **A334** (1989) 143.
14. M. Pignanelli, N. Blasi, J.A. Bordewijk et al., Nucl.Phys. **A559** (1993) 1.
15. V.G. Soloviev, A.V. Sushkov and N.Yu.Shirikova, Nucl.Phys. **A568** (1994) 244.
16. V.G. Soloviev, A.V.Sushkov and N.Yu.Shirikova (to be published).
17. V.G. Soloviev and A.V. Sushkov, Z.Phys. **A345** (1993) 155.
18. V.G. Soloviev and N.Yu.Shirikova, Nucl.Phys. **A542** (1992) 410.
19. V.G. Soloviev, A.V.Sushkov and N.Yu.Shirikova, (to be published).
20. V.G. Soloviev, A.V. Sushkov and N.Yu.Shirikova, J.Phys. **G20** (1994) 113.
21. Y. Akiyama, K.Heyde, A.Arima and N.Yoshinaga, Phys.Lett **B173** (1986) 1.
22. P.Piepenbring and M.K.Jammari, Nucl.Phys. **A481** (1988) 81.
23. R. Nojarov and A. Faessler, Nucl.Phys. **A484** (1988) 1.
24. C. DeCoster and K. Heyde, Nucl.Phys. **A524** (1991) 441.
25. W.F. Davidson and W.R.Dixon, J.Phys. **G17** (1991) 1683.
26. V.G. Soloviev and N.Yu. Shirikova, Z.Phys. **A334** (1989) 149.
27. V.G. Soloviev and N.Yu. Shirikova, Z.Phys. **A301** (1981) 263.
28. H.G. Börner et al., Phys.Rev.Lett. **66** (1991) 691.
29. R. Neu and F. Hoyler, Phys.Rev. **C46** (1992) 208.
30. M. Oshima et al., Nucl.Phys. **A557** (1993) 635c.
31. A. Aprahamian, Phys.Rev. **C46** (1992) 2093.
32. D.D. Warner e.a., Phys.Rev. **C27** (1989) 2292.
33. V.G. Soloviev and A.V. Sushkov, J.Phys. **G16** (1990) L57.
34. I.M. Govil, H.W. Fulbright e.a., Phys.Rev. **C33** (1986) 793.

373

NEW CHALLENGES FROM RADIOACTIVE BEAMS

DAVID D WARNER
DRAL Daresbury Laboratory, Daresbury
Warrington, WA4 4AD, United Kingdom

ABSTRACT

The status of several of the proposals and projects currently under way to build radioactive beam facilities is briefly reviewed. The potential for physics with radioactive beams is discussed with emphasis on the role of the Interacting Boson Model and its ability to address the broad range of nuclear structure problems which will soon become accessible to experimental study.

1. Introduction

The advent of radioactive beam accelerators offers the opportunity to study reactions involving nuclei very far from the line of stability. This capability will represent a step forward akin in importance to the original introduction of heavy-ion accelerators themselves and is likely to have a similar catalytic effect on the experimental study of nuclear structure. Although the realisation of this goal still poses considerable challenges, both technical and political, the past 2 or 3 years has seen remarkable progress made, driven by a worldwide consensus that this is indeed a crucial and essential development for the field and by an increasing degree of collaboration between the different groups working on the diverse methods of achieving the ultimate aim of a broad range of high intensity radioactive beams with energies spanning the Coulomb barrier and reaching up to the Fermi energy.

The number of proposals which have been made is already too large to discuss in any detail here. However, it is worth at least mentioning the major initiatives in Europe and North America which are under way and perhaps concentrating a little further on those which are either operating or fully funded and under construction. The point is to emphasise the fact that the goal that was little more than a pipe dream a few years ago is rapidly becoming a reality.

A list of projects and proposals is given below. Below each facility, an indication is given of the type of projectile used in the primary production process and of the primary and secondary accelerators used. It is important to note that the list here focuses on facilities using two accelerators, the first to provide the broadest possible range of exotic nuclei with maximal intensities, and the second to accelerate the unstable nuclei up to the required energy. It has by now been generally recognised that this is the preferred method to provide beams for nuclear structure studies. There are, of course, radioactive beams already available and in use at higher energies (50 MeV/u to 1 GeV/u) produced by the projectile fragmentation reaction.

a) Proposals
- **PIAFE – ILL-ISN, Grenoble France**
  - Thermal neutrons – Reactor – Two cyclotrons (SARA)
- **EXCYT – Catania, Italy**
  - Light and heavy ions – Superconducting cyclotron – Tandem
- **ISAC – TRIUMF, Vancouver, Canada**
  - Proton spallation – Cyclotron – RFQ + Linac
- **ISOLDE – CERN, Switzerland**
  - Proton spallation – PS booster – RFQ + Linac

b) Facilities operating or under construction
- **ARENAS[3] – Louvain-la-Neuve, Belgium**
  - Low energy protons – Cyclotron – Cyclotron
- **ORIB – Oak Ridge, U.S.A.**
  - Low energy protons, light-ions – Cyclotron – Tandem
- **SPIRAL – GANIL, Caen, France**
  - Heavy ion fragmentation – Two cyclotrons – Cyclotron
- **RIST – Rutherford Laboratory, U.K.**
  - Proton spallation – Synchrotron

The list of proposals given above summarises the cases where development work is actually in progress to achieve the particular aims. A more detailed discussion of the relative attributes of some of the projects can be found elsewhere.[1] The diversity of approaches is apparent, with the PIAFE project choosing thermal neutron fission at the Grenoble High Flux Reactor as the production mechanism, while in Catania, the construction of a superconducting cyclotron is under way to produce light and heavy ions for fragmentation. Work is continuing in Vancouver to upgrade the existing ISOL facility ISAC to allow 10 μA of high-energy protons to be used in the primary spallation process, while at ISOLDE, a proposal exists to couple an EBIS (Electron Beam Ion Stripper) to the existing beamlines to increase the charge state of the radioactive ions and hence greatly facilitate the subsequent acceleration process. This is a development which, if successful, could have crucial implications for all other facilities.

The second set of facilities shown are those which are fully funded and under construction or are actually operating. Thus they will represent the first generation of radioactive beam accelerators and, as such, deserve further comment.

2. The First Radioactive Beam Facilities and Experiments

Of the projects listed, only one involves a facility which already has a capability to accelerate radioactive beams, namely, ARENAS[3] at Louvain-la-Neuve, which has been operating since 1989.[2] However, the focus of this facility is centred on the study of reaction cross sections of interest in nuclear astrophysics and, as such, it is designed to produce energies well below the Coulomb barrier (< 1.5 MeV/u) for a rather limited number of beams. Nevertheless, the use of a cyclotron as the post accelerator coupled to an ECR ion source raises the possibility of using higher ionic charge states to reach the Coulomb barrier in a few selected cases and hence perform pilot experiments in the nuclear structure regime. Very recently, this has indeed become possible and the first fusion evaporation, gamma-ray spectroscopy experiment[3] has been performed with a radioactive beam.

The reaction studied was ^{40}Ca(^{19}Ne,xp.yn.zα) with a 70 MeV beam of ^{19}Ne in the 4+ charge state. A modified version of one of the TESSA family of gamma-ray arrays from Daresbury Laboratory was used, consisting of 7 BGO-shielded Ge detector assemblies mounted in the backward hemisphere and the LEDA charged particle detector positioned forward of the target. Each detector assembly was individually shielded with 2 cm. of Pb and additional shielding was used downstream of the target, both inside and outside the beam tube. One of the major points of interest up to now has been the likely problems associated with performing gamma-ray experiments in the intense background created by the beam itself. In this case, the beam intensity was in the range 10^8–10^9 particles/sec. and the ^{19}Ne decay is to the daughter ground state by β^+ so that the background is confined to 511 keV gamma-rays. Figure 1 shows a very preliminary singles spectrum which has been obtained by gating with the beam pulses and subtracting accidentals. Figure 1 represents $\sim$30% of the data and it is clear that the background from the radioactive beam poses no insurmountable problems. Essentially all of the peaks in the spectrum can be associated with fusion evaporation products.

The first radioactive beam facility with a significant capability for nuclear structure studies will be ORIB, which is due to come on-line in 1995. The Holifield Heavy Ion Research Facility (HHIRF) at Oak Ridge National Laboratory houses two accelerators, the Oak Ridge Isochronous Cyclotron and the 25 MV Tandem. These machines are being reconfigured into a radioactive beam facility to produce proton-rich beams for nuclear, atomic and astrophysical studies. The K=105 cyclotron will act as the primary accelerator, producing the radioactive species via (light-ion,xn) reactions, while the tandem will serve as the post-accelerator. In between the two, it will be necessary to construct a target/ion source and a charge exchange system to provide the negative ions for input to the tandem. The first beams are likely to be ^{17}F, ^{58}Cu, ^{69}As and ^{33}Cl. New beam lines and detection systems are being constructed and the recoil separator previously used at the Nuclear Structure Facility at Daresbury will be shipped to ORIB this summer to be used in astrophysical studies.

Towards the end of 1998, on current projections, the SPIRAL facility at GANIL should come on-line. This will be a major facility, dedicated to the production of unstable beams with energies of $\sim$2.7–25 MeV/u. The GANIL-PLUS scheme uses the existing coupled cyclotron facility for the primary beam and proposes an additional K=262 cyclotron for post-acceleration. The production mechanism is

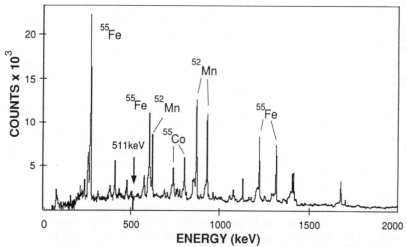

Fig. 1. γ-ray spectrum from 70 MeV ^{19}Ne radioactive beam on ^{40}Ca target.

heavy-ion fragmentation with primary beams of 50–100 MeV/u which provides a wide range of species with useful intensities. Technical aspects of the ECR ion sources used indicate that the first beams available are likely to be those involving the noble gaeses but further R&D on the ion sources is in progress in order to take full advantage of the enormous potential of this facility.

The last project, namely RIST (Radioactive Ion Source Test) at Rutherford Laboratory, differs from the others in that it is directed towards research into producing radioactive beams with primary beam intensities which are orders-of-magnitude greater than those used to date or foreseen in the other proposals. No post-acceleration is planned. The proposal is centred on the proton synchrotron at the ISIS spallation neutron facility, which can deliver up to 200 μA of 800 MeV protons. A schematic illustration of the project is shown in fig. 2. The aim is to prove the feasibility of using a thick, ISOLDE-style Ta target and ion source assembly in a primary beam intensity of up to 100 μA . (2 μA is available at ISOLDE). Success in these tests will pave the way for the eventual construction of a "second generation facility", as envisaged in recent reports from both the ISL[5] collaboration in the USA and NuPECC[1] in Europe, which would offer the broadest range of species and the highest intensities currently conceivable.

3. The IBM and Radioactive Beams

The Interacting Boson Model[6] has been broadened and developed steadily since its inception so that, by now, it encompasses a family of nuclear structure models linked by common underlying assumptions concerning their microscopic foundation and by the algebraic origins of their formalism. The various versions of the model

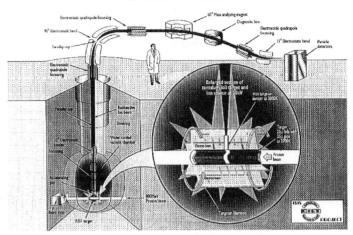

Fig. 2. Schematic illustration of the RIST facility.

are summarised in Table 1, along with their distinguishing characteristics, although many extensions have been made to these basic frameworks, particularly in the case of IBM-1.

The potential for physics with radioactive beams has been discussed at length over the past several years in many proposals and conferences. Some of the most relevant and recurring themes are listed below.

- Single particle transfer on unstable nuclei; direct empirical determination of shell structure across wide regions; new singly and doubly magic regions.
- Structure at the N=Z line up to A=100; role of isospin in collective nuclear structure; mirror nuclei at high mass and high spin.
- Coulomb excitation of unstable nuclei; lifetimes, low and high-lying vibrational modes.
- Complete spectroscopy.
- Structure of neutron-rich nuclei. Neutron halo and neutron skin.
- New collective modes; soft dipole, octupole, oblate, triaxial.
- Structure on and beyond the proton drip line. Particle radioactivity.
- Extension of studies in the actinide and transuranic regions; shell structure beyond Uranium.

The connection between the model and the experimental potential is reasonably obvious. The IBM-1 by now represents a wealth of knowledge of nuclear structure based on the symmetries of the model and the associated set of global predictions[7],

Table 1. Versions of the IBM

IBM-x	Features
IBM-1	Symmetries
	Global predictions
	Dependence on valence particle number
IBM-2	F-spin
	Quantitative dependence of params. on N,Z
	Predictions
IBM-3,4	Inclusion of isospin
	T=0 and 1, M_T=0 pair
IBFM-x	Structure of odd-A nuclei
	Pseudo L,S symmetry

many of which rely on the explicit inclusion of the valence particle number within the framework. These features now need to be tested at the very edges of stability and will in fact act as a sensitive monitor of the onset of unexpected structural changes. IBM-2[8] deals with the explicit recognition of the proton and neutron degrees of freedom via the inclusion of the F-spin quantum number. Moreover, its link to the underlying shell structure provides a powerful predictive capability which, as will be seen, can be tested as new regions are opened to experimental study.

The advantages of radioactive beams are not limited to the production of nuclei far from stability in nuclear reactions. The properties of the beams themselves can be studied in inverse kinematics, providing direct empirical information on single-particle structure, on giant resonance phenomena, on multipole matrix elements and so on, that hitherto has only been available for stable nuclei. Conversely, our knowledge of the high spin structure of stable nuclei is sparse and will be greatly enhanced by the use of neutron-rich, unstable beams. Thus the ideal of complete spectroscopy can be approached, where both low and high spin structure can be studied in detail in the same nucleus.

One of the particular foci of interest for radioactive beams has been the potential for studies along the N=Z line where the role of isospin in the collective structure of the nuclei in the 28-50 shell can be explored. Here, the IBM is already poised to provide the answers, in that the inclusion of the n-p pair has already been accomplished, within the bases of IBM-3 and 4.[9,10] This topic will be addressed more fully presently. Another area where radioactive beams have already made a major impact is in the study of very neutron-rich nuclei and the phenomena of the neutron halo and neutron skin; it will be demonstrated that the use of the algebraic approach can offer important insights to this problem also. Of course, all the attributes of the different versions of the IBM can equally well be applied to the IBFM and the study of odd-mass nuclei. Again the symmetry aspects are a crucial characteristic and, for example, the pseudo L,S symmetry associated with a large

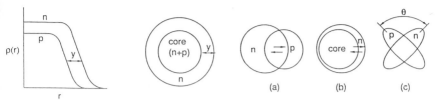

Fig. 3. Indication of dipole modes with a neutron skin.

family of IBFM group structures[11] could serve as a probe of the recently-proposed[12] breakdown in shell structure at extremes of one nucleon number.

4. The Neutron Skin and the Algebraic Approach

As mentioned earlier, experiments with radioactive beams from projectile fragmentation facilities have revealed[13] the presence of a neutron halo in several of the lightest nuclei on the neutron drip line. This is now understod as a tunneling effect where the last one or two neutrons are in low ℓ orbits very near the top of the well so that their wave functions have a very extended distribution which is manifest empirically in an anomalously large matter radius. There is, however, a distinctly different phenomenon which is predicted in Hartree Fock calculations[14] to occur in heavier nuclei in which an excess of several neutrons builds up so that the neutron density actually extends out significantly further than that of the protons, resulting in a mantle of dominantly neutron matter.

The situation is illustrated schematically in fig. 3. The presence of the skin may affect collective modes involving the out-of-phase motion of the neutrons against protons, such as the GDR and the scissors mode, as shown in a) and c). There is also the possibility of a new "soft" dipole mode b) in which the core nucleons, both neutrons and protons, move against the more weakly bound skin. Recently, the effect of an increasing skin thickness on the energy of these three modes was investigated[15] with a simple approach based on classical density oscillations, in which the change in the potential energy in each case was estimated from the density overlaps as a function of displacement. Not surprisingly, the effect on the modes a) and c) was minimal, while the soft mode b) dropped rapidly in energy, relative to the GDR. It is now interesting to pursue the behaviour of the scissors mode a little further, using the approach of the IBM.

The incorporation of both neutrons and protons in the IBM involves the algebra of the product $U_\nu(6) \otimes U_\pi(6)$. The starting point for the quadrupole modes of a nucleus with an additional neutron skin might therefore be taken as a triple product involving an additional group $U_s(6)$. The group structure could then be summarised as

$$U_\nu(6) \otimes U_\pi(6) \otimes U_s(6) \supset U_{\nu\pi}(6) \otimes U_s(6)$$

$$\supset \mathrm{U}_{\nu\pi s}(6)$$

$$\supset \left\{ \begin{array}{c} \mathrm{U}_{\nu\pi s}(5) \supset \mathrm{O}_{\nu\pi s}(5) \\ \mathrm{SU}_{\nu\pi s}(3) \\ \mathrm{O}_{\nu\pi s}(6) \supset \mathrm{O}_{\nu\pi s}(5) \end{array} \right\}$$

$$\supset \mathrm{O}_{\nu\pi s}(3) \supset \mathrm{O}_{\nu\pi s}(2) \tag{1}$$

where the weaker coupling of the skin is represented by coupling the corresponding U(6) group *after* those describing the core nucleons.

It is, of course, the group structure of $\mathrm{U}_{\nu\pi}(6)$ in IBM-2 which gives rise to the Majorana term and to the mixed symmetry states, the lowest of which in deformed nuclei represent the scissors mode. In the reduction (1), $\mathrm{U}_{\nu\pi}(6)$ is characterised by the irreducible representations $[N_c - f, f]$ where N_c is the number of nucleons in the core. The lowest irreps are then contained in the representation $[N_c, 0]$ which denotes the totally symmetric coupling while the lowest staes of mixed symmetry are in the next representation, $[N_c-1, 1]$. The group $\mathrm{U}_s(6)$ contains only $[N_s]$. The subsequent $\mathrm{U}_{\nu\pi s}(6)$ structure contains irreps with three rows, with the lowest couplings from $[N_c, 0]$ being $[N, 0, 0]$ and $[N-1, 1, 0]$, N denoting the total number of bosons. However, the $[N_c - 1, 1]$ irrep of $\mathrm{U}_{\nu\pi}(6)$ now also gives rise to an additional irrep of the triple sum group with labels $[N-1, 1, 0]$. The result is that, in deformed nuclei, there are now *two* scissors modes, one representing out-of-phase motion between the neutrons and protons in the core and the other denoting an angular oscillation between the core and the skin where, in this case, as in the soft dipole mode, the core protons presumably carry the core neutrons with them. The algebraic method thus reveals the possibility of a soft scissors mode. Moreover, a first estimate of the relative energies of the two modes can be obtained from the Casimir operators involved.

$$E = \lambda_1 C_2[\mathrm{U}_{\nu\pi}(6)] + \lambda_2 C_2[\mathrm{U}_{\nu\pi s}(6)] \tag{2}$$

There are now two Majorana terms with separate strengths and eq.(2) yields an energy for the soft scissors of $2\lambda_2 N$ versus $2\lambda_1 N_c + 2\lambda_2 N$ for the "normal" scissors states. Assuming that λ_1 and λ_2 have the same sign shows that the soft mode will be lower, whilst assuming that the two constants have equal magnitude, and that the number of neutrons in the skin is small compared to the total, suggests that the soft mode should appear at roughly half the energy of the normal one.

5. Predictions of IBM-2

Over the past 20 years, the understanding of the link between the IBM-2 parameters and the shell structure has developed steadily[16] and a large number of numerical calculations have been performed. Many of these calculations attempted to establish a dependence of parameters on valence neutron and proton number, either by empirical fitting or with the help of micrscopic understanding, and then went on to predict the structure of large chains of nuclei which, in many cases, included

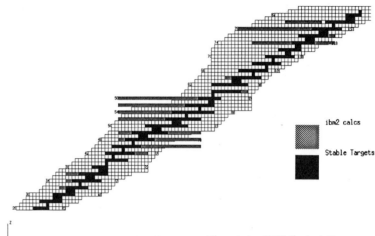

Fig. 4. Examples of regions covered by existing IBM-2 calculations.

regions which were totally unknown experimentally at that time. Examples of the extent of some of the calculations already in the literature are illustrated in fig. 4.

There are thus many detailed predictions awaiting the flow of new data from radioactive beam facilities. It is interesting to pick out a particular example from fig. 4, namely, the Ru isotopes for which predictions exist right out to N=82. In this case, the authors[17] used a standard Hamiltonian of the form

$$H = \epsilon(n_{d_\pi} + n_{d_\nu}) + \kappa Q_\pi \cdot Q_\nu + V_{\pi\pi} + V_{\nu\nu} + M_{\pi\nu} \qquad (3)$$

Where ϵ, κ and χ_ν (the parameter describing the structure of Q_ν) were varied as a function of the neutron boson number while χ_π was kept fixed, as was the Majorana term $M_{\pi\nu}$. The like-boson terms were negligible. The parameters values for the first half of the N=50-82 shell were extracted from fits to known Ru nuclei, but those for the second half of the shell had to be taken from a similar study[18] of the Xe, Ba nuclei.

Although it is too soon to test these results using radioactive beams, some very recent new data[19] has become available on neutron-rich Ru nuclei out to N=70, from experiments performed by a Manchester/Argonne/Strasbourg collaboration at Daresbury Laboratory. Partial decay schemes for $^{108-114}$Ru were determined from the study of prompt triple-γ coincidences in ^{248}Cm fission fragments, using the EUROGAM large detector array[20]. Levels with probable spin-values up to $10\hbar$ have been observed in the ground and quasi-γ bands and γ-ray branching ratios obtained.

The new data is compared with the IBM-2 predictions in fig. 5. The agreement for the ground bands is remarkably good, right up to the 10$^+$ states. For the γ-bands, the predictions work well for N=64 and 66 but, while the subsequent trend

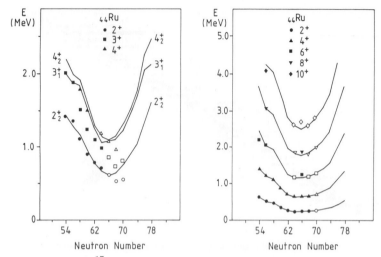

Fig. 5. IBM-2 calculations[17] for n-rich Ru nuclei; solid symbols denote data used in fit; open symbols denote new data[19].

is correct, the minimum of the bandhead energy is predicted to occur too early (by two neutrons) and the predicted minimum energy is not low enough. Furthermore, the calculations appear to predict a structure in the second half of the shell which may be too γ-soft, relative to the data. This question of triaxiality[19] is investigated further in Table 2 where the second column gives the value of the γ shape variable, extracted using the Davydov perscription[21] involving the ratio of energies of the 2+ states in ground and γ bands. The last column gives the value of the "staggering index" S_γ suggested by Zamfir and Casten[22] and defined as

$$S_\gamma = \frac{(E_{4_2} - E_{3_2}) - (E_{3_2} - E_{2_2})}{E_{2_1}} - 0.33 \qquad (4)$$

for the 3+ and 4+ γ-band states. S_γ provides one measure for the degree of softness in the γ degree of freedom in the nuclear potential. In general, distinguishing γ-soft from γ-rigid structure empirically is not easy since most quadrupole properties, such as B(E2) ratios, are the same if γ_{rms} in a soft potential is equal to γ_{rigid}. The quantity of eq. (4), however, takes the following values in the different limits.

$$S_\gamma(rigid) > 0; \ S_\gamma(rotor) = 0; \ S_\gamma(U(5)) = -1.33; \ S_\gamma(O(6)) = -2.33$$

Thus the values of Table 2 suggest that the nuclei are still γ-soft, albeit to a lesser extent than implied by either of the symmetry limits of the IBM. It seems likely that, as found elsewhere[22], addition of a cubic term in the IBM Hamiltonian is necessary to improve agreement with experiment. Such a term has been shown[23] to result in a shallow minimum in the otherwise flat γ potential of the IBM.

Table 2. Triaxiality in the neutron-rich Ru nuclei. See text for details.

A	γ	S_γ
108	22°	-0.57
110	24°	-0.42
112	26°	-0.33
114	27°	

6. The N=Z Line and the Role of Isospin In Collective Structure

The consequences of isospin symmetry in strongly deformed nuclei represent a facet of nuclear structure which has yet to be explored. The N=Z line crosses the proton drip line in the neighbourhood of the Ba and Ce isotopes but, currently, the limit of any detailed spectroscopic study with stable beams lies at ^{64}Ge. Beyond this the results have been limited to the observation of just one, or a few, states. This stems from the fact that, with stable beams, the 2n reaction must be used to reach the N=Z line in this region, with the result that the cross sections decrease rapidly with increasing mass, ranging from some 640μb for ^{64}Ge down to $\simeq$10μb for ^{84}Mo, which is the heaviest N=Z nucleus yet reached in spectroscopic studies and where only the first 2^+ state could be identified. In contrast, the cross sections with an unstable beam can be in the millibarn range and the channels of interest are now of the type (xα.yp) where the selectivity can be achieved with the much more efficient process of charged particle identification.

This greatly enhanced experimental capability has the potential to reveal a number of features. For instance, in light nuclei, isospin is a good quantum number because the Coulomb forces are small. In heavy nuclei, it remains good mainly because the neutrons and protons are filling different major shells, resulting in a large energy separation for the states of higher T. In nuclei where neutrons and protons are filling the same orbits in the 28-50 shell, neither condition holds and, in fact, a recent study of ^{64}Ge [24] suggests significant T-mixing in the low-lying states.

The interplay between Coulomb and nuclear forces can also be studied via the Coulomb energy differences in mirror nuclei. Most of the attention to date in such studies has been focussed on ground state energies, relatively little being known about high spin states. However, this situation has changed very recently, following an experiment[25] performed with the Daresbury recoil separator which succeeded in populating states in the A=49 and 47 mirror nuclei ^{49}Cr, ^{49}Mn and ^{47}Cr, ^{47}V up to spins in the region of the first band crossing and beyond. These data reveal clear collective rotational characteristics and a clear correlation between the behaviour of the Coulomb energy differences and the spin alignment as the rotational frequency increases through the band-crossing region.

The role of neutron-proton pairing must also be considered in this region. Earlier studies [26] suggest that the T=0 mode may dominate in N=Z nuclei. It has already been pointed[27] out that this may affect band-crossing phenomena and, indeed, the first backbend in a N=Z nucleus has recently been observed[25], albeit in the lower shell. The presence of T=0 pairing may also be manifest in the structure of odd-odd nuclei where a dominance of n-p pairing could result in the reappearance of a pairing gap. The nuclei around A=80 are known [28] to be highly deformed and, as such, have so far been treated in terms of collective frameworks which do not treat the isospin degree of freedom explicitly. However, the frameworks of IBM-3 and IBM-4 include the $M_T=0$ pair and have recently been developed to the stage where they are able to address many of these questions. Most importantly, a method to estimate the N (boson number) and T dependence of the parameters of IBM-3 has recently been determined[29] which should enable this framework to be applied on a similar footing to IBM-2. Moreover, a contribution to this conference[30] has shown that the IBM-3 formalism can now be given a geometric interpretation by being couched in the framework of intrinsic states.

When protons are particles and neutrons holes, or vice-cersa, the boson model does not incorporate isospin and so IBM-2 must be used. Nevertheless, it has been shown that the isospin mixing in the particle-hole system is small. It is large in the particle-particle or hole-hole cases, where IBM-3 must be used. To treat the T=0 nuclei themselves, IBM-4 may be necessary, and certainly so for odd-odd nuclei. Here the pseudo-L,S concept may be useful in regaining the degeneracies of the Wigner supermultiplet. Finally, the odd-A nuclei in this region may also be treated in these frameworks. Moreover, the bose-fermi symmetries stemming from a U(6/12) structure rely on a single particle space with j=1/2, 3/2 and 5/2 with the Fermi surface lying between the 3/2 and 5/2 orbits. This situation is likely to pertain to the negative parity structure in the neighbourhood of Z or N =35.

7. Conclusions

It seems evident that the prospect of radioactive beam accelerators is an extremely exciting one for the entire field of nuclear structure. The IBM is undoubtedly the model of choice in many crucial aspects of the new frontiers which will become accessible, not least because its explicit dependence on valence particle number renders it the ideal tool to extrapolate from our existing knowledge of low energy structure into hitherto unthought-of regimes and to predict or identify the onset of deviations from our current understanding.

8. References

1. *European Radioactive Beam Facilities. Statement by NuPECC. Report by Study Group* (May, 1993).
2. D. Darquennes *et al Proc. First Int. Conf. on Radioactive Nuclear Beams, Berkeley* ed. W. D. Myers *et al* (World Scientific, 1989) p. 3.
3. W. Catford *et al* , to be published.
4. H. J. Kluge *ISOLDE Users Guide* (CERN Report 86-05, 1986).

5. R. F. Casten et al *The IsoSpin (ISL), Research opportunities with Radioactive Nuclear Beams* (LANL Report LA-11964-C, 1991)
6. A. Arima and F. Iachello *Phys. Rev. Lett.* **35** (1975) 1069; *Ann. Phys. (N.Y.)* **99** (1976) 253.
7. R. F. Casten and D. D. Warner *Rev. Mod. Phys.* **60** (1988) 389.
8. A. Arima, T. Otsuka, F. Iachello and I. Talmi *Phys. Lett.* **B66** (1977) 205.
9. J. P. Elliott and A. P. White *Phys. Lett.* **B97** (1980) 169
10. J. P. Elliott *Prog. Part. Nucl. Phys.* **25** (1990) 325; contribution to these Proceedings.
11. F. Iachello and P. Van Isacker *The Interacting Boson-Fermion Model* (Cambridge: Cambridge University Press, 1991)
12. J. Dobaczewski, I Hamamoto W. Nazarewicz and J. A. Sheikh *Phys. Rev. Lett.* **72** (1994) 981.
13. Tanihata I et al *Phys. Rev. Lett.* **55** (1985) 2676; *Phys. Lett.* **B206** (1988) 592
14. N. Fukunishi, T. Otsuka and I. Tanihata *Phys. Rev.* **C48** (1993) 1648.
15. P. Van Isacker, M. A. Nagarajan and D. D. Warner *Phys. Rev.* **C45** (1992) R13
16. F. Iachello and I. Talmi *Rev. Mod. Phys.* **59** (1987) 339.
17. P. Van Isacker and G. Puddu *Nucl. Phys.* **A348** (1980) 125.
18. G. Puddu, O. Scholten and T. Otsuka *Nucl. Phys.* **A348** (1980) 109.
19. J. A. Shannon et al, to be published.
20. P. J. Nolan *Nucl. Phys.* **A520** (1990) 657.
21. A. S. Davydov and G. F. Fillippov *Nucl. Phys.* **8** (1958) 237.
22. N. V. Zamfir and R. F. Casten *Phys. Lett.* **B260** (1991) 265.
23. K. Heyde et al *Phys. Rev.* **C29** (1984) 1420.
24. Ennis P. J. et al *Nucl. Phys.* **A535** (1991) 392
25. Cameron J. A. et al *Phys. Lett.* **B235** (1990) 239 ; **B319** (1993) 58 ; *Phys. Rev.* **C49** (1994) 1347.
26. Goodman A L *Adv. Nucl. Phys.* **11** (1979) 263.
27. J. A. Sheikh, N. Rowley, M. A. Nagarajan and H. G. Price *Phys. Rev. Lett.* **64** (1990) 376.
28. Gelletly W et al *Phys. Lett.* **B253** (1991) 287 and references therein.
29. J. A. Evans, G. L. Long and J. P. Elliott *Nucl. Phys.* **A561** (1993) 201.
30. J. N. Ginocchio and A. Leviatan, contribution to these Proceedings.

GLOBAL SYSTEMATICS OF UNIQUE PARITY QUASIBANDS IN ODD-A NUCLEI

D.BUCURESCU, G.CATA-DANIL, M.IVASCU, L.STROE, C.A.UR
Institute of Atomic Physics, P.O.Box MG-6, Bucharest, Romania

ABSTRACT

Analysis of excitation energies in the favored and unfavored quasibands determined by the unique parity orbitals $g_{9/2}$, $h_{11/2}$ and $i_{13/2}$ reveals that all odd-A nuclei with Z =32-78 obey global systematics similar to those of the even-even ones: three general behaviours are found, of anharmonic vibrator, strong coupling and seniority, which are connected by narrow zones of rapid transition. The splitting between the two quasibands is found correlated with the P-factor.

Discovery of regularities in the vast amount of data on nuclei and their excited states helps to understand in a more unitary way the evolution of the nuclear structure over large mass ranges. Interesting correlations between the energies of states in the quasiground band of practically all known even-even nuclei have been recently evidenced[1,2], which led to a global view on the evolution of the nuclear structure, from the "shell model" nuclei to those with various collective behaviours.

In the present work, we undertake a similar study in odd-A nuclei, for which such global investigations are practically absent. It is expected that the physics observed in the even-even nuclei will repeat itself; on the other hand, since one always has an unpaired particle, additional complexity and phenomena may appear.

We focus on the quasiband structures resulted when the odd-particle occupies a unique parity orbital (u.p.o.). These are levels unambiguously assigned in most of the known nuclei and can therefore be followed over large regions of the nuclide chart; thus, with the orbitals $g_{9/2}$, $h_{11/2}$ and $i_{13/2}$ we span both neutron- and proton-odd nuclei with Z=30 to 80. For all nuclei from the NNDC collection where structures determined by a u.p.o. with spin j were available, we have followed the evolution of the favored (the sequence of levels of spins $j, j+2, j+4, ...$) and unfavored ($j-1, j+1, j+3, ...$) quasiband, respectively.

Fig. 1 shows the correlation between the energies of the states of spin $(j+4), (j+6)$ and $(j+2)$ within both the neutron and proton $g_{9/2}$ and $h_{11/2}$ favored quasibands (nuclei from Ge to Pt). The nuclei having even-even neighbours with $2.0 < R_{even}(4/2) = E(4_1^+)/E(2_1^+) < 3.0$, (therefore collective but not yet deformed) show a very well defined linear correlation $E(I) = \alpha E(j+2) + \beta$; the experimental values are $[\alpha, \beta]$=[1.99(3),182(19)], [2.84(12),564(52)], [3.94(23),920(114)] for $I = j+4, j+6, j+8$, respectively (β in keV). These relations are identical, within errors, with those from the gs band in the adjacent even-even nuclei[1], therefore we find the same behaviour of *anharmonic vibrator* (AHV) *with constant anharmonicity* ($\simeq 180$ keV) for the u.p.o. quasiband. The experimental data show a similar behaviour also for the unfavored band.

There are limiting situations, such as the weak coupling or the decoupling, when we automatically expect a behaviour similar to that of the even-even nuclei; however, the general relevance of an AHV behaviour for all non-deformed odd-A nuclei (which thus "duplicates" that of the even-even ones) has not been recognized until our recent work[3]. The deformed nuclei (with $R_{even}(4/2) > 3.0$) define a second branch, which is well described by the linear behaviour expected from the strong coupling (SC) formula. The way the transition from AHV to SC actually takes place is displayed in the more detailed Fig. 2. Both for the $h_{11/2}$ and $i_{13/2}$ u.p.o. one sees that by going towards deformed nuclei one follows first the AHV line up to a "turning point" where the curve changes abruptly its

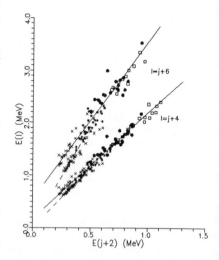

Fig. 1. Energy correlations within the favored u.p.o. quasibands $g_{9/2}$ and $h_{11/2}$.

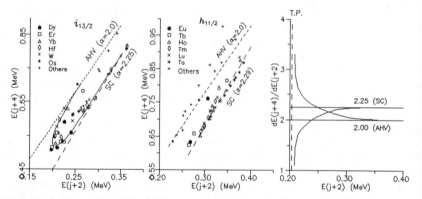

Fig. 2. Details of the AHV→SC transition. The full line is drawn by hand and its derivative is shown at the right.

direction and then approaches the SC line (slope $\simeq 2.25$). The turning point region corresponds, in the even-even nuclei, to the zone where a very rapid transition takes also place, from AHV to rotor[1]. That transition has been shown[1,4] to have the features of a critical phase transition, as known from macroscopic systems, such as the ferromagnets. The corresponding AHV → SC transition seen in the odd-A nuclei (Fig. 2) is also very rapid. The right hand side of Fig. 2 shows the derivative $dE(j + 4)/dE(j + 2)$ of the curve (full line) which describes empirically the $\nu i_{13/2}$

transition. The analogue curve $dE(4_1^+)/dE(2_1^+)$ in the e-e nuclei is discontinuous at the critical point $E_c(2_1^+)$; in our case the curve diverges at the turning point, its behaviour both below and above $E_c(2_1^+)$ being typical for a λ-anomaly (power law as a function of $E_{T.P.} - E(j+2)$). Nevertheless, this is a peculiar sort of λ-anomaly, with one of the branches inverted, which makes the turning point a unique phenomenon, not observed in other systems. Thus, the critical phase transition AHV→rotor(SC) in both e-e and odd-A nuclei is a challenging question for the microscopic structure theories.

Next, we add the "precollective" nuclei (with $R_{even}(4/2) < 2.0$, i.e., single magic or near closed shell nuclei (Fig. 3)). A third regime emerges: a straight line of slope 1.08(6) and intercept 549(72) keV, identical with that observed in the even-even case[2]. The slope 1.0 is characteristic of the "seniority" (SEN) regime, named so since it is a result of the (approximate) conservation of the seniority.

Fig. 3 summarizes nicely the results obtained until now: according to their u.p.o. structures, the odd-A nuclei fall into three classes: the seniority, anharmonic vibrator and strong coupling regimes, which are interconnected by narrow zones of rapid transitions.

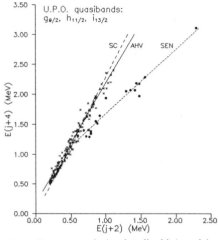

Fig. 3. Energy correlation for all odd-A nuclei with $Z = 32 - 78$;

Having determined global correlations for energies within the two (favored and unfavored) u.p.o. quasibands, we examine now the global evolution of their relative displacement, which is another very useful structure indicator. For this, we introduce the following "splitting index": $R_j{}^s = [E(j+2) - E(j+1)]/[E(j+2) - E(j)]$, which has a well defined positive value in the SC limit, is positive for partial deformation alignment and, in general, negative in the case of weak coupling and decoupling situations. We have found that this index has a remarkably compact evolution when represented as a function of the parameter $P[\equiv N_p N_n/(N_p + N_n)]$ of Casten (Fig. 4). Most of the nuclei fall into a narrow region which starts at negative $R_j{}^s$ values for small P, and increase with increasing P to positive values until one reaches the SC value. Nuclei with an almost closed shell show systematic deviations from this general pattern: Ag, Rh (for $\pi g_{9/2}$); Ir, Re ($\pi h_{11/2}$); Pt, Os, W ($\nu i_{13/2}$). An empirical explanation of this would be that these nuclei have different (larger) or "effective" P-values, which requires, in turn, larger effective values for either N_p or N_n. There is indeed a physical reason for that: in the e-e nuclei near closed shells there are quite low "intruder" structures based on 2p-2h excitations (across the major shell gap) - which play a role in the lowest nuclear excitations - and they have four valence particles of one type in addition to our usual counting.

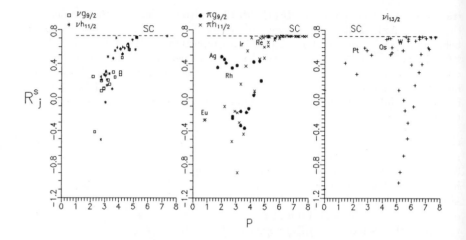

Fig. 4. Correlation between the splitting index (see text) and the P—factor.

Such configurations, known in Pd, Cd, Pt and Os nuclei, appear to be reflected also in the behaviour of our index $R_j{}^s$. Indeed, calculating P for the deviant nuclei with N_p (or N_n) larger by 4, would bring most of the points rather close to the global behaviour of all the other nuclei.

Conclusions. We have shown that the experimental energies of the u.p.o. quasibands in the medium heavy odd-A nuclei show remarkably compact correlations. Like the even-even nuclei, they fall into three classes, the AHV, the SC and the seniority regimes, which are conencted by zones of rapid transition. The AHV → SC transition is made through a turning point phenomenon which corresponds to the critical phase transition in the even-even nuclei. Finally, it has been shown that the splitting between the favored and unfavored u.p.o. quasibands has a compact evolution as a function of P.

These phenomenological correlations are useful for prediction of properties of nuclei in new regions far from stability. Also, they should allow to find a compact evolution, over large nuclear zones, for the phenomenological parameters of algebraic structure models, such as IBFM. Finally, the occurence of critical phase -like transitions between wide nuclear regions of "stable" behaviour rises challenging questions for the microscopic nuclear theories.

1. R.F.Casten, N.V.Zamfir and D.S.Brenner, *Phys. Rev. Lett.* **71** (1993) 227.
2. R.F.Casten, N.V.Zamfir and D.S.Brenner, in *Proc. of the Symposium "Capture Gamma-Ray Spectroscopy"*, Fribourg 1993.
3. D.Bucurescu, G.Cata-Danil, M.Ivascu, L.Stroe and C.A.Ur, *Phys. Rev.* **C49** (1994) R1759.
4. A.Wolf, R.F.Casten, N.V.Zamfir and D.S.Brenner, *Phys. Rev.* **C49** (1994) 802

393

ELECTRIC DIPOLE EXCITATIONS IN RARE EARTH NUCLEI

A. ZILGES, P. VON BRENTANO, AND R.-D. HERZBERG
Institut für Kernphysik, Universität zu Koeln
D-50937 Köln, Germany

and

U. KNEISSL, J. MARGRAF, AND H. H. PITZ
Institut für Strahlenphysik, Universität Stuttgart
D-70569 Stuttgart, Germany

ABSTRACT

A survey of recent photon scattering experiments for the investigation of the octupole degree of freedom in nuclei is given. The energies and strengths of Electric Dipole Excitations in spherical as well as in deformed nuclei are discussed.

1. Introduction

The elementary collective excitations of nuclei apart from the rotation of prolate nuclei are vibrations of the nuclear shape.[1] A vibration which occurs in spherical as well as in strongly deformed nuclei is the pear shaped 3^- octupole excitation. This vibration induces an electric dipole moment which leads to enhanced E1 transition rates. For a systematic survey of low lying electric dipole excitations one wants to use an experimental method which is strength selective. Nuclear Resonance Fluorescence (NRF) or Photon Scattering Experiments fulfil this requirement: Using a continous bremsstrahlung source one can populate *all* dipole excitations with groundstate transition widths above a certain detection limit in the energy interval below the endpoint energy of the bremsstrahlung spectrum. In addition to the strength information this electromagnetic probe yields the energies, spins, branching ratios and (using Compton polarimeters) parities of the dipole excitations.[2]

2. Octupole Vibrational States in Deformed Nuclei

A very interesting systematics has been obtained througout the last years concerning the low lying octupole vibrational states in deformed nuclei.[3,4] The octupole vibrational bands occur due to the coupling of the 3^- octupole vibration to the quadrupole deformed core. The bands are characterized by their K-quantum number, the bandheads with J=1 are populated in the photon scattering experiments. It has been shown that the mixing of the bands with different K-quantum numbers is a very important feature of these states,[5] the mixing matrix element can be calculated from the observables of the experiment and lies around 50 keV (which is relatively large compared to usual K-mixing matrix elements[6]). A calculation in the sdf-IBA shows excellent agreement with the experimentally observed branching ratios and absolute strengths.[7]

3. Multi Phonon Excitations in spherical nuclei

In spherical nuclei the octupole vibration couples to the quadrupole vibration creating a multiplet of five ($2^+ \otimes 3^-$) two phonon states. The 1^- member of this multiplet has been investigated in a number of nuclei via photon scattering.[8,9] In the N=82 region it lies around 3.5 MeV, a strong E1 groundstate transition with $B(E1) \simeq 5 \times 10^{-3}$ Weisskopf units is observed. Figure 1 shows as an example the photon scattering spectrum of the N=82 nucleus ^{140}Ce, the E1 groundstate transition is observed as a sharp peak near the sum energy of the octupole and quadrupole vibration.

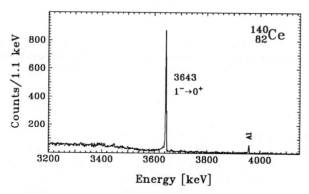

Fig. 1. Photon scattering spectrum of ^{140}Ce between 2.5 and 4.1 MeV; the line at 3643 keV arises due to the E1 groundstate transition from the two phonon 1^- state.

The intensive search for the other members of the two phonon multiplet with different experimental probes is still under way.[10] We learned more about the structure of these states by performing a photon scattering experiment on the odd A nucleus ^{143}Nd where the coupling of an additional neutron to the ($2^+ \otimes 3^-$) states is observed.[11] Due to level mixing it was possible to excite 15 out of the 31 two phonon particle levels in the experiment. A detailed calculation combining an IBA approach for the core with a simple core particle coupling model yields very good agreement with the experimental data.[11] Additional experiments on other odd A nuclei in the N=82 region have been performed and support the results for ^{143}Nd.

4. Evidence for γ-octupole states in deformed nuclei

Now one may ask if similar two phonon states, i.e. a band structure build on the coupling of two vibrations exists as well in deformed nuclei. As discussed in section 2 the coupling of the octupole vibration to the quadrupole deformed core leading to the octupole vibrational bands is well understood. A coupling of the 3^- excitation to the γ- or β-quadrupole vibration was hitherto unknown. Very recently we observed a strong dipole excitation at an energy of $\simeq 2.5$ MeV in the nucleus ^{162}Dy. Its branching ratio is deviating from the Alaga rule value for a pure K=1 or K=0

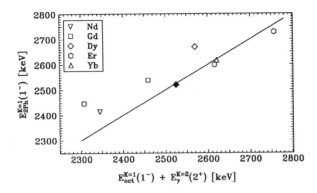

Fig. 2. Energy of the observed strong electric dipole excitations around 2.5 MeV versus the sum energy of the octupole vibration with K=1 and the γ-quadrupole vibration. The solid line marks a one to one correspondence.

state and an additional measurement with a Compton polarimeter yielded negative parity. The energy lies very close to the sum energy of the 3^- octupole vibration with K=1 and the 2^+ quadrupole γ vibration with K=2. Therefore we assumed a γ-octupole structure for this state, a picture that was proposed in a theoretical study of Donner and Greiner already in the sixties.[12] We investigated systematically other deformed rare earth nuclei and finally got the result shown in fig. 2. In all nuclei the branching ratio of this "two phonon" dipole excitation around 2.5 MeV points to a mixed K-quantum number. This may be explained because the K-quantum number of the γ-vibration is already impure. The figure displays the energy of this strong dipole excitation versus the sum energy of the K=1 octupole vibration and the K=2 quadrupole vibration.[13] The main problem in the construction of such a figure is the rare spectroscopic information on the K=1 octupole vibrational bandheads.

To support the γ-octupole structure for these low lying 1^- states we performed a calculation in the sdf–IBA model.[13] This calculation reproduces the observed two phonon state, nevertheless the calculated B(E1) strength is orders of magnitudes too small.

5. Conclusion and Outlook

Systematic photon scattering experiments have been performed at the Stuttgart Dynamitron accelerator for the investigation of low lying dipole excitations. The photon scattering or Nuclear Resonance Fluorescence (NRF) method is an excellent tool to achieve this goal.

In the *spherical* nuclei the energies and B(E1) strengths of the 1^- member of the $(2^+ \otimes 3^-)$ two phonon multiplet in the N=82 region have been observed. The coupling of an additional particle has been investigated in a number of neighbouring odd A nuclei. We hope to expand the systematics in the future by new (γ,γ') experiments in other regions of the chart of nuclides where the two phonon states lie at higher energies. For this purpose we will combine the NRF set up at the superconducting Darmstadt S–DALINAC[14] with the newly developed Euroball Cluster Ge–detectors. The Cluster detector combines very high efficiency and good energy resolution at high photon energies.[15] In addition it may be possible to use the Cluster detector as a Compton polarimeter for parity determination. Other studies in nuclei near closed shell will focus on the measurement of the two phonon to one phonon E1 transition strengths which are hitherto unknown and on systematic studies of electric dipole excitations in odd A nuclei.

In the *deformed* nuclei a short summary of the knowledge on octupole vibrational bands has been given. The E1 transitions to the bandheads are now well understood. The investigation of octupole vibrational bands in deformed nuclei will be expanded by a study of actinide nuclei. At energies around 2.5 MeV first evidence for the coupling of the octupole vibration to the γ-vibration with K=2 has been found recently. To clarify the nature of these two phonon γ-octupole states we plan a number of high precision experiments in the rare earth and actinide region. We hope that additional information will come from measurements with other experimental probes. Further theoretical studies are clearly warranted to gain a deeper insight into the structure of Multi Phonon Excitations in deformed nuclei.

The authors would like to thank their colleagues from the NRF collaboration Stuttgart–Giessen–Köln. We gratefully acknowledge valuable discussions with R. F. Casten, F. Dönau, A. Gelberg, K. Heyde, R. V. Jolos, T. Mizusaki, A. Oros, T. Otsuka, A. Richter, K. Sugaware–Tanabe, and N. V. Zamfir. This work was supported by the DFG under contracts Br 799/6–1 and Kn 154–21.

References

1. A. Bohr and B. R. Mottelson, *Nuclear Structure*, Benjamin, New York, 1975.
2. U. Kneissl, *Prog. Part. Nucl. Phys.* **24** (1990) 41 and **28** (1992) 331.
3. T. Guhr et al., *Nucl. Phys.* **A501** (1989) 95.
4. A. Zilges et al., *Z. Phys. A – Hadrons and Nuclei* **340** (1991) 155.
5. A. Zilges et al., *Phys. Rev.* **C42** (1990) 1945.
6. See e.g.: B. R. Barrett et al., *Phys. Rev.* **C45** (1992) R1417.
7. P. von Brentano et al., *Phys. Lett.* **B278** (1992) 221.
8. F. R. Metzger, *Phys Rev* **C14** (1976) 543.
9. H. H. Pitz et al., *Nucl. Phys.* **A509** (1990) 587.
10. See e.g.: E. Müller–Zanotti et al., *Phys. Rev* **C47** (1993) 2524.
11. A. Zilges et al., *Phys. Rev. Lett.* **70** (1993) 2880.
12. W. Donner and W. Greiner, *Z. Phys.* **197** (1966) 440.
13. U. Kneissl et al., *Phys. Rev. Lett.* **71** (1993) 2180.
14. W. Ziegler et al., *Nucl. Phys.* **A564** (1993) 366.
15. J. Eberth, *Nuclear Physics News* **3** (1993) 8.

IBA CALCULATION OF THE EFFECTIVE GAMMA DEFORMATION IN BARIUM NUCLEI

O. VOGEL, A. GELBERG, AND P. VON BRENTANO
Institut für Kernphysik, Universität zu Koeln
D-50937 Köln, Germany

AND

P. VAN ISACKER
GANIL, BP 5027, F-14021 Caen Cedex, France

ABSTRACT

The effective γ deformation of even–even Ba nuclei is derived from the IBA-1 by using a simple Hamiltonian $H = \kappa Q \cdot Q + \kappa' L \cdot L$ and calculating the matrix elements of $(Q \times Q)_{00}$ and $(Q \times Q \times Q)_{00}$ in the framework of the consistent-Q formalism.

1. Introduction

Many studies in recent years have shown that the transitional nuclei in the $A = 130$ mass region exhibit triaxiality.[1] They also showed that the triaxiality of these nuclei is soft in γ, that is, they do not have a fixed γ deformation. Generally, these nuclei are well described by the O(6) limit of the IBA.[1] One signature for the difference between γ soft and γ rigid is the staggering in the quasi-gamma band. Another question to be answered is how strong is the triaxiality in these nuclides.

In a recent approach the Rigid Triaxial Rotor Model (RTRM)[2] was used to extract the deformation parameter γ from experimental energies and transition ratios.[3] However, as stated above, this model cannot reproduce all spectroscopic features of the Ba isotopes. It thus seems appropriate to seek a more realistic description of these nuclei in the context of which one can also derive the deformation parameter γ. This is achieved in the following where a method to derive an effective γ deformation from IBA-1 is applied to the Ba nuclei.

2. Determination of Gamma

In the approach of Elliott *et al.*,[4] the following association between the invariants constructed from the quadrupole operator Q and the deformation parameters (β, γ) is proposed by analogy with the geometrical model[5]:

$$(Q \times Q)_{00} \rightarrow \beta^2/\sqrt{5}, \tag{1}$$

$$(Q \times Q \times Q)_{00} \rightarrow -\beta^3 \sqrt{\frac{2}{35}} \cos 3\gamma. \tag{2}$$

This allows to determine the expectation value of $\cos 3\gamma$, by the following equation:

$$\langle \cos 3\gamma \rangle = -\sqrt{\frac{7}{2\sqrt{5}}} \frac{\langle (Q \times Q \times Q)_{00} \rangle}{\langle (Q \times Q)_{00} \rangle^{3/2}}. \tag{3}$$

So the effective γ deformation of the ground state is given as

$$\gamma_{\text{eff}} = 1/3 \cdot \arccos \langle 0_1^+ | \cos 3\gamma | 0_1^+ \rangle. \tag{4}$$

In order to describe the Ba nuclides, we use the following simple Hamiltonian:

$$H = \kappa Q(\chi) \cdot Q(\chi) + \kappa' L \cdot L, \tag{5}$$

with

$$Q_\mu = d_\mu^\dagger s + s^\dagger \tilde{d}_\mu + \chi (d^\dagger d)_{2\mu}, \tag{6}$$

allowing us to move between the O(6) limit ($\chi = 0$) and the SU(3) limit ($\chi = -\sqrt{7}/2$) of the IBA. In the consistent-Q formalism[6] (CQF) the same value of χ is used in the Hamiltonian and in the E2 transition operator. The resulting wave functions and transition rates are thus totally determined by the parameter χ and the boson number N. Due to our choice of CQF Hamiltonian, the average γ deformation is a function of χ and N only.

In order to derive the parameter χ for a given nucleus we follow an approach similar to the one by Yan et al.[3] who use the RTRM[2] and an analytical expression for the B(E2) ratio

$$R_B = \frac{B(E2; 2_\gamma^+ \to 2_1^+)}{B(E2; 2_\gamma^+ \to 0_1^+)}. \tag{7}$$

R_B is a monotonous function of γ and experimentally it is very easily accessible from the branching ratios of the 2_γ^+ state.

In the case of the IBA, R_B is a monotonous function of χ (for fixed N) and can be used to determine the parameter χ. In table 1 the values for χ are given together with the effective γ deformation in the ground state from equation 4 and the γ from the RTRM. One can see that the deduced value of χ is nearly constant for large N, but the γ deformation, and therefore the triaxiality, decreases with increasing boson number. Also the agreement between the γ calculated with IBA and the γ from RTRM is very good.

To better understand the calculated values of γ_{eff}, it is useful to examine the energy surfaces $V(\beta, \gamma)$.[7] As an example, $V(\beta, \gamma)$ is shown for ^{130}Ba in figure 1, with the same parameters as in the energy fit below. Although $V(\beta, \gamma)$ has a minimum at the prolate side, the effective γ deformation is very large due to the softness in γ.

3. Comparison with Experimental Data

The question arises how well these nuclides can be described with our particular choice of Hamiltonian, since there are only two free parameters for the excitation

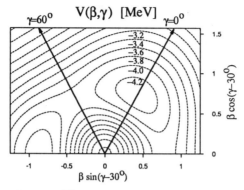

Fig. 1. Contour plot of $V(\beta,\gamma)$ for ^{130}Ba, calculated with the parameters $\kappa = -60.6$ keV, $\kappa' = 12.5$ keV, and $\chi = -0.181$.

Table 1. Experimental B(E2) ratio R_B,[8–10] boson number N, deduced parameter χ, the derived values of γ_{eff} in the IBA ground state, and the γ from the RTRM.

Nucleus	^{124}Ba	^{126}Ba	^{128}Ba	^{130}Ba	^{132}Ba	^{134}Ba
R_B	6.0(14)	7.7(12)	8.8(3)	16.1(18)	35.7(64)	176(9)
N	10	9	8	7	6	5
χ	-0.189(38)	-0.189(20)	-0.209(5)	-0.181(11)	-0.150(14)	-0.088(3)
γ_{eff}	19.4(15)°	20.4(8)°	20.7(2)°	22.8(4)°	24.8(5)°	27.3(1)°
γ_{RTRM}	20.6(14)°	21.8(8)°	22.3(2)°	24.4(4)°	26.2(4)°	28.3(1)°

energies and none for the relative B(E2) values. The parameters κ and κ' are fitted to excitation energies of the 6_1^+ state and the 4_γ^+ state. The result for ^{130}Ba is shown in figure 1. The agreement is quite good, although typically the staggering in the quasi-gamma band is too strong in the IBA and the 2_3^+ state lies too high. The first feature can be improved by including a cubic term $(d^\dagger d^\dagger d^\dagger)_3 \cdot (ddd)_3$ in the Hamiltonian.[11] For a better description of the 2_3^+ state, IBA-2 calculations are normally needed. A more sensitive test for the quality of a fit are the transition probabilities, which strongly depend on the nuclear structure. These are also very well reproduced, as can be verified from table 2.

4. Conclusions

1. The IBA calculations show a large effective γ deformation. In contrast to the RTRM, it is not characterized by a rigid γ, but by a softness in the γ direction.

2. The values of γ derived from the IBA are consistent with those obtained with the RTRM.

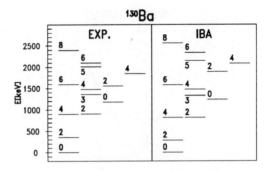

Fig. 2. Experimental[8] and IBA calculated low-lying positive-parity states in ^{130}Ba.

Table 2. Experimental[8] and IBA calculated B(E2) ratios for ^{130}Ba.

I_i	I_f	Exp.		IBA	I_i	I_f	Exp.		IBA	I_i	I_f	Exp.		IBA
2_2^+	2_1^+	100		100	4_2^+	2_2^+	100		100	2_3^+	0_2^+	100		100
	0_1^+	6.2	(7)	6.2		2_1^+	2.3	(4)	0.1		2_2^+	21	(4)	0.1
3_1^+	2_2^+	100		100		4_1^+	54	(10)	58		4_1^+	2.7	(5)	1.3
	2_1^+	4.5	(6)	6.1	0_2^+	2_2^+	100		100		2_1^+	3.3	(6)	0.001
	4_1^+	22	(3)	29		2_1^+	3.3	(2)	2.5		0_1^+	0.017	(3)	0.2

3. Simply fixing the parameter χ from the decay of the 2_2^+ state yields a very good fit to the low-lying states of the Ba isotopes in the framework of the CQF.

The authors would like to thank A. Dewald, R.V. Jolos, K. Kirch, T. Otsuka, P. Petkov, N. Pietralla, and N. Yoshinaga for stimulating discussions. This work was partly supported by BMFT under contract no. 06 OK 602 I.

References

1. R.F. Casten and P. von Brentano, *Phys. Lett.* **152** (1985) 22.
2. A.S. Davydov and G.F. Filippov, *Nucl. Phys.* **8** (1958) 237.
3. J. Yan, O. Vogel, P. von Brentano, and A. Gelberg, *Phys. Rev.* **C48** (1993) 1046.
4. J.P. Elliott, J.A. Evans, and P. Van Isacker, *Phys. Rev. Lett.* **57** (1986) 1124.
5. K. Kumar, *Phys. Rev. Lett.* **28** (1972) 249.
6. R.F. Casten and D.D. Warner, *Phys. Rev.* **C28** (1983) 1798.
7. J. Ginocchio and M.W. Kirson *Phys. Rev. Lett.* **44** (1980) 1744.
8. G. Siems et al., *Phys. Lett.* **A320** (1994) 1.
9. N. Idrissi et al. *Z. Phys.* **A341** (1992) 427.
10. NNDC Online Data Bank, National Nuclear Data Center, New York.
11. R.F. Casten et al., *Nucl. Phys.* **A439** (1985) 289.

ANOMALY IN ^{196}Pt: A CLEAN TEST?

M K HARDER
*Department of Mathematical Sciences, University of Brighton,
Brighton, E. Sussex, BN2 4GJ, U.K.*

B KRUSCHE
*Institute Laue-Langevin, Grenoble,
CEDEX F-38042, France*

ABSTRACT

The O(6) description of ^{196}Pt is reviewed in light of a strongly observed but forbidden E0 transition. The "cleanness" of the E0 test is compared to that of quadrupole-based tests. Options for retaining the O(6) description in spite of this anomaly are mentioned.

1. Introduction

Some nuclei become more famous than others by virtue of their generous display of characteristics predicted by nuclear models. The nucleus ^{196}Pt is one which has attained such a station. It was the first nucleus put forward[1] as an example of the O(6) limit of the Interacting Boson Model[2], assuring it everlasting fame. But new evidence[3] shows an anomaly in this description, demanding either acceptance that this nucleus is not pure O(6) or that the IBM description of E0 transitions is incomplete.

2. Quadrupole tests

The nucleus ^{196}Pt was first related to the O(6) symmetry via its level energies and B(E2) values[1]. A one-to-one correspondance was found between the levels observed experimentally (below the pairing energy of 1.8MeV) and those predicted by the IBM for a pure O(6) nucleus. These states have been labelled with the O(6) quantum numbers σ and τ, with values including σ =N, N-2 and N-4, where N is the number of bosons. All of the predicted E2 transitions were observed as strong branches, and those forbidden were either weak or not observed.

A serious discrepancy emerged when the quadrupole moment of the first 2$^+$ state was measured[4,5]. A pure O(6) nucleus is predicted to have zero quadrupole moments, yet a clear positive value was found for the first excited 2$^+$ state. A volley of papers then followed[5-8], with arguments for and against the O(6) description based on the relative importance of the B(E2) values to the quadrupole moment value. Clearly, neither quantity was a clean test of the symmetry, or the ambiguity would not exist.

As there was no reason to ignore the QM value, attempts were made to fit ^{196}Pt as a near-O(6) nucleus, with broken O(6) symmetry. In principle this could be done by either deviating from the pure O(6) hamiltonian and/ or from the pure O(6) E2 operator[4]:

$$H_{IBM} = \varepsilon"n_d + a_0 P.P + a_1 L.L + a_2 Q.Q + a_3 T_3.T_3 + a_4 T_4$$

with the O(6) limit having $\varepsilon"$=0, a_2=0, a_4=0;

$$T(E2) = \alpha Q = \alpha[(s^\dagger d + d^\dagger s)^{(2)} + \chi(d^\dagger d)^{(2)}]$$

with the O(6) limit having the value $\chi = 0$.
It was then shown[9] that the QM and B(E2) experimental values could not be simultaneously reproduced in the IBM using any value of χ with a pure O(6) hamiltonian. This rather implied that a non-O(6) term was needed in the hamiltonian. However, all attempts to reproduce the experimental data of the first two excited states by incorporating a SU(3) term have so far failed - including IBM-2 and CQF versions[7].

3. E0 tests

With the facility of the high-resolution electron spectrometer at the Institute Laue-Langevin[10], it was possible to examine the E0 transition strengths of ^{196}Pt. Although such measurements are less common than B(E2) measurements, they appear to be rather clean tests of nuclear models[11]. This is primarily due to the simple form of the E0 operator:

$$T(E0) = \alpha(s^\dagger s)^{(0)} + \beta(d^\dagger d)^{(0)}$$

which can be rewritten in terms of the total boson number $N = n_s + n_d$ as:

$$T(E0) = \alpha N + \beta' n_d$$

where $\beta' = \beta/(\sqrt{5}-\alpha)$.

Since N is constant for a given nucleus (N=6 for ^{196}Pt) and the basis states are orthonormal, the E0 transition matrix element is simply

$$\langle f| T(E0) |i\rangle = \beta' \langle f| n_d |i\rangle \quad \text{for } i \neq f.$$

If we expand the IBM wavefunction in terms of basis states where n_d is a good quantum number, the E0 transition strengths are related directly to the d-boson structure, providing clean test of the wavefunction:

$$\langle f| T(E0) |i\rangle = \sum_j a_j(i) \, a_j(f) \, n_{d_j}.$$

Table 1. O(6) wavefunctions in terms of the SU(5) basis states for the 0^+ states.

O(6) wave-function $\{\sigma,\tau,\nu\}$	Breakdown in terms of the SU(5) basis states $\{n_d, \eta_\beta, \eta_\Delta\}$						
	{0,0,0}	{2,1,0}	{3,0,1}	{4,2,0}	{5,1,1}	{6,0,2}	{6,3,0}
{6,0,0}	-0.433	-0.740	-	-0.491	-	-	-0.095
{6,3,1}	-	-	-0.886	-	-0.463	-	-
{4,0,0}	0.685	0.079	-	-0.673	-	-	-0.269
{4,3,1}	-	-	0.463	-	-0.886	-	-
{2,0,0}	0.559	-0.581	-	0.296	-	-	0.512
{6,6,2}	-	-	-	-	-	1.000	-
{0,0,0}	-0.177	0.306	-	-0.468	-	-	0.810

The basis states used are the ones found in standard IBM codes; the coefficients $a_j(i)$ and $a_j(f)$ are labelled for the initial and final states respectively, and j runs through each of the basis states in the set (which represents the symmetry SU(5)).

This is the general expression for E0 transitions within the IBM: in the case of U(5) symmetry the E0 strengths vanish since n_d is diagonal; none of the basis states overlap. Thus no E0 transitions are allowed in U(5). For the O(6) limit the wave functions are linear combinations of the U(5) basis states: these are given in table 1 for the 0^+ states. The strong selection rule $\Delta\tau = 0$ follows from the otherwise zero overlap of the wave functions. It can also be shown that the transition matrix elements vanish unless $\Delta\sigma = \pm 2$.

With this test in mind, an early study was made of some E0 transitions[12], but included only 0-0 transitions to the ground state and selected 2-2 transitions, up to 2.2 MeV. These agreed with O(6) predictions. The only anomalous transition was a strong E0 transition between the fifth and second 2^+ states. Transitions between excited 0^+ states were not mentioned.

4. Current Work

In the current work, the entire electron spectrum from 240keV to 1100keV was measured in order to identify any further E0 decays, especially between 0^+ states. The nucleus ^{196}Pt was produced by thermal neutron capture and the high-resolution BILL spectrometer was used to detect the electrons. The full results will be published[13], including all internal conversion data.

The transition of particular interest is a new one seen at 267.50keV, which links the $0_3^+(1402.7)$ and $0_2^+(1135.3)$ levels. Such a transition is forbidden by the strong $\Delta\tau=0$ selection rule within the O(6) limit. Although a limit exists for the lifetime of the $0^+(1402)$ state[14] it yields only limits for the E0 matrix elements of $\rho^2(E0: 0_3^+-0_1^+) < 0.018$, $\rho^2(E0: 0_3^+-0_2^+) < 0.0386$. Since the first transition is allowed and the second forbidden, obtaining two limits is not very helpful. It is far more instructive to obtain a precise ratio of the two transition probabilities directly from the internal conversion data:

$$\frac{\rho^2(E0:\ 267keV)}{\rho^2(E0:\ 1402keV)} = 21.0\ (28)$$

This result follows directly from the standard expressions of Alduschenkov & Voinova[15], where Ω is the electronic factor and I_e is the electron intensity of the transition. A full account of the experimental data used for this ratio is given elsewhere[16].

Clearly this result is at odds with the IBM O(6) description; the 267keV transition is forbidden not only within O(6) but also in SU(5). The transition as labelled would have $\Delta\tau=3$, violating the $\Delta\tau=0$ rule. Yet it is over 20 times the strength of the allowed 1402keV transition. This strength cannot be explained with small perturbations, and implies that the current O(6) description of the states cannot be correct. More seriously, the fact that both transitions originate from the same level make it impossible to assign the character of that level to any predicted 0^+ state in the strict O(6) limit as none of them depopulate via E0 transitions to both an excited state and the ground state.

Because the experimental value given is a ratio, it could be contended that the non-O(6) character observed is due to either or both of the 0^+ level wave functions. However, the wavefunction of the lower 0^+ level at 1135keV does appear to conform to O(6) requirements in that the forbidden transition to the ground state has not been seen. In fact the current experimental limit is as low as $\rho^2 < 7.2 \times 10^{-5}$. This suggests that the upper 0^+ level at 1402keV is responsible for the violation, and may be an intruder state. Indeed, both ^{192}Pt and ^{194}Pt have 0^+ states in this region which are not reproduced by the IBM[17]. The second-excited 0^+ state at 1481keV in ^{198}Pt has already been suggested as a candidate intruder state[18]. In the case of ^{196}Pt, the candidate "intruder" 0^+ state is known to have two 2^+ states feeding strongly into it. All three were previously interpreted as the third 0-2-2 band in the O(6) limit[3]. Thus, the character of this band is now not well understood, and throws into question the classification of this nucleus as an O(6) nucleus. More work now needs to be done to provide a better understanding of not only this band but subsequently the character of these heavier platinum nuclei.

The E0 transition between the 0_4^+ and 0_3^+ states was obscured by other intense lines, and its presence cannot be ascertained.

In conclusion, a strong IBM-forbidden E0 transition has been seen whose strength is 21.0 (28) times greater than an allowed transition from the same level. It appears that: a) the 0^+(1402keV) state lies outside the IBM space, or b) the IBM cannot predict E0 transitions well, or c) core interactions have affected the predicted E0 transition strength predicted by the IBM from this level or d) ^{196}Pt is not an O(6), or SU(5) nucleus.

A recent application of an "Effective Fermion SO(6) Dynamical Symmetry" to all of the platinum nuclei[18] has found good agreement with energy levels and transitions. It suggests that an accidental near-O(6) symmetry is present in ^{196}Pt. It would be very useful to have E0 predictions from this model.

1. J. A. Cizewski et al., *Nucl. Phys.* **A323** (1979)349.
2. A. Arima and F. Iachello, *Phys. Rev. Lett.* **40** (1978)385.
3. M. K. Harder et al., *Proc. 8th Conf. Gamma-Ray Spectroscopy*, (Fribourg) ed. J. Kern (World Scientific, Singapore, 1994) 983.
4. M. P. Fewell et al., *Phys. Lett.* **157B** 353 (1985),
5. C. S. Lim et al, *Nucl. Phys.* **A548** 308 (1992).
6. M. P. Fewell, *Phys. Lett.* **167B** 6 (1986).
7. G. J. Gyapong et al., *Nucl. Phys.* **A458** 165 (1986).
8. R. F. Casten and J. A. Cizewski, *Phys. Lett.* **B185** 293(1987).
9. P. Van Isacker, *Nucl. Phys.* **A465** 497 (1987).
10. W. Mampe et al., *Nucl. Inst.* **154** 127 (1978).
11. M. K. Harder et al., *Proc. 6th Conf. Gamma Spectroscopy*, (Leuven) ed. K. Abrahams and P. Van Assche (IOP Conference Series no. 88, London,1987) p. 529.
12. W. R. Kane et al., *Phys. Lett.* **117B** 15 (1982).
13. M. K. Harder, M. Veskovic, B. Krusche, to be published.
14. H. G. Borner et al., *Phys. Rev.* **C42** R2271 (1990).
15. A. V. Alduschenkov and N. A. Voinova, *Nucl. Data Tables* **11** 299(1971).
16. M. K. Harder and B. Krusche, to be published in *Phys. Letts B,* (1994.)
17. R. Bijker et al., *Nucl. Phys.* **A344** 207 (1980).
18. S. W. Yates et al., *Nucl Phys.* **A406** 519 (1983).
19. D. H. Feng et al., *Phys. Rev.* **C48** 1488 (1993).

EVOLUTION OF THE TRIAXIAL ASYMMETRY AT VARIATION OF THE SYMMETRIC QUADRUPOLE DEFORMATION

W. ANDREJTSCHEFF AND P. PETKOV

Bulgarian Academy of Sciences, Institute for Nuclear Research and Nuclear Energy, 1784 Sofia, Bulgaria

and

N. V. ZAMFIR

Institute of Atomic Physic, Bucharest Magurele, Romania

ABSTRACT

Shape parameters characterizing symmetric quadrupole deformation and triaxial asymmetry are model-independently deduced by the approximated sum-rule method (ASRM) for nearly seventy nuclei. For Barium isotopes with $A=124 \div 138$, available results obtained by the ASRM, the RTRM, the GCM and the IBA are shown.

1. Introduction

Triaxial shapes of atomic nuclei are under discussion[1] already for many years. A main problem thereby is to identify a clear-cut evidence for triaxiality. Investigations have been performed comparing ratios of experimental level energies and E2 transition rates to predictions of the Davydov rigid-triaxial-rotor model (RTRM). In this way, the asymmetry parameter γ could be recently deduced[2] for a series of Ba and Xe isotopes. The collective potential energy $V(\beta,\gamma)$ and thus the parameters β and γ (β: quadrupole deformation) can be obtained[3] by a fit in the framework of general collective model (GCM).

2. The Approximated Sum-Rule Method

Recently, we have shown[4] that ground states of even-even nuclei can be treated with an approximation of the model-independent Cline-Kumar sum-rule method. This approximation requires the knowledge of only four experimental E2 matrix elements which are available for a large body of nuclei. The explicit expressions for the mean values of the symmetric quadrupole deformation β_{rms}, the asymmetry angle δ_{eff} (closely related to the collective model parameter γ up to higher order terms) and the eccentricity of the nuclear ellipsoid are given[4] in those papers. Here, we recall the eccentricity e in the case when axis 3 is the symmetry axis. The semiaxes of the ellipsoid are R_κ ($\kappa=1,2,3$).

$$e = e_3 = \frac{R_1^2 - R_2^2}{R_1^2} \approx \sqrt{\frac{15}{\pi}} \beta_{rms} \sin\delta_{eff} \qquad \left(0° \le \delta \le 30°\right) \qquad (1)$$

We consider the eccentricity to be a suitable measure for deviation from axial symmetry and for comparison of nuclei with different β_{rms} and δ_{eff} (i.e. β and γ).

3. The β–γ Shape Distribution

In the figures, the shape parameters derived[4] as mentioned in sec. 2 are presented. Fig. 1 indicates a decrease of the asymmetry angle δ_{eff} with increasing deformation β_{rms}. A typical example is provided by the Barium isotopes with A=138, 136, 134 and 130 labeled by an asterisk. The data points of some nuclei constitute a separate track (framed) parallel to the main group.

In fig. 2, the eccentricities (Eq. (1)) are displayed versus mass number A. The overwhelming majority of these values belongs to the surprisingly narrow interval $0.14 \le e \le 0.21$. Below that are only values for some "classical" axial rotors ($150 \le A \le 190$) as well as for ^{138}Ba. For several of the nuclei with eccentricities within the above interval, (soft) triaxiality has been considered in the literature. For instance, $^{98-104}$Ru (Z=44) contribute to the small bump around A=100. Investigations[3] with the GCM found for $^{100-108}$Ru triaxial minima in the potential–energy surfaces.

A striking feature of fig. 2 are the eccentricities of $^{72-76}$Ge and $^{74-78}$Se (N=40, 42, 44) lying appreciably higher than all other values. Apparently, in these isotopes the shapes of the ground states reveal the most pronounced (effective) triaxiality among the nearly seventy nuclei in this systematics.

4. The Shape Evolution in Barium Isotopes

In table 1, shape parameters of the ground states of even Barium isotopes are shown derived within different approaches. The IBA calculation of E2 matrix elements was performed within the consistent Q formalism, the parameters of the general Hamiltonian were the same as used[5] previously. The value of the parameter χ is given in the last column. A correlation[1,6] between this value and the asymmetry parameter δ_{eff} is appreciable. In fact, this correlation is a general prediction of the IBA model and further studies could lead to new insights and understanding of the axially asymmetric degree of freedom of even-even nuclei.

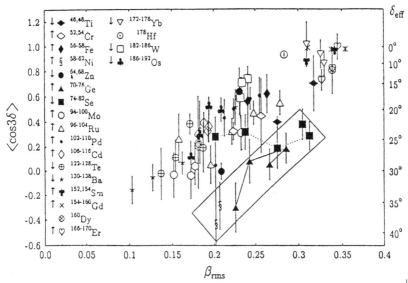

Fig.1. Asymmetry versus symmetric quadrupole deformation[4]

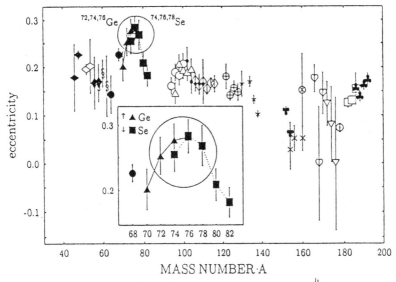

Fig.2. Eccentricity(eq.1) of the nuclei shown[4] in fig.1

Table 1. Shape parameters for Barium isotopes. To the far right, the IBA parameter χ as used in the calculations is given.

A	β_{rms}			δ_{eff} (deg)				χ^5
	ASRM⁴	RTRM²	GCM⁷	ASRM⁴	RTRM²	GCM⁷	IBA	
124		0.30	0.30		20	9		
126		0.28	0.26		21	11	16	-0.5
128	0.246	0.25	0.25		22	14	20	-0.5
130	0.226	(0.23)	0.215	20(3)	24	17	24	-0.35
132		0.2	0.21		26	23	27	-0.20
134	0.162	0.17		29(2)	29			
136	0.127			31(2)				
138	0.103			33(2)				

We are indebted to P. von Brentano, A. Gelberg and A. Dewald for fruitful discussions. This work has been supported by the Bulgarian National Research Foundation under contracts PH14 and PH31.

References

1. R. F. Casten, *Nuclear Physics from a Simple Perspective* (Oxford, 1990).
2. J. Yan *et al.*, *Phys. Rev.* **C48** (1993) 1046.
3. D. Troltenier *et al.*, *Z. Phys.* **A338** (1991) 261.
4. W. Andrejtscheff and P. Petkov, *Phys. Rev.* **C48** (1993) 2531 and *Phys. Lett.* **B** (1994).
5. A. Faessler *et al.*, *Nucl. Phys.* **A438** (1985) 78.
6. J. P. Elliot *et al.*, *Phys. Rev. Lett.* **57** (1986) 1124.
7. P. Petkov, A.Dewald, W.Andrejtscheff *et al.*, to be published

THE REVIVAL OF THE L-S COUPLING SCHEME AT SUPERDEFORMATION

K. Sugawara-Tanabe

Otsuma Women's University, Tama, Tokyo 206 Japan

Abstract

We found that the $L-S$ coupling scheme is restored not only for the parity doublet levels but also for some levels at superdeformation in a realistic calculation, which is caused from the strongly deformed quadrupole interaction. This real-spin mechanism includes the unique-parity level in contrast to the pseudo-spin mechanism. The contribution from the unique-parity level to the M1 transition becomes non-negligible order at superdeformation.

1. Real-spin mechanism

We showed in a simple model[1] that the expectation values of the spin-orbit force in the parity-doublet (P-D) levels at $\delta_{osc} \sim 0.6$ converge to the values by the asymptotic wave functions which is obtained by neglecting the residual ls and l^2 interactions in the axially symmetrically deformed Nilsson Potential[2]. The P-D levels are the pair levels with asymptotic quantum numbers $[N, n_z = N, \Lambda = 0]\frac{1}{2}$ and $[N-1, n_z = N-2, \Lambda = 1]\frac{1}{2}$, and they are degenerate around $\delta_{osc} \sim 0.6$. The former level of the P-D levels belongs to the unique parity level with $\Omega = \frac{1}{2}$ (unique-parity partner), and the latter level to the pseudo-spin family[3,4] with no pseudo-spin pair partner (pseudo-spin partner). Here in this work we extended more realistic calculations using Nicra code[5].

We use the Nilsson Hamiltonian[2] which is assumed to keep an axially symmetry with a rational ratio $a : b$ between the frequencies $\omega_\perp$ and ω_z. The deformation parameter δ_{osc} is given by $3(a-b)/(2a+b)$. An energy eigenvalue of H without the residual l^2 and ls interactions is described by $\hbar\omega_{sh}(N_{sh} + a + \frac{b}{2})$, where $\omega_{sh} = \omega_0(a^2b)^{-1/3}$, and $\hbar\omega_0 = 41A^{-1/3}MeV$. The corresponding eigenfunction is the so-called asymptotic wave function $|N, n_z, \Lambda, \Omega >$. The shell quantum number $N_{sh} = an_\perp + bn_z$, $n_\perp = n_+ + n_-$ and $\Lambda = n_+ - n_-$. In the superdeformed shape where $a : b = 2 : 1$, the P-D levels belong to the same $N_{sh} = 2n_\perp + n_z$ with the same shell energy $\hbar\omega_{sh}(N_{sh} + 5/2)$, which proves that the P-D levels are almost degenerate in the energy diagram. This energy degeneracy is not destroyed in the realistic calculation because the residual interaction is much smaller than the harmonic oscillator potential.

We evaluate the diagonal matrix element of the ls interaction using the wave function obtained numerically by the exact diagonalization of total H using the NICRA code[5]. The eigenvector of H is given by, $H|\sigma, \Omega, \alpha > = E(\sigma, \Omega, \alpha)|\sigma, \Omega, \alpha >$

where $\Omega = \Lambda + \Sigma$ ($= \frac{1}{2}$ for the P-D levels), α denotes the signature and σ denotes all the other quantum numbers except for Ω and α. Here energy eigenvalues are degenerate on signatures, as there is no cranking term. We calculated the expectation values of the angular momentum $< \sigma, \Omega, \alpha | \mathbf{J_x} | \sigma, \Omega, \alpha >$ and those of the spin-orbit force, $< \sigma, \Omega, \alpha | \mathrm{ls} | \sigma, \Omega, \alpha >$. Fig.1 shows these values for all the levels with $\Omega = \frac{1}{2}$ belonging to the same N_{sh} both for the proton levels (π) of $A = 152$ case. Inside of the figures the single particle levels are identified by the asymptotic quantum numbers $N\ n_z\ \Lambda$. It is very remarkable that all of the expectation values except for those by the unique-parity partner of the P-D levels approach to those by the asymptotic limit wavefunction, i.e. for the case of $< \sigma, \Omega, \alpha | \mathrm{ls} | \sigma, \Omega, \alpha >$ to $\Lambda\Sigma$ and for the case of $< \sigma, \Omega, \alpha | \mathbf{J_x} | \sigma, \Omega, \alpha >$ to $\pm\frac{1}{2}$. This indicates that the $L - S$ coupling scheme is recovered for these levels. The unique-parity partner of the P-D levels shows a tendency to reach to the asymptotic values, but not yet enough even at $\delta \sim 0.8$. We will come back to this point later. This feature of the small expectation value of the ls interaction in superdeformation is characteristic not only of the levels with $\Omega = \frac{1}{2}$, but also of some other single-particle levels in large deformation. We calculated the values of the $< \sigma, \Omega, \alpha | \mathrm{ls} | \sigma, \Omega, \alpha >$ for the $N_{sh} = 6$ of the proton shell and the $N_{sh} = 7$ of the neutron shell in ^{152}Dy case. All the levels show a tendency to converge to the asymptotic values, but $[541]\frac{3}{2}$ in π, $[651]\frac{3}{2}$ and $[532]\frac{5}{2}$ in ν show a gradual and not enough convergence. These levels belong to the unique parity levels with small Ω.

We calculated the effect of rotation on these expectation values. In this case $|\sigma, \Omega, \alpha >$ is the eigenstate of $H' = H - \omega_{rot}J_x$. So long as the rotational frequency ω_{rot} is small, the effect of the cranking is negligible especially in the expectation values of ls, which agrees with the prediction by the perturbation calculation[1]. As for the expectation values of $\mathbf{J_x}$, those by one signature partner of the unique-parity level show a gradual increase with an increasing $\hbar\omega_{rot}$, but those by the other signature partner come near to the asymptotic values. We will name this recovery of the $L - S$ coupling scheme at superdeformation as real-spin mechanism in contrast to the pseudo-spin mechanism.

Now we see the detailed character of the single-particle level at superdeformation using the spherical wave functions. The state $|\sigma, \Omega, \alpha >$ is expanded by the $L - S$ coupling spherical basis $|N, L, L_z = \Lambda, \Omega)$ through the formula

$$|\sigma, \Omega, \alpha >= \sum W^{\sigma\,\Omega}_{N\,L\,\Lambda\,\Omega} |N, L, \Lambda, \Omega). \qquad (1)$$

To make our observation clear we show these coefficients of the P-D levels, i.e. $W^{\sigma\,\frac{1}{2}}_{N\,L\,\Lambda\,\frac{1}{2}}$ as functions of δ_{osc} for the case of proton shell of ^{152}Dy in Fig.2. Inside of the figure, $|N, L, \Lambda, \Omega >$ is denoted by (L, Λ). At the spherical limit $\Lambda = 0$ and 1 components are mixed nearly equal order (0.54 to 0.46), which indicates $j-j$ coupling is more suitable. With increasing δ_{osc} the dominant spherical components decrease, while the other components increase and finally the amplitudes of the $\Lambda = 1$ ($\Lambda = 0$) components become dominant for the pseudo-spin partner (the unique-parity partner) of the P-D levels, respectively.

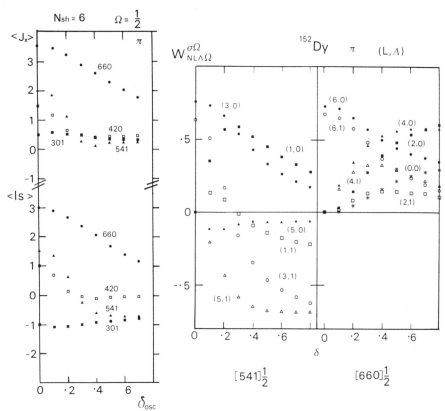

Fig.1: $\Omega = \frac{1}{2}$ levels in the same N_{sh} Fig.2: $W^{\sigma\Omega}_{NL\Lambda\Omega}$ in the P-D levels

In the $[541]\frac{1}{2}$ level of Fig.2, the ratio between the squares of $\Lambda = 1$ components to those of $\Lambda = 0$ is 0.23 to 0.77 at $\delta_{osc} = 0.2$, while at $\delta_{osc} = 0.6$ the ratio is 0.78 to 0.22. For the state $[660]\frac{1}{2}$, the ratio between the squares of $\Lambda = 1$ components to those of $\Lambda = 0$ is 0.58 to 0.42 at $\delta_{osc} = 0.2$, while at $\delta_{osc} = 0.6$ the ratio is 0.15 to 0.85. These results indicate the $L - S$ coupling scheme is restored at superdeformation, and this resurrection is much better for the unique-parity partner levels than the pseudo-spin partner levels of the P-D levels, i.e. 0.85 to 0.77 at $\delta_{osc} = 0.6$ for the proton shell and 0.81 to 0.79 for the neutron shell. Although Fig.1 show the convergence to the asymptotic limit is worse for the unique-parity partner level of the P-D levels, Fig.2 shows that the $L - S$ coupling scheme becomes much better for the unique-parity partner at $\delta_{osc} = 0.6$. The reason is the following; As is seen in Fig.2, the amplitudes of $\Lambda = 0$ and those of $\Lambda = 1$ are out of phase for the $[541]\frac{1}{2}$, while those are in phase for the $[660]\frac{1}{2}$. The out of phase in the pseudo-spin partner of the P-D levels causes the cancellation of the matrix element of ls, and helps to converge to the asymptotic values. For the unique-parity partner of the P-D levels, the in phase amplitude has no cancellation and the convergence to the asymptotic values looks very slow.

Now we will discuss the difference between the pseudo-spin mechanism and the real-spin one. At first real-spin mechanism works only around the superdeformation, as it is related with the shell structure. On the other hand the pseudo-spin works in any deformation. Next the pseudo-spin formalism do not include the unique-parity levels, while the real-spin can take into account of the unique-parity levels. In order to see the difference between the two mechanism, we calculated the effect of the unique-parity component on the M1 transition rate. For example in the diagonal matrix element between the levels $[541]\frac{1}{2}$, which belongs to the pseudo-spin family, the contribution of the $h\frac{11}{2}$ is 0 at $\delta = 0$, but at $\delta = 0.6$ it becomes 117 %. The non-diagonal matrix element between $[541]\frac{1}{2}$ and $[532]\frac{3}{2}$ has 20 % contribution of $h\frac{11}{2}$ at $\delta = 0.2$, while at $\delta = 0.6$ 52 %. Thus the contribution of the unique-parity level becomes very important at superdeformation.

2. Conclusion

We found that the $L - S$ coupling scheme is restored also in this realistic calculation not only for the P-D levels but also for the other levels in N_{sh}. Although the expectation values of the spin-orbit interaction by the unique-parity levels do not converge enough to the asymptotic values, their $L - S$ coupled wave function components become to fulfill 85% of the total. Once the $L - S$ coupling scheme works again, the quantization of alignment is explained by the real spin s. The differences between this real-spin scheme and the pseudo-spin one are followings; (1) real-spin works only around $\delta_{osc} \sim 0.6$ and small rotational frequency ($\hbar\omega_{rot} \leq 0.3$), while pseudo-spin in any deformation; (2) the contribution from the unique-parity level is included in this real-spin, but not in pseudo-spin. Therefore the quantization of alignment observed in the normal deformation and also those in $A \sim 130$ where $\hbar\omega_{rot} > 0.5$ are out of this mechanism and will be explained in the frame of psudo-spin mechanism.

References

1. K. Sugawara-Tanabe and A. Arima, *Phys. Lett.* **317B** (1993)1; *Nucl. Phys.* **A557**157C.
2. S.G. Nilsson, *Dan Mat. Fys. Medd.* **29**, no.16(1955).
3. A. Arima, M. Harvey and K. Shimizu, *Phys. Lett.* **30B**(1969)517.
4. K.T. Hecht and A. Adler, *Nucl. Phys.* **A137**(1969)129.
5. T. Bengtsson, I. Ragnarsson and S. Åberg, *The Cranked Nilsson Model* in *Computational NuclearPhysics* **1** (Springer Verlag, 1991)51.

ON THE RECONCILIATION OF MICROSCOPIC AND FITTED BOSON g-FACTORS*

E.D.DAVIS[a] and P.NAVRÁTIL[b,c]
[a] Fisika Departement, Universiteit van Stellenbosch, 7599 Stellenbosch, RSA
[b] Institut vir Teoretise Fisika, Universiteit van Stellenbosch, 7599 Stellenbosch, RSA
[c] The Institute of Nuclear Physics, Czech Academy of Sciences, 250 68 Řež near Prague, CR

ABSTRACT

We investigate the possibility that microscopic boson g-factor estimates may require revision and consider the impact on 2^+_1 g-factors of the difference ε_v between the neutron and proton d-boson energies found typically in microscopic evaluations of IBM parameters. We present arguments in support of the contention that microscopic estimates of ε_v (taken in conjunction with our microscopic boson g-factors) can account for the systematics of 2^+_1 g-factors in the Z=50-to-82 and N=82-to-126 shells. Our calculations also suggest a connection between the 2^+_1 g-factors and the position of mixed-symmetry states, in particular a mixed-symmetry β-band in deformed nuclei.

1. Introduction

The interacting boson model (IBM) provides a successful platform for the systematisation of the low-energy collective properties of medium-to-heavy nuclei [1]. However, there is a long-standing discrepancy between microscopic estimates for boson g-factors ($g_\pi \approx 1$ and $g_\nu \approx 0$) and the values extracted from fits to g-factors of 2^+_1 states [2] (under the assumption that they are of pure F-spin).

There are three possible explanations for the discrepancy: 1) the naive choices of $g_\pi = 1$ and $g_\nu = 0$ for boson g-factors are wrong; 2) the inclusion of the g-boson degree of freedom is required; 3) F-spin impurities and their effects are more substantial than previously thought. In fact, on the basis of detailed microscopic estimates of boson g-factors (which we substantiate below), we are lead to discard both of the first two options. Except near the beginning and end of shells, microscopic estimates do support the choices $g_\pi \approx 1$ and $g_\nu \approx 0$. As regards the g-boson degree of freedom, the difference between the microscopic g-factors of d- and g-bosons is not big enough to modify significantly sd-space results. Instead, we argue that an explanation in terms of F-spin impurities generated by a difference ε_v in neutron and proton d-boson energies may be microscopically viable.

2. Microscopic boson g-factors

To complement earlier microscopic studies of boson g-factors, we have calculated g-factors for both d- and g-bosons within the similarity-transformed Dyson boson

*Supported by the Stellenbosch 2000 Fund for support, grants from the South African FRD, and the grant GA CR No. 202/93/2472.

mapping [3,4]. In this formalism, the multipolarity λ boson g-factor [5]

$$g_\lambda = \sqrt{\frac{(2\lambda+1)}{\lambda(\lambda+1)}} \sum_{j_1,j_2,j} \langle l\tfrac{1}{2}j_1||g_l\hat{l}+g_s\hat{s}||l\tfrac{1}{2}j_2\rangle \sqrt{(1+\delta_{j,j_1})(1+\delta_{j,j_2})}\bar{\beta}^\lambda_{j_2,j}\beta^\lambda_{j,j_1} \begin{Bmatrix} j_1 & j_2 & 1 \\ J & J & j \end{Bmatrix}. \quad (1)$$

Here, $\bar{\beta}^\lambda_{j_2,j}$ and β^λ_{j,j_1} denote the bra and ket collective boson amplitudes, respectively. We evaluate these amplitudes as in Ref. [3], using the shell-model input of the global calculations of Refs. [3,6]. The resulting g-factors for both neutron and proton d- and g-bosons are shown in Fig. 1. We observe that the results support the claims made in the introduction and are compatible with other microscopic estimates of Refs. [7,8].

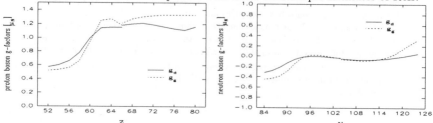

Fig. 1. Proton and neutron d- and g-boson g-factors for the Z=50-to-82 and N=82-to-126 shells within the similarity-transformed Dyson boson mapping approach.

3. Influence of ε_ν: general considerations

The predominantly numerical study of the influence of F-spin admixtures on 2_1^+ g-factors in Ref. [9] has drawn attention to the role played by a difference ε_ν in neutron and proton d-boson energies. The question is can we reproduce 2_1^+ g-factors when we confine ourselves to microscopic estimates of the magnitude of ε_ν?

Representative microscopic estimates of ε_ν for nuclei in Z = 50-to-82 and N = 82-to-126 shells can be inferred directly from Ref. [3] where global estimates of ε_π and ε_ν have been given. They have the advantage of being compatible with our earlier boson g-factor estimates.

Striking qualitative evidence in support of the viability of using microscopic estimates of ε_ν to account for the behaviour of 2_1^+ g-factors emerges if we compare on the sign of our ε_ν estimates and the sign of the discrepancies $\Delta g_{\exp}$ ($\equiv g^{\exp}_{2_1^+} - g^{(0)}$) between experimental 2_1^+ g-factors and $g^{(0)} = N_\pi/N$ (the F-spin symmetric prediction for $g_\pi = 1, g_\nu = 0$).

A perturbative treatment of the effect of ε_ν on 2_1^+ g-factors within the intrinsic state formalism yields

$$g_{g.s.b.} = g_\pi \frac{N_\pi}{N} + g_\nu \frac{N_\nu}{N} - 2\frac{\varepsilon_\pi-\varepsilon_\nu}{\Delta E}\frac{N_\pi N_\nu}{N^2} x_0^2 (g_\pi - g_\nu) , \quad (2)$$

where x_0^2 is the s-boson occupation probability of the groundstate and ΔE is the splitting between the bandheads of the groundstate band and a mixed-symmetry β-band. [Equation (2) also applies to vibrational nuclei if $x_0 = 1$ and ΔE identified as the splitting between the 2_1^+ state and lowest mixed-symmetry 2^+ state.] It follows from Eq. (2) that ε_ν and $\Delta g_{\exp}$ should be the same sign if the effect of ε_ν is dominant. We note that, just as $\Delta g_{\exp}$ is for the most part negative, so our estimates of ε_ν are negative for most nuclei. However, as the Z = 82 shell-closure is approached, our estimates for ε_ν approach zero, and, for Z = 78(80), ε_ν is positive for $N < 108(118)$. Were the role of ε_ν important, the implication would be that a change in the sign of $\Delta g_{\exp}$ from positive to negative should be observed (with increasing N) for a set of isotopes close to the Z = 82 shell-closure. Exactly such a change in sign is seen for the Pt isotopes (cf. Fig. 2c).

Another kind of correlation between our estimates for ε_ν and experimental 2_1^+ g-factors is made apparent in Figs. 2a and b. The rates of variation in $\Delta g_{\exp}/g^{(0)}$ are paralleled by the rates of variation in our microscopic estimates of ε_ν.

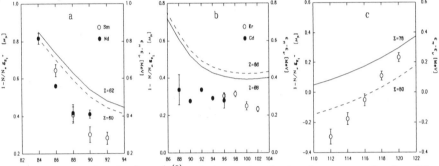

Fig. 2. Isotopic variation of $-\Delta g_{\exp}/g^{(0)}$ and microscopic estimates of $\varepsilon_\pi - \varepsilon_\nu$ as a function of the neutron number. Experimental values are taken from Refs.[10,11]. In part a, the Nd and Sm isotopes are considered; in part b, the Er and Gd isotopes; in part c, the Pt isotopes.

4. Influence of ε_ν: detailed calculations

We aim to show that with microscopically motivated choices for IBM parameters (among them ε_ν), we are able to reproduce 2_1^+ g-factor data quantitatively without forfeiting a reasonable description of other characteristics. We consider the ^{148}Sm and ^{154}Sm isotopes which have among the largest discrepancies between 2_1^+ g-factors and maximal F-spin predictions found (cf. Fig. 2a).

For the vibrational ^{148}Sm, we use microscopically calculated IBM-2 parameters without any adjustments. We obtain good agreement with the lowest energies (cf. Fig. 3) and reproduce the experimental 2_1^+ g-factor. We note that the boson g-factors are $g_\pi = 1.105\mu_N$ and $g_\nu = -0.136\mu_N$ and the difference of d-boson energies is $\varepsilon_\nu = -0.66$ MeV. Also, the ξ_2 parameter of the Majorana interaction is set to zero. (The other Majorana interaction parameters ξ_1 and ξ_3 do not influence the g-factor.)

In the case of the deformed ^{154}Sm, an IBM-2 fit is attempted. Reasonable agreement with the energies of the groundstate, β and γ bands is possible (Fig. 3). With $g_\pi = 1.05\mu_N$ and $g_\nu = -0.05\mu_N$, a difference of $\varepsilon_\nu = -0.4$ MeV is necessary to obtain the 2_1^+ g-factor correctly (again $\xi_2 = 0$). A mixed-symmetry β-bandhead at 1.34 MeV is predicted. Other choices of ε_ν and ξ_2 reproduce the 2_1^+ g-factor while shifting this β-bandhead upwards — e.g. $\varepsilon_\nu = -0.6$ and $\xi_2 = 0.05$ MeV (β-bandhead at 1.78

MeV). We note that $|\varepsilon_\nu|$ is unlikely to be more than 0.66 MeV (the ^{148}Sm value). In the calculations, the ε_ν term in the Hamiltonian is responsible for at least 75% of the discrepancy between the 2_1^+ g-factor and the maximal F-spin prediction.

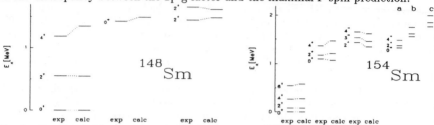

Fig. 3. Experimental and calculated energy spectrum of ^{148}Sm (left) and ^{154}Sm (right). The parameter sets are given in Ref. [5]. For ^{154}Sm, in part a the Majorana strength $\xi_2 = 0$, in parts b and c, the (calculated) mixed-symmetry bands are shown for non-zero values of ξ_2.

5. Conclusions

In this contribution, we have re-calculated microscopic boson g-factors and confirmed that for most isotopes $g_\pi \approx 1$ and $g_\nu \approx 0$. In Ref. [9], the sensitivity of 2_1^+ g-factors to a difference ε_ν in neutron and proton d-boson energies was noted. We have presented qualitative evidence that the systematics of 2_1^+ g-factors are governed by the systematics of microscopic estimates of ε_ν. Detailed calculations for ^{148}Sm and ^{154}Sm support this conclusion.

In closing, we note that reliance on F-spin admixtures to account for the behaviour of 2_1^+ g-factors introduces an interesting correlation with the location of mixed-symmetry states. If we take the magnitude of ε_ν to be constrained by our microscopic estimates, then, for example, we predict the presence of a mixed-symmetry β-bandhead in ^{154}Sm below 2 MeV.

6. References

1. A. Arima and F. Iachello, *The Interacting Boson Model* (Cambridge University Press, Cambridge, 1987).
2. P.O.Lipas, P.von Brentano and A.Gelberg, *Rep. Prog. Phys.* **53** (1990) 1355.
3. P. Navrátil and J. Dobeš, *Nucl. Phys.* **A533** (1991) 223.
4. J. Dobeš, P. Navrátil and O. Scholten, *Phys. Rev.* C **45** (1992) 2795.
5. E.D.Davis and P.Navrátil, *Phys. Rev.* C, submitted.
6. S.Pittel, P.D.Duval and B.R.Barrett, *Ann. Phys. (N.Y.)* **144** (1982) 275.
7. M.Sambataro, O.Scholten, A.E.L.Dieperink and G.Piccitto, *Nucl. Phys.* **A423** (1984) 333.
8. I.Morrison, P.von Bretano and A.Gelberg, *J. Phys. G: Nucl. Part. Phys.* **15** (1989) 801.
9. A.Wolf, O.Scholten and R.F.Casten, *Phys. Lett.* **B 312** (1993) 372.
10. P. Raghavan, *At. Data Nucl. Data Tables* **42** (1989) 189.
11. S.S. Anderssen and A.E. Stuchbery, to be published.

IBM Approach to the Rotational Damping

Takahiro Mizusaki and Takaharu Otsuka
Department of Physics, University of Tokyo,
Hongo, Tokyo 113, Japan

and

P. von Brentano
Department of Physics, University of Cologne, Germany

ABSTRACT

Within the framework of Interacting Boson Model (IBM), we discuss the rotational damping, which has a close relation to the transition from order to chaos at high spin states. We give an articulate elucidation on the importance of the higher multipole interactions and triaxiality, which cause the mixing among highly excited rotational bands.

1. Introduction

Recently much attention has been paid to the statistical feature of off-yrast nuclear structure of excited, rapidly rotating nuclei at high spin because of recent development of the experimental technique of the 4π detectors. The statistical analysis, which has recently been developed [1], plays a substantial role for clarifying the gross feature in off-yrast region. For instance, the $E_\gamma - E_\gamma$ coincidence is useful for the fluctuation analysis of the quasi-continuum spectra composed of mainly stretched E2 transitions. It shows that the rotational structure, which is typically shown near yrast region in rare-earth nuclei, is gradually damped as the excitation energy measured from the yrast line is increased. This smearing of the rotational structure is related to the coexistence of order and chaos in nuclei. For the theoretical counterpart, Lauritzen *et.al.* [2] advocated the concept of the damping of the rotational band. Matsuo *et.al.* [3] gave an extensive study by the extended cranked shell model.

We apply the Interacting Boson Model (IBM) [4] for describing the rotational damping. The first advantage of this approach is to carry out the calculation in the laboratory frame, *i.e.*, we can treat exactly the eigenstates with definite angular momentum, in contrast to the cranked shell model. The second one is that IBM has the group theoretical rotational limit and we can utilize it for the discussion of the damping of the rotational band. The third one is that recently the chaotic feature of IBM have been clarified by several authors [5,6] and we have a possibility of making a link between the chaotic feature of IBM and the rotational damping.

In the present work, we restrict ourselves to the boson degree of freedom. We examine the mixing among the rotational bands which IBM incorporates in its SU(3) limit. Needless to say, non-collective degrees of freedom are important and we can include their effect by the extended IBM, which allows the 2 quasi-particle excitation. The latter will be shown in a separate publication.

2. IBM and Chaos

The statistical properties of high-spin state have a close connection to chaos. Here we discuss the chaos of IBM. The IBM hamiltonian is written by,

$$H = \varepsilon_d n_d - \kappa_{(2)} Q^{(2)} \cdot Q^{(2)} - \kappa_{(3)} Q^{(3)} \cdot Q^{(3)} - \kappa_{(4)} Q^{(4)} \cdot Q^{(4)} + \kappa' L \cdot L \quad (1)$$

where $Q^{(2)} = (s^\dagger \tilde{d} + d^\dagger \tilde{s}) - \chi (d^\dagger \tilde{d})^{(2)}$, $Q^{(i)} = (d^\dagger \tilde{d})^{(i)}$ (i=3,4) and $L = \sqrt{10}(d^\dagger \tilde{d})^{(1)}$. It contains six parameters. When we consider the properties of low-lying states, we usually do not need the $\kappa_{(3)}$ and $\kappa_{(4)}$. At certain parameters, this hamiltonian expresses the U(5), O(6) and SU(3) symmetries, which correspond to the vibrational, γ unstable and rotational nuclei, respectively. As we consider the well-deformed nuclei, $\varepsilon_d = 0$ is assumed hereafter. Up to now, the chaos has been investigated in the classical and quantum systems in various ways. The chaos itself can be defined without any ambiguity only in the classical system. The classical equation of IBM can be defined in the limit of N $\to \infty$ by the coherent state [4]. In the classical system, the Lyapunov exponent and chaotic volume characterize chaos. In the quantum system, as the study of quantum manifestation of chaos, the nearest neighbor level spacing P(s) and the spectral rigidity Δ_3 are often used because their universal correspondence is well-known. As the measure of chaotic properties of wavefunctions, the statistics of transition matrix elements is also used. The previous works [5,6] show a strong correlation between the chaos of classical system and the one of quantum system and that there exists a strong chaotic region between the O(6) and SU(3) limits, i.e. $\varepsilon_d = 0$, $\chi \approx 0.5 \times \frac{\sqrt{7}}{2}$, which is shown in Fig.1.a. It shows the Brody parameter ω, by which P(s) optimally fits, as a function of χ parameter. Here, we have an interest in the high spin states. In Fig.1.a, we show the spin dependence of the Brody parameter ω, and that IBM rapidly looses its chaotic property as the angular momentum becomes larger. However, if we introduce the higher multipole components into IBM hamiltonian, we can get the very chaotic situation in the high spin state. This is shown in Fig.1.b, where we introduce the $Q^{(4)} \cdot Q^{(4)}$. We find that the $Q^{(3)} \cdot Q^{(3)}$ has similar effects. This result is similar to the one of Matsuo et.al.[3], who concluded that higher multipole components of the two-body residual interaction serves as a trigger to the damping. In the case of IBM, as the spin is increasing, the basis state contains less s bosons. Therefore, the matrix element of $s^\dagger \tilde{d} + d^\dagger \tilde{s}$ in the quadrupole interaction becomes smaller in the higher spin state. This makes the whole boson space to be dissolved into the direct sum of the subspaces which have a definite s boson number. The $(d^\dagger \tilde{d})^{(2)}$ behaves as the SU(3) generator approximately in high spin states. In the low-spin states, chaos occurs in the transition region between symmetries. In turn, in high spin state, chaos occurs by interference among the multipole interactions. Moreover, as triaxiality simply means the increase of the matrix elements of $s^\dagger \tilde{d} + d^\dagger \tilde{s}$, it has the tendency to break the direct sum of U(5) subspace, and brings about the enhancement of chaos. This is why the regularity appears in the high spin state only due to quadrupole interaction, and higher multipole interaction and triaxiality becomes more important. This is schematically shown in Fig.2.

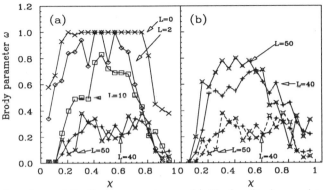

Fig. 1. Spin dependence of the Brody parameter ω. (a) The hamiltonian consists of only the $Q^{(2)} \cdot Q^{(2)}$ term. (b) The dotted lines are the same as those of fig.(a). The solid lines correspond to the results, for a hamiltonian comprising $Q^{(2)} \cdot Q^{(2)}$ and $Q^{(4)} \cdot Q^{(4)}$ terms.

3. Rotational Damping

The level statistics is independent of the over-all scale of hamiltonian. However, in order to discuss the rotational damping, it is important to adjust the over-all scale. In the case of low-lying states, we adjust the IBM parameters to fit the experimental data. In turn, in high spin state, there are no detailed data and they are meaningless for this purpose even if they exits. We adjust the over-all scale of the hamiltonian so as to fit the level density of the Fermi-gas [7]. The κ' is adjusted so as to reproduce the yrast energy of each angular momentum. Since the angular-momentum range where the quasi-continuum γ rays are emitted is around 30-60 $\hbar$, we calculate the transition $40\hbar \to 38\hbar$. In order to get better statistics, we use a boson number N=40.

By including the higher multipole interactions and triaxiality, we can reproduce the basic nature of rotational damping, i.e., level density and fragmentation of the E2 transitions. Note that, although too large strengths of the higher multipole interactions distort the level density unrealistically, we can successfully describe both of them well. The energy dependence of the mixing of the rotational bands is simply characterized by the branching number, $n_i = (\sum_j w_{ij}^2)^{-1}$ where w_{ij} is a transition probability from i state to j state. It shows the rapid increase of the mixing as a function of the excitation energy. However, in the present model, the damping width is quantitatively smaller than the experimental one. This is due to the lack of the non-collective degree of freedom. We are pursuing this issue.

In conclusion, we clarify the whole chaotic property of IBM and summarize it in the chaotic tetrahedron. We can give a simple explanation for the regularity of the quadrupole interaction and of the importance of higher multipole interactions in high

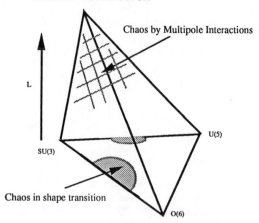

Fig. 2. Schematic explanation of chaos in IBM. Lower part of tetrahedron is Casten triangle. The vertical direction is the angular momentum.

spin states. These features are similar to the ones found in the analysis with the the cranked-shell model and the particle-rotor model. We conclude that the basic mechanism of chaos in high spin states is the same in these three models.

This work is supported in part by the International Joint Research Projects of the Japan Society for the Promotion of Sciences and the Deutsche Forschungsgemeinschaft.

4. References

1. B.Herskind, A.Bracco, R.A.Broglia, T.Døssing. A.Ikeda, S.Leoni, J.Lisle, M.Matsuo, and E.Vigezzi, *Phys. Rev. Lett.* **68**, (1992) 3008.
2. B.Lauritzen, T.Døssing and R.A.Brogilia, *Nucl. Phys.* **A457**, 61 (1986).
3. M.Matsuo, T.Døssing, B.Herskind, S.Frauendorf, *Nucl. Phys.* **A564**, 345 (1993), and references therein.
4. F.Iachello and A.Arima, *The Interacting Boson Model.* 1987, Cambridge: Oxford.
5. N.Whelan and Y.Alhassid, *Nucl. Phys.* **A556**, 42 (1993), and references therein.
6. T.Mizusaki, N.Yoshinaga, T.Shigehara and T.Cheon, *Phys. Lett.* **B269**, 6 (1991).
7. S. Åberg, *Nucl. Phys.* **A477**, (198) 18.

ALGEBRAIC AND GEOMETRIC APPROACHES TO THE COLLECTIVE ENHANCEMENT OF NUCLEAR LEVEL DENSITIES

A. MENGONI
ENEA, Dipartimento Innovazione, Settore Fisica Applicata, V.le G. B. Ercolani 8
40138 Bologna, Italy

G. MAINO, A.VENTURA
ENEA, Dipartimento Innovazione, Settore Fisica Applicata, V.le G. B. Ercolani 8
40138 Bologna, Italy
and INFN, Sezione di Firenze, Italy

and

Y. NAKAJIMA
JAERI, Nuclear Data Center, Tokai-mura
Ibaraki 319-11, Japan

ABSTRACT

The Interacting Boson Model (IBM) has been employed in the evaluation of the collective contribution to nuclear level density. Previous calculations have been extended to transitional nuclei in the framework of the neutron-proton version of the model (IBM-2). The results are shown in comparison with the predictions of geometric models.

1. Introduction

The nuclear motion at finite temperature is generated, in general, by contributions arising from quasiparticle as well as from collective excitation modes. In the low-temperature and low-spin region a variety of nuclear models, based on geometrical properties, shell structure, pair correlations and so forth, provide the description of discrete spectra. In a situation where the density of states is high enough to form a continuum, most of the nuclear structure models cease to be applicable and a description of nuclear excitation has to rely on relatively rough models: the Fermi-Gas model (FGM) or the like. This situation is already met at excitation energies corresponding to the neutron separation energy (≈ 8 MeV) for medium-mass and heavy nuclei. In spite of the fact that reasonable estimates of the nuclear density of states can be given if a good parametrization of the FGM is adopted, one is left out with the contribution of collective degrees of freedom which cannot be included in the FGM description itself.

Because the collective degrees of freedom play a fundamental role in the low temperature region, it is admissible to pose the question: up to what energy do the collective motion contributes to nuclear excitations? In the past, several attempts have been made to answer to this question, using phenomenological[1] as well as microscopic models[2].

Recently[3], we have proposed to adopt the Interacting Boson Model (IBM) to test excitation energy regions above the continuum threshold. This task has been accomplished in two steps: first, the IBM Hamiltonian is extended to finite temperature. Then, the contribution of the collective degrees of freedom is calculated and opportunely included into the level density formalism. As will be clarified below, the first step consists in evaluating the effective number of bosons, N, at finite temperature.

In our previous analysis we have considered only the three dynamical symmetry limits of the IBM-1. For the same symmetry cases and in the large-N limit of IBM-1, the relation of the collective enhancement with the results of geometric models has been recently investigated[4]. In that case, however, like for geometric models the assumption of a temperature-independent N, leads to an unphysical collective enhancement which is indefinitely increasing with nuclear temperature.

In a first extension[5] of these approaches, we have evaluated the collective contribution to the nuclear level density for transitional classes in the A $\approx$ 150 and A $\approx$ 190 mass regions, using the neutron-proton version of the IBM (IBM-2), but the calculations were performed only at fixed temperature, corresponding to the neutron binding energy. It was nevertheless found that the enhancement due to collective excitations increases drastically as the SU(3) dynamical symmetry limit of the IBM is approached.

Here, we will show the results obtained for the enhancement factor of the three samarium isotopes ^{148,150,152}Sm, describing a transitional class of nuclei going from vibrational (U(5) limit of the IBM) to rotational shapes (SU(3) limit).

2. The IBM at finite temperature

2.1. The effective number of bosons

The Talmi IBM-2 Hamiltonian[6]

$$\hat{H} = \hat{h}_0 + \epsilon(\hat{n}_{d\nu} + \hat{n}_{d\pi}) + \kappa \hat{Q}_\nu^{(2)} \cdot \hat{Q}_\pi^{(2)} + \hat{V}_{\nu\nu} + \hat{V}_{\pi\pi} + \hat{H}_M \qquad (1)$$

should have, in principle, all the parameters depending on the nuclear temperature (the definition of all the operators in Eq.(1) can be found in the literature[6]). It is, however, known[7] that the nucleon-nucleon effective interaction can be considered constant for low temperatures ($T \leq 5$ MeV). Then, the IBM parameters can be assumed to be constant and fixed to reproduce the low-lying spectra. On the other hand, the pairing interaction is known to be strongly temperature-dependent and therefore, the effective number of bosons N_τ too, can be expected to depend on the temperature.

The full description of the assumptions made in the evaluation of N_τ for neutrons ($\tau = \nu$) and protons ($\tau = \pi$), is given in reference[3]. Here, we will only recall the expressions for the average number of particle- and hole-pairs, in a given set of lsson orbits

$$N_{pp} = \sum_k [V_k^2(1-2f_k) + f_k^2], \qquad N_{hh} = \sum_k [U_k^2(1-2f_k) + f_k^2] \qquad (2)$$

here V_k^2 (U_k^2) and $f_k^2 = [1 + \exp(\beta E_k)]^{-2}$ are, respectively, the particle (hole) d quasiparticle occupation probabilities. As usual, the quasiparticle energies E_k e derived by solving[8] the BCS equations at finite temperature, $\beta \equiv 1/T$, with in McV. The summation in Eq.(2) runs over a set of levels around the Fermi ergy contained in an interval of half width $\sqrt{3}\Delta_0$, with Δ_0 given by the BCS lution at zero temperature. All the assumptions cited here are widely described the reference[3]. As usual, the effective number of bosons is defined as $N_\tau = in[N_{pp}, N_{hh}]$.

The results of the calculations for $N = N_\nu + N_\pi$ as a function of the nuclear mperature are shown in Fig. 1(a) for the three isotopes ^{148,150,152}Sm (a transitional ass U(5) → SU(3) of IBM).

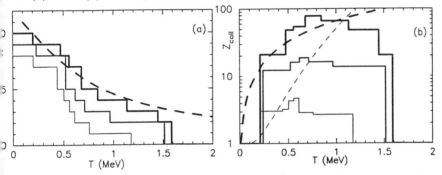

g. 1. (a) The effective number of bosons as a function of nuclear temperature for ^{148,150,152}Sm. hinner and thicker lines are for ^{148}Sm and ^{152}Sm respectively. (b) The collective enhancement ctors Z_{coll}: the thin dotted line is Z_{vib} for ^{148}Sm, the thick dotted line is Z_{rot} for ^{152}Sm. For rther explanations, see the text.

An analytical expression for N, above the critical temperature, can be derived, sulting in

$$N = 2\rho_0 T \ln \cosh(\frac{\sqrt{3}\Delta_0}{2T}) \qquad (3)$$

here ρ_0 is the density of single particle states around the Fermi energy. In Fig. (a), we show (dotted line) the values obtained using Eq.(3) with $\rho_0 = 6A/13\pi^2$ IeV^{-1} and $\Delta_0 = 12/\sqrt{A}$ MeV, for ^{152}Sm ($A = 152$).

.2. *The collective enhancement factor*

It is possible to show that if the coupling between the collective and the quasi- article degrees of freedom is weak, the total density of states at excitation U can e factorized into

$$\rho(U) = \rho_{qp}(U) \cdot Z_{coll}(\beta) \qquad (4)$$

where

$$Z_{coll}(\beta) = \sum_c \exp(-\beta E_c) \qquad (5)$$

is the canonical partition function for the collective spectrum. The set of collective states $\{E_c\}$ is calculated by diagonalizing the full Hamiltonian of Eq.(1) for each value of the nuclear temperature T.

The geometrical models give the very simple analytical expressions for the collective enhancement factor

$$Z_{rot}(\beta) = \begin{cases} 1 & \text{spherical} \\ \sigma_\perp^2 & \text{axisymmetric} \end{cases} \qquad Z_{vib}(\beta) = [\frac{1}{1-\exp(-\beta E_2)}]^5 \qquad (6)$$

where the commonly adopted notation for the parameters of spherical or axisymmetric rotors and quadrupole vibrators has been used.

The results of our calculations for the enhancement factor, Z_{coll}, are show in Fig. 1(b). There, we show the results for the three samarium isotopes obtained using the IBM-2 spectra together with the results given by the rotor and quadrupole vibrator for ^{148}Sm and ^{152}Sm, respectively. As can be seen, the geometric models do not predict any reduction in Z_{coll} as the nuclear temperature increases. However, for nuclei close to the SU(3) limit, the order of magnitude of the enhancement is similar for the two different approaches, in particular up to the region where the maximum value is reached.

3. Conclusions

The results show that reliable collective enhancement factors for nuclear level density can be estimated using IBM. In addition to the simple analytical expression which can be obtained for the dynamical symmetry limits of IBM, we have shown that transitional nuclei can be treated as well, using the full potentials of the IBM-2

4. References

1. S. Bjørnholm, A. Bohr and B. Mottelson, *Proceedings on the Symposium on the Physics and Chemistry of Fission* Vol. 1 (IAEA, Vienna, 1974) 367; A. V. Ignatyuk, K. K. Istekov, and G. N. Smirenkin, *Sov. J. Nucl. Phys.* **29** (1979) 450.
2. See for example: S. Shlomo and J. B. Natowitz, *Phys. Lett.* **252** (1990) 187
3. G. Maino, A. Mengoni and A. Ventura, *Phys. Rev.* **C42** (1990) 988.
4. V. K. B. Kota, *Europhys. Lett.* **23** (1993) 481.
5. A. Mengoni and G. Maino, *Proc. Nuclei in the Cosmos* (IOP, 1993), p. 411
6. F. Iachello and A. Arima, *The Interacting Boson Model* (Cambridge University Press, Cambridge, 1987).
7. M. Brack and P. Quentin, *Phys. Lett.* **52B** (1974) 159.
8. G. Maino and A. Ventura, *Comput. Phys. Commun.* **47** (1987) 303.

F-SPIN MULTIPLETS AND M1 TRANSITIONS

IN THE A=100 MASS REGION

A.GELBERG [1,2] and T.OTSUKA [2]

1.Institut für Kernphysik der Universität zu Köln,50937 Köln,Germany

2.Physics Department,University of Tokyo, Bunkyo-ku,Tokyo 113,Japan

ABSTRACT

F-spin multiplets in the A=100 region have been investigated.While the energy constancy inside these multiplets is bad in comparison with deformed rare earth nuclei, the M1 transitions between low lying collective levels are relatively strong.This shows that the energy variation inside a multiplet is at least partly due to F-spin mixing. IBM-2 calculations of excitation energies,E2 and M1 strengths in ^{102}Pd and ^{108}Pd have been carried out

1. INTRODUCTION

F-spin was introduced by A.Arima,F.Iachello, T.Otsuka and I.Talmi in the framework of the proton neutron Interacting Boson Model (IBM-2) [1,2,3]. A state is characterized by two quantum numbers: the F-spin F and the projection $F_0 = (N_p - N_n)/2$, where N_p and N_n are the proton and neutron boson numbers respectively. The values of F are

$$F_0 \leq F \leq (N_p + N_n)/2 = F_{max} \qquad (1)$$

The states with $F = F_{max}$ are totally symmetric and they have the lowest energy. States with $F < F_{max}$ are mixed symmetry states [2,6].

If F were a good quantum number and the IBM-2 parameters were the same for the nuclei with the same F_{max}, i.e. with the same total number of bosons N,their spectra would be identical.These nuclei form an F-spin multiplet, which is the analog of an isospin multiplet. The conditions which would lead to F-spin multiplets with constant excitation energies have been discuseed in refs [4,5] Let us consider the experimental low-energy spectra of the rare earth nuclei with $F_{max} = 6$ and $-2 \leq F_0 \leq 2$ from ^{156}Dy to ^{172}W [7] shown in Figure 1. The energy constancy in this multiplet is quite remarkable, although a slight F_0 dependence can be seen.

If we extend the search for F-spin multiplets to the A=100 region we notice that the energy differences are rather large. This can be noticed on the second multiplet of Figure 1 with F=4 (from ^{100}Mo to ^{112}Cd). The question arises whether this variation is due to F-spin mixing,or it is just caused by the variation of the IBM-2 parameters inside the multiplet.

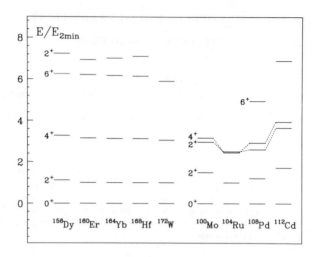

Fig. 1. F-spin multiplets

2. M1 TRANSITIONS

The magnetic moment is defined as [3]

$$\hat{\mu} = \mu_N(g_p\hat{L}_p + g_n\hat{L}_n) \qquad (2)$$

where $\hat{L}_p$ and $\hat{L}_n$ are the angular momenta of the proton and neutron boson respectively, and g_p and g_n are the proton boson and the neutron boson g-factors respectively. The general F-spin selection rule for M1 transitions is $\Delta F = 0, \pm 1$ and transitions between two totally symmetric states ($F = F_{max}$) are forbidden [3]. Therefore, if M1 transitions occur between low-lying collective states, these states must contain components with mixed symmetry ($F < F_{max}$). This allows to use the strength of M1 transitions between low-lying collective states as an indicator of F-spin mixing.

In order to characterize the relative M1 strength we can use the reduced mixing ratio Δ defined by

$$\Delta^2 = \delta^2/E_\gamma^2 \qquad (3)$$

where δ is the E2/M1 mixing ratio of a gamma transition and E_γ is its energy. We will consider transitions between the 2_2^+ and the 2_1^+ states, for which good experimental data are available [8] (see Table 1). One can notice that in the deformed region, where F-spin multiplets with nearly constant energy have been observed, Δ^2 is very large, indicating a relatively small M1 strength. On the contrary, in the A=100 region Δ^2 is much smaller, while these nuclei belong to multiplets which display a rather

Table 1. Mixing ratios of $2_2^+ \to 2_1^+$ transitions; energies in MeV.

Nucleus	E_γ	δ	Δ^2
^{100}Mo	.528	4.4	69.4
^{100}Ru	.822	3.2	15.15
^{102}Pd	.978	2.8	8.2
^{106}Pd	.616	9.4	232
^{108}Pd	.497	3.1	38.9
^{120}Xe	.491	12.6	658
^{134}Ba	.563	8.1	206
^{166}Er	.705	>25	>1257
^{168}Er	.741	28	1427
^{162}Dy	.808	>30	>1378

strong energy variation.In a qualitative way one can see that the energy constancy of the F-spin multiplet and the relative M1 strength of transitions between low-lying collective states are anticorrelated.This indicates that the energy variation is at least partly due to F-spin mixing.

3. IBM-2 CALCULATIONS OF ^{102}Pd and ^{108}Pd

It is of interest to see whether an IBM-2 calculation can reproduce the observed M1 strengths.We choose as examples the nuclei ^{102}Pd and ^{108}Pd for which reliable experimental data are available [9,10]. The following Hamiltonian was used [11]

$$H = \epsilon n_d + \kappa Q_p \cdot Q_n + \sum_{m=n,p} \sum_{k=0,2,4} c_{mk}((d^\dagger d^\dagger)^{(k)} \cdot (\tilde{d}\tilde{d})^{(k)})_m + \lambda M \quad (4)$$

M is the Majorana operator in a simplified form [4]

$$M = F_{max}(F_{max} + 1) - \hat{F}^2 \quad (5)$$

For ^{102}Pd $\epsilon = 0.88, \kappa = -0.18, c_{2n} = -0.05, \lambda = 0.24$, all in MeV; $\chi_p = -0.8$, $\chi_n = -0.55, e_p = 0.12, e_n = 0.1, g_p = 1.2, g_n = -0.17$; $e_{p(n)}$ are effective charges.For ^{108}Pd $\epsilon = 0.74, \kappa = -0.115, \chi_p = -0.8, \chi_n = -0.4, c_{2n} = -0.12, c_{4n} = 0.3, g_n = -0.05$.The effective charges and g_p are the same.These parameters are rather different from those of ref [13].

Only $\kappa, \chi_n, \lambda, c_{n2}$ and c_{n4} were treated as free parameters.The other parameters were extrapolated from the values for ^{112}Cd derived from microscopic calculations [12].As can be seen in Figure 2, the low-lying collective states of these nuclei can be satisfactorily reproduced by the calculation. The same can be said for the electromagnetic observables shown in Table 2.The fit of higher lying states, e.g. the 2_2^+ state, is less satisfactory.It has been assumed that those states may have intruder components [13]. It is extremely important to fit in a consistent way both g-factors and M1-strengths,without assuming good F-spin.

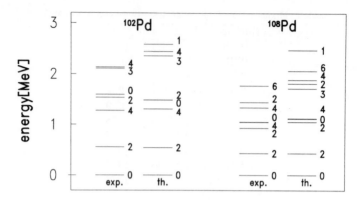

Fig. 2. Excited levels of ^{102}Pd and ^{108}Pd

Table 2. Electromagnetic observables; B(E2) in fm^4, B(M1) in μ_N^2

Observable	^{102}Pd		^{108}Pd	
	Experiment	Theory	Experiment	Theory
$B(E2; 2_1^+ \to 0_1^+)$	923(6)	924	1561(40)	1527
$B(E2; 2_2^+ \to 2_1^+)$	425(57)	693	2383(261)	1334
$B(M1; 2_2^+ \to 2_1^+)$	.0036(5)	.0037	.0024(6)	.0021
$g(2_1^+)$	.41(4)	.39	.367(20)	.367

4. CONCLUSIONS

The relatively strong M1 transitions in nuclei belonging to the A=100 region indicate F-spin mixing.This feature is correlated with the large energy variation inside F-spin multiplets.On the contrary, deformed nuclei in the rare earth region show very weak M1 transitions and the energy variation in F-spin multiplets is small.Electromagnetic transition strengths and excitation energies in ^{102,108}Pd have been correctly described by using the IBM-2.

5. ACKNOWLEDGEMENTS

The authors would like to thank Prof.P. von Brentano for stimulating discussions. This work has been partly supported by the Japan Society for Promotion of Science and by the Deutsche Forschungsgemeinschaft.

6. REFERENCES

1. A.Arima,T.Otsuka,F.Iachello and I.Talmi, *Phys.Lett.* **66B** (1977) 205.
2. T.Otsuka,A.Arima,F.Iachello and I.Talmi, *Phys.Lett.* **76B** (1978) 139; T.Otsuka, A.Arima and F.Iachello, *Nucl.Phys.* **A309** (1978) 1.
3. F.Iachello and A.Arima, *The Interacting Boson Model*, Cambridge University Press, Cambridge, 1987.
4. P.O.Lipas,P. von Brentano and A.Gelberg, *Rep.Progr.Phys.* **53** (1990) 1355.
5. P. von Brentano,A.Gelberg,H.Harter and P.Sala, *J.Phys.G.* **11** (1985) L85
6. T.Otsuka, in *Algebraic Approaches to Nuclear Structure*, ed. R.F.Casten, Harwood Academic, Langhorne,1993.
7. M.Sakai,*At. Data Nucl. Data Tables* **31** (1984) 399.
8. K.S.Krane, *At. Data Nucl. Data Tables* **25** (1980) 29
9. D. de Frennes and E.Jacobs, *Nucl. Data Sheets*, **63** (1991) 373;J.Blachot, *Nucl. Data Sheets*, **62** (1991) 803
10. P.Raghavan *At. Data Nucl. Data Tables* **42** (1989) 189
11. G.Puddu, O.Scholten and T.Otsuka, *Nucl. Phys.* **A348** (1980) 109
12. M.Deleze, S.Drissi, J.Kern, P.A.Tercier, J.P.Varlet, J.Rikovska, T.Otsuka, S.Judge and A.Williams, *Nucl. Phys.* **A551** (1993) 269
13. P.van Isacker and G.Puddu, *Nucl.Phys.* **A348** (1980) 125

Applications of IBM-3 to the $Z \sim N \sim 40$ Nuclei

Michiaki Sugita

Advanced Science Research Center, Japan Atomic Energy Research Institute Tokai, Ibaraki, 319-11, Japan

ABSTRACT

Using a complete diagonalization program for IBM-3 hamiltonians, the shape transitions from $A = 80$ to 90 are investigated. The splitting of the *scissors* motions by isospin is discussed.

1. Introduction

In nuclei with $Z \sim N \sim 40$, the valence protons and neutrons fill the same major shell. Thus isospin should be taken into account. We use the Interacting Boson Model with Isospin (IBM-3)[1,2] for describing these nuclei.

The nuclei at $A \sim 80$ are characterized by huge deformation ($\beta_2 \sim 0.4$) but rather low values of the ratio $\text{Ex}(4_1^+)/\text{Ex}(2_1^+) \sim 2.5$. Indeed, for the $^{80}_{38}\text{Sr}$ nucleus, the energy spectrum and the $B(E2; I \to I-2)$ values within the ground band ($I = 2, \ldots, 10$) can be fitted almost completely using the O(6) limit of IBM, as seen from Figs. 1 and 2.

2. IBM-3 Hamiltonian

We use the following IBM-3 hamiltonian:

$$H = e_d n_d - \sum_{t=0,1,2} \kappa(t) Q(t) : Q(t) + \alpha T(T+1), \tag{1}$$

where the symbol (:) means the scalar products with respect to both the angular momentum and the isospin, and Q is an O(6)-type quadrupole operator;

$$Q(t) = [s^\dagger \tilde{d} + d^\dagger \tilde{s}]^{(l=2,t)}. \tag{2}$$

This hamiltonian is O(5) invariant. The IBM-3 wavefunctions can be classified by the permutation symmetry of charge and spacial wavefunctions $[f] = [f_1 f_2 f_3]$ as $|[N]; [f]TT_z; [f]\beta LM >$; both the charge and spacial permutation symmetries should be the same because of boson symmetry. The hamiltonian in Eq. (1) contains the isovector and isotensor Q forces and thus does not conserve $[f]$. We assume that there is no quadrupole force between identical particles, which leads to

$$2\kappa(0) + 3\kappa(1) + \kappa(2) = 0. \tag{3}$$

Then the quadrupole force can be rewritten as

$$-x\{[s^\dagger d^\dagger]^{(t=0)} : [\tilde{d}\tilde{s}]^{(t=0)} + \sqrt{5/2}([s^\dagger s^\dagger]^{(t=0)} : [\tilde{d}\tilde{d}]^{(t=0)} + \text{h.c})\} + \xi_2[s^\dagger d^\dagger]^{(t=1)} : [\tilde{d}\tilde{s}]^{(t=1)}, \tag{4}$$

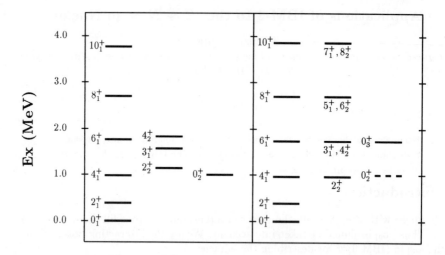

Figure 1: Comparison of energy levels of $^{80}_{38}$Sr between experiment (left) and the O(6) limit of the IBM (right). In the O(6) limit, we use the hamiltonian : $H = \kappa C(O(5)) + \kappa' C(O(6)) = \kappa \tau(\tau+3) + \kappa'\sigma(\sigma+4)$, with $\kappa = 0.0965$ MeV and $\kappa' = 0.0277$ MeV. The solid lines in the left figure show the $\sigma = N$ levels and the dashed line $\sigma = N - 2$ (0_2^+).

$\kappa' < 0$?

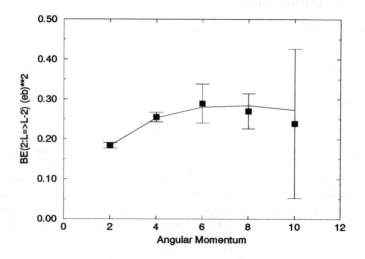

Figure 2: Ground band BE2 values of $^{80}_{38}$Sr. The symbols express experiment and the line the O(6) limit values.

where
$$x = 2(\kappa(0) + 2\kappa(2)) = 3(\kappa(2) - \kappa(1)), \quad (5)$$
and
$$\xi_2 = 2/3(x - 6\kappa(2)). \quad (6)$$

The second term is one of the so-called Majorana interactions and gives the energy splitting between $[f] - [N]$ and $[N-1,1]$ states. The values for α in Eq.(1) and ξ_2 are determined so as to reproduce the excitation energies (7 Mev, 9 Mev) of the $T = 1, 2$ states of the lighter $Z = N$ nuclei such as ^{44}Ti, ^{48}Cr and ^{52}Fe. The x value is determined by ^{80}Sr. The values for κ are decreased linearly from the values obtained at $A = 80$ to 0 at $A = 90$. The d-boson energy e_d is determined so as to reproduce Ex(2_1^+) for each nucleus. We got resonably good reproduction of the systematics of the energy levels of the ground bands and B(E2:$0_1^+ \to 2_1^+$).

3. Splitting of the scissors motion by isospin

In Figs. 3 and 4, we show an example of the results of the present study, where the excitation energies of the *scissors* states with $J^\pi = 1^+$ and the B(M1:$0_1^+ \to 1^+$) values to them from the ground states are plotted for the $_{40}$Zr isotopes. We use the the following M1 operator:

$$T(\text{M1}) = \sqrt{3/4\pi}(g_\pi L_\pi + g_\nu L_\nu + g_{\pi\nu} L_{\pi\nu}) \quad (7)$$

with
$$g_\pi = 1, g_\nu = 0, g_{\pi\nu} = 1/2(\mu_N). \quad (8)$$

In the $N = Z$ nucleus ^{80}Zr, the *scissors* mode is concentrated in a single state with $T = 1$ at the relatively high excitation energy ~ 7 MeV. However, in the $N \neq Z$ nuclei, they split into T and $T + 1$ states and the B(M1) values to $T_f = T$ is much bigger than the values to $T_f = T + 1$, where T is the isospin of the ground state. It would be interesting to search experimentally for the rather low-lying (~ 3 MeV) *scissors* states in the current region of nuclei.

4. Summary

It is interesting to use IBM-3 not just for light nuclei but rather for heavier deformed nuclei such as in the region of $Z \sim N \sim 40$.

The author would like to thank Dr. J.N. Ginocchio and T-5 group of Los Alamos National Laboratory for discussions and their kind hospitality.

References

1. J.P. Elliott and A.P. White, Phys.Lett., **B97**, 169 (1980).
2. P. Halse, J.P. Elliott and J.A. Evans, Nucl.Phys., **A417**, 301 (1984).

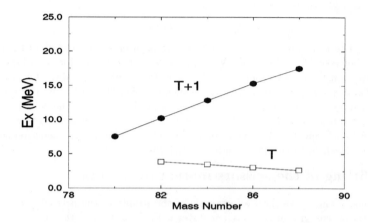

Figure 3: Excitation energies of the Scissors modes for the Zr isotopes.

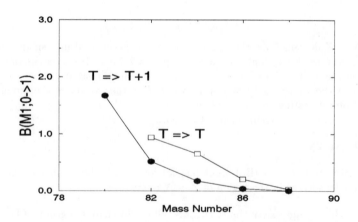

Figure 4: B(M1) values to the Scissors states for the Zr isotopes.

SUM RULES IN THE PROTON NEUTRON INTERACTING BOSON MODEL

C. DE COSTER[*] and K. HEYDE
*Institute for Theoretical Physics, Vakgroep Subatomaire en Stralingsfysica,
University of Gent, Proeftuinstraat 86, B-9000 Gent, Belgium*

ABSTRACT

We present results of sum rule calculations for M1,M3,E2 transitions in the IBM-2. In particular the M1 non-energy weighted and the M1,E2 and M3 energy-weighted sum rules are evaluated. We thereby study limiting situations using good F-spin and in the U(5), SU(3) and O(6) dynamical symmetries.

1. Introduction

Sum rules provide a useful tool for measuring quantitatively the degree of collectiveness of a given state, in particular since only the ground state expectation value of a single or double commutator needs to be evaluated.
In the light of the controversy about the nature of low-lying excitations in deformed rare-earth nuclei and in particular their degree of collectivity[1], we have used sum rule techniques to address some of these questions. We have studied both magnetic (M1, M3) and electric (E2) non-energy-weighted (NEWSR) and linear energy-weighted (LEWSR) sum rules within the interacting boson model framework. It is the aim of this paper to present some physically interesting results without going into detail on the calculations. For the latter and for applications to more realistic cases, we refer to other recent publications[1-3].

2. Magnetic Dipole Sum Rules

2.1. Non-Energy-Weighted Sum Rule

In the limit of good F-spin the NEWSR reads[1,3,4]

$$\sum_f B(M1; 0_1^+ \to 1_f^+) = \frac{9}{4\pi} \frac{(g_\pi - g_\nu)^2 P}{N-1} \langle 0_1^+ | \hat{n}_d | 0_1^+ \rangle \quad , \tag{1}$$

with $P = 2N_\pi N_\nu / N$ depending on the proton and neutron boson numbers, $\hat{n}_d$ the d-boson number operator and g_ρ the gyromagnetic factors ($\rho = \pi, \nu$).
Starting from the E0-operator within the IBM[5]

$$\hat{T}(E0) = \gamma_0 \hat{n}_s + \beta_0 \hat{n}_d = \gamma_0 \hat{N} + \beta_0' \hat{n}_d \quad , \tag{2}$$

---
[*] postdoctoral research associate N.F.W.O.

where $\hat{N}$ counts the total boson number and $\hat{n}_d$ the d-boson number content in the nuclear ground state, one can calculate the isotopic shift[5] (going from N to N+1 bosons)

$$\delta\langle r^2\rangle = \gamma_0 + \beta'_0 \delta\langle \hat{n}_d\rangle \quad . \qquad (3)$$

Therefore, nuclear properties like the isotope shift can be related to the magnetic dipole properties, i.e.,

$$\sum_f B(M1;0_1^+ \to 1_f^+) = \frac{9}{4\pi}(g_\pi - g_\nu)^2 \frac{1}{\beta'_0} \frac{P}{N-1}\left(\langle r^2\rangle - \gamma_0 N\right) \quad . \qquad (4)$$

Note that both quantities are shown to be related to nuclear deformation[5]. Indeed, in the classical limit of the IBM, $\langle \hat{n}_d\rangle$ is proportional to $\langle \beta^2\rangle$ while the nuclear isotope shift is found to depend on $\langle \beta^2\rangle$ within the liquid drop model. Based on these findings, Otsuka earlier proposed a relation between the isotope shift and the variation of mixed symmetry M1 strength, which is born out well by data for the Sm isotopes[6].
The above arguments might hint towards nuclear mass regions where large variations in the summed M1 strength could occur. In the very neutron deficient ^{195}Au region, large radius variations have been observed recently by Kluge et al.[7]. Subsequently, it might be interesting to search for a possible large increase in M1 strength in these nuclei and study its relation to the nuclear radius variation.

2.2. Linear Energy-Weighted Sum-Rule
In the limit of good F-spin, the LEWSR reads

$$\sum_f B(M1;0_1^+ \to 1_f^+)\left(E_x(1_f^+) - \lambda N\right) = \frac{-9}{4\pi}(g_\pi - g_\nu)^2 \kappa_{\pi\nu}\langle 0_1^+|\hat{Q}_\pi \cdot \hat{Q}_\nu|0_1^+\rangle \quad . \qquad (5)$$

From eq.(5) one observes a strong correlation between the LEWSR and the strength of the deformation driving quadrupole proton-neutron force. The expectation value in the ground state of the latter force is a measure of the quadrupole deformation energy which can also be related to the corresponding binding energy in a quadrupole deformed mean field such as the Nilsson model[8]. Similar results have been obtained using the shell model[9].
Introducing an intermediate set of 2+ states one finally obtains

$$\sum_f B(M1;0_1^+ \to 1_f^+)\left(E_x(1_f^+) - \lambda N\right) = c\sum_f B(E2;0_1^+ \to 2_f^+) \quad , \qquad (6)$$

with $\hat{T}(E2) = e_{\it eff}(\hat{Q}_\pi + \hat{Q}_\nu)$ an isoscalar operator,

$$e_{\it eff} = \left(-\frac{9}{4\pi}(g_\pi - g_\nu)^2 \kappa_{\pi\nu}\frac{N_\pi N_\nu}{N^2}\right)^{\frac{1}{2}} e.b. \qquad (7)$$

and c a conversion factor $\mu_N^2.MeV/(e.b.)^2$ to correct for the dimension mismatch. Note that λN is exactly the F-spin invariant Majorana contribution to the excitation energy of a mixed-symmetry state $F = F_{\max} - 1$.
In transitional and in strongly deformed nuclei, most of the summed E2 strength resides in the first 2+ state. So, within a good approximation the sum on the right-hand side of eq.(6) can be restricted to the 2_1^+ level only. In the rotational model[10], $B(E2;0_1^+ \to 2_1^+)$

depends on the electric quadrupole moment of the ground state from which one deduces empirically nuclear deformation, such that $B(E2;0_1^+ \to 2_1^+) \propto \langle \beta^2 \rangle$. Experimentally[11,12] as well as theoretically[2,13-16] it was shown that the summed M1 strength depends on nuclear deformation $\langle \beta^2 \rangle$.

This common dependence on nuclear deformation and the saturation of the latter when approaching mid-shell can explain the similar smooth behaviour of low-lying M1 and E2 strength as observed experimentally for the rare-earth nuclei[2,17]: a steep increase occurring in the transition from spherical vibrational-like towards strongly deformed nuclei, followed by a saturation of the strength for the region of well-deformed nuclei. Furthermore, based on numerical calculations, one can also introduce a few approximations, retaining the leading terms in the M1 and M3 LEWSR. One then obtains the interesting result

$$\frac{LEWSR(M1)}{LEWSR(M3)} \cong \left(\frac{6(g_\pi - g_\nu)}{7(\Omega_\pi - \Omega_\nu)} \right)^2 . \tag{8}$$

3. The linear energy-weighted electric quadrupole sum rule

We immediately give the results for the dynamical symmetries.

$$\sum_f B(E2;0_1^+ \to 2_f^+) E_x(2_f^+)$$

U(5)
$$= 5\varepsilon_d \left(e_\pi^2 N_\pi + e_\nu^2 N_\nu \right)$$

SU(3)
$$= \frac{-9}{8} \kappa_{\pi\nu} \left(e_\pi^2 N_\pi + e_\nu^2 N_\nu \right)(2N+3)$$
$$+ \frac{135}{32} \kappa_{\pi\nu} (e_\pi - e_\nu)^2 \frac{4}{3} \frac{N_\pi N_\nu}{2N-1}$$

O(6)
$$= \frac{-9}{8} \kappa_{\pi\nu} \left(e_\pi^2 N_\pi + e_\nu^2 N_\nu \right)(N+4)$$
$$+ \frac{135}{32} \kappa_{\pi\nu} (e_\pi - e_\nu)^2 \frac{N_\pi N_\nu}{2(N+1)} \tag{9}$$

From these equations, one immediately observes a strong dependence on valence nucleon pair number N. In this respect it is interesting to note that a general expression for the E2 sum rule was derived by Bohr and Mottelson[10] with $M(E2;\mu) = e \sum_i (1/2 - t_z(i)) r_i^2 Y_\mu^2(\hat{r}_i)$

$$\sum_f B(E2;0_1^+ \to 2_f^+) \propto \frac{Z}{M}\langle 0_1^+ | \sum_{i=1}^{Z} r_i^2 | 0_1^+ \rangle \quad . \tag{10}$$

This sum rule indicates a dependence on proton number Z. So, roughly, the IBM sum rule versus the classical one gives a ratio N/Z (N denoting the boson number), exhibiting the valence model properties of the IBM model. For ^{156}Gd ($e_\pi = e_v = 0.12$ e.b., $\varepsilon_d = 1$ MeV), e.w.E2(IBM-2)/e.w.E2(BM) $\cong$ 10%. Since it is known that low-lying collective E2 transitions exhaust about 10% of the classical e.w. sum rule, it is shown that the collective E2 strength within the IBM-2 is of the correct order of magnitude for a valence model space.

4. Conclusion

Calculations of NEWSR and LEWSR in the IBM-2, although elaborate, give in a few limiting cases physically interesting results pointing towards connections between different observables. From numerical tests it becomes clear that these relations remain valid in realistic cases, when introducing renormalization of IBM-2 parameters[1].

5. Acknowledgements

The authors are grateful to R.Casten, W.Nazarewicz, A.Richter, O.Scholten, J.Wood, L.Zamick for useful discussions and to NFWO, IIKW and NATO for financial support.

6. References

1. K. Heyde, C. De Coster, D. Ooms, *Phys. Rev.* **C49** (1994) 156 and refs. therein
2. K. Heyde, C. De Coster, A. Richter and H. Wörtche, *Nucl. Phys.* **A549** (1992) 103
3. K. Heyde, C. De Coster, D. Ooms, A. Richter, *Phys. Lett.* **B312** (1993) 267
4. J. Ginocchio, *Phys. Lett.* **B265** (1991) 6
5. F. Iachello, *Nucl. Phys.* **A358** (1981) 89c
6. T. Otsuka, *Hyperfine Interactions* **74** (1992) 93
7. K. Wallmeroth et al., *Nucl. Phys.* **A493** (1989) 224
8. V.V. Pal'chik and N.I. Pyatov, *Sov. J. Nucl. Phys.* **32** (1980) 476
9. L. Zamick and D. Zheng, *Phys. Rev.* **C44** (1991) 2522; *Phys. Rev.* **C46** (1992) 2106
10. A. Bohr and B. Mottelson, *Nuclear Structure* (W.A. Benjamin inc. 1975)
11. W. Ziegler et al., *Phys. Rev. Lett.* **65** (1990) 2515
12. J. Margraf et al., *Phys. Rev.* **C45** (1992) R521; **C47** (1993) 1474
13. C. De Coster and K. Heyde, *Phys. Rev. Lett.* **63** (1989) 2797
14. I. Hamamoto and C. Magnusson, *Phys. Lett.* **B260** (1991) 6
15. E. Garrido et al., *Phys. Rev.* **C44** (1991) R1250
16. I. Hamamoto and W. Nazarewicz, *Phys. Lett.* **B297** (1992) 25
17. C. Rangacharyulu et al., *Phys. Rev.* **C43** (1991) R949

EXPLORING THE VALIDITY OF THE Z=38 AND Z=50 PROTON CLOSED SHELLS IN EVEN-EVEN Mo ISOTOPES, AND ODD-A Tc ISOTOPES

Haydeh Dejbakhsh, and S. Shlomo
Cyclotron Institute, Texas A&M University
College Station, Texas 77843, USA

ABSTRACT

The neuron-rich even-even $^{94-104}Mo$ isotopes have been investigated in terms of the neutron-proton interacting boson model. Two different approaches were used. The first investigation is based on the validity of the Z=38 subshell closure considering ^{88}Sr as a doubly magic core. In the second calculation the Z=50 and N=50 are considered as valid closed shells leading to ^{100}Sn as a core. The results based on the validity of the Z=38 subshell closure are used to investigate odd-A Tc isotopes in the mass 100 region. The calculated results are compared with experimental data.

1. INTRODUCTION

The neutron-rich nuclei in the mass 100 region have been investigated extensively in the past decade due to the well known behaviour of the first 2^+ levels showing a sharp shape transition from spherical to deformed as a pair of neutrons are added beyond N=58. The presence of the proton Z=38 and neutron N=56 subshell closure strongly affects the shape and shape transition of nuclei in this mass region.

2. RESULTS

2.1. The even-even Mo isotopes

We investigated the even-even Mo in the IBM-2 (the interacting proton-neutron boson model) framework [1] in which proton and neutron degrees of freedom are included. The first calculation was based on the Z=38 proton and N=50 neutron shells. The number of proton bosons for Mo isotopes, outside the ^{88}Sr core, is 2. In our calculations we used different values for ϵ_π and ϵ_ν following the shell model structure of these nuclei. Figure 1a shows the first excited 2^+ states of ^{88}Sr (core nucleus for Mo isotopes), ^{90}Sr, and ^{90}Zr. The $E(2_1^+)$ value for ^{90}Sr corresponds to the excitation energy of the neutron d-bosons ($N_\nu=1$) from ^{88}Sr and $E(2_1^+)$ of ^{90}Zr corresponds to the excitation energy of the proton d boson ($N_\pi=1$). The energy of the first 2^+ state in ^{90}Zr is high relative to the energy of the 2^+ state of the nuclei with four or six protons added (^{92}Mo, ^{94}Ru, see Fig. 1a) to the ^{88}Sr core. Therefore, an effective excitation energy was considered for the proton d boson. The energy of the first excited states of the Sr nuclei are similar when two, four, or six neutrons are added to the ^{88}Sr core. In our calculations the value of ϵ_ν and ϵ_π are close to the value of the $E(2_1^+)$ for ^{90}Sr and ^{92}Mo, respectively. In order to investigate the effect of ϵ_ν differing from ϵ_π we performed two calculations in the U(5) limit. Figure 1b shows the results of the IBM-2 calculation for $N_\pi=1$ and $N_\nu=1$. In the first calculation $\epsilon_\nu=\epsilon_\pi=0.92$ MeV (shown on the left) and in the second calculation $\epsilon_\nu \neq \epsilon_\pi$ ($\epsilon_\nu=0.74$ MeV and $\epsilon_\pi=1.1$ MeV), all the other IBM-2 parameters are zero. This figure shows how some of the degeneracies are removed by making the ϵ_ν value

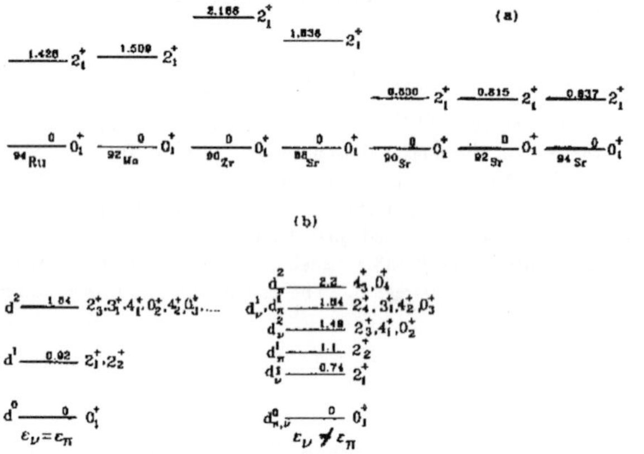

Figure 1: (a) Experimental excitation energy of the first 2^+ states for some of the nuclei in the mass 100 region. (b) The schematic spectra for a nucleus with one neutron and one proton bosons.

different from that of ϵ_π. This also explains why we are successful in reproducing the correct excitation energy levels of the Mo isotopes without introducing the mixing of two configurations. [2]

Our results for both calculations ($N_\pi=2$ and $N_\pi=4$) for Mo isotopes are shown in Figure 2a and 2b. In figure 2a we have plotted the excitation energies of the yrast levels for $J^\pi=2^+$-6^+ for $^{94-104}Mo$ isotopes. In Figure 2b the result of the calculation for the $J^\pi=0_2^+,2_2^+,2_3^+$, and 3_1^+ states is compared with the experimental data for $^{94-104}Mo$ isotopes. The solid lines are calculated excitation energies for $N_\pi=2$ and dashed lines for $N_\pi=4$. The experimental energies are given by points representing different angular momenta as shown in figure 2a and 2b. We obtained a good description of the Mo isotopes with the interacting proton-neutron boson model (IBM-2) assuming Z=38 or Z=50 as a closed shell and using different values for single boson energies for the proton and neutron bosons. [3] First these results indicate that the Z=38 is a valid subshell closure. Second this is an alternative approach and significantly simpler than one considered by Sambataro and Molnar [2] who considered the mixing of two boson configurations (of protons).

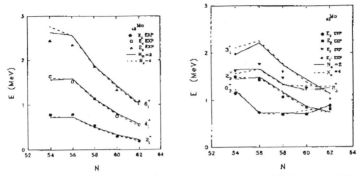

Figure 2: Excitation (a) energy of the yrast levels, and (b) for the *ith* levels with angular momentum J and parity π, (J^π).

2.2. The Structure of odd-A Tc isotopes

The odd-A nuclei have been investigated based on the interacting neutron-proton boson-fermion model (IBFM-2) [4] which couples a particle to an IBM-2 core. The even-even Mo isotopes are used as the core to investigate the Tc isotopes. We used our previous description of the Mo isotopes as a basis for this study. [3] The parameters for the boson-fermion interaction for positive and negative parity states are given in Ref. 5. In Fig. 3a we have plotted the experimental excitation energy of the positive-parity levels of $^{97-103}Tc$ isotopes along with the calculated values within the IBFM-2 using the parameters given in Ref. 5. The solid lines are

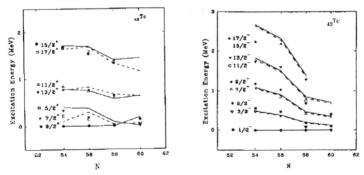

Figure 3: Excitation energy of (a) the positive-parity levels and of (b) the negative-parity levels of the odd-A Tc isotopes.

the calculated excitation energies for $9/2^+$, $5/2^+$, $13/2^+$, and $17/2^+$ states and the dashed lines are the calculated energies for $7/2^+$, $11/2^+$, and $15/2^+$ states. The experimental energies are given by points representing different angular momenta. The overall agreement between experimental and calculated values is good. It is interesting to note that our calculation reproduces the experimental data well for the three low-lying levels with $J^\pi = 9/2^+, 7/2^+$, and $5/2^+$, a significant improvement over the previous results of Refs. 6 and 7, especially for ^{101}Tc. In Fig. 3b we present the experimental results for the negative parity states for $^{97-103}Tc$ isotopes along with the result of the calculation using the parameters given in Ref. 5. The overall agreement between experimental and calculated values is reasonable. The calculated energy spectra and electromagnetic properties [5] are in good agreement with the experimental data. We have clarified the spin assignments of the energy levels in ^{101}Tc.

3. Conclusion

We have shown that the low-lying structure of the Mo and Tc isotopes can be reproduced equally well by the IBM-2 with two proton bosons or four proton hole bosons which indicates that the Z=38 subshell is as valid as the Z=50 closed shell.

We have also shown that the low-lying structure of Mo isotopes can be reproduced by the IBM-2 model without introducing mixing of two configurations by considering the single particle energy of the d proton boson to be different from that of the neutron boson. In this region where the n-p interaction [8] plays an important role in deformation deriving force, considering $\epsilon_\nu \neq \epsilon_\pi$ will remove some of the degeneracies.

4. Acknowledgements

This research was supported by NSF grant Nos. PHY-9110686, and PHY-9215014. We would like to thank Dr. J.D. Bronson for reading the manuscript.

5. References

1. F. Iachello and A. Arima, *The Interacting Boson Model* (Cambridge University Press, Cambridge, 1987).
2. M. Sambataro, G. Molnar, *Nucl. Phys.* **A376** (1982) 201.
3. H. Dejbakhsh, D. Latypov, G. Ajupova, S. Shlomo, *Phys. Rev.* **C46** (1992) 2326.
4. F. Iachello and P. Van Isacker, *The Interacting Boson-Fermion Model*, (Cambridge University Press, Cambridge, 1991).
5. H. Dejbakhsh, and S. Shlomo, *Phys. Rev.* **C48** (1993) 1695.
6. J.M. Arias, C.E. Alonso, and M. Lozano, *Nucl. Phys.* **A466** (1987) 295.
7. P. Degalder, *et. al.* Nucl. Phys. **A401**, 397 (1983).
8. R. F. Casten, Phys. Lett. **B152** (1985) 145; R. F. Casten, Phys. Rev. Lett. **54** (1985) 1991; R. F. Casten, Nucl. Phys. **A443**, 1 (1985).

SUPERSYMMETRY IN THE PT-AU REGION

S.M. FISCHER, A. APRAHAMIAN, X. WU[†]
Department of Physics, University of Notre Dame
Notre Dame, Indiana 46556, USA

and

J.X. SALADIN and M.P. METLAY[‡]
Department of Physics and Astronomy, University of Pittsburgh
Pittsburgh, Pennsylvania 15260, USA

ABSTRACT

The nucleus ^{195}Au has been studied via the reactions ^{196}Pt$(p,2n)$ at beam energies of 12 and 16 MeV, and ^{nat}Ir$(\alpha,2n)$ at 26 MeV. The positive parity energy levels show good agreement with U(6/4) predictions, while some discrepancy is observed between the experimental and theoretical E2/M1 mixing ratios. The data for ^{195}Au is combined with the existing data for ^{194}Pt and ^{195}Pt to present a modified set of parameters for the prediction of the negative parity states of ^{196}Au in the context of the $U_\nu(6/12) \otimes U_\pi(6/4)$ supersymmetry.

1. Introduction

Nuclei in the Pt mass region are well-established as γ-soft nuclei, with ^{196}Pt recognized as the first example[1] of a nucleus described by the O(6) limit of the Interacting Boson Model. Supersymmetries have been developed which are capable of providing a simultaneous description of neighboring even-even and odd-A nuclei. In particular, the positive parity states of the odd-A Au nuclei which result from coupling a single $2d_{3/2}$ proton to an O(6) core are good candidates for the single-j U(6/4) algebra. Similarly, the negative parity states of ^{195}Pt have been successfully interpreted[2] by the U(6/12) multi-j symmetry, which couples the odd neutron $3p_{1/2}$, $3p_{3/2}$ and $2f_{5/2}$ orbitals to an O(6) core.

A larger supersymmetry has been proposed by Van Isacker et al.[3] in which a quartet of nuclei consisting of an even-even, even-odd, odd-even and odd-odd nucleus, each with the same total number of bosons plus fermions, is simultaneously described by a single Hamiltonian. The proposed supersymmetry is given by $U_\nu(6/12) \otimes U_\pi(6/4)$, and the quartet of ^{194}Pt-^{195}Pt-^{195}Au-^{196}Au nuclei has been suggested[3] as the best example of the exact limit of this supersymmetry.

Several studies[4,5] have been undertaken to test the supersymmetry predictions, with attention being focused on the odd-odd member, ^{196}Au, for which very little data exists. In these studies, the experimental results are compared with the predicted negative parity states of ^{196}Au. These states are obtained from the energy eigenvalue equation for the supersymmetry for which the parameters may be determined completely by the other three

---

[†] Current address: University of California San Francisco Physics Research Laboratory, South San Francisco, CA 94080, USA
[‡] Current address: Physics Department, Florida State University, Tallahassee, FL 32306, USA

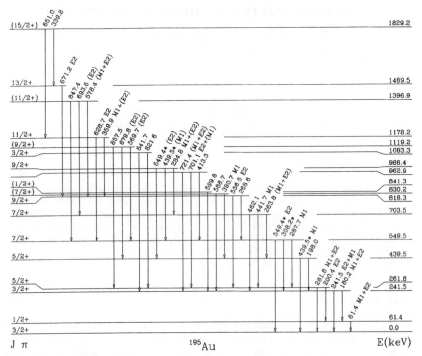

Figure 1. Partial level scheme of the positive parity states of ^{195}Au. Energies are given in keV, and an asterisk (*) indicates that the γ-ray is observed as a doublet.

members of the quartet, ^{194}Pt-^{195}Pt-^{195}Au. However, the experimental data for the odd-A nucleus ^{195}Au is also rather limited. We have therefore conducted in-beam studies of ^{195}Au to provide both a further test of the U(6/4) symmetry as well as to improve the predictions for ^{196}Au in the context of the larger supersymmetry.

2. Experimental Data

In-beam spectroscopic studies including γ-ray angular distributions and γ-γ coincidences and correlations have been performed for the ^{196}Pt$(p,2n)^{195}$Au reaction at beam energies of 12 and 16 MeV, and the ^{nat}Ir$(\alpha,2n)^{195}$Au reaction at 26 MeV. Conversion electrons have also been measured for the $(p,2n)$ reaction at 16 MeV using one element of the University of Pittsburgh ICEBall[6] array. All work was conducted at the University of Notre Dame FN Tandem facility and utilized the University of Pittsburgh multi-detector HPGe array.

Figure 1 shows a partial level scheme of the positive parity states of ^{195}Au relevant to the U(6/4) description. Two transitions are included which were not observed in this study, but which have been previously observed in radioactive decay measurements[7]. The 61.4 keV transition to the ground state is not observed due to poor detector efficiency at low γ-ray energies, and the 241.5 keV transition to ground is a weak branch of a low spin γ-ray cascade not strongly populated in the reaction studies. We note that while these reaction studies populate a significant *negative parity* structure not described in the U(6/4) symmetry, only one linking transition between the negative and positive parity structures was observed.

Several previously tentative spin assignments have been confirmed, including the 5/2+ level at 439.5 keV and the 7/2+ level at 549.5 keV. A notable discrepancy with the decay data is the previously identified cascade of transitions of energies 442(M1)→694(E2)→262 keV originating from the level at 1396.9 keV. This cascade has been misidentified, and is correctly given as the 693.6(E2)→441.7(M1)→261.8 keV cascade, thus replacing the (9/2+) level at 955.1 keV with the 7/2+ level at 703.5 keV. This error was noticed earlier by Zganjar et al.[8], but remains incorrect in the Nuclear Data Sheets compilation[7] and in previous supersymmetry interpretations. A new 9/2+ level at 988.4 keV is established by the 284.8 keV transition feeding into the 703.5 keV level, as well as an experimentally challenging pair of 439.5 and 549.4 keV doublet transitions. While the proposed sequence is a bit curious, it is substantiated by a set of analogous transitions observed in our (α,2n) study of ^{193}Au. Only a tentative assignment of (9/2+) can be made for the 1119.2 keV level, due largely to a difficulty in determining the internal conversion coefficient for the 679.8 keV transition which decays out of this level but also appears as a doublet with a known M1 transition in the negative parity band.

3. ^{195}Au in the U(6/4) and U(6/12)⊗U(6/4) Supersymmetries

Excitation energies of the positive parity levels below 1.4 MeV of ^{195}Au along with a

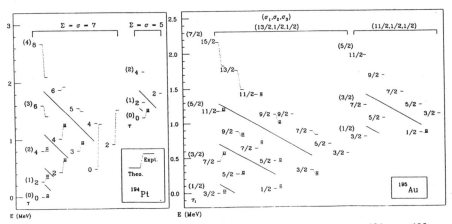

Figure 2. Comparison of the experimental and U(6/4) theoretical energy levels for ^{194}Pt and ^{195}Au. Levels which have been included in the fit are indicated by an asterisk (*). The diagonal lines separate states of different τ.

set of energy levels of ^{194}Pt below 1.5 MeV have been fit according to the U(6/4) eigenvalue equation[9] given by

$$E = A_1 \Sigma_1(\Sigma_1 + 4) + A_2[\sigma_1(\sigma_1 + 4) + \sigma_2(\sigma_2 + 2) + \sigma_3^2]$$
$$+ B[\tau_1(\tau_1 + 3) + \tau_2(\tau_2 + 1)] + CJ(J+1). \quad (1)$$

The parameters obtained are $A_1 = 6.9$, $A_2 = -50.1$, $B = 26.9$ and $C = 26.7$ keV. Figure 2 shows a comparison of the theoretical and experimental energy levels. Qualitative agreement for the energy levels is observed, and the fit displays an overall symmetry breaking of $\varphi \sim 9\%$ as defined by[9] $\varphi = \Sigma |E_i^{expt} - E_i^{theo}| / \Sigma E_i^{expt}$. The quantity φ has been calculated for the 19 levels included in the fit and compares favorably with previous values[9] of ~18% for the Os-Ir nuclei.

Both E2 and M1 transitions are governed by the selection rules $\Delta \sigma_i = 0$ and $\Delta \tau_1 = 0, \pm 1$. The $5/2_2^+ \rightarrow 3/2_1^+$ transition of energy 439.5 keV strongly violates the $\Delta \tau_1$ selection rule, while several other transitions which also violate this selection rule are observed to be relatively weak. Vervier et al.[10] have shown very good agreement between experimental and U(6/4) predictions for B(E2) ratios in the neighboring isotope of ^{197}Au, implying that the collective properties are calculated correctly within this model space.

The single particle properties may be tested by comparing the predicted and experimental E2/M1 mixing ratios, δ. Mixing ratio calculations require an explicit overall normalization factor, N, due to the inclusion of expressions for both E2 and M1 transition operators. Figure 3 shows a plot of the experimental |δ| versus the predicted U(6/4) values of δ/N, as calculated from Iachello et al.[11] A linear curve which passes through the origin may be used to determine the normalization. Mixing ratios for the 61 and 262 keV transitions are obtained from high resolution conversion electron measurements from the decay data in the literature[7], while the other mixing ratios are determined from angular distributions measured in this study. The

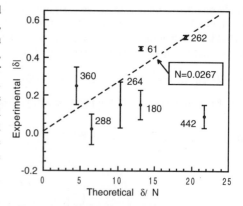

Figure 3. Experimental versus U(6/4) theoretical E2/M1 mixing ratios labeled by transition energy in keV. The dashed curve represents a sample normalization, N, to the experimental mixing ratio of the 262 keV transition.

dashed curve is a normalization to the measured mixing ratio for the 262 keV transition. It appears that no single normalization exists which will give a consistent representation of the data, and that the single particle description given by the U(6/4) symmetry is incomplete.

A likely source of this discrepancy, as well as of the symmetry-breaking 439.5 keV transition, is that single particle orbitals other than the $2d_{3/2}$ significantly contribute to the positive parity structure of this nucleus. For example, spectroscopic factors[12] show fairly large contributions of the $3s_{1/2}$ and $2d_{5/2}$ orbitals in the $1/2_1^+$ and $5/2_2^+$ levels, respectively. It would be quite interesting to explore in particular the B(M1) and δ(E2/M1) predictions of the U(6/20) multi-j symmetry in this mass region, which couples the $3s_{1/2}$, $2d_{3/2}$, $2d_{5/2}$ and $1g_{7/2}$ protons to the O(6) core.

While recognizing that some problems exist in the single particle description, it remains useful to combine the new data for ^{195}Au with the existing data for ^{194,195}Pt in the context of the $U_v(6/12) \otimes U_\pi(6/4)$ supersymmetry. A fit to a total of 41 levels for the three nuclei has been performed for the energy eigenvalue equation given in ref. 3. The parameters obtained are $A = 49.4$, $\bar{B} = -2.4$, $B = -45.9$, $C = 38.1$, $D = 5.7$, and $E = 12.5$ keV, producing an overall symmetry breaking of $\varphi \sim 13\%$ for the three members of the quartet. These parameters are then used to directly predict the energies of the negative parity states of ^{196}Au described by the supersymmetry.

In conclusion, the energy spectra for the three members ^{194}Pt-^{195}Pt-^{195}Au show quite good agreement with the $U_v(6/12) \otimes U_\pi(6/4)$ supersymmetry predictions. However, studies of ^{195}Au offer evidence for the contribution of single particle orbitals not included in the $U_\pi(6/4)$ limit, which implies that the resulting predictions for the negative parity states of ^{196}Au are also likely to be incomplete.

This work is supported by the National Science Foundation under Contracts Nos. PHY90-06246 and PHY90-22196.

References

1. J.A. Cizewski, R.F. Casten, G.J. Smith, M.L. Stelts, W.R. Kane H.G. Börner and W.F. Davidson, *Phys. Rev. Lett.* **40** (1978) 167.
2. A. Mauthofer, K. Stelzer, J. Gerl, Th.W. Elze, Th. Happ, G. Eckert, T. Faestermann, A. Frank and P. Van Isacker, *Phys. Rev.* **C34** (1986) 1958.
3. P. Van Isacker, J. Jolie, K. Heyde and A. Frank, *Phys. Rev. Lett.* **54** (1985) 653.
4. J. Jolie, U. Mayerhofer, T. von Egidy, H. Hiller, J. Klora, H. Lindner and H. Trieb, *Phys. Rev.* **C43** (1991) R16.
5. G. Rotbard, G. Berrier, M. Vergnes, S. Fortier, J. Kalifa, J.M. Maison, L. Rosier, J. Vernotte, P. Van Isacker and J. Jolie, *Phys. Rev.* **C47** (1993) 1921.
6. M.P. Metlay, J.X. Saladin, I.Y. Lee and O. Dietzsch, *Nucl. Instr. and Meth.* **A336** (1993) 162.
7. Z. Chunmei, Nucl. Data Sheets **57** (1989) 1.
8. E.F. Zganjar, J.L. Wood, R.W. Fink, L.L. Riedinger, C.R. Bingham, B.D. Kern, J.L. Weil, J.H. Hamilton, A.V. Ramayya, E.H. Spejewski, R.L. Mlekodaj, H.K. Carter and W.D. Schmidt-Ott, *Phys. Lett.* **58B** (1975) 159.
9. A.B. Balantekin, I. Bars and F. Iachello, *Nucl. Phys.* **A370** (1981) 284.
10. J. Vervier, R. Holzmann, R.V.F. Janssens, M. Loiselet and M.A. van Hove, *Phys. Lett.* **105B** (1981) 343.
11. F. Iachello and S. Kuyucak, *Ann. Phys.* **136** (1981) 19.
12. J.A. Cizewski, in *Interacting Boson-Boson and Boson-Fermion Systems*, ed. O. Scholten (World Scientific, Singapore, 1984), and private communication.

NEW EXPERIMENTAL SYSTEMATICS IN THE A ~ 100 REGION

T. Borello-Lewin, L.B. Horodynski-Matsushigue, J.L.M. Duarte, L.C. Gomes and G.M. Ukita
Instituto de Física, Universidade de São Paulo, CP 20516
São Paulo, SP, CEP 01452-990, Brasil

ABSTRACT: Hole and particle experimental valence and non-valence spectroscopic strengths for ^{99,101,103}Ru isotopes are presented. The peculiar excitation pattern disclosed could represent a stringent test on the IBM interpretation.

A survey of last year's publications shows that the tendency of predominantly focussing the IBM interpretation on even-even nuclei is still prevalent. On the other hand, the full power of the model will be better ascertained if the effect of the unpaired nucleon(s) on the nuclear properties can also be correctly predicted. The S. Paulo Pelletron nuclear spectroscopy (with light ions) group pursues a research program in the challenging A ~ 100 region, investigating the properties of low-lying levels in even-even cores and especially in their odd-even neighbours. Figures 1 and 2 display part of the results, referring to valence and non-valence spectroscopic strengths in ^{99,101,103}Ru. Levels with known spins and parities of $3/2^+$ or $9/2^+$ are marked and, exception made for those, the indicated strengths correspond, respectively, to spins 5/2 and 7/2 for $l = 2$ and $l = 4$ transfers and to 11/2 for $l = 5$. The hole spectroscopic components, represented to the right of the excitation energy of the corresponding states as full bars, result from the systematic ^{100,102,104}Ru(d,t) study of this group, to appear in a forthcoming paper[1]. These spectroscopic factors have been determined for the first time for ^{99}Ru and significantly extended for ^{101,103}Ru, covering >80% of the valence strength expected for each isotope. Besides this, important low-lying $l = 1$ and $l = 3$ hole components were detected for ^{101,103}Ru, but not for ^{99}Ru, up to 1.4 MeV (see Figure 2). The particle strengths are shown as open bars to the left of the level energies and were taken from a previous work of the group[2] and from the works of Fortune *et al.*[3] and Berg *et al.*[4], for ^{101,103}Ru respectively. Attention is to be called to the lack of information on particle strength for ^{99}Ru and to the relatively poorer sensitivity of the ^{103}Ru particle information, which is limited to less than 2 MeV of excitation. A ^{102}Ru(d,p)^{103}Ru work, designed to extend the older studies[3,4] is in progress in S. Paulo.

The even Ru isotopes have been a traditional testing ground for IBM and are supposed to undergo a structure transition from U(5) to O(6) in this mass region. To our knowledge, predictions for odd Ru are nevertheless restricted to calculations of energy levels for ^{99}Ru (Ref. 5) and ^{103}Ru (here also some wave function components[6]) and to the rather complete works of Arias *et al.*[7] and Maino *et al.*[8] Some rotor model calculations[9,10] have also been performed for the odd Ru isotopes of A ~ 100 and have consistently had to resort to deformations smaller by a factor of ~2 with respect to those of the even cores and the tendency of some B(E2) values published by Arias *et al.*[7] and also by Maino *et al.*[8] could be showing similar difficulties.

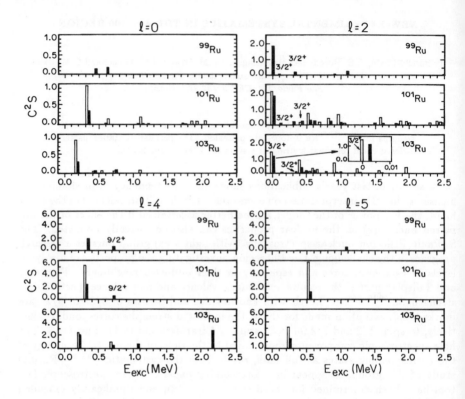

Fig. 1. Valence hole (full bars - right) and particle (open bars - left) spectroscopic strengths.

Inspection of Figures 1 and 2 does not indicate a clear filling pattern for neither orbital corresponding to valence even-l transfer, but rather an amazing overall similarity, being only the $l = 2$ orbitals appreciably fragmented. Although, in the excitation energy interval examined, there exist in all three isotopes many known levels which, in principle, could be excited by $l = 4$ transfer, only some of these are selected for very preferential population. The second intense $l = 4$ excitation at 2.2 MeV in ^{103}Ru is an interesting new feature disclosed in the systematic work[1] of the S. Paulo group and the relative constancy of excitation energy and strength of the first $7/2^+$ levels in the three isotopes is remarkable. As for the odd-l transfers, it is extremely interesting to note that no other than one, very low-lying and relatively intense, $l = 3$ excitation was observed in each ^{101,103}Ru, none having been located in ^{99}Ru above the $l = 3$ detection limit of $C^2S = 0.04$. Also no $l = 1$ excitations were seen in ^{99}Ru below 1.4 MeV and in ^{101}Ru, where hole and particle data of similar quality exist, these do not correspond to the same state. The $l = 3$ excitation is populated in both pickup and stripping and seems to go down in energy in a way similar to that of the known $11/2_1^-$ level.

The IBMF 2 works of Arias et al.[7] and Maino et al.[8] were able to appoint the general trends for the strength distributions corresponding to the valence orbitals. Unfortunately no systematic theoretical predictions are available, the published information being restricted to particle[7,8] and hole[8] states in ^{101}Ru and to hole[7] states in ^{103}Ru, up to 1.2 MeV. Maino et al.[8] do not present $l = 5$ spectroscopic strengths. Both calculations start from even Ru cores as described by parameters of Van Isacker and Puddu[11] and resort to fitted values for the boson-fermion interactions. The parameters adopted by Maino et al.[8] turn out quite different from those of Arias et al.[7], who

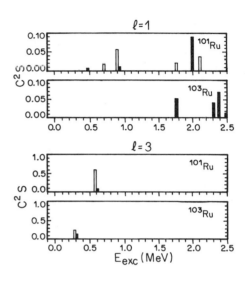

Fig. 2. Non-valence hole (full bars - right) and particle (open bars - left) spectroscopic strengths.

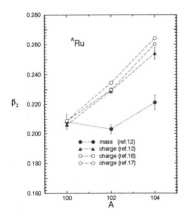

Fig. 3. Evolution of mass and charge deformation parameters in even Ru.

needed distinct ones to reproduce the excitation energies of positive and negative parity states in the odd isotopes. The spreading of the $l = 2$ strength could not be adequately reproduced and in ^{103}Ru Arias et al.[7] foresaw a doublet of $5/2^+$ states at about zero excitation energy, instead of the $3/2^+$ ground state populated in the stripping reaction[4] and the $5/2^+$ first excited state populated in pick-up.[1,4] Conspicuous discrepancies between predictions are noted for $l = 0$ and $l = 4$ transfers. The Maino et al.[8] calculations give better results for the low-lying $g_{7/2}$ transfers, while Arias et al.[7] agree better with the $s_{1/2}$ data. It is to regret that the challenging low-lying

$l = 3$ and $l = 1$ states could not have their spectroscopic properties confronted with theory. In Figure 3 some very recent results[12] on the 2_1^+ core states warn against a naïve utilization of the IBM-1 version for the heavier Ru isotopes. In fact, a small, but experimentally firmly established, difference between mass and charge deformation parameters was determined by coulomb-nuclear interference with alpha particles and deuterons. A similar trend, also favouring the proton component was observed for ^{94}Mo (Ref. 13), in contrast to the results for the ^{92}Zr isotope, where the 2_1^+ excitation appears dominated by the neutrons[14]. In view of the maintained interest in sdgIBM, attention is also to be recalled to the extremely strong $E4$ excitation disclosed in ^{100}Ru by this group[15].

Acknowledgements

This work was partially supported by Conselho Nacional de Desenvolvimento Científico e Tecnológico (CNPq), Coordenação do Aperfeiçoamento de Pessoal de Nível Superior (CAPES), Financiadora de Estudos e Projetos (FINEP) and Fundação de Amparo à Pesquisa do Estado de São Paulo (FAPESP).

References

1. J.L.M. Duarte et al., to be published in *Phys. Rev. C*
2. J.L.M. Duarte et al., *Phys. Rev. C* **36** (1988) 664.
3. H.T. Fortune et al., *Phys. Rev. C* **3** (1971) 337.
4. G.P.A. Berg et al., *Nucl. Phys.* **A379** (1982) 93.
5. E.H. du Marchie van Voorthuysen et al., *Nucl. Phys.* **A355** (1981) 93.
6. S. Brandt et al., *Phys. Rev. C* **34** (1986) 341.
7. J.M. Arias et al., *Nucl. Phys.* **A466** (1987) 295.
8. G. Maino et al., *Z. Phys.* **A340** (1991) 340.
9. C.S. Whisnant et al., *Phys. Rev. C* **34** (1986) 443.
10. J.Rekstad, *Nucl. Phys.* **A247** (1975) 7.
11. P. Van Isacker and G. Puddu, *Nucl. Phys.* **A348** (1980) 125.
12. L.C. Gomes, *Ph.D. Thesis*, Inst. Física, USP, 1993.
13. G.M. Ukita et al., *Workshop on Nucl. Phys.(Braz. Phys. Soc.)*. Abstracts, 1993. p. 60.
14. L.B. Horodynski-Matsushigue et al., *Workshop on Nucl. Phys.(Braz. Phys. Soc.)*. Abstracts, 1993. p. 61. L.B. Horodynski-Matsushigue et al., *Proc. Int. Nucl. Phys. Conf*, 1989. p. 307. D. Rychel et al., *Z. Phys.* **A326** (1987) 455.
15. S. Sirota et al., *Phys. Rev. C* **40** (1989) 1527.
16. J.H. Hirata, *Ph.D. Thesis*, Inst. Física, USP, 1984.
17. S. Landsberger et al., *Phys. Rev. C* **21** (1980) 588.

IBFM CALCULATION OF Xe AND Cs ISOTOPES

N. YOSHIDA[a], A. GELBERG[b,c], T. OTSUKA[b] and H. SAGAWA[d]

[a] *Faculty of Informatics, Kansai University*
Ryozenji-cho 2-1-1, Takatsuki-shi, 569 Japan

[b] *Department of Physics, University of Tokyo*
Hongo 7-3-1, Bunkyo-ku, Tokyo, 113 Japan

[c] *Institüt für Kernphysik, Univerisität zu Köln*
Zülpicherstraße 77, D-50937 Köln, Germany

[d] *Center for Mathematical Sciences, University of Aizu*
Ikkimachi, Aizuwakamatsu-shi, 965 Japan

ABSTRACT

We analyze the negative-parity states of ^{125,127}Xe and ^{127,129}Cs in the proton-neutron interacting boson-fermion model (IBFM2). We show that the IBFM2 describes quite well not only the energy levels and their signature dependence but also the branching ratios.

1. Introduction

The nuclei in the Xe-Ba region have been widely studied from the time of the invention of the interacting boson model (IBM). The even-even nuclei have attracted special interest there,[1-10] because they show typical features of the O(6) dynamic symmetry.[11] The study was extended to odd-A nuclei by using the interacting boson fermion model (IBFM).[12-19] In recent years, the number and the quality of experimental data on odd-A nuclei have increased greatly.[20] We therefore consider it worthwhile to study this region again taking into account these refined data. In the present report, we show the results for the negative-parity states in ^{125,127}Xe and ^{127,129}Cs.

2. Interacting Boson-Fermion Model 2

In the IBFM2, an odd particle is coupled to a core of the proton-neutron interacting boson model (IBM2). The total hamiltonian is $H = H^{\mathrm{B}} + H^{\mathrm{F}} + V^{\mathrm{BF}}$ where

$$
\begin{aligned}
H^{\mathrm{B}} &= \varepsilon_d n_d + \kappa \left(Q_\nu \cdot Q_\pi\right) + \frac{1}{2}\xi_2\left(d_\nu^\dagger s_\pi^\dagger - d_\pi^\dagger s_\nu^\dagger\right)\cdot\left(\tilde{d}_\nu s_\pi - \tilde{d}_\pi s_\nu\right) \\
&\quad + \sum_{K=1,3}\xi_K([d_\nu^\dagger d_\pi^\dagger]^{(K)}\cdot[\tilde{d}_\pi \tilde{d}_\nu]^{(K)}) + \frac{1}{2}\sum_{L=0,2,4}c_\nu^{(L)}([d_\nu^\dagger d_\nu^\dagger]^{(L)}\cdot[\tilde{d}_\nu \tilde{d}_\nu]^{(L)}), \quad (1) \\
V^{\mathrm{BF}} &= \sum_{i,j}\Gamma_{i,j}\left([a_i^\dagger \tilde{a}_j]^{(2)}\cdot Q_{\rho'}^{\mathrm{B}}\right) + \sum_j A_j[a_j^\dagger \tilde{a}_j]^{(0)}[d_{\rho'}^\dagger \tilde{d}_{\rho'}]^{(0)} \\
&\quad + BL\cdot J - \sum_{i,j,k}\frac{\Lambda_{i,j}^k}{\sqrt{2k+1}} : [[d_\rho^\dagger \tilde{a}_i]^{(k)}[a_j^\dagger \tilde{d}_\rho]^{(k)}]_0^{(0)} :, \quad (2)
\end{aligned}
$$

n_d is the d-boson number operator, $Q_\rho = d_\rho^\dagger s_\rho + s_\rho^\dagger \tilde{d}_\rho + \chi_\rho [d_\rho^\dagger \tilde{d}_\rho]^{(2)}$ is the boson quadrupole operator, L represents the boson total angular momentum, and J the fermion total angular momentum. The fermion hamiltonian H^F is the sum of the quasiparticle energies: $\sum \epsilon_j n_j$. The E2 operator is $T^{(E2)} = e_\nu^B Q_\nu + e_\pi^B Q_\pi + \sum_{i,j} e_{i,j}^F [a_i^\dagger \tilde{a}_j]^{(2)}$. The fermion quadrupole effective charges are determined by $e_{i,j}^F = e_{\text{eff}}(u_i u_j - v_i v_j) Q_{i,j}$ where u_j, v_j are the (un)occupation amplitudes and $Q_{i,j} = -(1/\sqrt{5}) < n_i l_i |r^2| n_j l_j > < i||Y^{(2)}||j>$. The M1 operator is $T^{(M1)} = \sqrt{3/(4\pi)} \, [g_\nu^B L_\nu^B + g_\pi^B L_\pi^B + \sum_j g_j J_j]$ where L_ν^B, L_π^B and J_j are angular momentum operators.

3. Xe Isotopes

The nuclei $^{125,127}_{54}\text{Xe}_{71,73}$ can be treated as IBM2 cores of $^{126,128}_{54}\text{Xe}_{72,74}$, respectively, plus an odd neutron (hole). We include the neutron $0h_{11/2}$, $1f_{7/2}$ and $0h_{9/2}$ orbitals. The values of $\chi_\nu = 0.4$ and $\chi_\pi = -0.4$ are fixed to the microscopically derived values (ref.[22]). Other parameters, shown in Table 1, are determined so as to give a good fit of energy levels. The parameters for the odd neutron are calculated

Table 1. The parameters of the boson hamiltonian. The unit of energies is MeV.

nucleus	ϵ_d	κ	ξ_2	ξ_1	ξ_3	$c_\nu^{(0)}$	$c_\nu^{(2)}$
^{126}Xe	0.67	−0.23	0.05	0.29	-0.40	0.1	−0.1
^{128}Xe	0.64	−0.28	0.05	0.29	-0.40	0.1	−0.1

from the Nilsson + BCS with deformation $\delta = 0$, using the the parameters taken from refs.[23,24] In V^{BF} (Eq. (2)), $\rho = \nu$ and $\rho' = \pi$. To reduce the number of free parameters, we assume the orbital dependence of ref.[21]: $\Gamma_{i,j} = \Gamma_0 (u_i u_j - v_i v_j) Q_{i,j}$. The values for $<|r^2|>$ are the harmonic oscillator values. The monopole and the exchange interactions are assumed only for $0h_{11/2}$. Table 2 shows the parameters fitted to the experimental energies. The parameters $A_j = 5.0$ MeV, $B = 0.013$ MeV, and Γ_0/κ and $\Lambda_{j,j}^j/\kappa$ are set constant. To calculate the electromagnetic transitions, the boson

Table 2. The parameters for the odd-A Xe isotopes. The unit of energies is MeV.

nucleus	ϵ_j			Γ_0	$u_j = \sqrt{1-v_j^2}$			Λ_{jj}^j
	$0h_{11/2}$	$1f_{7/2}$	$0h_{9/2}$		$0h_{11/2}$	$1f_{7/2}$	$0h_{9/2}$	$h_{11/2}$
^{125}Xe	1.160	5.469	6.120	10.0	0.7701	0.9945	0.9952	7.5
^{127}Xe	1.136	5.236	5.886	12.17	0.7003	0.9940	0.9953	9.13

charges $e_\nu^B = e_\pi^B = 0.108$ eb are determined from the experimental $B(E2; 0_1^+ \to 2_1^+)$ of ^{126}Xe. The effective charge of the odd neutron is set to $e_{\text{eff}} = 0.5\ e$. The boson

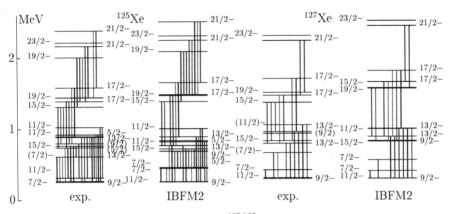

Fig. 1. Energy levels and branching ratios in ^{125,127}Xe. The branching ratios are indicated by the widths of the lines.

g-factors are fixed to $g_\nu^B = 0$, $g_\pi^B = 0.8$ μ_N. The fermion g-factors are the Schmidt values with the spin factors quenched by 0.7.

Figure 1 shows that the energy levels agree very with experimental data. Reasonable agreements are seen also for branching ratios. One notes that the 2nd and the 3rd observed $11/2^-$ levels in ^{125}Xe are interpreted as the 3rd and the 2nd calculated levels. This exchange gives better agreement for branching ratios to $9/2^-$ and $7/2^-$. The calculated branching ratios can be explained from the O(6) character of the core.[20]

4. Cs Isotopes

To study $^{127,129}_{55}$Cs$_{72}$, we couple an odd proton in $0h_{11/2}$ to the boson cores of $^{126,128}_{54}$Xe$_{72,74}$. Some adjustment of χ_ν and κ gives better features. The signature dependence is sensitive to χ_ν, while the absolute magnitude of excitation energies is more affected by κ. We adopt $\chi_\nu = 0.6$ for the both isotopes. We take $\kappa = -0.24$, -0.20 MeV for ^{126,128}Xe, respectively. In V^{BF}, $\rho = \nu$ and $\rho' = \pi$. We take $\Gamma = 0.25$, 0.3 MeV for ^{127,129}Cs. Other parameters including the exchange force are set equal to zero. In fact the $h_{11/2}$ shell is almost empty.

Figure 2 shows the results. The calculated branching ratios are in good agreement with the experiment.

5. Conclusions

We have anylized the negative-parity states in ^{125,127}Xe and ^{127,129}Cs in IBFM2. The branching ratios as well as the energy levels are described well by the calculation for the yrast and the yrare states. The O(6) property of the core explains the branching ratios Other isotopes are being studied.

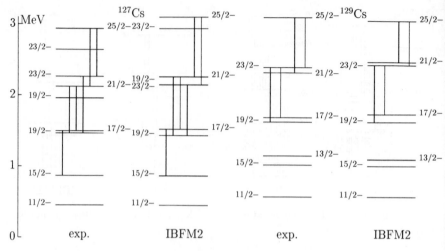

Fig. 2. Energy levels and branching ratios in ^{127,129}Cs.

References

1. G. Puddu, O. Scholten and T. Otsuka, *Nucl. Phys.* **A348** (1980) 109.
2. H. Kusakari and M. Sugawara, *Z. Phys.* **A317** (1984) 287.
3. R. F. Casten and P. von Brentano, *Phys. Lett.* **152B** (1985) 22.
4. A. Novoselsky and I. Talmi, *Phys. Lett.* **172B** (1986) 139.
5. H. Harter, P. von Brentano and A. Gelberg, *Phys. Rev.* **C34** (1986) 1472.
6. A. Sevrin, K. Heyde and J. Jolie, *Phys. Rev.* **C36** (1987) 2631.
7. P. von Brentano et al., *Phys. Rev.* **C38** (1988) 2386.
8. W. Krips et al., *Nucl. Phys.* **A529** (1991) 485.
9. P. F. Mantica et al., *Phys. Rev.* **C45** (1992) 1586.
10. F. Seiffert et al., *Nucl. Phys.* **A554** (1993) 287.
11. A. Arima and F. Iachello, *Phys. Rev. Lett.* **40** (1978) 385.
12. M. A. Cunningham, *Nucl. Phys.* **A385** (1982) 204, 221.
13. C. E. Alonso et al., *Phys. Lett.* **144B** (1984) 141.
14. J. M. Arias, C. E. Alonso and R. Bijker, *Nucl. Phys.* **A445** (1985) 333.
15. K. Fransson et al., *Nucl. Phys.* **A469** (1987) 323.
16. H. C. Chiang, S. T. Hsieh and D. S. Chuu, *Phys. Rev.* **C39** (1989) 2390.
17. S. T. Hsieh, H. C. Chiang and M. M. King Yen, *Phys. Rev.* **C41** (1990) 2898.
18. D. Bucurescu et al., *Phys. Rev.* **C43** (1991) 2610.
19. D. Bucurescu et al., *Z. Phys.* **A343** (1992) 139.
20. I. Windenhöver et al., submitted to *Nucl. Phys.* **A**.
21. O. Scholten, Ph. D. Thesis, Univ. of Groningen (1980).
22. T. Otsuka, *Nucl. Phys.* **A557** (1993) 531c.
23. D. Lieberz et al., *Nucl. Phys.* **A529** (1991) 1.
24. Jing-ye Zhang et al., *Phys. Rev.* **C39** (1989) 714.

SINGLE-NUCLEON TRANSFER REACTIONS AND THE IBFFM

P.E. GARRETT[a,b] and D.G. BURKE[b]

[a] Dept. of Physics, University of Fribourg, CH-1700, Fribourg, Switzerland
[b] Dept. of Physics and Astronomy, McMaster University, Hamilton, Ont., Canada

Abstract

Previous calculations performed with the IBFFM for Ir, Au, and Sb nuclei are compared with the results of single-nucleon transfer reactions. An overview of these comparisons is given, and possible reasons for discrepancies between theory and data are discussed.

1. Introduction

In order to describe odd-odd nuclei, the IBM was extended by coupling two fermions to the boson core. The resulting interacting boson fermion-fermion model (IBFFM) hamiltonian, which is typically solved numerically, contains many terms and also has many parameters. Most of these parameters can be determined by fits to the properties of adjacent even-even and odd-A nuclei, with the remaining ones found from fits to the various properties, such as the spin, parity, quadrupole and magnetic moments of the ground state, of the odd-odd nucleus in question. Once these parameters have been determined, the energy spectrum can be calculated, and the wave functions of the states used to determine transition probabilities, spectroscopic strengths for transfer reactions, etc. This procedure has been used in the calculations[1-3], for example, of a series of Sb and Ir isotopes.

Single-nucleon transfer reactions provide very sensitive tests of nuclear models since the cross sections can be directly related to the squares of the amplitudes of the microscopic components of the wave functions. It is therefore of interest to compare the predictions of the IBFFM with results of single-nucleon transfer. Surprisingly, there are very few examples where such comparisons have been made, and these are concentrated in the Ir-Au region[1,2] (^{192,194}Ir, ^{198}Au with $SO(6)$) cores and the Sb region[3] (^{120,122,124}Sb with $U(5)$ cores).

2. Calculation Procedures

The IBFFM hamiltonian is written[4] as

$$H_{IBFFM} = H_{IBFM}(\pi) + H_{IBFM}(\nu) - H_{IBM} + H_{RES}(\pi\nu) \qquad (1)$$

where $H_{IBFM}(\pi)$ and $H_{IBFM}(\nu)$ are the IBFM hamiltonians for the neighbouring odd-proton and odd-neutron nuclei, respectively, H_{IBM} is the IBM hamiltonian, and $H_{RES}(\pi\nu)$ is the residual interaction between the quasiparticles. In the calculations[5,2,6] for ^{192}Ir, ^{194}Ir, and ^{198}Au, the numbers of core bosons were taken as $N = 4$, and thus the boson spaces for ^{192}Ir and ^{194}Ir were truncated from $N = 7$ and

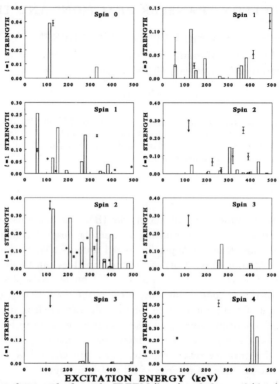

Figure 1: Predicted strengths from the IBFFM (bars) and observed (dots) for the ^{193}Ir(d,t) reaction (from ref.[1]).

$N = 6$, respectively. The even-even boson cores were fitted in the $SO(6)$ limit, which is applicable in the Pt region. In the calculation for ^{198}Au, the $\pi s_{\frac{1}{2}}, \pi d_{\frac{3}{2}}, \pi h_{\frac{11}{2}}$ protons and the $\nu p_{\frac{1}{2}}, \nu p_{\frac{3}{2}}, \nu f_{\frac{5}{2}}, \nu i_{\frac{13}{2}}$ neutrons were included, while the calculations for ^{192}Ir and ^{194}Ir included the above quasiparticles and also the $\pi d_{\frac{5}{2}}$ and $\nu h_{\frac{9}{2}}$ quasiparticles. For ^{192,194}Ir, the H_{IBFM} hamiltonians were fitted to the neighbouring odd-A nuclei, and the residual interaction parameters were found by fitting the properties of the odd-odd ground states, whereas in the calculations for ^{198}Au, the parameters were fine tuned by fitting the properties of the excited states of ^{198}Au. For the Sb isotopes[3], the even-even boson core was fitted in the $U(5)$ limit, and the maximum number of d-bosons was restricted to 2. The fermion model spaced spanned the $s_{\frac{1}{2}}, d_{\frac{3}{2}}, d_{\frac{5}{2}}, g_{\frac{7}{2}}$, and $h_{\frac{11}{2}}$ orbitals for both protons and neutrons. The proton was assumed to be a particle, while the neutrons were taken as quasiparticles. The IBFM hamiltonian parameters were found by fitting the neighbouring odd-A nuclei.

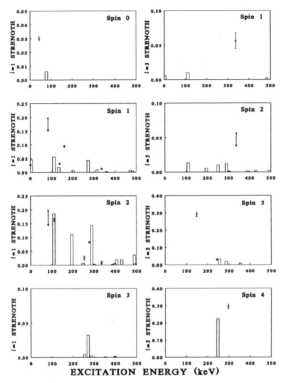

Figure 2: Predicted strengths from the IBFFM (bars) and observed (dots) for the ^{193}Ir(d,p) reaction (from ref.[2]).

3. Discussion

In Fig.1, experimental strength distributions[1] for $l = 1$ and $l = 3$ transfer for the ^{193}Ir(d,t) reaction are shown. As can be seen, the predicted $l = 1$ (d,t) strength to some of the low-lying states is much greater than observed. The major discrepancies for the $l = 3$ (d,t) strength occur for the 4⁻ state at 66.3 and the 4⁻ or 5⁻ state at 256.8 keV, which have large $l = 3$ strengths, and are predicted to have an energy greater than 400 keV. While the IBFFM cannot correctly predict the energy of states with large $l = 3$ strengths in ^{192}Ir, it reproduces the correct magnitude of total (d,t) $l = 3$ strength to low-lying levels. The calculated ^{193}Ir(d,p) strengths for both $l = 1$ and $l = 3$ transfer are shown in Fig.2 along with the experimental values[2]. There is only one strong $l = 3$ transition predicted, contrary to experiment. The 3⁻ state at 148.7 keV in ^{194}Ir has one of the strongest $l = 3$ transitions populating it, but no 3⁻ states are predicted to have large amounts of $l = 3$ strength in the IBFFM.

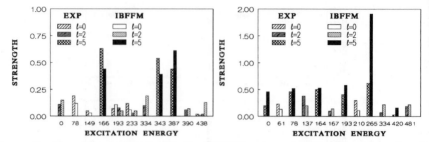

Figure 3: Strengths predicted with the IBFFM compared with observations for the ^{121}Sb(p,d)^{120}Sb reaction (left) and the ^{123}Sb(p,d)^{122}Sb reaction (right) (from ref.[3]).

The discrepancy of the $l = 3$ strength may be related to the exclusion of $f_{\frac{7}{2}}$ neutrons in the calculations. The $f_{\frac{7}{2}}$ states were observed with significant strength in the single-neutron transfer studies on the Pt nuclei[7]. Overall, the agreement between theory and experiment is much better for ^{194}Ir than for ^{192}Ir, and this may be a reflection of the effects of boson truncation, being more serious for ^{192}Ir than for ^{194}Ir, and also that the boson core for ^{194}Ir is a better $SO(6)$ nucleus.

In ^{198}Au, the IBFFM predictions for the relative (d,p) cross sections are in reasonable agreement with the experimental results[6]. It must be noted, however, that the relative cross sections are not as sensitive a test as the absolute strengths.

In the odd-odd Sb isotopes, ^{120,122}Sb, the IBFFM reproduces the spectroscopic strengths rather well (see Fig.3), except for the $l = 5$ strength for a triplet of levels observed at $\simeq$ 265 keV in the ^{123}Sb(p,d)^{122}Sb reaction (greatly overestimated) and to a level at 283 keV observed in the ^{121}Sb(d,p)^{122}Sb reaction (greatly underestimated)[3]. There may be a problem with the experimental results of the latter, however, since the observed strength exceeds the sum rule limit.

4. Conclusions

The IBFFM in some cases seems to provide an adequate description of odd-odd nuclei. Serious problems can easily arise, however, if the model parameters are not fine-tuned for each calculation. Also, the truncation of the boson space, needed in order to reduce the length of the computations, can introduce a bias in the results.

This work was supported in part by the Swiss National Science Foundation.

References
1. P.E. Garrett and D.G. Burke, *Nucl. Phys.* **A568** (1994) 445
2. P.E. Garrett et al., *Nucl. Phys. A* submitted for publication
3. Zs. Dombrádi, S. Brant and V. Paar, *Phys. Rev. C* **47** (1993) 1539
4. S. Brant, V. Paar and D. Vretenar, *Z. Phys.* **A319** (1984) 255
5. J. Kern et al., *Nucl. Phys.* **A534** (1991) 77
6. U. Mayerhofer et al., *Nucl. Phys.* **A492** (1989) 1
7. D.G. Burke and G. Kajrys, *Nucl. Phys.* **A517** (1991) 1

SURFACE PLASMONS OF METAL CLUSTERS IN THE INTERACTING BOSON MODEL

E. Lipparini
*Dipartimento di Fisica dell' Universita' di Trento
and Istituto Nazionale di Fisica Nucleare, I-38050 Povo, Italy*

and

A. Ventura
*ENEA, Dipartimento Innovazione, Settore Fisica Applicata,
Centro Ezio Clementel, I-40138 Bologna, Italy*

ABSTRACT

We study the excitation of surface plasmons in the absorption of visible light by charged silver clusters within the framework of the Interacting Boson Model.

1. Introduction

The excitation of surface dipole plasmons in photoabsorption by medium-size sodium clusters has already been analysed in the frame of the interacting boson model (IBM), by phenomenological simulation of quadrupole surface oscillations with d bosons and of dipole plasmons with p bosons[1].

Aiming at a microscopic foundation of the model, we extend the analysis to charged silver clusters, by investigating whether the experimental photon absorption cross sections[2] are consistent with the following assumptions: (i) s and d bosons are collective electron pairs in the valence shell; (ii) the total dipole strength, calculated in random-phase approximation in the frame of a spherical jellium model, is fragmented by coupling high-energy p bosons, of particle-hole type, to low-energy s-d bosons, of particle-particle, or hole-hole type.

2. Model and Calculations

The model Hamiltonian describing the excitation of dipole plasmons on the surface of a cluster and their interaction with low-energy modes of the cloud of valence electrons is written in the form

$$H = H_B + H_P + V_{B,P} , \qquad (1)$$

where H_B describes the low-energy modes of the open-shell cluster, H_P the plasma resonance and $V_{B,P}$ the interaction between the plasma resonance and the low-energy modes.

On the assumption that the observed odd-even staggering of ionization potentials of neutral Ag_N clusters[3], $\Delta \simeq 0.5$ eV, is due to an attractive residual interaction between valence electrons, the correlation is stronger for $L=0$ and $L=2$ pairs, in analogy with the nuclear case, and the boson images of these pairs interact according to the simplified IBM Hamiltonian

$$H_B = E_0 + \varepsilon \hat{n}_d + k \hat{Q} \cdot \hat{Q} \quad , \tag{2a}$$

$$\hat{n}_d = d^+ \cdot \tilde{d} \quad , \tag{2b}$$

$$\hat{Q}_v = [d^+ \times \tilde{s} + s^+ \times \tilde{d}]_v^{(2)} + \chi [d^+ \times \tilde{d}]_v^{(2)} \quad , \quad (v = -2, \ldots, +2) \tag{2c}$$

where $\tilde{d}_\mu = (-1)^\mu d_\mu$ ($\mu = -2, \ldots, +2$), $\tilde{s} = s$ and $n_s + n_d \equiv n_B = n/2$ is the total number of electron pairs.

In eq. (2a) E_0 is a c-number, such that the ground state has zero energy, and the coefficients ε, k and χ can be obtained, in principle, from knowledge of single-particle energies and strengths of effective residual interactions of the electrons in the valence shell.

The Hamiltonian describing the dipole states is written in terms of p bosons, with $L^\pi = 1^-$, as

$$H_P = \Sigma_i \varepsilon_{p_i} \hat{n}_{p_i} \quad . \tag{3}$$

While the s- and d degrees of freedom are boson images of particle pairs, or hole pairs in a valence shell, the p bosons represent particle-hole excitations across shell closures. That is why the energy, ε_{p_i}, and the number, n_{p_i}, of the p bosons of ith type are determined by performing an RPA calculation in the open-shell clusters in the frame of the spherical jellium model.

Finally, the Hamiltonian describing the coupling of s-d bosons and p bosons has been taken in the form of a quadrupole-quadrupole interaction

$$V_{BP} = k' \Sigma_i [\hat{Q} \cdot (p_i^+ \times \tilde{p}_i)^{(2)}] \quad , \tag{4}$$

where $\tilde{p}_{i\mu} = (-1)^{1+\mu} p_{i-\mu}$. This coupling is the leading term appropriate in the limit of small amplitude oscillations.

Diagonalization of the Hamiltonian provides us with energies, E_n, and wave functions $|1_i^-\rangle$ of the dipole states. The reduced matrix elements of the electric dipole operator yield the transition probabilities between ground state, $|0_1^+\rangle$, and dipole states, to be inserted in the formula of generalized polarizability at frequency ω, from which the photoabsorption cross section is obtained through the optical theorem[1].

The width of the dipole states is assumed to depend on $E = \hbar \omega$ according to the power law

$$\Gamma(E) = \gamma \left(\frac{E}{\varepsilon_p}\right)^\delta \quad , \tag{5}$$

where $\gamma \simeq 0.4 - 0.5$ eV and $\delta \simeq 2$ are adjusted to the experimental photoabsorption profile. The increase of Γ with E takes phenomenologically into account the increase with E of the density of states to which the dipole excitations are coupled.

The parameters of the s-d Hamiltonian (2) cannot be uniquely fixed on the basis of photoabsorption data; we have chosen ε and k in such a way that the (unobserved) 2_1^+ state is lower than $2\Delta \simeq 1$ eV by almost an order of magnitude. The p-boson energies of the Hamiltonian (3) have been estimated from the corresponding energies of an RPA calculation in the frame of the spherical jellium model, lowered by about 10 %, in order to take into account the effect of core electrons. The same RPA calculation provides us with the total fraction of energy-weighted sum rule exhausted by the surface plasmons. The k' and χ parameters of the s-d-p interaction (4), as well as the γ and δ parameters of formula (5) have been adjusted to the experimental photoabsorption data. For the open-shell clusters considered in the present work, two p bosons are sufficient for good data fitting, as shown in Fig. 1, while the relevant parameters are listed in table 1.

Table 1. IBM parameters for $N = 11 \div 19$ Ag cluster ions.

Cluster	Ag_{11}^+	Ag_{13}^+	Ag_{15}^+	Ag_{17}^+	Ag_{19}^+
n_B	1	2	3	2	1
ε (eV)	0.01	0.01	0.01	0.01	0.01
k (eV)	-0.05	-0.05	-0.05	-0.05	-0.05
χ	-1.323	-1.000	0.000	0.000	0.100
ε_{p1} (eV)	3.80	3.80	3.80	3.75	3.85
S_1 (%EWSR)	82.0	81.0	81.0	81.0	82.0
ε_{p2} (eV)	5.50	5.40	5.40	5.30	5.50
S_2 (%EWSR)	8.0	9.0	10.0	10.0	9.6
k' (eV)	0.27	0.16	0.16	0.16	0.18
γ_1 (eV)	0.408	0.408	0.408	0.462	0.544
δ_1	2	2	2	2	2
γ_2 (eV)	0.408	0.408	0.408	0.462	0.544
δ_2	2	2	2	2	2

Summing up, the photoabsorption data of silver cluster ions appear to be consistent with the hypothesis of a correlation between valence electrons, possibly due to the phonon field of the ions, which deserves further investigation.

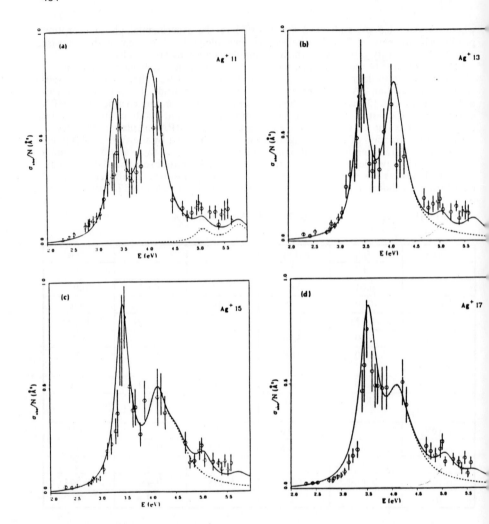

Fig.1. Photoabsorption cross sections of silver cluster ions ($Å^2$), divided by the number of valence electrons and compared with experimental data[2].

3. References

1. F. Iachello, E. Lipparini and A. Ventura, Phys. Rev. **B45**(1992) 4431.
2. J. Tiggesbäumker, L. Köller, H. O. Lutz and K. H. Meiwes-Broer, Chem. Phys. Lett. **190** (1992) 42.
3. G. Alameddin, J. Hunter, D. Cameron, and M. M. Kappes, Chem. Phys. Lett. **192** (1992) 122.

EFFECTIVE E1–OPERATOR IN IBFM

D. N. DOJNIKOV

Cyclotron Laboratory, Ioffe Physical–Technical Institute,
194021 St.–Petersburg, Russia

and

V. M. MIKHAJLOV

Physical Institute of St.–Petersburg State University,
198904 St.–Petersburg, Russia

ABSTRACT

A quasiparticle version of the microscopic foundation of IBFM is presented where pair quasiparticle quadrupole phonons are mapped onto the d–bosons. New terms simultaneously changing fermion and boson states are introduced in effective E2– and M1–transition operators. The E1–operator in IBFM is constructed from analogous terms. Calculations are performed for transitional nuclei of Se and Kr.

The investigation of E1–transitions between lowlying states in odd nuclei can give additional information concerning structure of these states. Indeed, E1–transitions among single–particle states inside valence shells are strongly forbidden, therefore, their probabilities are entirely caused by small components of wave functions involving single–particle states far from the Fermi surface.

In transitional nuclei which are the subject of our presentation the calculation of such small admixtures becomes especially complicated because of many quasiparticle structure of states under consideration containing an odd fermion and several pairs which can be represented as phonons. Some properties of these states such as energies, probabilities of E2– and M1–transitions are rather successfully described by IBFM[1] the space of which consists of s– and d–bosons and an ideal fermion inside valence shells. In order to explain the existence of E1–transitions between states embraced by IBFM one has to apply for the microscopic foundation of the model and take into account fermion structure of bosons and a wider fermion basis including high energy quasiparticles and various types of pair–fermion excitations which do not explicitly enter into the IBFM–space, that can be performed in the framework of a general microscopic approach a version of which was developed in our works[2].

In transitional nuclei with $A \sim 70 \div 80$ for which our calculation were performed the pairing correlations play an important role. The most appropriate way to take them into account is the quasiparticle formalism which was first employed for the foundation of IBM in ref.[3]. Therefore, in our approach the basic part of the full quasiparticle space consists of $a_i^+|c\rangle_F$ (a^+, a are Bogolubov quasiparticles) i.e. includes quasiparticle state inside valence shells (i) and collective excitations $|c\rangle_F$, which are build up from the most collective low energy quadrupole phonons of the Tamm-Dankoff type D_μ^+, $(\mu = 0, \pm 1, \pm 2)$

$$D_\mu^+ = \frac{1}{\sqrt{2}} \sum_{12} \psi_{12}(a_1^+ a_2^+)_\mu^{(2)}; \quad D_\mu^+ \to d_\mu^+ \sqrt{\Omega - \hat{n}_d} \to \sqrt{\Omega} d_\mu^+ s; \quad (1)$$

Extrabasic states involve functions $a_k^+|c\rangle_F$ where k labels particle states outside valence shells and $a_j^+ B_n^+|c\rangle_F$ where B_n^+ is any phonon with the exception of D's. If the interaction between quasiparticles is absent the basic and extrabasic states are separated by an energy interval which is not less then 2Δ (Δ is the pairing gap). Approximately treating the Pauli exclusion principle we suppose that extrabasic functions are orthogonal to each other and to basic functions while the orthogonality of the basic ones is considered especially.

In order to obtain the effective Hamiltonian (H_{eff}) and transition operators in the basic space we use a simplified version of Feschbach method. For this purpose we omit the interaction between quasiparticles in extrabasic states. In this approximation operators can be expressed in terms of the initial fermion operators and series over intermediate extrabasic states.

The exact calculation of elements of H_{eff} is a complicated task as collective states $|c\rangle_F$ can involve a number of the D–phonons. In order to approximately calculate $_F\langle c'|a_f H_{\text{eff}} a_i^+|c\rangle_F$ we use the procedure[4] for mapping the operator $a_f H_{\text{eff}} a_i^+$ with even number of fermions onto the boson space. Then we cut off the boson series retaining only the d–bosons and replacing $|c\rangle_F$ by boson states $|c\rangle$.

Coefficients appearing at the transformation of $a_f H_{\text{eff}} a_i^+$ depend on the initial (i) and final (f) states. In order to formally take off this dependence we introduce the ideal fermion operators (α^+, α) commuting with boson ones. Thus, for a boson operators $\hat{m}(i,f)$ we have

$$\langle c'|\hat{m}(i,f)|c\rangle \equiv \langle c'|\alpha_f \sum_{12} m(1,2)\alpha_1^+ \alpha_2 \alpha_i^+|c\rangle, \quad \langle c'|\alpha_f \alpha_i^+|c\rangle = \delta_{i,f}\delta_{c,c'}. \quad (2)$$

The transition to the boson Hamiltonian was considered for negative parity states in even nuclei after reducing fermion operators to the normal product in ref.[4]. For odd system we have found it more convenient in some cases to expand $a^+ a$ in bosons independently. Such a trick turns out to be of use for the more correct treatment of Pauli principle revealing itself in the nonorthogonality of the basic states:

$$_F\langle c'|a_f a_i^+|c\rangle_F = \langle c'|\alpha_f \hat{N}\alpha_i^+|c\rangle, \hat{N} = 1 - \sum_{L12} N'_{L12}(\alpha_s^+ \bar{\alpha}_t)^{(L)} \cdot (d^+ d)^{(L)}, N' \sim (\psi)^2 \quad (3)$$

Our microscopic approach results in a more complicated form of operators in comparison with their usual form in IBFM. Most important alterations take place for transitions operators. In addition to standard IBFM terms in E2– and M1–operators we include components which can simultaneously change fermion and boson states[2]. Thus they now contain $\alpha^+\alpha d^+ s$, $\alpha^+\alpha d^+ d$, $\alpha^+\alpha d^+ d^+ ss$ and h.c. terms.

The microscopic calculation of parameters of boson–fermion interaction and E2–, M1–transition operators[2] gives results that are in quite reasonable agreement with experimental data. The calculation is performed with factorized forces[4]. Their strength constants are close to Bohr–Mottelson estimations. As an example in the figure the probabilities of E2($L \to L - 2$) and M1($L \to L - 1$) transitions between positive parity states in ^{77}Kr are represented.

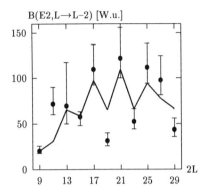
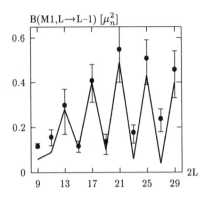

An analogous approach is employed for the establishment of the effective E1-operator inside the basic space. It does not contain collective and single-particle parts as it takes place for E2- and M1-operators.

$$\hat{T}_\mu(E1) = \sum_{\lambda st} \left\{ (\alpha_s^+ \alpha_t)^{(\lambda)} \left[q_1 s^+ d + \sum_\kappa (q_2^\kappa (d^+ d)^{(\kappa)} + q_3^\kappa (d^+ d^+)^{(\kappa)}) \right] + h.c.\&t.r. \right\}, \quad (4)$$

$q_i \equiv q_i(s, t, \lambda)$, h.c.&t.r. is hermitian conjugation and time reversal.

A great number of microscopic processes is determined parameters q_i. Among them we take firstly into account E1-transitions caused by those small components of the D-phonon which include one of quasiparticles on a level outside valence shells. Then we consider the appearance of an odd quasiparticle in intermediate states on a level of neighbour shells. This process is accompanied by the creation or annihilation of D-phonon. It can proceed also with scattering the D-phonon. At last a contribution in parameters of Eq. (4) is brought by processes leading to intermediate states with a B-phonon. All calculations have been carried out with standard effective charges $e_\nu^* = -(Z/A)e$, $e_\pi^* = (N/A)e$. The order of magnitude of q-parameters ($10^{-2} A^{1/6} e \times fm$) calculated is caused mainly by large energy denominators $\sim \hbar\omega = (41/A^{1/3})$ MeV both in ψ-amplitudes of the D-phonons and in admixtures of B-phonons.

Our theoretical and experimental values of B(E1) (in 10^{-5}W.u.) for ^{77}Se and ^{79}Kr are compared in the table. One can see that the theoretical results reproduce not only the order of magnitude of B(E1)$_{\rm exp}$ but also the tendency in their change from ^{77}Se to ^{79}Kr. We believe that the agreement obtained is quite reasonable particularly in view of the fact that we use no special parameters to fit B(E1).

	^{77}Se				^{79}Kr		
L_i^π	L_f^π	$B(E1)$ exp[5]	th	L_i^π	L_f^π	$B(E1)$ exp[6]	th
$5/2_1^-$	$7/2_1^+$	$1.74(3)$	2.57	$5/2_1^+$	$5/2_1^-$	12_{-2}^{+3}	1.8
$7/2_1^-$	$5/2_1^+$	$0.54_{-0.20}^{+0.29}$	0.62	$7/2_1^-$	$5/2_1^+$	$2.8_{-0.5}^{+0.8}$	5.3
$7/2_1^-$	$9/2_1^+$	$0.79_{-0.23}^{+0.27}$	0.20	$7/2_1^-$	$7/2_1^+$	$5.9_{-1.0}^{+1.5}$	3.3
$7/2_1^-$	$7/2_1^+$	$2.2_{-0.6}^{+0.7}$	1.76	$7/2_1^-$	$9/2_1^+$	$2.1_{-0.4}^{+0.6}$	3.8
$9/2_1^-$	$9/2_1^+$	$2.5_{-1.3}^{+2.9}$	0.22	$9/2_1^-$	$7/2_1^+$	$2.6_{-0.6}^{+1.4}$	1.3
$9/2_1^-$	$7/2_1^+$	$5.2_{-1.9}^{+3.3}$	3.72	$11/2_1^-$	$9/2_1^+$	$3.1_{-0.6}^{+1.0}$	1.9
$5/2_2^-$	$7/2_1^+$	$0.54_{-0.06}^{+0.07}$	0.25	$13/2_1^-$	$11/2_1^+$	$5.2_{-2.6}^{+5.1}$	2.7
$7/2_2^-$	$9/2_1^+$	$0.13_{-0.06}^{+0.13}$	0.95	$15/2_1^-$	$13/2_1^+$	$4.0_{-0.7}^{+1.1}$	1.7
$5/2_3^-$	$5/2_1^+$	$5.1_{-2.0}^{+4.2}$	4.61	$17/2_1^-$	$15/2_1^+$	$6.2_{-1.3}^{+2.0}$	4.4
$5/2_3^-$	$7/2_1^+$	$5.0_{-1.8}^{+3.7}$	1.89	$19/2_1^-$	$17/2_1^+$	$4.8_{-1.1}^{+2.1}$	4.0
$13/2_1^-$	$11/2_1^+$	13_{-4}^{+5}	5.16				

In conclusion we developed a quasiparticle version of the microscopic approach for the foundation IBFM and calculated parameters of the model Hamiltonian and E2−, M1−transition operators. We constructed an effective E1−transition operator acting inside the basic space and simultaneously changing fermion and boson states that allows to find $B(E1)$ without a special fit of parameters.

The authors wish to thank I. Lemberg and A. Pasternak for valuable discussions. The work was partially supported by the International Science Foundation.

References

1. *Interacting Bose–Fermi Systems in Nuclei,* ed. F. Iachello (Plenum Press. N.Y. and London, 1981).
2. D. N. Dojnikov and V. M. Mikhajlov, *Izvestiya Akad. Nauk. ser. fiz.*[1] **57** (1993) n.1, p.80, ibid n.9, p.180.
3. F. Dönau, D. Janssen, R. V. Jolos, *Nucl. Phys.* **A224** (1974) 93.
4. D. N. Dojnikov et al., *Nucl. Phys.* **A531** (1991) 326.
5. M. F. Kudojarov et al., in *Proc. of Intern. Conf. on Nucl. Spectroscopy and Nucl. Structure* (Nauka, Leningrad, 1990), p.60, 61.
6. R. Schwengner et al., *Nucl. Phys.* **A509** (1990) 550.

---

[1]The English translation of Izvestiya Akad. Nauk (ser.fiz.) is available as Bull. Acad. Sci. (phys.ser). Volume numbers are the same as those in the Russian edition.

STRUCTURE OF ODD-ODD Sb NUCLEI

ZS. DOMBRÁDI, T. FÉNYES, Z. GÁCSI and J. GULYÁS
Institute of Nuclear Research, H-4001, Debrecen, P.O.Box 51, Hungary

S. BRANT and V. PAAR
Department of Physics, University of Zagreb, 41000 Zagreb, Croatia

The low-lying states of the odd-odd $^{114-118}$Sb nuclei have been investigated through (p,nγ) and (α,nγ) reactions using gamma and electron spectroscopic methods. Nuclear structure calculations were performed for the $^{114-124}$Sb nuclei within the framework of the interacting boson-fermion-fermion model. The experimentally observed states were classified into proton-neutron multiplets on the basis of the measured and calculated level energies, electromagnetic properties and spectroscopic factors. From the splitting of the $\pi d_{5/2}\nu h_{11/2}$ and $\pi g_{7/2}\nu h_{11/2}$ multiplets, as well as of the $\pi d_{5/2}\nu d_{3/2}$ and the $\pi g_{7/2}\nu d_{3/2}$ multiplets we determined how the state of the single proton affects the occupation probabilities of the neutron orbits.

1. Experimental methods

In the last few years we studied the structure of odd-odd Sb nuclei.[1-4] The method of the experimental work was in-beam spectroscopy using the proton and α particle beams of the Debrecen cyclotron. In order to obtain complete spectroscopic information on odd-odd antimony nuclei γ-ray, $\gamma\gamma$-coincidence and internal conversion electron spectra as well as γ-ray angular distributions and relative cross sections were measured at different bombarding energies. New levels, spins, parities, γ-branching and mixing ratios have been determined. For spin determination we used three different methods: Hauser Feshbach analysis of the (p,n) reaction cross sections, analysis of the internal conversion coefficients of the transitions and of the γ-ray angular distributions. The large amount of the new experimental data made the reanalysis[1-6] of the structure of odd-odd Sb nuclei meaningful. In contrast to the previous attempts[7-9], we used the interacting-boson-fermion-fermion model (IBFFM)[10] for the interpretation of the experimental data, and calculated a whole chain of isotopes.

2. Proton-neutron multiplet states

In order to determine the dominant component of the wave functions of the low-lying states in odd-odd Sb nuclei we calculated the level energies, electromagnetic properties and spectroscopic factors in the framework of IBFFM. The model parameters used were close to the parameters determined from the neighbouring odd tin and antimony nuclei. The boson core was approximated with the SU(5) limit of IBM and the d-boson energy was taken form the corresponding even-even tin isotope.

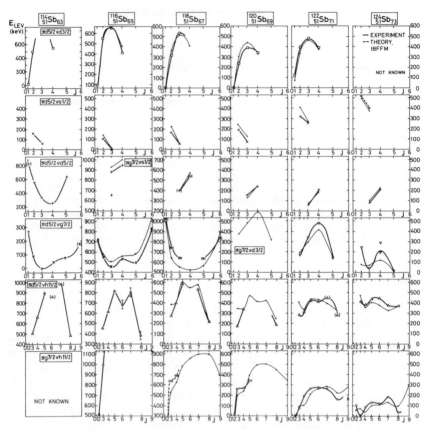

Fig. 1. Systematics of the energy splitting of the proton-neutron multiplets in $^{114-124}$Sb nuclei.[1-6] The abscissa is scaled according to $J(J+1)$, where J is the spin of the state.

After fine tuning of the quasiparticle energies the position of the states could be predicted with about 100 keV uncertainty. The close lying states of the same spin were distinguished on the basis of their decay properties, as the strong ($\approx$1 W.u.) M1 transitions connecting the $\Delta J = 1$ members of the multiplets were observed in most cases. Weaker M1 transitions helped to determine the mixing of the states. The transition rates were described with about 0.05 W.u. reliability, the magnetic moments were reproduced with a few percent accuracy, while the spectroscopic factors agreed with the experimental value within a factor of three. We could make new assignments for more than 50 states in the $^{114-124}$Sb nuclei on the basis of the theoretical analysis, which are shown in Fig.1.

Proton-neutron multiplet-like components were the most significant ones in the wave functions of the low-lying states, typically with 50-60% weight. 35-45%

was fragmented over the one d-boson states and an additional 4–8% over the two d-boson states. The wave functions were based mainly on a single multiplet, suggesting that even a diagonal quasiparticle description should give a reasonable result. In spite of this, a striking deviation from the quasiparticle predictions was found.

The parabolic like splitting of the $\pi g_{7/2}^{-1}\nu h_{11/2}$ and $\pi d_{5/2}^{-1}\nu h_{11/2}$ multiplets is not inverted through smoothing out of the splitting, as expected from the quasiparticle model, but through the kinking of the ends of the parabola. This kind of inversion is well described by the IBFFM as a result of the neutron-core exchange interaction.[6]

3. Rearrangement of the core nucleons

According to Federman and Pittel,[12] as a result of the proton-neutron correlation, when protons start to occupy an orbit, they pull valence neutrons onto the strongly overlapping orbitals, too, resulting in rearrangement of the valence neutrons.

Studying the odd-odd nuclei we found that the energy splitting of the proton-neutron multiplets strongly depends on the occupation probabilities. When a particle or a hole is coupled to a particle, the splitting of the multiplets can be approximated with an open down or open up parabola[11], respectively, while when a quasiparticle in a nearly half-filled orbit is coupled to the particle, the multiplet splitting has a significant fourth order component, too.[6] Using this V^2 dependency, the occupation probabilities can be determined from the multiplet splittings.

Numerous, or all of the members of the $\pi d_{5/2}\nu h_{11/2}$ and $\pi g_{7/2}\nu h_{11/2}$ multiplets, as well of the $\pi d_{5/2}\nu d_{3/2}$ and the $\pi g_{7/2}\nu d_{3/2}$ multiplets are known in ^{120,122,124}Sb. The energy splitting of these multiplets can be used to investigate the change of the occupation probabilities of the $h_{11/2}$ and $d_{3/2}$ neutron states, depending on whether the proton is in the $d_{5/2}$ or $g_{7/2}$ orbit. In this case the core is the same, and the neutron state is the same, so the uncertainty of the parameters describing the core and the neutron-core interaction will not contribute to the uncertainty of the change of occupation probabilities. The proton-core interaction is a simple quadrupole-quadrupole interaction, where the ℓ dependence of the quadrupole moment of a single particle is well known. The change of the quadrupole field of the core due to a normal excitation of a single valence proton is expected to be negligible, so the only parameter responsible for the change of the splitting of the above multiplets is the V^2 value.

We have fitted the occupation probability of the neutron states having the proton either in $d_{5/2}$ or in $g_{7/2}$ state, while fixing all the other parameters of the IBFFM at some reasonable value. The results of the fitting are shown in Fig. 2. for the case of ^{122}Sb. It is clearly seen that when the proton is in $d_{5/2}$ state, then the occupation of the $\nu d_{3/2}$ state gets larger, and when the proton is in $g_{7/2}$ state, the $\nu h_{11/2}$ state, overlapping more with the $g_{7/2}$ state, gets more occupied. The latter situation was observed also in ^{120}Sb and ^{124}Sb nuclei. The difference in the occupation probabilities was $\Delta V^2_{h_{11/2}} \approx 0.07$ in those cases, too.

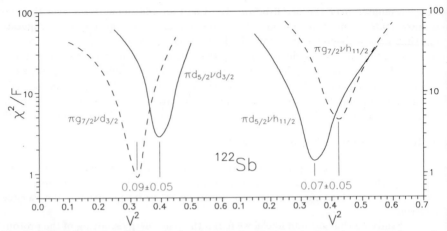

Fig. 2. IBFFM fit to the splitting of $\pi d_{5/2}\nu h_{11/2}$, $\pi g_{7/2}\nu h_{11/2}$ and $\pi d_{5/2}\nu d_{3/2}$, $\pi g_{7/2}\nu d_{3/2}$ multiplets of ^{122}Sb as a function of the neutron occupation probabilities. The experimental data are from Ref.[13], the parameters of the IBFFM from Ref.[6]

This work was supported in part by the Hungarian Scientific Research Foundation (OTKA), Contract No. 3004.

References

1. Z. Gácsi, T. Fényes, Zs. Dombrádi, Phys.Rev. C **44** (1991) 626
2. Z. Gácsi, Zs. Dombrádi, T. Fényes, S. Brant and V. Paar, Phys. Rev. C **44** (1991) 642
3. J. Gulyás, T. Fényes, M. F. F. M. Hassan and Zs. Dombrádi, Phys. Rev. C **46** (1992) 1218
4. Z. Gácsi and Zs. Dombrádi, Phys. Rev. C, submitted for publication
5. T. Fényes and Zs. Dombrádi, Phys. Lett. **B 275** (1992) 7
6. Zs. Dombrádi, S. Brant and V. Paar, Phys. Rev. C **47** (1993) 1539
7. V. L. Alexeev, B. A. Emelianov, D. M. Kaminker, Yu. L. Khazov, I. A. Kondurov, Yu. E. Loginov, V. L. Rumiantsev, S. L. Sakharov and A. I. Smirnov, Nucl. Phys. **A262** (1976) 19
8. W. F. Van Gunsteren and D. Rabenstein, Z. Phys. A **282** (1977) 55
9. S. A. Artamonov and V. I. Isakov, Izv. Akad. Nauk SSSR, Ser. Fiz. **43** (1979) 2071
10. V. Paar, in *In-Beam Nuclear Spectroscopy*, eds.: Zs. Dombrádi and T. Fényes, (Akadémiai Kiadó, Budapest, 1984), Vol. 2, p. 675
11. V. Paar, Nucl. Phys. **A331**, 16 (1979).
12. F. Federman and S. Pittel, Phys. Rev. C **20** (1978) 820
13. V. L. Alexeev, B. A. Emelianov, A. I. Egorov, L. P. Kabina, D. M. Kaminker, Yu. L. Khazov, I. A. Kondurov, E. K. Leushkin, Yu. E. Loginov, V. V. Martynov, V. L. Rumiantsev, S. L. Sakharov, P. A. Sushkov, H. G. Börner, W. F. Davidson, J. A. Pinston and K. Schreckenbach, Nucl. Phys. **A297** (1978) 373

473

ROTATIONAL BANDS IN THE sp(6) ⊃ u(3) ALGEBRA
OF THE FERMION DYNAMICAL SYMMETRY MODEL

Sudha R. Swaminathan and K. T. Hecht
*Department of Physics, The University of Michigan, 500 E. University Drive
Ann Arbor, Michigan 48109, U.S.A*

ABSTRACT

An overview of recent work on the sp(6) ⊃ u(3) algebra of the fermion dynamical symmetry model (FDSM) is presented. An assessment of the FDSM-SU(3) symmetry in the individual proton and neutron spaces is made. The special conditions under which rotational spectra are seen for the coupled proton-neutron system in the strong coupling limit are discussed.

The interacting boson model (IBM) exemplifies the successful application of a mathematically-dictated truncation of the shell model basis for describing nuclear spectra throughout the periodic table.[1] Much research has also been devoted towards unearthing the exact mapping from the shell model fermion pairs of even nuclei coupled to a total angular momentum of $J = 0$ and $J = 2$, to the s and d bosons of the IBM. The fermion dynamical symmetry model (FDSM), whose development was motivated by such studies, is based on the favoured S ($J = 0$) and D ($J = 2$) fermion-pair algebra.[2] The phenomenological successes of the FDSM are not entirely unexpected, given the similarity of the various group chains that arise in the FDSM to those present in the IBM. The question is whether the construction of a pure fermion favoured-pair basis in the FDSM has simplified the renormalisation procedure necessary for shell model effective interactions to be used in the model space. Our objective has been to assess the microscopic validity of the SU(3) limit for the Sp(6) ⊃ U(3) group chain of the FDSM. This group chain is the appropriate one for studying valence nucleons in the actinide region. We present a brief overview of some of our recent results and conclusions.

In actual applications of the FDSM, the shell model space for even nuclei is truncated to the generalised seniority $u = 0$ subspace. The model Hamiltonian is built purely from group generators of the sp(6) algebra. These are the S, D favoured-pair creation operators, and the one-body operators P of ranks 0, 1, and 2, which generate the u(3) sub-algebra. The assumption made in the FDSM is that the lowest-lying states of even nuclei are states with the lowest generalised seniority, $u = 0$, and the largest SU(3) quantum numbers allowed in the model. Our calculations have tested this assumption by using modified FDSM Hamiltonians, which include the microscopically based quadrupole-quadrupole interaction built from the real quadrupole moment operator Q. In addition to being a semi-realistic interaction for studying low-lying rotational spectra of deformed actinide nuclei, it provides a quantifiable assessment of generalised-seniority mixing in the FDSM.

This is because Q, which lies outside the group algebra, mixes different u states, unlike the rank 2 generator P of the FDSM, which cannot. Analytical expressions for the matrix elements of Q have been derived by vector coherent state theory,[3] making such tests of the model feasible.

Actinide nuclei with many active protons and neutrons filling the $N = 5$ and $N = 6$ shells respectively, exhibit beautifully rotational spectra. Both protons and neutrons are classified with $k = 1$ in the k-active $\mathrm{sp}(6) \supset \mathrm{u}(3)$ algebra of the FDSM. Matrix elements of the quadrupole moment operator in this algebra are written in terms of easily available SU(3) coefficients. The goodness of the FDSM-SU(3) symmetry can now be subjected to the same test as that of Elliott's SU(3) symmetry and its pseudo-SU(3) analogue, namely that the dominance of a strong $Q \cdot Q$ interaction should automatically lead to observed low-lying rotational spectra.

With this aim, a complete calculation was performed separately for ten valence protons and fourteen valence neutrons, the configuration with the largest Pauli-allowed FDSM-SU(3) quantum numbers in the actinide region. The Hamiltonian which was diagonalised in the pure $u = 0$ space is

$$\frac{H}{V_0} = -\left[(1 - gr)Q \cdot Q + gr S^\dagger S\right] \quad (1)$$

where V_0 is an unspecified value of the potential introduced to make the Hamiltonian dimensionless, $S^\dagger S$ is the $J = 0$ monopole pairing term, and gr, which ranges from 0.0 to 1.0, is its coupling strength. The interaction in Eq. (1) simulates the traditional Kumar-Baranger interaction,[4] which was used to study the competition between the shell model and the collective rotational model for nuclei with increasing numbers of valence protons and neutrons. By comparing the coupling strengths of the operators in the interaction of Eq. (1) with those used by Kumar and Baranger for actinide nuclei, values of $1 - gr$ equal to 0.11 and 0.13 were obtained for the protons and neutrons respectively. The diagonalisation was performed explicitly for these values of $1 - gr$ as well.

Our first major result demonstrates the existence of a multiplicity of FDSM bases for a particular oscillator shell, two for the $N = 5$ shell and four for the $N = 6$ shell.[5] This multiplicity of phase choices arises in the transformation of the quadrupole moment operator Q from the shell model basis to the coupled k-spin, i-spin basis of the FDSM. The expansion coefficients of Q depend on the relative signs of the radial wave functions of the harmonic oscillator shell model with different values of the orbital angular momentum l. Since the construction of the favoured D pair in the FDSM basis does not depend on l or on the relative signs of the radial wave functions, the overlap between a state vector in the shell model basis with specific values of l and a state vector in the FDSM $k - i$ basis is affected by the phase choice. For calculations performed in a truncated space, the compositions of the eigenstates vary drastically from one phase choice to another, and even the eigenvalues of physical operators such as $Q \cdot Q$ are no longer independent of the phase choices as expected. We urge the founders of the model, who have overlooked this important issue, to address the sensitivity to these phase choices evident in our calculations.

A quadrupole pairing term $D^\dagger D$ has been included in some of our calculations. With $D^\dagger D$ and $S^\dagger S$ terms of equal coupling strengths gr, strong generalised-seniority mixing was observed for values of gr less than 0.6 in a two-particle calculation.[5] With $gr = 0.6$, the lowest $J = 2$ eigenstate in the complete two-identical-particle basis contains only 59% of the $u = 0$ $J = 2$ state for the best possible phase choice for two protons, and only 50% for the best possible phase choice for two neutrons, with other phase choices giving results as low as 15%. Therefore, to assess the nature of the FDSM-SU(3) symmetry, the pure $S^\dagger S$ plus $Q \cdot Q$ interaction of Eq. (1) is chosen.[5]

Since the S and D pairs of the FDSM scheme are identical nucleon pairs, an assessment was first made of the nature of the FDSM-SU(3) scheme in configurations of identical particles only. Calculations were first performed for the $u = 0$ subspace of ten protons with the possible SU(3) irreps (10 0), (62), (24), (40) and (02), and separately for the $u = 0$ subspace of fourteen neutrons with SU(3) irreps (14 0), (10 2), (64), (26), (80), (42), (20) and (04). In both cases, the $u = 0$ subspace contains both large and small SU(3) quantum numbers, so the simple interaction of Eq. (1) is sufficient for determining whether the FDSM-SU(3) symmetry is a good one for collective states. If it is, then a large $Q \cdot Q$ component in Eq. (1) should drive the most collective SU(3) irreps (10 0) and (14 0) to be the dominant components of the lowest eigenstates of the ten-proton and fourteen-neutron systems. Our calculations show that the contribution of the basis state labelled (10 0) is close to 0% for all values of $1-gr$, for both phase choices for the protons.[5] The state labelled (24) was the dominant component of the lowest ten-proton $J = 0$ eigenstate, with a contribution of about 30-45% for values of $1 - gr$ greater than 0.2, for both phase choices.[5] Similarly, the contribution of the state labelled (14 0) was close to 0% for all values of $1 - gr$ and all four phase choices for the neutrons.[5] The state labelled (42) was the dominant component of the lowest fourteen-neutron $J = 0$ eigenstate, with a contribution of about 25-50% for values of $1-gr$ greater than 0.2, for the four different phase choices.[5] The results quoted above indicate that the SU(3) content of these eigenstates is very different from that of Elliott's SU(3) scheme which is known to have a large overlap with shell model eigenstates associated with observed rotational bands. The rotational feature of the FDSM-SU(3) symmetry for both protons and neutrons of actinide nuclei is thus subject to question.

Our next calculation incorporated the co-operative effects of both protons and neutrons, which is necessary for studying quadrupole collectivity in nuclei. We diagonalised the interaction of Eq. (1) which included a proton-neutron quadrupole interaction in a strongly coupled proton-neutron basis. The smallness of the overlap between the truncated FDSM basis and the full shell model basis for n-identical particle calculations necessitates a complicated renormalisation of shell model effective interactions in the individual proton and neutron spaces. Assuming that such a renormalisation can be achieved, we forced the basis to include only the pure proton (10 0) and the pure neutron (14 0) states as its SU(3) irreps. The question now is whether a strong proton-neutron $Q \cdot Q$ interaction singles out the largest possible irrep in the coupled proton-neutron system and leads to rotational bands, with the ground state band dominated by the SU(3) irrep (24 0). The answer is yes, provided the phase choice made in the FDSM is the favoured one out of the eight possibilities. In this favoured basis, a proton-neutron $Q \cdot Q$ interaction with coupling strength $1 - gr$ greater than 0.2 leads to rotational spectra for the lowest band (24 0), followed by those for the (22 1) and (20 2) bands, as expected for an Elliott collective SU(3) structure.[5] The purity of the (24 0) $J = 0$ state for this case, for example, is greater than 99%.[5] This calculation dramatically illustrates our earlier contention regarding the significance of the phase choices in an extremely truncated space, because the rotational band structure for all J values in the strong coupling limit arises only for one of the eight possible basis choices.

An extension of the calculation in the coupled proton-neutron basis was performed for $J = 0$, including all the five $u = 0$ SU(3) irreps for the protons, and all eight $u = 0$ SU(3) irreps for the neutrons, with the favoured FDSM phase choice. We find that in order for the state with (24 0) to dominate the lowest eigenstate of a $J = 0$ pairing plus $Q \cdot Q$ interaction, the strength of the $Q_\pi \cdot Q_\nu$ term must be at least fifty times that of the $J = 0$ pairing term, the pure proton $Q_\pi \cdot Q_\pi$ term, and the pure neutron $Q_\nu \cdot Q_\nu$ term.[5] (Here π and ν refer to the proton and neutron components.) There is no reasonable justification for enhancing the strength of $Q_\pi \cdot Q_\nu$ relative to $Q_\pi \cdot Q_\pi$ or $Q_\nu \cdot Q_\nu$ by such a large factor for obtaining rotational bands. This result adds to our doubts regarding the applicability of the FDSM in the individual proton and neutron spaces without elaborate renormalisation procedures.

The advantage of the pure proton or pure neutron S and D pairs of the FDSM, originally postulated to gain insights into the s and d boson pairs of the IBM is partially lost. We conclude that the mapping from the shell model fermion space to the FDSM fermion space is just as complicated as the mapping to the IBM boson space.

Acknowledgements

One of us (S.R.S) would like to thank Prof. Jean Krisch at the University of Michigan for her support, and the Director and staff at the Institute of Mathematical Sciences, Madras for their hospitality during the preparation of this manuscript.

References

1. A. Arima and F. Iachello, *Annals of Physics* **99** (1976) 253.
2. C. L. Wu, D. H. Feng, X. G. Chen, J. Q. Chen, and M. W. Guidry, *Phys. Lett.* **168B** (1986) 313 ; *Phys. Rev.* **C36** (1987) 1157.
3. K. T. Hecht and J. Q. Chen, *Nucl. Phys.* **A512** (1990) 365.
4. K. Kumar and M. Baranger, *Nucl. Phys.* **A62** (1965) 113.
5. Sudha R. Swaminathan and K. T. Hecht, *Phys. Rev.* **C**, to be published (1994).

On possible equidistance of some groups of levels in deformed nuclei at excitation energies up to 5 MeV

A.V. Voinov

*Laboratory of Neutron Physics, Joint Institute for Nuclear Research,
Dubna, Moscow region, Russia*

Abstract

The frequency analysis of the energy dependence of two-step γ-cascade intensities for heavy deformed nuclei is performed. The harmonic frequencies are shown to be present with a high probability in many nuclei. The existence of phonon excitations in nuclei at an excitation energy of up to 5 MeV is assumed.

The low energy region of nuclear excitations is studied intensively using various types of nuclear reactions. However, there is still no concerted understanding, and no theoretical description of the levels structure in this region of nuclei excitations exists yet. Evidently it is connected with a lack of experimental possibilities for studying nuclei levels over large energy and spin intervals. The majority of experimental methods using various types of nuclear reactions orients to studying low - lying levels (up to ~ 2 MeV) of nuclei. The levels at higher excitation energies are poorly known. The method of measuring two-step γ-cascades in the thermal neutron beam [1] of the IBR-30 reactor developed in Dubna compensates partly this gap to allow one to identify nuclear levels at energies up to the neutron binding energy $B_n \approx 5 \div 9$ MeV. By using this method a series of new, interesting results was obtained [2].

The two-quanta cascades populating low-lying levels of nuclei up to ~ 1 MeV were observed experimentally. In this case the energy of the intermediate levels populated by the primary transitions changes from 0 to B_n (within the limits of method and apparatus used). The spin values for intermediate levels lie in the interval from 1 to 5 for even-even, and from 1/2 to 3/2 for odd-even nuclei in dependence on the combination of spins of the initial (compound state) and final levels of the cascade. For all cases dipole multipolarity of primary transitions is assumed as the most probable.

In the figure the distributions of cascade intensities for low-lying levels of some deformed nuclei are shown. In these distributions groups of equally spaced intermediate levels of intense cascades can be seen clearly. In the figure possible equidistant level groups are shown by arrows.

For the mathematical analysis of the observed effect an expansion of the cascade intensity distribution in terms of harmonic frequencies (periods of equidistance) was performed by using the Fourier transformation in the form:

$$F(T) = \sum_{k=1}^{n} S_k \cdot \exp(-i \cdot 2\pi \cdot E_k / T)$$

where S_k, E_k are the intensity and the energy of the k-th cascade, n is the number of the cascades in the distribution, T is the period of equidistance. The analysis has shown that for many nuclei one can observe a harmonic components with a large amplitude in the intensity distributions of two-quanta cascades. The figure shows the F(T) function for the corresponding distributions calculated for some even-even and odd-even nuclei. The maxima in the F(T) correspond to the large harmonic component shown in the intensity distributions by arrows.

The method of statistical simulation was used for estimating the statistical confidence level of the observed maxima in the F(T). The estimated parameters were: the peak amplitude and the peak width at half maximum in the F(T). The values of the cascade intensities and energies of intermediate levels were simulated for each experimental intensity distribution. The form of the corresponding statistical distribution was fitted to an experimental one averaged over all measured nuclei.

72 experimental intensity distributions of cascades for 22 even-even and odd-even nuclei have been analysed. In 41 of them the harmonic component was found with a statistical confidence of >95%. This is a strong argument in favour of nonaccidental equidistance effect. The period of the harmonic component changed basically in the range of $T \sim 300 \div 1000$ keV.

Unfortunately there is no theoretical indication of the existence of the equidistance effect for the energy excitations of about several MeV. However, it should be pointed out that the energy of quadrupole phonons for heavy deformed nuclei is in the interval of the observed period T. Therefore, the equidistance effect can be interpreted in terms of the phonon excitations in nuclei (maybe not necessarily quadrupole). If one assumes a weak coupling between the collective and internal motions, a vibrational band can be considered to be built on any of quasiparticle states. Then, if the intensity of a cascade depends on a quasiparticle component of the wave function, there is no dependence on the phonon number in it, and the observation of the effect of equidistance becomes possible.

As is known, the existence of multiphonon excitations in deformed nuclei is a crucial contradiction between the theoretical models. The Quasiparticle - Phonon Nuclear Model claims that the energy centroid of two-phonon states is shifted towards energies larger than 3 MeV, where they lose they collective character. On the other extreme, the Interacting Boson Model admits existence of multiphonon excitations [3] but question about harmonicity or anharmonicity of a vibrational spectra is still open especially for higher excitation energies.

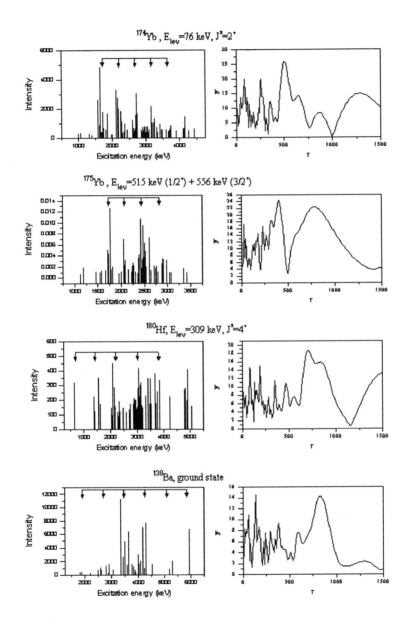

Fig. The cascade intensity distributions for some heavy deformed nuclei and the F(T) dependence. Possible equidistant groups are shown by the arrows.

Of course, the above, being just a qualitative reasoning, calls for a close theoretical examination, and I would like to hope that these results will attract the attention of the theorists.

References

1. A.A. Bogdzel et al., JINR Report No. P15-82-706, Dubna, 1982
2. S.T. Boneva et al., Sov. J. Part. Nucl. 22(2), March-April 1991.
3. R.Piepenbring, in Proceedings of the International Conference on Serlected Topics in Nuclear Structure, Dubna, June 1989, p.76

483

THE NT-DEPENDENCE OF THE IBM-3 HAMILTONIAN

J P ELLIOTT, J A EVANS, G L LONG‡ and V-S LAC

School of Mathematical and Physical Sciences
University of Sussex, Brighton, BN1 9QH, U.K.

ABSTRACT

A seniority mapping from the isospin invariant form of the interacting boson model IBM-3 into the shell-model with isospin leads to a dependence of the IBM-3 hamiltonian on isospin T and boson number N.

Since the early days of the interacting boson model it has been clear that there are two alternative physical foundations for the model. The first relates the bosons to the quadrupole phonon operators in the macroscopic collective model of Bohr and Mottelson[1], built on the five degrees of freedom of a quadrupole nuclear shape. In this interpretation, the boson number N simply determines the truncation point for the maximum number of phonons allowed in the calculation and has no physical significance. There is also no physical significance to the s-boson and the precise relation between the phonon operator $c^\dagger$ and the bosons is

$$d^\dagger s = c^\dagger (N_s)^{1/2} \tag{1}$$

where N_s is the number of s-bosons. In this interpretation the IBM contains no new physics and is nothing more than an algebraic device for calculations in a truncated quadrupole phonon model.

The second interpretation relates the s- and d-bosons to nucleon pairs with $J = 0$ and 2 in the microscopic shell-model. The boson number N is now fixed by $2N = n$ where n is the number of valence nucleons and both bosons have clear physical significance. In this picture the IBM contains new physics through the assumption of the dominance of the s and d pairs and the generation of collective motion microscopically.

The empirical use of the IBM by fitting hamiltonian parameters separately in each nucleus cannot distinguish between these two interpretations. Given the undoubted prevalence of quadrupole shapes in nuclei, it is most natural to conclude that any success of the model is due to the first interpretation. It is therefore important to study the second, microscopic, interpretation to see if any of its

‡ Permanent address : Department of Physics, Tsinghua University, Beijing, China.

consequences, not present in the first, can be tested. These consequences relate to the role of the boson number N and the relation between the boson hamiltonian and the shell- model hamiltonian. In this talk I shall concentrate on the role of both N and the isospin quantum number T.

An important step in this direction was taken[2,3] early in the development of the model by mapping the IBM states into a seniority basis of shell-model states with the equation $v = 2N_d$ to relate the seniority v to the number of d-bosons. In simple terms, the seniority is the number of nucleons not in $J = 0$ pairs. Thus the mapping is carried out according to

$$|s^{N_s} d^{N_d} J\rangle \to (S^\dagger)^{(n-v)/2} |j^v J\rangle \qquad (2)$$

where $S^\dagger = (a^\dagger a^\dagger)^0$ is the $J = 0$ pair state and $N_s = (n - v)/2$. In group-theoretical language, seniority is described by the quasi-spin group SU(2) with generators $S^\dagger$, S and $(\hat{n} - \Omega)/2$ where $\Omega = j + 1/2$. The shell-model interaction V can be analysed into components which belong to definite irreducible representations of SU(2) and the use of SU(2) Wigner coefficients gives closed formulae for the n-dependence of shell-model matrix elements of each component of V, in the seniority basis. By equating corresponding matrix elements, it is possible to deduce a one and two-body boson hamiltonian which reproduces the shell-model matrix elements, up to seniority $v = 4$. The parameters in the boson hamiltonian are then functions of $N = n/2$ although this dependence vanishes for large j. The N-dependence compensates, to some extent, for the absence of the Pauli-principle in the IBM.

The mapping described above refers to a single kind of nucleon, i.e. neutrons only and IBM-1, but the IBM-2 extension[2,3] to neutrons and protons is immediate, with two kinds of boson ν and π and with the mapping carried out separately in neutron and proton spaces. In IBM-2 it is convenient to classify states by their separate permutation symmetries in the orbital (sd) and charge $(\nu\pi)$ spaces, using the partition labels $[N]$, $[N-1,1]$, ... etc. into at most two parts $N = N_1 + N_2$ to describe each space. These partitions describe, at the same time, the U(6) representation in sd-space and the SU(2) or F-spin representation in $\nu\pi$-space through $\text{F}=(N_1 - N_2)/2$.

It can be shown that the products of neutron and proton shell-model states used in the IBM-2 mapping will have good isospin if the valence neutrons and protons occupy different shells, but in lighter nuclei, where the valence nucleons occupy the same shell this is no longer the case. IBM-2 then violates the isospin symmetry which is known to be valid for nuclear forces. Isospin symmetry may however be incorporated into the boson model by introducing a third kind of boson δ which, together with the ν- and π-bosons, forms a $T = 1$ triplet. These three kinds of boson, all with s or d orbital motion, form the building blocks for an isospin invariant version[4] of the boson model IBM-3. It is again convenient to use partition labels to describe the separate orbital and charge spaces but now the partitions may have three parts $N = N_1 + N_2 + N_3$ and the group for the charge space is SU(3). The isospin SU(2) group is a subgroup of SU(3). Although the addition of a third kind of boson increases the number of states, much of the increase is accounted for by the inclusion of all $2T + 1$ members of each multiplet T.

Seniority may again be used to establish a mapping[5] between IBM-3 and the shell-model with isospin. Since isospin is present in both IBM-3 and the shell-model the same label T can be used. In this respect, T is treated like the angular momentum. The quasi-spin group is O(5) with ten generators made up of the three $T = 1$, $J = 0$ pair creation operators $S_q^\dagger$, the corresponding three destruction operators S_q, the three isospin operators T_q and the number operator $\hat{n} - 2\Omega$. The representations of O(5) require two labels (v, t) and reduce in a rather complicated way to a set[6] of n, T combinations. The symbol t denotes the total isospin of the "unpaired" nucleons, usually referred to as the "reduced isospin". The complete set of shell-model states belonging to the representation (v, t) is generated by taking all possible T_p and T in the expression

$$\{(S^\dagger)_{T_p}^{(n-v)/2} |j^v Jt\rangle\}_T \tag{3}$$

where T_p denotes the total isospin of the $T = 1$, $J = 0$ pairs. The structure of the shell-model states Eq.(3) clearly suggests a mapping onto boson states

$$|s_{T_p}^{N_s} d_t^{N_d} JT\rangle \tag{4}$$

where, as before, $n = 2N$, $v = 2N_d$ and $N_s = (n - v)/2$. In addition, T_p is equated to the total isospin of the s-bosons while t is equated to the isospin of the d-bosons.

Although boson states Eq.(4) with different T_p are independent and orthogonal this is not true of the shell-model states Eq.(3). However, some recent developments in coherent state theory[7,8,9] show how to construct an orthonormal basis which is close to the basis Eq.(3) and may still carry the label T_p. This enables the mapping to be made between Eq.(4) and the orthonormal version of Eq.(3). The coherent state method also leads to formulae[9] giving the N and T dependence of shell-model matrix elements in a (v, t) basis and hence from the mapping we can deduce the N and T dependence of the corresponding IBM-3 hamiltonian. Our derivation of the shell-model formulae[9] is equivalent to the use of Wigner coefficients for the group O(5) but is more direct in this case since the necessary Wigner coefficients for O(5) were not known.

Time prevents any significant discussion of the method but the essence is the coherent state

$$e^{z^*\cdot S^\dagger}|j^v Jt\rangle \qquad (5)$$

introducing the complex variable z, which is an isovector, the exponent in Eq.(5) being a scalar product in isospin. Each shell model state $|\psi\rangle$ in the representation (v,t) then has a z-space representative

$$\langle j^v Jt|e^{z\cdot S}|\psi\rangle. \qquad (6)$$

The label T_p in Eq.(3) then acquires orthogonality in z-space. Matrix elements may then be calculated simply in z-space before a rather complicated transformation back into the shell-model.

There is an interesting connection between the label T_p and the boson U(6) label. This is illustrated in the table below for $n = 6$, i.e. $N = 3$. For seniority zero, T_p is redundant since it is the same as T, the boson configuration contains no d-bosons and the U(6) symmetry is totally symmetric. For seniority two, there are two s-bosons so that $T_p = 0$ or 2 giving the T values shown. The SU(3) symmetry (in the ν, π, δ space) must be the same as the U(6) symmetry and, since the SU(3) reductions to SU(2) are [3] $\to T = 1$ and 3, and [21] $\to T = 1$ and 2, it follows that the $T = 3$ and $T = 2$ states have unique U(6) labels. However, for $T = 1$ there are two possible SU(3) labels and two possible T_p labels. In fact there is a precise transformation between these two boson bases, given by the SU(3) Wigner coefficients shown in the table. With this transformation, the mapping enables us to carry the SU(3), or U(6), label into the shell-model. Because of the lack of orthogonality in the shell-model basis Eq.(3), these groups have no precise role in the shell-model but numerical work[5,9] shows that the eigenstates of a typical interaction are much closer to the SU(3) basis than to the T_p basis.

Table 1. An illustration of IBM-3 bases for some states with $N = 3$.

(v,t)	T_p	T	Configuration	U(6)	Transformation
(0,0)	3	3	s^3	[3]	-
	1	1			
(2,1)	2	3	s^2d	[3]	-
		2		[21]	-
	1	1		[3]	$(5/9)^{1/2}|T_p=0\rangle + (4/9)^{1/2}|T_p=2\rangle$
	0	1		[21]	$(4/9)^{1/2}|T_p=0\rangle - (5/9)^{1/2}|T_p=2\rangle$

The most general isoscalar IBM-3 hamiltonian may be written[10] as

$$H = H_0 + \epsilon N_d + H_2 \qquad (7)$$

where H_0 and ϵ are numbers and the two-boson part H_2 is given by

$$H_2 = a_0\big((d^\dagger s^\dagger)_{20} \cdot (\tilde{s}\tilde{d})_{20}\big) + a_1\big((d^\dagger s^\dagger)_{21} \cdot (\tilde{s}\tilde{d})_{21}\big) - x a_0 \hat{N}_s \hat{N}_d$$

$$+ \frac{1}{2} \sum_{L_2 T_2} c_{L_2 T_2}\big((d^\dagger d^\dagger)_{L_2 T_2} \cdot (\tilde{d}\tilde{d})_{L_2 T_2}\big)$$

$$+ \frac{1}{2} \sum_{T_2} b_{T_2}\Big\{ \big((s^\dagger s^\dagger)_{0T_2} \cdot (\tilde{d}\tilde{d})_{0T_2}\big) + \text{h.c}\Big\}$$

$$+ \sqrt{\frac{1}{2}} \sum_{T_2} d_{T_2}\Big\{ \big((d^\dagger s^\dagger)_{2T_2} \cdot (\tilde{d}\tilde{d})_{2T_2}\big) + \text{h.c}\Big\} \qquad (8)$$

where $x = (N-T)(N+T+1)/3N(N-1)$. The subscript T_2 on c takes the values 0 and 2 for $L_2 = 0, 2$ and 4 while $T_2 = 1$ for $L_2 = 1$ and 3. For b and d, $T_2 = 0$ and 2. The dots in Eq.(8) denote scalar products in both angular momentum and isospin. This is the reduced form of H obtained by eliminating four trivial operators such as N_s, which can be written as $N - N_d$. There is some freedom in the manner of elimination. We have done it in such a way that the parameters a_0 and a_1 make no contribution in states of full U(6) symmetry $[N]$. All parameters are allowed to be functions of N and T.

Starting from some shell-model interaction it is now possible to deduce the parameters in the boson hamiltonian Eq.(8) by equating corresponding matrix elements for states with seniority $v \leq 4$ and any N and T. The functional dependence of the parameters on N and T is generally complicated and is best treated numerically although H_0 is a simple linear combination of N, N^2 and $T(T+1)$, a familiar shell-model result. Some approximate simple forms can be discerned in other parameters.

Some results, from ref.[10], are shown in figures 1 to 5 for a typical effective shell-model interaction[11] and a large $j = 21/2$. The physical significance of ϵ is the excitation of the fully symmetric $[N]$ 2^+ state from the configuration $s^{N-1}d$ relative to the s^N state. Each curve is labelled by a value for $N-T$ so that the top curve labelled 0 corresponds to the all-neutron case of IBM-1. It shows the familiar constancy with N. As T decreases ϵ is seen to decrease and to have a greater variation with N. Figures 2 and 3 show the combinations $\bar{c}_L = x c_{L0} + (1-x) c_{L2}$ for $L = 0$ and 2. This is the combination which is relevant in the fully symmetric U(6) state. Again there is little variation with N for the all-neutron case $N = T$ but a strong variation for other T. The behaviour of the two L-values is quite different although $L = 4$ is similar to $L = 2$. Figures 4 and 5 show the matrix elements b_0 and d_0 which are off-diagonal in

N_d. The off-diagonal elements b_2 and d_2 are very small, consistent with small mixing of seniority in the all-neutron case.

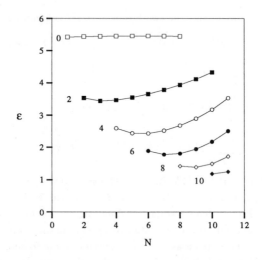

Figure 1. In this figure, and in figures 2-8 IBM-3 parameters are shown as a function of N and T. Each curve is tagged by the value of $N - T$.

The method described above for a single j-shell has been generalised[12] to a system of several j shells and numerical work for the realistic system $j = 3/2$, $1/2$ and $5/2$, relevant to the region between the closed shells at 28 and 50 shows curves similar to those above except for an additional linear N-dependence in some parameters arising from the shell-model single-particle energies.

From a practical point of view, it is not expected that the precise values of the IBM parameters deduced from the mapping would lead to a successful description of data, because of the significant mixing between boson and non-boson states in a shell-model framework. However it might be hoped that the derived N, T dependence, with suitable parameters, could be used in a fit to data. A single set of parameters would then apply to a range of nuclei with different N and T. Such a calculation is under way for the nuclei between the closed shells at 28 and 50 where neutrons and protons are filling the same orbits.

Finally, it is of some interest to compare the IBM-3 results with those from IBM-2. Although IBM-2 states do not have good isospin, the main isospin component of IBM-2 states with boson numbers N_ν and N_π will generally be $T = |N_\nu - N_\pi|$. (There

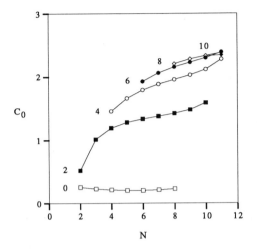

Figure 2. See caption to figure 1.

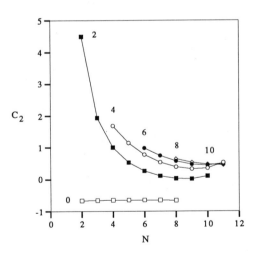

Figure 3. See caption to figure 1.

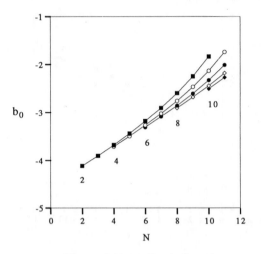

Figure 4. See caption to figure 1.

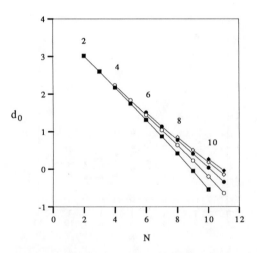

Figure 5. See caption to figure 1.

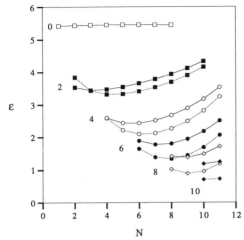

Figure 6. See caption to figure 1. In addition, in this figure and in figures 7-8, the IBM-2 results (dotted lines) are superimposed on the previous IBM-3 results (solid lines).

are a few exceptions to this, such as the mixed symmetry states with $N_\nu = N_\pi = 1$ for which $T = 1$.) For $N_\nu > N_\pi$ this gives the relations

$$N_\nu = (N + T)/2, \qquad N_\pi = (N - T)/2. \qquad (9)$$

It is then possible to use the two separate $SU(2)$ mappings for neutrons and protons to find the dependence of the IBM-2 hamiltonian on N_ν and N_π, and then to transform to an N, T dependence using Eqs.(9). A detailed comparison between full calculations in the two models is difficult to interpret but, in each case we may project onto the IBM-1 subspace of states of full $U(6)$ symmetry $[N]$. We may then start from a given shell-model interaction and proceed to an N, T dependent IBM-1 hamiltonian via either IBM-2 or IBM-3. Note that there are two factors contributing to the N, T dependence in each case. One factor comes from the quasi-spin group and the other from the projection. A comparison of this kind has been carried out[13] for the same shell-model system as used in figures 1 to 5 above. Figures 6 to 8 show the IBM-2 values for some of the IBM-1 projected parameters superimposed, as dotted lines, on the previous IBM-3 results. The curves must be identical for the all-neutron case $T = N$ and they are qualitatively similar for other T. However this is only a very

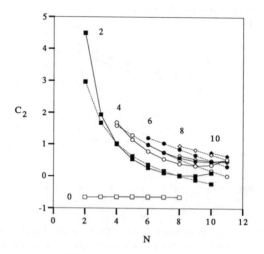

Figure 7. See caption to figure 6.

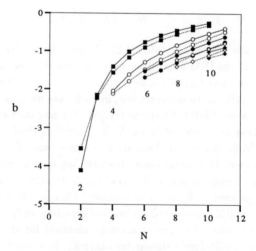

Figure 8. See caption to figure 6.

restricted comparison, in the IBM-1 subspaces and it remains to be seen how the two models differ in a complete calculation with the inclusion of all the states of mixed symmetry. The mixing of isospin in IBM-2 states is not small, for example[14] with $N_\nu = 2$, $N_\pi = 1$, states with F=3/2 have 80%($T = 1$) with 20%($T = 3$) while states with F=1/2 have 50% each of $T = 1$ and $T = 2$.

In conclusion, we have used the neutron-proton quasi-spin group O(5) in the shell-model to deduce the dependence of the IBM-3 hamiltonian on boson number and isospin. This will be used to guide the empirical fitting of the IBM-3 hamiltonian to a range of nuclei between the closed shells at 28 and 50.

References

[1] A. Bohr, *Mat. Fys. Medd. Dan. Vid. Selsk.* **26** No.14 (1952).
[2] A. Arima, T. Otsuka, F. Iachello and I. Talmi, *Phys. Lett.* **B66** (1977) 205.
[3] T. Otsuka, A. Arima and F. Iachello, *Nucl. Phys.* **A309** (1978) 1.
[4] J.P. Elliott and A.P. White, *Phys. Lett.* **B97** (1980) 169.
[5] J.A. Evans, J.P. Elliott and S. Szpikowski, *Nucl. Phys.* **A435** (1985) 317.
[6] K.T. Hecht, *Nucl. Phys.* **A102** (1967) 11.
[7] K.T. Hecht, *The Vector Coherent State Method and its Application to Problems of Higher Symmetries (Lecture Notes in Physics 290)* (Springer,Berlin 1987).
[8] K.T. Hecht and J.P. Elliott, *Nucl. Phys.* **A438** (1985) 29.
[9] J.P. Elliott, J.A. Evans and G.L. Long, *J. Phys.* **A25** (1992) 4633.
[10] J.A. Evans, G.L. Long and J.P.Elliott, *Nucl. Phys.* **A561** (1993) 201.
[11] W.W. Daehnick, *Phys. Reports* **96** (1983) 317.
[12] J.P. Elliott, J.A. Evans, G.L. Long and V-S. Lac, *J. Phys.* **A** (1993) (To be published).
[13] V-S. Lac, J.P. Elliott, J.A. Evans and G.L. Long, (To be published).
[14] J.P.Elliott, *Prog. Part. Nucl. Phys.* **25** (1990) 325.

NUCLEAR COLLECTIVE MOTION WITH ISOSPIN

JOSEPH N. GINOCCHIO
Theoretical Division, Los Alamos National Laboratory
Los Alamos, NM 87545

and

AMIRAM LEVIATAN
Racah Institute of Physics, The Hebrew University,
Jerusalem 91904, Israel

ABSTRACT

We study intrinsic aspects of quadrupole collectivity with conserved isospin in the framework of the interacting boson model (IBM-3) of nuclei. A geometric visualization is achieved by means of a novel type of intrinsic states which are deformed in angular momentum yet have well defined isospin. The energy surface of the general IBM-3 Hamiltonian is derived and the normal modes are identified for prolate deformations.

1. Introduction

The original interacting boson model[1] (IBM-1), based on interacting monopole and quadrupole bosons, was initially interpreted as a quantization of the Bohr-Mottelson quadrupole liquid drop model. However, the geometric connection to a quadrupole shape was not made until an intrinsic state was introduced[2-4]. Later a distinction was made between neutron and proton bosons which were interpreted as correlated pairs of neutrons and pairs of protons[5] (IBM-2). This paved the way for a microscopic interpretation of the boson model by establishing its link with the nuclear shell model. The IBM-1 states were shown to correspond to the subset of states of IBM-2 which are completely symmetric in the neutron and proton degrees of freedom and belong to the highest representation of the F-spin group[6] (an $SU_F(2)$ group of transformations between proton and neutron bosons). Like IBM-1, the connection of IBM-2 to the Bohr-Mottelson-like quadrupole deformations is made by introducing an intrinsic state[7-8] for the boson system. With this intrinsic state, collective motion in the IBM can then be given a geometrical visualization in terms of quadrupole deformation parameters.

IBM-2 is valid for nuclear states of valence protons (neutrons) filling a major shell which is closed with respect to neutrons (protons). This is the situation encountered in heavy nuclei near the valley of stability due to the neutron excess beyond neutron number 82. It follows that the subset of nuclear states which correspond to the IBM-2 states all have good isospin and so for these states the isospin label is trivial and of no interest. However, when the valence neutrons and protons are filling the same major shell such as in light nuclei, then isospin must be introduced[9] as discussed by J. P. Elliott in this Proceedings. To take account of isospin conservation in the framework of boson models, a neutron-proton monopole and quadrupole pair must be included in addition to the proton-proton and neutron-neutron pairs. These pairs are represented in the IBM-3 by three types of monopole $(s_\tau^\dagger)$ and quadrupole $(d_{m,\tau}^\dagger)$ bosons, where $m = (-2,-1,0,1,2)$ is the angular momentum projection (for $L = 2$) and $\tau = (1,0,-1)$ is the isospin projection indicating neutron, proton-neutron, and proton bosons respectively.[9-10] Although IBM-3 was introduced for light nuclei, there are some heavy nuclei which also have neutrons (ν) and protons (π) filling the same major shell, for example, the lighter isotopes of Tellurium, Xenon, Barium, and the isotopes in the $n_\pi \sim n_\nu \sim 40$ region. Furthermore, with the advent of radioactive beams, more nuclei may be studied which fall into this category. Quantitative calculations of energy levels and transition rates by matrix diagonalization in IBM-3 are discussed by M. Sugita in this Proceedings. However, since these heavy nuclei have collective low-lying states, it is illuminating to determine the intrinsic motion which produces collectivity in a similar manner as that was done for IBM-1 and IBM-2.

2. Intrinsic State with Isospin

Because isobaric analog states are at high excitation energy, we want to construct an IBM-3 intrinsic state with a definite isospin for each isospin multiplet.[11-12] The basic ingredient in this construction is a deformed (in angular momentum space) condensate boson for each isospin projection $\tau = (1,0,-1)$,

$$b_{c,\tau}^\dagger = (1+\beta^2)^{-1/2}\left[s_\tau^\dagger + \beta\cos\gamma\, d_{0,\tau}^\dagger + \beta\sin\gamma\frac{1}{\sqrt{2}}\left(d_{2,\tau}^\dagger + d_{-2,\tau}^\dagger\right)\right] \quad (1)$$

The deformed bosons in (1) have the same quadrupole deformations β, γ for all isospin projections. The intrinsic state is then a deformed state (in angular momentum space) of N of these bosons (representing one-half of the total number of valence nucleons) coupled to good isospin T

$$\Psi_c(N, T, T_z = T, \beta, \gamma) = \mathcal{N}_c (b_c^\dagger \cdot b_c^\dagger)^{(N-T)/2}\left(b_{c,1}^\dagger\right)^T |0> \quad (2a)$$

$$\mathcal{N}_c(N,T) = \sqrt{\frac{(2T+1)!!}{(N-T)!!(N+T+1)!!\,T!}} \quad (2b)$$

Here the dot · means scalar product only in isospin space and $|0>$ is the boson vacuum representing the doubly-closed shell. The intrinsic state involves a product of $(N-T)/2$ deformed boson-pairs with isospin zero, and a condensate of T deformed

$b_{c,1}^\dagger$ bosons. Intrinsic states with other isospin projection (T_z) can be obtained by repeated application of the isospin lowering operator on the state in (2).

3. Neutron, Proton, Neutron-Proton Symmetry

In going from IBM-2 to IBM-3, the F-spin $SU_F(2)$ group is enlarged to an $SU_f(3)$ group which classifies the states of IBM-3 with respect to interchange of 3 flavors (neutron, proton, neutron-proton) of bosons. The generators of $SU_f(3)$ are given by

$$G_\tau^{(t)} = \left(s^\dagger \tilde{s}\right)_{0,\tau}^{(0,t)} + \sqrt{5}\left(d^\dagger \tilde{d}\right)_{0,\tau}^{(0,t)}, \quad t = 1,2 \tag{3}$$

where $\tilde{s}_\tau = (-1)^\tau s_{-\tau}$, $\tilde{d}_{m,\tau} = (-1)^{m+\tau} d_{-m,-\tau}$ and the superscript (l,t) means coupling to angular momentum and isospin respectively. These generators are scalars under rotations ; i.e., commute with the angular momentum operators $\hat{L}_m = -\sqrt{30}$ $(d^\dagger \tilde{d})_{m,0}^{(1,0)}$. The $SU_T(2)$ isospin subgroup of $SU_f(3)$ is generated by $\hat{T}_\tau = -\sqrt{2} G_\tau^{(1)}$.

The bosons $s_\tau^\dagger, d_\tau^\dagger,$ and $b_\tau^\dagger$ all transform under the three-dimensional (1,0) irreducible representation of $SU_f(3)$ and the set of intrinsic states in (2) have good (N,T,T_z), $T = N, N-2, ..., 0$ or 1, and observe the highest $SU_f(3)$ symmetry $(N, 0)$. Realistic calculations reported by M. Sugita in this Proceedings confirm the fact that the ground state has almost pure $(N, 0)$ $SU_f(3)$ symmetry. This result parallels the discovery in IBM-2 that the ground state has very good F-spin even though the Hamiltonian is not F-spin invariant[13].

4. Energy Surface with Conserved Isospin

The most general isospin-invariant Hamiltonian with one and two-boson interactions which conserves the total number of bosons, hence is in the enveloping algebra of $U(18)$, is given by $H = H_0 + V$, where

$$H_0 = \epsilon_s s^\dagger : \tilde{s} + \epsilon_d d^\dagger : \tilde{d} \tag{4a}$$

$$V = \sum_{T=0,2} u_0^{(T)} \left(s^\dagger s^\dagger\right)^{(0,T)} : \left(\tilde{s}\tilde{s}\right)^{(0,T)} + \sum_{T=0,1,2} u_2^{(T)} \left(s^\dagger d^\dagger\right)^{(2,T)} : \left(\tilde{d}\tilde{s}\right)^{(2,T)}$$

$$+ \sum_{T=0,2} \left[v_0^{(T)} \left(s^\dagger s^\dagger\right)^{(0,T)} : \left(\tilde{d}\tilde{d}\right)^{(0,T)} + v_2^{(T)} \left(s^\dagger d^\dagger\right)^{(2,T)} : \left(\tilde{d}\tilde{d}\right)^{(2,T)} + \text{H. c.}\right]$$

$$+ \sum_{L=0,2,4}\sum_{T=0,2} c_L^{(T)} \left(d^\dagger d^\dagger\right)^{(L,T)} : \left(\tilde{d}\tilde{d}\right)^{(L,T)} + \sum_{L=1,3} c_L^{(1)} \left(d^\dagger d^\dagger\right)^{(L,1)} : \left(\tilde{d}\tilde{d}\right)^{(L,1)} \tag{4b}$$

The symbol : denotes the scalar product in both angular momentum and isospin space and H. c. stands for Hermitian conjugate. The superscript (subscript) in any of the coefficients of the two-body terms in V (e.g. $u_L^{(T)}$) indicates the isospin (angular momentum) of the boson pairs.

The expectation value of the Hamiltonian in the intrinsic states (2) defines an energy surface $E_{N,T}(\beta,\gamma)$ given by

$$E_{N,T}(\beta,\gamma) = K_{N,T} + (1+\beta^2)^{-2}\beta^2 \left[a_{N,T} - b_{N,T}\beta\cos 3\gamma + c_{N,T}\beta^2\right] \quad (5a)$$

with

$$K_{N,T} = \epsilon_s N + u_0^{(2)} N(N-1) + \delta u_0 A_{N,T} \quad (5b)$$

$$a_{N,T} = \epsilon N + f(u_2^{(2)}, v_0^{(2)}, u_0^{(2)}) N(N-1) + f(\delta u_2, \delta v_0, \delta u_0) A_{N,T} \quad (5c)$$

$$b_{N,T} = 2\sqrt{\frac{2}{7}}\left[v_2^{(2)} N(N-1) + \delta v_2 A_{N,T}\right] \quad (5d)$$

$$c_{N,T} = \epsilon N + g(c_0^{(2)}, c_2^{(2)}, c_4^{(2)}, u_0^{(2)}) N(N-1) + g(\delta c_0, \delta c_2, \delta c_4, \delta u_0) A_{N,T} \quad (5e)$$

$$A_{N,T} = \frac{(N-T)(N+T+1)}{3} \quad (5f)$$

where $f(x,y,z) = x + 2y/\sqrt{5} - 2z$, $g(w,x,y,z) = w/5 + 2x/7 + 18y/35 - z$, $\epsilon = \epsilon_d - \epsilon_s$, and $\delta x_L = x_L^{(T=0)} - x_L^{(T=2)}$. Although the IBM-3 Hamiltonian in (4) depends on 18 parameters, the shape of the energy surface (i.e. its β-γ dependence) is determined by only 3 combinations of parameters: $a_{N,T}$, $b_{N,T}$, $c_{N,T}$. The N and T dependence in the energy surface are manifested in three functional forms: N, $N(N-1)$ and $A_{N,T}$ with the one-boson terms contributing only to the N dependence, only the isotensor interactions contributing to the $N(N-1)$ dependence, and only the difference of isoscalar and isotensor interactions contributing to the $A_{N,T}$ dependence. This means that, if the corresponding strengths of the isoscalar and isotensor interactions are equal, there is no isospin dependence in the energy surface. This (unrealistic) situation would be the case for an $SU_f(3)$ invariant Hamiltonian. The $A_{N,T}$ dependence is essential to reproduce the excitation energies of the isobaric analog states. It should be noted, however, that the mapping from fermions to bosons can induce number and isospin dependence into the parameters of the IBM-3 Hamiltonian.[14] The three isovector interactions in eq. (4), $u_2^{(1)}$, $c_1^{(1)}$ and $c_3^{(1)}$, do not affect the energy surface at all.

The energy surface in (5) has the same functional dependence on β,γ as the corresponding energy surface[4] for IBM-1. Hence it follows that the only types of ground state shapes allowed for an isospin invariant IBM-3 Hamiltonian with at most two-boson interactions are spherical shapes ($\beta = 0$) or deformed shapes ($\beta > 0$) which can be either gamma unstable or axially symmetric ($\gamma = 0$ prolate, $\gamma = \pi/3$ oblate). However, because of the explicit isospin dependence of the energy surface, the deformation parameters will depend on isospin T as well as on the total numbers of valence pairs N, and hence different types of quadrupole shapes can co-exist in the same nucleus. The conditions on the parameters $a_{N,T}, b_{N,T}$, $c_{N,T}$ to obtain a local/global, spherical/deformed minima are the same as those given[15] for IBM-1.

Nuclei with $N = T$ (all neutrons or all protons) are spherical. In this subspace only the isotensor interactions contribute to the dynamics since they are the only interactions which have terms that can destroy two neutron or two proton bosons. Hence we expect the choice of parameters in the Hamiltonian to be such that the energy surface will have a global spherical minimum for $N = T$. Such considerations have been used[11] to determine some of the parameters of the IBM-3 Hamiltonian for the Sn, Te, Xe and Ba isotopes by calculating two-nucleon separation energies and comparing them with their measured values.

5. Normal Modes

Given an IBM-3 Hamiltonian one determines the equilibrium shape for a prescribed isospin T by minimizing its energy surface (5). Each such shape is characterized by (N, T, β, γ) and is represented in the model by a condensate wave-function (2). Projected states of good angular momentum (and isospin) form rotational bands. Henceforth we confine the discussion to axially deformed prolate shapes ($\beta > 0, \gamma = 0$) with a symmetry z-axis. Excitations of these shapes are obtained by replacing condensate bosons in (2) by orthogonal deformed bosons

$$b^\dagger_{c,\tau} = (1+\beta^2)^{-1/2}\left[s^\dagger_\tau + \beta\, d^\dagger_{0,\tau}\right] \quad , \quad b^\dagger_{\gamma,\pm 2,\tau} = d^\dagger_{\pm 2,\tau} ,$$
$$b^\dagger_{\beta,\tau} = (1+\beta^2)^{-1/2}\left[d^\dagger_{0,\tau} - \beta\, s^\dagger_\tau\right] \quad , \quad b^\dagger_{sc,\pm 1,\tau} = d^\dagger_{\pm 1,\tau} , \qquad (6)$$

representing β, γ and rotational (scissors) modes. Intrinsic states, with good isospin and good $SU_f(3)$ symmetry, involving single boson-excitations of the condensate are given by

$$\Psi^i_1 = \mathcal{N}_1\left[2\alpha\left(b^\dagger_i \cdot b^\dagger_c\right)\left(b^\dagger_c \cdot b^\dagger_c\right)^{\alpha-1}\left(b^\dagger_{c,1}\right)^T + T\, b^\dagger_{i,1}\left(b^\dagger_c \cdot b^\dagger_c\right)^\alpha\left(b^\dagger_{c,1}\right)^{T-1}\right]|0> \qquad (7a)$$

$$\Psi^i_2 = \mathcal{N}_2\left[\left(b^\dagger_i \cdot b^\dagger_c\right)\left(b^\dagger_c \cdot b^\dagger_c\right)^{\alpha-1}\left(b^\dagger_{c,1}\right)^T - b^\dagger_{i,1}\left(b^\dagger_c \cdot b^\dagger_c\right)^\alpha\left(b^\dagger_{c,1}\right)^{T-1}\right]|0> \qquad (7b)$$

$$\Psi^i_3 = \mathcal{N}_3\left(b^\dagger_i b^\dagger_c\right)^{(1)}_1\left(b^\dagger_c \cdot b^\dagger_c\right)^{\alpha-1}\left(b^\dagger_{c,1}\right)^T|0> \qquad (7c)$$

where $\alpha = (N-T)/2$ and $i = \beta, \gamma, sc$. The respective isospin, $SU_f(3)$ assignments and normalizations are shown in Table 1.

Table 1: $SU_f(3)$ and isospin content of IBM-3 intrinsic states. Total isospin $T' = T$ for Ψ_1^i, Ψ_2^i and $T' = T+1$ for Ψ_3^i, with T the isospin of Ψ_c. Norms are given in terms of $\mathcal{N}_c(N,T)$, eq. (2). In Ψ_k^i, superscript $i = \beta, \gamma, sc$; subscript $k = c, 1, 2, 3$.

Intrinsic state	$SU_f(3)$	Total isospin	$\mathcal{N}_k/\mathcal{N}_c(N,T)$
$\Psi_c(T)$	$(N,0)$	$N, N-2, \ldots, 0$ or 1	1
$\Psi_1^i(T')$	$(N,0)$	$N, N-2, \ldots, 0$ or 1	$\frac{1}{\sqrt{N}}$
$\Psi_2^i(T')$	$(N-2,1)$	$N-2, N-4 \ldots, 2$ or 1	$\sqrt{\frac{(N-T)(N+T+1)T}{(T+1)N}}$
$\Psi_3^i(T')$	$(N-2,1)$	$N-1, N-3 \ldots, 1$ or 2	$\sqrt{\frac{(N-T)(2T+3)}{(T+2)}}$

Ψ_1^i is an isoscalar excitation of the condensate, and has the same isospin as the ground state condensate $\Psi_c(T)$ in eq. (2). These β, γ, and scissors excitations are completely symmetric under interchange of neutrons and protons and therefore transform like (N, 0) under $SU_f(3)$, just like the ground state condensate. However the scissors mode is actually a Goldstone boson and hence represents a spurious excitation. This is due to the fact that it is obtained by a global rotation of the condensate (i.e. $\Psi_1^{sc} \propto \hat{L}_{m=\pm 1} \Psi_c(T)$) about directions perpendicular to the symmetry axis. Consequently, the two ($K = \pm 1$) isoscalar scissors excitations are Goldstone modes not genuine excitations.

Ψ_2^i corresponds to a particular combination of isoscalar and isotensor excitations while Ψ_3^i is an isovector excitation of the condensate. Both Ψ_2^i and Ψ_3^i transform as $(N-2, 1)$ under $SU_f(3)$ but Ψ_2^i has the same isospin as $\Psi_c(T)$ (no $T = 0$ or N) whereas Ψ_3^i has isospin $T + 1$. Hence Ψ_3^i ($i = \beta, \gamma, sc$) are the analog of excited collective states in the neighboring odd-odd nucleus. As far as we know, such collective β, γ and scissors modes in odd-odd nuclei have not been considered previously.

In Figure 1, we illustrate schematically the isospin and flavor symmetry of the condensate and the band heads for the different types of excitation as a function of the isospin projection $T_z = |n_\nu - n_\pi|/2 = |N_\nu - N_\pi|$ for a fixed (even) number of pairs, $N = (n_\nu + n_\pi)/2 = N_\nu + N_\pi$. Here $N_\pi(N_\nu)$ is the number of $\pi(\nu)$ bosons.

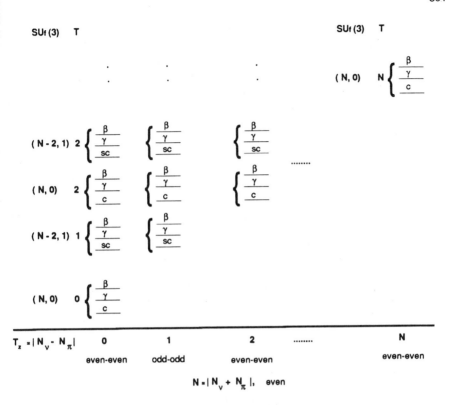

Fig. 1. A schematic view of the flavor symmetry and isospin of the condensate and the excitation modes as a function of the difference in the number of neutron and proton pairs (T_z) and for a fixed (even) number of total pairs (N).

To distinguish the intrinsic vibrational (genuine) modes from the spurious rotational (Goldstone) modes, it is advantageous to separate the Hamiltonian into intrinsic and collective parts,[15-16] $H = H_{\text{int}} + H_{\text{coll}}$. The intrinsic part of the Hamiltonian (H_{int}) annihilates the condensate wave-function (2) and can be transcribed as

$$H_{\text{int}} = \sum_{T=0,2} \xi_0^{(T)} \left[\sqrt{5} \left(d^\dagger d^\dagger \right)^{(0,T)} - \beta^2 \left(s^\dagger s^\dagger \right)^{(0,T)} \right] : \left[\text{H.c.} \right]^{(0,T)}$$

$$\sum_{T=0,2} \xi_2^{(T)} \left[\beta \left(s^\dagger d^\dagger \right)^{(2,T)} + \sqrt{\frac{7}{2}} \left(d^\dagger d^\dagger \right)^{(2,T)} \right] : \left[\text{H.c.} \right]^{(2,T)}$$

$$\sum_{L=1,3} \xi_L^{(1)} \left(d^\dagger d^\dagger \right)^{(L,1)} : \left(\tilde{d}\tilde{d} \right)^{(L,1)} + \xi_2^{(1)} \left(s^\dagger d^\dagger \right)^{(2,1)} : \left(\tilde{d}\tilde{s} \right)^{(2,1)} \qquad (8)$$

The collective part (H_{coll}) does not affect the shape of the energy surface and its rotational terms (e.g. $\hat{L} \cdot \hat{L}$) bring about intraband splitting. The presence of a $\hat{T} \cdot \hat{T}$ term in H_{coll} ensures that the lowest-energy bands have the lowest possible isospin (i.e. $T = T_0 = |n_\nu - n_\pi|/2$) with a $T(T+1)$ splitting among the different allowed isospin multiplets ($T = T_0, T_0 + 1, \ldots$). To identify the excitation modes associated with a given Hamiltonian and to estimate their energies, we diagonalize the Bogoliubov image of the intrinsic Hamiltonian (H_{int}^B). The latter is composed of components with maximal number of condensate bosons, in an expansion of H_{int} in the deformed basis of eq. (6). It has the form

$$H_{\text{int}}^B = \sum_{T=0,1,2} \left\{ \lambda_\beta^{(T)} \left(b_\beta^\dagger b_c^\dagger\right)^{(T)} \cdot \left(\tilde{b}_c \tilde{b}_\beta\right)^{(T)} + \lambda_\gamma^{(T)} \sum_{K=\pm 2} \left(b_{\gamma,K}^\dagger b_c^\dagger\right)^{(T)} \cdot \left(\tilde{b}_c \tilde{b}_{\gamma,K}\right)^{(T)} \right\}$$
$$+ \lambda_{sc}^{(1)} \sum_{K=\pm 1} \left(b_{sc,K}^\dagger b_c^\dagger\right)^{(1)} \cdot \left(\tilde{b}_c \tilde{b}_{sc,K}\right)^{(1)} \tag{9}$$

where $\tilde{b}_{i,\tau} = (-1)^\tau b_{i,-\tau}$ ($i = c, \beta, \gamma, sc$) and

$$\lambda_\beta^{(T)} = \beta^2 \left[4\xi_0^{(T)} + \xi_2^{(T)}\right]\left(\delta_{T,0} + \delta_{T,2}\right) + \xi_2^{(1)} \delta_{T,1} \tag{10a}$$
$$\lambda_\gamma^{(T)} = (1+\beta^2)^{-1}\left\{9\beta^2 \xi_2^{(T)}\left(\delta_{T,0} + \delta_{T,2}\right) + \left[\xi_2^{(1)} + 2\beta^2 \xi_3^{(1)}\right]\delta_{T,1}\right\} \tag{10b}$$
$$\lambda_{sc}^{(1)} = (1+\beta^2)^{-1}\left[\beta^2\left(6\xi_1^{(1)} + 4\xi_3^{(1)}\right)/5 + \xi_2^{(1)}\right] \tag{10c}$$

By construction H_{int}^B is isospin invariant (but not rotational invariant) and its action on the intrinsic states in (7) can be read from

$$H_{\text{int}}^B \begin{pmatrix} \Psi_1^i \\ \Psi_2^i \\ \Psi_3^i \end{pmatrix} = \begin{pmatrix} m_{11}^i & m_{12}^i & 0 \\ m_{21}^i & m_{22}^i & 0 \\ 0 & 0 & m_{33}^i \end{pmatrix} \begin{pmatrix} \Psi_1^i \\ \Psi_2^i \\ \Psi_3^i \end{pmatrix} \tag{11}$$

The matrix elements are ($i = \beta, \gamma, sc$)

$$m_{11}^i = \lambda_i^{(2)}(N-1) + \left(\lambda_i^{(0)} - \lambda_i^{(2)}\right) A_{N,T}/N \tag{12a}$$
$$m_{12}^i = m_{21}^i = \left(\lambda_i^{(0)} - \lambda_i^{(2)}\right)\left(\sqrt{3T(T+1)A_{N,T}}\right)/3N \tag{12b}$$
$$m_{22}^i = m_{33}^i + \left(\lambda_i^{(0)} - \lambda_i^{(2)}\right) T(T+1)/3N \tag{12c}$$
$$m_{33}^i = \lambda_i^{(2)}(N/2 - 1) + \lambda_i^{(1)} N/2 \tag{12d}$$

In the $SU_f(3)$ dynamical symmetry limit $\lambda_i^{(0)} = \lambda_i^{(2)}$ and the states Ψ_1^i and Ψ_2^i do not admix. However, as noted earlier, this is not a realistic limit and in general the resulting eigenstates are a mixture of Ψ_1^i and Ψ_2^i with a mixing proportional to $\left(\lambda_i^{(0)} - \lambda_i^{(2)}\right)$. For $T = 0$ or N there are no admixtures, because Ψ_2^i does not exist (see Table 1). In-between these values, the admixtures increase and, for large N, will

503

peak at $T \sim N/\sqrt{2}$, because of the particular N-T dependence of the off-diagonal matrix element (12b). The two orthonormal combinations represent symmetric and antisymmetric β and γ excitations. The $\beta(K = 0)$ mode is one-dimensional while the $\gamma(K = \pm 2)$ mode is two-dimensional, where K denotes the projection of the angular momentum on the symmetry z-axis. For the scissors mode, since $\lambda_{sc}^{(0)} = \lambda_{sc}^{(2)} = 0$, it follows that Ψ_1^{sc} has eigenvalue zero, i.e. these excitations are not intrinsic excitations but are Goldstone bosons as noted earlier. The second eigenstate Ψ_2^{sc} corresponds to a genuine two-dimensional ($K = \pm 1$) antisymmetric scissors mode and has definite $SU_f(3)$ symmetry. All isovector (β, γ and scissors) excitations are genuine and are represented by Ψ_3^i and have good $(N-2, 1)$ $SU_f(3)$ symmetry and total isospin even (odd) for N odd (even). The eigenvalues of the matrix in (11) provide an estimate, for large N, of the position of the bandheads corresponding to these genuine excitations (there are also additional shifts from terms in H_{coll}, e.g. from $\hat{T} \cdot \hat{T}$).

In Figure 2 we illustrate schematically the realistic excitation spectrum with and without $SU_f(3)$ breaking. We see that the excitations with isospin $T + 1$ remain unchanged with $SU_f(3)$ breaking as well as the scissors mode with isospin T.

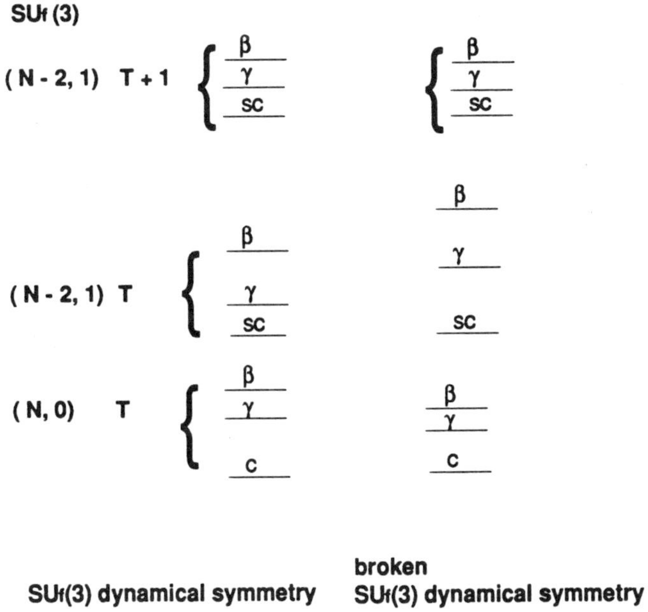

Fig. 2. Schematic Spectrum with and without $SU_f(3)$ breaking.

6. Summary

We have shown that quadrupole collective motion with isospin as a conserved quantum number can be given a geometric interpretation in terms of quadrupole deformation parameters by introducing intrinsic states which are made up of deformed IBM-3 bosons coupled to a definite isospin. This is a novel type of intrinsic states in the sense that they break the rotation symmetry (i.e. deformed in angular momentum space) but simultaneously preserve an internal symmetry (i.e. have good isospin). The intrinsic matrix element of the Hamiltonian defines an energy surface whose minima represent the different possible shapes. Intrinsic modes corresponding to small oscillations about the minima in the energy surface have been identified for prolate deformations. They involve isoscalar, isovector and isotensor excitations of the condensate and correspond to β, γ and scissors type of vibrations. The isoscalar scissors modes are found to be Goldstone modes associated with global rotations of the equilibrium shape. In addition we have predicted the occurrence of excited β, γ and scissors collective modes in odd-odd nuclei.

New nuclear physics facilities are exploring new regions of nuclei far from the valley of stability, whose description requires an extrapolation of existing models beyond their domain of validity. Since there is an element of uncertainty in any extrapolation, it makes sense to preserve in the process those symmetries and systematics which are known to be observed in ordinary nuclei. In the framework of boson models several extrapolation schemes have been suggested, e.g. the $N_\pi N_\nu$ scheme[17] or F-spin multiplets.[18] In the current work we have presented an additional scheme based on combining isospin with quadrupole collectivity in a boson description of nuclei for which protons and neutrons fill the same major shell. We suggest to use the version of the boson model with isospin (IBM-3) not just for light nuclei, but rather for heavier and possibly deformed nuclei like those which will be accessed by radioactive beams.

7. Acknowledgements

This work is supported in part by the United States Department of Energy.

References

1. F. Iachello and A. Arima, *The Interacting Boson Model* (Cambridge Univ., Press, Cambridge, 1987).

2. J.N. Ginocchio and M.W. Kirson, *Phys. Rev. Lett.* **44**, (1980) 1744;

3. A.E.L. Dieperink, O. Scholten and F. Iachello, *Phys. Rev. Lett.* **44**, (1980) 1747.

4. J.N. Ginocchio and M.W. Kirson, *Nucl. Phys.* **A350**, (1980) 31.

5. T. Otsuka, A. Arima, F. Iachello and I. Talmi, *Phys. Lett.* **B76** (1978) 139.

6. A. Arima, T. Otsuka, F. Iachello and I. Talmi, *Phys. Lett.* **B66** (1977) 205.

7. A. Leviatan and M.W. Kirson, *Ann. Phys.* **201**, (1990) 13.

8. J.N. Ginocchio and A. Leviatan, *Ann. Phys.* **216**, (1992) 152.

9. J.P. Elliott and A.P. White, *Phys. Lett.* **B97**, (1980) 169.

10. P. Halse J.P. Elliott and J.A. Evans, *Nucl. Phys.* **A417**, (1984) 301.

11. J.N. Ginocchio and A. Leviatan, in *Symmetries in Science VII* (B. Gruber and T. Otsuka Eds.), (Plenum Press, NY, 1994), p. 201.

12. J.N. Ginocchio and A. Leviatan, submitted to the *Phys. Rev. Lett.* (1994).

13. P. Van Isacker, et al., *Nucl. Phys.* **A476** (1988) 301.

14. J.A. Evans, G.L. Long and J.P. Elliott, *Nucl. Phys.* **A561**, (1993) 201.

15. A. Leviatan, *Ann. Phys.* **179**, (1987) 201.

16. M.W. Kirson and A. Leviatan, *Phys. Rev. Lett.* **55**, (1985) 2846.

17. R.F. Casten, *Phys. Lett.* **B152**, (1985) 145; *Phys. Rev. Lett.* **54**, (1985) 1991; *Nucl. Phys.* **A443**, (1986) 1.

18. N.V. Zamfir, R.F. Casten, P. von Brentano and W.-T. Chou, *Phys. Rev.* **C46**, (1992) R393.

MULTISTEP PROCESSES IN MEDIUM ENERGY SCATTERING

R. BIJKER

R.J. Van de Graaff Laboratory, University of Utrecht
P.O. Box 80000, 3508 TA Utrecht, The Netherlands

ABSTRACT

Multistep processes in medium and high energy scattering to collective states are treated by combining eikonal scattering methods and an algebraic treatment of the target dynamics. First the application to proton scattering from odd-mass nuclei is discussed and applied to ^{195}Pt, with particular emphasis on the relative contributions of nonspherical and channel coupling effects. Multistep processes are found to be particularly important in the application to electron scattering from polar molecules, such as LiF.

1. Introduction

The scattering of a high energy projectile from a target with a set of strongly coupled collective states generally involves many partial waves and the virtual excitation of many intermediate states. This situation occurs both in proton scattering from collective nuclei, in which the effects of channel coupling become important especially for larger momentum transfer [1], and in electron scattering from polar molecules, in which the forward angle scattering is dominated by the long range dipole interaction [2]. In both cases the scattering cannot be treated in (distorted wave) Born approximation, but must be calculated to all orders in the channel coupling, involving fully the rotational and vibrational degrees of freedom of the target. The standard approach is that of a coupled channels (or close coupling) calculation, which becomes very complicated when the number of channels that have to be included is large.

In this contribution I discuss an alternative approach in which the channnel coupling can be solved in closed form to all orders in the coupling between the projectile and the target. This method, called the algebraic-eikonal approach, is a combination of an algebraic description of the target dynamics (*e.g.* the interacting boson model [3] for collective even-even nuclei and the vibron model [4] for molecules) with eikonal scattering methods in the adiabatic approximation. It was originally developed in nuclear physics and applied to proton-nucleus scattering [1,5,6]. More recently it has also found useful applications in electron scattering from polar molecules [7,8].

After a brief review of the eikonal approximation we first extend the algebraic-eikonal approach to odd-mass nuclei, and discuss the application to ^{195}Pt, which provides one of the best examples of dynamic symmetries in odd-mass nuclei. We discuss in some detail the relative importance of nonspherical and channel coupling effects. In a second application to electron scattering from LiF we find that multistep processes are far more important than in the nuclear case. The eikonal transition matrix elements are derived in closed analytic form either exactly or in mean field and provide a clear physical understanding of the scattering process.

2. Eikonal Approximation

For medium and high energy scattering the eikonal approximation is a good approximation for elastic and inelastic scattering. The hamiltonian is in general given by

$$H = \frac{\hbar^2 k^2}{2m} + H_t(\xi) + V(\vec{r}, \xi) , \qquad (1)$$

where H_t describes the target dynamics and $V(\vec{r}, \xi)$ represents the interaction between the projectile and the target. The projectile coordinate $\vec{r}$ is measured from the center of mass of the target. The internal coordinates of the target are collectively denoted by ξ. If the kinetic energy of the projectile is much larger than the interaction strength V, and is also sufficiently large that the projectile wavelength is small compared with the range of variation of the potential, one may use the eikonal approximation to describe the scattering. If, in addition, the projectile energy is large compared with the target excitation energies, one can neglect H_t (adiabatic limit). Under these approximations the scattering amplitude for scattering a projectile with initial momentum $\vec{k}$ from an initial state $|i\rangle$ to final momentum $\vec{k}'$ and a final state $|f\rangle$ is given by

$$A_{fi}(\vec{q}) = \frac{k}{2\pi i} \int d^2 b \, e^{i\vec{q}\cdot\vec{b}} \langle f | e^{i\chi(\vec{b},\xi)} - 1 | i \rangle , \qquad (2)$$

where $\vec{q} = \vec{k}' - \vec{k}$ is the momentum transfer and $\chi(\vec{b}, \xi)$ is the eikonal phase that the projectile acquires as it goes by the target

$$\chi(\vec{b}, \xi) = -\frac{m}{\hbar^2 k} \int_{-\infty}^{\infty} dz \, V(\vec{r}, \xi) . \qquad (3)$$

In the derivation of the scattering amplitude the projectile coordinate is written as $\vec{r} = \vec{b} + \vec{z}$, where the impact parameter $\vec{b}$ is perpendicular to the z-axis, which is chosen along $\hat{z} = (\vec{k}+\vec{k}')/|\vec{k}+\vec{k}'|$. In the eikonal approximation the scattering amplitude is expressed in terms of an integral over a two-dimensional impact parameter rather than as a sum over partial waves.

3. Proton-Nucleus Scattering

For medium energy proton-nucleus scattering the eikonal approximation has been applied successfully to elastic and inelastic scattering to forward angles [1]. The target hamiltonian H_t describes the nuclear excitations and $V(\vec{r}, \xi)$ represents the interaction between the incoming proton and the target nucleus. If the range of the projectile-nucleus interaction is short compared to the size of the nucleus, the potential can be expressed in terms of the projectile-nucleon forward scattering amplitude f,

$$V(\vec{r}, \xi) = -\frac{2\pi\hbar^2 f}{m} \left[\rho(r) + [a_B(r)Q_B(\xi) + a_F(r)Q_F(\xi)] \cdot Y_2(\hat{r}) \right]. \quad (4)$$

Here $\rho(r)$ is the nuclear density for the general distorting or optical potential, and $a_B(r)$ and $a_F(r)$ are the collective (bosonic) and single-particle (fermionic) transition densities for the quadrupole coupling of the projectile to the nuclear degrees of freedom.

For a strongly absorbing probe the scattering is dominated by peripheral collisions, which allows one to keep only the leading order term in the expansion of the spherical harmonic around $\theta = \pi/2$,

$$Y_{2\mu}(\hat{r}) = Y_{2\mu}(\hat{b}) + \mathcal{O}(z/r). \quad (5)$$

The effects of the peripheral approximation can most easily be studied in Born approximation, which arises by expanding the eikonal transition operator to first order in the coupling potential

$$e^{i\chi(\vec{b},\xi)} = 1 + i\chi(\vec{b},\xi) + \cdots, \quad (6)$$

and keeping the first two terms.

The calculation of the nuclear matrix elements to all orders in χ is a complicated task. The combination with algebraic models to describe the nuclear excitations makes such a calculation feasible.

3.1. The Interacting Boson Fermion Model

In the IBFM [9] the collective and single-particle quadrupole operators are given by

$$\begin{aligned} \hat{Q}_{B,\mu} &= (s^\dagger \tilde{d} + d^\dagger s)^{(2)}_\mu + \chi_Q (d^\dagger \tilde{d})^{(2)}_\mu, \\ \hat{Q}_{F,\mu} &= \sum_{jj'} t^{(2)}_{jj'} (a^\dagger_j \tilde{a}_{j'})^{(2)}_\mu, \end{aligned} \quad (7)$$

with $\tilde{d}_\mu = (-1)^{2-\mu} d_{-\mu}$ and $\tilde{a}_{j,\mu} = (-1)^{j-\mu} a_{j,-\mu}$. Since the quadrupole operators are linear in the generators of the symmetry group of the IBFM $G = U_B(6) \otimes U_F(m)$ (with $m = \sum_j (2j+1)$), the eikonal transition matrix elements can be interpreted as group elements of G. They are a generalization of the Wigner $\mathcal{D}$-matrices for $SU(2)$. These representation matrix elements can be calculated exactly (albeit numerically) to all orders in the projectile-nucleus coupling strength, either with or without the peripheral approximation. This holds for any collective nucleus, either spherical, deformed, γ unstable or an intermediate situation between any of them. The general result can be derived in terms of a five-dimensional integral for the collective part [10] and a contribution from the single-particle part, which is easily obtained for a single nucleon.

In the peripheral approximation the expression for the scattering amplitude for scattering from an initial state $|i\rangle = |\alpha, J, M\rangle$ to a final state $|f\rangle = |\alpha', J', M'\rangle$ reduces to a one-dimensional integral over the impact parameter

$$A_{fi}(\vec{q}) = \frac{k}{i} i^{M-M'} \int_0^\infty b db \, J_{M-M'}(qb) \left[e^{i\chi_{opt}(b)} \sum_{M''} \mathcal{D}_{M'M''}^{(J')}(\hat{q}) \right.$$
$$\left. \times \langle \alpha', J', M''| e^{i[\epsilon_B(b)\hat{Q}_{B,z} + \epsilon_F(b)\hat{Q}_{F,z}]} |\alpha, J, M''\rangle \mathcal{D}_{M''M}^{(J)}(-\hat{q}) - \delta_{fi} \right] . \quad (8)$$

The projectile distorted wave is given by

$$\chi_{opt}(b) = \frac{2\pi f}{k} \int_{-\infty}^\infty dz \, \rho(r) , \quad (9)$$

and the quadrupole eikonal profile functions by

$$\epsilon_B(b) = \frac{2\pi f}{k} \int_{-\infty}^\infty dz \, a_B(r) \sqrt{\frac{5}{4\pi}} ,$$
$$\epsilon_F(b) = \frac{2\pi f}{k} \int_{-\infty}^\infty dz \, a_F(r) \sqrt{\frac{5}{4\pi}} . \quad (10)$$

Hence the only representation matrix elements that are needed are those that depend on the z-component of the quadrupole operator. Without the peripheral approximation the other components have to be included as well. In the special case of a dynamic symmetry the matrix elements appearing in Eq. (8) can be derived in closed analytic form. Following the study of dynamic symmetries in even-even nuclei [5], I present in this contribution the first results for odd-mass nuclei.

One of the best examples of dynamic symmetries in odd-even nuclei is provided by the lowlying negative parity states of ^{195}Pt, which have been analyzed successfully in terms of the $U(6) \otimes SU(2) \supset SO(6) \otimes SU(2)$ limit of the IBFM [11]. The odd neutron in ^{195}Pt occupies the $3p_{1/2}$, $3p_{3/2}$ and $2f_{5/2}$ shell model orbits with $n = 5$, which are treated here in a pseudo-spin coupling scheme as the $3\tilde{s}_{1/2}$, $2\tilde{d}_{3/2}$ and $2\tilde{d}_{5/2}$ pseudo-orbits with $\tilde{n} = n - 1 = 4$. The quadrupole operators in the $SO(6) \otimes SU(2)$ limit are given by

$$\hat{Q}_{B,\mu} = (s^\dagger \tilde{d} + d^\dagger s)_\mu^{(2)} ,$$
$$\hat{Q}_{F,\mu} = -\sqrt{\frac{4}{5}}(a_{1/2}^\dagger \tilde{a}_{3/2} - a_{1/2}^\dagger \tilde{a}_{3/2})_\mu^{(2)} - \sqrt{\frac{6}{5}}(a_{1/2}^\dagger \tilde{a}_{5/2} + a_{1/2}^\dagger \tilde{a}_{5/2})_\mu^{(2)} . \quad (11)$$

The classification scheme and the structure of the wave functions are discussed in detail in [11]. The eigenstates are labeled by $|[N_1, N_2], (\sigma_1, \sigma_2, \sigma_3), (\tau_1, \tau_2), L, J^P, M\rangle$. Some lowlying excitations are listed in Table 1 together with their $B(E2)$ values to the ground state, $|[7, 0], (7, 0, 0), (0, 0), 0, 1/2^-, M\rangle$.

Table 1: $B(E2)$ values leading to the ground state in ^{195}Pt, calculated with $e_B = 0.184$ (eb) and $e_F = -0.257$ (eb). The number of bosons is $N = 6$.

Initial state	$B(E2)$ th	J^P(keV)	$B(E2)(e^2b^2)$ exp [12]	calc
$[7,0],(7,0,0),(1,0),2_1,J^P$	$(Ne_B + e_F)^2 \frac{N+5}{5(N+1)}$	$3/2^-(211)$	0.240(25)	0.225
		$5/2^-(239)$	0.210(23)	0.225
$[6,1],(6,1,0),(1,0),2_2,J^P$	$(e_B - e_F)^2 \frac{2N(N+3)}{5(N+1)(N+2)}$	$3/2^-(99)$	0.085(20)	0.075
		$5/2^-(130)$	0.066(10)	0.075
$[6,1],(6,1,0),(1,1),1,J^P$	0	$1/2^-$		0
		$3/2^-(199)$	0.019(5)	0

The eikonal transition matrix elements in the $SO(6) \otimes SU(2)$ limit can all be expressed in terms of Gegenbauer polynomials. For example, for elastic transitions we find

$$U_{\text{el}}(b) = \frac{4!N!}{(N+2)(N+4)!} \left[(N+2) \cos \epsilon_F \, C_N^{(3)}(\cos \epsilon_B) \right.$$
$$\left. - ((N+4)\cos(\epsilon_B - \epsilon_F) - 2) \, C_{N-1}^{(3)}(\cos \epsilon_B) \right]$$
$$\rightarrow (2 + 3\cos \epsilon_F)\frac{3j_1(\bar{\epsilon}_B)}{5\bar{\epsilon}_B} + (1 - \cos \epsilon_F)\frac{6j_3(\bar{\epsilon}_B)}{5\bar{\epsilon}_B} - \sin \epsilon_F \frac{3j_2(\bar{\epsilon}_B)}{\bar{\epsilon}_B} \, . \quad (12)$$

In the last step we have taken the large N limit, which is such that $\bar{\epsilon}_B = \epsilon_B N$ remains finite. The matrix elements for transitions to excited states can be derived in a similar way. It is interesting to note, that in the peripheral approximation the states with odd values of the (pseudo-orbital) angular momentum L (such as the pseudo-spin doublet with $L = 1$ in Table 1) cannot be excited.

3.2. Application to ^{195}Pt

In Figure 1 we show the differential cross section (d.c.s.) for scattering 800 MeV protons from an odd-mass nucleus with $SO(6) \otimes SU(2)$ symmetry, in particular for elastic scattering and for the excitation of two lowlying pseudo-spin doublets with $L = 2$ (see Table 1). The first one belongs to the ground state band with $(\sigma_1, \sigma_2, \sigma_3) = (7,0,0)$ and the second one to the excited band with $(6,1,0)$.

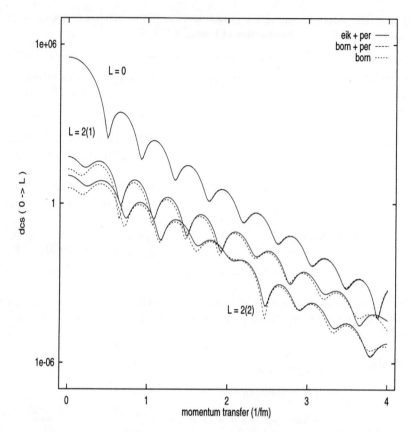

Figure 1: Differential cross section in mb/sr for the excitation of the pseudo-orbital doublets with $L = 0$, 2_1 and 2_2 in ^{195}Pt by 800 MeV protons.

In the calculations we take a Woods-Saxon form for the nuclear density with a nuclear radius of $1.2A^{1/3} = 6.96$ fm and a diffusivity of 0.75 fm, normalized to the total number of nucleons $A = 195$. For the collective transition density we take the derivative of the nuclear density (Tassie form) and for the single-particle transition density a product of radial wave functions for the pseudo-oscillator orbits. We note, that in the pseudo-spin coupling scheme there is only a single transition density for the fermion quadrupole operator of Eq. (11). The transition densities are normalized to the $B(E2)$ values in Table 1. The forward proton-nucleon scattering amplitude is $f = \frac{ik}{4\pi}\sigma$, in which we take the isospin averaged proton-nucleon cross section as $\sigma = 46(1 + 0.38i)$ mb [5].

In Figure 1 we present the results of three different calculations: the eikonal with peripheral approximation (solid), and the DWBA with and without the peripheral appoximation (dashed and dotted, respectively). To the accuracy of plot the solid and the dashed curves coincide, indicating that for these transitions and for momentum transfer $q \leq 4$ fm^{-1} the effects of channel coupling are very small. On the other hand, isospin doublets with $L = 4$ can only be excited in eikonal, and not in DWBA, since with quadrupole couplings one needs at least a two step process. A similar conclusion was reached in [5], in which 'the $B(E2)$ values were chosen rather large to make the coupling effects visible for the lowest states'. In the present calculation we use $B(E2)$ values fitted to experiment.

By comparing the two DWBA calculations for the $L = 2$ isospin doublets with and without the peripheral approximation (dashed and dotted) we see that the nonperipheral effects are quite substantial, especially at small angles. In these calculations they are far more important than the effects of the channel coupling. This is confirmed by a study of nonperipheral contributions in the eikonal approximation [13]. In deformed nuclei the $B(E2)$ values are larger and hence the multistep processes become relatively more important, so that the two effects may become comparable in magnitude. This indicates that one has to be very careful in drawing conclusions about the importance of channel couplings and/or other effects, if nonperipheral effects are not included in the calculations.

4. Electron-Molecule Scattering

Another application of the algebraic-eikonal approach is to electron scattering from polar molecules. In this case, multistep processes are much more important than in the nuclear case. Due to the dipole nature of the coupling there is no need to invoke the peripheral approximation to simplify the calculations.

Unlike the situation in proton-nucleus scattering, in electron-molecule scattering it is in general not possible to resolve the excitation of individual rotational states experimentally. In order to compare with experiment one therefore has to sum over the final rotational states and average over the initial magnetic substates. This summed and averaged differential cross section for scattering from an initial vibrational state v to a final vibrational state v' is given by

$$\frac{d\sigma(v \to v'|q)}{d\omega} = \frac{1}{2l+1} \sum_m \sum_{l'm'} |A_{fi}(\vec{q})|^2 . \tag{13}$$

It can be obtained in a straightforward way by calculating the contribution for each individual rotational excitation and then summing and averaging. This is not only time consuming, but also rather wasteful, since it first introduces a great amount of detailed information, which is subsequently lost in the summing and averaging.

It is however not necessary to first calculate the individual rotational transitions. The d.c.s. of Eq. (13) can be expressed in terms of intrinsic (or body-fixed) T-matrix elements. The intrinsic wave functions are expressed in terms of geometric variables which characterize the equilibrium shape of the molecule and its orientation. The basic assumption made in medium energy electron-molecule scattering is that the electron goes by the molecule in a time short compared with the rotational time. This is equivalent to the assumption that the T-matrix is diagonal in the geometric variables. Assuming that the rotational motion is independent of the vibrational state, the completeness of the rotational states allows one to do the sums involved in summing and averaging by closure. Under these approximations, the rotationally summed and averaged d.c.s. of Eq. (13) reduces to [8]

$$\frac{d\sigma(v \to v'|q)}{d\omega} = \frac{1}{8\pi^2} \int d\Omega \left|\langle \vec{k}', v', \Omega |T| \vec{k}, v, \Omega\rangle\right|^2 . \tag{14}$$

It represents the d.c.s. for a vibrational excitation, which is averaged over the orientation Ω of the corresponding intrinsic states. This form is quite general. It depends only on the completeness of the rotational states, and on the assumption that during the scattering process the orientation Ω of the intrinsic state does not change. Eq. (14) also shows explicitly that the summed and averaged d.c.s. does not depend on the initial angular momentum, which allows one, if one were to calculate this cross section according to Eq. (13), to take $l_i = 0$ which simplifies the calculations considerably.

The intrinsic T-matrix element appearing in the r.h.s. of Eq. (14) can be calculated directly with mean-field methods. Since the eikonal approximation is a short wavelength approximation, it shares the same domain of validity as the approximation that makes the T-matrix element diagonal in Ω. In the eikonal approximation the T-matrix element for vibrational transitions $v \to v'$ can be written as

$$\begin{aligned} T_{v'v}(\vec{q}, \Omega) &= \langle \vec{k}', v', \Omega |T| \vec{k}, v, \Omega\rangle \\ &= \frac{k}{2\pi i} \int d^2b \, e^{i\vec{q}\cdot\vec{b}} \langle v', \Omega | e^{i\chi(\vec{b},\xi)} - 1 | v, \Omega\rangle . \end{aligned} \tag{15}$$

Within the approximations made to solve the scattering problem the above expression for the T-matrix element is valid in general, *i.e.* irrespective of the model used to describe the molecular dynamics. In the next section I discuss how these T-matrix elements can be calculated in the vibron model [4].

In electron scattering from strongly polar molecules the forward angle scattering is determined by a long range dipole interaction

$$\begin{aligned} V(\vec{r}, \xi) &= a(r) \, \vec{T}(\xi) \cdot \hat{r} = a(r) \, \vec{T}(\xi) \cdot \sqrt{\frac{4\pi}{3}} Y_1(\hat{r}) , \\ a(r) &= -\frac{e}{r^2 + R_0^2} . \end{aligned} \tag{16}$$

Here $\vec{r} = \vec{b} + \vec{z}$ is the projectile coordinate and $\vec{T}(\xi)$ is the molecular dipole operator which depends on the internal degrees of freedom of the molecule. In this case the transition eikonal phase

$$\chi(\vec{b}, \xi) = g(b) \, \vec{T}(\xi) \cdot \hat{b} = g(b) \, \vec{T}(\xi) \cdot \sqrt{\frac{4\pi}{3}} Y_1(\hat{b}) ,$$

$$g(b) = -\frac{m}{\hbar^2 k} \int_{-\infty}^{\infty} dz\, a(r)\frac{b}{r}, \tag{17}$$

depends on $Y_1(\hat{b})$. Unlike the situation in proton-nucleus scattering here this is a consequence of the symmetry properties of the dipole interaction and not of using the peripheral approximation.

Just as in the nuclear case, the scattering amplitude for a rotational transition from an initial state $|i\rangle = |v,l,m\rangle$ to a final state $|f\rangle = |v',l',m'\rangle$ can be expressed in terms of a one-dimensional integral over the impact parameter

$$A_{fi}(\vec{q}) = \frac{k}{i} i^{m-m'} \int_0^\infty b\,db\, J_{m-m'}(qb) \left[\sum_{m''} \mathcal{D}_{m'm''}^{(l')}(\hat{q})\right.$$
$$\left. \times \langle v',l',m''|e^{ig(b)T_z(\xi)}|v,l,m''\rangle \mathcal{D}_{m''m}^{(l)}(-\hat{q}) - \delta_{fi}\right]. \tag{18}$$

The eikonal transition matrix elements in the intrinsic frame, Eq. (15), and in the laboratory frame, Eq. (18), contain the channel coupling to all orders in $g(b)$.

4.1. The Vibron Model for Diatomic Molecules

In a study of infrared intensities in hydrogen halide molecules in the vibron model it was shown [14] that vibrational transitions are well described by the dipole operator

$$\hat{T}_\mu = d_0 \hat{D}_\mu + \frac{1}{2} d_1 \left(e^{\lambda \hat{n}_p} \hat{D}_\mu + \hat{D}_\mu e^{\lambda \hat{n}_p}\right),$$
$$\hat{n}_p = \sqrt{3}\,(p^\dagger \tilde{p})^{(0)},$$
$$\hat{D}_\mu = (p^\dagger s - s^\dagger \tilde{p})_\mu^{(1)}, \tag{19}$$

with $\tilde{p}_\mu = (-1)^{1-\mu} p_{-\mu}$. The calculation of the matrix elements of the eikonal operator with this complicated form for $\hat{T}$ is a formidable challenge, since the dipole operator of Eq. (19) is not linear in the generators of the symmetry group. Since for molecules the number of vibrons is large (typically $N \sim 100$), one can calculate matrix elements of $\hat{T}$ to a good approximation by replacing the boson operators by their expectation values. The method is illustrated for vibrational excitation of diatomic molecules in electron scattering, but we emphasize its general validity. The extension to more complicated molecules and to other dynamical schemes is straightforward.

In the vibron model for diatomic molecules the vibrational wave functions in the intrinsic frame can be parameterized in terms of three geometric variables, a coordinate β and two Euler angles $(\Omega) = (\theta, \phi, 0)$,

$$|N,v,\beta,\Omega\rangle = R(\Omega)\,|N,v,\beta\rangle, \tag{20}$$

with

$$|N,v,\beta\rangle = \frac{1}{\sqrt{v!(N-v)!}} \left(-\beta s^\dagger + \sqrt{1-\beta^2}\, p_0^\dagger\right)^v \left(\sqrt{1-\beta^2}\, s^\dagger + \beta p_0^\dagger\right)^{N-v} |0\rangle \,. \quad (21)$$

The value of β can be determined by minimizing the expectation value of the hamiltonian in the ground state $|N, v=0, \beta\rangle$, or equivalently, by solving the hamiltonian in Hartree approximation.

The intrinsic T-matrix elements of Eq. (15) can be evaluated simply in the mean-field approximation by replacing the operators by their expectation values in the intrinsic states of the vibron model. As a result the T-matrix element for vibrationally elastic scattering $v = 0 \to v' = 0$ is given by [8]

$$T_{00}(\vec{q},\Omega) = \frac{k}{2\pi i} \int d^2 b\, e^{i\vec{q}\cdot\vec{b}} \left(e^{i\xi_0 g(b)\cos(\phi_b-\phi)\sin\theta} - 1\right) \,. \quad (22)$$

The coefficient ξ_0 is the intrinsic matrix element of the dipole operator

$$\begin{aligned}
\xi_0 &= \langle N, v=0, \beta | \hat{T} | N, v=0, \beta \rangle \\
&= 2N\beta\sqrt{1-\beta^2}\left(d_0 + d_1 e^{\lambda N \beta^2}\right) \,, \quad (23)
\end{aligned}$$

which can be normalized to the static dipole moment of the molecule $\xi_0 = d_{exp}$. Similarly, the T-matrix element for vibrationally inelastic scattering $v = 0 \to v' = 1$ is given by

$$T_{10}(\vec{q},\Omega) = \frac{k}{2\pi i} \int d^2 b\, e^{i\vec{q}\cdot\vec{b}}\, i\xi_1 g(b) \cos(\phi_b - \phi) \sin\theta\, e^{i\xi_0 g(b)\cos(\phi_b-\phi)\sin\theta} \,, \quad (24)$$

where ξ_1 denotes the intrinsic transition matrix element

$$\begin{aligned}
\xi_1 &= \langle N, v=1, \beta | \hat{T} | N, v=0, \beta \rangle \\
&= (1-2\beta^2)\sqrt{N}(d_0 + d_1 e^{\lambda N \beta^2}) + 2\beta^2(1-\beta^2)N\sqrt{N}\,\lambda d_1 e^{\lambda N \beta^2} \,. \quad (25)
\end{aligned}$$

The first term in ξ_1 has the same dependence on the parameters in the dipole operator as ξ_0, whereas the second term arises from the exponential part of the dipole operator. ξ_1 can be normalized to the dipole transition matrix element $\xi_1 = R_{01}$, either taken from experiment or from *ab initio* calculations. The angular part of the integral over the two-dimensional impact parameter can be done analytically by expanding the exponentials in terms of Bessel functions. The remaining integral over the impact parameter can be solved numerically.

4.2. Application to LiF

In Figure 2 we show the summed and averaged d.c.s. for vibrationally excitation ($v = 0 \to v' = 0, 1$) of LiF by 5.44 eV electrons. The d.c.s. are calculated using the

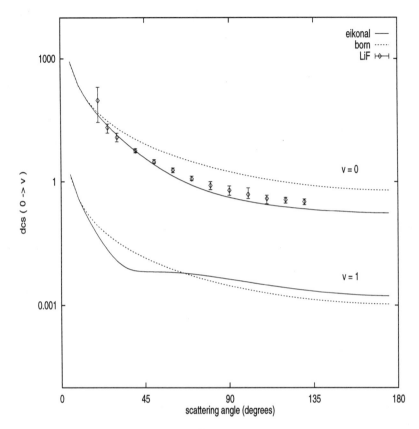

Figure 2: Differential cross section in 10^{-20} m^2/sr for vibrational excitation of LiF by 5.44 eV electrons. The experimental data [15] are normalized to our results at 40 degrees.

intrinsic T-matrix elements which according to Eqs. (22) and (24) are completely determined by ξ_0 and ξ_1. These coefficients are normalized to the static dipole moment of LiF, $\xi_0 = d_{exp} = 6.58$ Debye, and the transition matrix element, $\xi_1 = R_{01} = 0.2718$ Debye, which was obtained from an *ab initio* calculation [16]. For the dipole cut-off radius we take $R_0 = 0.5$ Å. A comparison between the eikonal (solid) and the Born approximation (dotted) shows that in this case the channel coupling effects are very important. The cross section for excitation of the first vibrational excitation ($v = 0 \rightarrow v' = 1$) is about two orders of magnitude smaller than the elastic one. Similar results for inelastic scattering were obtained in [16,17].

In Figure 3 we show the contribution of the individual rotational transitions. Although at present rotational excitations cannot be resolved experimentally, it

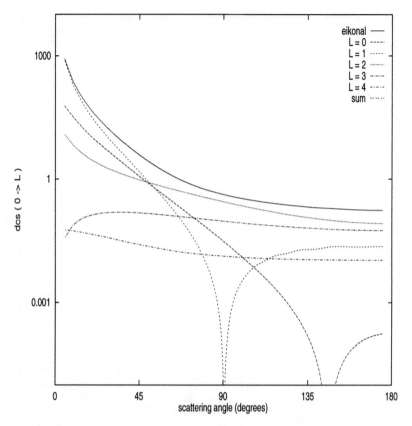

Figure 3: Differential cross section in 10^{-20} m^2/sr for vibrationally elastic rotational excitation of LiF by 5.44 eV electrons.

may be relevant for future use to present some results for rotational scattering as well. The matrix elements for rotational transitions can be evaluated by projecting states with good angular momentum and parity from the intrinsic states. In the mean-field approximation we find for vibrationally elastic transitions

$$U_{00}(b) = \langle v'=0, l', m'' | e^{ig(b)\mathcal{T}_z} | v=0, l, m'' \rangle$$

$$\rightarrow \sqrt{\frac{2l+1}{2l'+1}} \sum_\lambda i^\lambda (2\lambda+1) j_\lambda(\xi_0 g(b)) \langle l, 0, \lambda, 0 | l', 0 \rangle \langle l, m'', \lambda, 0 | l' m'' \rangle \ , (26)$$

in agreement with the classical rotor model [18], and for vibrationally inelastic transitions

$$U_{10}(b) = \langle v'=1, l', m'' | e^{ig(b)\mathcal{T}_z} | v=0, l, m'' \rangle$$

$$\rightarrow i\xi_1 g(b)\sqrt{\frac{2l+1}{2l'+1}} \sum_\lambda i^{\lambda+1}[(\lambda+1)j_{\lambda+1}(\xi_0 g(b)) - \lambda j_{\lambda-1}(\xi_0 g(b))]$$
$$\times \langle l,0,\lambda,0|l',0\rangle\langle l,m'',\lambda,0|l'm''\rangle \ . \tag{27}$$

The summed and averaged d.c.s. of Eq. (13) is to the accuracy of the plot identical to that of Figure 2 which was calculated directly in the intrinsic frame. This once again shows that the d.c.s. measured in electron scattering does not depend on any detailed information about the rotational states, but is determined solely by the static dipole moment of the molecule and a typical dipole transition matrix element.

5. Summary and Conclusions

In this contribution multistep processes in medium and high energy scattering were treated by combining the eikonal approximation for the scattering with algebraic models for the target dynamics. In the ensuing algebraic-eikonal approach the transition matrix elements that contain the channel coupling to all orders can be derived in closed analytic form either exactly or, to a very good approximation, in mean field. This is equivalent to solving the coupled channels equations in the adiabatic limit.

The algebraic-eikonal approach was applied to two totally different areas, namely proton-nucleus and electron-molecule scattering. We first discussed the extension to odd-mass nuclei, and derived the eikonal transition matrix elements using the $SO(6) \otimes SU(2)$ limit of the IBFM and the peripheral approximation. By comparing the d.c.s. in the eikonal approximation and the DWBA with and without the peripheral approximation, it was found that in this case the channel coupling effects are much smaller than the nonperipheral effects. This indicates that one has to be very careful in drawing conclusions about the importance of channel couplings and/or other effects, if nonperipheral effects have not been taken into account.

In electron scattering from polar molecules multistep processes are much more important. We showed that in order to calculate the summed and averaged differential cross section which is needed to compare with experiment, it is not necessary to calculate the individual rotational transitions separately. This d.c.s. can be calculated directly in the intrinsic frame. The intrinsic T-matrix elements are completely determined by the static dipole moment of the molecule and a transition dipole matrix element. No detailed knowledge of the various rotational states is needed, nor of the vibron hamiltonian, nor of the dipole transition operator (with the exception of its selection rules).

In the last application the eikonal T-matrix elements were derived in meanfield approximation. This is a very powerful method, that can be used to evaluate the matrix elements of an arbitrary function of the algebraic operators. It not only simplifies the eikonal treatment, but can also be used in other more accurate dynamic schemes. These include a direct calculation of the scattering in the intrinsic (or body-fixed) frame for vibrational excitation of both diatomic and more complex molecules, and, in the nuclear case, the calculation of the scattering matrix elements in the neutron-proton IBM.

In summary, the algebraic-eikonal approach presents an alternative way to treat multistep processes in which the effects of channel coupling are taken into account in an elegant, transparent and numerically very efficient way.

6. Acknowledgements

This work is supported in part by the Stichting voor Fundamenteel Onderzoek der Materie (FOM) with financial support from the Nederlandse Organisatie voor Wetenschappelijk Onderzoek (NWO).

7. References

1. R.D. Amado, J.A. McNeil and D.A. Sparrow, *Phys. Rev.* **C25** (1982) 13.
2. L.A. Collins and D.W. Norcross, *Phys. Rev.* **A18** (1978) 467.
3. A. Arima and F. Iachello, *Phys. Rev. Lett.* **35** (1975) 1069.
4. F. Iachello, *Chem. Phys. Lett.* **78** (1981) 581.
5. J.N. Ginocchio, T. Otsuka, R.D. Amado and D.A. Sparrow, *Phys. Rev.* **C33** (1986) 247.
6. G. Wenes, J.N. Ginocchio, A.E.L. Dieperink and B. van der Cammen, *Nucl. Phys.* **A459** (1986) 631.
7. R. Bijker, R.D. Amado and D.A. Sparrow, *Phys. Rev.* **A33** (1986) 871; R. Bijker and R.D. Amado, *Phys. Rev.* **A34** (1986) 71; *ibid.* **A37** (1988) 1425; R. Bijker, R.D. Amado and L.A. Collins, *Phys. Rev.* **A42** (1990) 6414.
8. R. Bijker and R.D. Amado, *Phys. Rev.* **A46** (1992) 1388.
9. F. Iachello and O. Scholten, *Phys. Rev. Lett.* **43** (1979) 679.
10. G. Wenes, A.E.L. Dieperink and O.S. van Roosmalen, *Nucl. Phys.* **A424** (1984) 81.
11. R. Bijker and F. Iachello, *Ann. Phys. (N.Y.)* **161** (1985) 360.
12. E.J. Bruton, J.A. Cameron, A.W. Gibbs, D.B. Kenyon and L. Keszthelyi, *Nucl. Phys.* **A152** (1970) 495.
13. B.I.M. van der Cammen, A.E.L. Dieperink, O. Scholten and G. Wenes, *Phys. Rev.* **C37** (1988) 1624.
14. F. Iachello, A. Leviatan and A. Mengoni, *J. Chem. Phys.* **95** (1991) 1449.
15. L. Vuskovic, S.K. Srivastava and S. Trajmar, *J. Phys.* **B11** (1978) 1643.
16. Y. Alhassid and B. Shao, *Phys. Rev.* **A46** (1992) 3978.
17. A. Mengoni and T. Shirai, *Phys. Rev.* **A44** (1991) 7258.
18. O. Ashihara, I. Shimamura and K. Takayanagi, *J. Phys. Soc. Jpn.* **38** (1975) 1732.

The Algebraic Scattering Theory and its application to heavy-ions reactions

Rubens Lichtenthäler Filho*
Departamento de Física Nuclear, Universidade de São Paulo
C.P. 20516, CEP 01498 São Paulo SP, Brasil

Abstract

We present the most conspicuous aspects of the Algebraic Scattering Theory as well as recent developments and applications to heavy ion collisions near the Coulomb barrier.

1 The $SO(3,1)$ S-matrix

Algebraic Scattering Theory[1] has been proposed by Alhassid and Iachello as an alternative approach to the study of low energy heavy ion collisions. In the algebraic approach, the Schroedinger equation is replaced by recursion relations for the algebraic Jost functions $A_l(k)$ and $B_l(k)$ and one obtains the S-matrix for elastic scattering:

$$S_l(k) = exp(il\pi)B_l(k)/A_l(k) \tag{1}$$

For the SO(3,1) symmetry the recursion relations take the form:

$$- N(l,v)A_{l+1}(k) = (l+1-iv)A_l(k) \tag{2}$$

$$+ N(l,v)B_{l+1}(k) = (l+1+iv)B_l(k) \tag{3}$$

where $N(l,v)$ is a normalization factor which drops out when we calculate the S-matrix. The function v, the algebraic potential, contains all the dynamic information about the scattering process. Solution of the recursion relations yields the SO(3,1) S-matrix as a ratio of two Euler Gamma functions

$$S_l = \frac{\Gamma(l+1+iv)}{\Gamma(l+1-iv)}. \tag{4}$$

---
*Supported by FAPESP

In the case of pure Coulomb scattering the algebraic potential is the Sommerfeld parameter:
$$v = \eta = \mu Z_1 Z_2 e^2 / \hbar^2 k \qquad (5)$$
and the $SO(3,1)$ symmetry is exact. In order to take into account the strong interaction, the algebraic potential must be generalized to be dependent on the angular momentum l:
$$v(l,k) = \eta(k) + v_s(l,k) \qquad (6)$$
It is important to mention that the unitary bound for the S-matrix, $|S| \leq 1$, imposes that the imaginary part of the algebraic potential must be positive.

A few models have been proposed for the l dependence of the real and imaginary parts of the algebraic potential[1,2]. Also theoretical investigations based on semiclassical approaches give some insight about its shape for the large l region[3,4]. In the next section we will present an inversion procedure which will permits the investigation of the relation between the optical model and algebraic potential.

2 The Inverse Problem

One important feature of SO(3,1) S-matrix as defined in Eq. (4) is the simplicity of the inverse problem.
$$S_l(k) \longrightarrow v(l,k). \qquad (7)$$
This problem is always solvable for any given set of S-matrix elements by expanding the SO(3,1) S-matrix in terms of $v(l,k)$ around a given v^0.

$$v(l,k) = v^0 + \frac{-i\log(S/S^0)}{\psi(l+1+i\eta+iv^0) + \psi(l+1-i\eta-iv^0)}, \qquad (8)$$

where
$$S^0 = \frac{\Gamma(l+1+i\eta+iv^0)}{\Gamma(l+1-i\eta-iv^0)}, \qquad (9)$$

and ψ is the Digamma function. For $v^0 \approx 0$ and using the asymptotic form for the Digamma function $\Psi \to \ln$ one obtains:

$$v(l,k) = \frac{\delta_r(l,k) + i\delta_i(l,k)}{\ln[(l+1)^2 + \eta^2(k)]^{1/2}}. \qquad (10)$$

Eq. (10) shows that the algebraic potential is related to the phase shifts a straightforward way.

We applied the inversion method to determine the algebraic potential corresponding to a realistic Woods- Saxon optical potential for the system $^{16}O + ^{63}Cu$ at energies in the neighbourhood of the Coulomb barrier. In figure 1 we present the real and imaginary parts of the algebraic potential obtained by this procedure at laboratory energies 42MeV, 46MeV, 56MeV, and 64MeV. We can observe that in spite of the fact that the optical potential is constant for all energies, this is not the case for the algebraic one. As the imaginary part increases, the real one bends down to negative values in the region $l \leq l_{grazing}$. This is an indication that there is a repulsive term in the potential which causes the deflection function $2d\delta_r/dl$ to be positive in this region. The decreasing of the real algebraic potential with energy also confirms the presence of a repulsive potential. For higher values of the angular momentum $l \geq l_g$, the algebraic potential becomes positive again with an exponential decrease with l as expected for an attractive nuclear potential[3]. This repulsion for small values of l has the simple optical interpretation as a reflection in the imaginary well which contributes with a real negative phase shift and therefore a real negative algebraic potential.

In order to describe this effect we suggest the following form for the algebraic potential:

$$v_s(l) = v_r f_r(l) - v_i^2 f_i^2(l) + iv_i f_i(l) \qquad (11)$$

where v_r and v_i are the real and imaginary strengths and f_r and f_i the corresponding form factors taken as a Woods-Saxon[1] or modified Woods-Saxon[2] shapes. The first and third terms in the right hand side of Eq. (11) are the usual form factors that describe the refractive and absorptive scattering. The second term is supposed to take into account the contribution due to the reflection in the imaginary well. A justification to the use of the square of the imaginary form factor in the second term can be found in a paper by S.K.Kauffmann[5,6] where the author investigates the relation between the optical model parameters and the phase shifts for very heavy ion systems. In principle we could also include an additional term in Eq. (11) due to the second order scattering in the real and imaginary potentials which would give a positive imaginary contribution to the algebraic potential. However for the sake of simplicity we prefer to keep Eq. (11) and argue that positive contributions to the imaginary potential can be taken into consideration by an adequate choice of the parameters. This is not the case for the second order term due to reflection in the

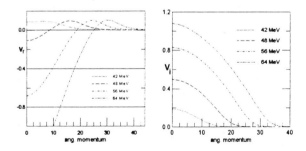

Figure 1: The real and imaginary parts of the algebraic potential.

imaginary well since it produces a negative contribution to the algebraic potential which cannot be simulated by reasonable variation of the parameters.

3 Analysis of the Elastic Scattering

Using Eq. (11) we analyzed the elastic angular distributions for the $^{16}O +^{63}Cu$ at energies ranging from $E_{lab} = 39 MeV$ up to $E_{lab} = 64 MeV$ measured by Chamon and Pereira[7] For $f(l)$ we used the Wood-Saxon shape. The six parameters to be known are the strengths the grazing angular momentum and the diffuseness for the real and imaginary potentials. They have been freely varied to adjust the experimental data. In Fig. 2 we show our results. The reduced Chi-square of the fits are about unity or even lower at some energies.

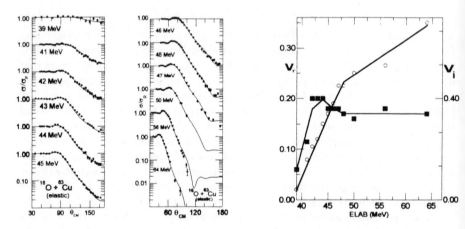

Figure 2: On the left the fits of the elastic scattering angular distributions of the $^{16}O+^{63}Cu$ system. On the right the strengths of the real and imaginary algebraic potential.

The values of the grazing angular momentum obtained from the fits follow approximately the relation $l_g = kR$ where $k = \sqrt{2\mu(E - E_b)}/\hbar$ and E_b is the Coulomb barrier. By adjusting R and E_b to reproduce the imaginary grazing angular momentum one obtains $E_b = 40.4 Mev$ and $R = 7.6 fm$ which agrees very well with those obtained from optical model calculations and fusion measurements for this system[7].

In Fig. 2 we plot the strengths of the real and imaginary potential as a function of the energy. The solid curves are a guide to the eyes. An interesting phenomena occurs at the energies around 43 MeV where the strength of the real potential presents a maximum. This behavior is necessary to reproduce the principal maximum of the diffraction pattern at forward angles. If we do not increase the real potential at these energies, the calculated angular distributions become flat in this region unlike the experimental data which present still a maximum for 42, 43 and 44 MeV. For lower energies the real potential goes to zero since we are below the Coulomb barrier. Above 46 MeV the real potential seems to become constant. The imaginary strength increases with energy although apparently in a less pronounced way for higher energies. This is expected to be due to the opening of reaction channels, mainly inelastic. As can be seen in Fig. 2, the real and imaginary strengths of $v_s(l)$ exhibit in the neighborhood of the Coulomb barrier a variation with energy similar to the one associated with the threshold anomaly[7].

4 Coupled Channels Formalism

When many reaction channels are explicitly taken into account the S-matrix assumes the form:

$$S = \frac{\Gamma(L + 1 + iV)}{\Gamma(L + 1 - iV)} \qquad (12)$$

where V is the algebraic potential matrix which contains the dynamics of the scattering in different channels. In the spinless case L is a multiple of the unity matrix and commutes with V what allows to solve the problem by means of a simple algorithm: (i) diagonalize V (ii) write the S-matrix in the representation where both S and V are diagonal

$$S_\alpha = \frac{\Gamma(l + 1 + iv_\alpha)}{\Gamma(l + 1 - iv_\alpha)} \qquad (13)$$

where v_α are the α-eigenvalues of V. (iii) go back to the original representation by means of the transformation $S = ZS_\alpha Z^{-1}$ where Z is the matrix of the eigenvectors of V.

The introduction of spin makes the algebraic approach more difficult to treat because L is no longer a multiple of the unitary matrix and does not commute with V. A possible approximation[8] is to replace L by J in the argument of the gamma function:

$$S = \frac{\Gamma(J + 1 + iV)}{\Gamma(J + 1 + iV)}. \qquad (14)$$

As J and V commute, we can use the same algorithm as used above. We do not make approximations in the potential matrix V which remains the same as in the exact treatment. In fact for the spinless case the algebraic potential V is an $n \times n$ matrix where n is the number of channels considered in the calculation. However when we take into account the spin of the target S_i in the different channels, the dimension of the V matrix is enlarged to $N = \sum_{i=1}^{n}(2S_i + 1)$ for each J^π. The coupling matrix elements are determined by[9]:

$$V_{\alpha L,\alpha' L'} = (4\pi)^{-1/2}\sum_{\lambda}(-1)^{J-S_\alpha}i^{L'-L-\lambda}\hat{L}\hat{L}'\hat{S}_{\alpha'}(LL'00\mid \lambda 0) \quad (15)$$
$$\times W(L,L',S_\alpha,S_{\alpha'}\mid \lambda,J)V_{\alpha\alpha'}^{\lambda}(l)(S_{\alpha'}\parallel V \parallel S_\alpha),$$

where α and α' are the channel masses and charges, L and L' are the possible values of the initial and final orbital angular momentum respectively. The inelastic coupling form factors $V_{\alpha,\alpha'}^{\lambda}(l)$ are composed of the nuclear (which is usually taken as the derivative of the elastic form factor) and the Coulomb excitation terms. for the latter we used the expression derived by Broglia and Winther[10]:

$$V_{coul}(l) = V_0 \frac{exp(-\xi(E,\Delta E,\eta,l))}{\left(1+\sqrt{1+(l/\eta)^2}\right)^{\lambda}} \quad (16)$$

where ξ is given by:

$$\xi = C_\lambda \eta \Delta E/(4E)(1+\sqrt{1+(l/\eta)^2}); \quad (17)$$

the parameters C_λ and V_0 are adjustable. The reduced matrix elements $(S_{\alpha'}\parallel V \parallel S_\alpha)$ contain the nuclear structure information. They can be treated as free parameters or being determined from nuclear models such as *Rotational Model, Vibrational model, Interacting Boson Model(IBM),* or *Interacting Boson Fermion Model(IBMF)*.

5 Nuclear Structure Calculations and Results

Inelastic scattering of ^{16}O by ^{63}Cu with the excitation of the low energy levels of the target ^{63}Cu can be described with good accuracy within the framework of the Interacting Boson-Fermion Model(IBMF)[11]. In the preliminary calculations of the present work, we have made the simplifying assumption that the low-lying states of the $^{64}_{30}Zn_{34}$ are generated by the excitation of two valence protons above the $Z=28$ shell closure and six valence neutrons above the $N=28$ closure without mixing with other configurations that are important at higher energies. This yields $N_p=1$ proton boson and $N_n=3$ neutron bosons interacting in the $U^{(b)}(5)$ Hamiltonian.

$$H_b = \epsilon\hat{n}_d + 10a_1 B^{(1)}.B^{(1)} + a_3 B^{(3)}.B^{(3)} \quad (18)$$

where $\hat{n}_d = \sqrt{(5)}(d^+ \times \tilde{d})^{(0)}$, $B_\mu^{(\lambda)} = (d^+ \times \tilde{d})_\mu^{(\lambda)}$, $(\mu = -\lambda,...,+\lambda)$ are spherical tensors made of the usual d-boson creators and annihilators. Formula (18) with $\epsilon = 1.3 MeV$, $a_1 = 0.01 MeV$ and $a_3 = -0.205$ MeV reproduces the low-lying levels of ^{64}Zn with $n_d \leq 3$ which form the spectrum of an anharmonic vibrator. The spectrum of the ^{63}Cu is obtained from the ^{64}Zn by replacing the proton boson with an unpaired proton whose valence space is limited to the first accessible orbital above the $Z = 28$ i.e. $2_{p_{3/2}}$, thus giving an $SU^{(f)}(4)$ symmetry to the fermion Hamiltonian

$$H_f = E_{3/2} a^+_{3/2} \cdot \tilde{a}_{3/2}, \qquad (19)$$

where E_j is the quasiparticle energy and a^+_{jm} creates a proton in the m substate of the j orbital. The modified annihilator is $\tilde{a}_{jm} = (-1)^{j-m} a_{j,-m}$.
the boson-fermion interaction is chosen of the form

$$H_{BF} = b_1 (B^{(1)} \times A^{(1)})^{(0)} + b_3 (B^{(3)} \times A^{(3)})^{(0)}, \qquad (20)$$

where $A^{(\lambda)}_\mu = (a^+_{3/2} \times \tilde{a}_{3/2})^{(\lambda)}_\mu$, $(\lambda = 1, 3, \mu = -\lambda, ..., +\lambda)$.
The low energy levels of the ^{63}Cu considered in the coupled-channels calculations of the present work are obtained from the total Hamiltonian $H = H_f + H_B + H_{BF}$ with $b_1 = -0.07$ MeV and $b_3 = -2.65$ MeV. It may be shown that this choice of the boson-fermion interaction Eq.(20) is consistent with a Spin(5) Bose-Fermi symmetry[11] of the Hamiltonian.
In Fig. 3 we present the results of the analysis of the elastic and inelastic scattering angular distributions for the system $^{16}O + ^{63}Cu$.

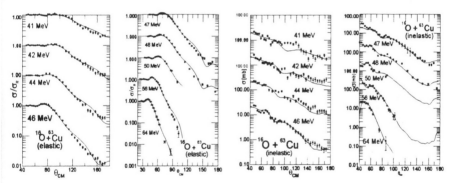

Figure 3: The fits of the elastic and inelastic angular distributions for the $^{16}O + ^{63}Cu$ system as indicated.

We considered the coupling with the five low lying states of the ^{63}Cu. The relative strengths of the reduced nuclear matrix elements used in this calculations have been determined in the context of the IBMF model as described above. It reduces the number of parameters to practically just an overall strength for the Coulomb excitation and a complex one for the nuclear excitation. The diagonal coupling elements were basically those determined from the analysis of the elastic angular distributions.

6 conclusions

We discuss some recent developments of the Algebraic Scattering Theory concerning the properties of the algebraic potential for the one channel elastic scattering. We also present a formalism which permits the inclusion of the spin in the algebraic description when many reaction channels are involved. The formalism is applied to the elastic and inelastic angular distributions for the ^{16}O $+^{63}Cu$ system at energies in the neighborhood of the Coulomb barrier.

7 references

1. Y. Alhassid, F. Iachello, *Nucl. Phys.* **A501** (1989) 585.

2. R. Lichtenthäler Filho, A. C. C. Villari, L. C. Gomes and P. Carrilho Soares, *Phys. Lett.* **B269** (1991) 49.

3. M. S. Hussein, M. P. Pato and F. Iachello, *Phys. Rev.* **C38** (1988)1072.

4. K. Amos, L. Berge, L. J. Allen and H. Fiedeldey, *Phys. Rev.* **C47** (1993) 2827.

5. S.K.Kauffmann *Relation of Phase Shifts to Potential Parameters in the Elastic Scattering of Very Heavy Ions* (unpublished) (1976).

6. W. E. Frahn in *Treatise on Heavy Ion Sciences*, (D. A. Bromley Ed.,Plenum Press, N. Y. 1984), vol. 1, p 216.

7. D. Pereira, G. R. Razeto, O. Sala, L. C. Chamon, C. A. Rocha, J. C. Acquadro and C. Tenreiro, *Phys. Lett.* **B220** (1989) 347.

8. R. Lichtenthäler Filho, A. Ventura, L. Zuffi, *Phys.Rev.* **C46** (1992) 707.

9. G.R.Satchler, *Direct Nuclear Reactions*(Oxford University Press, New York, 1983),p161.

10. R. A. Broglia and A. Winther, *Heavy Ion Reactions*, Addison-Wesley,1990.

11. F.Iachello and P.Van Isacker, *The Interacting Boson-Fermion Model* (Cambridge University Press, 1991).

INTERACTING BOSON TECHNIQUES IN CLUSTER STUDIES

J. CSEH, G. LÉVAI

Institute of Nuclear Research of the Hungarian Academy of Sciences
Debrecen, Pf. 51, H-4001, Hungary

and

W. SCHEID

Institut für Theoretische Physik der Justus-Liebig-Universität Giessen
Heinrich-Buff-Ring 16, D-35392 Giessen, Germany

ABSTRACT

We apply algebraic techniques, similar to those of the IBM in nuclear cluster studies. In the semimicroscopic algebraic cluster model a microscopically constructed model space, which is free from the Pauli forbidden states and from the spurious center of mass excitations, is combined with phenomenological forces. The cluster–cluster interactions, written in terms of group generators, can be related to effective two-nucleon forces. Low– and high–lying cluster states of light nuclei are described simultaneously in this approach. The extension of the model to heavy nuclei seems to be possible based on the pseudo SU(3) scheme. The use of more general algebraic structures is discussed.

1. Introduction

The algebraic techniques of the Interacting Boson Model[1] proved to be successful not only in treating the quadrupole collective motion, but also in connection with other degrees of freedom. Group theoretical descriptions of dipole rotations and vibrations[2] have been applied to nuclear molecular states[3], as well as to systems relevant to molecular and hadronic spectroscopy.

In order to describe cluster states of nuclei, one has to take into account the internal structure of the clusters in addition to their relative motion. The first attempt to this problem within the algebraic approach has been made on the phenomenological level of the Nuclear Vibron Model[4], and Vibron–Fermion Model[5]. It turns out, however, that the applicability of these treatments is rather limited, due to the essential role of the antisymmetrization, missing from these descriptions[6,7]. In order to overcome this difficulty, we have recently proposed[8] and developed[9] a semimicroscopic algebraic cluster model.

2. The Semimicroscopic Algebraic Cluster Model

2.1. Concept and Formalism

In the semimicroscopic algebraic cluster model the internal structure of a cluster is described in terms of the $SU(3)$ shell model[10], therefore its wavefunction is characterized by the $U_C^{ST}(4) \otimes U_C(3)$ symmetry, where C refers to cluster, and $U^{ST}(4)$

is Wigner's spin-isospin group[11]. The relative motion of the clusters is accounted for by the vibron model with $U_R(4)$ group structure[2]. The representation labels of the group chain

$$U_{C_1}^{ST}(4) \otimes U_{C_1}(3) \otimes U_{C_2}^{ST}(4) \otimes U_{C_2}(3) \otimes U_R(4)$$
$$\supset U_C^{ST}(4) \otimes U_C(3) \otimes U_R(3) \supset U_C^S(2) \otimes U(3)$$
$$\supset U_C^S(2) \otimes O(3) \supset U(2) \supset O(2) \quad (1)$$

provide us with the quantum numbers for the basis states of a two-cluster system. From this set we have to skip those states that are Pauli forbidden, or that correspond to spurious excitations of the center of mass. A simple recipe for eliminating these states is applying a matching requirement between the quantum numbers of the shell model basis of the whole nucleus and its cluster model basis[8,9]. This recipe is based on the connection between the harmonic oscillator shell model and harmonic oscillator cluster model[12]. This procedure corresponds to a special truncation of the extensive shell model basis in the sense, that only those states survive, which are Pauli–allowed, do not contain contribution from the spurious center of mass excitation, and are relevant to the cluster structure under study.

When the internal structure of each cluster is described by a single $U_C^{ST}(4) \otimes U_C(3)$ representation, then the physical operators of the system can be obtained in terms of the generators of the $U_{C_1}^{ST}(4) \otimes U_{C_1}(3) \otimes U_{C_2}^{ST}(4) \otimes U_{C_2}(3) \otimes U_R(4)$ group. In such a case the description is algebraically closed, i.e. the matrix elements can be deduced by means of group theoretical techniques. In the limiting case when the Hamiltonian is given by the invariant operators of (1), then the eigenvalue problem has an analytical solution, and a $U(3)$ dynamical symmetry is said to hold.

The problem can be simplified further if one or both of the clusters are $U_C^{ST}(4)$ and/or $U_C(3)$ scalars. In this case these groups and the quantum numbers associated with them do not appear explicitly in the formulas. In Ref.[9] the formalism is presented in detail for the $U_C(3) \otimes U_R(4)$ and $U_{C_1}(3) \otimes U_{C_2}(3) \otimes U_R(4)$ models, as well as for the restricted $U_C^{ST}(4) \otimes U_C(3) \otimes U_R(4)$ model. In this latter case the restriction implies that only spin- and isospin-free interactions and a single $U_C^{ST}(4)$ representation are considered. If both of the clusters are $U_C^{ST}(4)$ and $U_C(3)$ scalars, the model reduces to that of the simple vibron model with a basis truncation corresponding to the Wildermuth condition[6].

Here we give a brief account of the $U_C(3) \otimes U_R(4)$ model, which is able to describe two–cluster systems in which one of the clusters is a closed–shell nucleus (e.g. ^4He, ^{16}O, or ^{40}Ca), while the other one is an even–even nucleus with spin and isospin zero. In this simple case the basis states can be labeled without explicit reference to the $U^{ST}(4)$ group, (unless some higher excitations of the non–closed–shell nucleus are also considered), and the cluster model basis states are characterized by the representation labels of the group chain:

$$U_C(3) \otimes U_R(4) \supset U_C(3) \otimes U_R(3) \supset SU_C(3) \otimes SU_R(3) \supset SU(3) \supset O(3) \supset O(2)$$
$$|[n_1^C, n_2^C, n_3^C], [N, 0, 0, 0], \quad [n_\pi, 0, 0,], \ (\lambda_C, \mu_C) \ , \ (n_\pi, 0) \ , \ (\lambda, \mu), K_L, L \quad , M \ \rangle. \tag{2}$$

The irreducible representations (λ, μ) of $SU(3)$ are obtained by taking the outer product of $(\lambda_C, \mu_C) \otimes (n_\pi, 0)$. N stands for the maximal number of the excitation quanta assigned to the relative motion, and it determines the size of the model space. The angular momentum content of a (λ, μ) representation is given by the usual relations of the Elliott model[10]. For technical reasons, however, it is more convenient to use the orthonormal $SU(3)$ basis of Draayer and Akiyama[13], rather than the Elliott basis, which is not orthogonal. The parity of the basis states is determined by the parity assigned to the relative motion: $P_R = (-1)^{n_\pi}$. (The internal states of the non–$U(3)$–scalar cluster carry positive parity $P_C = (-1)^{n_1^C + n_2^C + n_3^C}$, unless major shell excitations of the clusters are also considered.)

The coupled wavefunction can be expressed in terms of $SU(3) \supset O(3)$ Wigner coefficients:

$$|(\lambda_C, \mu_C), N(n_\pi, 0); (\lambda, \mu)\chi LM\rangle$$
$$= \sum_{\chi_C L_C M_C} \sum_{L_R M_R} \langle (\lambda_C, \mu_C)\chi_C L_C M_C; N(n_\pi, 0)L_R M_R | (\lambda, \mu)\chi LM \rangle$$
$$\times |(\lambda_C, \mu_C)\chi_C L_C M_C\rangle |N(n_\pi, 0)L_R M_R\rangle. \tag{3}$$

The physical operators can be constructed from the generators of the groups present in group chain (2). In particular, the most general form of the Hamiltonian can be obtained in terms of a series expansion of these generators. In the simplest case, however, when we use the $SU(3)$ dynamical symmetry approximation, and consider only one $U_C(3)$ representation to describe the structure of the non–closed–shell even–even cluster, the energy eigenvalues can be obtained in a closed form. When written up to quadratic terms it reads:

$$E = \epsilon + \gamma n_\pi + \delta n_\pi^2 + \eta C_2(\lambda, \mu) + \beta L(L+1). \tag{4}$$

In this approximation the energy levels can easily be assigned to rotational bands labeled by the quantum numbers $n_\pi(\lambda, \mu)\chi$. (See Eqs. (3) and (4).)

The electromagnetic transition operators are also constructed from the group generators, which automatically implies selection rules in the dynamical symmetry approximation. The electric quadrupole transition operator, for example, is written as the sum of the rank–2 generators of the $U_C(3)$ and the $U_R(3)$ groups:

$$T^{(E2)} = q_R Q_R^{(2)} + q_C Q_C^{(2)}. \tag{5}$$

The matrix elements of the operators with the basis states (3) are calculated using tensor algebraic techniques[14].

The cluster spectroscopic factors can be calculated microscopically, by applying a $SU(3)$ basis[15]. In practice, however, this calculation is very complicated except for systems of small nucleon numbers. A phenomenological method, that is in line with the microscopic content, is discussed in Refs.[7,9].

The $U_{C_1}(3) \otimes U_{C_2}(3) \otimes U_R(4)$ and $U_C^{ST}(4) \otimes U_C(3) \otimes U_R(4)$ models can be formulated via a straightforward generalization of the results presented here.

The applications of these models away from the $SU(3)$ dynamical symmetry limit require numerical diagonalization of the Hamiltonian containing symmetry breaking terms as well.

Although the interactions applied in this approach are phenomenological ones, they can be related to the effective two-nucleon forces, due to the use of the microscopic $SU(3)$ cluster model basis. The connection is established by equating the corresponding matrix elements. See Ref.[16] for the details.

2.2. Applications

Applications of the semimicroscopic algebraic cluster model have been carried out so far within the $SU(3)$ dynamical symmetry approximation. This approximation allows exact analytical expression of the energy eigenvalues and electromagnetic transition rates in terms of reduced matrix elements, Clebsch–Gordan coefficients, etc. obtained from the algebraic description. Its validity can be estimated from the comparison of the results with the corresponding experimental data.

The $T = 1$ states of the ^{18}O nucleus have been described in terms of core–plus–alpha–particle configuration[17] by means of the restricted $U_C^{ST}(4) \otimes U_C(3) \otimes U_R(4)$ model. 34 energy levels of the experimental spectrum were identified with model states, and were assigned to 11 bands, some of which were new (Fig. 1). This spectrum is more complete, than those of the previous calculations. Reduced $E2$ and $E1$ transition probabilities were also determined. Their agreement with the experimental data is similar to those of the best microscopic or semimicroscopic calculations. Our results also showed strong correlation with those of fully microscopic calculations, which seems to indicate that the semimicroscopic algebraic cluster model approximates certain microscopic effects reasonably well.

In Ref.[18] the $U_{C_1}(3) \otimes U_{C_2}(3) \otimes U_R(4)$ model is used to treat the low–lying $T = 0$ states and the molecular resonances of the ^{24}Mg nucleus in a uniform $^{12}C + ^{12}C$ cluster description. We have analyzed about 150 experimental levels in the energy range of 0 to 40 MeV, and nearly 100 electric quadrupole transition probabilities in our study, which is a more complete account of the energy spectrum and $E2$ transitions of the ^{24}Mg nucleus than any previous model calculation. The description is reasonably good. This example demonstrates that a large number of experimental data, including the ground–state region as well as the highly excited molecular resonances can be reconciled in terms of relatively straightforward calculations, which is one of the major advantages of the semimicroscopic algebraic cluster model.

The extension of the model to heavy nuclei seems to be possible based on the pseudo $SU(3)$ scheme[19]. The $^{210}Pb + ^{14}C$ clusterization was found to be present in the ground state wavefunction of the ^{224}Ra nucleus. In addition, the low–energy spectrum could also be reproduced in terms of this cluster configuration.

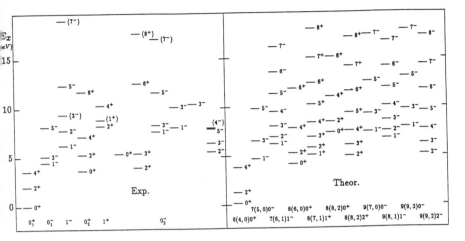

FIG. 1. Experimental and model spectra of ^{18}O from Ref.[17]. In the experimental part the K^π assignment indicate bands which were identified previously.

3. The Multichannel Character of the U(3) Dynamical Symmetry

The $U(3)$ dynamical symmetry of this approach has an interesting new feature: it is related to several cluster configurations. This is a consequence of the antisymmetry of the total wavefunction.

Let us now consider the case of states that have $U(3)$ quantum numbers with a single multiplicity both in the shell model description of the whole nucleus and in different cluster model bases. Due to the fact that the Hamiltonian of the harmonic oscillator shell model and the Hamiltonian of the harmonic oscillator cluster model have exactly the same eigenvalue spectrum[12], the relation of the $U(3)$ bases of these two description is very simple. Each basis state of a cluster configuration is a linear combination of those basis states of the shell model, that belong to the same energy. Since the basis states of different $U(3)$ labels are orthogonal, these wavefunctions are identical with each other. The general effect of the antisymmetrization, namely that seemingly different cluster configurations can overlap to a large extent, in the special case of the dynamical symmetry (and single multiplicity) results in complete identity of their wavefunctions. The consequences are interesting from different respects. For special nuclear states we obtain a selection rule for cold fission and fusion, while for the energy spectra we end up with correlations between the distribution of different cluster configurations and between the interactions of different reaction channels.

The link between the superdeformed and cluster states of alpha–like nuclei ($N = Z = even$) was explored in Ref.[20]. A spheroidal shape is called superdeformed here, when the ratio of its main axes can be expressed as ratios of natural numbers. They were obtained from Nilsson–Strutinsky calculations[21], and we addressed the question: what kind of alpha-like cluster configurations correspond to these states? It turns out, that in most cases several different clusterizations have the same wavefunction and in addition, the same cluster structure appears at different excitation energies of the same nucleus, depending on the relative orientations of the deformed clusters (Fig. 2).

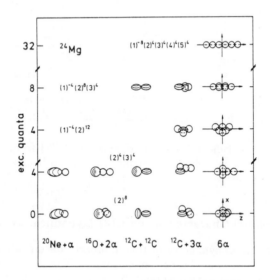

FIG. 2. Possible cluster configurations[20] of some highly deformed states of the ^{24}Mg nucleus. Only the directions of the symmetry axes and the directions of amalgamations of the clusters carry physical content, while the relative distances do not. The y axis is perpendicular and coming out of the plain.

The allowed and forbidden binary fission modes of ground–state–like configurations in sd shell nuclei were determined[22]. Similar investigations can be of great interest in the region of heavy nuclei, where different cluster emissions have been observed experimentally from the same states.

The fact that different clusterizations have a common part of their models spaces gives rise to well defined relation between the phenomenological cluster-cluster interactions, and as a consequence, there is a strong correlation between the distributions of different cluster configurations at low and high energies[23]. The high-lying cluster states are usually populated as resonances in heavy–ion reactions. Their analysis in terms of multichannel dynamical symmetry involves much more constraints, then in terms of the phenomenological models (algebraic or geometric), therefore we have much less ambiguities in this description than usual.

4. A Note on the Possible Use of Quantum Algebras

Quantum algebras are more general algebraic structures then the Lie algebras, that serve as a mathematical basis for the present as well as for many other models in nuclear physics. Therefore the question of how useful the new algebras can be in nuclear structure studies arises naturally. So far the investigations concentrated mainly on the possible extensions of the concept of dynamical symmetry, motivated by the following considerations. When a dynamical symmetry is based on a more general algebraic structure, it may give a better description of the experimental data, or it can cover the physically relevant territory of symmetry breaking (which is treated in Lie algebraic models by numerical diagonalization). Recently we have studied this question in simple, but non–trivial models. In Ref.[24] the relative motion of the $^{16}O+\alpha$ system is described in terms of an $SU_q(3)$ anharmonic oscillator model[25]. It turns out that the description of the energy spectrum can be improved when we deform the underlying (originally Lie) algebraic structure, but this deformation destroyes the description of the spectroscopic factors. Similar conclusion was found with a two–dimensional IBM model[26]: different physical quantities would require different group deformation for their best description. These examples indicate that for problems of nuclear structure the validity of the quantum algebraic dynamical symmetry does not go far beyond those of the Lie algebraic models, e.g. it can not take us from the vibrational limit to the rotational one.

5. Conclusions

Here we have presented a semimicroscopic algebraic cluster model, and some of its applications. In this approach the model space is constructed microscopically, and the interactions we apply are phenomenological ones. Nevertheless, their relation to effective two–nucleon–forces can be established. The model has been developed so far for two–cluster systems, and the applications have been carried out within the dynamical symmetry approach.

Extensions of this model to several directions are possible and desirable. E.g. symmetry breaking interactions could be applied, spin– and isospin–dependent forces can be introduced, multicluster systems should be investigated, etc. The relative advantages and disadvantages of the present approach can be deduced only after more systematic applications. Nevertheless, based on the first studies mentioned here, it seems to be promising for the description of complex cluster systems.

6. Acknowledgements

This work was supported by the OTKA (Contr. No. T14321) and by the PHARE ACCORD Programme (Contr. No. H9112-0694).

7. References

1. F. Iachello and A. Arima, *The Interacting Boson Model* (Cambridge University Press, Cambridge, 1987).
2. F. Iachello and R. D. Levine, J. Chem. Phys. **77** (1982) 3046.
3. K. A. Erb and D. A. Bromley, Phys. Rev. **C23** (1981) 2781.
   J. Cseh, Phys. Rev. **C31** (1985) 692.
4. H. J. Daley and F. Iachello, Ann. Phys. (N.Y.) **167** (1986) 73.
5. G. Lévai and J. Cseh, Phys. Rev. **C44** (1991) 152 and 166.
6. J. Cseh and G. Lévai, Phys. Rev. **C38** (1988) 972.
   J. Cseh, J. Phys. Soc. Jpn. Suppl. **58** (1989) 604.
7. J. Cseh, G. Lévai, and K. Katō, Phys. Rev. **C43** (1991) 165.
8. J. Cseh, Phys. Lett. **B281** (1992) 173.
9. J. Cseh and G. Lévai, Ann. Phys. (N.Y.), **230** (1994) 165.
10. J. P. Elliott, Proc. Roy. Soc. **A245** (1958) 128 and 562.
11. E. P. Wigner, Phys. Rev. **51** (1937) 106.
12. K. Wildermuth and Th. Kanellopoulos, Nucl. Phys. **7** (1958) 150.
13. J. P. Draayer, and Y. Akiyama, J. Math. Phys. **14** (1973) 1904.
14. B. G. Wybourne, *Classical Groups for Physicists*, (Wiley, New York, 1974).
15. H. Horiuchi and K. Ikeda, *Cluster Models and Other Topics*, (World Scientific, Singapore, 1986).
16. K. Varga and J. Cseh, Phys. Rev. **C48** (1993) 602.
17. G. Lévai, J. Cseh, and W. Scheid, Phys. Rev. **C46** (1992) 548.
18. J. Cseh, G. Lévai, and W. Scheid, Phys. Rev. **C48** (1993) 1724.
19. J. Cseh, R. K. Gupta, and W. Scheid, Phys. Lett. **B299** (1993) 205.
20. J. Cseh and W. Scheid, J. Phys. **G18** (1992) 1419.
21. G. Leander and S. E. Larsson, Nucl. Phys. **A239** (1975) 93.
22. J. Cseh, J. Phys. **G19** (1993) L97.
23. J. Cseh, Proc. Int. Conf. on Nuclear Reaction Mechanism, Varenna, 1994, in press.
24. J. Cseh, J. Phys. **G19** (1993) L63.
25. A. Del Sol Mesa, G. Loyola, M. Moshinsky and V. Velázquez, J. Phys **A26** (1993) 1147.
26. J. Cseh, J. Phys. **A25** (1992) L1225.

INTERACTING BOSONS IN MOLECULAR STRUCTURE

ALEJANDRO FRANK
Instituto de Ciencias Nucleares, UNAM
Apdo. Postal 70-543, México, D.F., 04510 México and
Instituto de Física, Laboratorio de Cuernavaca, UNAM
Apdo. Postal 139-B, Cuernavaca, Mor., México.

and

RENATO LEMUS
Instituto de Ciencias Nucleares, UNAM
Apdo. Postal 70-543, México, D.F., 04510 México.

ABSTRACT

We present some recent applications of algebraic techniques in molecular structure.

1. Introduction

The introduction of the interacting boson model twenty years ago gave rise to a remarkably fruitful period of research into nuclear structure physics.[1,2] Although by then symmetry methods had already proved to be of great value in different branches of physics, from solid state to nuclear and particle physics,[3] the IBM represented a radical proposal, where the language and methodology of group theory play a truly fundamental role. This is not only due to the elegance and simplification these techniques bring about in calculations, but because of the physical insights they provide for dealing with more complex situations, as illustrated by the numerous applications and new concepts that have arisen in the last two decades.

As a result of the success of these methods in nuclear structure physics, similar techniques have been proposed for the description of other physical systems, including molecules,[4] hadrons[5] and scattering processes,[6] where a common objective of the investigations is the establishment of algebraic models following the IBM paradigm. In the case of molecules, however, a new ingredient is required, namely the introduction of point symmetries into the mathematical framework. The algebraic approach to molecular structure should bring together the techniques of both continuous and discrete groups.[7]

In its original formulation, the vibron model associates a U(4) algebra to each molecular bond, in close analogy to the IBM case, but with S and P bosons as building blocks, the latter reflecting the dipole character of the local modes in each bond.[4] The model successfully describes the rotation-vibration spectra of small linear molecules, including infrared intensities.[8] Later on, the electronic degrees of freedom in diatomic molecules were incorporated through the introduction of fermionic operators, following in this case the IBFM paradigm, giving rise to a complete description of molecular excitations in the hydride molecules.[9] These studies included the evaluation of electronic potential energy surfaces, through the use of coherent states.[10]

Although these were important works that demonstrated the power of the algebraic approach, they were not yet capable of having an impact in the case of large molecules, where it is necessary to incorporate the discrete symmetries inherent to these systems.[11] The subsequent introduction of the vibronic U(2) models paved the way towards solving this problem, although they were originally restricted to the description of stretching vibrations.[12] In this formulation, vibrations and rotations are considered separately in a first approximation, and a purely vibronic Hamiltonian constructed. In this paper we present a description of this approach and an application to the complete vibrational spectrum of methane.[13]

2. The U(2) Model

The U(2) model is based on the isomorphism between a one-dimensional Morse oscillator and a two-dimensional harmonic oscillator, classified by the $U(2) \supset O(2)$ chain of groups.[14] In the algebraic approach, the Morse Hamiltonian has the realization

$$\hat{H}_M = -\frac{\hbar^2}{2\mu d^2} \hat{J}_z^2 \ , \qquad (1)$$

where $\hat{J}_z$ is the O(2) generator, while the potential depth is fixed by the total boson number N, which classifies the U(2) representations.[14,15] For more than one bond repeated U(2) couplings are necessary, a simple task given the well known nature of the SU(2) angular momentum algebra. All interactions can then be expressed in terms of the $\hat{J}_i(k)$ generators, where k refers to the bond number and i to the projection component. The great virtue of this procedure is the simplicity with which we can incorporate the point symmetries into the Hamiltonian and other operators.[12,15] In particular, for triatomic molecules one can readily identify bases which correspond to local and normal modes. These correspond to the chain of groups

$$U^1(2) \times U^2(2) \supset$$
$$O^1(2) \times O^2(2) \supset O^{12}(2) \qquad (2)$$

and

$$U^{12}(2) \supset O^{12}(2) \ , \qquad (3)$$

respectively.[16(libro)] The chains (2) and (3) imply phonon-number conservation (Dennison-Darling resonances), but phonon non-conservation (Fermi resonances) can be incorporated simply through the introduction of the $U(2) \supset U(1)$ chians. A very general procedure has been devised to generate the correct molecular symmetry.[15] For stretching vibrations this model leads to a unique and straightforward prescription. If we now wish to include the bending vibrations, however, there are complications due to the appearance of spurious states. Iachello and Oss have proposed to associate a U(2) algebra to each atomic degree of freedom, and later eliminate the (translational and rotational) spurious states through projection operators.[14] Using this technique they have been able to study the spectrum of benzene up to high overtones and combinations.[17] On the other hand, we have recently proposed a generalization of the model in which a U(2) algebra is associated to every interatomic potential. In this case one can systematically incorporate group theoretical techniques to construct a normal basis, identify the spurious states and eliminate them from the outset. Consequently, the method requires an appropriate combination of continuous and discrete group techniques.[13] We have carried out a preliminary test of these ideas in the context of non-linear triatomic molecules[18] and we report here a study of methane incorporating its full methodology.[13,19]

3. Tetrahedral Molecules

For tetrahedral molecules we assign a $U^i(2)$ algebra to each interaction present, as shown in Fig. 1. The first four algebras are chosen to correspond to the $A - B$

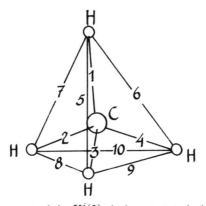

FIG. 1. Assignment of the $U^i(2)$ algebras to tetrahedral molecules.

interactions, while the other six represent the $B - B$ couplings. The molecular dynamical group is then given by the product $U^1(2) \times \ldots \times U^{10}(2)$, and the most general Hamiltonian, up to two body interactions, conserving the total number of quanta and invariant under the tetrahedral group $\mathcal{T}_d$, can be written as

$$\hat{\mathcal{H}} = \hat{\mathcal{H}}^S + \hat{\mathcal{H}}^B + \hat{V}^{S-B} \ . \qquad (4)$$

The term $\hat{\mathcal{H}}^S$ describes the stretching degrees of freedom and has the form

$$\hat{\mathcal{H}}^S = A_1 \sum_{i=1}^{4} \hat{C}_{2O^i(2)} + B_{12} \sum_{i=1}^{3} \sum_{j=i+1}^{4} \hat{C}_{2O^{ij}(2)} + \lambda_{12} \sum_{i=1}^{3} \sum_{j=i+1}^{4} \hat{\mathcal{M}}_{ij} ,$$

while $\hat{\mathcal{H}}^B$ is the bending contribution, given by

$$\hat{\mathcal{H}}^B = A_5 \sum_{i=5}^{10} \hat{C}_{O^i(2)} + B_{5,6} \left\{ \sum_{j=6}^{9} \sum_{i=5,10} \hat{C}_{2O^{ij}(2)} + \sum_{j=7,9} \sum_{i=6,8} \hat{C}_{2O^{ij}(2)} \right\}$$

$$+ B_{5,10} \left\{ \hat{C}_{2O^{5,10}(2)} + \hat{C}_{2O^{6,8}(2)} + \hat{C}_{2O^{7,9}(2)} \right\}$$

$$+ \lambda_{5,6} \left\{ \sum_{j=6}^{9} \sum_{i=5,10} \hat{\mathcal{M}}_{ij} + \sum_{j=7,9} \sum_{i=6,8} \hat{\mathcal{M}}_{ij} \right\}$$

$$+ \lambda_{5,10} \left\{ \hat{\mathcal{M}}_{5,10} + \hat{\mathcal{M}}_{6,8} + \hat{\mathcal{M}}_{7,9} \right\} .$$

The last operator, $\hat{V}^{S-B}$, represents the stretching-bending interactions, which will be neglected as a first approximation. In these expressions $\hat{C}_{2O^{ij}(2)}$ corresponds to the $O^{ij}(2)$ Casimir invariant, while $\hat{\mathcal{M}}_{ij}$ is the Majorana operator, which is related to the $U^{ij}(2)$ Casimir operator.[12,15]

The simplest basis to diagonalize the Hamiltonian is the one associated to the local-mode chain[12,15]

$$\begin{array}{ccccccc}
U^{(1)}(2) \times \ldots \times U^{(4)}(2) \times U^{(5)}(2) \times \ldots \times U^{(10)}(2) & \supset & O^{(1)}(2) \times \ldots \times O^{(10)}(2) & \supset & O(2) \\
\downarrow \quad \quad \quad \downarrow \quad \quad \downarrow \quad \quad \quad \downarrow & & \downarrow \quad \quad \quad \downarrow & & \downarrow \\
|[N_1] , \ldots [N_1] \quad [N_2] , \ldots , [N_2] & & v_1 , \ldots , v_{10}; & & V >,
\end{array}$$
(5)

where below each group we have indicated the quantum numbers characterizing the eigenvalue of the corresponding invariant operator. The two boson numbers N_1 and N_2, are related to the two sets of physical modes (stretching and bending). The quantum numbers v_i correspond to the number of phonons in each oscillator ($v_i = \frac{N_i}{2} - m_i$), while $V = \sum_{i=1}^{10} v_i$.

A simple analysis of an AB_4 tetrahedral molecule[11] shows that it presents 9 vibrational degrees of freedom, four of them corresponding to the fundamental stretching modes ($A_1 \oplus F_2$) and the other five to the fundamental bending modes ($E \oplus F_2$). Comparing this result with the local basis (2), we deduce that an unphysical bending mode is present in the algebraic formalism. We thus proceed to eliminate this spurious state both from the Hamiltonian and the basis.

To accomplish this goal we first transform, for the one phonon case, the local basis to a normal one, which carries the irreducible representations (irreps) of the

T_d group. With this change of basis we obtain the decomposition $A_1 \oplus F_2$ for the stretches and $A_1 \oplus E \oplus F_2$ for the bends. From this result we readily identify the A_1 bending mode as the spurious state. We now eliminate this spurious state from the space and proceed to construct the higher phonon basis from the physical one-phonon set by means of the coupling coefficients $C(\ ;\ )$

$$^{v_1+v_2}\Psi_\gamma^\Gamma = \sum_{\gamma_1 \gamma_2} C(\Gamma_1 \Gamma_2 \Gamma; \gamma_1 \gamma_2 \gamma)\ ^{v_1}\Psi_{\gamma_1}^{\Gamma_1}\ ^{v_2}\Psi_{\gamma_2}^{\Gamma_2} ,\qquad (6)$$

where Γ and γ label the irreps of T_d and its components, respectively.

To eliminate the spurious contributions from the Hamiltonian we demand its null expectation value with respect to the one-phonon spurious functions[20]

$$< {}^1\Psi_{bending}^{A_1} |\hat{\mathcal{H}}| {}^1\Psi_{bending}^{A_1} > = 0 , \qquad (7)$$

which leads to a constraint

$$4(1-N_2)A_5 + 16(1-2N_2)B_{5,6} + 4(1-2N_2)B_{5,10} = 0$$

between the interaction parameters.

The vibrational energies are obtained by diagonalizing the Hamiltonian (4) with respect to the normal basis (6), constructed from the projected one-phonon functions (A_1, F_2)—stretching and (E, F_2)— bending, taking into account the constraint (7).

4. Methane

We now apply this approach to describe the vibrational levels of methane. According to the Hamiltonian (4) the number of parameters is eight, plus the boson numbers N_1 and N_2. The vibron number N_1 can be fixed from the anharmonicity of the $C - H$ bond, while for the bending vibrations we have taken N_2 from the $H - H$ interaction in H_2O given in reference [16]. From these considerations, the number of free parameters is seven, taking into account the constraint (7).

The Hamiltonian (4) is diagonalized in the normal basis, built by repeated couplings of the form (6). Since by construction this basis is symmetry adapted, the Hamiltonian matrix separates into blocks corresponding to the irreps of T_d. In Table I we present the least square fit for methane up to three quanta. Following Herzberg's notation,[11] the four fundamental energies for A_1, F_2 (stretching) and E, F_2 (bending) have been denoted by ν_1, ν_3, ν_2 and ν_4, respectively. The final parameters are given in Table II. The model in its simplest form (without including the V^{B-S} interaction or higher order terms) provides a good description of 19 experimental energy levels with an rms deviation of 12.16 cm^{-1} and theoretical predictions for many unobserved levels.[19]

TABLE I. Experimental [11,20] and calculated energies (cm^{-1}) for methane.

V	Normal label	Γ	Expt.	Calc.	V	Normal label	Γ	Expt.	Calc.
1	ν_4	F_2	1310.0	1303.7	3	$\nu_1+\nu_2+\nu_4$	F_1		5745.6
	ν_2	E	1533.0	1520.4			F_2	5775.0	5759.9
	ν_1	A_1	2916.5	2918.4			A_2		5854.4
	ν_3	F_2	3019.4	3027.2			E		5854.4
2		A_1		2474.5			F_1		5854.4
	$2\nu_4$	E		2476.4		$\nu_2+\nu_3+\nu_4$	F_2		5854.4
		F_2	2614.0	2610.5			A_1		5868.7
	$\nu_2+\nu_4$	F_1		2827.2			E		5868.7
		F_2	2830.4	2841.5			F_1		5868.7
	$2\nu_2$	A_1		3003.7			F_2	5861.0	5868.7
		E		3026.3		$\nu_1+2\nu_2$	A_1		5922.0
	$\nu_1+\nu_4$	F_2	4223.0	4222.0			E		5944.7
		A_1		4330.9			F_2		6030.9
	$\nu_3+\nu_4$	E		4330.9		$2\nu_2+\nu_3$	F_1		6053.5
		F_1		4330.9			F_2		6053.5
		F_2	4319.0	4330.9		$2\nu_1+\nu_4$	F_2		7091.7
	$\nu_2+\nu_3$	F_1		4547.7			A_1		7160.4
		F_2	4549.0	4547.7		$\nu_1+\nu_3+\nu_4$	E		7160.4
	$\nu_1+\nu_2$	E		4438.8			F_1		7160.4
	$2\nu_1$	A_1		5788.0			F_2		7160.4
	$\nu_1+\nu_3$	F_2	5861.0	5856.7			F_2		7278.1
		A_1		5974.4			E		7308.4
	$2\nu_3$	F_2	6004.7	6014.5			A_1		7318.2
		E		6047.7		$2\nu_3+\nu_4$	F_1		7318.2
3		F_2		3624.3			F_2		7318.2
	$3\nu_4$	F_1		3778.3			F_2		7351.4
		F_2		3779.4			F_1		7351.4
		A_1		3920.4		$2\nu_1+\nu_2$	E		7318.2
		A_1		3925.7		$\nu_1+\nu_2+\nu_3$	F_1		7377.1
		E		3935.6			F_2		7377.1
	$\nu_2+2\nu_4$	E		3987.9			E		7494.8
		A_2		4017.6			F_1		7534.9
		F_2	4123.0	4123.9		$\nu_2+2\nu_3$	F_2	7514.0	7534.9
		F_1		4260.4			E		7568.1
		F_1		4425.5			A_2		7568.1
	$2\nu_2+\nu_4$	F_2		4317.4			A_1		7568.1
		F_2		4387.6		$3\nu_1$	A_1		8581.1

V	Normal label		Γ	Expt.	Calc.	V	Normal label		Γ	Expt.	Calc.
	$3\nu_2$	$\Big\{$	A_1		4495.3		$2\nu_1 + \nu_3$	$\Big\{$	F_2	8604.0	8603.0
			E		4510.9				A_1		8725.5
			A_2		4575.7		$\nu_1 + 2\nu_3$	$\Big\{$	F_2	8807	8794.1
	$\nu_1 + 2\nu_4$	$\Big\{$	A_1		5392.8				E		8838.5
			E		5394.7				F_2	8900.0	8910.0
			F_2		5501.7		$3\nu_3$	$\Big\{$	F_1		8944.8
	$\nu_3 + 2\nu_4$	$\Big\{$	F_1		5503.6				A_1		8982.1
			F_2		5503.6				F_2	9045.0	9034.5
			F_2		5528.8						
			F_2		5637.7						
			A_1		5637.7						
			E		5637.7						
			F_1		5637.7						

TABLE II. Parameters of the Hamiltonian obtained in the least square fitting (cm^{-1}). The numbers of bosons are taken to be N$_1$=43 and N$_2$=28.

Stretching			Bending				
A_1	B_{12}	λ_{12}	A_5	$B_{5,6}$	$B_{5,10}$	$\lambda_{5,6}$	$\lambda_{5,10}$
−13.2125	−0.6850	0.6328	35.4844	2.6492	−28.0164	9.0501	5.1799

5. Conclusions

We have discussed the application of algebraic techniques to molecular structure calculations and presented a new method, which applied to an algebraic model of coupled anharmonic oscillators is able to describe the complete vibrational spectrum of polyatomic molecules. We emphasize that the method systematically incorporates group theoretical techniques which simplify the diagonalization of the Hamiltonian and provide a clear methodological procedure that can be applied to other molecules.[19] Although we have used the model in its simplest form, it can be improved by the inclusion of the stretching-bending interactions $\hat{V}^{S-B}$, by introducing in the Hamiltonian higher order terms and by incorporating interactions which do not conserve the total number of quanta.

Acknowledgments

We are grateful to F. Iachello and P. Van Isacker for their continuous interest and useful comments. This work was supported in part by CONACyT, Mexico, under project 400340-5-3401E.

References

1. F. Iachello and A. Arima, *The Interacting Boson Model*, Cambridge University Press, Cambridge, 1987.
2. F. Iachello and P. Van Isacker, *The Interacting Boson-Fermion Model*, Cambridge University, Press, Cambridge, 1991.
3. J. P. Elliott and P. G. Dawber, *Symmetry in Physics*, Oxford University Press, New York, 1979.
4. F. Iachello, *Chem. Phys. Lett.* **78** (1981) 581.
5. F. Iachello, *Nucl. Phys.* **A560** (1993) 23.
6. Y. Alhassid, F. Gürsey and F. Iachello, *Ann. Phys. (NY)* **167** (1986) 181.
7. M. Hamermesh, *Group Theory and its Application to Physical Problems*, Addison Wesley, Reading, MA, 1962.
8. O. S. van Roosmalen, *et.al.*, *J. Chem. Phys.* **79** (1983) 2515; F. Iachello, S. Oss and R. Lemus, *J. Mol. Spectr.* **146** (1991) 56, Ibid., **149** (1991) 132.
9. R. Lemus and A. Frank, *Ann. Phys. (NY)* **206** (1991) 122, Ibid., *Phys. Rev.* **A47** (1993) 4920; Ibid., *Phys. Rev. Lett.* **66** (1991) 2863.
10. R. Lemus, A. Leviatan and A. Frank, *Chem. Phys. Lett.* **194** (1992) 327.
11. G. Herzberg, *Molecular Spectra and Molecular Structure*, van Nostrand, New York, 1945.
12. F. Iachello and S. Oss, *Phys. Rev. Lett.* **66** (1991) 2976.
13. R. Lemus and A. Frank, *Phys. Rev. Lett.* to be published.
14. A. Frank in *Nuclear Physics at the Borderlines*, Springer Verlag 111, (1992).
15. A. Frank and R. Lemus, *Phys. Rev. Lett.* **68** (1992) 413.
16. A. Frank ad P. Van Isacker, *Algebraic Methods in Molecular and Nuclear Structure Physics*, Wiley, N.Y., (1994).
17. F. Iachello and S. Oss, *Chem. Phys. Lett.* **205** (1993) 205; Ibid., *J. Chem. Phys.* **99** (1993) 7337.
18. J. M. Arias, A. Frank, R. Lemus and F. Pérez-Bernal, *et.al. J. Molec. Spectr.*, to be published.
19. R. Lemus and A. Frank, *J. Chem. Phys.*, to be published.
20. J. C. Hilico, *J. Phys. Paris* **31** (1970) 289; B. Bobin and G. Guelachvili, *J. Phys. Paris* **39**, (1978) 33.

age 545

Description of Nuclear Structure Effects in Subbarrier Fusion by the Interacting Boson Model

A.B. Balantekin
Department of Physics, University of Wisconsin,
Madison, Wisconsin 53706 USA

Abstract

Recent theoretical developments in using the Interacting Boson Model to describe nuclear structure effects in fusion reactions below the Coulomb barrier are reviewed. It is shown that including higher order coupling effects between the nuclear excitations and the translational motion leads to improved agreement between calculated barrier distributions and the data as compared with the linear coupling. Including only s and d bosons are shown to be sufficient to calculate physically observable quantities.

1 Introduction

Many researchers have established that although a simple barrier penetration model of two colliding spherical nuclei describes well the fusion of light nuclei, for heavier systems one needs to include coupling to other degrees of freedom [1, 2]. These extra degrees of freedom yield a distribution of barriers [3] and consequently enhance the cross section below the barrier [4, 5]. Enhancement of fusion cross sections due to coupling of levels in colliding nuclei to the relative motion, has opened a new avenue for testing nuclear structure models. The recent experimental determination of average angular momenta [6] and the barrier distributions [7] have provided even more stringent tests for such models.

The natural language to study multidimensional barrier penetration is the coupled channels formalism, which was widely used in investigating subbarrier fusion phenomena [2]. An alternative formulation of the multidimensional quantum tunnelling is given by the path integral formalism [4]. In all these studies either the geometrical model of Bohr and Mottelson [8] or its simplifications were used to describe the nuclear structure effects. Especially in the path integral formulation of the problem, an algebraic nuclear structure model significantly simplifies evaluation of the path integral. The Interacting Boson Model (IBM) of Arima and Iachello [9] is one such model which has been successfully employed to describe the properties of low-lying collective states in medium heavy nuclei. Here attempts to use IBM in describing nuclear structure effects are reviewed.

Path integral formulation of this problem, as sketched in the next section, requires analytic solutions for the nuclear wave functions. In our first attempt in using IBM to describe nuclear structure effects in subbarrier fusion, we employed the SU(3) limit of IBM [10]. However, the SU(3) limit corresponds to a rigid nucleus with a particular quadrupole deformation and no hexadecapole deformation, a situation which is not realized in most deformed nuclei. Thus analytic solutions away from the limiting symmetries of the IBM are needed for realistic calculations of subbarrier fusion cross sections.

In a parallel development, a $1/N$ expansion was investigated [11] for the IBM which provided analytic solutions for a general Hamiltonian with arbitrary kinds of bosons. This technique proved useful in a variety of nuclear structure problems where direct numerical calculations are prohibitively difficult. Later it was applied to medium energy proton scattering from collective nuclei [12] in the Glauber approximation, generalizing the earlier work done using the SU(3) limit [13]. As we briefly describe in the next section, using the $1/N$ expansion technique in the path integral formulation of the fusion problem [14] makes it possible to go away from the three symmetry limits of IBM, in particular arbitrary quadrupole and hexadecapole couplings can be introduced.

Recently a series of recent high-precision measurements were carried out at the Australian National University where distributions of fusion barriers were determined directly [7]. IBM based description of fusion cross sections describes the Australian National University data well [14, 15, 16, 17] especially when the higher order coupling effects are included.

In the next section first the influence functional method is briefly summarized and its application to the linear coupling of translational motion to the structure of target nuclei is described. The significance of nonlinear couplings is discussed and a Green's function method to describe such nonlinear effects is discussed. Section 3 includes attempts to describe subbarrier fusion data with these techniques.

2 Algebraic Models in Subbarrier Fusion

The Hamiltonian for the multidimensional quantum tunnelling problem relevant to subbarrier fusion is

$$H = H_k + H_0(\xi) + H_{\text{int}}(\mathbf{r}, \xi_i) \quad (1)$$

with the term H_k representing the kinetic energy

$$H_k = -\frac{\hbar^2}{2\mu}\nabla^2, \quad (2)$$

where **r** is the relative coordinate of the target and projectile and ξ represents any internal degrees of freedom of the target. In this equation the term $H_0(\xi)$ represents the internal structure of the target nucleus and the propagator to go from an initial state characterized by relative radial coordinate (the magnitude of **r**) r_i and internal quantum numbers n_i to a final state characterized by the radial position r_f and the internal quantum numbers n_f may be written as

$$K(r_f, n_f, T; r_i, n_i, 0) = \int \mathcal{D}[r(t)] e^{\frac{i}{\hbar}S(r,T)} W_{n_f n_i}(r(t), T), \qquad (3)$$

where $S(r,T)$ is the action for the translational motion and $W_{n_f n_i}$ is the propagator for the internal system:

$$W_{n_f n_i}(r, T) = \left\langle n_f \left| \hat{U}_{\text{int}}(r(t), T) \right| n_i \right\rangle. \qquad (4)$$

$\hat{U}_{\text{int}}$ satisfies the differential equation

$$i\hbar \frac{\partial \hat{U}_{\text{int}}}{\partial t} = [H_0 + H_{\text{int}}] \hat{U}_{\text{int}}, \qquad (5)$$

$$\hat{U}_{\text{int}}(t=0) = 1. \qquad (6)$$

We want to consider the case where r_i and r_f are on opposite sides of the barrier. In the limit when the initial and final states are far away from the barrier, the transition amplitude is given by the S-matrix element, which can be expressed in terms of the propagator as [4]

$$S_{n_f, n_i}(E) = -\frac{1}{i\hbar} \lim_{\substack{r_i \to \infty \\ r_f \to -\infty}} \left(\frac{p_i p_f}{\mu^2}\right)^{\frac{1}{2}} \exp\left[\frac{i}{\hbar}(p_f r_f - p_i r_i)\right]$$

$$\int_0^\infty dT e^{+iET/\hbar} K(r_f, n_f, T; r_i, n_i, 0), \qquad (7)$$

where p_i and p_f are the classical momenta associated with r_i and r_f. In heavy ion fusion we are interested in the transition probability in which the internal system emerges in any final state. For the ℓth partial wave, this is

$$T_\ell = \sum_{n_f} |S_{n_f, n_i}(E)|^2, \qquad (8)$$

which becomes, upon substituting Eqs. (3) and (7),

$$T_\ell = \lim_{\substack{r_i \to \infty \\ r_f \to -\infty}} \left(\frac{p_i p_f}{\mu^2}\right) \int_0^\infty dT e^{\frac{i}{\hbar}ET} \int_0^\infty \tilde{T} e^{-\frac{i}{\hbar}E\tilde{T}}$$

$$\int \mathcal{D}[r(t)] \int \mathcal{D}[\tilde{r}(\tilde{t})] \exp^{\frac{i}{\hbar}(S(r,T) - S(\tilde{r},\tilde{T}))} \rho_M(\tilde{r}(\tilde{t}), \tilde{T}; r(t), T). \qquad (9)$$

Here we have assumed that the energy dissipated to the internal system is small compared to the total energy and taken p_f outside the sum over final states. We identified the two-time influence functional as

$$\rho_M(\tilde{r}(\tilde{t}),\tilde{T};r(t),T) = \sum_{n_f} W^*_{n_f,n_i}(\tilde{r}(\tilde{t});\tilde{T},0)W_{n_f,n_i}(r(t);T,0). \tag{10}$$

Using the completeness of final states, we can simplify this expression to write

$$\rho_M(\tilde{r}(\tilde{t}),\tilde{T};r(t),T) = \left\langle n_i \left| \hat{U}^\dagger_{\text{int}}(\tilde{r}(\tilde{t}),\tilde{T})\hat{U}_{\text{int}}(r(t),T) \right| n_i \right\rangle. \tag{11}$$

Eq. (11) displays the utility of the influence functional method when the internal system has symmetry properties. If the Hamiltonian in Eq. (5) has a dynamical or spectrum generating symmetry, i.e. if it can be written in terms of the Casimir operators or generators of a given Lie algebra, then the solution of Eq. (5) is an element of the corresponding Lie group [4]. Consequently the two time influence functional of Eq.(11) is simply a diagonal group matrix element for the lowest-weight state and it can be evaluated using standard group-theoretical methods. This is exactly the reason why the path integral method is very convenient when the internal structure is represented by an algebraic model such as IBM.

We take H_{int} to be of the form of the most general one-body transition operator for IBM,

$$H_{\text{int}} = V(r) + \sum_{kj\ell} a_{kj\ell}(r) \left[b^\dagger_j \tilde{b}_\ell \right]^{(k)} \cdot Y^{(k)}(\hat{\mathbf{r}}), \tag{12}$$

where the boson operators are denoted by b_ℓ and $b^\dagger_j$. The k sum runs over $k = 2, 4, \ldots 2\ell_{\text{max}}$. Odd values of k are excluded as a consequence of the reflection symmetry of the nuclear shape; the $k = 0$ term is already included in the bare potential $V(r)$. The form factors $\alpha_{kj\ell}(r)$ represent the spatial dependence of the coupling between the intrinsic and translational motions. The interaction term given in Eq.(12) is an element of the $SU(6)$ algebra for the original form of the Interaction Boson Model with s and d bosons and is an element of the $SU(15)$ algebra when g bosons are included as well [9].

To simplify the calculation of the influence functional, we can perform a rotation at each instant to a frame in which the z-axis points along the direction of relative motion. Neglecting the resulting centrifugal and Coriolis terms in this rotating frame is equivalent to ignoring the angular dependence of the original Hamiltonian. In this approximation, the coupling form factors become independent of ℓ and only $m = 0$ magnetic substates of the target are excited [18]. For heavy systems the neglected centrifugal and Coriolis forces are small. We take the scattering to be in the x-y plane. Then making a rotation through the Euler angles $\hat{\mathbf{b}} = (\phi, \pi/2, 0)$, we can write the Hamiltonian as the rotation of a simpler Hamiltonian depending only on the magnitude of $\mathbf{r}$

$$H = R(\hat{\mathbf{b}})H^{(0)}(r)R^\dagger(\hat{\mathbf{b}}). \tag{13}$$

Since in Eq.(1) H_0 and H_k are rotationally invariant, H_{int} is the only term whose form is affected by the transformation. Hence we introduce the rotated interaction Hamiltonian $H_{int}^{(0)}(r)$, given by

$$H_{int} = R(\hat{b})H_{int}^{(0)}(r)R^{\dagger}(\hat{b}), \tag{14}$$

$$H_{int}^{(0)}(r) = \sum_{j\ell m} \phi_{j\ell m}(r) b_{jm}^{\dagger} b_{\ell m}, \tag{15}$$

$$\phi_{j\ell m}(r) = (-)^m \sum_k \sqrt{\frac{2k+1}{4\pi}} \langle jm\ell - m|k0\rangle \alpha_{kj\ell}(r). \tag{16}$$

If we assume now that the form factors $\alpha_{kj\ell}(r)$ are all proportional to the same function of r then the Hamiltonian $H_{int}^{(0)}$ commutes with itself at different times and hence we can write the two-time influence functional as

$$\rho_M = \left\langle n_i \left| e^{\frac{i}{\hbar}\int_0^{\overline{T}} dt H_{int}^{(0)}(\bar{r}(t))} e^{-\frac{i}{\hbar}\int_0^T dt H_{int}^{(0)}(r(t))} \right| n_i \right\rangle. \tag{17}$$

Eq. (17) illustrates the utility of using an algebraic model such as the Interacting Boson Model to describe nuclear structure effects within the path integral formalism. Since the exponents of the two operators in the influence functional commute, ρ_M becomes the matrix element of an SU(6) transformation between SU(6) basis states, in other words it is a representation matrix element for this group and can easily be calculated using standard techniques. The two-time influence functional for the sd-version of IBM was calculated in Ref. [14] and, for the particular case of SU(3) limit, in Ref. [10].

Up to this point we utilized only a first-order coupling between nuclear states and translational motion. One can include the effects of coupling to all orders. This can be achieved by exploiting the symmetry properties of the resolvent operator directly without utilizing its path integral representation. Such a Green's function approach has also been used to study quantum tunnelling in a heat bath [19].

To include the effects of couplings to all orders, the interaction Hamiltonian in Eq. (1) is written as

$$H_{int}(\mathbf{r},\xi) = V_{\text{Coul}}(\mathbf{r},\xi) + V_{\text{nuc}}(\mathbf{r},\xi), \tag{18}$$

where the Coulomb part is

$$V_{\text{Coul}}(\mathbf{r},\xi) = \frac{Z_1 Z_2 e^2}{r}(1 + \frac{3}{5}\frac{R_1^2}{r^2}\hat{O}) \quad (r > R_1),$$
$$= \frac{Z_1 Z_2 e^2}{r}(1 + \frac{3}{5}\frac{r^2}{R_1^2}\hat{O}) \quad (r < R_1). \tag{19}$$

The nuclear part is taken to have Woods-Saxon form,

$$V_{\text{nuc}}(\mathbf{r},\xi) = -V_0 \left(1 + exp\frac{r - R_0 - R_1\hat{O}(\hat{r},\xi)}{a}\right)^{-1}. \tag{20}$$

In Eqs. (19) and (20), R_0 is the sum of the target and projectile radii and R_1 is the mean radius of the deformed target. $\hat{O}$ is a general coupling operator between the internal coordinates and the relative motion

$$\hat{O} = \sum_k v_k T^{(k)}(\xi) \cdot Y^{(k)}(\hat{r}). \tag{21}$$

The v_k represent the strengths of the various multipole transitions in the target nucleus. In the standard IBM with s and d bosons, the only possible transition operators have $k = 0, 2, 4$, odd values being excluded as a consequence of the reflection symmetry of the nuclear shape. The monopole contribution is already included in the Woods-Saxon parameterization and so is not needed. The quadrupole and hexadecapole operators are given by

$$\begin{aligned} T^{(2)} &= [s^\dagger \tilde{d} + d^\dagger s]^{(2)} + \chi [d^\dagger \tilde{d}]^{(2)}, &(22)\\ T^{(4)} &= [d^\dagger \tilde{d}]^{(4)}. &(23) \end{aligned}$$

We adopt the "consistent-Q" formalism of Casten and Warner [20], in which χ in Eq. (22) is taken to be the same as in H_{IBM} (fitted to reproduce the energy level scheme and the electromagnetic transition rates of the target nucleus) and is thus not a free parameter.

In the previous section, we used the usual approximation in which the nuclear potential of Eq. (20) is expanded in powers of the coupling, keeping only the linear term (cf. Eq. (12)). In order to calculate the fusion cross section to all orders we consider the resolvent operator for the system

$$G^+(E) = \frac{1}{E^+ - H_k - H_{IBM}(\xi) - H_{\text{int}}(r, \hat{O})}. \tag{24}$$

The basic idea is to identify the unitary transformation which diagonalizes the operator $\hat{O}$

$$\hat{O}_d = \mathcal{U} \hat{O} \mathcal{U}^\dagger \tag{25}$$

in order to calculate its eigenvalues and eigenfunctions

$$\hat{O}_d |n\rangle = h_n |n\rangle. \tag{26}$$

Assuming the completeness of these eigenfunctions

$$\sum_n |n\rangle \langle n| = 1 \tag{27}$$

one can write the matrix element of the resolvent as

$$\begin{aligned} &\langle \xi_f, r_f | G^+(E) | \xi_i, r_i \rangle \\ &= \langle \xi_f, r_f | \mathcal{U}^\dagger \mathcal{U} \left[E^+ - H_k - H_{\text{int}}(r, \hat{O}) \right]^{-1} \mathcal{U}^\dagger \mathcal{U} | \xi_i, r_i \rangle \\ &= \langle \xi_f, r_f | \mathcal{U}^\dagger \left[E^+ - H_k - H_{\text{int}}(r, \hat{O}_d) \right]^{-1} \sum_n |n\rangle \langle n| \mathcal{U} | \xi_i, r_i \rangle \\ &= \sum_n \langle \xi_f | \mathcal{U}^\dagger |n\rangle \langle n| \mathcal{U} | \xi_i \rangle \langle r_f | G_n^+ | r_i \rangle \end{aligned} \tag{28}$$

where
$$G_n^+(E) = \frac{1}{E^+ - H_k - H_{\text{int}}(r,h_n)}. \quad (29)$$

Since we ignored the excitation energies in the target nucleus, i.e. the term H_{IBM}, then $G_n^+(E)$ given in Eq. (20) is the resolvent operator for one-dimensional motion in the potential $H_{\text{int}}(r,h_n)$, the fusion cross section of which can easily be calculated within the standard WKB approximation. The total cross section can be calculated by multiplying these eigenchannel cross sections by the weight factors indicated in Eq. (28). The calculation of the matrix element $\langle n|\mathcal{U}|\xi_i\rangle$ within the Interacting Boson Model is straightforward and is given in Refs. [15, 16, 17], to which the reader is referred for further details.

3 A Systematic Study of Subbarrier Fusion in Rare Earth Nuclei

The inclusion of higher order effects is expected to have a large effect on subbarrier fusion leading to orders of magnitude enhancement of the cross sections compared to the linear couplings. We have shown the difference between linear and higher order coupling effects for the ^{16}O + ^{154}Sm system in Figure 1 [15]. In this figure the parameters of the Woods-Saxon potential and the coupling strengths v_k are taken to be $V_0 = 63.93$ MeV, $R_0 = 8.35$ fm, $a = 1.17$ fm, $v_2 = 0.20$, $v_4 = 0.17$, leading to a potential barrier with height 59.2 MeV, radius 10.5 fm and curvature at the top of the barrier 2.8 MeV. Full coupling results for the fusion cross section and distribution of barriers are shown with the solid lines and linear coupling results with dashed lines. One sees that including couplings to all orders gives a much larger cross section at subbarrier energies, and a barrier distribution of rather different shape.

Figure 1 indicates the need to renormalize the interaction strengths in order to describe the data. Even after renormalization of the interaction strengths so that cross sections calculated with linear and nonlinear couplings roughly agree, one expects the barrier distributions to differ in details.

Having established the importance of nonlinear couplings, we made a systematic study of subbarrier fusion data, accumulated over the last few years, for rare-earth nuclei [16]. We first show in Figure 2 the barrier distributions in the three limits of the IBM, namely the SU(5) (dotted) line), SU(3) (dashed line), and O(6) (solid line) which represent the vibrational, rotational, and γ-unstable nuclei. We next fitted the existing data on vibrational and rotational nuclei with a consistent set of parameters. To illustrate the quality of this global fit, in Figure 3 we compare our theoretical result for the ^{16}O + ^{154}Sm system with the data from the Australian National University. There is excellent agreement between IBM-based model and the data.

Having extracted a consistent set of parameters we can predict fusion cross sections and barrier distributions in transitional nuclei. One expects that barrier distributions for transitional nuclei exhibit sharp changes due to shape phase transitions, contrary

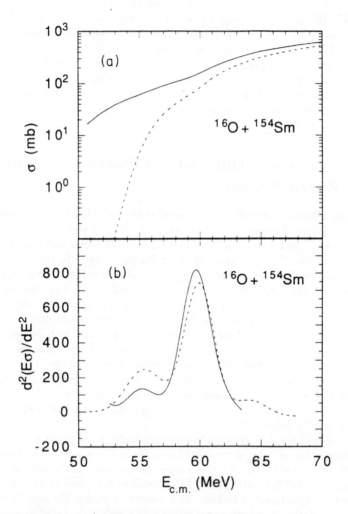

Figure 1. A comparison between the effect of coupling to all orders (solid line) and linear coupling (dashed line) on fusion cross section (a) and the distribution of barriers (b).

to vibrational and rotational nuclei where the cross sections increase smoothly with increasing deformation or mass number [16]. An accurate measurement of subbarrier fusion cross sections for transitional nuclei such as Pt and Os could provide a sensitive test for competing models in this region and possibly point to new directions in research.

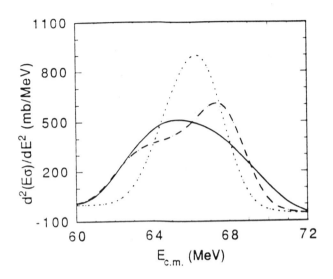

Figure 2. Barrier distributions in the three limits of the IBM. The dotted, dashed and solid lines correspond to the SU(5), SU(3), and the O(6) dynamical symmetries.

The Xe-Ba nuclei exhibit characteristics similar to the transitional Os-Pt region. These nuclei are not as well studied as the Os-Pt isotopes, and very basic information, such as whether they are prolate or oblate, is still missing. Subbarrier fusion could be an effective experimental tool in learning more about the shape transitions in the Xe-Ba region.

It is also possible to generalize the previous formalism to include arbitrary kinds of bosons in the target nucleus and investigate whether g bosons have any discernible effects on subbarrier fusion reactions. We found that [17] except for slight differences in the barrier distributions (which can be made even smaller by fine tuning the coupling strengths), there are no visible differences between the sd and sdg model predictions. The similarity of the results implies that subbarrier fusion probes the overall coupling strength in nuclei, but otherwise is not sensitive to the details of the nuclear wavefunctions. In this sense, subbarrier fusion is in the same category as other static quantities (energy levels, electromagnetic transition rates), and does not seem to constitute a dynamic probe of nuclei, in contrast to proton scattering.

Another experimental test of our model would be to study angular momentum distributions in subbarrier fusion, which can be determined reasonably accurately from the gamma-ray multiplicities data [6]. Vandenbosch and his collaborators at the

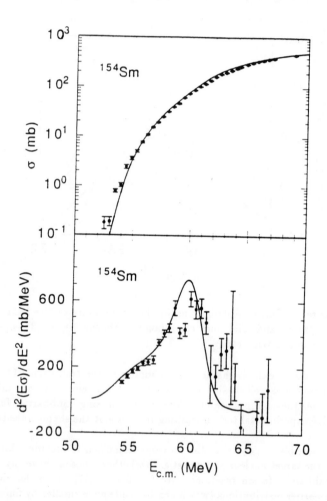

Figure 3. Comparison of predicted fusion cross section and barrier distribution with the data for the reaction $^{16}O+^{144}Sm$. The parameters are given in Ref. [16].

University of Washington, Seattle devoted considerable time to the measurement of angular momentum distributions [6, 21]. Work on the calculation of average angular

momenta in subbarrier fusion reactions, in particular the effects of the shape phase transition on angular momentum distributions, is currently in progress [22].
I would like express my gratitude to my collaborators, J. Bennett, S. Kuyucak, and N. Takigawa, without whom the work reported here would not be possible. This research was supported in part by the U.S. National Science Foundation Grant No. PHY-9314131 and in part by the University of Wisconsin Research Committee with funds granted by the Wisconsin Alumni Research Foundation.

References

[1] A.B. Balantekin, S.E. Koonin, and J.W. Negele, Phys. Rev. **C28**, 1565 (1983).

[2] S.G. Steadman and M.J. Rhoades-Brown, Ann. Rev. Nucl. Sci. **36**, 649 (1986); M. Beckerman, rep. Prog. Phys. **51**, 1047 (1988).

[3] N. Rowley, G.R. Satchler, and P.H. Stelson, Phys. Lett. **B254**, 25 (1991).

[4] A.B. Balantekin and N. Takigawa, Ann. Phys. (NY) **160**, 441 (1985).

[5] H. Esbensen, Nucl. Phys. **A352**, 147 (1981).

[6] R. Vandenbosch, Ann. Rev. Part. Nucl. Sci. **42**, 447 (1992).

[7] J.X. Wei *et al.*, Phys. Rev. Lett. **67**, 3368 (1991); J.R. Leigh *et al.*, Phys. Rev. **C47**, R437 (1993).

[8] A. Bohr and B. Mottelson, Danske Videnskab. Selskab Mat. Fys. Medd. **27**, 16 (1953).

[9] F. Iachello and A. Arima, *The Interacting Boson Model*, (Cambridge, 1987).

[10] A.B. Balantekin, J. Bennett and N. Takigawa, Phys. Rev. **C44**, 145 (1991).

[11] S. Kuyucak and I. Morrison, Ann. Phys. (NY) **181**, 79 (1988); **195**, 126 (1989).

[12] S. Kuyucak and I. Morrison, Phys. Rev. C, in press.

[13] J.N. Ginocchio *et al.*, Phys. Rev. **C33**, 247 (1986).

[14] A.B. Balantekin, J.R. Bennett, A.J. DeWeerd, and S. Kuyucak, Phys. Rev. **C46** (1992) 2019.

[15] A.B. Balantekin, J.R. Bennett, and S. Kuyucak, Phys. Rev. **C48**, 1269 (1993).

[16] A.B. Balantekin, J.R. Bennett, and S. Kuyucak, Phys. Rev. **C49**, 1079 (1994).

[17] A.B. Balantekin, J.R. Bennett, and S. Kuyucak, Phys. Rev. **C49**, 1294 (1994).

[18] N. Takigawa and K. Ikeda, *Proc. Symp. on Many Facets of Heavy Ion Fusion Reactions*, Ed. W. Henning et al., (Argonne : ANL-PHY-86-1) (1986), pp. 613-620; H. Esbensen, S. Landowne, and C. Price, Phys. Rev. **C36**, 1216 (1987); O. Tanimura, Phys. Rev. **C35**, 1600 (1987).

[19] N. Takigawa, Y. Alhassid, and A.B. Balantekin, Phys. Rev. **C45**, 1850 (1992); A.B. Balantekin, J.R. Bennett, N. Takigawa, and Y. Alhassid, Japanese J. Appl. Phys. Series **9**, 90 (1993).

[20] R.F. Casten and D.D. Warner, Rev. Mod. Phys. **60**, 389 (1988).

[21] S. Gil et al, Phys. Rev. **C43**, 701 (1991); Phys. Rev. Lett. **65**, 3100 (1990).

[22] A.B. Balantekin, J.R. Bennett, and S. Kuyucak, in preparation.

QCD AND THE NUCLEAR PHYSICS IBM

Yuval Ne'eman * #

Raymond and Beverly Sackler Faculty of Exact Sciences, Tel-Aviv University

69978 Tel-Aviv, Israel

and

Djordje Sijacki * #

Institute of Physics, P.O.Box 57, Belgrade, Yugoslavia

Abstract

QCD has been applied perturbatively to the UV region of strong interactions. However, it has not been useful in the description of hadron physics and the IR limit. We introduce "Chromogravity", reviewing the proof that the non-quark (gluonic) component of interhadron interactions produces a "strong" gravity-like interaction in the IR limit. This reproduces Regge trajectories, the Pomeron etc.. Using a Bethe-Salpeter approach, we show that the $J = 2, 0$ basic boson of the Nuclear Physics IBM could arise as a bound state of two nucleons in a spinless, I-spin symmetric combination, with the $J = 2, 0$ chromogravitons.

---

* Wolfson Distinguished Chair of Theoretical Physics
# Also on leave from Centre for Particle Physics, University of Texas, Austin, Texas 78712, USA
* Suppported in part by the Science Foundation (Belgrade)
# Supported by the Wolfson Distinguished Chair of Theoretical Physics, Tel-Aviv University.

QCD and the Nuclear Physics IBM

Introduction: QCD after 20 years.

This conference commemorates the 20th anniversary of the very successful Arima-Iachello "IBM" phenomenological symmetry (and systematics) in nuclei [1]; this presentation points to a possible theoretical foundation for it at the Particle Physics level.

Nuclear Physics was for many years considered as a "2nd level theory", with protons, neutrons and mesons as the 1st level, at which the fundamental forces operate; symmetries and structure at the nuclear level, though formulated directly, were also supposed to be derivable - at least in principle or qualitatively - from a set of "fundamental" fields and their interactions. As a matter of fact, the parametrization of nuclear forces, in terms of the exchange of the many species of meson fields discovered in high energy experiments, continued to be a useful exercise throughout the Sixties and Seventies.

However, around the time the IBM was conceived, the Physics of Particles and Fields itself went through a transition, from a "1st level theory" to a "2nd level" one. Particles, like nuclei, were, from now on, a higher order structure. The phenomenological theory of (in modern nomenclature) "flavor"- $SU(3)$ (the Eightfold Way), which we had conceived of early in 1961, had quickly led us to the discovery of a "deeper" layer, that of quarks. There was, however, a complication. The"constituent" quark model (CQM) had led to $SU(6)$ and to a wealth of additional systematics, at the hadron level, but with non-relativistic ad hoc dynamics, hard to reconcile with a regime of high-energy physics. At the same time, many new weak and electromagnetic results were derived from a rather different reflection of the existence of quarks, namely from a quark field model (FQM), sometime also described as "current" quarks. As a constituent, the (charge -1/3) "d" quark has a mass of roughly 300 MeV; as a field or as a "current" quark its mass (as appearing in propagators) is around 5 MeV; moreover, the overlap between the FQM and CQM d quark is only within $cos^2\theta_{Cabibbo} = .96$, i.e. between the two pictures there is a mixing and rediagonalization of the quark states. How all of this occurs and the relationship between the two quark pictures is still an unresolved issue in our understanding of the Strong Interactions.

The Sixties were otherwise a period of soul-searching in Particle Physics, because of difficulties with Relativistic Quantum Field Theory (RQFT); many of the new results were therefore expressed through on-mass- shell physics: Dispersion Relations, Regge Trajectories, Dual Models and Hadron Strings. Then, in 1970-73, with the renewed faith in RQFT as the basic medium for the description of both fundamental and phenomenological levels (after the Veltman-'t Hooft success in the renormalization of Yang-Mills gauge theories), came the discovery of QCD [2]. Its adoption and incorporation in a new grand synthesis, the Standard Model (SM), were the outcome of the success of asymptotic freedom (AF) in fitting the (then fresh) scaling results of deep inelastic electron-nucleon scattering, as interpreted by Feynman- Bjorken parton phenomenology. These were high-energy weak-coupling results, similar to those of the FQM or "current" quarks.

At the same time, QCD does incorporate two very general features relating to the hadron spectrum, i.e. to the CQM: the resolution of the problem of the "wrong" spin-statistics of the baryon (56 in $SU(6)$) ground state, and the zero-triality of the entire $SU(3)$ (Eightfold-Way) physical spectrum. Both questions are answered by the presence of color-$SU(3)$. This was part of the motivation in Nambu's original introduction of this gauge group, in its first (integral charges) version, and it holds for the fractional set.

As a field theory, however, QCD is utilized at the weak-coupling end. There, AF provides a successful perturbative treatment for the "ultraviolet" (UV) region, including both high-energy strong interactions (jets, etc.) and high-energy electro-weak hadronic interactions, corresponding to the current-quarks aspects of FQM.

Nothing of the sort, however, has emerged to date at the energy antipodes, the "infrared" (IR) region. After almost a quarter of a century, we still lack a complete proof of color-confinement, beyond Wilson's original lattice calculations. Issues related to the IR region continue to be treated by using pre-QCD models or hard-to-relate approaches: "bag" models, the Skyrme-Witten, Kazakov-Migdal and other models, in which QCD is replaced by a scalar force and/or reduced dimensionalities, or calculations based on lattice methods. At best, we have effective potentials containing oscillator-like components with positive powers in r, partially demonstrated to correspond to a static approximation of QCD in this region. Very little has been achieved in the understanding of the precise structure of the hadron spectrum, in reproducing the CQM or Regge systematics.

1. QCD-induced Chromogravity Explains the IR Sector

The authors of this paper have conjectured [3] (note: these statements should be taken to apply non-perturbatively, in this strong coupling limit),

(a) that (the entire system of) gluon exchange forces (with the gluons in color-neutral combinations) make up an important component of inter- hadron interactions, including the dominating diffractive contribution;

(b) that the physical role of this component is to produce a longer-range force, with many of the characteristics of gravity, starting with the basic mathematical foundation, namely invariance under diffeomorphisms;

(c) that the simplest such n-gluon exchange, that of the two-gluon system

$$G_{\mu\nu}(x) = (\kappa)^{-2} \ g_{ab} \ B^a_\mu(x) \ B^b_\nu(x) \tag{1.1}$$

(κ has the dimensions of mass, $< \mu, \nu, \cdots >$ are Lorentz 4-vector indices, $< a, b, \cdots >$ are $SU(3)$ adjoint representation (octet) indices, g_{ab} is the Cartan metric for the $SU(3)$ octet, B^a_μ is a gluon field) fulfills the role of an effective (pseudo) metric, gauging these (pseudo) diffeomorphisms. These assumptions were then proven in ref. [4,5] and we shall give in section 2 a brief summary of the proofs.

QCD thus contains a gravity-like component, which we have labelled "chromogravity"; in hadron physics this explains a large number of facts: 1) The Pomeranchuk trajectory, in its Dual Model version, is thereby reproduced from QCD. The mechanism precisely justifies the 1969-70 Harari-Rosner Duality quark diagram method and the Harari-Freund conjecture, namely that it is s-channel elastic scattering which produces the t-channel Pomeron. 2) The dynamics produce p^{-4} propagators, the closest thing to a proof of confinement, within the context of RQFT. 3) The same dynamics predict Regge trajectories with $M^2 \sim J$, fitting observations and hard to obtain in conventional field theory models. 4) The system of hadron resonances is given a relativistic classification [6] in terms of $SL(4, R)$. Moreover, the role played by quadrupolar excitations, pointed out in the sixties and seemingly out of place in a strongly interacting system, can now be understood as relating to chromogravity rather than to gravity itself. 5) Interference between the input (UV) masses (or Yukawa Interactions) of the "current" quarks (FQM) and the $J = 0$ component of chromogravity, produces the "constituent" quarks' masses (CQM) and breaks chiral symmetry. The common origin of the two basic excitations, $J = 2$ and $J = 0$, implies degenerate excitation energies between this process and the Regge excitations of the previous item - indeed validated by experiment. 6) Salam and others throughout the Seventies and Eighties had concluded that QCD is insufficient for the description of the Strong Interactions and had postulated an independent "strong gravity", to be added to QCD. Our work thus satisfies Salam's intuitive identification of a need – though finding the missing component within QCD itself. 7) We provide an explanation of the fact that the string, now considered as a theory of Quantum Gravity, could also provide a theory of hadrons and their interactions, especially successful in reproducing the key features of the hadron spectrum. 8) The central role played by the two-gluon term as a "chromograviton" is also a reflection of the fact that closed strings (i.e. the contraction of two open strings) have gravity as their RQFT limit, while open strings reduce in that limit to gluon-like massless Yang-Mills fields. 9) Chromogravity has its analog in QED, where the two-photon exchange contribution (with J=2,0) is the basic constituent of Van der Waals forces. 10) Lastly – and in the context of this meeting – we have suggested [7] that the relatively long range of chromogravity, together with the spins $J = 2, 0$ of its quanta, could produce (by binding to $J = 0$ isospin-symmetric nucleon pairs) the basic $J = 2, 0$ six excited defining states of IBM systematics in even-even nuclei. The degenerate energies represent a scaled down version of the same in our points 4-5 above.

2. Effective Diffeomorphisms

The gluon color-$SU(3)$ gauge field transforms under an infinitesimal local $SU(3)$ variation according to

$$\delta_\epsilon B_\mu^a = \partial_\mu \epsilon^a + B_\mu^b \{\lambda_b\}_c^a \epsilon^c = \partial_\mu \epsilon^a + i f_{bc}^a B^b \epsilon^c \tag{2.1}$$

(we use the adjoint representation $\{\lambda_b\}_c^a = -i f_b^{ac} = i f_{bc}^a$). To deal with the non-perturbative IR region, we expand the gauge field operator around a constant global vacuum solution N_μ^a,

$$\partial_\mu N_\nu^a - \partial_\nu N_\mu^a = i f_{bc}^a N_\mu^b N_\nu^c \tag{2.2a}$$

$$B^a_\mu = N^a_\mu + A^a_\mu \qquad (2.2b)$$

Such a vacuum solution might be of the instanton type [8], for instance. Consider, e.g. the first non-trivial class, with Pontryagin index $n = 0$. Expand around this classical configuration, working, as always for instantons, in a Euclidean metric (i.e. a tunneling solution in Minkowski spacetime). At large distances the instanton field is required to approach a constant value

$$g_{ab}\, N^a_\mu\, \partial_\nu \epsilon^b = \partial_\nu (g_{ab}\, N^a_\mu\, \epsilon^b) \qquad (2.2c)$$

Instantons are constructed through the mapping of an $SU(2)$ subgroup of the gauge group onto the "great sphere" S^3, namely $SU(2) \to S^3$. The full $SU(3)$ mapping is then induced over the various $SU(2)$ subgroups, following Bott's theorem. To gain some additional insight we choose, instead, to replace the instanton by an alternative vacuum solution, constructed by mapping $SU(3) \to S^4$, namely directly onto the complete and completed by a point at infinity – Euclidean manifold.

In the $SU(3)$ Cartan basis, H^1 and H^2 are the diagonal operators spanning the Cartan subalgebra's space; E^{+i} and E^{-i}, $i = 1,..3$ are respectively the three raising and three lowering operators of I, U, V spins, using Lipkin's original flavor-inspired terminology. In this basis, the Cartan metric takes the form

$$g_{ab} = \sigma^0 \oplus \sigma^{1I} \oplus \sigma^{1U} \oplus \sigma^{1V} \qquad (2.3a)$$

where σ^0 is the 2×2 identity matrix, operating in the Cartan subalgebra subspace. σ^1 is that Pauli matrix, operating in each of the three non-diagonal subspaces constituted by paired positive and negative roots. We now select for the vacuum solution a gauge

$$N^{h1}_\mu = \frac{1}{\sqrt{2}}\begin{vmatrix}1\\0\\0\\0\end{vmatrix},\quad N^{h2}_\mu = \frac{1}{\sqrt{2}}\begin{vmatrix}1\\0\\0\\0\end{vmatrix},\quad N^{I+}_\mu = \frac{1}{\sqrt{2}}\begin{vmatrix}0\\1\\0\\0\end{vmatrix},\quad N^{I-}_\mu = \frac{1}{\sqrt{2}}\begin{vmatrix}0\\1\\0\\0\end{vmatrix}$$

$$N^{U+}_\mu = \frac{1}{\sqrt{2}}\begin{vmatrix}0\\0\\1\\0\end{vmatrix},\quad N^{U-}_\mu = \frac{1}{\sqrt{2}}\begin{vmatrix}0\\0\\1\\0\end{vmatrix},\quad N^{V+}_\mu = \frac{1}{\sqrt{2}}\begin{vmatrix}0\\0\\0\\1\end{vmatrix},\quad N^{V-}_\mu = \frac{1}{\sqrt{2}}\begin{vmatrix}0\\0\\0\\1\end{vmatrix} \qquad (2.3b)$$

and obtain the (flat) Euclidean metric from

$$\eta_{\mu\nu} = N_\mu^a \, g_{ab} \, N_\nu^b \tag{2.3c}$$

writing the two non-perturbative field solutions respectively as 4×8 and 8×4 matrices; the octet 8-dimensionality is contracted by the action of the 8×8 Cartan metric matrix. Note that in the instanton sytem, such a constant solution would arise from a null solution $B_\mu^a = 0$ through the application of a local gauge transformation involving a gauge function $\epsilon^a(x)$ linear in x,

$$\epsilon^1(x^\mu) = \frac{x^0}{\sqrt{2}}, \quad \epsilon^2(x^\mu) = \frac{x^0}{\sqrt{2}}, \quad \epsilon^{I+}(x^\mu) = \frac{x^1}{\sqrt{2}}, \quad \epsilon^{I-}(x^\mu) = \frac{x^1}{\sqrt{2}}$$

$$\epsilon^{U+}(x^\mu) = \frac{x^2}{\sqrt{2}}, \quad \epsilon^{U-}(x^\mu) = \frac{x^2}{\sqrt{2}}, \quad \epsilon^{V+}(x^\mu) = \frac{x^3}{\sqrt{2}}, \quad \epsilon^{V-}(x^\mu) = \frac{x^3}{\sqrt{2}} \tag{2.3d}$$

the N_μ^a then being given by the constants $\partial_\mu \epsilon^a(x)$ and thus defining an instanton with Pontryagin index $n = 1$, i.e. a topologically non-trivial object.

In what follows, we preserve the definition (2.2c) and the gauge (2.3); should a gauge transformation adjoin an x-dependent variation, we choose to include it in the $A_\mu^a(x)$ component of (2.2b). Returning to the "curved" pseudo-metric (1.1), we can now replace κ by the "flat" density,

$$G_{\mu\nu} = \frac{g_{ab} \, B_\mu^a \, B_\nu^b}{(\det(g_{ab} \, N_\mu^a \, N_\nu^b))^{1/4}} \tag{2.3e}$$

thus yielding a non-singular dimensionless Euclido-Riemannian "metric". Its color-$SU(3)$ infinitesimal gauge variation is given by

$$\delta_\epsilon G_{\mu\nu} = \delta_\epsilon \left\{ g_{ab}(N_\mu^a + A_\mu^a)(N_\nu^b + A_\nu^b) \right\}$$
$$= g_{ab}(\partial_\mu \epsilon^a \, N_\nu^b + N_\mu^a \, \partial_\nu \epsilon^b \, \partial_\mu \epsilon^a \, A_\nu^b + A_\mu^a \, \partial_\nu \epsilon^b) \tag{2.4a}$$
$$= i g_{ab} \left\{ f_{cd}^a \, B_\mu^c \, \epsilon^d \, B_\nu^b + f_{cd}^b \, B_\mu^a \, B_\nu^c \, \epsilon^d \right\}$$

The last bracket vanishes, since it represents the homogeneous $SU(3)$ transformation of the $SU(3)$ scalar expression in (1.1)

$$i f_{bcd} \, (B_\mu^b \, B_\nu^c + B_\mu^c \, B_\nu^b) \, \epsilon^d \tag{2.4b}$$

(or, more technically, due to the total antisymmetry of f_{abc} in a compact group). With N_μ^a, N_ν^b, representing constant fields, we rewrite the terms in which they appear as a new infinitesimal variation,

$$\xi_\mu = \eta_{ab}\, \epsilon^a\, N_\mu^b \tag{2.4c}$$

Integrating by parts the terms in A_μ^a, A_ν^b we get

$$g_{ab}\left(\epsilon^a\, \partial_\mu A_\nu^b + \partial_\nu A_\mu^a\, \epsilon^b\right) \tag{2.4d}$$

an expression whose Fourier transform vanishes for $k \to 0$, i.e. *in the infrared sector*.

As a result, we can write in this limit,

$$\delta_\xi G_{\mu\nu} = \partial_\mu \xi_\nu + \partial_\nu \xi_\mu = \partial_\mu(\xi^\sigma\, G_{\sigma\nu}) + \partial_\nu(\xi^\sigma\, G_{\mu\sigma}) \tag{2.5}$$

where we have changed over to the ξ^σ variable of (2.4c), and where we can reidentify δ_ξ as a variation under a formal diffeomorphism of the R^4 manifold. Eq. (2.5) simulates the infinitesimal variation of a "world tensor" $G_{\mu\nu}$ under Einstein's covariance group, $x^\sigma \to x^\sigma \to \xi^\sigma$. ξ^σ thus has to be defined as a contravariant vector; $G_{\mu\nu}$ of (1.1) is invertible, thanks to the constant part N_μ^a, in (2.2b), using a Taylor expansion to evaluate the inverse $G^{\mu\nu}(x)$. Note that as the μ, ν indices are "true" Lorentz indices, acted upon by the physical Lorentz group, the manifold has to be Riemannian: only Riemannian manifolds – with or without torsion – have tangents with orthogonal or pseudo-orthogonal symmetry. Thus

$$D_\sigma G_{\mu\nu} = 0 \tag{2.6}$$

$G_{\mu\nu}$ with its 10 components reduces to a sum of $J = 2, 1, 0, 0$. As a result of (2.6), the $J = 1, 0$ are cancelled and the physically operative components are $J = 2, 0$. Evaluating the commutator of two such variations reproduces another diffeomorphism, which serves to complete the proof (5).

3. n-gluon fields.

Our treatment is non-perturbative; we use, however, the formal expansion provided by the generating functional of the Green's functions, for a classification of the contributions making up the overall non-quark component. In this, we parallel the on-mass-shell considerations which had led Harari and Freund [9] to a precise definition of the s- channel processes composing (in the Finite Energy Sum Rules) the t-channel Pomeranchuk trajectory: elastic scattering, as represented by duality Harari-Rosner diagrams [10] in which no resonance is formed and no Susskind "rubber-bands" can be stretched into the t-channel. The expansion involves all possible color-singlet configurations of gluon fields. We rearrange the sum by lumping together contributions from n-gluon irreducible parts, $n = 2, 3, ..., \infty$ and with the same Lorentz quantum numbers. Thus, QCD "gluon-made" operators which mutually connect various hadron states are characterized by color-singlet quanta. The corresponding color-singlet n-gluon field operator has the following form

$$G^{(n)}_{\mu_1\mu_2\cdots\mu_n} = d^{(n)}_{a_1a_2\cdots a_n} B^{a_1}_{\mu_1} B^{a_2}_{\mu_2} \cdots B^{a_n}_{\mu_n} \qquad (3.1)$$

The set of all $d^{(n)}_{a_1a_2\cdots a_n}$ tensors, $n = 1, 2, \ldots$, can be used to form, together with the group generators, a basis of all $SU(3)$ invariant operators. In this case all such higher rank operators can be expressed in terms of two invariant operators. In our case, the set of all $G^{(n)}_{\mu_1\mu_2\cdots\mu_n}$ operators, $n = 1, 2, \ldots$, forms a basis of a vector space of colorless purely gluonic configurations. Moreover, in our case, in contradistinction to the ordinary group theoretical situation, these field operators are also all functionally independent.

The QCD variation of the $G^{(n)}_{\mu_1\mu_2\cdots\mu_n}$ field can be shown to be expressible in terms of effective pseudo-diffeomorphisms,

$$\delta_\epsilon G^{(n)}_{\mu_1\mu_2\cdots\mu_n} = \partial_{\{\mu_1} \xi^{(n-1)}_{\mu_2\mu_3\cdots\mu_n\}} \equiv \delta_\xi G^{(n)}_{\mu_1\mu_2\cdots\mu_n}, \qquad (3.2)$$

where $\{\mu_1\mu_2 \cdots \mu_n\}$ denotes symmetrization of indices, and

$$\xi^{(n-1)}_{\mu_1\mu_2\cdots\mu_{n-1}} \equiv d^{(n)}_{a_1a_2\cdots a_n} N^{a_1}_{\mu_1} N^{a_2}_{\mu_2} \cdots N^{a_{n-1}}_{\mu_{n-1}} \epsilon^{a_n} \qquad (3.3)$$

generalizing our results as derived for $G^{(2)}_{\mu\nu} = G_{\mu\nu}$.

A subsequent application of two $SU(3)$-induced variations implies

$$[\delta_{\epsilon_1}, \delta_{\epsilon_2}] G^{(n)}_{\mu_1\mu_2\cdots\mu_n} = \delta_{\epsilon_3} G^{(n)}_{\mu_1\mu_2\cdots\mu_n} \quad i.e. \quad [\delta_{\xi_1}, \delta_{\xi_2}] G^{(n)}_{\mu_1\mu_2\cdots\mu_n} = \delta_{\xi_3} G^{(n)}_{\mu_1\mu_2\cdots\mu_n} \qquad (3.4)$$

generalizing the $n = 2$ case, i.e. an infinitesimal nonlinear realization of the $Diff(4,R)$ group in the space of fields $\left\{ G^{(n)}_{\mu_1\mu_2\cdots\mu_n} \mid n = 2, 3, \ldots \right\}$.

4. $L^{(m)}$ operators.

Let us consider an ∞-dimensional vector space over the field operators $\left\{ G^{(n)} \mid n = 2, 3, \ldots \right\}$, i.e.,

$$V(G^{(2)}, G^{(3)}, \ldots) = V(G^{(2)}_{\mu_1 \mu_2}, G^{(3)}_{\mu_1 \mu_2 \mu_3}, \ldots).$$

We can now define an infinite set of field-dependent operators $\left\{ L^{(m)} \mid m = 0, 1, 2, \ldots \right\}$ as follows

$$\begin{aligned} L^{(0)\rho}_{\nu_1} &= d^{(2)}_{a_1 a_2} B^{a_1}_{\nu_1} \frac{\delta}{\delta(g_{a_2 b} B^b_\rho)} \equiv g_{a_1 a_2} B^{a_1}_{\nu_1} \frac{\delta}{\delta(g_{a_2 b} B^b_\rho)}, \\ L^{(1)\rho}_{\nu_1 \nu_2} &= d^{(3)}_{a_1 a_2 a_3} B^{a_1}_{\nu_1} B^{a_2}_{\nu_2} \frac{\delta}{\delta(g_{a_3 b} B^b_\rho)} \equiv d_{a_1 a_2 a_3} B^{a_1}_{\nu_1} B^{a_2}_{\nu_2} \frac{\delta}{\delta(g_{a_3 b} B^b_\rho)}, \\ &\cdots \cdots \\ L^{(m)\rho}_{\nu_1 \nu_2 \cdots \nu_{m+1}} &= d^{(m+2)}_{a_1 a_2 \cdots a_{m+2}} B^{a_1}_{\nu_1} B^{a_2}_{\nu_2} \cdots B^{a_{m+1}}_{\nu_{m+1}} \frac{\delta}{\delta(g_{a_{m+2} b} B^b_\rho)}. \end{aligned} \quad (4.1)$$

. . . .

In the general case, $L^{(m)\rho}_{\nu_1 \nu_2 \cdots \nu_{m+1}}$, $m = 0, 1, 2, \ldots$ action on the field operators $\left\{ G^{(n)} \mid n = 2, 3, \ldots \right\}$ reads

$$\begin{aligned} L^{(m)\rho}_{\nu_1 \nu_2 \cdots \nu_{m+1}} G^{(2)}_{\mu_1 \mu_2} &= \delta^\rho_{\mu_1} G^{(2+m)}_{\nu_1 \nu_2 \cdots \nu_{m+1} \mu_2} + \delta^\rho_{\mu_2} G^{(2+m)}_{\mu_1 \nu_1 \nu_2 \cdots \nu_{m+1}}, \\ L^{(m)\rho}_{\nu_1 \nu_2 \cdots \nu_{m+1}} G^{(3)}_{\mu_1 \mu_2 \mu_3} &= \delta^\rho_{\mu_1} G^{(3+m)}_{\nu_1 \nu_2 \cdots \nu_{m+1} \mu_2 \mu_3} + \delta^\rho_{\mu_2} G^{(3+m)}_{\mu_1 \nu_1 \nu_2 \cdots \nu_{m+1} \mu_3} + \delta^\rho_{\mu_3} G^{(3+m)}_{\mu_1 \mu_2 \nu_1 \nu_2 \cdots \nu_{m+1}}, \\ &\cdots \cdots \\ L^{(m)\rho}_{\nu_1 \nu_2 \cdots \nu_{m+1}} G^{(n)}_{\mu_1 \mu_2 \cdots \mu_n} &= \delta^\rho_{\mu_1} G^{(n+m)}_{\nu_1 \nu_2 \cdots \nu_{m+1} \mu_2 \cdots \mu_n} + \delta^\rho_{\mu_2} G^{(n+m)}_{\mu_1 \nu_1 \nu_2 \cdots \nu_{m+1} \mu_3 \cdots \mu_n} \\ &\quad + \cdots + \delta^\rho_{\mu_n} G^{(n+m)}_{\mu_1 \mu_2 \cdots \mu_{n-1} \nu_1 \nu_2 \cdots \nu_{m+1}}, \end{aligned}$$

. . . .
$$(4.2)$$

Let us now consider the algebraic structure defined by the $\left\{ L^{(m)} \mid m = 0, 1, 2, \ldots \right\}$ operators Lie brackets. In the most general case, for the brackets of $L^{(l)}$ and $L^{(m)}$ we find

$$[L^{(l)}, L^{(m)}] \subset L^{(l+m)}, \quad (4.3)$$

and more specifically,

$$[L^{(l)\rho_1}_{\nu_1 \nu_2 \cdots \nu_{l+1}}, L^{(m)\rho_2}_{\sigma_1 \sigma_2 \cdots \sigma_{m+1}}] = \sum_{i=1}^{m+1} \delta^{\rho_1}_{\sigma_i} L^{(l+m)\rho_2}_{\sigma_1 \sigma_2 \cdots \sigma_{i-1} \nu_1 \nu_2 \cdots \nu_{l+1} \sigma_{i+1} \cdots \sigma_{m+1}} - \sum_{j=1}^{l+1} \delta^{\rho_2}_{\nu_j} L^{(l+m)\rho_1}_{\nu_1 \nu_2 \cdots \nu_{j-1} \sigma_1 \sigma_2 \cdots \sigma_{m+1} \nu_{j+1} \cdots \nu_{l+1}}. \quad (4.4)$$

We have constructed an ∞-component vector space, $V = V(G^{(2)}_{\mu_1\mu_2}, G^{(3)}_{\mu_1\mu_2\mu_3}, \ldots)$, over the n-gluon field operators, as well as the corresponding algebra of homogeneous diffeomorphisms, $diff_0(4, R) = \left\{ L^{(m)\rho}_{\nu_1\nu_2\cdots\nu_{m+1}} \middle| m = 0, 1, 2, \ldots \right\}$; the vector space V is invariant under the action of the $diff_0(4, R)$ algebra. We thus have an induced "covariance group" gauged at the IR limit, with the "chromometric" $G_{\mu\nu}$ as the "chromogravitational field". The conclusions are listed in section 2. Note that the "chromograviton" is (in the IR limit in which it exists) a massless field, due to the diffeomorphic gauge. Lattice calculations predict the appearance of di-gluonic $J = 0, 2$ "glueballs", with masses around $1.5 - 2.0 GeV$. Quantum Mechanics will cause a mixing between these states and the chromograviton, which will also extend to the $J = 0, 2$ quark-antiquark bound states, a subject for further study.

5. IBM Bosons as $NH_{\mu\nu}N$ Bound States

QCD has not modified Yukawa's original ansatz, namely that the strong binding of nucleons in nuclei is produced by pions (and other, heavier mesons). We now consider these mesons as quark-antiquark bound states; their dominance is due to their relatively small masses, the lightest in hadron systematics. This is produced in QCD by the very small UV masses of the u,d quarks (an input parameter), least amplified by QCD (acting through interference of the $J = 0$ component of chromogravity - a topic we discuss elsewhere [11]) and is reflected in the spontaneously broken chiral $SU(2) \times SU(2)$, with the pion as pseudo-Goldstone particle. In our present context, this is the force which - aside from binding nucleons in the nuclear shells - also "almost binds" the first isospin-symmetric nucleon pair after a closed shell, into a spin-antisymmetric $J = 0$ state, similar to the analogous virtual state in the deuteron (or two-nucleon) system.

This $J = 0$ pseudo-particle is then enacted upon by the chromogravity $J = 2, 0$ potential, emitted by the inner nucleus, a quasi-long range force. We can treat this dynamical situation by either using a Schroedinger equation with a chromogravitational potential, placing the $A = 2, I = 1, J = 0$ pseudo- particle in it; or, alternatively, study the interaction between the two particles, the di-nucleon and the (digluon) chromograviton, in a

Bethe- Salpeter treatment. We shall use the latter approach in this work.

We apply a phenomenological effective field theory, with chromogravity treated like the conventional "weak field approximation" of gravity, in which we subtract the Minkowski (or Euclidean) metric,

$$G_{\mu\nu}(x) = \eta_{\mu\nu} + H_{\mu\nu(x)} \tag{5.1a}$$

The Feynman diagrams are given by the pure chromogravity action, bilinear in $H_{\mu\nu}$, supplemented by the matter-field action and the following expansion

$$\sqrt{-G} \longrightarrow 1 + \tfrac{1}{2} H^\alpha_\alpha - \tfrac{1}{4} H^\alpha_\beta H^\beta_\alpha + \tfrac{1}{8}(H^\alpha_\alpha)^2 + \cdots. \tag{5.1b}$$

The simplest relevant diagrams are illustrated in Fig. #1.

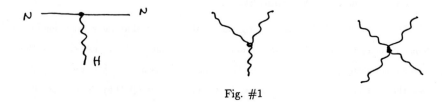

Fig. #1

Consider the two-nucleon – chromograviton scattering amplitude. For nucleons N_1, N_2 and $H_{\mu\nu}$, it is given by (cf. Fig. #2)

$$\begin{aligned}
\langle N_1 H N_2 | S | N_1 H N_2 \rangle &= (-i)^3 g_N^2 g_H \int d^4 x_1 d^4 x_2 d^4 x_3 \\
&\times \langle N_1 H N_2 | \bar\Psi H^{\mu\nu} \bar\Psi | 0 \rangle D_F(x_1 - x_2) D_F(x_2 - x_3) \langle 0 | \Psi H_{\mu\nu} \Psi | N_1 H_{\mu\nu} N_2 \rangle \\
&+ (-i)^6 g_N^4 g_H^2 \int d^4 x_1 d^4 x_2 d^4 x_3 d^4 x_4 d^4 x_5 d^4 x_6 \langle N_1 H N_2 | \bar\Psi H^{\mu\nu} \bar\Psi | 0 \rangle \\
&\times D_F(x_1 - x_2) D_F(x_2 - x_3) S_F(x_1 - x_4) D_F(x_2 - x_5) S_F(x_3 - x_6) D_F(x_4 - x_5) D_F(x_5 - x_6) \\
&\times \langle 0 | \Psi H_{\mu\nu} \Psi | N_1 H N_2 \rangle + \cdots
\end{aligned} \tag{5.2}$$

where we have suppressed the spinorial indices as well as the tensorial indices in $D_{F\mu\nu,\rho\sigma}$ and in the vertices.

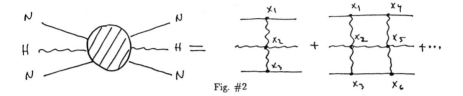

Fig. #2

Define a function

$$F_{\mu\nu}(x_1, x_2, x_3; N_1 H N_2) = \langle 0|\Psi(x_1)H_{\mu\nu}(x_2)\Psi(x_3)|N_1 H N_2\rangle +$$
$$(-i)^3 g_N^2 g_H \int d^4x_4 d^4x_5 d^4x_6 S_F(x_1-x_4) D_F(x_2-x_5) S_F(x_3-x_6)$$
$$\times \langle 0|\Psi H_{\mu\nu}\Psi|N_1 H N_2\rangle + \cdots \tag{5.3a}$$

The inhomogeneous term is of the following form

$$F_{0\mu\nu}(x_1, x_2, x_3; N_1 H N_2) \sim F_0(x_1; N_1) F_{0\mu\nu}(x_2; H) F_0(x_3; N_2) \equiv$$
$$\langle 0|\Psi|N_1\rangle \langle 0|H_{\mu\nu}|H\rangle \langle 0|\Psi|N_2\rangle . \tag{5.3b}$$

The scattering amplitude becomes

$$\langle N_1 H N_2|S|N_1 H N_2\rangle = \int d^4x_1 d^4x_2 d^4x_3 F_0^{*\mu\nu}(x_1, x_2, x_3; N_1 H N_2)$$
$$\times \left[(-i)^3 g_N^2 g_H D_F(x_1-x_2) D_F(x_2-x_3)\right] F_{\mu\nu}(x_1, x_2, x_3; N_1 H N_2), \tag{5.4a}$$

where $F_{\mu\nu}(x_1, x_2, x_3; N_1 H N_2)$ satisfies the following integral equation

$$F_{\mu\nu}(x_1, x_2, x_3; N_1 H N_2) = F_{0\mu\nu}(x_1, x_2, x_3; N_1 H N_2) +$$
$$\int d^4x_4 d^4x_5 d^4x_6 S_F(x_1-x_4) D_F(x_2-x_5) S_F(x_3-x_6) \tag{5.4b}$$
$$\times (-i)^3 g_N^2 g_H D_F(x_4-x_5) D_F(x_5-x_6) F_{\mu\nu}(x_4, x_5, x_6; N_1 H N_2).$$

The (2-gluon) chromometric 1-particle state $|H_{ph}\rangle$ corresponds to the physical modes of the $H_{\mu\nu}$-field wave equation

$$\left[\tfrac{1}{4}a\Box^2 - \tfrac{1}{2}l_S^{-2}\Sigma_{\alpha\beta}{}^\gamma \Sigma^{\alpha\beta}{}_\gamma - \tfrac{1}{2}l_Q^{-2}\Delta_{\alpha\beta}{}^\gamma \Delta^{\alpha\beta}{}_\gamma\right]\left(P_{(2+0)}H\right)_{\mu\nu}(x) = 0. \tag{5.5}$$

This state describes the $J^P = 2^+ \oplus 0^+$ physical di-gluon modes, i.e. $|H_{ph}\rangle \sim P_{(2+0)}|H\rangle$. Moreover, the two-nucleon wave function is totally antisymmetric and thus the $B = (N_1HN_2)_{ph}$ bound state is given as follows

$$|B\rangle = \frac{1}{\sqrt{2}}\epsilon^{ij}|N_i\rangle|H_{ph}\rangle|N_j\rangle. \tag{5.6}$$

The relativistic equation for the bound states (dropping the inhomogeneous term), with $F_{\mu\nu} \to B_{\mu\nu}(x_1,x_2,x_3;NHN) = \frac{1}{\sqrt{2}}\epsilon^{ij}\langle 0|\Psi(x_1)|N_i\rangle\langle 0|H_{\mu\nu}(x_2)|H_{ph}\rangle\langle 0|\Psi(x_3)|N_j\rangle + \cdots$
finally reads

$$\begin{aligned}B_{\mu\nu}(x_1,x_2,x_3;NHN) &= \int d^4x_4 d^4x_5 d^4x_6 S_F(x_1-x_4)D_F(x_2-x_5)S_F(x_3-x_6) \\ &\times (-i)^3 g_N^2 g_H D_F(x_4-x_5)D_F(x_5-x_6)B_{\mu\nu}(x_4,x_5,x_6;NHN).\end{aligned} \tag{5.7}$$

This equation is diagrammatically illustrated in Fig. #3.

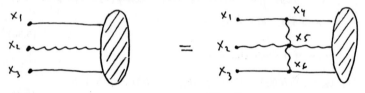

Fig. #3

To summarize, we have constructed explicitly a two-nucleon – (di- gluon) chromograviton relativistic 3-particle bound state. Owing to the Pauli principle, we find for a two-proton and/or two-neutron ($J_{NN} = 0$, $I_{NN} = 1$), excited by a chromograviton, a bound state characterized by the total $J^P = 2^+ \oplus 0^+$. We have discussed elsewhere [7] the quantitative aspect: the kinematical factors reduce the energy increment from about 1 GeV in single hadrons to about $.5MeV$ in nuclei. Finally, we can define the IBM boson quantum-mechanical operators as

$$b^+_{\mu\nu}(p)|0\rangle \equiv |B_{\mu\nu}(p)\rangle \tag{5.8}$$

Acknowledgements

I would like to thank Professors R. Orava and M. Chaichian for the hospitality of SEFT in Helsinki, where this work was completed.

References:

1. A. Arima and F. Iachello, *Phys. Rev. Lett.* **35** (1975) 1069.
2. M. Han and Y. Nambu, *Phys. Rev.* **B139** (1965) 1006; D.J. Gross and F. Wilczek, *Phys. Rev. Lett.* **30** (1973) 1323; H.D. Politzer, *Phys. Rev. Lett.* **30** (1973) 1346; H. Fritzsch, M. Gell Mann and H. Leutwyler, *Phys. Lett.* **B47** (1973) 365; S. Weinberg, *Phys. Rev. Lett.* **31** (1973) 494.
3. Dj. Šijački and Y. Ne'eman, *Phys. Lett.* **B247** (1990) 571.
4. Y. Ne'eman and Dj. Šijački, *Phys. Lett.* **B276** (1992) 173.
5. Y. Ne'eman and Dj. Šijački, "QCD-Induced Diffeomorphisms and the Duality Pomeron", TAU preprint (May 1994).
6. Dj. Šijački and Y. Ne'eman, *Phys. Rev.* **D47** (1993) 4133.
7. Dj. Šijački and Y. Ne'eman, *Phys. Lett.* **B250** (1990) 1.
8. A. Belavin, A. Polyakov, A. Schwartz and Y. Tyupkin, *Phys. Lett.* **59B** (1975) 85.
9. P.G.O. Freund, *Phys. Rev. Lett.* **20** (1968) 235; H. Harari, *Phys. Rev. Lett.* **20** (1968) 1395.
10. H. Harari, *Phys. Rev. Lett.* **22** (1969) 562; J. L. Rosner, *Phys. Rev. Lett.* **22** (1969) 689.
11. Y. Ne'eman and Dj. Šijački , paper to be read at 1994 Como QCD Conference.

q-deformed vibron model for diatomic molecules

R. N. Alvarez [1], Dennis Bonatsos [2] and Yu. F. Smirnov [3]*

[1] Institute of Nuclear Physics, Moscow State University
117234 Moscow, Russia
[2] Institute of Nuclear Physics, NCSR Demokritos
GR-15310 Aghia Paraskevi, Attiki, Greece
[3] Instituto de Fisica, Universidad Nacional Autonoma de Mexico
Apartado Postal 20-364, 01000 Mexico, D.F., Mexico

Abstract

A deformed version of the vibron model for diatomic molecules is constructed. Both of the O(4) and U(3) dynamical symmetries of the model are rewritten, using the concept of complementary subalgebras, in a more convenient form, which is subsequently deformed. The present model unifies the so far independent successful quantum algebraic approaches to rotational and to vibrational spectra of diatomic molecules. In addition, the method can be used for the construction of deformed versions of the U(5) and O(6) limits of the Interacting Boson Model of nuclear structure.

PACS numbers: 33.10.Cs, 31.15.+q, 02.20.Sv

1. Introduction

The mathematical structure of quantum algebras (quantum groups) [1–4] is attracting recently much attention. They are deformed versions of the usual Lie algebras, to which they reduce when the deformation parameter q is set equal to 1. In parallel, applications of quantum algebras in physics have begun to develope, in particular in cases in which Lie algebras are known to describe approximately the symmetries of a physical system. The quantum algebra $SU_q(2)$ has been successfully used for describing rotational spectra of diatomic molecules [5–7], deformed nuclei [8–10], and superdeformed nuclei [11]. Vibrational spectra of diatomic molecules have been described in terms of deformed oscillators [12–16], as well as in terms of an $SU_q(1,1)$ symmetry [17,18]. Potentials giving spectra equivalent to those of the deformed oscillators just mentioned have been constructed [19,20] and found to be deformed versions of the modified Pöschl–Teller potential, or, equivalently, the Morse potential.

On the other hand, the vibron model [21–23], having an overall U(4) symmetry, is known to provide a unified description of molecular rotations and vibrations through the use of algebraic techniques, in a way similar to the description of collective nuclei in terms of the Interacting Boson Model (IBM) [24]. The O(4) limiting symmetry of the vibron model has been found to be appropriate for diatomic molecules, while the U(3) limiting symmetry has been used for the description of clustering effects in nuclei, as well as for the quasi-molecular description of heavy-ion resonances (see [25] for lists of references).

The question is therefore created if a deformed version of the vibron model can accommodate in a unified framework the improved descriptions of rotational and vibrational molecular spectra obtained so far in terms of separate quantum algebras. The problem of constructing the deformed version of the vibron model (or of the IBM) is not a simple one, since the construction of the reduction chains of $U_q(4)$ and $U_q(6)$ has not been achieved yet. It suffices to be mentioned that the reduction from $SU_q(3)$ to $SO_q(3)$ has been carried out only for fully symmetric irreducible representations (irreps) of $SU_q(3)$ [26]. However, a few efforts towards constructing deformed versions of the vibron model [27,28] and the IBM [29,30] already exist.

In this paper a deformed version of both the O(4) and U(3) dynamical symmetries of the U(4) vibron model will be constructed, taking advantage of the techniques of complementary algebras, introduced by Moshinsky and Quesne [31–33], which by-pass the difficulties in the construction of reduction chains of quantum algebras. In addition to unifying the existing independent quantum algebraic descriptions of rotational and of vibrational molecular spectra, the present approach allows in a simple way for the introduction of cross-terms describing the coupling between these two excitation mechanisms.

A brief account of the vibron model for diatomic molecules will be given in section 2, while in section 3 the model will be formulated in a new way, using the techniques of complementary algebras. In section 4 the q-deformed version of the complementary analogues of both the O(4) and U(3) dynamical symmetries of the vibron model will be given, while section 5 will contain discussion of the present results and plans for further work.

2. The vibron model for diatomic molecules

In this section the briefest possible account of the vibron model [21–23] is given in its usual form. In the vibron model the rotations and vibrations of a diatomic molecule are described in terms of 4 bosons: a scalar boson of positive parity and angular momentum $l=0$, denoted by s^+, and the three components of a vector boson of negative parity and $l = 1$, denoted by p_μ^+, $\mu = 0, \pm 1$. The corresponding annihilation operators transforming as spherical tensors are $\tilde{s} = s$ and $\tilde{p}_\mu = (-1)^{1-\mu} p_{-\mu}$. Denoting these bosons by $b_{l,\mu}^+$, $l = 0,1$, $-l \leq \mu \leq l$, and $\tilde{b}_{l,\mu} = (-1)^{l-\mu} b_{l,-\mu}$, and defining the tensor product of two operators $T_{u_1}^{k_1}$ and $T_{u_2}^{k_2}$ as

$$[T^{k_1} \otimes T^{k_2}]^{k_3} = \sum_{u_1 u_2} (k_1 u_1 k_2 u_2 | k_3 u_3) T_{u_1}^{k_1} T_{u_2}^{k_2}, \quad (2.1)$$

one observes that the 16 possible bilinear quantities $[b_l^+ \otimes \tilde{b}_{l'}]^L$ generate the algebra U(4), which is, therefore, the overall symmetry of the vibron model.

There are two chains of subalgebras of U(4) containing the angular momentum algebra SO(3) as a subalgebra. These are

I. $\quad$ U(4) $\supset$ O(4) $\supset$ SO(3) $\supset$ SO(2), $\quad$ (2.2)

II. $\quad$ U(4) $\supset$ U(3) $\supset$ SO(3) $\supset$ SO(2). $\quad$ (2.3)

In the case of chain I the basis has the form $|N\omega L M>$, where the various quantum numbers are:

i) N is the total number of bosons. It characterizes the irreducible representations (irreps) of U(4), which are fully symmetric, since we are dealing with a system of bosons.

ii) ω is the seniority quantum number, characterizing the irreps of O(4) and obtaining the values $\omega = N, N - 2, \ldots, 1$ or 0.

iii) L is the angular momentum quantum number, labelling the irreps of SO(3) and taking the values $L = \omega, \omega - 1, \ldots, 1, 0$.

iv) M denotes the z-component of the angular momentum, labelling the irreps of SO(2) and having the values $-L \leq M \leq L$.

When the Hamiltonian is characterized by the dynamical symmetry of chain I, it can be written in terms of the Casimir operators of the algebras appearing in this chain:

$$H_I = \epsilon_0 + \epsilon_1 C_1(\text{U}(4)) + \epsilon_2 C_2(\text{U}(4)) + A C_2(\text{O}(4)) + B C_2(\text{SO}(3)), \quad (2.4)$$

where N and N^2 are related to the first and second order Casimirs of U(4). The eigenvalues of the Hamiltonian in the basis given above are then

$$E(N, \omega, L) = \epsilon_0 + \epsilon_1 N + \epsilon_2 N(N + 3) + A\omega(\omega + 2) + BL(L + 1). \quad (2.5)$$

Usually the vibrational quantum number

$$v = \frac{N - \omega}{2} \quad (2.6)$$

is introduced, and the energy eigenvalues are rewritten as

$$E(N, v, L) = \epsilon_0' + \epsilon_1' N + \epsilon_2' N^2 - 4A(N+2)(v+\frac{1}{2}) + 4A(v+\frac{1}{2})^2 + BL(L+1), \quad (2.7)$$

where ϵ'_0, ϵ'_1, ϵ'_2 are related to ϵ_0, ϵ_1, ϵ_2, A. It should be noticed that the 4th and 5th term in the rhs correspond to the spectrum of the Morse potential [34].

In the case of chain II the basis is $|Nn_pLM>$, where N is again the total number of bosons, while the other quantum numbers are:

i) n_p is the number of p-bosons, labelling the irreps of U(3) and obtaining the values $n_p = 0, 1, \ldots, N$.

ii) L is labelling the irreps of SO(3), obtaining the values $L = n_p, n_p - 2, \ldots, 1$ or 0.

iii) M is labelling the irreps of SO(2), with values $-L \leq M \leq L$.

When the Hamiltonian is characterized by the dynamical symmetry of chain II, it can be written in terms of the Casimir operators of the algebras appearing in it:

$$H_{II} = \epsilon_0 + \epsilon_1 C_1(U(4)) + \epsilon_2 C_2(U(4)) + \epsilon C_1(U(3)) + \alpha C_2(U(3)) + \beta C_2(SO(3)). \quad (2.8)$$

The eigenvalues of the Hamiltonian in the basis given above are

$$E(N, n_p, L) = \epsilon_0 + \epsilon_1 N + \epsilon_2 N(N+3) + \epsilon n_p + \alpha n_p(n_p + 2) + \beta L(L+1). \quad (2.9)$$

3. Alternative formulation of the vibron model

An alternative formulation of the vibron model can be achieved in terms of complementary algebras. The notion of complementary algebras was introduced by Moshinsky and Quesne [31–33]. It is especially fruitful in the case of multidimensional harmonic oscillators or many particle systems of few kinds of bosons. In the present case of 4 kinds of bosons (s^+, p^+_μ, $\mu = 0, \pm 1$) the host algebra is Sp(8,R). Two chains of subalgebras are

$$\text{Sp}(8, \text{R}) \supset \text{U}(4) \supset \text{O}(4) \supset \text{SO}(3) \supset \text{SO}(2), \quad (3.1)$$

$$\text{Sp}(8, \text{R}) \supset \text{Sp}(2, \text{R}) \supset \text{U}(1). \quad (3.2)$$

The quantum numbers N, ω, L, M, labelling the irreps of the subalgebras of the first chain, have been described in section 2. Sp(2,R) is isomorphic to SU(1,1). The irreps of SU(1,1) and U(1) are labelled by the quantum numbers j and m, respectively. Two subalgebras A_1 and A_2 of a larger algebra A are complementary within a definite irrep of A, if there is an one-to-one correspondence between all the irreps of A_1 and of A_2 contained in this irrep of A [31]. In the example given above, the only irreps of the host algebra Sp(8,R) which can be realized in a Fock boson space are the even irrep $[\dot{0}]$, including the vectors $|N\omega LM>$ with N=even, and the odd irrep $[\dot{1}]$, including the vectors with N=odd. It can then be proved that O(4) and Sp(2,R) (and thus also O(4) and SU(1,1)) are complementary. The same holds for U(4) and U(1).

For convenience let us denote the 4 kinds of bosons introduced in section 2 by b^+_ν, $\nu = 1, 2, 3, 4$, corresponding to p^+_{+1}, p^+_{-1}, p^+_0, s^+ respectively. To each kind ν of bosons corresponds an algebra $\text{Sp}^\nu(2, \text{R})$, generated by

$$K^\nu_+ = \frac{1}{2} b^+_\nu b^+_\nu, \qquad K^\nu_- = \frac{1}{2} b_\nu b_\nu, \qquad K^\nu_0 = \frac{1}{2}(N_\nu + \frac{1}{2}), \quad (3.3)$$

where $N_\nu = b^+_\nu b_\nu$. These generators satisfy the commutation relations

$$[K^\nu_0, K^\nu_\pm] = \pm K^\nu_\pm, \qquad [K^\nu_+, K^\nu_-] = -2K^\nu_0. \quad (3.4)$$

The Sp(2,R) $\approx$ SU(1,1) algebra, mentioned above, is realized in the space of 4 kinds of bosons. Therefore we are going to use for it the symbol Sp$^{(1234)}$(2,R) $\approx$ SU$^{(1234)}$(1,1). This algebra is generated by

$$K_+ = \frac{1}{2}\sum_\nu b_\nu^+ b_\nu^+, \qquad K_- = \frac{1}{2}\sum_\nu b_\nu b_\nu, \qquad K_0 = \frac{1}{2}(N+2), \tag{3.5}$$

where $N = \sum_\nu b_\nu^+ b_\nu$. These generators satisfy the commutation relations

$$[K_0, K_\pm] = \pm K_\pm, \qquad [K_+, K_-] = -2K_0. \tag{3.6}$$

The Casimir operator is

$$C_2(\text{Sp}^{(1234)}(2,\text{R})) = -K_+ K_- + K_0(K_0 - 1), \tag{3.7}$$

with eigenvalue $j(j+1)$. It is known that when O(n) and SU(1,1) are complementary, the quantum numbers ω and j characterizing their irreps are connected by [35]

$$j = \frac{1}{2}(\omega + \frac{n-4}{2}). \tag{3.8}$$

In the present case SU$^{(1234)}$(1,1) is complementary to O(4), so that

$$j = \frac{\omega}{2} = \frac{N-v}{4}. \tag{3.9}$$

It is also known that the Casimir operators of two algebras complementary to each other are connected by a simple, usually linear, relation of the type

$$C_2(A_1) = c_1 C_2(A_2) + c_2. \tag{3.10}$$

In the case of O(n) and SU(1,1) this relation is

$$C_2(\text{SU}(1,1)) = \frac{1}{4}C_2(\text{O}(n)) + \frac{n(n-4)}{16}, \tag{3.11}$$

which in the present case of O(4) reduces to

$$C_2(\text{SU}^{(1234)}(1,1)) = \frac{1}{4}C_2(\text{O}(4)). \tag{3.12}$$

The U(1) subalgebra of SU$^{(1234)}$(1,1) is generated by the operator K_0 alone, the eigenvalues of which we label by m. In the general case of the complementary algebras U(n) and U(1), the quantum numbers N and m characterizing their irreps are connected by [35]

$$m = \frac{1}{2}(N + \frac{n}{2}), \tag{3.13}$$

which in the present special case of U(4) and U(1) reduces to

$$m = \frac{1}{2}(N+2). \tag{3.14}$$

The chain of eq. (3.1) already studied is of interest in the case of the chain I of the vibron model. It implies that in studying chain I, one can replace in the basis $|N\omega LM>$ the quantum numbers $N\omega$ by the quantum numbers jm of the complementary subalgebras. Furthermore, in the Hamiltonian of eq. (2.4) one is entitled to replace the second order Casimir of O(4) by the second order Casimir of SU$^{(1234)}$(1,1), and the first and second order Casimirs of U(4) (N and $N(N+3)$), by the first and second order Casimirs of U(1) (K_0 and K_0^2).

In the case of chain II, the U(3) subalgebra of U(4) involves only the p-bosons. The host algebra is then Sp(6,R), having the two chains of subalgebras

$$\text{Sp}(6,\text{R}) \supset \text{U}(3) \supset \text{SO}(3) \supset \text{SO}(2), \tag{3.15}$$

$$\text{Sp}(6,\text{R}) \supset \text{SU}^{(123)}(1,1) \supset U(1), \tag{3.16}$$

where the superscript (123) means that only the bosons b_1^+, b_2^+, b_3^+ are involved in the formation of SU$^{(123)}$(1,1) $\approx$ Sp$^{(123)}$(2,R). Further details on these chains are given below (see eqs (3.26) – (3.28)).

The building up of the bases related to the two limiting symmetries of the vibron model can then be achieved as follows. To each kind of boson b_ν^+, an Sp$^\nu$(2,R) algebra corresponds, as already mentioned, generated by the operators given in eq. (3.3). The states with N_ν=even correspond to the irrep $D^{-3/4}$, while the states with N_ν=odd correspond to the irrep $D^{-1/4}$. Thus to each boson state

$$|N_1 N_2 N_3 N_4> = \frac{1}{\sqrt{N_1! N_2! N_3! N_4!}} (b_1^+)^{N_1} (b_2^+)^{N_2} (b_3^+)^{N_3} (b_4^+)^{N_4} |0>, \tag{3.17}$$

one can correspond a set of 4 noncompact "angular momenta" $j_\nu = -3/4$ or $-1/4$, $\nu = 1, 2, 3, 4$, the value of each angular momentum j_ν depending on the parity of the corresponding boson number N_ν.

We can now proceed to the vector coupling of the first two "angular momenta" j_1 and j_2, using the SU(1,1) Clebsch–Gordan coefficients [36,37]

$$|j_1 j_2 : j_{12} m_{12}> = \sum_{m_1 m_2} <j_1 m_1 j_2 m_2 | j_{12} m_{12}>_{\text{SU}(1,1)} |j_1 m_1> |j_2 m_2>. \tag{3.18}$$

This means that the intermediate SU$^{(12)}$(1,1) algebra has been introduced, generated by

$$K_\mu^{12} = K_\mu^1 + K_\mu^2, \tag{3.19}$$

with $\mu = 0, \pm 1$. The host algebra of this space of two kinds of bosons is Sp(4,R). The following two chains of subalgebras exist

$$\text{Sp}(4,\text{R}) \supset \text{U}(2) \supset \text{SO}(2), \tag{3.20}$$

$$\text{Sp}(4,\text{R}) \supset \text{SU}^{(12)}(1,1) \supset U(1), \tag{3.21}$$

where the irreps of U(2) are labelled by the total number of bosons $N_{12} = N_1 + N_2$, while the irreps of SO(2) are labelled by $M = N_1 - N_2$. SO(2) is complementary with SU$^{(12)}$(1,1) $\approx$ Sp$^{(12)}$(2,R), the irreps of which are labelled by

$$j_{12} = \frac{1}{2}(M - 1), \tag{3.22}$$

according to eq. (3.8), while U(2) is complementary to U(1), the irreps of which are labelled by
$$m_{12} = \frac{1}{2}(N_{12} + 1), \tag{3.23}$$
according to eq. (3.13).

The next step along this line is to couple j_{12} with j_3. In this case 3 kinds of bosons are involved, so that the host algebra is Sp(6,R). The relevant chains for this case have been given in eqs (3.15) – (3.16), SO(3) being complementary to $\mathrm{SU}^{(123)}(1,1)$, which is generated by
$$K_\mu^{123} = K_\mu^{12} + K_\mu^3, \tag{3.24}$$
with $\mu = 0, \pm 1$. The resulting eigenvectors are

$$|j_1 j_2 (j_{12}) j_3 : j_{123} m_{123}>$$
$$= \sum_{m_{12} m_3} <j_{12} m_{12} j_3 m_3 | j_{123} m_{123}>_{\mathrm{SU}(1,1)} |j_1 j_2 : j_{12} m_{12}> |j_3 m_3>. \tag{3.25}$$

The Casimir operators of SO(3) and $\mathrm{SU}^{(123)}(1,1)$ are connected by
$$C_2(\mathrm{SO}(3)) = 4 C_2(\mathrm{SU}^{(123)}(1,1)) + \frac{3}{4}, \tag{3.26}$$
according to eq. (3.11), while the quantum numbers labelling their irreps, L and j_{123} respectively, are connected by
$$j_{123} = \frac{1}{2}(L - \frac{1}{2}), \tag{3.27}$$
according to eq. (3.8). The eigenvalues of $C_2(\mathrm{SU}^{(123)}(1,1))$ in the above mentioned basis are given by $j_{123}(j_{123} + 1)$. Furthermore, in the chains of eqs (3.15) – (3.16), U(3) and U(1) are complementary, the quantum numbers n_p and m_{123} labelling respectively their irreps being connected by
$$m_{123} = \frac{1}{2}(n_p + \frac{3}{2}), \tag{3.28}$$
according to eq. (3.13).

The coupling of the two "angular momenta" j_1 and j_2, performed above, can be avoided, by noticing that the $\mathrm{SU}^{(12)}(1,1)$ can be generated by
$$K_+^{12} = b_1^+ b_2^+, \qquad K_-^{12} = b_1 b_2, \qquad K_0^{12} = \frac{1}{2}(N_1 + N_2 + 1). \tag{3.29}$$

The vectors
$$|j_{12} m_{12}> = \frac{1}{\sqrt{N_1! N_2!}} (b_1^+)^{N_1} (b_2^+)^{N_2} |0>, \tag{3.30}$$
with
$$j_{12} = \frac{1}{2}(N_1 - N_2 - 1), \qquad m_{12} = \frac{1}{2}(N_1 + N_2 + 1), \tag{3.31}$$
(which are in agreement with eqs (3.22) – (3.23)) are eigenvectors of the Casimir operator $C_2(\mathrm{SU}^{(12)}(1,1))$, with eigenvalues $j_{12}(j_{12} + 1)$. In this way one can avoid the

coupling of j_1 and j_2 to j_{12}. The coupling of j_{12} and j_3 cannot be avoided, however. The resulting vectors in this case we denote by $|j_{12}j_3 : j_{123}m_{123}>$.

In the last step the coupling of j_{123} to j_4 is performed. The host algebra in this case is Sp(8,R), the relevant chains having been given in eqs (3.31) – (3.32). The basis vectors, denoted by $|j_1j_2(j_{12})j_3(j_{123})j_4 : jm>$, or by $|j_{12}j_3(j_{123})j_4 : jm>$ in the case in which the shortened version of eq. (3.30) is used for the vectors with "angular momentum" j_{12}, correspond to the irreps of the $\text{Sp}^{1234}(2,R) \approx \text{SU}^{1234}(1,1)$ algebra, generated by

$$K_\mu^{1234} = K_\mu^{123} + K_\mu^4, \tag{3.32}$$

with $\mu = 0, \pm 1$. Here K_μ^4 are the generators of the $\text{SU}^4(1,1)$ algebra, associated with the s-bosons. The total noncompact "angular momentum" j, characterizing the irreps of $\text{SU}^{(1234)}(1,1)$, is connected to the seniority quantum number ω, characterizing the irreps of O(4), by eq. (3.8), while the Casimir operators of these two complementary algebras are connected by eq. (3.12).

Given the above, it is clear that for the chain I of the vibron model, instead of the basis $|N\omega LM>$, the basis $|j_{12}j_3(j_{123})j_4 : jm>$ can be used. Furthermore, in the case of chain II, instead of the basis $|Nn_pLM>$, the basis $|j_{12}j_3 : j_{123}m_{123}> |j_4m_4>$ can be used. It is clear that the connection between the two new bases for the dynamical symmetries of the vibron model is

$$|j_{12}j_3(j_{123})j_4 : jm> = \sum_{m_{123}m_4} <j_{123}m_{123}j_4m_4|jm>_{\text{SU}(1,1)} |j_{12}j_3 : j_{123}m_{123}> |j_4m_4>. \tag{3.33}$$

The Hamiltonian of chain I, given in eq. (2.4), can be rewritten using the complementarity relations described in eqs (3.1), (3.2), (3.15), (3.16), as

$$H_I = \epsilon'_0 + \epsilon'_1 C_1(\text{U}(1)) + \epsilon'_2 C_2(\text{U}(1)) + 4AC_2(\text{SU}^{(1234)}(1,1)) + 4BC_2(\text{SU}^{(123)}(1,1)). \tag{3.34}$$

The eigenvalues of this Hamiltonian are

$$E(m, j, j_{123}) = \epsilon'_0 + \epsilon'_1 m + \epsilon'_2 m^2 + 4Aj(j+1) + 4Bj_{123}(j_{123}+1). \tag{3.35}$$

Using eqs. (3.14), (3.9), (3.27), which connect the quantum numbers m, j, j_{123} to the previous ones (N, ω, L), it is easily verified that eq. (3.35) is an alternative way of writing eq. (2.5).

Similarly the Hamiltonian of chain II, given in eq. (2.8), can be rewritten, taking into account the complementarity relations given in eqs (3.1), (3.2), (3.15), (3.16), as

$$H_{II} = \epsilon'_0 + \epsilon'_1 m + \epsilon'_2 m^2 + \epsilon' m_{123} + \alpha' m_{123}^2 + 4\beta' C_2(\text{SU}^{(123)}(1,1)), \tag{3.36}$$

where in the rhs the 2nd and 3rd term correspond to the first and second order Casimir of the U(1) algebra of the chain of eq. (3.2), while the 4th and 5th terms correspond to the first and second order Casimir operators of the U(1) algebra appearing in the chain of eq. (3.16). The eigenvalues of this Hamiltonian are

$$E(m, m_{123}, j_{123}) = \epsilon'_0 + \epsilon'_1 m + \epsilon'_2 m^2 + \epsilon' m_{123} + \alpha' m_{123}^2 + 4\beta' j_{123}(j_{123}+1). \tag{3.37}$$

Using eqs (3.14), (3.28), (3.27), which connect m, m_{123}, j_{123} to N, n_p, L, it is easily verified that eq. (3.37) is an alternative way of writing eq. (2.9).

In this section we have therefore rewritten the bases and the Hamiltonians corresponding to the two dynamical symmetries of the vibron model in terms of complementary subalgebras. This formulation is useful because it can be q-deformed in a very simple way.

4. q-deformation of the vibron model

In the previous section the subalgebra chains of the vibron model were reduced to equivalent chains of complementary subalgebras

$$SU^{(1234)}(1,1) \supset SU^{(123)}(1,1) \supset SU^{(12)}(1,1) \supset U(1). \tag{4.1}$$

The corresponding Hamiltonians were then written in terms of the Casimir operators of the new reduction chains. An evident possibility for q-deforming these Hamiltonians is to substitute the SU(1,1) algebras of eq. (4.1) by their q-deformed counterparts, $SU_q(1,1)$ [38–40]

$$SU_q^{(1234)}(1,1) \supset SU_q^{(123)}(1,1) \supset SU_q^{(12)}(1,1) \supset U_q(1). \tag{4.2}$$

In this section we shall explain how this can be achieved, after giving a brief account of the necessary mathematical details.

q-numbers are defined as

$$[x]_q = \frac{q^x - q^{-x}}{q - q^{-1}}. \tag{4.3}$$

For q real ($q = e^\tau$ with τ real), they can be written as

$$[x]_q = \frac{\sinh \tau x}{\sinh \tau}, \tag{4.4}$$

while in the case of q being a phase ($q = e^{i\tau}$ with τ real), they obtain the form

$$[x]_q = \frac{\sin \tau x}{\sin \tau}. \tag{4.5}$$

In the limit $q \to 1$ ($\tau \to 0$), q-numbers reduce to usual numbers.

q-deformed oscillators [41,42] are introduced through the relations

$$aa^+ - q^{\pm 1}a^+a = q^{\mp N}, \qquad [N, a^+] = a^+, \qquad [N, a] = -a, \tag{4.6}$$

where a^+, a are the q-deformed boson creation and annihilation operators and N the relevant number operator. Using eq. (4.6) one can easily show that

$$a^+a = [N]_q, \qquad aa^+ = [N+1]_q. \tag{4.7}$$

q-deformed algebras can be expressed in terms of q-deformed bosons. Introducing a_i^+ as the q-deformed analogues of b_i^+ ($i = 1, 2, 3, 4$), with the properties

$$[a_\nu^+, a_\mu^+] = [a_\nu, a_\mu] = [a_\nu, a_\mu^+] = 0 \quad \nu \neq \mu, \tag{4.8}$$

one can prove that $SU_q^{(12)}(1,1)$ is generated by [38–40]

$$K_+^{(12)} = a_1^+ a_2^+, \qquad K_-^{(12)} = a_1 a_2, \qquad K_0^{(12)} = \frac{1}{2}(N_1 + N_2 + 1), \tag{4.9}$$

the relevant commutation relations being

$$[K_0^{(12)}, K_{\pm}^{(12)}] = \pm K_{\pm}^{(12)}, \qquad [K_+^{(12)}, K_-^{(12)}] = -[2K_0^{(12)}]_q. \qquad (4.10)$$

The vectors
$$|j_{12}m_{12}>_q = \frac{1}{\sqrt{[N_1]_q![N_2]_q!}}(a_1^+)^{N_1}(a_2^+)^{N_2}|0>, \qquad (4.11)$$

with j_{12}, m_{12} still given by eq. (3.31), are eigenvectors of the deformed Casimir operator

$$C_2(SU_q^{(12)}(1,1)) = -K_+^{(12)}K_-^{(12)} + [K_0^{(12)}]_q[K_0^{(12)} - 1]_q, \qquad (4.12)$$

with eigenvalues $[j_{12}]_q[j_{12}+1]_q$.

For the $SU^\nu(1,1)$ algebras one has the boson realization [38]

$$K_+^\nu = \frac{1}{[2]_q}a^+a^+, \qquad K_-^\nu = \frac{1}{[2]_q}aa, \qquad K_0^\nu = \frac{1}{2}(N_\nu + \frac{1}{2}), \qquad (4.13)$$

where N_ν is the number of ν-bosons. These generators satisfy the commutation relations

$$[K_0^\nu, K_{\pm}^\nu] = \pm K_{\pm}^\nu, \qquad [K_+^\nu, K_-^\nu] = -[2K_0^\nu]_{q^2}. \qquad (4.14)$$

The $SU_q^1(1,1)$ and $SU_q^2(1,1)$ algebras we are not going to use explicitly in couplings, since for the $SU_q^{(12)}(1,1)$ algebra we already have the form given in eq. (4.9), which avoids the direct coupling. In order to be able to couple $SU_q^3(1,1)$ and $SU_q^4(1,1)$ to $SU_q^{(12)}(1,1)$, it is useful to have the same deformation parameter in all of these algebras, i.e. it is useful to have the same deformation parameter in the commutation relations of eq.(4.10) and eq. (4.14). In order to achieve that, we replace in eqs. (4.13), (4.14) q^2 by q. As a result, for $\nu = 3, 4$, eq. (4.6) is meant from now on with q replaced by $\sqrt{q}$. Then one also has

$$a_\nu^+ a_\nu = [N_\nu]_{\sqrt{q}}, \qquad a_\nu a_\nu^+ = [N_\nu + 1]_{\sqrt{q}}. \qquad (4.15)$$

Eq. (4.12), giving the Casimir operator, is therefore valid in this case with the usual q-numbers.

The $SU_q^{(123)}(1,1)$ algebra is generated by the operators

$$K_{\pm}^{(123)} = K_{\pm}^{(12)} \; q^{K_0^3} + K_{\pm}^3 \; q^{-K_0^{(12)}}, \qquad K_0^{(123)} = K_0^{(12)} + K_0^3, \qquad (4.16)$$

i.e. it is a standard coproduct of the irreps $D^{j_{12}+}$ and D^{j_3+} of the $SU_q(1,1)$ algebra. Therefore the basis vectors, in analogy to eq. (3.25), are of the form

$$|j_{12}j_3 : j_{123}m_{123}>_q = \sum_{m_{12}m_3} <j_{12}m_{12}j_3m_3|j_{123}m_{123}>_{SU_q(1,1)} |j_{12}m_{12}>_q |j_3m_3>_q, \qquad (4.17)$$

where $<j_1m_1j_2m_2|jm>_{SU_q(1,1)}$ are Clebsch–Gordan coefficients for the tensor product of two $SU_q(1,1)$ irreps. Explicit analytical formulae for these coefficients, as well as for the relevant $SU_q(2)$ coefficients, can be found in [43–47].

The $SU_q^{(1234)}(1,1)$ algebra is generated by the operators

$$K_{\pm}^{(1234)} = K_{\pm}^{(123)} \; q^{K_0^4} + K_{\pm}^4 \; q^{-K_0^{(123)}}, \qquad (4.18)$$

while the vectors analogous to eq. (3.33) are

$$|j_{12}j_3(j_{123})j_4 : jm>_q$$
$$= \sum_{m_{123}m_4} <j_{123}m_{123}j_4m_4|jm>_{SU_q(1,1)} |j_{12}j_3 : j_{123}m_{123}>_q |j_4m_4>_q. \quad (4.19)$$

These vectors are the q-analogues of the eigenvectors of the dynamical symmetry I of the vibron model. Similarly the vectors

$$|j_{12}j_3 : j_{123}m_{123}>_q |j_4m_4>_q \quad (4.20)$$

are the q-analogues of the eigenvectors of the dynamical symmetry II of the vibron model. Therefore eq. (4.19) connects the eigenvectors of the two dynamical symmetries, as eq. (3.33) does in the classical case.

In the case of dynamical symmetry I the Hamiltonian reads

$$H_I = \epsilon_0 + \epsilon_1[m]_q + \epsilon_2[m]_q^2 + 4AC_2(SU_q^{(1234)}(1,1)) + 4BC_2(SU_q^{(123)}(1,1)), \quad (4.21)$$

which is the q-analogue of eq. (3.34) (with the primes of the coefficients dropped). The eigenvalues of this Hamiltonian are

$$E(m,j,j_{123}) = \epsilon_0 + \epsilon_1[m]_q + \epsilon_2[m]_q^2 + 4A[j]_q[j+1]_q + 4B[j_{123}]_q[j_{123}+1]_q. \quad (4.22)$$

In the limit $q \to 1$, eq. (3.35) is obtained. Assuming that m, j, j_{123} are still connected to quantum numbers N, ω, L, through eqs (3.14), (3.9), (3.27), the last equation can be rewritten in a way resembling its classical counterpart, eq. (2.5), as

$$E(N,\omega,L) = \epsilon_0' + \epsilon_1'[N+2]_{\sqrt{q}} + \epsilon_2'[N+2]_{\sqrt{q}}^2 + A'[\omega]_{\sqrt{q}}[\omega+2]_{\sqrt{q}} + B'[L]_{\sqrt{q}}[L+1]_{\sqrt{q}}. \quad (4.23)$$

In producing the last equation, identities like

$$\left[\frac{\omega}{2}\right]_q = [\omega]_{\sqrt{q}}(q^{1/2}+q^{-1/2})^{-1}, \quad (4.24)$$

$$\left[L-\frac{1}{2}\right]_q\left[L+\frac{3}{2}\right]_q = [L]_q[L+1]_q - \left[\frac{1}{2}\right]_q\left[\frac{3}{2}\right]_q, \quad (4.25)$$

have been used.

Using eq. (2.6), eq. (4.23) can be rewritten as

$$E(N,v,L) = \epsilon_0' + \epsilon_1'[N+2]_{\sqrt{q}} + \epsilon_2'[N+2]_{\sqrt{q}}^2 + A\left[v-\frac{N}{2}\right]_q\left[v-1-\frac{N}{2}\right]_q + B'[L]_{\sqrt{q}}[L+1]_{\sqrt{q}}, \quad (4.26)$$

whice reduces to eq. (2.7) in the limit $q \to 1$, up to a redefinition of ϵ_0', ϵ_1', ϵ_2'.

In the case of the dynamical symmetry II the Hamiltonian can be written as

$$H_{II} = \epsilon_0 + \epsilon_1[m]_q + \epsilon_2[m]_q^2 + \epsilon\ [m_{123}]_q + \alpha\ [m_{123}]_q^2 + \beta C_2(SU_q^{(123)}(1,1)), \quad (4.27)$$

which is the q-analogue of eq. (3.36) (with the primes of the coefficients dropped). The eigenvalues of this Hamiltonian are

$$E(m,m_{123},L) = \epsilon_0 + \epsilon_1[m]_q + \epsilon_2[m]_q^2 + \epsilon\ [m_{123}]_q + \alpha\ [m_{123}]_q^2 + \beta[j_{123}]_q[j_{123}+1]_q. \quad (4.28)$$

In the limit $q \to 1$, eq. (3.37) is obtained. Assuming that m, m_{123}, j_{123} are connected to N, n_p, L, through eqs (3.14), (3.28), (3.27), the last equation can be rewritten in a way resembling its classical counterpart, eq. (2.9), as

$$E(N, n_p, L) = \epsilon'_0 + \epsilon'_1 [N+2]_{\sqrt{q}} + \epsilon'_2 [N+2]^2_{\sqrt{q}}$$

$$+ \epsilon' \left[n_p + \frac{3}{2} \right]_{\sqrt{q}} + \alpha' \left[n_p + \frac{3}{2} \right]^2_{\sqrt{q}} + \beta'[L]_{\sqrt{q}}[L+1]_{\sqrt{q}}. \qquad (4.29)$$

The results obtained in this section call for the following comments:

i) Rotational–vibrational spectra of diatomic molecules are described empirically by the Dunham expansion [48]

$$E(v, L) = \sum_{ik} Y_{ik} \left(v + \frac{1}{2} \right)^i (L(L+1))^k, \qquad (4.30)$$

where v is the vibrational quantum number, L the angular momentum, and Y_{ik} the Dunham coefficients, fitted to experiment. It is clear that the Dunham expansion contains powers of $(v+1/2)$, powers of $L(L+1)$, as well as cross terms. Eqs (4.23), (4.29) contain no cross terms. This is due to the fact that in the Hamiltonians of eqs (4.21), (4.27), only terms up to quadratic in the generators are included, as in the case of the classical vibron model. Cross terms can be taken into account in the dynamical symmetry I, for example, by modifying eq. (4.21) as follows

$$H'_I = H_I + DC_2(SU_q^{(1234)}(1,1))C_2(SU_q^{(123)}(1,1)). \qquad (4.31)$$

Then eq. (4.26) is modified as

$$E'(N, v, L) = E(N, v, L) + D' \left[v - \frac{N}{2} \right]_q \left[v - 1 - \frac{N}{2} \right]_q [L]_{\sqrt{q}}[L+1]_{\sqrt{q}}. \qquad (4.32)$$

ii) Rotational spectra in both dynamical symmetries (eqs (4.26), (4.29)) are described by the term $[L]_{\sqrt{q}}[L+1]_{\sqrt{q}}$. This is known to be the Casimir operator of $SU_{\sqrt{q}}(2)$. The $SU_q(2)$ model has been extensively used for the description of rotational spectra of diatomic molecules [5–7], deformed [8–10] and superdeformed [11] nuclei. It has been found [9] that this term is equivalent to an expansion in terms of powers of $L(L+1)$

$$[L]_q[L+1]_q = \frac{1}{(j_0(\tau))^2} (j_0(\tau)L(L+1) - \tau j_1(\tau)(L(L+1))^2 + \frac{2}{3}\tau^2 j_2(\tau)(L(L+1))^3$$

$$- \frac{1}{3}\tau^3 j_3(\tau)(L(L+1))^4 + \frac{2}{15}\tau^4 j_4(\tau)(L(L+1))^5 - \ldots), \qquad (4.33)$$

where $j_n(\tau)$ are the spherical Bessel functions of the first kind and $q = e^{i\tau}$. This expansion is similar to the one contained in the Dunham expansion. In the case of $SU_q(2)$, however, all the expansion coefficiens are related to powers of τ, thus resulting in economy of parameters. Notice that the decreasing of the coefficients of increasing powers of $L(L+1)$, as well as the alternating signs of the terms, facts that are known empirically to hold, occur in eq. (4.33) automatically, since τ is known [5–11] to obtain small positive values. Furthermore, it has been proved [9] that the $SU_q(2)$ model is equivalent to the Variable Moment of Inertia (VMI) model, which describes rotational stretching effects. The q parameter has been found [9] to correspond to the softness parameter of the VMI model. The implications of the $SU_q(2)$ model on the electromagnetic transition probabilities connecting the rotational levels of nuclei have been considered [10].

iii) The 4th term in eq. (4.26) corresponds to the Casimir operator of $SU_q(1,1)$, already used [17] for the description of vibrational spectra of diatomic molecules. It has been proved [17] that this term, for $q = e^{i\tau}$, can be expanded as

$$\left[v - \frac{N}{2}\right]_q \left[v - 1 - \frac{N}{2}\right]_q = \frac{1}{\sin(\tau)^2}(\frac{1}{2}(\cos(\tau) - \cos(\tau(N+2))) - \tau \sin(\tau(N+2))(v+\frac{1}{2})$$

$$+\tau^2 \cos(\tau(N+2))(v+\frac{1}{2})^2 + \frac{2}{3}\tau^3 \sin(\tau(N+2))(v+\frac{1}{2})^3 - \frac{1}{3}\tau^4 \cos(\tau(N+2))(v+\frac{1}{2})^4 + \ldots),$$
(4.34)

We remark that a series of powers of $(v + 1/2)$ is obtained, similar to the one contained in the Dunham expansion. In the present case, however, the expansion coefficients are all related to τ (and N, which in the vibron model is a constant for a given molecule), thus resulting in economy in parameters.

iv) The anharmonicity constant (i.e. the ratio Y_{20}/Y_{10}) in the classical case (eq. (2.7)) is fixed to $-1/(N+2)$. In the deformed case of eq. (4.26), however, is equal to $-\tau/\tan(\tau(N+2))$, as it is easily seen from the expansion of eq. (4.34). The extra freedom gained this way has been found [17] to improve the fits of vibrational molecular spectra.

v) Since N is fixed for a given molecule (related to the maximum number of bound states below the dissociation limit), the first three terms in eqs (4.26), (4.29) have no influence on the spectrum.

vi) In eq. (4.26) it is clear that the deformation parameter for the vibrational part of the spectrum is τ, while for the rotational part it is $\frac{\tau}{2}$. Therefore a relation is implied between the rotational stretching and the anharmonicity corrections. Careful empirical fits are needed in order to decide if this is a restriction or an advantage of the present model. There is no *a priori* reason, however, that these two physically different mechanisms be described by the same parameter. A more general version of the model, allowing for these two deformation parameters to be independent from each other, might give better results.

5. Discussion

In this paper a deformed version of the O(4) and U(3) dynamical symmetries of the vibron model for diatomic molecules has been constructed. This has been achieved by first rewriting, through use of the concept of complementary subalgebras, the model in a more convenient form, which is subsequently deformed. The present approach unifies into a common framework the so far separate quantum algebraic approaches to rotational and to vibrational spectra of diatomic molecules.

For the O(4) limit of the present model, fittings to experimental data for diatomic molecules are required. Its U(3) limit can be used for the description of clustering phenomena in nuclei [49], as well as for the quasi-molecular description of heavy-ion resonances [50]. The present work can be extended to the study of triatomic molecules

[23]. The method of complementary subalgebras can also be used in constructing [30] the deformed versions of the U(5) and O(6) dynamical symmetries of the Interacting Boson Model (IBM) [24] of nuclear structure.

References

* On leave of absence from the Institute of Nuclear Physics, Moscow State University.
[1] P. P. Kulish and N. Yu. Reshetikhin, Zapiski Semenarov LOMI **101**, 101 (1981).
[2] E. K. Sklyanin, Funct. Anal. Appl. **16**, 262 (1982).
[3] V. G. Drinfeld, in *Proceedings of the International Congress of Mathematicians*, edited by A. M. Gleason (American Mathematical Society, Providence, RI, 1987) p. 798.
[4] M. Jimbo, Lett. Math. Phys. **11**, 247 (1986).
[5] D. Bonatsos, P. P. Raychev, R. P. Roussev and Yu. F. Smirnov, Chem. Phys. Lett. **175**, 300 (1990).
[6] Z. Chang and H. Yan, Phys. Lett. A **154**, 254 (1991).
[7] J. G. Esteve, C. Tejel and B. E. Villaroya, J. Chem. Phys. **96**, 5614 (1992).
[8] P. P. Raychev, R. P. Roussev and Yu. F. Smirnov, J. Phys. G **16**, L137 (1990).
[9] D. Bonatsos, E. N. Argyres, S. B. Drenska, P. P. Raychev, R. P. Roussev and Yu. F. Smirnov, Phys. Lett. B **251**, 477 (1990).
[10] D. Bonatsos, A. Faessler, P. P. Raychev, R. P. Roussev and Yu. F. Smirnov, J. Phys. A **25**, 3275 (1992).
[11] D. Bonatsos, S. B. Drenska, P. P. Raychev, R. P. Roussev and Yu. F. Smirnov, J. Phys. G **17**, L67 (1991).
[12] Z. Chang, H. Y. Guo and H. Yan, Phys. Lett. A **156**, 192 (1991).
[13] Z. Chang and H. Yan, Phys. Rev. A **44**, 7405 (1991).
[14] Z. Chang, H. Y. Guo and H. Yan, Commun. Theor. Phys. **17**, 183 (1992).
[15] D. Bonatsos and C. Daskaloyannis, Phys. Rev. A **46**, 75 (1992).
[16] D. Bonatsos and C. Daskaloyannis, Chem. Phys. Lett. **203**, 150 (1993).
[17] D. Bonatsos, E. N. Argyres and P. P. Raychev, J. Phys. A **24**, L403 (1991).
[18] D. Bonatsos, P. P. Raychev and A. Faessler, Chem. Phys. Lett. **178**, 221 (1991).
[19] D. Bonatsos, C. Daskaloyannis and K. Kokkotas, Phys. Rev. A **45**, R6153 (1992).
[20] D. Bonatsos, C. Daskaloyannis and K. Kokkotas, J. Math. Phys. **33**, 2958 (1992).
[21] F. Iachello, Chem. Phys. Lett. **78**, 581 (1981).
[22] F. Iachello and R. D. Levine, J. Chem. Phys. **77**, 3046 (1982).
[23] O. S. van Roosmalen, F. Iachello, R. D. Levine and A. E. L. Dieperink, J. Chem. Phys. **79**, 2515 (1983).
[24] F. Iachello and A. Arima, *The Interacting Boson Model* (Cambridge University Press, Cambridge, 1987).
[25] D. Bonatsos, *Interacting Boson Models of Nuclear Structure* (Clarendon, Oxford, 1988).
[26] J. Van der Jeugt, J. Phys. A **25**, L213 (1992).
[27] Y. F. Cao and H. L. Lin, Chem. Phys. Lett. **207**, 317 (1993).

[28] F. Pan, *Quantum deformation of the $U(4) \supset SO(4) \supset SO(3)$ chain and the description of rotation-vibration spectra of diatomic molecules*, ICTP Trieste preprint (1992).
[29] D. Bonatsos, A. Faessler, P. P. Raychev, R. P. Roussev and Yu. F. Smirnov, J. Phys. A **25**, L267 (1992).
[30] F. Pan, *q-deformations in the Interacting Boson Model for nuclei*, ICTP Trieste preprint (1992).
[31] M. Moshinsky and C. Quesne, J. Math. Phys. **11**, 1631 (1970).
[32] G. Couvreur, J. Deenen and C. Quesne, J. Math. Phys. **24**, 779 (1983).
[33] C. Quesne, J. Phys. A **18**, 2675 (1985).
[34] P. M. Morse, Phys. Rev. **34**, 57 (1929).
[35] F. Pan and Y. F. Cao, J. Math. Phys. **29**, 2384 (1988).
[36] H. Ui, Ann. Phys. (NY) **49**, 69 (1968).
[37] W. J. Holman, III, and L. C. Biedenharn, Jr, Ann. Phys. (NY) **39**, 1 (1966).
[38] P. P. Kulish and E. V. Damaskinsky, J. Phys. A **23**, L415 (1990).
[39] H. Ui and N. Aizawa, Mod. Phys. Lett. A **5**, 237 (1990).
[40] N. Aizawa, J. Phys. A **26**, 1115 (1993).
[41] L. C. Biedenharn, J. Phys. A **22**, L873 (1989).
[42] A. J. Macfarlane, J. Phys. A **22**, 4581 (1989).
[43] T. Maekawa, J. Math. Phys. **32**, 2598 (1991).
[44] Yu. F. Smirnov, V. N. Tolstoy and Yu. I. Kharitonov, Yad. Fiz. **53**, 959 (1991). [Sov. J. Nucl. Phys. **53**, 593 (1991).]
[45] Yu. F. Smirnov, V. N. Tolstoy and Yu. I. Kharitonov, Yad. Fiz. **53**, 1746 (1991). [Sov. J. Nucl. Phys. **53**, 1068 (1991).]
[46] N. Aizawa, J. Math. Phys. **34**, 1937 (1993).
[47] R. N. Alvarez, Diploma Thesis (Moscow State University, 1992).
[48] J. L. Dunham, Phys. Rev. **41**, 721 (1932).
[49] J. Cseh, Atomki preprint 9-1992-P.
[50] Z. Chang and H. Yan, Academia Sinica preprint ASITP-91-25.

CHAOS IN THE COLLECTIVE DYNAMICS OF NUCLEI

Y. Alhassid
*Center for Theoretical Physics, Sloane Physics Laboratory,
Yale University, New Haven, CT 06511*

Abstract

The onset of chaos in the collective dynamics of nuclei is investigated within the framework of the interacting boson model. The appropriate "classical" limit of the nuclear many-body system is a time-dependent mean-field approximation and is obtained through the use of coherent states. It is shown that the quantal fluctuations of spectra and B(E2) intensities are strongly correlated with the classical onset of chaos. A new quasi-regular region is found within the Casten triangle.

1. Introduction

The idea underlying quantum chaos originated in nuclear physics over thirty years ago when a series of neutron resonances at threshold (~ 6 MeV) was observed. In this energy region the nuclear level density is very high ($\sim 10^6$ MeV^{-1}) and it is beyond the scope of any theory to predict the detailed level scheme. The spacings between neighboring resonances showed large fluctuations and their distribution was in good agreement with the Wigner distribution. The latter displays level repulsion and is rather different from a Poisson distribution which would correspond to a random sequence of energy levels. This was found to be a universal behavior in a large number of nuclei and was explained by random matrix theory (RMT) [1] developed by Wigner, Dyson, Mehta and others. Although it explains statistical properties of spectra, the RMT approach is quite different from that of statistical mechanics. While in statistical mechanics all states at a given energy, spin and any other conserved quantities are assumed to be equally probable, in RMT all interaction laws that are consistent with the given symmetries of the system are assumed to be equally probable. For systems with time reversal symmetry, the proper ensemble is the gaussian orthogonal ensemble (GOE). The use of RMT for the neutron resonances was justified by the complexity of the compound nucleus [2] and by lack of knowledge of the exact effective nuclear interaction. However, it was conjectured more recently that the GOE is the proper ensemble to describe systems whose classical limit is chaotic. This followed a study of a simple quantum mechanical system in two degrees of freedom, the Sinai's billiard, where the spectral statistics were found to follow that of the GOE [3].

An interesting question is whether chaos could prevail in the low-lying collective part of the nuclear spectrum, where the number of degrees of freedom is larger than two but smaller than that of the compound nucleus. Weidenmuller et al [4] analyzed experimental low-lying levels by grouping together levels of similar properties in different nuclei. Shriner et al [5] analyzed the statistical properties of low lying levels in ^{26}Al, and Raman et al [6] analyzed a complete set of levels in ^{116}Sn. Levels from a large number of nuclei were also studied in Ref. [7]

From a theoretical point of view it is necessary to have a realistic and tractable model of collective motion. The advantage in a theoretical analysis is that one can analyze a given Hamiltonian and states with given spin/parity and still obtain reasonable statistics. In addition, the nucleus is a many-body system with strong interactions, and one must find the appropriate "classical" limit where the concept of chaotic motion could be investigated.

The interacting boson model [8, 9], has been very successful in describing the low-lying collective levels and electromagnetic transition intensities of heavy nuclei. In its simplest version, the IBM-1, the nuclear Hamiltonian is described with few parameters and possesses three physical limits where it is exactly solvable. Even a general IBM-1 Hamiltonian (with no symmetries) is relatively easy to solve due to the moderate size of its basis.

These same features make the model very attractive for the study of the onset of chaos in the nuclear collective motion [10]. One can investigate both the statistical fluctuation properties of the quantal Hamiltonian and the dynamics of its classical limit. The quantal analysis requires an accurate and complete set of energy levels and transition intensities. The finite Hilbert space of the model allows an exact solution with no truncation errors. The classical (mean-field) limit is easily obtained through the use of coherent boson states in the limit of large numbers of bosons. Moreover, the completely integrable limits of the model are easily identified without any detailed calculations.

2. Dynamical Symmetries and Integrability

In our studies we used the IBM-1, where the degrees of freedom are one monopole s boson and five quadrupole bosons d_μ. The total number of bosons $N = s^\dagger s + \sum_\mu d_\mu^\dagger d_\mu$ is conserved. The Hamiltonian is a rotational scalar built for one- and two-body terms in the bosons. To study the dynamics of the IBM throughout its parameter space, it is best to use its most economical parametrization, the consistent Q formalism [11].

The Hamiltonian is

$$H = E_0 + c_0 \hat{n}_d + c_2 Q^\chi \cdot Q^\chi + c_1 L^2 \quad , \tag{1}$$

where $n_d = d^\dagger d$ is the number of d bosons, Q^χ is a quadrupole operator

$$Q^\chi = (d^\dagger \times \tilde{s} + s^\dagger \times \tilde{d})^{(2)} + \chi (d^\dagger \times \tilde{d})^{(2)} \quad , \tag{2}$$

and $\vec{L}$ is the angular momentum. We have used the notation $\tilde{d}_\mu = (-)^\mu d_{-\mu}$ so that $\tilde{d}_\mu$ transforms like $d_\mu^\dagger$ under rotations. The same operator as in (2) is used to describe the E2 transitions

$$T(E2) = \alpha_2 Q^\chi \quad . \tag{3}$$

A dynamical symmetry occurs in an algebraic model with an algebra G when the Hamiltonian can be written as a function of the Casimir invariants of a chain of subalgebras of G

$$G \supset G' \supset G'' \supset \cdots \quad . \tag{4}$$

The eigenstates of H can then be labeled by the eigenvalues of the Casimir invariants C(G), C(G'), C(G''), $\cdots$ and the energy levels are given analytically. A dynamical symmetry implies integrability [12, 13] since the above set of Casimir invariant are constants of motion in involution (i.e. commute among themselves).

$$[H, C(G^{(i)})] = 0 \quad ; \quad [C(G^{(i)}), C(G^{(j)})] = 0 \quad . \tag{5}$$

If the set is not complete, there are missing labels. If a missing label occurs, for example, in the reduction $G' \supset G''$, then it is possible [14] to construct from the generators of G' an invariant of G''. With these additional invariants we obtain a complete set of constants and hence integrability. One can therefore identify apriori the completely integrable limits of an algebraic model by simply identifying its dynamical symmetries.

It is well known that the IBM has three such limits

$$U(6) \supset \left\{ \begin{array}{c} U(5) \supset O(5) \\ SU(3) \\ O(6) \supset O(5) \end{array} \right\} \supset O(3) \quad \begin{array}{c} (I) \\ (II) \\ (III) \end{array} . \tag{6}$$

which are denoted by (I), (II), and (III), respectively [8]. (I) describes vibrational nuclei, (II) rotational nuclei and (III) γ-unstable nuclei. (I) is obtained in (1) for $c_2 = 0$, (II) for $c_0 = 0$ and $\chi = -\sqrt{7}/2$, and (III) for $c_0 = 0$ and $\chi = 0$.

We remark that $N, \vec{L}^2, L_z$ are always constants in involution. To get a complete set we need three more (since the problem is in six degrees of freedom). In the above three limits they are the following:
(i) $U(5)$; two are the number of d bosons, $\hat{n}_d$, and the quadratic Casimir invariant of $O(5)$. The third is an $O(3)$ invariant built from the $O(5)$ generators.
(ii) $SU(3)$; the quadratic and cubic Casimir invariants of $SU(3)$ and the $O(3)$ invariant $(\vec{L}\times Q)^{(1)}\cdot\vec{L}$ where Q is as in (2) with $\chi = -\sqrt{7}/2$.
(iii) $O(6)$; two constants are the quadratic $O(6)$ and $O(5)$ Casimir invariants and the third is as discussed in (i).

We note that in all the cases above the Hamiltonian does not depend on the invariants which are associated with the missing labels. This will cause exact degeneracies which are beyond what the Poisson statistic predicts. These situations will be termed "overintegrable".

3. Statistical tests: spectral and intensity fluctuations

To study the signatures of chaos in the quantal IBM Hamiltonians we have investigated fluctuation properties of the spectrum and the E2 transition intensities [15]. A study of spectral fluctuations alone was also carried out in Ref. [16] and [17].

To separate the average part of the spectrum from its fluctuating part, we calculate the staircase function N(E) and find $N_{av}(E)$ by fitting to it a sixth order polynomial. The unfolded energy levels $\tilde{E}_i$ are then defined by $\tilde{E}_i = N_{av}(E_i)$. The nearest neighbor level spacing distribution is calculated and fitted to a Brody distribution

$$P_\omega(S) = AS^\omega exp(-aS^{1+\omega}) \quad , \tag{7}$$

where $\alpha = \Gamma[(2+\omega)/(1+\omega)]^{1/2}$ and $A = (1+\omega)\alpha$ are determined by the conditions that $< S > = 1$ and P is normalized to 1. The Brody distribution interpolates between the Poisson distribution

($\omega = 0$) and Wigner distribution ($\omega = 1$). The latter is very close to the one obtained from the GOE.

Another statistical measure of the spectrum is the Δ_3 statistics of Dyson and Metha

$$\Delta_3(\alpha, L) = \min_{A,B} \frac{1}{L} \int_\alpha^{\alpha+L} [N(\tilde{E}) - (A\tilde{E} + B)]^2 d\tilde{E} \quad . \tag{8}$$

It measures the deviation of the staircase function (of the unfolded spectrum) from a straight line. To obtain a smoother function $\bar{\Delta}_3(L)$, we average $\Delta_3(\alpha, L)$ over n_α intervals $(\alpha, \alpha + L)$, $\bar{\Delta}_3(L) = \sum_\alpha \Delta_3(\alpha, L)/n_\alpha$. The successive intervals are taken to overlap by $L/2$. A more rigid spectrum has smaller values of Δ_3, so we except $\bar{\Delta}_3$ in the GOE limit to be smaller than $\bar{\Delta}_3$ for a Poisson spectrum. For a Poisson statistics $\bar{\Delta}_3(L) = L/15$ and the asymptotic result for the GOE is $\bar{\Delta}_3(L) \approx \ln L/\pi^2 - 0.007$ ($L >> 1$). In general, we fit $\bar{\Delta}_3$ to a function which depends on a parameter q and interpolates between the Poisson statistics (q=0) and the GOE (q=1)

$$\Delta_3^q(L) = \Delta_3^{Poisson}((1-q)L) + \Delta_3^{GOE}(qL) \quad . \tag{9}$$

We also analyze the distribution of the E2 transition intensities y, where

$$y \equiv B(E2; i \to f) = \frac{1}{2J_i + 1} \mid (f \parallel T(E2) \parallel i) \mid^2 \quad , \tag{10}$$

and $T(E2)$ is given by (3). To separate the secular variation of $B(E2)$ with energy we first divide (10) by an average intensity calculated by broadening each level with a Gaussian of width γ. The choice for γ is discussed in Ref. [15]. We then construct the distribution $P(y)$ such that $P(y)dy$ is the probability to find an intensity y in the interval dy around y.

The histogram $P(y)$ is fitted to the following distribution [18]

$$P_\nu(y) = \frac{(\nu/2 < y >)^\nu/2}{\Gamma(\nu/2)} y^{\nu/2-1} \exp(-\nu y/2 < y >) \quad , \tag{11}$$

which is a χ^2 distribution in ν degrees of freedom. For $\nu=1$, this is the Porter-Thomas distribution obtained from GOE in the limit where the dimension of the matrices is large [2]. This should correspond to a chaotic behavior. When the dynamics becomes more regular, we expect selection rules to set in so that ν will decrease monotonically [18].

Fig. 1 shows an example of the three quantal measures described above ($P(S), \Delta_3(L)$ and $P(y)$) for the $J = 6$ states of the IBM Hamiltonian with $c_0 = 0$ and $-\sqrt{7}/2 \leq \chi \leq 0$ describing the $SU(3) \to O(6)$ transition. Near the dynamical symmetry limits the spectral fluctuations are close to Poisson and ν is small. For intermediate values of χ the spectral fluctuations are similar to those of the GOE and ν is largest. These results suggest the onset of chaos in the transition between rotational and γ-unstable nuclei.

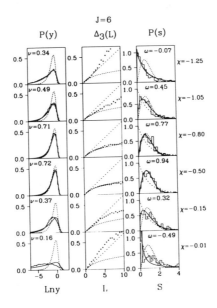

FIG. 1. Spectral and intensity fluctuations for the $J = 6^+$ levels of the Hamiltonian (1) with $N = 25$ bosons, $c_0 = 0$ and various values of χ between $-\sqrt{7}/2$ ($SU(3)$) and 0 ($O(6)$). Right: Level spacing distribution $P(S)$ where the solid line is the best fitted Brody distribution. Middle: Δ_3 statistics denoted by the + symbols. Left: The $E2$ intensity distribution $P(y)$ where the solid line is a χ^2 distribution in ν degrees of freedom. In all columns the dotted-dashed lines are the Poisson statistics and the dashed lines are the GOE statistics.

The negative values of ω at the dynamical symmetry limits reflect the overintegrability explained in Section 2. Due to missing labels we have exact degeneracies beyond what the Poisson statistic predicts. It is possible to restore a generic behavior without breaking the dynamical symmetry by adding to the Hamiltonian invariant terms which are associated with the missing labels [19].

4. Classical Limit

To establish that the GOE statistic, observed in the transition between $SU(3)$ and $O(6)$ nuclei, is indeed a signature of chaos, it is important to study a "semi-classical" limit of the system. The nucleus is a many-body system and we would argue that the appropriate "classical" limit to consider is the time-dependent mean-field approximation.

The many-body propagator can be described as a functional integral over coherent states. For bosons we use (in the IBM) [20]

$$| \alpha > \equiv \exp(-|\alpha|^2/2) \exp(\alpha_s s^\dagger + \sum_\mu \alpha_\mu d^\dagger_\mu) | 0 > \quad . \quad (12)$$

where $\alpha \equiv \{\alpha_s, \alpha_\mu; \mu = -2, \cdots, 2\}$ are six complex parameters. The action in the functional integral is constructed from a Hamiltonian that is obtained from the quantal one through the substitution

$$s^\dagger, d^\dagger_\mu \rightarrow \alpha^*_s, \alpha^*_\mu \quad (13)$$
$$s, d_\mu \rightarrow \alpha_s, \alpha_\mu \quad . \quad (14)$$

Rescaling $\alpha \rightarrow \alpha/\sqrt{N}$, the action becomes proportional to N. Therefore, $1/N$ plays the role of $\hbar$ so that the classical limit is obtained for $N \rightarrow \infty$ through a mean field approximation. In practice, we use the stationary phase approximation to obtain Hamilton-like equations of motion [21]

$$i\dot{\alpha}_j = \frac{\partial h}{\partial \alpha^*_j}$$
$$i\dot{\alpha}^*_j = -\frac{\partial h}{\partial \alpha_j}, \quad (15)$$

where $h = < \alpha | H | \alpha > /N$ is given by

$$h(\alpha, \alpha^*) = \epsilon_0 + \bar{c}[\eta n_d - (1-\eta)q^\chi \cdot q^\chi] + \bar{c}_1 \vec{\ell}^2 \quad . \quad (16)$$

n_d, q^χ_μ and $\vec{\ell}$ are c numbers obtained from $\hat{n}_d, Q^\chi_\mu$ and $\vec{L}$, respectively, by the substitution (13),(14). The parameters in (16) are related to those in the quantal Hamitonian (1) by

$$\epsilon_0 = E_0/N, \; \bar{c}_1 = Nc_1$$
$$\bar{c} = c_0/\eta, \; \frac{\eta}{1-\eta} = -\frac{c_0}{Nc_2} \quad . \quad (17)$$

$\vec{\ell}$ in (5) is the angular momentum per boson. Since $\vec{\ell}^2$ is a constant we can set $\bar{c}_1 = 0$ without loss of generality. Furthermore with proper scaling of the Hamiltonian we can take $\bar{c} = 1$. This leave us with two parameters: η $(0 \leq \eta \leq 1)$ and χ $(-\sqrt{7}/2 \leq \chi \leq 0)$.

The family of Hamiltonians (16) can be described by Casten triangle whose three vertices correspond to the three dynamical symmetry limits. A point inside the triangle is parametrized by coordinates (χ, η) such that η is the height above the base and χ is the intersection with the base of a line originating at the upper vertex and passing through the given point. The base describes the transition between the $SU(3)$ ($\chi = -\sqrt{7}/2$ and $O(6)$ ($\chi = 0$).

The Hamiltonian (16) is in six degrees of freedom. It is possible to eliminate one degree of freedom by factoring out a total phase $exp(i\theta)$ from all six α_j. The phase is chosen such that α_s becomes real. The action variable conjugate to θ is the number of bosons N. After the scaling $\alpha \to \alpha/\sqrt{N}$, the constraint on the boson number becomes

$$\alpha_s^* \alpha_s + \sum_\mu \alpha_\mu^* \alpha_\mu = 1 \quad . \tag{18}$$

By substituting $\alpha_s = \sqrt{1 - \sum_\mu \alpha_\mu^* \alpha_\mu}$ in (16) we obtain a problem in five degrees of freedom. The remaining α_μ (after factoring out $exp(i\theta)$) can be rewritten as

$$\begin{aligned} \alpha_\mu &= [(-)^\mu q_{-\mu} + i p_\mu]/\sqrt{2} \\ \alpha_\mu^* &= [q_\mu - (-)^\mu i p_{-\mu}]/\sqrt{2} \quad . \end{aligned} \tag{19}$$

The real tensors q_μ, p_μ play the role of the quadrupole deformation variables and its conjugate momenta, respectively. It is possible to transform q_μ to the intrinsic variables β, γ and Euler angles Ω, and p_μ to the respective conjugate momenta.

5. Classical Chaos

In the limit of dynamical symmetry the classical dynamics is completely integrable. The substitution (13),(14) provides the classical constants which satisfy the same relations as in (5) but with the Poisson brackets replacing the commutator. As we break the symmetry, chaotic dynamics may set in part of the phase space. This can be detected by examining the Lyapunov exponents. A Lyapunov exponent is defined as the rate of exponential separation D(t) between two neighboring trajectories $\alpha_j(t)$ and $\alpha_j(t) + \Delta\alpha_j(t)$

$$\lambda = \lim_{t \to \infty} \frac{1}{t} ln\left[\frac{D(t)}{D(0)}\right] \quad . \tag{20}$$

Depending on the choice of $\Delta\alpha_j(0)$ there are several such exponents whose number is equal to the dimension of phase space. Since the time evolution has a symplectic structure, they come in pairs $\pm\lambda$. Since N, L^2, L_z and the Hamiltonian H are always constants of the motion, there are at most two non-zero pairs of Lyapunov exponents. The motion is thus effectively in three degrees of freedom so that the method of Poincare sections to demonstrate chaos is impractical. Instead we observe the maximal λ and call a point in phase space chaotic when this λ is positive. For a regular trajectory $\lambda = 0$.

We define two measures of classical chaos in the subspace of given energy and angular momentum: the fraction σ of chaotic volume and the average maximal Lyapunov exponent. Instead of the latter we may calculate an average of all the non-negative Lyapunov exponents which is equal to the Kolmogorov entropy. In practice we solve Hamilton's equations in the full twelve-dimensional phase-space. The initial points are chosen uniformly (by Monte Carlo) on a twelve-dimensional sphere $\sum_j |\alpha_j|^2 \equiv 1$ and their maximal Lyapunov exponent is determined by solving the classical equations of motion for two neighboring trajectories. By choosing initial points which are within narrow energy and angular momentum windows, we can determine σ and λ as a function of energy and angular momentum as well.

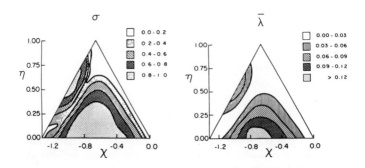

FIG. 2. Classical diagrams of Casten's triangle.
Left: the chaotic fraction σ of phase space for $0.16 < \ell < 0.80$. Right: the average maximal Lyapunov exponent $\bar{\lambda}$.

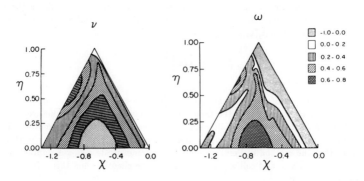

FIG. 3. Quantal diagrams of Casten's triangle for states with spin $4 \leq J \leq 20\hbar$ ($N = 25$ bosons). Right: the level spacing parameter ω. Left: the $B(E2)$ distribution parameter ν.

Fig. 2 shows the classical results in Casten's triangle where σ and $\bar{\lambda}$ are averaged over energy and over angular momentum (per boson) in the range $0.16 \leq \ell \leq 0.80$. The unshaded regions ($0 \leq \sigma \leq 0.3$) are highly regular while the dotted areas ($0.7 \leq \sigma \leq 1$) are highly chaotic. In addition to the expected regularity of the three vertices, notice that the transition between $U(5)$ and $O(6)$ is regular. It is in fact completely integrable since $O(5)$ is a common subalgebra. The quadratic Casimir invariant of $O(5)$ and the invariant which corresponds to that $O(5) \supset O(3)$ reductions, together with N, L^2, L_z and H form a complete set of constants in involution. In the transition from $SU(3)$ to $O(6)$ we see highly chaotic behavior in the intermediate regions which is consistent with the spectral and intensities statistics of Fig. 1. A chaotic region is also observed in the transition between $SU(3)$ and $U(5)$. An unexpected result is the existence of an almost regular narrow region inside the triangle connecting the $SU(3)$ and $U(5)$ vertices [12]. This region is probably associated with an approximate symmetry of the model that was previously unknown.

To demonstrate the strong correlation between the classical results (Fig. 2) and the quantal fluctuations, we show in Fig. 3 a contour diagram for the quantal measures ω and ν in the triangle.

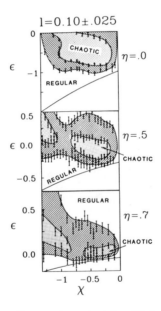

FIG. 4. Classical phase diagram at constant $\ell = 0.1$ in the ϵ (energy per boson) - χ plane and for several values of η. The regular, intermediate and chaotic regions correspond to $\sigma < 0.3$, $0.3 < \sigma < 0.7$ and $0.7 < \sigma < 1$, respectively.

To demonstrate the energy dependence of the dynamics we show in Fig. 4 an energy $(\epsilon) - \chi$

diagram at constant angular momentum and for three values of η. ϵ_{min} denotes the lowest possible energy for a given χ. The immediate region above ϵ_{min} is always regular so that for each χ (and η) there is a critical energy for the onset of chaos.

6. The case of zero spin

For the special case $\ell = 0$, the β, γ variables decouple from the Euler angles and it is possible to rewrite the classical equation of motion (5) in the intrinsic variables alone using the Hamiltonian

$$\begin{aligned} h &= \eta n_d - (1-\eta) q^\chi \cdot q^\chi \\ &= \frac{1}{2}[\eta + 2(1-\eta)\beta^2](\beta^2 + T) - 2(1-\eta)\beta^2 \\ &\quad + -\frac{\chi(1-\eta)}{\sqrt{7}/2}\sqrt{1-(\beta^2+T)/2}[(p_\gamma^2/\beta - \beta p_\beta^2 - \beta^3)cos3\gamma + 2p_\beta p_\gamma sin3\gamma] \\ &\quad + -\frac{\chi^2}{7/4}(1-\eta)[(\beta^2+T)^2/8 - p_\gamma^2/2] \quad , \end{aligned} \qquad (21)$$

where $T = p_\beta^2 + p_\gamma^2/\beta^2$.

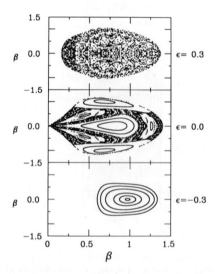

FIG. 5. Poincare Sections $\gamma = 0$ in the $p_\beta - \beta$ plane for the Hamiltonian (21) with $\eta = 0.5$, $\chi = -\sqrt{7}/2$ at three different energies. This case is intermediate between $SU(3)$ and $U(5)$.

In this case there is only one non-zero pair of Lyapunov exponents and we can construct two dimensional Poincare sections. For the $SU(3) \to O(6)$ transition we observe that the γ-vibration becomes unstable at a lower energy than the β-vibration, while in the $SU(3) \to U(5)$ transition, it is the β-vibration which is first to become unstable. An example of the latter transition is shown in Fig. 5 where Poincare sections of (21) for an intermediate case between $SU(3)$ and $U(5)$ are drawn. In the quasi-regular region, both the β- and γ- vibrations have similar stability properties that are closed to being marginal at all energies. For more details see [22].

7. Non-generic nuclear spectral fluctuations

Recently Paar et al [23] observed spectral fluctuations in the $SU(3)$ limit which are GOE-like for states with $J = 0$. In that limit the Hamiltonian must be integrable. In fact, it is possible to construct explicitly action-angle variables, and express H analytically in terms of the action variables I_1, I_2. In Fig. 6 we show contours of constant energy in the $I_1 - I_2$ plane [24]. We see that they are approximately "flat", so that $H \approx f(\vec{\omega} \cdot \vec{I})$. This situation is similar to the case of the harmonic oscillator which was shown to be non-generic[25]. Thus the present case also has non-generic spectral fluctuations. A closer look suggests that the level spacing distribution has no limit and is not a Wigner distribution. It is possible to restore generic fluctuations without breaking the $SU(3)$ dynamical symmetry by adding the cubic Casimir of $SU(3)$ to the Hamiltonian[19].

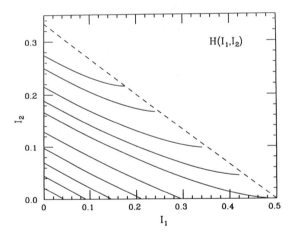

FIG. 6 Contour lines H =const. for the classical $SU(3)$ Hamiltonian at $\ell = 0$. Notice that the contours are almost straight lines.

8. Conclusions and additional studies

We have studied the onset of chaos in the collective dynamics of nuclei, using the framework of the interacting boson model. Strong correlations are found between spectral and transition intensity statistics on one hand and the character of the classical dynamics on the other hand.

A surprising result is the existence of a quasi-regular region that connects the $SU(3)$ and $U(5)$ dynamical symmetries inside the triangle. It is associated with an approximate symmetry, and it would be interesting to find the experimental signatures of that symmetry.

In many experimental situations, we cannot fully resolve the spectrum and this might be an obstacle to a precise determination of the statistical properties. It is thus important to find signatures of quantum chaos that are less sensitive to experimental resolution. Such a probe, the spectral autocorrelation function and its Fourier transform, the survival probability, are discussed in Refs. [26, 27, 28, 29]. An exact analytical formula for that quantity is derived in Ref. [28] using random matrix theory.

We have argued that dynamical symmetry always leads to complete integrability. A novel concept of partial dynamical symmetry was recently introduced in Ref. [30], where the Hamiltonian does not have the given symmetry but a subset of its eigenstates does. We have shown that such a partial symmetry leads to the suppression of chaos even though in the example we analyzed the number of exactly solvable states is vanishingly small in the classical limit $N \to \infty$ [31].

At higher spin and/or energy, the bosons break into pairs of fermions and it is then important to include these additional fermionic degrees of freedom to obtain a realistic description. A first step towards such a study can be found in Ref. [32]. Another interesting question is the role played by the neutron-proton interaction in the onset of chaotic collective motion, and this can be studied in the framework of IBM-2.

I would like to thank A. Novoselsky and N. Whelan for their collaboration on various parts of this work. This work was supported in part by DOE Grant No. DE-FG-0291ER-40608.

References

[1] M.L. Metha, Random Matrices, *Academic Press* (1990).

[2] C.E. Porter, Statistical Theories of Spectra: Fluctuations, *Academic Press* (1965).

[3] O. Bohigas, M.J. Giannoni and C. Schmit, Phys. Rev. Lett. **52**, 1 (1984).

[4] A.Y. Abul-Magd and H.A. Weidenmuller, Phys. Lett. **B162**, 223 (1985).

[5] J.F. Shriner, Jr. et al., Z. Phys. **A335**, 393 (1990).

[6] S. Raman et al, Phys. Rev. **C43**, 521 (1991).

[7] J.F. Shriner, Jr. et al., Z. Phys. **A338**, 309 (1991).

[8] F. Iachello and A. Arima, "The Interacting Boson Model", *Cambridge University Press*, (1987).

[9] A.Arima and F. Iachello, *Ann. Phys.* **99**, 253 (1976); **111**, 201 (1978); **123**, 468 (1979).

[10] Y. Alhassid, A. Novoselsky and N. Whelan, *Phys. Rev. Lett.* **65**, 2971 (1990).

[11] R.F. Casten and D.D. Warner, *Rev. Mod. Phys.* **60**, 389 (1988).

[12] Y. Alhassid and N. Whelan, *Phys. Rev.* **C67**, 816 (1991).

[13] W. Zhang, C.C. Martens, D.H. Feng and J. Yuan, *Phys. Rev. Lett.* **61**, 2167 (1988).

[14] G. Draayer and R. Gilmore, *J. Math Phys.* **26**, 3053 (1986).

[15] Y. Alhassid and A. Novoselsky, *Phys. Rev.* **C45**, 1677 (1992).

[16] V. Paar and V. Vorkapic, *Phys. Lett.* **B205**, 7 (1988); *Phys. Rev.* **C41**, 2379 (1990).

[17] T. Mizusaki, N. Yoshinaga, T. Shigenhara and T. Cheon, *Phys. Lett.* **B269**, 6 (1991).

[18] Y. Alhassid and R.D. Levine, *Phys. Rev. Lett.* **57**, 2789 (1986); Y. Alhassid and M. Feingold, *Phys. Rev.* **A34**, 374 (1989).

[19] N. Whelan and Y. Alhassid, *Nucl. Phys.* **A556**, 42 (1993).

[20] R.L. Hatch and S. Levit, *Phys. Rev.* **C25**, 614 (1982).

[21] Y. Alhassid and N. Whelan, *Phys. Rev. Lett.* **67**, 816 (1991).

[22] N. Whelan, Ph.D. thesis, Yale University (1993).

[23] V. Paar, D. Vorkapic and A. Dieperink and , *Phys. Rev. Lett.* **69**, 2184 (1992).

[24] B. Lauritzen, Y. Alhassid and N. Whelan, *Phys. Rev. Lett.* **72**, 2809 (1994).

[25] M.V. Berry and M. Tabor, *Proc. Roy. Soc. London* **A 356**, 375 (1977).

[26] L. Leviander, M. Lombardi, R. Just and J.P. Pique, *Phys. Rev. Lett.* **56**, 2449 (1986).

[27] Y. Alhassid and R.D. Levine, *Phys. Rev.* **A46**, 4650 (1992).

[28] Y. Alhassid and N. Whelan, *Phys. Rev. Lett.* **70**, 572 (1993).

[29] T. Guhr and H.A. Weidenmuller, *Chem. Phys.* **146**, 21 (1990).

[30] Y. Alhassid and A. Leviatan, *J. Phys.* **A25**, L1265 (1992).

[31] N. Whelan, Y. Alhassid and A. Leviatan, *Phys. Rev. Lett.* **71**, 2208 (1993).

[32] Y. Alhassid and D. Vretenar, *Phys. Rev.* **C46**, 1334 (1992).

605

BARYON MAPPING OF CONSTITUENT QUARK MODELS [*]

S. PITTEL
Bartol Research Institute, University of Delaware
Newark, DE 19716, USA

J. M. ARIAS
Departamento de Física Atómica y Nuclear
Universidad de Sevilla
Apdo. 1065, 41080 Sevilla, Spain

J. DUKELSKY
Grupo de Física Nuclear – Facultad de Ciencias
Universidad de Salamanca
37008 Salamanca, Spain

and

A. FRANK
Instituto de Ciencias Nucleares and
Laboratorio de Cuernavaca, Instituto de Física, UNAM
Apdo. Postal 70-543, 045100 México, D.F., México

ABSTRACT

A new and consistent mapping of colorless three-quark clusters onto colorless baryons is presented and then tested in the context of a three-color extension of the Lipkin model. The results suggest that this new mapping may provide a practical means of incorporating dynamical multi-quark correlations in a description of nuclei built out of quarks.

1. Introduction

Traditionally, the nucleus is treated as a quantum liquid of pointlike structureless objects, the nucleons. We now know, however, that nucleons are neither pointlike nor structureless. They are composite objects, built up collectively from the constituents of QCD. As such, they are not exact fermions. In the standard view of nuclear structure, we simply ignore this complication and treat them as real fermions, with effective interactions generated either semi-microscopically or from phenomenology. This procedure has been enormously successful in describing the main spectroscopic properties of nuclei.

Recent experiments carried out by the EMC collaboration, however, suggest that at a deeper level the constituent structure of the nucleon may play a role. These experiments indicate that the internal structure of a nucleon is modified in a nuclear medium, much the same as the structure of collective S and D pairs in nuclei

---

[*]Talk presented by S. Pittel at the International Conference on Perspectives of the Interacting Boson Model on the Occasion of its 20th Anniversary

changes in the presence of other collective pairs. If we wish to incorporate this in our description of nuclei, and in doing so perhaps make more evident where to look for explicit quark effects, we need to treat simultaneously the QCD interactions that build nucleons and the residual interactions between nucleons that lead to nuclear structure.

Ideally, such a treatment should start from QCD, but to date not much progress has been achieved in this direction. On the other hand, significant progress has been made in the use of nonrelativistic constituent quark models[1], which in principle can be obtained from QCD by eliminating the antiquark and gluon degrees of freedom. Such models have been applied with considerable success to nuclear systems with few particles. Despite conceptual questions inherent to such models, e.g. their precise link to QCD and the related problems of quark confinement and chiral invariance, on the basis of their success, they seem to provide an attractive starting point.

At present, such models have been directly applied to one- and two-baryon systems only. Efforts to apply them to many-nucleon systems have not been as successful. The reason is that in a nuclear environment quark triplets cluster into spatially-localized nucleons, a scenario that cannot be described with existing many-body methods[2].

A possible way to overcome this problem is to follow the analogy provided by the Interacting Boson Model (IBM), whose twentieth anniversary is being celebrated at this conference. The IBM starts from the observation that nuclear collective motion at low energies is dominated by a few collective degrees of freedom, namely S and D pairs of identical nucleons. There too it is not feasible to describe the resulting dynamics directly in terms of the collective degrees of freedom, since they are not real bosons. The IBM proceeds by replacing the collective pairs by real bosons, which at the microscopic level is done via a boson mapping[3]. The result of such a mapping (at least in principle) is a boson hamiltonian that reproduces all of the dynamics of the original system of nucleons. The complications arising from nucleon exchange, which are the reason that collective fermion pairs do not behave as exact bosons, are transferred to the effective boson hamiltonian.

Here we propose an analogous treatment of quark clustering in nuclei, namely to carry out a mapping from colorless quark triplets onto baryons, real fermions satisfying exact fermion anticommutation relations. Assuming that such a mapping can be carried out, it will lead from the original quark hamiltonian to an effective hamiltonian for these baryons that rigorously incorporates all quark exchange effects. Most importantly, since the baryons are exact fermions, we should be able to treat their dynamics using familiar fermion many-body techniques.

The idea that baryon mappings may provide a practical means of deriving nuclei from quarks was first proposed a few years ago by several independent groups[4-6]. Until recently, however, the development of a consistent mapping, capable of treating in a practical way both the two- and three-quark interactions that are suggested by QCD, has proven elusive. We have now succeeded in formulating a mapping[7] that seems to satisfy all of the necessary requirements.

The remainder of the paper is structured as follows. In Section 2, we review some general features of baryon mappings and then present the new and improved version we have recently formulated. In Section 3, we describe a simple model that we have developed to test our mapping. The model we have chosen is a three-color extension of the well-known Lipkin model[8], which in its traditional version has been used extensively to test various nuclear many-body techniques. In Section 4, we apply our mapping to this model for a system with two baryons and present the results. The bottom line is that it seems to work perfectly.

2. Baryon Mapping of Quark Systems

2.1. Preliminaries

Our starting point is a nonrelativistic model of constituent quarks. We denote the quark creation and annihilation operators by $q_{1a}^\dagger$ and q_{1a}, respectively. The first subscript denotes the color quantum number and the second all the rest. These operators satisfy the usual fermion anticommutation relation

$$\{q_{1a}, q_{2b}^\dagger\} = \delta_{1a,2b} \equiv \delta_{12}\,\delta_{ab}\,. \tag{1}$$

QCD considerations suggest that a realistic quark hamiltonian may include up to three-body interactions, all of which are color scalars. Such a hamiltonian can always be expressed in terms of the following colorless one-, two- and three-body operators, respectively:

$$A_{ab} = \sum_{1} q_{1a}^\dagger q_{1b}\,; \tag{2}$$

$$B_{abcd} = \sum_{12345} \epsilon_{123}\,\epsilon_{145}\, q_{2a}^\dagger q_{3b}^\dagger q_{5d} q_{4c}\,; \tag{3}$$

and

$$C_{abcdef} = \sum_{123456} \epsilon_{123}\,\epsilon_{456}\, q_{1a}^\dagger q_{2b}^\dagger q_{3c}^\dagger q_{6f} q_{5e} q_{4d}\,. \tag{4}$$

The ϵ_{123} quantities are antisymmetric tensors that guarantee the colorless nature of these operators.

The idea of a baryon mapping is to replace this problem by an *equivalent* one involving baryons. We denote the creation and annihilation operators of the baryon space by $\Lambda_{1a2b3c}^\dagger$ and Λ_{1a2b3c}, respectively. They, by definition, satisfy the multi-index anticommutation relation

$$\{\Lambda_{1a2b3c}, \Lambda_{4d5e6f}^\dagger\} = \delta(1a2b3c, 4d5e6f)\,, \tag{5}$$

where

$$\delta(1a2b3c, 4d5e6f) \equiv \delta_{1a,4d}\delta_{2b,5e}\delta_{3c,6f} + \delta_{1a,5e}\delta_{2b,6f}\delta_{3c,4d} + \delta_{1a,6f}\delta_{2b,4d}\delta_{3c,5e}$$
$$- \delta_{1a,4d}\delta_{2b,6f}\delta_{3c,5e} - \delta_{1a,5e}\delta_{2b,4d}\delta_{3c,6f} - \delta_{1a,6f}\delta_{2b,5e}\delta_{3c,4d}\,. \tag{6}$$

Furthermore, by assumption, they are antisymmetric under the interchange of their quark indices, e.g., $\Lambda^\dagger_{1a2b3c} = -\Lambda^\dagger_{2b1a3c}$, etc.

The space generated by these baryon operators is in fact larger than the original quark space. It includes a subset of states that are fully antisymmetric under the interchange of quark indices and in one-to-one correspondence with the original states of the quark space; this is referred to as the *physical subspace*. But it also contains states that are not fully antisymmetric under quark interchange. These states, called *unphysical*, have no counterparts in the original quark space. They are a pure artifact of the mapping and provide the principal difficulty in developing a *practical* baryon mapping, as they also do for boson mappings.

As we will discuss shortly, there are several possible mappings that can be developed, all of which exactly preserve the physics of the original problem *in the physical-baryon subspace*. Where they differ is in their predictions for unphysical states. Clearly, for a mapping to be of practical use, the unphysical states must lie high in energy relative to the physical states. Otherwise, it will be difficult to disentangle the physical states of interest from those that are unphysical, particularly in the presence of variational approximations.

So far, we have not discussed how to guarantee that the physics of the original problem is preserved by the mapping. Here we follow the Belyaev-Zelevinski prescription, whereby the mapping is defined so as to exactly preserve the commutation (and/or anticommutation) relations for physical operators in the original quark space.

There is a simple way to accomplish this. Consider the colorless one-body operator A_{ab} defined in (2). The commutator $[A_{ab}, A_{cd}]$ is exactly preserved if this operator is mapped according to

$$A_{ab} = \sum_1 q^\dagger_{1a} q_{1b} \to \frac{1}{2} \sum_{123cd} \Lambda^\dagger_{1a2c3d} \Lambda_{1b2c3d} . \tag{7}$$

Since the operators in (3–4) can be rewritten in terms of colorless one-body operators, it would seem that we could also apply (7) to them and achieve our goal. Unfortunately, this is not the case.

Since a one-body operator contains just one creation and one annihilation operator, it cannot incorporate information on the quark Pauli principle. Thus, when we apply this simple mapping, we find that (*i.*) it reproduces all quark dynamics in the physical subspace, *but* (*ii.*) it invariably leads to unphyical states lower in energy than the physical states of interest.

To incorporate quark Pauli effects in a practical way, we *must* map the multiquark creation and annihilation operators that appear in the two- and three-body interactions directly. What this means is that we must find a mapping that preserves the commutation relations between the colorless one-, two- and three-body operators *simultaneously*.

2.2. Colorless Baryons

The baryon operators introduced in the previous subsection all have color. We know, however, that it is possible to describe all of the relevant physics solely in terms of colorless baryons. The relevant operators for colorless baryons can be defined according to

$$\Lambda^\dagger_{abc} = \frac{1}{6} \sum_{123} \epsilon_{123} \Lambda^\dagger_{1a2b3c} ,\qquad (8)$$

and

$$\Lambda_{abc} = \frac{1}{6} \sum_{123} \epsilon_{123} \Lambda_{1a2b3c} .\qquad (9)$$

They are fully symmetric under the interchange of their indices and satisfy the anticommutation relation

$$\{\Lambda_{abc}, \Lambda^\dagger_{def}\} = \frac{1}{6} S(abc, def) ,\qquad (10)$$

where

$$S(abc, def) \equiv \delta_{ad}\delta_{be}\delta_{cf} + \delta_{ae}\delta_{bf}\delta_{cd} + \delta_{af}\delta_{bd}\delta_{ce} + \delta_{ad}\delta_{bf}\delta_{ce} + \delta_{ae}\delta_{bd}\delta_{cf} + \delta_{af}\delta_{be}\delta_{cd} .\qquad (11)$$

The mappings that we present in the next two subsections go directly from the space of colorless quark triplets to the space of colorless baryons.

2.3. The Hermitean Mapping

First we present the results obtained using the simple mapping described earlier, in which the colorless quark operators are first expressed as products of colorless one-body operators A_{ab} and then mapped according to (7). A subsequent truncation to colorless baryons is then achieved by making the replacements

$$\begin{aligned}\Lambda^\dagger_{1a2b3c} &\to \epsilon_{123}\Lambda^\dagger_{abc} , \\ \Lambda_{1a2b3c} &\to \epsilon_{123}\Lambda_{abc} ,\end{aligned}\qquad (12)$$

as discussed in Ref. 4.

Implementing the above prescription leads to the following results:

$$A_{ab} \to A^h_{ab} = 3 \sum_{cd} \Lambda^\dagger_{acd} \Lambda_{bcd} ;\qquad (13)$$

$$B_{abcd} \to B^h_{abcd} = 12 \sum_f \Lambda^\dagger_{abf}\Lambda_{cdf} - 9 \sum_{efgh} \Lambda^\dagger_{aef}\Lambda^\dagger_{bgh}(\Lambda_{cef}\Lambda_{dgh} + \Lambda_{def}\Lambda_{cgh}) ;\qquad (14)$$

$$\begin{aligned}C_{abcdef} \to C^h_{abcdef} &= 36\,\Lambda^\dagger_{abc}\Lambda_{def} \\ &- 36 \sum_{ghi}(\Lambda^\dagger_{gha}\Lambda^\dagger_{bci} + \Lambda^\dagger_{ghb}\Lambda^\dagger_{cai} + \Lambda^\dagger_{ghc}\Lambda^\dagger_{abi}) \\ &\quad (\Lambda_{dgh}\Lambda_{efi} + \Lambda_{egh}\Lambda_{dfi} + \Lambda_{fgh}\Lambda_{dei}) .\end{aligned}\qquad (15)$$

We refer to this as the *hermitean baryon mapping* and include a superscript h to distinguish it from the improved *nonhermitean baryon mapping* to follow.

2.4. The Nonhermitean Mapping

The nonhermitean baryon mapping arises when we require the simultaneous preservation of *all* (anti)commutation relations. As noted earlier, we expect it to provide a more satisfactory description of unphysical states, through implicit incorporation of quark Pauli effects. A detailed discussion on how this mapping was obtained can be found in Ref. 7. Here, we simply present the final results:

$$A_{ab} \to A_{ab}^{nh} = 3 \sum_{cd} \Lambda_{acd}^\dagger \Lambda_{bcd} ; \qquad (16)$$

$$B_{abcd} \to B_{abcd}^{nh} = 12 \sum_e \Lambda_{abe}^\dagger \Lambda_{cde} + 9 \sum_{efgh} \Lambda_{aef}^\dagger \Lambda_{bgh}^\dagger (\Lambda_{cde}\Lambda_{fgh} + \Lambda_{efg}\Lambda_{cdh}) ; \qquad (17)$$

$$\begin{aligned}C_{abcdef} \to C_{abcdef}^{nh} &= 36\, \Lambda_{abc}^\dagger \Lambda_{def} \\ &+ 36 \sum_{ghi} (\Lambda_{gha}^\dagger \Lambda_{bci}^\dagger + \Lambda_{ghb}^\dagger \Lambda_{cai}^\dagger + \Lambda_{ghc}^\dagger \Lambda_{abi}^\dagger)\Lambda_{ghi}\Lambda_{def}\,.\end{aligned} \qquad (18)$$

The superscript nh denotes that these are the nonhermitean baryon images.

We should emphasize here that the sets of mapping equations given in this and the preceding subsections can be applied to *any* colorless constituent quark hamiltonian written in uncoupled form.

3. The Three–Color Lipkin Model

The three-color Lipkin model is based on the well-known Lipkin model[8]. Since many of the characteristics of the three-color Lipkin model are already in the original one, we first devote a few lines to reviewing it.

The Lipkin model has two levels, each Ω-fold degenerate, separated by an energy Δ. It is assumed that in the unperturbed ground state, $N=\Omega$ particles occupy *all* the single-particle states of the lower level. The fermion creation and annihilation operators of the model are written as $q_{\sigma m}^\dagger$ and $q_{\sigma m}$ respectively, where σ characterizes whether the particle is in the lower level, $\sigma = -$, or in the upper one, $\sigma = +$, and m denotes which of the Ω degenerate states of that level the particle occupies.

The hamiltonian of the model can be expressed as

$$H = H_1 + H_2 , \qquad (19)$$

$$H_1 = \frac{\Delta}{2} \sum_m (q_{+m}^\dagger q_{+m} - q_{-m}^\dagger q_{-m}) , \qquad (20)$$

$$H_2 = -\frac{\chi}{\Omega} \sum_{m_1 m_2} \{q_{+m_1}^\dagger q_{+m_2}^\dagger q_{-m_2} q_{-m_1} + q_{-m_1}^\dagger q_{-m_2}^\dagger q_{+m_2} q_{+m_1}\}. \qquad (21)$$

In addition to the one-body term, it contains a two-body interaction that scatters pairs of particles among the two levels, without changing their m values.

This model can be solved exactly using group theoretical techniques for any value of Ω and any values of the parameters Δ and χ. The set of all possible one-body operators built from its creation and annihilation operators generates the Lie algebra $U(2\Omega)$. The structure of the problem suggests a decomposition

$$U(2\Omega) \supset U(\Omega) \otimes U(2) . \qquad (22)$$

The Lipkin hamiltonian can be rewritten solely in terms of the $U(2)$ generators and all the states belong to a definite irreducible representation of $U(2)$ (or $SU(2)$). This in turn implies that the hamiltonian matrix can be analytically evaluated using the well known $SU(2)$ angular momentum algebra.

The three-color model involves three sets, one for each color, of standard two-level Lipkin models. Again the lower levels are assumed to be completely filled in the unperturbed ground state, which in this case contains $N = 3\Omega$ particles. The creation and annihilation operators now include a label 1 that represents the color quantum number and are thus written as $q_{l\sigma m}^\dagger$ and $q_{l\sigma m}$, respectively. The model hamiltonian includes one-body, two-body *and* three-body interactions, all colorless, which scatter particles coherently among the levels:

$$H = H_1 + H_2 + H_3 , \qquad (23)$$

$$H_1 = \frac{\Delta}{2} \sum_{1m} (q_{1+m}^\dagger q_{1+m} - q_{1-m}^\dagger q_{1-m}) , \qquad (24)$$

$$H_2 = -\frac{\chi_2}{\Omega} \sum_{12345, m_1 m_2} \epsilon_{123}\epsilon_{145} \{ q_{2+m_1}^\dagger q_{3+m_2}^\dagger q_{5-m_2} q_{4-m_1} + q_{4-m_1}^\dagger q_{5-m_2}^\dagger q_{3+m_2} q_{2+m_1} \} , \qquad (25)$$

and

$$H_3 = -\frac{\chi_3}{\Omega^2} \sum_{123456, m_1 m_2 m_3} \epsilon_{123}\epsilon_{456} \{ q_{1+m_1}^\dagger q_{2+m_2}^\dagger q_{3+m_3}^\dagger q_{6-m_3} q_{5-m_2} q_{4-m_1}$$
$$+ q_{4-m_1}^\dagger q_{5-m_2}^\dagger q_{6-m_3}^\dagger q_{3+m_3} q_{2+m_2} q_{1+m_1} \} . \qquad (26)$$

The model contains three parameters, one for each of the terms in the hamiltonian. Whenever $\chi_2 \gg \chi_3$ *and* Δ, the system will be dominated by two-quark correlations. Whenever $\chi_3 \gg \chi_2$ *and* Δ, it will be dominated by three-quark correlations. Thus, it has a rich enough structure to make it useful as a test of our proposed mapping.

The group structure of this model is significantly more complex than for the usual Lipkin model. Now the set of one-body operators generates a Lie algebra $U(6\Omega)$, and the structure of the model suggests a classification of states in terms of the chain

$$U(6\Omega) \supset U(\Omega) \otimes U(6) \supset U(\Omega) \otimes U(3) \otimes U(2) . \qquad (27)$$

The group $U(3)$ is essential in this classification, since all physically admissible states should be colorless, i.e., they should belong to the (Ω, Ω, Ω) representation of $U(3)$ (or equivalently the $(\lambda, \mu) = (0, 0)$ scalar representation in Elliott's $SU(3)$ notation). The complication is that different $U(6)$ representations can contain these states and, moreover, for each of them several $U(2)$ representations are connected by the hamiltonian. Nevertheless, we have succeeded in generating algebraic solutions for this model for both $\Omega = 2$ and 3, by appropriate commutator manipulations.

4. Test of Baryon Mappings on the Three–Color Lipkin Model

Here, we apply the colorless baryon mappings given in Section 2 to the three-color Lipkin model. We carry out the analysis for $\Omega = 2$ only, for which the number of baryons is likewise 2. Diagonalization of the effective baryon hamiltonian can be carried out exactly for this case, leading to a direct test of the mapping.

The colorless states of the model, following the mapping, can be expressed as

$$\Lambda^\dagger_{\sigma_1 m_1 \sigma_2 m_2 \sigma_3 m_3} \Lambda^\dagger_{\sigma_4 m_4 \sigma_5 m_5 \sigma_6 m_6} |0> .$$

The total number of two-baryon states that can be formed is 52. In contrast, the total number of colorless six-quark states in the Lipkin model is 20. That the two-baryon space is larger than the six-quark space was anticipated in our earlier discussion. The two-baryon space includes not only physical states (in one-to-one correspondence with the states of the quark model) but unphysical states as well.

The general three-color Lipkin hamiltonian (23–26) can be mapped in either nonhermitean or hermitean form, using the results of the previous section. We will be particularly interested in the nonhermitean mapping, since it is expected to provide a more practical incorporation of quark Pauli effects. However, in the results that follow, we consider both, to see whether our expectations are realized.

In order to assess the feasibility of "pushing up" unphysical states with respect to physical states with the nonhermitean mapping, it is important to have a criterion for distinguishing one from the other. This can be done by introducing a Majorana operator[9], analogous to the one used in boson mappings.

In Figures 1–2, we present some representative results of our test calculations for two different choices of the model parameters. In both, we show the algebraic results obtained prior to the mapping (denoted *exact*) and the results obtained after the nonhermitean (nh) and hermitean (h) mappings. In the spectra that refer to diagonalization after the mapping, we distinguish physical from unphysical states by using the Majorana operator mentioned above. Physical states are indicated by solid lines and unphysical states by dashed lines. We use a heavy solid line to denote degenerate (or nearly degenerate) solutions, and indicate to the right the number of physical (P) and unphysical (U) states at that energy. We only show the low-energy portions of the spectra corresponding to $E < 0$.

Figure 1 shows our results for the choice $\Delta = 0$, $\chi_2 = 1$ and $\chi_3 = 0$, for which the system is dominated by two-quark correlations. Both the hermitean and nonhermitean mappings exactly reproduce the spectrum of states obtained by exact diagonalization of the quark model. Following the hermitean mapping, however, the lowest eigenvalues are unphysical. In contrast, when the nonhermitean mapping is used, the unphysical states are pushed up in energy, and the lowest four eigenvalues are physical. This is precisely what we had hoped would occur.

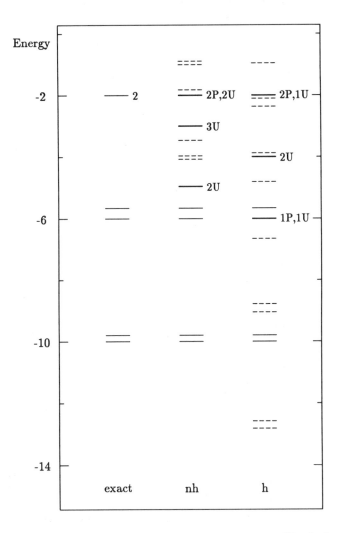

Figure 1. Calculated spectra of the three-color Lipkin model for $\Omega = 2$, $\Delta = 0$, $\chi_2 = 1$ and $\chi_3 = 0$. The notation is described in the text.

It is important, however, to see whether this also occurs in the presence of three-quark correlations. Thus, in Figure 2, we show results obtained for $\Delta = 0$, $\chi_2 = 0$ and $\chi_3 = 1$, namely for a system dominated by three-quark correlations. Exactly the same conclusions apply. Both mappings exactly reproduce the spectrum of physical states. The hermitean mapping, however, leads to unphysical states very low in energy, whereas the nonhermitean mapping yields them significantly raised.

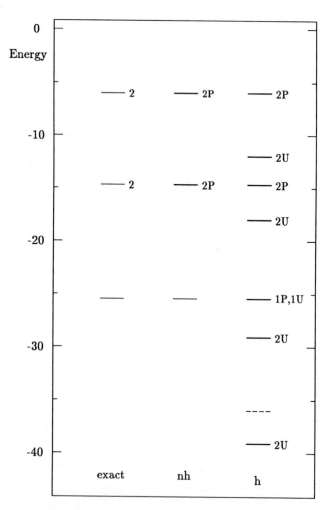

Figure 2. The same as Figure 1 except that the hamiltonian parameters used are $\Delta = 0$, $\chi_2 = 0$ and $\chi_3 = 1$.

We have also carried out calculations for mixed scenarios in which all three terms in the quark hamiltonian are active. All such calculations lead to the same general conclusion; our nonhermitean baryon mapping seems to provide a practical means of incorporating dynamical many-body correlations in multi-quark systems.

5. Outlook

Despite the promising results of our test calculations, much work is still needed.

- We must still demonstrate the usefulness of our mapping in the presence of variational approximations or truncations, as will no doubt be required in real problems.

- We also need to show that our mapping can properly describe spatial three-quark correlations, as are certainly present in real nuclei but not in the Lipkin model, since it has no spatial degrees of freedom.

Further tests in these directions are now underway.
Our ultimate goal of course is the description of real nuclei using real constituent quark models. Work along these lines is likewise now in progress.

6. Acknowledgments

This work was supported by the National Science Foundation under grant #s PHY-9303041 and INT-9314535, by NATO under grant # CRG.900466, by the Spanish DGICYT under project # PB92-0663 and by CONACYT, Mexico, under project # E-120.3475.

7. References

1. M. Oka, K. Shimizu and Y. Yazaki, *Nucl. Phys.* **A464** (1987) 700; K. Brauer, A. Faessler, F. Fernandez and K. Shimizu, *Nucl. Phys.* **A507** (1990) 599.
2. P. Ring and P. Schuck, *The Nuclear Many-Body Problem* (Springer, Berlin, 1980).
3. Abraham Klein and E. R. Marshalek, *Rev. Mod. Phys.* **632** (1991) 375.
4. S. Pittel, J. Engel, J. Dukelsky and P. Ring, *Phys. Lett.* **B247** (1990) 185.
5. E. G. Nadjakov, *J. Phys. G: Nucl. Part. Phys.* **16** (1990) 1473.
6. J. Meyer, *J. Math. Phys.* **328** (1991) 2142.
7. S. Pittel, J. M. Arias, J. Dukelsky and A. Frank, to be published in Physical Review C (1994).
8. H. J. Lipkin, N. Meshkov and A. Glick, *Nucl. Phys.* **62** (1965) 188.
9. P. Park, *Phys. Rev.* **C35** (1987) 807.

WEAK INTERACTIONS IN THE INTERACTING BOSON-FERMION MODEL

Giuseppe Maino
*ENEA, Dipartimento Innovazione, Settore Fisica Applicata,
viale Ercolani 8, I-40138 Bologna, Italy, and
INFN, Sezione di Firenze*

ABSTRACT

The theory of nuclear β decay is discussed within the framework of the interacting boson-fermion model (IBFM) and results of astrophysical interest are shown. We then review the role played by the isospin degree of freedom in the proton-neutron boson model as far as β decay in light-mass nuclei and negative muon capture are concerned. Measurements to check the supersymmetry scheme in the mass region around A≈30 are suggested.

1. Introduction

After the classical paper of Fermi where he firstly attempted a theoretical description of the β decay[1], the phenomenological approach to the weak interaction developed fast resulting in the understanding of both allowed[2] and forbidden[3] β-decay processes and attaining its definite vector-axial vector (V-A) form in the fifties with the prediction and the consequent observation of the maximal parity breaking[1,4]. The Puppi triangle of the *universal* weak interaction, shown in fig. 1, summarizes these basic results. The three sides of the triangle refer, respectively, to the nuclear muon capture, the β decay and the decay of a free muon into an electron and two neutrinos.

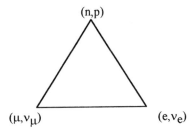

Fig. 1 - Puppi triangle of the weak interaction.

Finally, in the sixties a fully renormalizable gauge theory was introduced, leading to the unification of weak and electromagnetic interactions[1].

As for the treatment of weak-interaction processes in nuclei, their quantitative description requires a detailed knowledge of the wavefunctions of both the initial and final states since the β decays, for instance, are very sensitive even to small configuration mixing between the two states. Moreover, the existence of Gamow-Teller giant resonances, as revealed in charge-exchange (p,n) reactions, points out the need of the inclusion of collective degrees of freedom too, in order to achieve a consistent treatment of weak interactions in nuclei. Complete shell-model calculations with realistic forces and large enough configuration spaces are feasible only for light-mass nuclei. β decays in heavier nuclei can be described by means of a truncated shell-model approach or a quasi-particle random phase approximation (QRPA), extending the old Tamm-Dancoff and RPA techniques to nuclei far from closed-shell configurations. Shortcomings of the QRPA description can be found in ref.[5], where applications of the Migdal theory of finite Fermi systems to the nuclear β decay are also mentioned.

In fact, with increasing number of valence nucleons outside the major shell closures, one has to introduce more and more single-particle orbitals in order to describe all possible decay branches and the shell-model configuration space has to be truncated in a suitable way. This is a very serious problem in doubly-odd nuclei whose β-decay properties have been hardly dealt with in theoretical papers; recently, a simplified approach has been proposed[6] and applied to the calculations of first-forbidden β- and β+/EC (electron conversion) transitions from odd-odd to even-even isobars with mass A = 136.

The interacting boson-fermion model[7] (IBFM) in its version 2, where neutron and proton degrees of freedom are explicitly introduced, seems a convenient framework to deal with weak-interaction processes since it allows us simple yet reliable calculations of nuclear wavefunctions, including collective and single-particle excitations. It is worth mentioning the interacting boson model[8] (IBM) description[9] of the double β decay where the effects of nuclear deformation are properly taken into account in the calculations of ^{128,130}Te to Xe decays.

2. Beta decay in the IBFM

2.1 Medium- and heavy-mass nuclei

In the IBFM-2 the odd (unpaired) proton and/or neutron is coupled to a bosonic IBM-2 core by means of a suitable fermion-boson interaction[7,10] whose form has been derived in the frame of a generalized seniority scheme. The complete Hamiltonian has the following expression:

$$\hat{H} = \hat{H}_B(\pi, \nu) + \hat{H}_F(p,n) + \hat{V}_{BF}(\pi, \nu, p, n), \quad (1)$$

where π and ν refer, respectively, to the proton and neutron boson degrees of freedom and $\hat{H}_B(\pi, \nu)$ is the usual IBM-2 Hamiltonian in the Talmi form[8]. The fermionic part of the Hamiltonian (1) is[7]

$$\hat{H}_F(p,n) = \sum_{j_p} E_{j_p} \hat{n}_{j_p} + \sum_{j_n} E_{j_n} \hat{n}_{j_n} + \hat{V}_F(p,n), \quad (2)$$

where E_j is the energy of a quasiparticle in the single-particle orbital with angular momentum j, evaluated in the usual BCS approximation, and $\hat{n}$ the quasiparticle number operator. When one deals with doubly-odd nuclei, the two-body residual interaction between the unpaired neutron and proton has to be considered and its form is discussed, for instance, in ref.[11]. Finally, the boson-fermion interaction has the following form[7,10]:

$$\hat{V}_{BF}(\pi,\nu,p,n) = \hat{V}_{BF}^m + \hat{V}_{BF}^q + \hat{V}_{BF}^{exc}, \qquad (3)$$

consisting, as usual, in a monopole-monopole interaction, a quadrupole-quadrupole part and an exchange term which takes into account the Pauli principle, since bosons approximate correlated nucleon pairs.

By diagonalizing Hamiltonian (1), it is possible to calculate eigenvalues and eigenfunctions of non-magic nuclei with an odd number of neutrons and/or protons. It is then a simple matter to evaluate the $\beta^\pm$ matrix elements introducing[12] the following one-nucleon transfer operators that are the images in the IBFM space of the fermion creation and annihilation operators:

$$P_\rho^{+(j)} = \zeta_j a_j^+ + \sum_{j'} \zeta_{jj'} (s^+ \tilde{d} a_{j'}^+)^{(j)} \qquad (4)$$

$$P_\rho^{+(j)} = \vartheta_j (s^+ a_j)^{(j)} + \sum_{j'} \vartheta_{jj'} (d^+ \tilde{a}_{j'})^{(j)}, \quad (\rho = \pi, \nu), \qquad (5)$$

where the coefficients, ζ and θ, can be deduced in the generalized seniority scheme and read as[10,12]

$$\zeta_j = b_1 u_j, \quad \zeta_{jj'} = -b_2 v_j \beta_{jj'} \sqrt{\frac{10}{N_\rho (2j+1)}}, \qquad (6)$$

$$\theta_j = b_1' \frac{v_j}{\sqrt{N_\rho + 1}}, \quad \vartheta_{jj'} = b_2' u_j \beta_{jj'} \sqrt{\frac{10}{(2j+1)}}, \qquad (7)$$

where $\{b_i\}$ are suitable normalization coefficients[7,12]. Therefore, the calculations of β-decay rates in the IBFM do not involve adjustable parameters other than those which occur in the Hamiltonian (1). It is worth noticing that operator (4) conserves the boson number between the initial and final states and operator (5) does not. Introducing the operators, $P_\rho^{-(j)}$, Hermitian conjugates of operators (4,5), the β-decay matrix elements for Fermi ($\Delta J=0$) and Gamow-Teller ($\Delta J=1$) transitions can be written, respectively, as

$$\langle J_f | \hat{F} | J_i \rangle = -\sum_j \sqrt{(2j+1)} \langle J_f | (P_\pi^{\pm(j)} x P_\nu^{\pm(j)})^{(0)} | J_i \rangle, \qquad (8)$$

$$\left\langle J_f \left| \hat{GT} \right| J_i \right\rangle = -\frac{1}{\sqrt{3}} \sum_{jj'} \left\langle l \frac{1}{2} j \|\bar{s}\| l' \frac{1}{2} j' \right\rangle \left\langle J_f \left| (P_\pi^{\pm(j)} x P_\nu^{\pm(j')})^{(1)} \right| J_i \right\rangle \delta_{ll'}.$$

(9)

For instance, the odd-mass Tc isotopes have 3 (hole) proton bosons and one unpaired proton coupled to neutron bosons of particle type. Their β⁻ decays populate states belonging to Ru isotopes with the same number of (hole) proton bosons, while one neutron boson is annihilated and one unpaired neutron created. Therefore, the conjugate operators of eqs. (4) and (5) are appropriate for the proton and neutron part, respectively.

The β-decay rate, λ_β, is given by

$$\lambda_\beta = \frac{\ln 2}{t_{\beta,1/2}} = \frac{1}{2\pi^3} \sum G_i^2 |M_i|^2 f,$$

(10)

where $t_{\beta,1/2}$ is the half-life, G_i is the vector (or axial vector) coupling constant of the weak interaction depending on the matrix element, M_i, eq.(8) or (9), and f is a known function describing the lepton phase volume[4]. In general, the weak interaction transition rates are expressed in terms of the product $(ft_{\beta,1/2})$ - the so-called ft value - in order to stress the nuclear structure effects. We then have the following expression[12]:

$$ft = \frac{6163}{\left|\left\langle J_f |\hat{F}| J_i \right\rangle\right|^2 + \left(\frac{G_A}{G_V}\right)^2 \left|\left\langle J_f |\hat{GT}| J_i \right\rangle\right|^2} \text{(sec)},$$

(11)

where the ratio of the axial-vector to the vector coupling constant is reduced with respect to the free value, 1.59±0.02, in the nuclear environment due to the coupling with non-nucleonic degrees of freedom, etc.; a quenced value, $(G_A/G_V)^2 = 1.2$, is assumed[12].

With these ingredients, some calculations of β-decay strengths have been performed for medium- and heavy-mass nuclei. In ref.[13,14] the Bose-Fermi symmetries are exploited to investigate the first nonunique forbidden transitions between states of Ir, Au, Pt and Hg isotopes with mass number 195 and 197 with satisfactory results. An analysis of Gamow-Teller transitions between I, Te and Xe isobars is presented in ref.[12], where a discrepancy of a factor 3.5 is found between observed and calculated values, much smaller than that springing from the pairing model calculations (about 70).

Assuming the IBFM parametrization of ref.[15], the β⁻ decay of odd-mass Tc isotopes has been investigated[16]; results for selected transitions are shown in Table I and compared with the experimental data taken from *Nuclear Data Sheets*, being in quite reasonable agreement on the whole.

Table I. β decay rates for odd-mass Tc → Ru transitions.

Initial nucleus	Jπ (E)	Final nucleus	Jπ (E)	log(ft) expt.	log(ft) IBFM
^{101}Tc	9/2+ (0.)	^{101}Ru	7/2+ (306.)	4.8	4.30
			7/2+ (545.)	5.6	6.63
			7/2+ (843.)	5.6	4.83
^{103}Tc	5/2+ (0.)	^{103}Ru	3/2+ (0.)	5.23	7.20
			5/2+ (136.)	5.66	7.32
			7/2+ (214.)	6.21	6.47
			3/2+ (346.)	5.03	5.49
^{105}Tc	5/2+ (0.)	^{105}Ru	3/2+ (0.)	> 7.3	8.62
			7/2+ (229.)	8.28	7.90

Energies of the states are within brackets and in keV units.

Theoretical calculations of β-decay rates are often essential in astrophysical applications[17] when direct measurements are impossible. In fact, at the high temperatures of stellar interiors nuclear excited states can be populated, in addition to the ground state, according to the Boltzmann distribution,

$$g_i = (2J_i + 1)\exp\left(-\frac{E_i}{k_B T}\right), \quad (12)$$

where J_i and E_i are the spin and the excitation energy of a level i, respectively, while k_B is the Boltzmann constant and T the stellar temperature. Of course, more and more levels are thermally populated as the temperature increases. In the late stages of stellar evolution, T can be of the order of 5×10^9 °K. The effective β-decay rate of a nucleus at finite-temperature environment can be then expressed in terms of the nuclear partition function, $g = \sum_{i=1}^{N} g_i$, as $\lambda_{eff} = \frac{1}{g}\sum_{i=1}^{N} \lambda_i g_i$, where λ_i is the β-decay rate of the i-th level at thermal equilibrium, when N levels are populated on the whole. Therefore, even at relatively low temperatures, as those involved in the s process, the contributions from the excited states could lead to a value of the effective β-decay rate considerably different from that of the ground state. This fact is very important when the ground state has a very low decay probability whereas fast transitions are expected from low-lying excited states, as in the case of ^{99}Tc, which is nearly stable at room temperature. Fig. 2 shows the predictions of the present IBFM approach in comparison with the truncated shell-model calculations of ref.[18] for Gamow-Teller transitions from excited states of ^{99}Tc. The present calculations lead to a half-life of about 2 years for a typical stellar temperature of 3×10^8 °K (see fig. 3), in agreement with previous estimates[19,20] but in contradiction with the existing astrophysical scenarios for s processes[18].

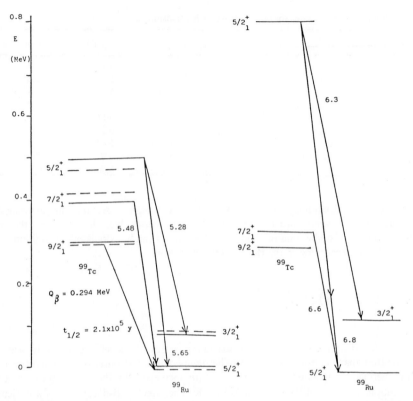

Fig. 2 - ^{99}Tc β⁻ decays of astrophysical interest. On the left, IBFM results (energy spectra and log(ft) values); experimental levels are shown by dashed lines. On the right, shell-model calculations of ref.[18].

2.2 Light-mass nuclei

In light-mass nuclei, where neutrons and protons occupy the same major shells, neutron-proton pairing correlations may play an important role, analogously to the pairing force leading to correlated neutron-neutron and proton-proton pairs in the heavy- and medium-mass nuclei. Moreover, the (approximate) isospin symmetry of nuclear forces must be taken explicitly into account. Therefore, in this mass region, the IBFM-4 applies which represents a suitable generalization of the usual IBFM-2 approximation[8].

The relevant formalism for the description of the β-decay has been studied in refs.[21,22]. In particular, in ref.[22] a Gamow-Teller operator for IBM-4 is derived microscopically and found to be proportional to the SU(4) generator given by the product of spin and isospin-raising operators. Ref.[22] deals with the role played by neutron-proton bosons.

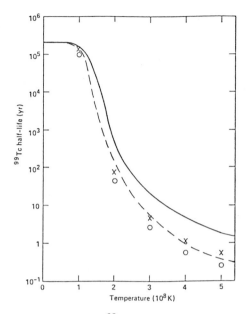

Fig. 3 - Calculated β⁻ decay half-life of ^{99}Tc versus temperature of stellar interiors. Dashed line: IBFM calculations; solid line: ref.[18]; open circles, ref.[19]; crosses, ref.[20].

The basic ingredients of the IBM-4 are bosons characterized by isospin T and intrinsic spin S, namely,

$$\begin{cases} \vartheta \ (T=0, S=1), \\ \pi \ (T=1, S=0), \\ \delta \ (T=1, S=0), \\ \nu \ (T=1, S=0), \end{cases}$$

where each boson may have total spin J=0 or 2 and positive parity; therefore, π and ν bosons are the usual correlated neutron-neutron and proton-proton pairs occuring in IBM-2. The 36 bilinear products of the relevant boson creation and annihilation operators span the $U^B(36) \supset U^B(6) \otimes U^B_{ST}(6)$ algebra, where the spin and isospin degrees of freedom are explicitly introduced. It is worth stressing that odd-odd nuclei can be treated on the same foot than even-even nuclei within this algebraic approach incorporating neutron-proton bosons.

The β-decay is then described[21] in terms of boson transitions where, for instance, one ν boson is transformed into one δ boson (Fermi transition) or one θ boson (Gamow-Teller transition). Therefore, transitions corresponding to *allowed* boson changes are enhanced with respect to those involving different boson configurations. As an example, fig. 4 shows the low-energy levels of A = 42 nuclei. The low log(*ft*) values correspond to favourite boson transitions, while β^+ decays requiring a change of the orbital angular momentum are retarded, as in the case of the decay from the ground state of ^{42}Ti to the second 1^+ level of ^{42}Sc. A similar situation occurs in the A = 18 nuclei and is extensively discussed in ref.[21], thus supporting the validity of the IBM-4 boson scheme. For a comparison with extended shell-model calculations, already feasible for these nuclei, see for instance ref.[23].

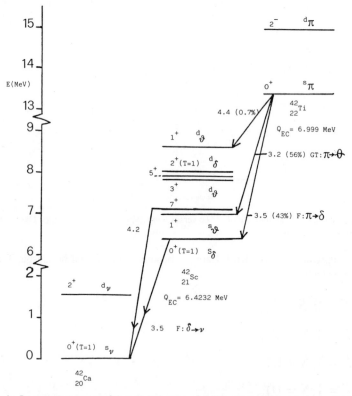

Fig. 4 - Low-energy spectra of A = 42 isobaric nuclei. Observed log(*ft*) values are shown together with the IBM-4 classification of the states.

3. Nuclear muon capture

The last side of the Puppi triangle is left, referring to the nuclear muon capture reactions. The relevant theory dates back to the pioneering paper[24] of Primakoff and is reviewed in refs.[25,26]. As far as the vector part of the weak current is concerned, the dipole transitions are particularly important in muon capture and rule over the other components both for doubly closed-shell nuclei and N > Z nuclei. In this latter case, a correspondence was pointed out[27,28] between the upper component produced by the

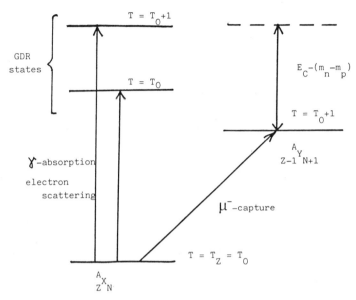

Fig. 5 - Schematic representation of negative muon capture, photoabsorption and electron scattering processes by N > Z nuclei.

isospin splitting of the E1 giant dipole resonance and the parent dipole mode excited by muon capture. The situation is shown schematically in fig. 5. The muon capture corresponds to a transition from the $|0, T_0, T_z = T_0\rangle$ ground state to $|n, T_0 + 1, T_z + 1\rangle$ states of the adjacent (Z-1) isobaric nucleus; the giant dipole resonance (GDR) components excited by photoabsorption or elastic scattering reactions have isospin values $T = T_0$ and T_0+1, respectively, the third component, T_Z, being unchanged with respect to the ground state.

Within the IBM approach to giant resonances[7], the well-known phenomenon of the GDR isospin splitting has been described[29] on the assumption of a $SU_B(6) \otimes SU_T(2)$ symmetry of IBM-3; a remarkable agreement with the measured photoabsorption cross sections has been found both in the s-d and f-p shells. As an example, fig. 6 shows the calculated[30] and observed[31] total photoabsorption cross section of ^{64}Zn. The double-humped structure, clearly seen in the ^{63}Cu(p,γ_0)^{64}Zn reaction[32], has been interpreted as arising from the isospin fragmentation and this interpretation is nicely confirmed by the present IBM analysis.

One is thus tempted to apply the same approach to a rough first-order approximation description of muon capture rates in N > Z nuclei. Following the sum-rule technique developed in ref.[33], the isospin effects can be simply introduced in an analytical formula for the muon capture rate, Λ_μ, also including dynamical correlations. Unlike ref.[33], where the isospin is taken into account assuming the volume symmetry energy coefficient of the semiempirical mass formula, the same expression of ref.[29] for

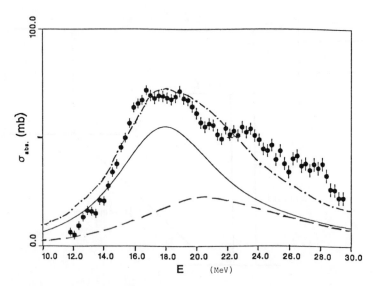

Fig. 6 - Experimental[31] and calculated (dotted-dashed line) photoabsortion cross section of ^{64}Zn. Solid line: T=2 component; dashed line: T=3 component.

the GDR isospin splitting is adopted (used in the calculations shown in fig. 6, too) and the following formula holds:

$$\Lambda_\mu = 20 Z^3 A R \left[1 - 2.60 \left(\frac{N-Z+2}{A} \right) \right]^2 + \Lambda'_\mu, \qquad (13)$$

with A, Z and N, respectively, mass, proton and neutron number of the concerned nucleus. Λ'_μ is the recoil correction to the capture rate and R is given by

$$R = \frac{\pi |\varphi_\mu|^2}{(Z \alpha m_\mu)^3}, \qquad (14)$$

where φ_μ is the 1s muonic atomic wave function averaged on the nucleus, m_μ is the mass of the muon and α the fine structure constant. Assuming the values of R and Λ'_μ given in the literature[34], eq.(13) gives the total capture rates shown in Table II, in comparison with the corresponding experimental data. Once again, a satisfactory agreement is found even for those doubly closed-shell nuclei to which the IBM approach would not apply directly.

Since muon capture induces $\Delta T = 1$ transitions and the relevant operators are generators of the su(4) Wigner algebra, selection rules exist[25,26] for light-mass nuclei

Table II. Total nuclear capture rates, Λ_μ, for negative muons.

Nucleus	R [34]	Λ'_μ (10^{-6} s^{-1})	Λ_μ expt.[35] (10^{-6} s^{-1})	Λ_μ IBM (10^{-6} s^{-1})
^{40}Ca	0.4400	0.36	2.51±0.03	2.49
^{48}Ca	0.4700	0.05	-	0.81
^{58}Ni	0.3203	0.16	6.11±0.10	5.65
^{60}Ni	0.3203	0.16	5.56±0.10	4.78
^{62}Ni	0.3203	0.16	4.72±0.10	4.01
^{90}Zr	0.1754	0.31	8.62±0.08	8.93
^{140}Ce	0.0751	0.38	11.51±0.13	11.35
^{208}Pb	0.0292	0.59	13.17±0.39	12.69

such as those belonging to the 1p shell, confirmed by the experiments to a large extent. The IBM-4 approach recalled in sect.2.2 can be extended to odd-mass isotopes, once the unpaired nucleon is described by means of suitable creation and annihilation fermion operators in the s-l coupling, thus spannning a $\sum_{l_F}(2l_F + 1)$ - dimensional space. For instance, in the s-d shell the single-particle levels have $l_F = 0$ and 2 and the relevant representation space is $U^F(24) \supset U^F(6) \otimes U^F_{ST}(4)$.

These Bose-Fermi algebras can be embedded into a superalgebra, $U(36|24)$, whose supermultiplets are characterized by the total number of particles, $N=N+M$ with N number of bosons and M number of fermions (M=0,1,2). Odd-odd nuclei, in particular, may have states consisting of neutron-proton collective excitations (δ and θ bosons) and states with two unpaired nucleons (neutron and proton). The boson and fermion algebras can be combined together in a suitable Bose-Fermi lattice of algebras[7] to obtain a convenient basis of states as well as its classification scheme by means of the labels characterizing the representations of the lattice algebras.

A particular scheme[36] has been considered for nuclei in the s-d shell, where a boson-fermion basis can be constructed encompassing the Wigner supermultiplet scheme. In fact, one has

$$U^B(36) \otimes U^F(24) \supset U^B(6) \otimes U^B_{ST}(6) \otimes U^F(6) \otimes U^F_{ST}(4)$$
$$\supset U^B(6) \otimes O^B_{ST}(6) \otimes U^F(6) \otimes U^F_{ST}(4) \supset U^{BF}(6) \otimes SU^{BF}_{ST}(4),$$

since $O(6)$ and $SU(4)$ are isomorphic. Therefore, approximate selection rules for muon capture hold analogously to those valid for 1p-shell nuclei. In particular, muon captures where both the initial and final nucleus belongs to the same supermultiplet should be enhanced with respect to those between nuclei with different numbers, N, of bosons plus fermions. It is the case of ^{30,31}S $\to$ ^{30,31}P $\to$ ^{30,31}Si transitions, while the

$^{32}S \to {}^{32}P$ muon capture is forbidden in the supersymmetry scheme. Measurements of inclusive muon capture by these isotopes would be very useful to check the validity of the IBFM-4 supersymmetry limit.

4. Concluding remarks

The interacting boson-fermion model is a quite simple, yet reliable tool that allows us to perform many analyses of weak interaction processes for isotopes even far from the stability valley. The extension to the $\beta^\pm$ decay of doubly-odd nuclei is straighforward and will be object of future works. Moreover, within this algebraic framework, decay rates from nuclear excited states can be easily evaluated, of interest in astrophysical problems, and approximate formulae derived in order to deal with negative muon captures in nuclei.

5. Acknowledgements

Useful discussions with F. Iachello are gratefully acknowledged. Part of this work has been done in collaboration with L. Zuffi.

6. References

1. *Fifty Years of Weak-Interaction Physics*, eds. A. Bertin, R. A. Ricci and A. Vitale (Italian Physical Society, Bologna, 1984).
2. G. Gamow and E. Teller, *Phys. Rev.* **49** (1936) 895.
3. E. J. Konopinski and G. E. Uhlenbeck, *Phys. Rev.* **60** (1941) 308.
4. A. DeShalit and H. Feshbach, *Theoretical Nuclear Physics*, vol. I, chapt. IX (John Wiley & Sons, New York, 1974).
5. K. Takahashi and W. Hillebrandt, in Proceed. of the 2nd Int. Symp. on Nuclear Astrophysics, *Nuclei in the Cosmos*, eds. F.Käppeler and K.Wisshak (Institute of Physics Publishing, Bristol and Philadelphia, 1993), p. 381.
6. J. Suhonen, *Nucl. Phys.* **A563** (1993) 205.
7. F. Iachello and P. Van Isacker, *The Interacting Boson-Fermion Model* (Cambridge University Press, Cambridge, 1991).
8. F. Iachello and A. Arima, *The Interacting Boson Model* (Cambridge, Cambridge University Press, 1987).
9. O. Scholten and Z. R. Yu, *Phys. Lett.* **B161** (1985) 13.
10. O. Scholten and A. E. L. Dieperink, in *Interacting Bose-Fermi Systems in Nuclei*, ed. F. Iachello (Plenum Press, New York and London, 1981), p. 343.
11. A. Ventura, G. Maino, A. M. Bizzeti-Sona, P. Blasi and A. A. Stefanini, contribution to this conference.
12. F. Dellagiacoma and F. Iachello, *Phys. Lett.* **B218** (1989) 399.
13. P. Navrátil and J. Dobeš, in *Symmetries and Nuclear Structure*, eds. R.A.Meyer and V.Paar, Nuclear Science Research Conf. Series vol. 13 (Harwood Academic Publ., Chur, Switzerland, 1987) p. 195.

14. P. Navrátil and J. Dobeš, *Phys. Rev.* **C37** (1988) 2126.
15. G. Maino, A. Ventura, A. M. Bizzeti-Sona and P. Blasi, *Z.Phys.* **A340** (1991) 241.
16. G. Maino and L. Zuffi, in Proceed. of 8th Int. Conf. on *Nuclear Reaction Mechanisms*, Varenna, June 6-11, 1994.
17. T. Mizusaki and T. Otsuka, in Proceed. of 8th Int. Symp. on Capture Gamma-Ray Spectroscopy, ed. J. Kern (World Scientific, Singapore, 1994), p. 748
18. K. Takahashi, G. J. Mathews and S. D. Bloom, *Phys. Rev.* **C33** (1986) 296.
19. K. Cosner and J. W. Truran, *Astrophys. Space Sci.* **78** (1981) 85.
20. K. Yokoi and K. Takahashi, report KfK-3849 (Karlsruhe, 1985).
21. F. Iachello, in *Shell Model and Nuclear Structure*, ed. A. Covello (World Scientific, Singapore, 1989).
22. P. Halse and B. R. Barrett, *Ann. Phys. (N.Y.)* **192** (1989) 204.
23. A. Arima, K. Shimizu, W. Bentz and H. Hyuga, *Adv. Nucl. Phys.* **18** (1987) 1.
24. H. Primakoff, *Rev. Mod. Phys.* **31** (1959) 802.
25. N. C. Mukhopadhyay, *Phys. Rep.* **C30** (1977) 1.
26. F. Cannata, R. Graves and H. Uberall, *Riv. Nuovo Cim.* **7** (1977) 133.
27. B. Goulard, J. Joseph and F. Ledoyen, *Phys. Rev. Lett.* **27** (1971) 1238.
28. O. Nalcioglu, D. J. Rowe and C. Ngo-Trong, *Nucl. Phys.* **A218** (1974) 495.
29. G. Maino, A. Ventura and L. Zuffi, *Phys. Rev.* **C37** (1988) 1379.
30. G. Maino and M. Rosetti, to be published.
31. P. Carlos, H. Beil, R. Bergère, J. Fagot, A. Leprêtre, A. Veyssière and V. G. Solodukhov, *Nucl. Phys.* **A258** (1976) 365.
32. P. Paul, J. F. Amann and K. A. Snover, *Phys. Rev. Lett.* **27** (1971) 1013.
33. E. Lipparini, S. Stringari and R. Leonardi, *Phys. Lett.* **B212** (1988) 6.
34. N. Auerbach and A. Klein, *Phys. Lett.* **B114** (1982) 95.
35. T. Suzuki, D. F. Measday and J. P. Roalsvig, *Phys. Rev.* **C35** (1987) 2212.
36. S. Szpikowski, P. Klosowski and L. Próchniak, *Nucl. Phys.* **A487** (1988) 301.

TWENTY YEARS OF IBA AND TRANSIENT FIELDS

N. BENCZER-KOLLER, G. KUMBARTZKI, A. MOUNTFORD, T. VASS
M. SATTESON and N. MATT
Department of Physics and Astronomy, Rutgers University
New Brunswick, NJ 08903 USA

R. TANCZYN
Department of Physical Science, Kutztown University

C.L. LISTER
Argonne National Laboratory

P. CHOWDHURY
Wellesley College

ABSTRACT

The magnetic moments of the low-lying yrast states in ^{86}Zr, ^{150}Sm and ^{152,154}Gd have been measured by the transient field technique. The results are discussed in terms of collective, single particle and IBA models.

1. Introduction

Twenty years ago the Interacting Boson Approximation model made its appearance in the realm of nuclear structure physics [1]. At about the same time a new technique, the transient field technique, particularly effective for precision measurements of magnetic moments of short-lived nuclear states was perfected [2]; the lucky conjunction on the firmament of these two "constellations" led to a wealth of measurements, discoveries and revealing interpretations of nuclear structure. In the conventional collective model, the g factors of all states are expected to be equal to $\approx Z/A$, corresponding to the fact that both neutrons and protons in the nucleus participate equally in the collective motion. Even though the existing measurements at the time were not particularly precise, it was obvious already in the late sixties that the collective description was not an adequate representation of nature. Greiner [3] noted that differences in proton and neutron quadrupole deformations could account for the discrepancies between model and experiment. However, subsequent calculations showed that the effect was too small to reconcile the disagreement. The proton-neutron interacting boson approximation IBA-II [4], however, does provide a natural framework for studying proton-neutron degrees of freedom in collective states. The calculations that ensued were extraordinarily successful in calculating the magnetic moments of the 2^+ states of medium weight and heavy nuclei in the regions [5] $44 < Z < 56$ with $54 < N < 78$ and $56 < Z < 78$ with $88 < N < 120$. IBA-II yields a simple expression for $g(2_1^+)$,

$$g(2_1^+) = g_\pi(N_\pi)[N_\pi/(N_\pi + N_\nu)] + g_\nu(N_\nu)[N_\nu/(N_\pi + N_\nu)] \quad (1)$$

where $g_\pi(N_\pi)$ and $g_\nu(N_\nu)$ are the g factors of the correlated proton or neutron pairs and the N_π and N_ν represent the number of "active" proton and neutron pairs.

The detailed microscopic calculations of these parameters yielded considerable insight into the underpinning assumptions of the model. As so often found in physics, once systematics were accurately described by the model, deviations from the predictions could be interpreted [6] in terms of new structure phenomena, such as the effect of proton subshell closure at $Z = 64$ for $N < 88$.

As the years went on, both the theoretical models and the experimental techniques progressed. The interest shifted to new frontiers such as very high spins and deformation, and nuclei far-from-stability with unusual neutron to proton ratios. In order to explain the structure of high spin states, g-bosons had to be introduced [7], as well as contributions of fermions-boson interactions [8]. Measurements were extended to high excitation energies and continuum spectra.

Magnetic moments of higher spin states along the yrast band have been measured in recent times in rotational rare earth nuclei with $64 < Z < 70$ and $N > 90$. In general, the moments remain fairly constant at low spin, *decrease* at higher spin and, in some cases, have been noted to *increase* at yet higher spins. These observations were first explained in the framework of the cranked shell model approach as being due to rotational alignment of $i_{13/2}$ neutrons followed by alignment of protons [9]. Self-consistent cranked Hartree-Fock-Bogoliubov calculations [10] have also been very successful in describing the trends observed in several nuclei.

For transitional rare-earth nuclei with neutron number close to 90, a simplified model calculation was performed in the Nilsson-Strutinsky plus BCS approach by calculating the spin dependence of the deformation and pairing gaps [11]. These calculations predict a gradual *increase* of g factors as the angular momentum increases to $10\hbar$ for ^{160}Yb, and a more dramatic rise for ^{154}Dy; similar effects are expected for ^{152}Gd and ^{150}Sm. The magnetic moments of the yrast states in ^{154}Dy have been recently measured [12]. Whereas the decrease in the moments of states above spin 20 has indeed been observed, the predicted rise of the magnetic moments of the low-lying states was not confirmed. Thus it becomes interesting to further test the model in the $N = 88$ isotones ^{150}Sm and ^{152}Gd, as well as in ^{154}Gd.

This paper describes recent experiments on the low-lying states of ^{86}Zr, ^{150}Sm and ^{152}Gd and ^{154}Gd. The results are evaluated in the framework of cranking calculations, shell model results and recent IBA calculations. The ^{150}Sm data have been published [13]. The ^{86}Zr experiments have been completed and a paper is being prepared for publication [14]. The ^{152,154}Gd experiment is on-going and preliminary results are presented here.

2. Experimental Procedures

The magnetic moments of excited states in ^{86}Zr, ^{150}Sm, ^{152}Gd and ^{154}Gd were measured by the transient field technique. The technique is relatively straightforward when applied to the measurement of states excited by one- or two-step Coulomb excitation. However, whenever the interesting states lie at higher excitations and are populated by fusion-evaporation reactions or multistep Coulomb excitation, the procedures for application of the tansient field technique become much more complex, and the analysis of the results is consequently more difficult. In the experiments reported here, both these reactions were used. The ^{86}Zr states were populated by the ^{12}C(^{77}Se,3n)^{86}Zr reaction; data acquisition was done essentially in the "singles" mode. The states in ^{150}Sm and ^{152,154}Gd were Coulomb-excited by ^{58}Ni beams; recoil-particle gamma-ray coincidences were recorded. In all cases, the angular distribution of the decaying gamma-rays was measured.

The experiments were performed at the A. W. Wright Nuclear Structure Laboratory at Yale University. Heavy ion beams of ^{58}Ni or ^{32}S with energies ranging from 80 to 230 MeV were used to Coulomb excite either the 2_1^+, or the 2_1^+ and 4_1^+, or the 2_1^+, 4_1^+ and 6_1^+ states of ^{150}Sm and a 180 MeV ^{58}Ni beam was used for the ^{152}Gd experiment. The ^{86}Zr states were produced with 240 Mev ^{77}Se beams.

Seven different experiments—involving different targets, ferromagnetic foils, beams, beam energies and detectors—were performed in the case of the ^{150}Sm. Experiments on ^{86}Zr and ^{152,154}Gd were carried out with a single target/ferromagnetic foil/beam combination. The experimental details of and a description of the very different data analysis used in the ^{150}Sm/^{152,154}Gd and the ^{86}Zr experiments are extensively described in Refs. 13, 14, and 15.

3. Results

3.1 ^{86}Zr

^{86}Zr is a transitional nucleus that lies between the collective deformed ^{80,82,84}Zr nuclei and ^{88,90}Zr nuclei best described by single particle excitations. Transitional nuclei lie far enough from shell closures to prohibit full microscopic calculations. Furthermore they have not developed sufficient collectivity to be reliably interpreted through deformed cranking models. The energies involved in single particle excitations, in collective vibrations and in collective rotations are similar, so states associated with each of these degrees of freedom tend to mix and have complicated wave functions. Recently these nuclei have been analyzed in terms of the extended IBA where the collectivity, both vibrational and rotational, is described by the IBA and single particle contributions are incorporated through the coupling of unpaired nucleons [16]. This approach has been successful in interpreting bands in ^{84,86}Zr, which have unusual properties [17]. The model predicts transitional matrix elements very well, but falls short of predicting the exact nature of the states, particularly their proton or neutron parentage. These details can be obtained however from the magnetic moment measurements.

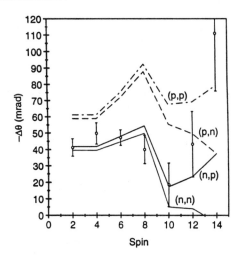

Fig.1. Precession of gamma-rays de-exciting ^{86}Zr states, calculated for various schemes of n and p alignments. The data points are experimental.

Table 1: Measured g factors in ^{86}Zr.

I	8_1^+	10_1^+	12_1^+	14_1^+	$< 8_2^+, 10_2^+ >$	$< 5_1^-, 7_1^-, 9_1^- >$
g	-1.1(6)	-0.5(10)	-0.4(8)	1.9(6)	1.1(2)	0.5(2)

In ^{86}Zr three I = 8 states are anticipated. One is related to the 2_1^+-4_1^+-6_1^+ yrast sequence and has a complex wave function. The other two have a rather pure configuration corresponding to either a pair of neutrons or a pair of protons in the $g_{9/2}$ shell, which decouple and align to their maximum spin. It is expected that in a deformed shell model potential with prolate deformation, the proton alignment is seen first. For an oblate potential with similar deformation, the neutron alignment occurs first [18].

The experimental results are displayed in Table I and Fig. 1. The negative g factor of the 8_1^+ state confirms a previous result obtained in a static hyperfine field [19]. This result clearly corresponds to a neutron alignment and an oblate nucleus. The 8_2^+ and 10_2^+ states have precessions consistent with g factors larger than the collective value suggesting a proton-aligned structure. The 8_3^+ state was not seen in this experiment.

Above 3.5 MeV there is sufficient energy available for two pairs of particles to decouple and align to form states with I > 12 $\hbar$. Here four sets of states each with four quasiparticles—all neutrons, all protons or mixed neutrons and protons—are anticipated. Only one of these bands is known from experiment. In the IBA + quasiparticles model the "all neutron" band is predicted to lie lowest, further reflecting the lower position of the first neutron alignment. Both four quasi-protons and four quasi-neutrons are needed to reproduce the pattern of transitional matrix elements that have been measured. Unfortunatly, the properties of the mixed proton-neutron configuration is beyond the scope of the first IBA version of the model, so its position and properties have not been calculated yet. The mixed configuration is the one predicted lowest in deformed calculations.

The measurements at the highest spins show a swing from negative to positive values of the g factor. This result can only arise from proton contributions to the wavefunctions. The value of $g(14_1^+) = +1.8(6)$ clearly rules out a pure four quasi-neutron scenario as well as the "collective" state. Two possibilities remain, the four aligned proton scenario or one of a mixed proton-neutron configuration. To bring the proton single particle states lower than the neutron ones requires deformation of the potential to a prolate shape, which would result in considerable enhancement of the B(E2) matrix elements in contradiction to the data. Thus, the mixed proton-neutron scenario with modest static deformation of the potential to the highest spin seems the most likely.

The negative parity states exhibit a precession corresponding to positive g factors, and thus support a mainly proton configuration such as $g_{9/2}$, $p_{3/2}$. However, the measured average value of $< g(5^-, 7^-, 9^-) > = +0.5(2)$ is much smaller than expected for a pure single particle aligned configuration where the g factor should be 1.4. Thus it may well be that the negative parity states arise from a mix of neutron- and proton-aligned configurations.

3.2 ^{150}Sm

The magnetic moment of the 2_1^+ state in ^{150}Sm, g = 0.385(27) has been measured by different techniques. In order to prevent uncertainties in the calibration of the transient magnetic field, and the increasing complex data analysis corresponding to the lower spin states from marring the present measurements of the moments of the higher states, only ratios of moments $g(4_1^+)/g(2_1^+)$ and $g(6_1^+)/g(2_1^+)$ will be quoted here.

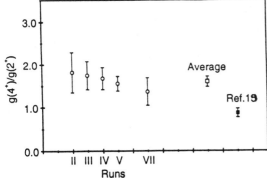

Fig.2. $g(4_1^+)/g(2_1^+)$ in ^{150}Sm measured in different runs under very different beam target/ferromagnetic foil/detector arrangements.

The seven ^{150}Sm experiments yielded results in very good agreement with each other as shown in Fig. 2. The following g factor ratios were obtained: $g(4_1^+)/g(2_1^+)$ = 1.60(12) and $g(6_1^+)/g(2_1^+)$ = 1.14(34). These data, however, disagree with the theoretical prediction for magnetic moments within a collective band, $g(I+2)/g(I) \approx 1$, and are significantly larger than the ratios obtained in a similar experiment [20]. This discrepancy was extensively discussed in Ref. 13, but no understanding of its cause was achieved.

The resulting $g(4_1^+)/g(2_1^+)$ is unexpectedly large. This large ratio cannot be explained through purely collective excitations nor through IBA models with either a restricted sd-boson or an extended sdg-boson basis [21].

How can such a large ratio of $g(4_1^+)/g(2_1^+)$ be understood? A number of alternative explanations are possible.

a. The experiment might be wrong. This assumption is always possible, but unlikely in view of the great care with which the experiment was carried out and different procedures were checked.

b. Pairing is strong for neutrons but breaks down for protons at higher spins. Then, as $g(2_1^+) \approx Z/A \approx 0.41$ and $g(4_1^+) \approx \mathcal{J}_{rigid} / (\mathcal{J}_{\pi,rigid} + 2/5\, \mathcal{J}_{\nu,rigid}) = Z / (Z + (2/5)N) = 0.61$, $g(2_1^+)/g(4_1^+) = 1.5$. This scenario is also possible, but not likely either [22]!

c. Configuration mixing plays a part. In the most naive shell model, one could construct the low-lying states with $d_{5/2}$ protons and $h_{9/2}$ neutrons for which g = 1.13 [22]. Thus, even a small admixture of such configurations could indeed result in magnetic moments that deviate from the predictions presented above. Again this model might be possible, but is also unlikely.

d. The simplified model calculation in the Nilsson-Strutinsky + BCS approach [11] predicts that changes in the nuclear deformation and pair fields occur with increasing spin in N = 88 nuclei. However, the effect is small. In addition, the ratio $g(2_1^+)/g(4_1^+) \approx 1$ has been measured recently for ^{154}Dy [12].

e. The 4_1^+ is more deformed than the 2_1^+ state [23]. This occurrence is unlikely but definitely possible. The measurement of the $g(2_1^+)$ state of Nd, Sm, and Gd isotopes confirmed the closure of the sub-shell at Z = 64. This sub-shell was shown to vanish for N = 90, an effect strongly correlated with the onset of deformation. Can it be that the phase transition is more complete for the 4_1^+ state than for the 2_1^+ state? Calculations [24] show that backbend does indeed occur stronger and sooner for the 4_1^+ state. In addition, the ratio $B(E2; 4_1^+ \rightarrow 2_1^+)/B(E2; 2_1^+ \rightarrow 0_1^+)$ peaks for ^{150}Sm suggesting indeed a larger deformation for that state (Fig.3).

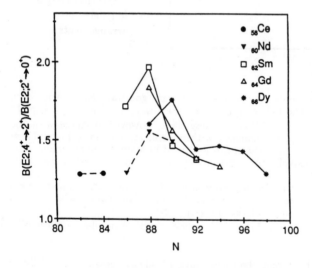

Fig.3. Ratio of $B(E2; 4_1^+ \rightarrow 2_1^+)/B(E2; 2_1^+ \rightarrow 0_1^+)$ for nuclei with 82<N<98.

A more specific explanation will have to wait for new single particle calculations, an examination of the structure of this vibrational nucleus within the framework of the IBA model, more detailed calculations of the effects of deformation on pairing gaps and alignments or another experimental test.

As a test of the above hypotheses, the measurement of the magnetic moments of the 2_1^+ and 4_1^+ states of ^{152,154}Gd was undertaken.

3.3 ^{152,154}Gd

As hinted by the experimental results on ^{150}Sm, and by the theoretical calculations described in Ref.10, the large ratio $g(4_1^+)/g(2_1^+)$ could be related to the nuclear structure of N = 88 nuclei, or could be a reflection of higher collectivity in the 4_1^+ state. These ideas are being tested through an experiment on ^{152,154}Gd isotopes. The first of these isotopes has the required number of neutrons, while the second has a higher $B(E2; 4_1^+ \rightarrow 2_1^+)/B(E2; 2_1^+ \rightarrow 0_1^+)$ ratio than the neighbouring Gd isotopes. The preliminary results yield $g(2_1^+) = 0.42(2)$ and $0.5(1)$, and $g(4_1^+)/g(2_1^+) = 1.1(2)$

and 1.0(3) for ^{152}Gd and ^{154}Gd respectively. The experiment will be repeated at higher energies. Similar measurements in ^{156}Dy could help understand this problem; however, this particular experiment presents formidable challenges.

4. Conclusions

The magnetic moments of states in the yrast band in ^{86}Zr were measured over the spin region I = 8–14$\hbar$. The results demonstrate that the first crossing at the 8_1^+ state involves the alignment of neutron $g_{9/2}$ quasiparticles, and that the proton $g_{9/2}$ quasiparticles align at higher spin.

The magnetic moment ratios $g(4_1^+)/g(2_1^+)$ and $g(6_1^+)/g(2_1^+)$ in ^{150}Sm have been measured by the transient field technique in a variety of experiments involving different targets mounted on different magnetic materials, different beams at different energies and different detectors. While individual moments may be very sensitive to the details of the transient field velocity dependence and target inhomogeneities, ratios of moments ought to be free from such systematic errors. The results are consistent with each other from run to run but yield ratios that are too large compared with previous experiments as well as with current theoretical predictions. Similar measurements in ^{152}Gd and ^{154}Gd yield g factor ratios ≈ 1, in agreement with the collective description.

5. Acknowledgments

We wish to thank A. Lipski at the Stony Brook Nuclear Physics Laboratory for his very generous assistance in target preparation. We also acknowledge the help and support of the students and technical staff of the Wright Nuclear Structure Laboratory. This work was supported in part by the National Science Foundation.

References

[1] F. Iachello and A. Arima, Phys. Lett. **53B**, 309 (1974); A. Arima, T. Otsuka, F. Iachello,and I. Talmi, Phys.Lett. **66**,205 (1977).

[2] B. Herskind, R. Borchers, J. D. Bronson, D. E. Murnick, L. Grodzins, and R. Kalish, Phys. Rev. Lett. **20**, 424 (1968); M. Forterre, J. Gerber, J. P. Vivien, M. P. Goldberg and K.-H. Speidel, Phys. Lett. **55B**, 56 (1975); J. L. Eberhardt, G. van Middelkoop, R. E. Horstman and H. A. Doubt, Phys. Lett. **56B**, 329 (1975); R. Kalish, M. Hass, M. Brennan and H. T. King, Nucl. Phys. **A276**,339 (1977)

[3] W. Greiner, Nucl.Phys **80**, 417 (1966).

[4] T. Otsuka, A. Arima, F. Iachello, and I. Talmi, Phys. Lett. **76B**, 139 (1978)

[5] M. Sambataro and A. E. L. Dieperink, Phys. Lett. **107B**, 249 (1981).

[6] A. Wolf, D. D. Warner and N. Benczer-Koller, Phys.Lett. **B158**, 7 (1985) and A. Wolf, R. F. Casten, D. D. Warner, Phys. Lett. **B190**, 19 (1987).

[7] B. Barrett and K. A. Sage, Interacting Bose-Fermi Systems in Nuclei, ed. F. Iachello, Plenum Press (1981).

[8] O. Scholten and A. E. L. Dieperink, Interacting Bose-Fermi Systems in Nuclei, ed. F. Iachello, Plenum Press (1981).

[9] S. Frauendorf, Phys. Lett. **B100**, 219 (1981); Y. S. Chen and S. Frauendorf, Nucl. Phys. **A393**, 135 (1983).

[10] A. Ansari, E. Wüst and K. Mühlhans, Nucl. Phys. **A415**, 215 (1984).

[11] R. Bengtsson and S. Åberg, Phys. Lett. **172**, 277 (1986).

[12] H. Hübel, S. Heppner, U. Birkental, G. Baldsiefen, A. P. Byrne, W. Schmitz, M. Bentley, P. Fallon, P. D. Forsyth, D. Howe, J. R. Roberts, H. Kluge, G. Goldring, A. Dewald, G. Siems and E. Lubkiewicz, Prog. Part. Nucl. Phys. **28**, 295 (1992); U. Birkental, A. P. Byrne, S. Heppner, H. Hübel, W. Schmitz, P. Fallon, P. D. Forsyth, J. W. Roberts, H. Kluge, E. Lubkiewicz, and G. Goldring, Nucl. Phys. **A555**, 643 (1993).

[13] T. Vass, A. W. Mountford, G. Kumbartzki, N. Benczer-Koller, and R. Tanczyn, Phys. Rev. **C48**, 2640 (1993).

[14] A. W. Mountford, T. Vass, G. Kumbartzki, L. A. Bernstein, N. Benczer-Koller, R. Tanczyn, C. J. Lister, P. Chowdhury, and S. J. Freeman, (to be published in Phys. Rev. C).

[15] A. W. Mountford, J. Billowes, W. Gellety, H. G. Price, and D. D. Warner, Phys. Lett **B279**, 228 (1992).

[16] A. A. Chishti, P. Chowdhury, D. J. Blumenthal, P. J. Ennis, C. J. Lister, Ch. Winter, D. Vretenar, G. Bonsignori, and M. Savoia, Phys. Rev. **C48**, 2607 (1993).

[17] P. Chowdhury, C. J. Lister, D. Vretenar, Ch. Winter, V. P. Janzen, H. R. Andrews, D. J. Blumenthal,B. Crowell, T. Drake, P. J. Ennis, A. Galindo-Uribarri, D. Horn, J. K. Johansson, A. Omar, S. Pilotte, D. Prevost, D. Radford, J. C. Waddington and D. Ward, Phys. Rev. Lett. **67**, 2950 (1991).

[18] J. Billowes, F. Cristancho, H. Grawe, C. J. Gross, J. Heese, A. W. Mountford, and M. Weiszflog, Phys. Rev. **C47**, R917 (1993).

[19] M. Weiszflog, C. J. Gross, A. Harder, M. K. Kabadiyski, K. P. Lieb, A. Raguse, D. Rudolph, J. Billowes, T. Burkhardt, J. Eberth, T. Myläus, and S. Skoda, Manchester University, Annual Report (1993).

[20] A. P. Byrne, A. E. Stuchbery, H. H. Bolotin, C. E. Doran and G. J. Lampard, Nucl. Phys. **A466**, 419 (1987).

[21] I. Morrison, Phys. Lett. **B175**,1 (1986).

[22] L. Zamick, private communication.

[23] R. Casten, private communication.

[24] J. Y. Zhang, Nucl. Phys. **A421**, 353c (1984).

639

MAGNETIC MOMENTS IN TRANSITIONAL NUCLEI: PROBING F-SPIN SYMMETRY AND SUPERSYMMETRY

ANDREW E. STUCHBERY
Department of Nuclear Physics, Research School of Physical Sciences and Engineering
Australian National University, Canberra, ACT 0200, Australia

ABSTRACT

Measurements of $M1$ properties in the transitional Os and Pt nuclei are presented and discussed in terms of their sensitivity to (i) F-spin mixing, (ii) the effective proton boson number, and (iii), in odd-A nuclei, the single-particle distribution required by $U(6/12)$ supersymmetry.

1. Introduction

The $M1$ properties of nuclei can provide insights into nuclear structure not gleaned from other properties. For example, in the proton-neutron Interacting Boson Model (IBM-2), g-factors and $M1$ transitions are sensitive to the breaking of F-spin symmetry, while the $E2$ properties are only slightly affected by it. The g-factors are more sensitive to the effective proton boson number than $E2$ transitions within an intruder band. Likewise, the $M1$ properties are sensitive to the occupation of single particle orbits in odd-A nuclei and therefore provide a critical test of the supersymmetry limits of the Interacting Boson-Fermion Model (IBFM). After reviewing new experimental data, I shall discuss these aspects of the $M1$ properties in the transitional Os and Pt nuclei.

2. New Measurements of $M1$ Properties in Transitional Nuclei

2.1. Mixing Ratios of $2_2^+ \to 2_1^+$ Transitions in ^{188,190,192}Os

Some time ago, the g-factors of the 2_1^+, 2_2^+ and 4_1^+ states in ^{188,190,192}Os were measured by the transient field (TF) technique, following Coulomb excitation by beams of 220 MeV ^{58}Ni from the ANU 14UD Pelletron accelerator[1]. As a concomitant, angular correlations between back-scattered beam ions and de-exciting γ rays were measured. Experimental details and the angular correlations for the pure $E2$ transitions ($2_1^+ \to 0_1^+$, $4_1^+ \to 2_1^+$, $2_2^+ \to 0_1^+$) have been presented elsewhere[1]. The angular correlations for the mixed $2_2^+ \to 2_1^+$ transitions are shown in Fig. 1. From these we obtained mixing ratios for ^{190,192}Os that agree well with the values in the literature, but for ^{188}Os, the present result, $\delta(E2/M1) = +7.2 \pm 1.1$, has the opposite sign of that reported previously[2]. The importance of this change in sign is discussed in Sect. 3.

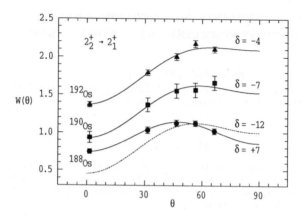

Fig. 1. Measured (data points) and fitted (solid lines) angular correlations for the $2_2^+ \to 2_1^+$ transitions in ^{188,190,192}Os, following Coulomb excitation by 220 MeV ^{58}Ni beams. For clarity the angular correlations for the different isotopes are offset from each other. The dashed line shows the angular correlation implied by the previously published mixing ratio for ^{188}Os.

2.2. Transient Field Measurement of $g(2_1^+)$ in ^{190}Pt

The g-factor of the first 2^+ state in ^{190}Pt has been measured by the TF technique. A platinum foil 2.2 mg/cm^2 thick, enriched to 4.19% in ^{190}Pt (natural abundance 0.01%), was pressed onto an annealed, Cu backed Gd foil 3.85 mg/cm^2 thick using a flashing of In (0.2 mg/cm^2) as adhesive. The 2_1^+ states of the even Pt isotopes present in the target (2.02% ^{192}Pt, 49.86% ^{194}Pt, 14.45% ^{196}Pt, 3.18% ^{198}Pt, 26.30% ^{195}Pt) were excited using 160 MeV ^{58}Ni beams from the ANU Pelletron accelerator. Angular correlations and TF precessions were measured. Full details will appear elsewhere[3]. Results are summarized in Table 1. The present relative g-factors were normalized to the weighted average of previously reported values for ^{194}Pt and ^{196}Pt.

Table 1. TF precession angles and $g(2_1^+)$ values in ^{190,192,194,196,198}Pt.

Nucleus	$\Delta\theta_{TF}$ [mrad]	$g/g(^{194}\text{Pt})$	$g(2_1^+)$		
			previous[a]	present	adopted
^{190}Pt	−40.6(1.1)	0.96(3)	-	0.287(12)	0.287(12)
^{192}Pt	−42.2(2.0)	1.00(5)	0.296(12)	0.298(17)	0.297(10)
^{194}Pt	−42.2(0.8)	1.00	0.300(13)	[0.300(13)][b]	0.300(13)
^{196}Pt	−42.3(1.0)	1.00(3)	0.297(10)	[0.297(10)][b]	0.297(10)
^{198}Pt	−44.8(1.8)	1.06(5)	0.294(19)	0.316(16)	0.307(12)

[a] Weighted average values from Refs.[4,5,6,7], and references therein.
[b] Calibration values.

2.3. Implantation-Decay Measurements of $g(2_1^+)$ in $^{184,186,188}Pt$

As the measurement of magnetic moments of short-lived excited states in neutron deficient, unstable nuclei is an experimental challenge, we have employed a novel implantation-decay technique to extend the $g(2_1^+)$ systematics for even platinum isotopes to include ^{184}Pt, ^{186}Pt and ^{188}Pt. A detailed description of the technique and results is being prepared for publication[8]. Briefly, heavy ion reactions were used to recoil-implant iron hosts with nuclei that β-decay to the nuclei of interest. This irradiation phase took place at some distance from the ANU γ ray detector array CAESAR[9]. After an appropriate irradiation period, the implanted Fe foil was transported to the centre of the γ-ray detector array (using a "rabbit"); it was polarized perpendicular to the plane of the detectors, and the perturbed γ-γ angular correlations measured. Irradiation and counting cycles were repeated until sufficient counts were accumulated.

Table 2. Measured precession angles and derived g-factor ratios for ^{184,186,188,192}Pt.

Nucleus	Reaction	Energy (MeV)	$\omega_L\tau^a$ (mrad)	$\tau(2_1^+)$ (ps)	$g(2_1^+)/g(192)$
^{184}Pt	^{165}Ho(^{24}Mg,5n)^{184}Au(β^+)	128	896(65)	542(20)	0.99(8)
^{186}Pt	^{173}Yb(^{19}F,6n)^{186}Au(β^+)	112	585(40)	375(14)	0.94(7)
^{188}Pt	^{169}Tm(^{24}Mg,5n)^{188}Tl(β^+)	128	162(13)	91(8)	1.07(13)
^{192}Pt	^{181}Ta(^{16}O,5n)^{192}Tl(β^+)	100	106(8)	64(1)	1.00(8)

^{a} Average precession measured from the perturbation of the $0_2^+ \to 2_1^+ \to 0_1^+$ cascade and either the $2_2^+ \to 2_1^+ \to 0_1^+$ or the $4_1^+ \to 2_1^+ \to 0_1^+$ cascade.

Results for the 2_1^+ state g-factors in ^{184,186,188}Pt, relative to that of the first excited state in ^{192}Pt, are summarized in Table 2. The magnetic moment data for $^{184-198}$Pt now span most of the upper half of the valence neutron shell. As can be seen from Tables 1 and 2, the g-factors of the 2_1^+ states in the Pt isotopes remain remarkably constant.

3. F-spin mixing and $M1$ properties of even-even transitional nuclei

There is a long-standing problem concerning the g-factors of the 2_1^+ and 2_2^+ states in the isotopes of in Os and Pt, as is indicated in Fig.2. The dotted lines show the IBM-2 predictions for the case of exact F-spin symmetry, namely

$$g(L) = (g_\pi N_\pi + g_\nu N_\nu)/N, \qquad (1)$$

where the boson g-factors have their bare values $g_\pi = 1$, $g_\nu = 0$. As g_π and g_ν are not expected to depart far from these values, the g-factors in the Os and Pt isotopes imply significant departures from F-spin symmetry. Specifically, to fit $g(2_1^+)$ for $^{190-198}$Pt requires $g_\pi \sim g_\nu \sim 0.3$, while for the Os isotopes, $g(2_1^+)$ is reasonably well reproduced with $g_\pi = 1$ and $g_\nu = 0$, but $g(2_2^+)$ would require the absurd values $g_\pi \sim -0.9$ and $g_\nu \sim 1$. The first attempt[1], to explain the $g(2_2^+)/g(2_1^+)$ variations by invoking F-spin mixing met with disaster. As can be seen in Fig. 2 (solid and dashed lines), with the hamiltonian of Bijker et al.[10], it appears that F-spin mixing makes the disparity between $g(2_2^+)$ and $g(2_1^+)$ worse.

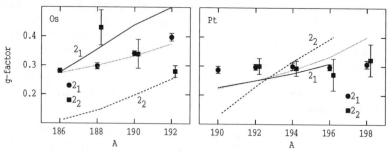

Fig. 2. Experimental g-factors and IBM-2 predictions in the F-spin symmetric limit (dotted), and with the hamiltonian of Bijker et al.[10] (solid: 2_1^+; dashed: 2_2^+); $g_\pi = 1$ and $g_\nu = 0$.

Important insights towards the solution of this problem have been obtained recently. Initially, the effects of F-spin mixing on g-factors were examined using analytic formulae derived in the intrinsic state formalism[11,12]. Then, these effects were investigated through exact diagonalizations of the IBM-2 hamiltonian using the code NPBOS[13]. We consider the hamiltonian

$$H = \epsilon_\pi \hat{n}_{d\pi} + \epsilon_\nu \hat{n}_{d\nu} + \kappa Q_\pi \cdot Q_\nu + \kappa_\pi Q_\pi \cdot Q_\pi + \kappa_\nu Q_\nu \cdot Q_\nu + \xi M, \qquad (2)$$

where $\hat{n}_{d\rho}$ are the d_ρ-boson number operators, M is the Majorana operator and Q_ρ are the quadrupole operators given by

$$Q_\rho = [d_\rho^\dagger s_\rho + s_\rho^\dagger \tilde{d}_\rho] + \chi_\rho [d_\rho^\dagger \tilde{d}_\rho]^{(2)}. \qquad (3)$$

It is convenient to define $\chi_S = (\chi_\pi + \chi_\nu)/2$ and $\chi_V = (\chi_\pi - \chi_\nu)/2$. The hamiltonian is F-spin symmetric when $\kappa = 2\kappa_\pi = 2\kappa_\nu$, $\chi_V = 0$ and $\Delta\epsilon = \epsilon_\pi - \epsilon_\nu = 0$.

As the mechanisms chosen to break F-spin symmetry are to some extent a matter of taste, we set $\kappa_\pi = \kappa_\nu = 0$ in the following calculations and study the effects of F-spin breaking in ^{190}Os by varying (i) χ_V, and (ii), $\Delta\epsilon$ (with $2\bar{\epsilon} = \epsilon_\pi + \epsilon_\nu$ fixed). From the results shown in Fig. 3 it is evident that: (a) $\Delta\epsilon$ breaking moves the g-factors of the states together producing very weak $M1$ transitions. (b) χ_V breaking produces sufficiently strong $M1$ transitions and differences between $g(2_2^+)$ and $g(2_1^+)$, with the condition that the sign of χ_V determines both the sign of the mixing ratio and the the sign of $g(2_2^+) - g(2_1^+)$. The import of the mixing ratio measurement described above is now clear. If $\delta(2_2^+ \to 2_1^+)$ in ^{188}Os were negative, as previously reported, there would be an inherent contradiction between the IBM-2 and the experimental $M1$ properties of ^{188}Os.

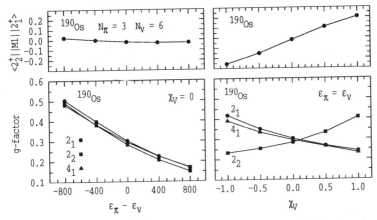

Fig. 3. Effects on $M1$ properties of ^{190}Os when χ_V and $\epsilon_\pi - \epsilon_\nu$ are varied. Fixed parameters are: $\epsilon_\pi + \epsilon_\nu = 900$ keV, $\kappa = 150$ keV, $\xi = 170$ keV and $\chi_S = -0.25$.

With this obstacle removed, a set of hamiltonian parameters was sought for $^{186-192}$Os and $^{190-198}$Pt, invoking both χ_V and $\Delta\epsilon$ to break F-spin and explain the g-factors. A comprehensive search for the best possible hamiltonian parameters was not made; rather the chosen parameters (which may not be unique) give a reasonable description of the energy spectra and electromagnetic properties. The resultant g-factors and M1 transition matrix elements are shown in Fig. 4. The F-spin purity of the low-excitation states ranges between $\sim 96\%$ and $\sim 98\%$. While the prediction of the $M1$ transition rates in the Os isotopes is not perfect, these may be improved by allowing g_π and g_ν to depart from their bare values.

Thus a solution has been found to the long-standing problem for collective theories posed by the g-factor variations in the transitional nuclei. One does not have to invoke single particle behaviour, assume spin contributions to the boson g-factors, or appeal to shape effects which are not supported by the measured E2 matrix elements. Rather, the solution is found in differences between the behaviour of the proton and neutron "fluids" and the interactions between them. While it is plausible that the protons and neutrons in these soft, transitional nuclei behave differently, a more quantitative microscopic understanding would be very interesting.

4. Magnetic Moments and Shape Coexistence in Unstable Isotopes of Pt

The spectra of the neutron-deficient Pt isotopes show features which indicate the coexistence of two sets of states based on different shapes. A proper treatment of these nuclei therefore requires configuration mixing calculations, along the lines of those performed for the mercury isotopes by Barfield et al.[14] In the absence of such calculations, the importance of shape coexistence for the g-factors will be demonstrated through a simplified semi-empirical approach.

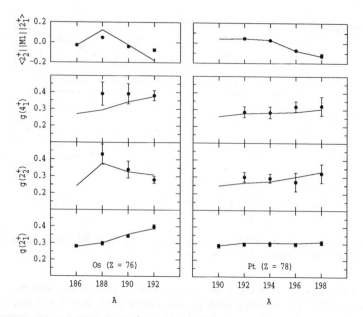

Fig. 4. Experimental g-factors and $2_2^+ \to 2_1^+$ $M1$ transition rates compared with the present calculations.

The prolate (deformed) ground-state bands in these nuclei are believed to be based on a configuration in which a pair of protons is excited across the $Z = 82$ shell gap, which, in effect, increases the proton boson number by two. Taking g-factors in the limit of F-spin symmetry as a first estimate, we have

$$g_{normal} \approx N_\pi/N \quad \text{and} \quad g_{deformed} \approx (N_\pi + 2)/(N + 2). \tag{4}$$

These estimates of the g-factors are compared with experiment in Fig. 5, along with an empirical estimate of the mixing, obtained from the observed excitation energies and band structures assuming a spin-independent interaction (see e.g. Ref.[15]).

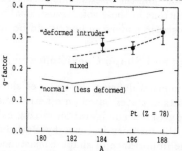

Fig. 5. g-factors in neutron-deficient Pt isotopes

Given the simplicity of the calculations, the agreement between theory and experiment is satisfactory. We have somewhat arbitrarily grouped ^{188}Pt with the neutron-deficient nuclei which exhibit shape coexistence, rather than with the oblate, stable nuclei considered in the previous section. It lies at the transition point. As the observed near constancy of $g(2_1^+)$ in the even isotopes from ^{184}Pt to ^{198}Pt appears to arise from an interplay of F-spin mixing and shape coexistence, it would be of considerable interest to model both processes through the shape transition and then seek a microscopic picture of the development and interaction between them.

5. Magnetic moments in ^{195}Pt and $U(6/12)$ supersymmetry

The nucleus ^{195}Pt, with its partner ^{194}Pt, has drawn considerable attention as an example of multi-j supersymmetry (see Refs.[16,17] for extensive references). As the supersymmetry schemes require very specific couplings between the fermion and boson states, and magnetic moments are sensitive to the occupation of the fermion orbits, they provide one of the most critical tests for multi-j supersymmetry.

We have studied ^{195}Pt experimentally and theoretically. Magnetic moments of several excited states in ^{195}Pt were measured using the transient field technique[16], and analytic expressions were derived for the magnetic moments in the $SO^{(BF)}(6) \times SU^{(F)}(2)$ limit of $U(6/12)$ multi-j supersymmetry[17]. In contrast with some properties, the magnetic moment formulae contain no adjustable parameters (aside from the usual quenching of the spin g-factor).

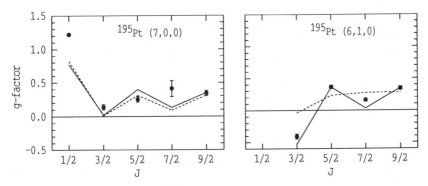

Fig. 6. Comparison of g-factors in ^{195}Pt for the stretched states in the $(N+1,0,0)$ and $(N,1,0)$ representations ($N = 6$). The solid lines are the supersymmetry model; broken lines show particle-rotor calculations. For details see Refs.[16,17].

Fig. 6 shows the experimental and theoretical g-factors for the stretched states $((\tau,0), L = 2\tau)$ in the $(N + 1,0,0)$ and $(N,1,0)$ representations, for which the data are complete up to spin $\frac{9}{2}^-$. Particle-rotor model calculations are shown for comparison. In view of the fact that there are no adjustable parameters in the supersymmetry model, the agreement with experiment is remarkable. Both models have difficulty explaining the g-factor of the ground-state, possibly because it has a larger $p_{3/2}$ contribution than predicted by either model[17,16]. Even so, the supersymmetry model provides a simple and successful description of the negative parity states in ^{195}Pt.

6. Conclusions

Magnetic moments provide important tests for the IBM and IBFM descriptions of the transitional nuclei. Recent advances include the description of the g-factor variations in the stable isotopes of Os and Pt through appropriate F-spin mixing; the measurement of $g(2_1^+)$ in the neutron-deficient Pt nuclei, which are sensitive to the boson number of the intruder configuration; and the demonstration that $^{194-195}$Pt provide one of the best examples of supersymmetry yet found in nature.

7. Acknowledgements

The assistance of my colleagues, particularly S.S. Anderssen, with the experiments, and the contributions of S. Kuyucak in the theoretical work are gratefully acknowledged. This work is supported in part by the Australian Research Council.

8. References

1. A. E. Stuchbery, et al., *Nucl. Phys.* **A435** (1985) 635.
2. B. Singh, *Nucl. Data Sheets* **59** (1990) 133.
3. S.S. Anderssen and A.E. Stuchbery, to be published.
4. A.E. Stuchbery, G.J. Lampard, and H.H. Bolotin, *Nucl. Phys.* **A528** (1991) 447.
5. F. Brandolini et al., *Nucl. Phys.* **A536** (1992) 366.
6. E. Bodenstedt et al., *Z. Phys.* **A342** (1992) 249.
7. R. Tanczyn et al., *Phys. Rev.* **C48** (1993) 140.
8. S.S. Anderssen, et al., to be published.
9. G.D. Dracoulis, to be published.
10. R. Bijker et al., *Nucl. Phys.* **A344** (1980) 207.
11. J.N. Ginocchio, W. Frank and P. von Brentano, *Nucl. Phys.* **A541** (1992) 211.
12. S. Kuyucak, in *Proc. 8th Int. Symp. on Capture γ-ray Spectroscopy, Fribourg, Switzerland*, ed. J. Kern, (World Scientific, Singapore, 1994), p. 251.
13. T. Otsuka and N. Yoshida, program NPBOS, report JAERI-M 85-094.
14. A.F. Barfield et al., *Z. Phys.* **A311** (1983) 205.
15. G.D. Dracoulis et al., *J. Phys. G: Nucl. Phys.* **12** (1986) L97.
16. G.J. Lampard, A.E. Stuchbery, H.H. Bolotin, and S. Kuyucak, *Nucl. Phys.* **A568** (1994) 617.
17. S. Kuyucak and A.E. Stuchbery, *Phys. Rev.* **C48** (1993) R13

G-FACTOR MEASUREMENTS IN SOME STABLE EVEN-EVEN HEAVY NUCLEI AS A SEVERE TEST OF IBM AND OTHER THEORETICAL MODELS

F. BRANDOLINI

Dipartimento di Fisica and Sezione INFN, Padova
35131 Padova, Italy

ABSTRACT

A review is given for Transient Field g-factor measurements with Gd done recently at LNL using Coulomb excitation induced by ^{58}Ni ions. For a direct comparison with IBM some stable even-even nuclei are discussed: ^{148,150}Sm, ^{156,158}Gd, ^{162,164}Dy, ^{164,166,168}Er, ^{192,194,196}Pt, ^{198,200,202}Hg.

1. Introduction

The measurement of excited state g factors is important in order to clarify the mode of nuclear excitation, in particular in even-even nuclei it can monitorize the accuracy of a collective description, since the g-factor value reflects the microscopical composition of the state. The most interesting aspect is the case of short lived states (0.1-100 ps), for measuring them hyperfine interaction techniques are mostly necessary. We used the Transient magnetic Field (TF) method, as it is, in many circumstances, superior to the other ones. The technique is well established in its main aspects both in Coulomb Excitation (C.E.) and in fusion reactions[1], however we have introduced some improvements[2], since the precision of experimental data is a crucial point.

The nuclei which we have studied with the C.E. TF technique are the stable ^{148,150}Sm, ^{156,158}Gd, ^{162,164}Dy, ^{164,166,168}Er, ^{192,194,196}Pt and ^{198,200,202}Hg isotopes. Sm isotopes are rather vibrational, Pt and Hg rather γ-soft and Gd, Dy, Er isotopes are well deformed. To describe these different aspects is a difficult task for a nuclear model.

2. Experimental procedure

The states on study were populated using Coulomb excitation induced by ^{58}Ni at the LNL XTU Tandem. Various energies in the range 117-225 MeV were employed in order to get selective populations of levels. The beam current was typically of 2-4 pnA. The target consisted in general of three main layers: the proper target, the ferromagnetic layer of Gd and a metallic backing. The foils were attached each other with a thin 0.2 mg/cm^2 layer of In. In the case of Hg, the compound HgS was evaporated with a special technique[3]. Gd has been used as ferromagnet since gives rise to about a two times bigger effect than Fe, but needs to be kept at liquid nitrogen temperature. The external field of about 0.04 T was periodically inverted every few minutes. The Gd annealing was done following ref.1 and the magnetization was checked with a double coil induction magnetometer. This was found important since the Gd foil preparation is critical in order to get the best magnetization. In some cases isotopically enriched Gd has been used.

Four Ge detectors were used located at ± 68 and ± 112 degrees or nearby angles, since quadrupole transitions have been mainly detected. The backscattered projectiles were detected with a 4 cm × 8 cm Parallel Plate Avalanche Counter, covering a 2 sr solid angle. The Gd thickness was chosen such that the average exit velocity was bigger than 2 v_0 in order to avoid any stopping in the ferromagnetic layer. A typical sensitivity was 150-200 mr/g. The effect has been deduced with the double ratio method and is related to the precession angle by the relation $\varepsilon = S \Delta \Theta$. The slope S was determined by rotating the Ge assembly of ±3 degrees. In order to avoid systematic errors when studying isotopic chains multiisotopic targets were used. The adherence between the layers was checked monitoring the γ-line shapes. In fact even a small detachment may result in a remarkable attenuation of the anisotropy. The experimental slopes were then systematically compared with those calculated with the code Coulex. A summary of used target is reported in table 1. In general each target corresponds to a different run. Due to shortage of beam time each run was of the order of one day.

Table 1. Targets used in the experiments

target ident.	isotopes (ratios)	target (mg/cm^2)	ferromagnet (mg/cm^2)		backing (mg/cm^2)		^{58}Ni energies (MeV)
Sm1	^{148,150}Sm (4:1)	1.9	5.3	Gd	6	Ag	217
Sm2		1.9	5.0	Gd	6	Ag	217
Gd1	^{156,158}Gd (1:1)	1.3	3.4	^{160}Gd	6	Ag	117
Gd2		1.2	2.5	Fe	6	Ag	130
Gd3		1.2	3.5	Fe	6	Ag	180
Gd4		1.2	4.7	^{160}Gd	25	Ag	217
Dy1	^{162}Dy	1.5	5.3	^{156}Gd	25	Ag	160,217
Dy2	^{164}Dy	1.5	5.5	^{156}Gd	20	Cu	217
Er1	^{166,168}Er (1:1)	1.5	5.5	^{160}Gd	25	Ag	210
Er2	^{166,168}Er (1:1)	1.5	5.7	^{160}Gd	25	Ag	225
Er3	^{164,166}Er (1:1)	1.5	5.3	^{160}Gd	25	Ag	160,210,225
Pt	^{192,194,196}Pt (1:1:1)	2.0	4.6	Gd	25	Ag	180,210
Hg1	^{199,200}Hg (1:1)	1.0	5.8	Gd	6	Ag	217
Hg2	^{198,200}Hg (1:1)	2.0	5.6	Gd	6	Ag	217
Hg3	^{200,202}Hg (1:1)	2.0	5.4	Gd	6	Ag	217
Hg4	^{198,200,202}Hg(1:1:1)	2.0	4.3	Gd	20	Cu	217
Hg5	^{198,200,202}Hg(1:1:2)	2.0	5.5	Gd	20	Cu	217

3. Experimental results

An example of accumulated spectra is shown in fig.1 for ^{164}Er and ^{166}Er nuclei. Spectra are very clean and the possibility of systematic errors is little. A summary of all the experimental results we have obtained in the nuclei on study is reported in table 2. In Gd the Chalk River (CR) parametrization[4] $B_{TF} = 27.5 v/v_0 Z \exp(-0.135 v/v_0)$ kT was employed. In Sm, Gd, Pt and Hg an internal calibration has been used, which shows that the parametrization is correct within 10 percent. Data were analysed with the code MAGMO[5], which provides the best values for g-factors, when giving as input the measured effects and all kinematic informations. Our data are compared with the most relevant references. For other references one should look at the compilation of ref. 6.

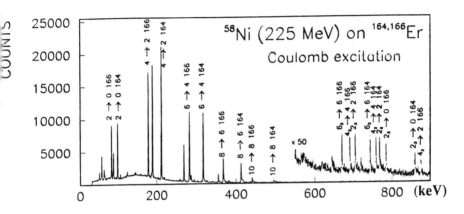

Fig.1. Typical coincidence γ-spectrum in the case of ^{164}Er and ^{166}Er measurement.

In general there is agreement between our data and those of Canberra, where the less efficient Fe was generally employed as ferromagnetic medium. Our data are however clearly more precise. For example for the 6$^+$ → 4$^+$ transition in ^{166}Er we get an effect of 0.0512(22) in one of the three runs, while they report 0.0142(11). The discrepancy in the case of ^{166}Er is mainly apparent, due to the large error bar of their values.

Table 2. Summary of experimental g-factors

nucleus g(2+)	I$^\pi$	lifetime (ps)	LNL[a]	Canberra[b]	Bonn[c]	others
^{148}Sm	2+	10.5	0.261(17)			0.301(33)[6]
	4+	3.3	0.247(29)			
^{150}Sm	2+	71.7	0.396	0.407(10)		0.385(27)
0.396(22)	4+	9.4	0.410(28)	0.35(5)		0.616(50)[19]
	6+	8.0	0.403(61)	0.38(8)		
^{156}Gd	2+	3200	0.386[d]			
0.386(4)	4+	160	0.415(25)	0.387(35)	0.327(19)	
	6+	23	0.401(37)	0.360(70)		
	8+	6.2	0.342(41)			
^{158}Gd	2+	3650	0.387(27)	0.420(100)		
0.381(14)	4+	230	0.395(28)	0.387(31)	0.409(15)	
	6+	23	0.406(37)	0.380(50)		
	8+	7.4	0.415(45)			

Table 2. Continued

^{162}Dy	4+	190	0.360(30)		
0.343(14)	6+	26.5	0.364(16)		
	8+	6.0	0.390(18)		
	10+	2.3	0.364(35)		
	2_γ^+	2.9	0.480(38)		
^{164}Dy	4+	288	0.310(20)	0.37(12)	
0.348(9)	6+	38	0.345(15)	0.28(8)	
	8+	10	0.292(20)	0.27(9)	
	10+	3.3	0.312(35)	0.35(13)	
	2_γ^+	6.6	0.382(30)	0.31(10)	
^{164}Er	4+	124	0.341(20)		
0.343(8)	6+		0.314(15)		
	8+	3.7	0.340(18)		
	10+	1.4	0.318(34)		
	2_γ^+	2.7	0.404(30)		
^{166}Er	4+	170	0.285(20)		0.315(16)
0.320(4)	6+		0.290(12)	0.267(32)e	0.258(11)
	8+	8.6	0.282(15)	0.236(46)e	0.262(47)
	10+	2.5	0.275(25)	0.203(74)e	
	2_γ^+	4.6	0.371(24)	0.280(50)e	
^{168}Er	4+	164	0.304(30)		
0.329(7)	6+		0.325(15)	0.328(46)	
	8+	4.9	0.301(16)	0.321(63)	
	10+	2.0	0.302(31)	0.322(80)	
	2_γ^+	4.9	0.387(30)	0.361(69)	
^{192}Pt	2+	62	0.318(17)		
	4+	6.0	0.281(30)		
	2_2^+	38	0.278(46)		
^{194}Pt	2+	60	0.295(10)	0.296(22)	
	4+	5.3	0.279(31)		
	2_2^+	50	0.281(55)		
^{196}Pt	2+	48	0.297(10)	0.294(23)	
	4+	6.1	0.345(40)	0.271(45)	
^{198}Hg	2+	33	0.380(30)		0.52(10)[20]
	4+	2.6	0.410(59)		
^{200}Hg	2+	67	0.326(26)		0.29(6)
	4+	4.7	0.254(43)		
^{202}Hg	2+	39	0.392(31)		0.44(9)
	4+	2.9	0.341(68)		

a) ^{148}Sm: ref 7. Gd and Er: ref. 8, 9. Pt: ref. 10. Hg: ref 11. In Er and Dy the values were obtained from C.R. parametrization and the quoted errors are statistic. b)^{150}Sm: ref.12. Gd: ref 13. Dy, Er: ref 14, 15. Pt: ref. 16. c) Gd: ref. 17. Er: ref. 18. d) Reference value[6]. e) Renormalized to the best data of the 4+ state.

4. Discussion

4.1. 2^+ g-factor in even-even stable heavy nuclei

This has been one of the first subjects extensively studied with the C. E. TF technique, because of the high anisotropy and cross section. The previous measurements of short lived states suffered of poor precision[6]. In fig. 2 our experimental data for Pt and Hg are shown togheter with the values known for W and Os. Pt isotopes have a very similar g factor of about 0.3. On the contrary Os and W stable isotopes show increasing g factors with N. Experimentally is known[6] that the Pt isotopes are oblate, while Os and W ones are prolate. These features were described with the Pairing plus Quadrupole Model by Kumar and Baranger[21] (KB), which also predicted the g factor trend, even if the transition prolate-oblate is somewhat anticipated in Os and results in an underestimation of the g-factor for high N.

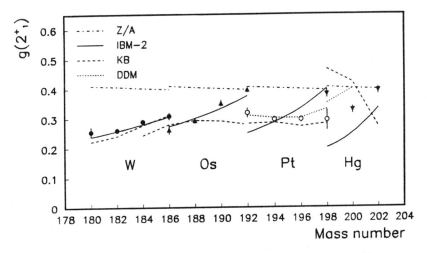

Fig. 2. Experimental and theoretical $g(2^+)$ values of stable even-even W, Os, Pt and Hg nuclides. KB: Kumar and Baranger[21], DDM: Dynamical Deformation Model[22].

This problem was reconsidered, as IBM provided the formula $g = (g_\pi N_\pi + g_\nu N_\nu)/(N_\pi + N_\nu)$, which seemed to reproduce the general features, when assuming $g_\pi \approx 1$ and $g_\nu \approx 0$ as suggested by the model[23]. Subshell effect were taken into account and estimated to contribute with $|\Delta g| \leq 0.1$. In this description the Pt isotopes were an evident exception and were then asked to be remeasured with better accuracy. In recent years two measurements were done: one at Canberra[16] and one at LNL[10]. In the first one ^{194,196,198}Pt were studied, using natural Pt, while we investigated ^{192,194,198}Pt, using enriched isotopes. The flat behaviour was confirmed with high accuracy. One may think that it is related to the oblate shape in Pt, however a similar situation of constancy occurs in the ^{164,166,168,170}Er isotopes and nearby nuclei, which are well prolate. According to Mössbauer measurements[6] $g(2^+)$ are nearly equal, and this has been checked again carefully by us along the gs band for the three first isotopes. The KB model works again well, apart for a normalization factor. Since IBM is in general quite successful, it has been tried to find an adjustment. Empirically, effective g_π and g_V were used, which are 0.3-0.4 for both proton and neutron bosons in Pt and Er isotopes[24]. Microscopic calculations have been made to calculate the effective values in Gd, Dy, Er, but the obtained values don't differ much from the standard values of 1 and 0. A different view has been to assume quenched effective values g_π =0.7 and g_V=0 and to attribute the deviation to the effect of partially closed shells, which gives rise to effective N_π and N_V[26]. Finally the quenching of the g_π was attributed possibly to F-spin impurity[27]. However microscopic predictions of effective N_π and N_V are so far not available for these nuclei. Such empirical approachs, using effective parameters, may appear not satisfactory and others more basic in term of mixed symmetry would be valuable.

The measurement of the Hg isotopes[14] has been the most difficult one among that selected here. In particular an absolute g-factor value of 0.326(36) has been obtained in ^{200}Hg, by comparison with ^{199}Hg. Also Hg isotopes have an oblate shape[6] and their g factors are rather similar, excepting that of ^{200}Hg which is somewhat smaller. ^{200}Hg has also a higher level density and is known as an exotic case. It has been studied with the DDM model[28] and it has been found to be rather vibrational. In the IBM language Hg isotopes are described with the O(6) symmetry as the Pt isotopes, while ^{200}Hg seems to be better described with a mixed U(5)-O(6) symmetry. This observation however doesn't help to predict the g factor. Because of computing limitations, shell model calculations could only be done for ^{204}Hg where the 2+ g factor was reasonably reproduced[29]

Concerning Sm our value of 0.261(17) for ^{148}Sm is somewhat smaller than the previous one[6] and much smaller than in ^{150}Sm. It has been discussed that at N<88, as is in the case of ^{148}Sm, the nucleus becomes quite spherical, and at the same time the closure at Z=64 becomes effective. The description of Sm and Nd isotopes was made empirically with effective N_π and N_V[26], but for example N_π = 0.9(4) was obtained for ^{144}Nd and not 2, as it would be predicted for a complete Z=64 closure. It can be observed that a complete shell closure is unrealistic and that microscopic calculations[30] predict a value higher than 2. An explanation of the smaller value in ^{148}Sm , based on shell model, would be a major role of the $\nu f_{7/2}$ orbital for N=84 and 86 , as shown by the fact that the first 2+ state is much lower in ^{148}Gd, roughly described with a $\nu f_{7/2}^2$ configuration, than in N=82 nuclei.

4. 2. *Higher levels in Sm, Pt, Hg isotopes*

The increased sensitivity allowed also to measure the g-factors of higher levels, which is particularly difficult in not much deformed nuclei because of the small cross section and anisotropy. Our interest was stimulated by the observed g factor disparity in ^{184,186}W and ^{188,192}Os , which could not be explained by IBM[31]. We therefore collected analogous information in Pt and Hg, but we didn't point out disparity with the first 2+ g factor.

Concerning the Pt isotopes, when included in the comparison with IBM2[31], the M1 transitions are poorly described. In Sm isotopes g(4) ≅ g(2) as one likely expects. Our data for the 4+ and 6+ states in ^{150}Sm are in good agreement with previous data[12], but in disagreement with a recently reported[19] g(4)/g(2) = 1.60(12).

4. 3. *Gd, Dy, Er isotopes. The gs band*

These isotopes are prototypes of deformed nuclei, where backbending has been first observed. This is attributed to the alignment of two $\nu(i_{13/2})$, which should affect the value of the g-factors. A second backbending is expected to occur at a somewhat higher energy as due to two $\pi(h_{11/2})$.

The first measurement of a g-factor after the first backbending was made in ^{158}Dy, where a greatly reduced value has been found[33], which has been explained with the contribution of the aligned neutrons. A similar measurement[34] was done in ^{156}Dy, followed by some others. One question which still deserves much attention, is to ascertain if the g-factor varies gradually or just at the backbending energies. This can be explored with good accuracy with C.E. TF and in fact has been done rather systematically at Chalk River about 10 years ago[4]. The result was that a smooth reduction of the order of 10%-20% at the 10$^+$ state was observed, in agreement with Cranked Model predictions[35]. However this has been put in discussion because of two works done at Bonn[17,18], which reported g(4)/g(2) =0.80(5) in ^{156}Gd and g(6)/g(2)= 0.816(35) in ^{166}Er. Therefore we have paid a special attention to a systematic study of Gd, Dy and Er nuclei. The results for the gs band confirm with much greater accuracy the Chalk River data. The situation seems to recall that of the Pt isotopes. As in that case, after some intermediary controversies, we are back at the original understanding. Our experimental data are plotted in fig. 3. They are not in contrast with a generally constant value. The displayed Dy and Er values are referred to the CR parametrization. This calibration causes a reduction (≤ 10%) in ^{164}Dy and Er isotopes and a small overestimation in ^{162}Dy.

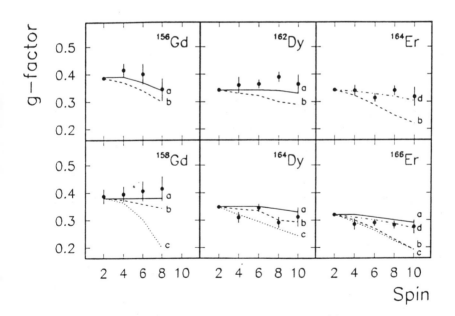

Fig. 3. Experimental and theoretical g-factors in ^{156,158}Gd, ^{162,164}Dy, ^{166,168}Er gs bands.
a) Ref. 35 (Cranked Model). b) Ref. 37. c) Ref. 36. d) Ref. 38. 2$^+$ g-factors in Dy and Er isotopes: ref.6.

Concerning the comparison with theoretical predictions, several HFBC calculations have been done, but they differ very much among them and are invoked selectively by experimentalists in other to justify their experimental data. For example calculations[36] of Ansari et al. could reproduce the attenuation found at Bonn for ^{166}Er, but predict attenuation in too many cases. Recent calculations, using the same approach as Ansari et al., predict remarkable attenuations[37]. However elsewhere[38] a gentle decrease consistent with our data was obtained. In the case of Dy even a slight increase was calculated[39]. Also IBM doesn't predict a sensible attenuation at high spins[40]. Constancy in the range of spins investigated by us is predicted by the Geometrical Collective Model[41] (GCM).
In our opinion the experimental situation is now clear, but still some contradictory signals continue. Very recently again a paper of Bonn has been published[42] which reports $g(6)/g(2)=0.57(11)$ in ^{158}Dy and which again cannot be explained theoretically even with Ansari curves. We note that C.E. TF measurements are very clean and that the possibility of systematic error when comparing excited states is small. This may be not the case in IMPAC measurement with static hyperfine field, which has been used in Bonn.

4.4 g-factor of the 2_γ^+ state in Dy and Er isotopes

The γ-band was reasonably populated in Dy and Er. The ratio $g(2_\gamma)/g(2_{gs})$ resulted to be generally noticeable bigger than 1: 1.25(13), 1.18(11) for ^{162,164}Dy and 1.27(10), 1.30(11), 1.25(11) for ^{164,166,168}Er. This finds a natural explanation in the frame of the spectroscopic properties of these nuclei. Experimentally[43] for the γ-band M1 transitions in these nuclei $|g_K-g_R| \approx 0.1$. If $g_K-g_R \approx 0.1$ from the formula $g=g_R+(g_K-g_R)K^2/I(I+1)$ one gets $g(2_\gamma)/g(2_{gs}) \approx 1.25$, in accordance with our data. The fact that these γ-bands are not entirely collective is indicated by the strong dependence on Z and N of their excitation energy, due to the particular particle-hole configurations. A value $g(2_\gamma)/g(2_{gs})= 1.06$ is predicted by the GCM model[42], which is bigger than 1, even if much smaller than observed. On the contrary a reduction of the g-factor is expected[44] by IBM-2 in contradiction with our data. In the frame of IBM the present measurement was considered of particular importance in order to establish the contribution of F-spin impurity in these nuclei[44], but it is not clear whether our data can be reconciled with the model.

5. Conclusions

A good precision has been achieved at LNL in C.E. TF g-factor measurements of short-lived states. Our data confirm that precise g-factor values can make a clear discrimination among theoretical models in even-even heavy nuclei.

Acknowledgements

Thanks are due : to all the coworkers in the measurements here reported (D. Bazzacco, M. De Poli, A. M. I. Haque, K. Löwenich, P. Pavan, R. V. Ribas, C. Rossi-Alvarez and M. Ionescu-Bujor), to A. Buscemi, S. Martini and R. Zanon for the help in the experiment preparation, to dr. R. Pengo and G. Manente for the assistance in the target preparation.

References

1. O.Häusser, Hyperfine Interactions 9 (1981) 19
2. D. Bazzacco et al., Phys. Rev. C 33 (1986) 1785
3. P. Pavan and F. Brandolini. LNL Annual Report 1988, 020/89, pg. 178.
4. O. Häusser et al. Nucl. Phys. A 412 (1984)141
5. R. V. Ribas, Nucl. Instr. and Meth in Phys. Res. 328 (1993) 553
6. P. Raghavan. Atomic Data and Nuclear Data Tables 42 (1989) 190
7. D. Bazzacco et al., Hyperfine Interactions 59 (1990) 133
8. D. Bazzacco et al., Hyperfine Interactions 59 (1990)125
9. F. Brandolini et al., Phys. Rev. C 45 (1992) 1549
10. F. Brandolini et al., Nucl. Phys. A 536 (1992) 366
11. D. Bazzacco et al., Hyperfine Interactions 59 (1990) 129 and to be published
12. A. P. Byrne et al., Nucl. Phys. A 466 (1987) 419
13. A. E. Stuchbery et al., Z.Phys. A 338 (1991) 135
14. C. E. Doran et al., Phys. Rev. C 40 (1989) 2035
15. C. E. Doran et al., Z. Phys. A 325 (1986) 285
16. A. E. Stuchbery, G.J. Lampard and H.H. Bolotin, Nucl. Phys. A 528 (1991) 447
17. A. Alzner et al., Z. Phys. A 322 (1985) 467
18. A. Alzner et al., Z. Phys. A 331 (1988) 277
19. T. Vass et al., Phys. Rev. C 48 (1993)2640
20. W.R.Kölbl et al., Nucl. Phys. A498(1986) 123
21. K. Kumar and M. Baranger, Nucl. Phys. A110 (1968) 490
22. Th. J. Köppel, A. Fässler and U. Gotz, Nucl. Phys. A 403 (1983) 263
23. M. Sambataro et al., Nucl. Phys. A 423 (1984) 333
24. A.Wolf, R.F.Casten, D. D. Warner. Phys. Lett. B 190(1987) 19
25. P.O. Lipas et al., Nucl. Phys.A 509 (1990) 509
26. A.Wolf, R.F.Casten, Phys. Rev. C 36 (1987) 851
27. R. F. Casten and A. Wolf, Phys. Lett. B 312 (1993) 372
28. S. T. Ahmad et al. J. Phys, G 15 (1989) 93
29. E. Maglione, private communication
30. O. Scholten. Phys. Lett. B 127 (1983) 144
31. A. E. Stuchbery et al., Nucl. Phys. A 435 (1985) 635
32. Yu Xin Liu, Gui Lu Long, Hong Zhou Sim, J. Phys. G 17 (1991) 877
33. G. Seiler-Clark et al., Nucl. Phys. A 399 (1983) 211
34. P. Taras et al., Nucl. Phys. A 345 (1985) 294
35. Y.S. Chen and S. Frauendorf, Nucl. Phys. A 393 (1983) 135
36. A. Ansari, E. Wüst and K. Mühlhans, Nucl. Phys. A 415 (1984) 215
37. M. Saha and S. Sen, Nucl. Phys. A 552 (1993) 37
38. K. Sugawara-Tanabe and K.Tanabe, Phys.Lett. 207B (1988)1 243
39. M. I. Cescato et al., Nucl. Phys. A 533 (1991) 455
40. L. D. Wood, I. Morrison, J. Phys. G 11 (1987) 315
41. D. Troltenier, J.A. Mahrun, W. Greiner, Z. Phys. A 348 (1994) 1
42. I. Alfter et al., Z. Phys. A 345 (1993) 273
43. K.Schreckenbach and W. Gelletly, Phys. Lett. 94B (1980) 298
44. J. N. Ginocchio et al., Nucl. Phys. A 541 (1992) 211

CONSISTENT TREATMENT OF M1 OBSERVABLES WITHIN THE IBA

A. WOLF

Nuclear Research Center Negev, Beer-Sheva 84190, Israel

ABSTRACT

It is shown that under certain assumptions, a consistent description can be obtained for both magnetic moments and M1 transition rates within IBA-2, using the same boson g-factors. The relation of this description to F-spin purity is discussed. A procedure is presented by which the F-spin purity of 2_1^+ states in even-even nuclei can be estimated from the experimental values of the magnetic moments.

1. Introduction

One of the main goals of the Interacting Boson Approximation is to provide a realistic description of low-energy collectivity in nuclei. The importance of pairing and the fact that quadrupole deformation is one of the main features of a considerable number of nuclei that are not near closed shells, are the main reasons for introducing the s and d bosons in the standard versions of the IBA. This is why this model is especially suited to describe E2 properties of nuclei. The interest in the study of M1 properties, stimulated by the discovery of the strong M1 transitions to the 1^+ states around $E_x = 3$ MeV, and by the fact that many M1 transitions between low-lying levels have collective character [1], has prompted investigations of M1 observables using the IBA. However, in IBA-1 M1 transitions are actually forbidden in first order, due to the fact that the M1 operator is proportional to the total angular momentum. Only introduction of higher order terms to the M1 operator allows one to calculate M1 transition rates in IBA-1. On the other hand, in the proton-neutron version of the IBA the M1 operator is no longer proportional to the total angular momentum, and therefore M1 transitions are allowed. Many detailed calculations of M1 properties using both IBA-1 (with higher order terms added to the M1 operator) and IBA-2 have been performed to date, and in general reasonable description of the experimental data has been obtained [2,3,4]. The basic parameters involved in the IBA-2 calculations are the usual IBA Hamiltonian parameters, and the boson g-factors, g_π, g_ν, which appear in the definition of the M1 operator. The "bare" values of g_π, g_ν, corresponding to the case in which only orbital contributions from the fermions are considered, are 1, 0 respectively. From the analysis of a considerable amount of experimental data, an inconsistency becomes evident as to the use of the boson g-factors: while for the description of magnetic moments values of g_π, g_ν considerably different from 1, 0 are needed, numerical calculations of M1 transition rates using the "bare" values yield reasonable agreement with experimental data. The purpose of this paper is to discuss the treatment of magnetic observables within IBA-2, and to show that under certain

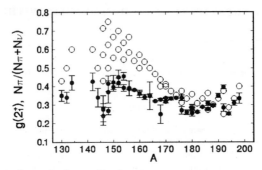

Fig. 1. Experimental g-factors (dots) and values calculated from eq. (2) (circles)

assumptions, by an appropriate choice of the Hamiltonian parameters, a consistent description of both magnetic moments and M1 transition rates can be obtained, using the "bare" values of the boson magnetic g-factors. In this approach, the need to use g_π, g_ν values different from 1, 0 for the description of the magnetic moments is explained as being related to the F-spin purity of the 2_1^+ states of the respective nuclei. Part of the results presented here have been published elsewhere [5,6].

2. Magnetic moments of 2_1^+ states in IBA-2

An important feature of algebraic models such as the IBA, is that they provide, under certain assumptions and for limiting cases, simple analytic formulas for many nuclear observables. In particular, in IBA-2 a simple relation for the g-factors of 2_1^+ states is obtained:

$$g(2_1^+) = (g_\pi N_\pi + g_\nu N_\nu)/(N_\pi + N_\nu) \qquad (1)$$

This relation is valid for any even-even nucleus, the only assumption being that its 2_1^+ state is completely symmetrical in the proton and neutron degrees of freedom, i.e., F-spin is a good quantum number and the respective state has maximal F-spin, $F = F_{max}$. N_π, N_ν are the numbers of proton, neutron bosons respectively. If we now use the "bare" values 1, 0 for the boson g-factors g_π, g_ν in eq. (1), we obtain :

$$g(2_1^+) = N_\pi/(N_\pi + N_\nu) \qquad (2)$$

Eq. (2) is very similar to the Z/A prediction of the geometrical model. However, it is significantly different from Z/A since it depends only on the valence number of particles, and therefore it predicts that in the first half of a shell the g-factors should increase as a function of Z and A, while in the second half of the shell, where the bosons are counted as holes, the g-factors should decrease when approaching the next major shell. From a systematic analysis of experimental g(2_1^+) values, Sambataro et al.[7] have shown that the data indeed follows such a trend, thus confirming the advantage of using eq. (2) over Z/A or other geometrical models. However, a more

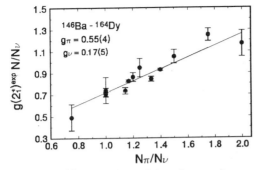

Fig. 2. Fit of eq. (1) to experimental data, for several rare-earth isotopes

detailed analysis of the experimental data shows that although eq. (2) correctly predicts the main trends of the data, significant deviations occur. For example, in Fig. 1 we present experimental g(2_1^+) values and the prediction of eq. (2) for a number of even-even isotopes with A=130-200. A possible way of understanding deviations of eq.(2) from experimental data such as those in Fig. 1, is to test the linear relationship in eq. (1) against the data. In Fig. 2 we present a fit of (N_π + N_ν)g(2_1^+) /N_ν vs. N_π /N_ν, for several rare-earth nuclei. The linear relationship is remarkably well obeyed by the data. The boson g-factors obtained from the fit, g_π = 0.55(4), g_ν = 0.17(5), are considerably different from the "bare" values. Similar fits were obtained for six different groups of even-even nuclei in the range A = 70 - 200 [5]. In all cases g_π, g_ν different from 1, 0 were obtained, and the sum g_π + g_ν was found to be approximately constant and equal to about 0.7. A systematic behavior was observed, namely, that g_π tends to increase and g_ν tends to decrease with A. The inherent assumption in this kind of analysis is that g_π, g_ν are constant for the isotopes considered in each group. Although a priori there is no theoretical reason for this assumption, the experimental data seems to confirm it quite well. It is interesting to note that microscopic calculations performed by Lipas et al. [8] for Sm, Gd, and Dy isotopes yielded g_π values in the range 0.87÷ 0.92, g_ν in the range -0.18 ÷ -0.11, and g_π + g_ν in the range 0.70 ÷ 0.81. Thus, there is some theoretical support for the constancy of g_π, to within about 5 %, although the calculated values are quite far from those obtained by the fit to the experimental data presented in Fig. 2, and, in fact, are close to the "bare" value of 1.

3. M1 Transition Rates

M1 transitions have been extensively studied both within IBA-1 and IBA-2 [2,3,4]. Since the neutron-proton degrees of freedom are essential for the description of magnetic properties, there is a clear advantage in using IBA-2 for the calculation of M1 transition rates. Moreover, M1 transitions between states with maximal F-spin vanish, and therefore, in order to describe such transitions between low-lying levels, one needs to use a Hamiltonian that has some amount of F-spin symmetry breaking. Detailed numerical calculations of various M1 matrix elements have been performed by Van Isacker et al., [4] using χ symmetry breaking, i.e., using $\chi_\pi \neq \chi_\nu$ for the parameter of the quadrupole operator in the IBA-2 Hamiltonian. They obtained reduced matrix elements in quite good agreement with experiment for many transitions in and between the γ and g.s. band in rare earth elements. For g_π, g_ν in the M1 operator they use the "bare" values 1, 0. In the next section we attempt to explain the apparent inconsistency regarding the use of g_π, g_ν in the treatment of magnetic moments and M1 transitions, and show that it can be resolved by a rather simple modification of the Hamiltonian.

4. Consistent description of magnetic moments and M1 transitions

The explanation we propose for the apparent inconsistency mentioned above is related to the fact that eq. (1) is valid only for pure F_{max} states. However, as mentioned in the previous section, some amount of F-spin symmetry breaking in the low-lying states is necessary in order to account for the existence of non-zero M1 transition rates. Therefore, it is possible that the inconsistency we want to explain is related to the F-spin purity of the low-lying states. A possible way to investigate this point is to make numerical calculations using an appropriate symmetry-breaking term in the IBA-2 Hamiltonian, and try to fit both the magnetic moments and the M1 transition rates. We performed such calculations using a modified version of NPBOS [9] for 17 even-even isotopes, namely: ^{110}Pd, ^{130,132}Ba, ^{146,148}Ce, ^{150}Nd, ^{152}Sm, ^{154}Gd, ^{168}Er, ^{184}W, ^{170}Yb, ^{180}Hf, ^{186,192}Os, and ^{192,194,196}Pt. Two types of Hamiltonians were used :

$$H_1 = n_\pi(\epsilon_d + \epsilon_{d\pi}) + n_\nu(\epsilon_d + \epsilon_{d\nu}) + \kappa Q_\pi \cdot Q_\nu + \lambda M \quad (3)$$

$$H_2 = n_\pi(\epsilon_d + \epsilon_{d\pi}) + n_\nu(\epsilon_d + \epsilon_{d\nu}) + \kappa (Q_\pi + Q_\nu)^2 + \lambda M \quad (4)$$

where

$$Q_\rho = (d_\rho^+ s_\rho + s_\rho^+ \tilde{d}_\rho)^{(2)} + \chi_\rho (d_\rho^+ \tilde{d}_\rho)^{(2)} \rho = \pi, \nu$$

and M is the Majorana operator [4]. In some cases an additional boson-boson interaction term was used in either H_1 or H_2. The Majorana parameter λ was taken as [10] $\lambda = 3.4\delta/\sqrt{(N_\pi N_\nu)}$, where δ is the mass deformation parameter. The results reported here were not found to be very sensitive to the particular choice of the Hamiltonian or its parameters. The values of $\epsilon_{d\pi}$, $\epsilon_{d\nu}$, are usually taken as $\epsilon_{d\pi} = \epsilon_{d\nu} = 0.0$. However, we found that the calculated values of $g(2_1^+)$ are quite sensitive to $\epsilon_{d\pi}$, $\epsilon_{d\nu}$, and used these parameters to fit the g-factors. In order to minimize the change in the calculated energy levels we used the condition:

$$N_\pi \epsilon_{d\pi} + N_\nu \epsilon_{d\nu} = 0 \quad (5)$$

Thus, only one of the parameters $\epsilon_{d\pi}$, $\epsilon_{d\nu}$ is free to vary, so we define $\epsilon_V \equiv \epsilon_{d\nu} - \epsilon_{d\pi}$ and use ϵ_V as a free parameter. In order to avoid negative values of the boson energies, we also required that $\epsilon_\rho \equiv \epsilon_d + \epsilon_{d\rho} \geq 0$, and in cases when the

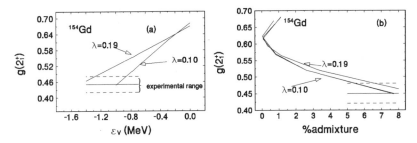

Fig. 3. Calculated g(2_1^+) vs. ϵ_V and vs. the % admixture of $F < F_{max}$ states into the 2_1^+ state for ^{154}Gd.

fits for the g-factors gave negative values for ϵ_ρ, the value of ϵ_d was modified in order to keep ϵ_ρ positive. For each isotope studied in this work, we began by using a set of parameters and a particular Hamiltonian that were previously shown to provide energy levels, E2, and M1 transition rates in reasonable agreement with the respective experimental values. Then we changed ϵ_V until the calculated g(2_1^+) was within the error of the experimental value. In Fig. 3a we show as an example the sensitivity of the calculated g(2_1^+) to ϵ_V for two different values of λ, for ^{154}Gd. We see that in both cases the experimental value can be fitted, albeit with different values of ϵ_V. In general, the variation of ϵ_V did not affect very much the energy levels and E2 transition rates. When needed, some "fine tuning" of the other parameters (in particular χ_ρ) was done so that the calculated observables were still in agreement with experiment. We found that a set of parameters could be found for each isotope, so that all observables, including g(2_1^+), are reasonably well predicted by the calculations. In Table 1 we show for example the results of applying this procedure to ^{154}Gd. We see that the calculated values of the M1 reduced matrix elements are not very much affected by the change in ϵ_V while g(2_1^+) is strongly changed when ϵ_V changes from 0.0 to -1.0 MeV. We conclude that the procedure presented here provides a consistent description of M1 observables, without affecting significantly the other results of the calculation. It is important to emphasize that ϵ_V is used here as a generic parameter in order to fit the magnetic moments, and therefore it does not have a special physical significance. However, we note that for most cases studied negative and quite large values of ϵ_V were needed to fit the data, meaning that in general $\epsilon_{d\pi} \geq \epsilon_{d\nu}$. This relation is consistent with the fact that proton 2^+ excitations generally lie higher than neutron 2^+ excitations. An inspection of the systematics of 2^+ energies illustrates this point: for nuclei with Z = 50 the values of E(2_1^+) are all around 1.2 MeV, while for isotopes with N = 82, the E(2_1^+) values are between 1.2 and 2 MeV.

Table 1. Calculated and experimental M1 reduced matrix elements and $g(2_1^+)$ (in μ_N) for ^{154}Gd, from refs. [4,6]

J_i	J_f	$\epsilon_V = 0.0$	$\epsilon_V = -1.0$ MeV	Experiment
2γ	2g	0.045	0.042	0.044(3)
3γ	2g	0.053	0.054	0.066(4)
3γ	4g	0.075	0.080	0.072(3)
4γ	4g	0.120	0.072	0.16(2)
3γ	2γ	0.152	0.110	0.20(5)
$g(2_1^+)$		0.670	0.517	0.45(3)

5. F-spin purity of 2_1^+ states

The use of $\epsilon_V \neq 0$ introduces separate proton and neutron boson energies and also has the effect of breaking the F-spin symmetry and introducing $F < F_{max}$ admixtures in the low-lying states. Since for any given ϵ_V in a particular calculation there corresponds a certain admixture of $F < F_{max}$ states in the 2_1^+ state, we can use the experimental values of $g(2_1^+)$ to investigate the F-spin purity of the respective 2_1^+ states. In order to do this, we used projections of the wave functions from the numerical calculations onto states with good F-spin ($F_{max}, F_{max} - 1$, etc.). The amplitudes of the projections were calculated using the modified form of NPBOS [9]. An example of this procedure is shown in Fig. 3b, where the calculated $g(2_1^+)$ is plotted against the % admixture of $F < F_{max}$ states into the 2_1^+ state. An important feature of Fig. 3 is that although for the two values of the Majorana parameter ($\lambda = 0.10, 0.19$) two completely different ϵ_V 's are obtained (Fig. 3a), the deduced admixtures are about the same (Fig. 3b). Thus, as mentioned above, the use of $\epsilon_V \neq 0.0$ is just a convenient way of breaking F-spin purity, and the particular values of this parameter do not necessarily have a quantitative meaning. Another example of the present procedure is shown in Fig. 4 for ^{186,192}Os. We found that the F-spin admixtures obtained from the experimental values of $g(2_1^+)$ are only weakly dependent on the particular choice of the Hamiltonian and therefore provide meaningful information about F-spin purity. In Fig. 5a we present the % admixtures of $F < F_{max}$ states into the 2_1^+ states, obtained using the same procedure for the isotopes studied in this work. It is important to emphasize that the values in Fig. 5a should only be regarded as upper limits estimates, because the deviations of the experimental $g(2_1^+)$'s from eq. (2), attributed here to the F-spin purity of the respective states, could also be due to other mechanisms [7,8]. The results in Fig. 5a show a systematic pattern, with admixtures of up to about 10%. This pattern is consistent with the systematics presented in Fig. 1, where the largest deviations of $g(2_1^+)$ from eq. (2) occur around A = 150 and Z = 64. Moreover, a similar pattern has been observed [11] for the systematics of the ratio e_ν/e_π of neutron, proton boson effective charges (Fig. 5b). A correlation between large (close to unity) values of e_ν/e_π and F-spin admixtures has been predicted by Ginocchio and Kuyucak [12], and the results of this study provide evidence for such a correlation.

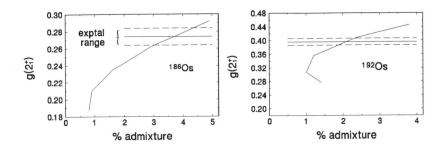

Fig. 4. Calculated $g(2_1^+)$ vs. the % admixture of $F < F_{max}$ states into the 2_1^+ state for ^{186,192}Os

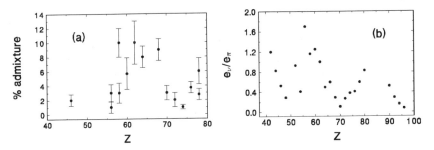

Fig. 5. (a) Upper limits for the % admixtures of $F < F_{max}$ states into the 2_1^+ state for the isotopes studied in this work; (b) The ratio of effective charges e_ν/e_π vs. Z, as determined from ref [11]

In conclusion, we have shown that by a particular choice of the IBA-2 parameters, a consistent description of both $g(2_1^+)$ and M1 transition rates can be obtained. The deviations of the effective boson g-factors g_π, g_ν from their "bare" values of 1, 0, observed previously [5], are explained here as being due to the F-spin impurity of the 2_1^+ states. Furthermore, estimates of upper limits for the F-spin admixtures of $F < F_{max}$ states in the 2_1^+ states were obtained for 17 isotopes, and found to vary up to about 10%. The observed correlation of these admixtures with the effective boson charges is of particular interest, and warrants further investigation.

6. Acknowledgements

Work partly supported by the US DOE under contracts DE-AC02-76CH00016 and DE-FG02-88ER40417. Thanks are due to Professor P. von Brentano for helpful discussions and for providing the modified version of NPBOS, and to I. Wiedenhover for help in implementing the code.

7. References

1. K. Schrekenbach and W. Gelletly, *Phys. Letters* **B190** (1980) 298.
2. P. O. Lipas, P. Toivonen, and E. Hammaren, *Nucl. Physics* **A469** (1987) 348.
3. A. E. L. Dieperink, O. Scholten, and W. W. Warner, *Nucl. Physics* **A469** (1987) 173.
4. P. Van Isacker et al., *Nucl. Physics* **A476** (1988) 301.
5. A. Wolf, R. F. Casten, and D. D. Warner, *Phys. Letters* **B190** (1987) 19.
6. A. Wolf, O. Scholten, and R. F. Casten, *Phys. Letters* **B312** (1993) 372.
7. M. Sambataro, O. Scholten, A. E. L. Dieperink, and G. Piccito, *Nucl. Physics* **A423** (1984) 333.
8. P. O. Lipas et al., *Nucl. Physics* **A509** (1990) 509.
9. H. Harter, unpublished.
10. H. Harter et al., *Phys. Letters* **B205** (1988) 174.
11. A. Wolf and R. F. Casten, *Phys. Rev.* **C46** (1992) 1323.
12. J. N. Ginocchio and S. Kuyucak, *Phys. Rev.* **C45** 867.

665

TRANSITIONAL ODD-ODD NUCLEI IN IBFM-2

A. Ventura and G. Maino
*ENEA, Dipartimento Innovazione, Settore Fisica Applicata, Bologna,
and INFN, Sezione di Firenze, Florence, Italy*

and

A. M. Bizzeti-Sona, P. Blasi and A. A. Stefanini
*Dipartimento di Fisica dell' Universita' di Firenze
and INFN, Sezione di Firenze, Florence, Italy*

ABSTRACT

Recent experimental data on high-spin states of odd-odd technetium isotopes are analysed in the frame of version 2 of the Interacting Boson-Fermion Model.

1. Introduction

Nuclei in the A = 100 mass region display a rapid transition from spherical to moderately deformed shapes, resulting in a variety of collective spectra of the even-even isotopes, accurately described in the Interacting Boson Model-2 (IBM-2) either as a U(5) → O(6) transition for the Ru chain[1], or as a U(5) → SU(3) transition for the Mo chain[2]. Energy levels and electromagnetic transitions of odd-mass Tc isotopes have been reproduced in the Interacting Boson-Fermion Model-2 (IBFM-2) either by assuming that the unpaired proton is a hole in a Ru core[3], below the Z = 50 shell closure, or a particle added to a Mo core[4], above the Z = 38 closure.

Scope of the present work is to extend the IBFM-2 analysis to the odd-odd isotopes ^{98}Tc and ^{100}Tc, with particular reference to recent measurements of high-spin states of negative parity. This preliminary analysis assumes the validity of the Z = 50 closure, but a comparison with results obtained with the Z = 38 closure is also planned for a more extended paper.

2. Experimental data on high-spin states of $^{98,100}Tc$

High-spin states in ^{98}Tc have been investigated via the reaction $^{94}Zr(^7Li, 3n)^{98}Tc$ by in-beam γ spectroscopy and conversion-electron measurements[5]. The ^7Li beam was provided by the 16 MV XTU Tandem accelerator of Laboratori Nazionali di Legnaro. The target consisted of a self-supporting foil of ^{94}Zr (618 μg/c m² thick and 91.2 % enriched). Gamma-ray excitation functions, γ − γ coincidences and $I(0°)/I(90°)$ angular anisotropies have been measured in the energy interval $E(^7Li) = 23 ÷ 31$ MeV.

Coincidence spectra have been recorded at $E(^7Li) = 26.5$ MeV, using three hyperpure Ge counters ($\simeq 18$ % efficiency) placed at $\theta = +90°, 0, -90°$ to the beam direction. Conversion electron measurements have been carried out with the solenoidal SPEL spectrometer[6] of L.N.L. More details on this part of the experiment are given in ref.[7].

From coincidence measurements, the level scheme of ref[5], partly reproduced in Fig. 1(a), has been obtained. Spin and parity have been assigned to the levels mainly on the basis of γ-ray angular anisotropies and electron conversion coefficients. High-spin states have been observed up to $J = 14$ and $E^x = 3200$ KeV. In particular, a negative parity band with $\Delta J = 1$ transitions, based on the $J^\pi = 8^-$, 1091 KeV state, has been identified. This band, analysed in the frame of the two-quasiparticle-plus-rotor model[5], has been associated to the configuration $(\pi g_{9/2} \otimes \nu h_{11/2})$, with the $g_{9/2}$ proton strongly coupled to the rotating core and the $h_{11/2}$ neutron aligned to the rotation.

A similar experiment[8] has been carried out to investigate the yrast states of ^{100}Tc by means of the reaction $^{96}Zr(^7Li, 3n)^{100}Tc$. Also in this case, γ-ray excitation functions, $\gamma - \gamma$ coincidences, angular asymmetries and electron conversion measurements have been performed. $\gamma - \gamma$ coincidences have been recorded at $E(^7Li) = 27$ MeV, using the MIPAD array consisting of six Compton suppressed Ge detectors, placed in the horizontal plane at angles $\pm 40°, \pm 90°, \pm 140°$ to the beam direction. The linear polarization, P_γ, of several γ transitions has also been investigated with a four-sector Ge polarimeter, placed in the horizontal plane of the MIPAD array at 90° to the beam direction. In addition, half-life measurements in the ns region have been carried out with the ^7Li beam pulsed at 5 MHz. In these experiments the target was a 2.5 mg/cm² self-supporting foil of enriched ^{96}Zr, apart from the conversion electron measurements, where a thin target (815 μg/cm²) has been used.

The level scheme obtained, with spin and parity assignments, is reported in part in Fig. 1(b). It is very similar to that of ^{98}Tc. In fact, like in that case, the lower part of the scheme shows a multiplet of positive parity states, with energies within 460 KeV and $J \leq 7$, which has been interpreted as belonging to the $(\pi g_{9/2} \otimes \nu d_{5/2})$ and $(\pi g_{9/2} \otimes \nu g_{7/2})$ configurations. Moreover, at high excitation energy, a negative parity band is present, which extends from $J^\pi = 6^-$ to $J^\pi = 14^-$ and to $E^x = 2693$ keV.

The negative parity bands observed in $^{98,100}Tc$ are very similar to those observed in the N = 55 and N = 57 isotones $^{100,102}Rh$ and have been interpreted[8] in terms of collective excitations of the ^{100}Ru core, based on a two-quasiparticle state (8^-) with configuration $\pi g_{9/2} \otimes \nu h_{11/2}$.

3. The Model

Odd-odd nuclei in the A = 100 mass region have already been treated in the interacting boson-fermion-fermion model[9] (IBFFM), where the unpaired proton and neutron are coupled to an IBM-1 core through the usual boson-fermion interaction and to each other by a residual interaction.

Our approach is close to the IBFFM, but replaces version 1 of IBM and IBFM with version 2, being based on the total Hamiltonian

$$H = H_B(\pi, \nu) + H_F(n) + H_F(p) + H_{BF}(\pi, n) + H_{BF}(\nu, p) + H_{FF}(n,p) \ . \tag{1}$$

Here, $H_B(\pi, \nu)$ is the standard Talmi Hamiltonian[10]

$$H_B(\pi, \nu) = \varepsilon_\nu \hat{n}_{d\nu} + \varepsilon_\pi \hat{n}_{d\pi} + k\hat{Q}_\rho(\chi_\nu) \cdot \hat{Q}_\pi(\chi_\pi) + \lambda \hat{M}_{\pi\nu} + \hat{V}_{\nu\nu} + \hat{V}_{\pi\pi}, \qquad (2)$$

where the boson quadrupole operators, $\hat{Q}_\rho(\chi_\rho)$, the Majorana operator $\hat{M}_{\pi\nu}$, and the residual boson interactions $\hat{V}_{\rho\rho}$ have the usual definition[10].

The free-fermion term, $H_F(k)$, with $k = n$, or p, reads

$$H_F(k) = \sum_j E_j(k) a_j^+(k) \cdot \tilde{a}_j(k), \qquad (3)$$

where $a_{j,m}^+(k)$ creates a nucleon of kind k in the j,m orbital, while the modified annihilator is $\tilde{a}_{j,m}(k) = (-1)^{j-m} a_{j,-m}$ and E_j is the corresponding quasiparticle energy.

The boson-fermion interaction is limited, for the time being, to bosons of one type with nucleons of the other type and is the sum of the standard terms, quadrupole-quadrupole, monopole-monopole and exchange[11]; for instance, in the case of an odd proton:

$$H_{BF}(\nu, p) = \Gamma_\nu \hat{Q}_\nu \cdot \hat{q}_p + A_\nu \hat{n}_{d\nu} \sum_{jp} \hat{n}_{jp} + \Lambda_\nu \hat{F}_{\nu p}. \qquad (4)$$

Here, $\hat{q}_p$ is the single-particle quadrupole operator of the odd proton

$$\hat{q}_p = \sum_{j,j'} \left[(u_j u_{j'} - v_j v_{j'}) Q_{j,j'} (a_j^+ \times \tilde{a}_{j'})^{(2)} \right]_p, \qquad (5)$$

where $Q_{jj'} = <l\,1/2\,j\,||\,Y^{(2)}\,||\,l'\,1/2\,j'>$ and u_j and v_j are the usual BCS coefficients, with $u_j^2 + v_j^2 = 1$.

The exchange term differs from the IBFM-1 definition and is given by

$$\hat{F}_{\pi\nu} = -(s_\nu^+ \times \tilde{d}_\nu)^{(2)} \cdot \{ \sum_{j,j',j''} \left[\frac{10}{N_\pi (2j+1)} \right]^{1/2} Q_{j,j'} \beta_{j'',j} (u_j v_{j'} + v_j u_{j'}) \cdot$$

$$:[(d^+ \times \tilde{a}_{j''})^{(j)} \times (a_{j'}^+ \times \tilde{s})^{(j')}]^{(2)}:\}_\pi + h.c., \qquad (6)$$

where : : means normal ordering, and

$$\beta_{j,j'} = \frac{1}{E_j + E_{j'} - \hbar\omega} (u_j v_{j'} + v_j u_{j'}) Q_{j,j'} \qquad (7)$$

Here, $\hbar\omega$ is the energy of a $|D>$ pair relative to the $|S>$ pair, obtained from the energy of the 2_1^+ state of a semimagic nucleus in the same mass region.

Finally, the residual proton-neutron interaction is the sum of a surface delta, with spin-spin coupling, a long-range quadrupole-quadrupole interaction and a short-range tensor interaction:

$$H_{FF}(n,p) = (\alpha_1 + \alpha_2 \vec{\sigma}_n \cdot \vec{\sigma}_p)\delta(\vec{r}_n - \vec{r}_p)\,\delta(r_p - R_0) + \alpha_3 r_n^2 r_p^2 Y^{(2)}(\Omega_n) \cdot Y^{(2)}(\Omega_p) \\ + \alpha_4 \frac{e^{-(r/r_0)^2}}{r^2}[3(\vec{\sigma}_n \cdot \vec{r})(\vec{\sigma}_p \cdot \vec{r}) - (\vec{\sigma}_p \cdot \vec{\sigma}_n)r^2]\,, \tag{8}$$

where $\vec{r} = \vec{r}_n - \vec{r}_p$.

As for electromagnetic transitions, the usual IBFM assumption is made, namely that the relevant operators are satisfactorily approximated by one-body terms only:

$$\hat{T}(L) = \hat{T}_B(L) + \hat{T}_p(L) + \hat{T}_n(L)\,, \tag{9}$$

where the boson operators are defined in ref.[10] and the fermion operators have the general structure:

$$\hat{T}_k(L) = \sum_{j \leq j'} \frac{e_{jj'}^{(L)}}{1+\delta_{jj'}} [(a_j^+ \times \tilde{a}_{j'})^{(L)} + (-1)^{j-j'}(a_{j'}^+ \times \tilde{a}_j)^{(L)}]\,, \quad (k = n, p) \tag{10}$$

where the single-particle matrix elements are, for $L = E2$:

$$e_{jj'}^{(2)} = -e_k \frac{1}{\sqrt{5}}(u_j u_{j'} - v_j v_{j'}) < l\,1/2\,j\,||\,r^2 Y^{(2)}\,||\,l'\,1/2\,j' > \tag{11}$$

and for $L = M1$:

$$e_{jj'}^{(1)} = -\frac{1}{\sqrt{3}} < l\,1/2\,j\,||\,g_l \vec{l} + g_s \vec{s}\,||\,l'\,1/2\,j' > \delta_{ll'}(u_j u_{j'} + v_j v_{j'})\,. \tag{12}$$

Analogous expressions hold for higher multipolarities.

4. Numerical Results

The IBM-2 Hamiltonian (2) is fully diagonalized in a spherical $U_\pi(5) \otimes U_\nu(5)$ basis, the five lowest levels for each value of the angular momentum of the core are subsequently stored and coupled to the odd nucleons. The complete IBFM-2 Hamiltonian (1) is thus diagonalized in a suitable spherical basis, whose fermion part contains up to five single-particle levels for each kind of nucleon.

Since, in the present calculation, protons are holes below the $Z = 50$ closure, and neutrons are particles above $N = 50$, the even-even cores of $^{98,100}Tc$ are $^{98,100}Ru$, with $N_\pi = 3$, $N_\nu = 2$ and 3, respectively. The relevant IBM-2 parameters are taken from ref.[1].

The single-particle energies of valence protons and neutrons have been evaluated by means of a spherical Woods-Saxon potential and the corresponding quasiparticle energies and occupation probabilities have been estimated in a standard BCS approximation. It has to be pointed out, however, that the ground states are not spherical, but exhibit a moderate quadrupole deformation ($\beta \simeq 0.2$). In a Nilsson diagram appropriate for this mass region[12], two branches of the neutron orbitals 2d5/2 and 1g7/2 cross at the above mentioned deformation; as a consequence, the ground state of the less deformed nucleus ^{98}Tc ($J^\pi = 6^+$) seems to belong to the ($\pi 1g9/2 \otimes \nu 2d5/2$) configuration, the more deformed ^{100}Tc ($J^\pi = 1^+$) to ($\pi 1g9/2 \otimes \nu 1g7/2$). That is why the corresponding quasiparticle energies and occupation probabilities were slightly modified with respect to the Woods-Saxon-BCS predictions; the adopted values are listed in table 1.

$\alpha \equiv (\rho nlj)$	Quasiparticle Energy (MeV)		Occupation Probability, v_α^2	
------	^{98}Tc	^{100}Tc	^{98}Tc	^{100}Tc
$\nu 2d5/2$	1.341	1.443	0.425	0.354
$\nu 1g7/2$	1.657	1.184	0.237	0.538
$\nu 1h11/2$	4.858	4.376	0.022	0.026
$\pi 1g9/2$	1.497	1.488	0.331	0.331
$\pi 2p1/2$	2.469	2.864	0.910	0.912
$\pi 2p3/2$	3.618	2.993	0.960	0.961

Table 1. Quasiparticle Energies and Occupation Probabilities.

The parameters of the boson-fermion interaction (4) for the odd proton, Γ_ν, Λ_ν, and A_ν have been adjusted to the low-energy spectra of both parities of ^{97}Tc in the case of ^{98}Tc, and ^{99}Tc in the case of ^{100}Tc. In the same way, the boson-fermion interaction parameters for the odd neutron, Γ_π, Λ_π and A_π, have been adjusted to the spectra of ^{99}Ru for ^{98}Tc and ^{101}Ru for ^{100}Tc. The IBFM calculations for the odd-mass Ru isotopes are described in detail in ref.[13]. Finally, the proton-neutron residual interaction is limited to a combination of a spin-spin delta plus a quadrupole-quadrupole interaction and the relevant parameters are listed in table 2.

ρ	Γ_ρ (MeV)		Λ_ρ (MeV)		A_ρ (MeV)	
---	^{98}Tc	^{100}Tc	^{98}Tc	^{100}Tc	^{98}Tc	^{100}Tc
ν	0.50	0.50	1.60	1.60	-0.40	-0.40
π	0.50	0.50	3.00	3.00	-0.45	-0.45
α_1 (MeV fm^4)		α_2 (MeV fm^4)		α_3 (MeV fm^{-4})		α_4 (MeV)	
^{98}Tc	^{100}Tc	^{98}Tc	^{100}Tc	^{98}Tc	^{100}Tc	^{98}Tc	^{100}Tc
0.0	0.0	-200.	-200.	-0.0003	-0.0002	0.0	0.0

Table 2. IBFM-2 Parameters.

Calculated energy levels of both parities are compared with experimental data in Fig. 1(a) for ^{98}Tc and in Fig. 1(b) for ^{100}Tc, with special emphasis on high-spin states of negative parity.

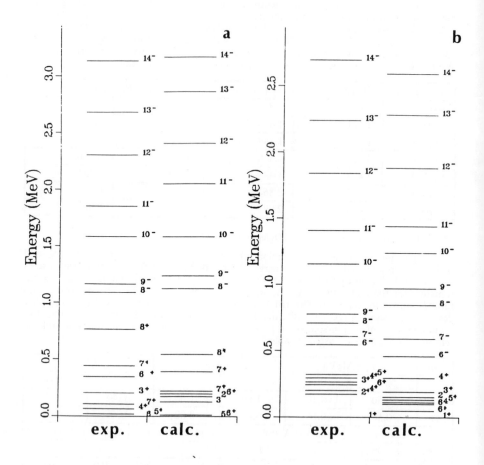

Fig.1. (a) Experimental[5] and calculated levels of ^{98}Tc; (b) experimental[8] and calculated levels of ^{100}Tc.

As far as the negative parity levels of $^{98,100}Tc$ are concerned, the present calculations confirm that the states of highest spin ($J^\pi = 13^-, 14^-$) are pure $(\pi 1g9/2 \otimes \nu 1h11/2)$ configurations coupled to low-spin states of the boson core of $^{98,100}Ru$. The states ranging from $J^\pi = 12^-$ to 6^- contain contributions from configurations where the proton occupies a negative parity orbital $\pi 2p1/2$, or $\pi 2p3/2$, while the odd neutron is found in the $\nu 2d5/2$, or in the $\nu 2g7/2$ orbital, coupled to core states of relatively high spin: the importance of these contributions increases with decreasing spin. The content of $(\pi 1g9/2 \otimes \nu 1h11/2)$ in the negative parity states depends on the quasiparticle energy of the $\nu 1h11/2$ orbital, assumed to be close to the spherical estimate in the present calculations; if one reduces it by about 1 MeV, to bring it to better agreement with the prediction[12] based on a Nilsson scheme at $\beta \simeq 0.2$, the $(\pi 1g9/2 \otimes \nu 1h11/2)$ configuration is largely dominant in all the negative parity states, with the exception of the lowest spins, $J^\pi = 6^-$ and 7^-.

The yrast states of positive parity are mainly built on two configurations, namely $(\pi 1g9/2 \otimes \nu 2d5/2)$ and $(\pi 1g9/2 \otimes \nu 1g7/2)$, coupled to the ruthenium core. The former contains the ground state of ^{98}Tc, with $J^\pi = 6^+$, almost degenerate with the first excited state, $J^\pi = 5^+$, which are inverted in the IBFM calculation. This inversion, occurring also in the analysis based on the two-quasiparticle-plus-rotor model[7], suggests the necessity of further improvements of the residual interaction. The latter configuration contains the ground state of ^{100}Tc, with $J^\pi = 1^+$, reproduced in the IBFM calculation, provided the $\nu 1g7/2$ quasiparticle state is lower in energy than the $\nu 2d5/2$ state, as a possible deformation effect. The agreement between experimental and calculated positive parity levels of both nuclei is poorer than in the case of negative parity and suggests deeper analysis of the fermionic part of the model.

5. References

1. P. Van Isacker and G. Puddu, *Nucl. Phys.* **A348** (1980) 125.
2. H. Dejbakhsh, D. Latypov, G. Ajupova and S. Shlomo, *Phys. Rev.* **C46** (1992) 2326.
3. J. M. Arias, C. E. Alonso and M. Lozano, *Nucl. Phys.* **A466** (1987) 295.
4. H. Dejbakhsh and S. Shlomo, *Phys. Rev.* **C48** (1993) 1695.
5. A. M. Bizzeti-Sona, P. Blasi, A. A. Stefanini and A. J. Kreiner, *Phys. Rev.* **C36** (1987) 2330.
6. G. Galeazzi, Annual Report L.N.L. (1985), p. 163.
7. A. M. Bizzeti-Sona, P. Blasi, A. A. Stefanini, G. Galeazzi and A. J. Kreiner, *Europhys. Lett.* **3** (1987) 163.
8. A. M. Bizzeti-Sona, P. Blasi, and A. A. Stefanini, Proc. of the 4th Int. Conf. on Nucleus-Nucleus Collisions, Kanazawa (Japan), June 10-14, 1991, p. 331.
9. W. Andrejtscheff, L. K. Kostov, P. Petkov, S. Brant, V. Paar, V. Lopac, G. Boehm, J. Eberth, R. Wirowski and K. O. Zell, *Nucl. Phys.* **A516** (1990) 157.
10. F. Iachello and A. Arima, *The Interacting Boson Model* (Cambridge University Press, 1987).
11. F. Iachello and P. Van Isacker, *The Interacting Boson-Fermion Model* (Cambridge University Press, 1991).
12. T. Seo, *Z. Phys.* **A324** (1986) 43.
13. G. Maino, A. Ventura, A. M. Bizzeti-Sona and P. Blasi, *Z. Phys.* **A340** (1991) 241.

STRUCTURE OF ODD-ODD Ga AND As NUCLEI, DYNAMICAL AND SUPERSYMMETRIES

T. FÉNYES, A. ALGORA, ZS. PODOLYÁK, D. SOHLER, J. TIMÁR
Institute of Nuclear Research, 4001 Debrecen, Hungary
and
V. PAAR, S. BRANT, Lj. ŠIMIČIĆ
Dep. of Phys.,Fac. of Sci., Univ. of Zagreb, Zagreb, 41000 Croatia

ABSTRACT

γ-ray, $\gamma\gamma$-coincidence and internal conversion electron spectra of the ^{68}Zn(p,nγ)^{68}Ga, ^{65}Cu(α,nγ)^{68}Ga, ^{66}Zn(p,nγ)^{66}Ga, ^{74}Ge(p,nγ)^{74}As, ^{73}Ge(p,nγ)^{73}As, ^{72}Ge(p,nγ)^{72}As and ^{70}Ge(p,nγ)^{70}As reactions were measured with Ge(HP) γ and combined superconducting magnetic lens plus Si(Li) electron spectrometers at different bombarding proton energies between 3.3 and 8.7 MeV and at 14.5 MeV α-particle energy. The $\sigma(E_{LEV},E_p)$ relative cross sections (with a subsequent Hauser-Feshbach analysis) and (in the case of ^{70}As product nucleus) γ-ray angular distribution spectra were also measured. Energies and relative intensities of about 810 (among them 440 new) γ-rays, as well as 250 (including 210 new) internal conversion coefficients were determined. The proposed new level schemes contain 260 (among them 70 new) levels. γ-ray branching and mixing ratios, level spin and parity values have been deduced.

In the framework of interacting boson(-fermion-fermion) model we have calculated the energy spectra, electromagnetic moments, reduced transition probabilities, γ-ray branching ratios, spectroscopic factors for 19 nuclei in the Ga-As region ($^{64-67}$Zn, $^{65-68}$Ga, $^{68-73}$Ge, $^{70-74}$As). The odd-odd nuclei have been described with consistent parametrization for the even-even core and for the neighbouring odd-A nuclei. Reasonable description of experimental data have been obtained. The energy spectra of ^{74}Se, ^{75}Se, ^{73}As and ^{74}As were calculated also using the formulae derived by Van Isacker and Jolie for the U(5) limit of the $U_\pi(6/12)xU_\nu(6/12)$ supersymmetry.

1. Introduction

The main intention of the present work was a detailed in-beam γ and electron spectroscopic study of the odd-odd Ga and As isotopes and a consistent description of the structure of Zn, Ga, Ge and As nuclei in the framework of the interacting boson (-fermion-fermion) model. The nuclei investigated in this program are shown in fig. 1.

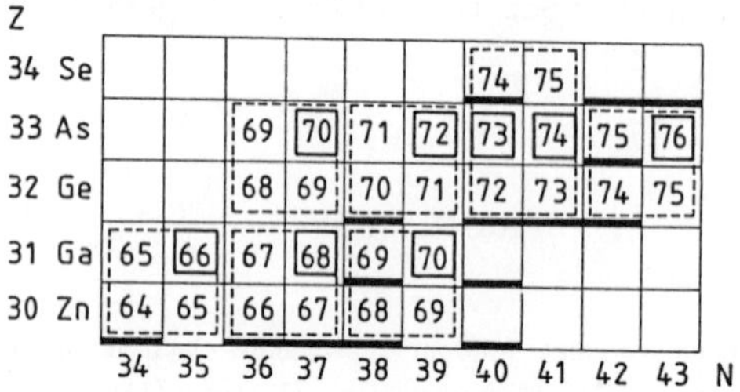

Z										
34 Se						74	75			
33 As		69	70	71	72	73	74	75	76	
32 Ge			68	69	70	71	72	73	74	75
31 Ga	65	66	67	68	69	70				
30 Zn	64	65	66	67	68	69				
	34	35	36	37	38	39	40	41	42	43 N

Fig. 1. Part of the chart of nuclides. Eight mass numbers in solid frames indicate experimental investigations. Mass numbers in dotted frames: theoretical calculations. Mass numbers underlined with thick lines: stable nuclei.

2. Experimental methods and results

We have studied the excited states of the Ga and As nuclei in the proton and α-particle beams of the Debrecen 103-cm (and in some experiments of the Jyväskylä 90-cm) isochronous cyclotrons via (p,nγ) and (α,nγ) reactions. The targets were prepared from enriched Cu, Zn, and Ge isotopes.

Different high resolution Ge(HP and LEPS) detectors were used for γ and a superconducting magnetic lens plus Si(Li) spectrometer for electron spectroscopic measurements. This latter provided unique possibility for in-beam electron spectroscopic studies (high transmission, good energy resolution, low background).

In order to obtain "complete" spectroscopic information, γ-ray (E_γ,I_γ), $\gamma\gamma$-coincidence and internal conversion electron spectra, as well as γ-ray angular distributions and relative cross sections $\sigma_{rel}(E_{LEV})$ were measured at different bombarding particle energies. Level schemes, spin and parity values, γ-branching and γ-mixing ratios have been deduced. Great attention was paid to the reliability and consistency of the obtained data. For example the level spins have been determined with three different methods: a) from internal conversion coefficients of transitions, b) from Hauser-Feshbach analysis, and c) from γ-ray angular distributions. Configuration of levels has been determined from all available data:

from nucleon transfer reactions, parabolic rule calculations, log ft values of the β-decay, electromagnetic moments, transition probabilities, γ-branching ratios, etc.

The results obtained up to now have been published in the following papers (or reports): ^{68}Ga [1,2], ^{66}Ga [3], ^{74}As [4], ^{72}As [5], ^{70}As [6]. We have studied ^{70}Ga [7] and ^{76}As [8] in our earlier works.

3. Interacting boson (-fermion-fermion) model calculations

The hamiltonian of the model is (refs. see in [3])

$$H_{IBFFM} = H_{IBFM}(p) + H_{IBFM}(n) - H_{IBM} + H_{RES}, \qquad (1)$$

where H_{IBM} denotes the IBM hamiltonian for the even-even core nucleus, $H_{IBFM}(p)$ and $H_{IBFM}(n)$ are the IBFM hamiltonians for the neighbouring odd-even nuclei, with an odd proton and odd neutron respectively. H_{RES} is the hamiltonian of the residual interaction.

The hamiltonian of the core has the following form

$$H_{IBM} = h_1 \hat{N} + h_2 \{(d^+d^+)_0 \sqrt{(N-\hat{N})(N-\hat{N}-1)} + h.c.\} +$$
$$+ h_3 \{(d^+d^+\tilde{d})_0 \sqrt{N-\hat{N}} + h.c.\} + \sum_{L=0,2,4} h_{4L} \{(d^+d^+)_L (\tilde{d}\tilde{d})_L\}_0,$$

where $\hat{N}$ is the d-boson number operator and N is the total number of s and d bosons.

The IBFM hamiltonian employed here is in the form:

$$H_{IBFM}(\alpha) = H_{IBM} + \sum_j \tilde{\epsilon}_j(\alpha) + H_{PVI}(\alpha),$$

where α stands for odd proton (α=π) or neutron (α=ν). The second and third terms are the quasiparticle and boson-fermion interaction hamiltonians, respectively.

$$H_{PVI}(\alpha) = \sum_j A_j [(d^+\tilde{d})_0 (c_j^+(\alpha)\tilde{c}_j(\alpha))_0]_0 + \sum_{j_1,j_2} \Gamma_{j_1,j_2} [Q_2(c_{j_1}^+(\alpha)\tilde{c}_{j_2}(\alpha))_2]_0 +$$
$$+ \sum_{j_1,j_2,j_3} \Lambda_{j_1,j_2,j_3} [(c_{j_1}^+(\alpha)\tilde{d})_{j_3} (\tilde{c}_{j_2}(\alpha)d^+)_{j_3}]_0;$$

with $A_j = A_0 \sqrt{5(2j+1)}$, $\Gamma_{j_1,j_2} = \Gamma_0 \sqrt{5} (u_{j_1}u_{j_2} - v_{j_1}v_{j_2}) <j_1\|Y_2\|j_2>$,
$\Lambda_{j_1,j_2,j_3} = -2\Lambda_0 \frac{\sqrt{5}}{\sqrt{2j_3+1}} (u_{j_1}v_{j_3} + v_{j_1}u_{j_3})(u_{j_3}v_{j_2} + v_{j_3}u_{j_2}) \cdot <j_3\|Y_2\|j_1> <j_3\|Y_2\|j_2>$,
$Q_{2\mu} = d_\mu^+ \sqrt{(N-\hat{N})} + \sqrt{(N-\hat{N})}\tilde{d}_\mu + \chi(d^+\tilde{d})_{2\mu}$.

The hamiltonian of the residual interaction:

$$H_{RES} = 4\pi V_\delta \delta(r_\pi - r_\nu) \delta(r_\pi - R_0) - \sqrt{3} V_{\sigma\sigma}(\boldsymbol{\sigma}_\pi \cdot \boldsymbol{\sigma}_\nu) + V_{TENS} \left[\frac{3(\boldsymbol{\sigma}_\pi \cdot \mathbf{r}_{\pi\nu})(\boldsymbol{\sigma}_\nu \cdot \mathbf{r}_{\pi\nu})}{r_{\pi\nu}^2} - (\boldsymbol{\sigma}_\pi \cdot \boldsymbol{\sigma}_\nu) \right],$$

where $r_{\pi\nu} = r_\pi - r_\nu$, $R_0 = 1.2 \sqrt[3]{A}$ fm.

Hamiltonian (1) was diagonalized in the proton-neutron-boson basis: $|(j_p j_n)I, NR; J>$, where j_p and j_n stand for proton and neutron angular moments coupled to I, N is the number of d bosons, R is their angular momentum, and J is the spin of the state. The computer codes, used in the calculations, were written by Brant, Paar and Vretenar [9].

First the $\{h_i\}$, χ and vibrational charge parameters were fitted to the energy spectrum, reduced transition probabilities and electromagnetic moments of the corresponding even-even nucleus. The quasiparticle energies and occupation probabilities were obtained mainly from BCS calculations with some modifications, to have better agreement with experimental level energies. The strength parameters of the boson-fermion interaction (A_o, Γ_o and Λ_o) and effective gyromagnetic ratios were determined mainly by fitting to the energy spectra and electromagnetic properties of the neighbouring odd-A nuclei. Finally the parameters of the residual interaction were fitted to the properties of the odd-odd nucleus. In this way a consistent description was obtained for the corresponding quartet of nuclei.

The experimental and calculated IB(FF)M energy spectra of ^{66}Ga, ^{68}Ga, ^{72}As, ^{74}As and neighbouring nuclei are compared in figs. 2-4, respectively.

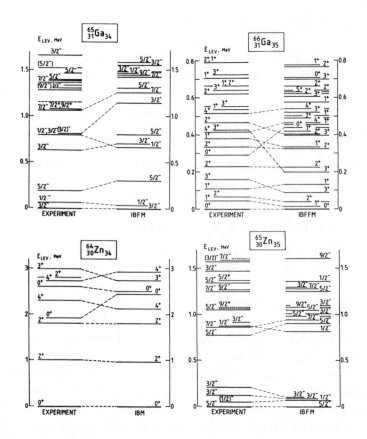

Fig. 2. Experimental and calculated IB(FF)M energy spectra of ^{64}Zn, ^{65}Zn, ^{65}Ga, and ^{66}Ga nuclei [3].

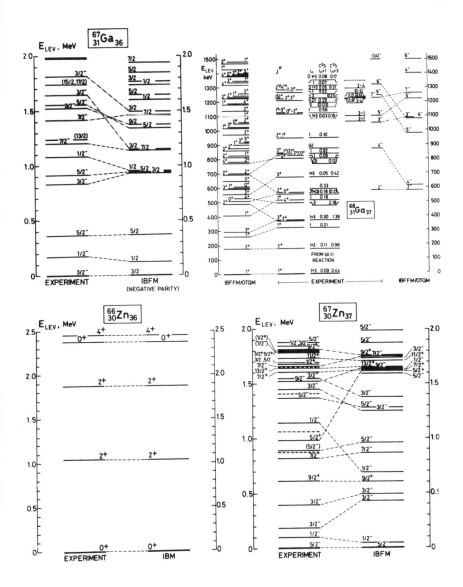

Fig. 3. Experimental and calculated IB(FF)M energy spectra of ^{66}Zn, ^{67}Zn, ^{67}Ga, and ^{68}Ga nuclei [2].

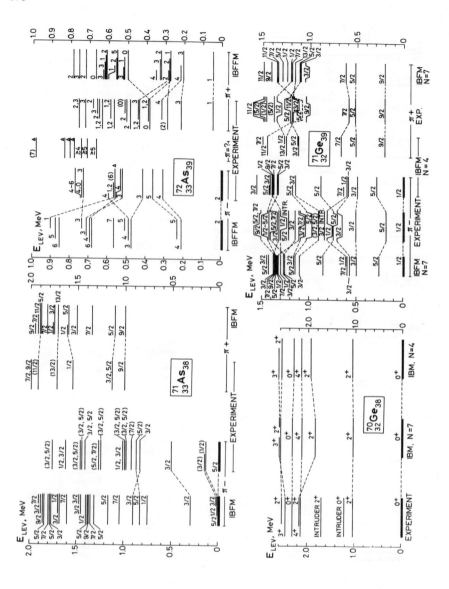

Fig. 4. Experimental and calculated IB(FF)M energy spectra of ^{70}Ge, ^{71}Ge, ^{71}As, and ^{72}As nuclei [5].

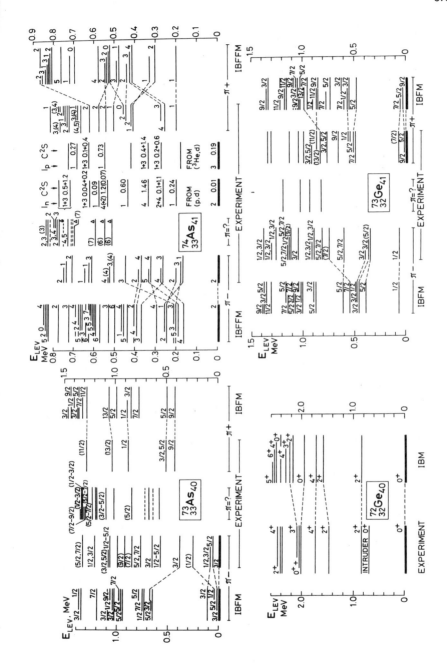

Fig. 5. Experimental and calculated IB(FF)M energy spectra of ^{72}Ge, ^{73}Ge, ^{73}As, and ^{74}As nuclei [4].

The calculated <u>wave functions</u>, B(E2) and B(M1) <u>reduced transition probabilities, γ-baranching ratios, magnetic dipole and electric quadrupole moments</u> of states, and <u>spectroscopic factors</u> are given and compared with experimental data in refs. [2,3,4,5] for ^{68}Ga, ^{66}Ga, ^{74}As, and ^{72}As and neighbouring 12 nuclei, respectively. For example in the case of the ^{64}Zn, ^{65}Zn, ^{65}Ga, and ^{66}Ga quartet using only ≲25 more or less freely fitted parameters about 400 nuclear data have been calculated and compared with available experimental data. With a few exceptions reasonable description of experimental data have been obtained for all investigated quartets. As the parameters usually vary slowly as we proceed from a quartet to the neighbouring one, the method allows consistent description of a larger group of nuclei. We remark that the present calculations gave the first theoretical description of the structure of ^{66}Ga and ^{68}Ga and the first IBFFM interpretation of ^{72}As and ^{74}As nuclei.

4. Supersymmetry calculations around ^{74}As

The more complete level scheme of ^{74}As, obtained in our work [4], offers a new possibility to check the predictions of the supersymmetric scheme. For the calculation of energy spectra we have used the (2,4a) and (2.6) formulae of ref. [10]. The formulae were derived by Van Isacker and Jolie for the U(5) limit of the $U_\pi(6/2) \times U_\nu(6/12)$ supersymmetry (proton-particle, neutron-hole case). The parameters of the formulae were fitted first to the levels of ^{74}Se, ^{75}Se, and ^{73}As by a least squares method. Then with the obtained $A_\pi + A_{\pi\nu} = 55$, $A_\nu + A_{\pi\nu} = 26$, $B_1 = 525$, $B_2 = 0$, $C = 4$, $D = -28$, a high negative value for E, and $F = 41$ (all in keV) parameters we have calculated the level spectrum of ^{74}As. The $A_{\pi\nu}$ parameter was taken zero, after testing its role in the generated ^{74}As spectra. The experimental and theoretical level spectra are compared in fig. 6. Although the generated spectrum does not reproduce the experimental ^{74}As levels perfectly (see differences in the 1^+ and 4^+ levels), the supersymmetry calculations describe 44 levels of four different nuclei reasonably well, using 7 fitted parameters only.

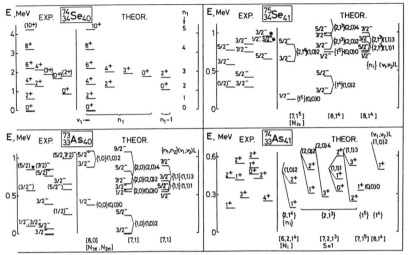

Fig. 6. Comparison of the experimental energy spectra of ^{74}Se, ^{75}Se, ^{73}As, and ^{74}As with the calculated ones. The levels are labelled with [N$_j$], [N$_{j\pi}$], [N$_{j\nu}$], {n$_i$}, (v$_1$, v$_2$), L, S, J, the irreducible representations of the groups U$_{\pi\nu}^{BF}$(6), U$_{\pi}^{BF}$(6), U$_{\nu}^{BF}$(6), U$_{\pi\nu}^{BF}$(5), O$_{\pi\nu}^{BF}$(5), O$_{\pi\nu}^{BF}$(3), SU$_{\pi\nu}^{BF}$(2) and Spin(3).

5. References

1. J. Timár, T. X. Quang, T. Fényes, Zs. Dombrádi, A. Krasznahorkay, J.Kumpulainen and R. Julin, *Nucl. Phys.* **A552** (1993)149.
2. J. Timár, T. X. Quang, Zs. Dombrádi, T. Fényes, A. Krasznahorkay, S. Brant, V. Paar and Lj. Šimičić, *Nucl. Phys.* **A552** (1993) 170.
3. J. Timár, T. X. Quang, T. Fényes, Zs. Dombrádi, A. Krasznahorkay, J. Kumpulainen, R. Julin, S. Brant, V. Paar, Lj. Šimičić, *Nucl. Phys.* **A...**(1994), in print.
4. A. Algora, D. Sohler, T. Fényes, Z. Gácsi, S. Brant, V. Paar, Lj. Šimičić, *ATOMKI Ann. Rep. 1993*, Debrecen, p. 16, and to be published.
5. D. Sohler, A. Algora, T. Fényes, Z. Gácsi, S. Brant, V. Paar, Lj. Šimičić, *ATOMKI Ann. Rep. 1993*, Debrecen, p. 14 and to be published.
6. Zs. Podolyák, T. Fényes and J. Timár, *ATOMKI Ann. Rep. 1993*, Debrecen, p. 12 and to be published.
7. T. Fényes, J. Gulyás, T. Kibédi, A. Krasznahorkay, J. Timár, S. Brant, and V. Paar, *Nucl. Phys.* **A419** (1984) 557.
8. Z. Gácsi, J. Gulyás, T. Kibédi, E. Koltay, A. Krasznahorkay, T. Fényes, *Izv. AN SSSR, ser. fiz.* **47** (1983) 45.
9. S. Brant, V. Paar and D. Vretenar, Computer code IBFFM/OTQM, *IKP Jülich (1985)*, unpublished.
10. P. Van Isacker and J. Jolie, *Nucl. Phys.* **A503** (1989) 429.

IBFFM CALCULATIONS ON ^{126}Cs

N. BLASI

Istituto Nazionale di Fisica Nucleare, sez. Milano, 20133 Milano, Italy

ABSTRACT

The extension of the IBA model to odd-odd nuclei is applied to the transitional nucleus ^{126}Cs. The parameteres are determined by fitting the neigbouring nuclei. Although ^{127}Cs and ^{125}Xe are only qualitatively reproduced, the results are rather satisfactory.

1. Introduction

Many interesting aspects trigger the study of odd-odd nuclei, both theoretically and experimentally, in spite of the difficulty of dealing with very complex spectra. In general, informations on the proton-neutron interaction V_{pn} can be deduced from the low-lying part of the spectra. At higher excitation energies, the interplay of collective and single-particle degrees of freedom plays an important role: several experimental studies in different mass regions have evidenced the presence of relatively high spin collective bands, although only in few cases the connection to the low-lying non-collective part of the spectrum could be observed. In the Cs region (A=120-130), the γ-softness of the even-even core nuclei, and the different driving shape effect of the coupled neutrons or protons triggered a number of recent experimental studies[1]. Collective bands have been observed, but in most cases spins and parities could be assigned only on the basis of crancking shell model calculations.

The extension of the IBA model to odd-odd nuclei has been applied to study the proton-neutron interaction V_{pn} by Brant et al.[2] for ^{134}Cs, and by Dombradi et al.[3] for several In and Sb isotopes. Since in this case the low-lying non collective levels of nuclei close to shell closure were investigated, the boson space was taken in an exact limit (O(6) or SU(5)) and truncated to two bosons. Collective bands in some well deformed odd-odd Os isotopes were described within the model by Chou et al.[4], in the SU(3) limit. One of the appealing features of the model is that it may be applied to nuclei from spherical to deformed or γ-unstable. Calculations in a transitional region, where other models are less appropriate, are of particular interest, also because in these regions spectra are more complex and experimental informations are more difficult to obtain. The possibility to fix the model parameters on the neighbouring even-even and odd-A nuclei may give a useful predictive power to the model, that could be used together with crancking shell model calculations for spin and parity assignments to collective bands in odd-odd nuclei. In order to investigate whether the IBFFA model can be used to predict collective energy levels of odd-odd nuclei, we performed calculations on the transitional nucleus ^{126}Cs.

2. Extension of the IBA model to odd-odd nuclei

Odd-odd nuclei can be thought as systems in which one proton and one neutron are coupled to a core. In general, the Hamiltonian can be written as

$$H = H_c + H_p + H_n + V_{cp} + V_{cn} + V_{pn}$$

The IBA model, and its extension to odd-A nuclei, has the appealing feature of a great flexibility in describing nuclei from spherical to deformed within the same framework, using a limited number of parameters. It is therefore interesting to extend the model to odd-odd systems, expecialy in transitional regions, were other models are less appropriate. Furthermore, since all the model parameters can be fixed on the neighbouring even-even and odd-A nuclei, we may hope on one hand to gain informations on the residual proton-neutron interaction, V_{pn}, and on the other to find a predictive power for nuclei where experimental informations are scarce.

Within the IBA model, the core is expressed in terms of interacting s and d bosons and the Hamiltonian is the usual IBA Hamiltonian[5] H_B. Each fermion interacts with the boson core according to the IBFA model, and the form of the Hamiltonian $H_{p(n)} + V_{cp(n)}$ is the same as in ref[6,7]. In order to perform real calculations, one needs to have enough experimental informations on the core and the two odd-A neighbouring nuclei. Among the Cesium isotopes, ^{126}Cs represents the best case, since the two odd-A neigbouring nuclei ^{127}Cs and ^{125}Xe were recently studied[8,9].

3. The parameters

The even-even Xe isotopes were studied by Puddu et al.[10] within the framework of the IBA-2 model, as an example of chain transition between the rotational (SU(3)) and γ-soft (O(6)) limits. We started from the IBA-2 parameters reported for the ^{126}Xe nucleus, projected the calculations into IBA-1[7], and sligtly adjusted some of the parameters in order to account for the space truncation (see table 1).

To the even-even core ^{126}Xe, we coupled separately one proton and one neutron hole, in order to reproduce ^{127}Cs and ^{125}Xe, respectively. For both nuclei, the unpaired particle moves in the 50-82 major shell, and the involved orbitals are: $g_{7/2}$, $d_{5/2}$, $s_{1/2}$, $d_{3/2}$ for the positive parity states, and $h_{11/2}$ for the negative parity ones.

In the case of ^{127}Cs, there are only five valence protons, which could be thought to occupy mainly the $g_{7/2}$ and the $d_{5/2}$ orbitals. Nevertheless, the experimental spectrum[8] shows the need of including all the orbitals in the calculations: in fact, three positive parity bands are observed at low excitation energy, of which one is interpreted as based mainly on the $d_{5/2}$ configuration, and the remaining two as based mainly on $g_{7/2}$, but the ground state $1/2^+$ and a low-lying $3/2^+$ not belonging to any known band, cannot be reproduced by the calculations without the explicit inclusion of the $s_{1/2}$ and $d_{3/2}$ orbitals. A negative parity decoupled band, based on $h_{11/2}$ configuration, is also observed at low excitation energy. In performing the calculations, the formulas derived by Scholten[7] were used. These formulas, which relate the parameters to the occupation number of each orbital, limit the number of parameters to three, but require an estimation of the quasi-particle energies ϵ_j and the occupation probabilities v_j^2. Usually, these are derived via a simple BCS calculation from single-particle energies taken from experimental data on double magic nuclei[7], or from various shell model calculations. The uncertainty of this procedure always leads to a readjustment in order to reproduce the data. Since we want to reduce as much as possible the number of free parameters in our calculations, and therefore to avoid an ad hoc riadjustment of the quantities ϵ_j and v_j^2, we have taken the single particle energies from a study[11] of the isotone ^{123}Sb via the reaction $^{122}Sn(^3He,d)^{123}Sb$. We then checked our choice by comparing the resulting quasi-particle energies and v_j^2 values with the experimental strength distributions[12] in ^{125}I. The quasi-particle energies and occupation probabilities obtained for ^{127}Cs were then kept fixed, and only the three remaining parameters A, Γ, Λ were varied. The resulting spectrum is shown in fig 1, and the final paremeters are listed in table 1. Note that although positive and negative parity states were calculated independently, the resulting parameters are quite similar.

Table 1. Model parameters used in the calculation of ^{126}Cs. χ and v^2 are dimensionless, all the others are in MeV.

	ϵ	κ	$c_{0\nu}$	$c_{2\nu}$	χ_ν	χ_ν
^{126}Xe	0.74	-0.155	0.1	-0.25	0.2	-0.8

	$\epsilon_{s_{1/2}}$	$\epsilon_{d_{3/2}}$	$\epsilon_{d_{5/2}}$	$\epsilon_{g_{7/2}}$	$v^2_{s_{1/2}}$	$v^2_{d_{3/2}}$	$v^2_{d_{5/2}}$	$v^2_{g_{7/2}}$	A	Γ	Λ
^{127}Cs	2.52	2.43	1.42	1.10	0.05	0.05	0.17	0.37	-0.35	0.65	2.20
^{125}Xe	1.08	1.25	1.70	1.65	0.39	0.31	0.78	0.82	-0.15	0.12	0.60

	$\epsilon_{h_{11/2}}$	$v^2_{h_{11/2}}$	A	Γ	Λ
^{127}Cs	2.20	0.06	-0.33	0.66	2.20
^{125}Xe	1.07	0.65	-0.13	0.85	1.95

In calculating the energy spectrum of ^{125}Xe we had again to include all valence orbitals: two decoupled positive parity bands are observed at low excitation energies, and interpreted as having main components $g_{7/2}$ and a mixture of $s_{1/2}$ and $d_{3/2}$ configurations, respectively. A negative parity decoupled band based on the $h_{11/2}$ configuration is also observed[9]. In this case, however, we cannot deduce the neutron single-particle energies from ^{135}Xe, since no good neutron transfer studies can be performed, Xe being a gas. We have then chosen to take the same quasi-particle energies and occupation probabilities as used by Cunningham[13] in his study of the odd Xe and Ba isotopes, and to adjust them to improve the fit to the data. The resulting energy spectrum is shown in fig.2. The calculations for positive and negative parity states were performed indipendently: in this case, the parameters A, Γ and Λ obtained in the two calculations are quite different. For the negative parity states, several authors[14,15] have investigated the importance of including the $h_{9/2}$ and/or $f_{7/2}$ orbitals. It might be that the restriction of the model space to only the $h_{11/2}$ orbital bring to an effective increase of the value of the parameters. However, we cannot exclude that a different but still reasonable choice of the ϵ_j and v_j^2 values for the positive parity orbitals would lead to different model parameters, closer to the ones found for the negative parity states. Further experimental informations are required to improve the calculations.

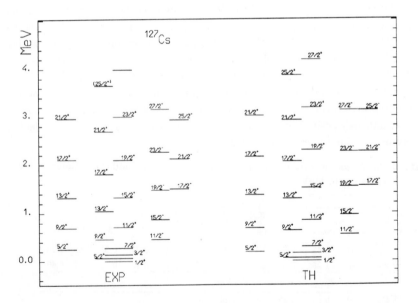

Figure 1. Experimental and calculated energy spectrum of ^{127}Cs.

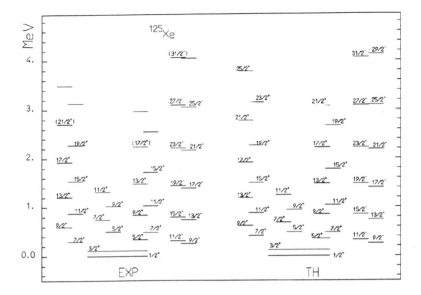

Figure 2. Experimental and calculated energy spectrum of ^{125}Xe. The negative parity states are normalised to the $11/2^-$ level.

In both cases, the energy spectra for the odd-A nuclei are reproduced only in a qualitative way. A more detailed analysis on the validity of the model and/or Scholten's formulas in this mass region would be auspicable: these formulas were in fact derived in the spherical limit, and it is not obvious that their use can be extended to other regions.

At this point we have determined all the parameteres needed in the calculation of the odd-odd nucleus, with the exception of the proton-neutron interaction. For sake of simplicity, we chose a simple modified surface delta interaction, whose explicit expression can be found in ref[16]. The strength k of the interaction is then the only free parameter left.

4. The calculations for ^{126}Cs

Experimentally, the odd-odd cesium isotopes were recently investigated by several groups, and many collective bands were found[1]. However, in most cases no connection between these bands and the low-lying non collective states were found. Even when the connection was observed, the difficulty of dealing with highly complicated spectra makes extremely hard to deduce the spin and parity of the states. Usually, spin assignments are based on systematics and cranking shell model calculations. The most intense collective bands observed in in-beam γ spectroscopy studies in the odd-odd cesium isotopes from ^{120}Cs to ^{128}Cs show strong similarities, and one would think that tyey are based on the same single particle configuration. However, the experimental groups working on these isotopes have assigned different spin and parity, and consequently different configurations[1].

In principle, we might perform the calculation for ^{126}Cs, and compare then the results with the experimental data. The model would have a predictive power, which would be of great use not only in this case, but in general for odd-odd nuclei, where the experimental informations are often scarce. However, the fact that all five orbitals had to be included in the calculations both for the unpaired neutron and proton produces an enormous amount of levels and bands when coupled together. Furthermore, the poor quality of the fit to ^{127}Cs and ^{125}Xe on one hand and the uncertainty in the quasi particle energies on the other make less reliable the model predictions for bands based on different configurations and their relative energies.

It appears then extremely important to obtain more experimental informations. The group of Milano performed a study on ^{126}Cs at the National Laboratory of Legnaro, using a system of six germanium detectors (MIPAD), two of which were planars, in order to study the connection of the yrast collective band to the ground state[17]. The use of planar detectors was essential, since the bandhead decays via very low energy transitions. It was concluded that, contrary to what presented in ref[1], the parity of the band is negative, and the spin of the bandhead is 5. A 4^- state is observed below the 5^- level, but the level spacing and the decay mode seems to indicate that it does not belong to the collective band. The configuration of such a band can then be based on π $h_{11/2}$ coupled to the neutron positive parity orbitals, or viceversa. The comparison of signature splittings of the neighbouring odd-A nuclei on one hand, and cranking shell model calculations on the other, point to the configuration with the neutron in the $h_{11/2}$ orbital[17].

IBFFA calculations including the neutron $h_{11/2}$ and the proton $g_{7/2}$, $d_{5/2}$, $d_{3/2}$, $s_{1/2}$ orbitals were then performed. In order to reduce the dimensions of the calculations, the core was prediagonalised, and only the levels belonging to the lowest four bands up to 3.2 MeV were considered. The resulting yrast band is shown in fig.3, in the case of no V_{pn} interaction (a), or an interaction strength k=0.3 (b). The energy of the calculated levels is normalised to the experimental 5^- state. The effect of the interaction is to expand the band without changing the relative spacing. The 4^- state is calculated few keV above the 5^- level, indicating that our choice for the form of V_{pn} might be oversimplified. Nevertheless, the agreement with the data is quite reasonable. The band turns out to be based on a mixture of all quasi particle configurations.

In order to verify that the agreement is not accidental, we calculated the yrast bands based on different configurations: in figure 3c the yrast negative parity band based on the proton $h_{11/2}$ and the neutron positive parity orbitals is shown. In this case the most contributing neutron configuration is $g_{7/2}$, which becomes predominant for spins higher than 8^-. The band turns out to be much more compressed, in complete disagreement with the experimental one. The positive parity band based on the ($\pi\, h_{11/2}$,$\nu\, h_{11/2}$) configuration can be seen in figure 3d. This was the configuration assumed by Komatsubara et al.[1], who assigned positive parity to the band: in this case, the high spin of the calculated bandhead and the signature splitting would not agree with the data.

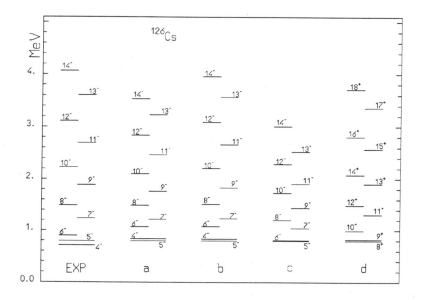

Figure 3. Experimental and calculated energy levels of ^{126}Cs.

5. Conclusions

The yrast collective band of ^{126}Cs has been reproduced by IBFFA calculations in which all parameters were fixed by fitting the neighbouring nuclei, with the exception of the proton-neutron interaction for which a simple modified surface delta interaction was chosen. Although the neighbouring odd-A nuclei were only qualitatively reproduced, the agreement can be considered satisfactory. The presence of the proton-neutron interaction turns out to be important to improve the fit to the data. However, since the members of the band have similar single particle components, no statement on the validity of our choice for V_{pn} can be drawn. The comparison of the collective bands calculated using different configurations indicate that this model could have a predictive power and could be used in addition to the usual crancking shell model in the understanding of the experimental data. However, some problems, mainly concerning the coupling of the fermion to the boson core, have still to be solved: Scholten's formulas, relating the parameters to the occupation probabilities of the single particle orbitals, should be verified in this transitional region, and a more realistic way of calculating quasi-particle energies and occupation numbers should be found. At the same time, more experimental information on odd-A nuclei (B(E2) values, spectroscopic factors, etc.) are necessary for a good test of the model. With an improvement in the description of odd-A nuclei, the extension of the model to odd-odd nuclei looks very promising. Also in this case, however, a good test of the model requires an effort on the experimental side to gather more experimental data of good quality.

6. References

1. T.Komatsubara, *Nucl.Phys.* A557 (1993) 419, and ref. therein.
2. S.Brant et al., *Phys.Lett.* B195 (1987) 111.
3. Zs. Dombradi et al., it Phys.Rev. C44 (1991) 1701, and ref. therein.
4. W.T.Chou et al. *Phys.Rev.* C37 (1988) 2834.
5. A.Arima and F.Iachello, *Phys.Rev.Lett.* 35 (1975) 1069.
6. F.Iachello and O.Scholten, *Phys.Rev.Lett.* 43 (1979) 679.
7. O.Scholten, PhD thesis, University of Gronigen, 1980, unpublished.
8. Y. Liang et al., *Phys.Rev.* C42 (1990) 890.
9. A. Granderath et al., *Nucl.Phys.* A524 (1991) 153.
10. G.Puddu et al., *Nucl.Phys.* A348 (1980) 109.
11. M.Conjeaud et al., *Nucl.Phys.* A117 (1968) 449.
12. J.R. Lien et al., *Nucl.Phys.* A281 (1977) 443.
13. M.A.Cunningham, *Nucl.Phys.* A385 (1982) 221.
14. M.A.Cunningham, *Nucl.Phys.* A385 (1982) 204.
15. S.T.Hsieh et al., *Phys.Rev.* C41 (1990) 2898.
16. H.Toki et al., *Z.Phys.* A292 (1977) 79.
17. N.Blasi et al., submitted to *Nucl.Phys.* A

Conference Summary

International Conference on Perspectives for the
Interacting Boson Model on the Occasion of
Its 20th Anniversary
— Concluding Remarks — *

Herman Feshbach

Center for Theoretical Physics
Laboratory for Nuclear Science
and Department of Physics
Massachusetts Institute of Technology
Cambridge, Massachusetts 02139 U.S.A.

It is appropriate for this celebration of the 20th anniversary of the IBM model to recall some of the early history. The title, Interacting Boson Model, was first used in two papers by Iachello and myself[1]. Iachello had arrived at MIT at the recommendation of Sergio Fubini together with the suggestion that he work with me, which very fortunately for me Iachello adopted. I remember suggesting to him to consider the possible use of the idea of the doorway state to bound state systems — still an excellent idea in my opinion. What this developed into is a model of ^{16}O, which was the subject of the two papers. In the first step of this model the particle–hole interaction was diagonalized. Two collective levels, the well known $T = 0$, 1^- and 3^- relatively far below the other particle states, emerged. These states were taken as the building blocks of the states of ^{16}O and for this reason it was called an interacting boson model; interacting because the next step was to take into account the effect of particle–particle and hole–hole interaction. In addition, the effect of two phonon annihilation represented by the figure was also calculated. Note no explicit use of the boson character of the 1^- and 3^- state was made. The interaction was taken to be the Soper mixture with radial dependence given by the $\delta(r_{ij})$. The matrix element of the two boson states was calculated directly from the particle–hole components of these states. Diagonalization was the next step. The center of mass spurious state was eliminated and a rough treatment of Pauli blocking was used. The results for $T = 0$ states are shown in the figure. Identification of the levels used the α-particle widths. Agreement is pretty good and we were proud of the fact that the low lying 6.05 level was primarily a four boson 3^- state (that is four particle–four hole state as proposed by Brown and Green). Actually the amplitude of this state was 0.771; there was an additional -0.468 contribution from the two 3^- boson state. Excellent agreement with the BE(2)'s was obtained using an effective charge. Note that these microscopic calculations were in the form generally used in the late sixties and early seventies.

---

* This work is supported in part by funds provided by the U. S. Department of Energy (D.O.E.) under cooperative agreement DE-FC02-94ER40818.

The revolutionary changes introduced by Iachello and Arima are now clear. Of primary value was the introduction of group theory. The s and d bosons were postulated and as a consequence one could deduce a U(6) symmetry. (In the ^{16}O case the bosons would be of the $p(1^-)$ and $f(3^-)$ variety and the group would be U(10).) Moreover, through the use of sub-group chains leading to O(3) special cases could be treated analytically. Finally and most importantly, a phenomenological Hamiltonian expressed in terms of boson operators was used. Pragmatically this is a much more effective approach than starting with the nucleon–nucleon force. This theory IBM-1 was immediately successful. It had the merit of being accessible to the experimentalists, which was not the case for the ^{16}O calculation.

This theory was, however, not well received by the Copenhagen group. It would serve no purpose to describe their reaction. But it is worthwhile to realize that Arima and Iachello had faith in their model and persisted in the face of this opposition. They were helped by Talmi and by the experimentalists, many of whom are here today, who kept producing results which agreed with the structure of IBM-1.

The success of IBM-1 has been followed by developments which (1) looked into elaborations which would include degrees of freedom beyond those which were included in IBM-1, and (2) examined the connection of the theory with the shell-model — or, more generally, with the many body problem of a finite number of fermions, which we shall refer to as IBA.

I will digress here to point out that nuclear physics is a unique subject at the frontiers of science because it deals with the structure and reactions of a finite number of interacting particles. It therefore has much in common with molecular physics (it is no accident that the IBM methodology finds applications there as Iachello has told us), with metallic clusters, with the conductivity of small bits of metal, *etc.* This area is referred to as "mesoscopic physics" — not macroscopic, with which the condensed matter physics is concerned, nor particle physics, which is the concern of particle physics, although it has become apparent that some elementary particles such as the nucleon are really composite systems.

To return to our categories. It will be observed that the papers presented this week belong to one or the other. Taking the first category, which I shall label IBM-1, we have (1) IBM-2, which took into account the neutron–proton composition of nuclei. This included the introduction of F spin by Talmi. (2) SUSY, the supersymmetric theory which extended the IBM to odd nuclei. (3) The s' and d' bosons introduced by Arima to understand back-bending. (4) The addition of g bosons. (5) The inclusion of isospin by Elliott. (6) The extension to superdeformed nuclei in this conference. (7) The discovery of the scissors mode. (8) Broken pairs. (9) Particle–hole excitation. (10) The extension to octupole deformed nuclei. (11) Application to high energy scattering. (12) Application to molecular structure. (13) QCD. (14) For the future: Radioactive beams, weak interactions, spin. (15) Three body forces:

Because of the severe truncation there will be an impact of the emitted states — for example, intruder states. The effect of truncation is reduced by inclusion of g, i, etc. bosons. Some of the effect can be reduced by renormalization. Another effect will be the presence of three body forces. This could be tested at the points where one has exact solutions, e.g. SU(3), by the addition of a term proportional to g^3, the eigenvalue of the trilinear Casimir operator.

We see the return of many of the shell model degrees of freedom. But there still remains the overriding approximation of the truncation leading principally to the s and d bosons.

We turn now to the second phase of this report — namely the connection with the many-body theory. This has two sub-sections.

The first, to which we shall now turn, compares the Arima–Iachello formulation with the "geometric theory." This I believe has been resolved — leading to the introduction of the intrinsic state. Ginocchio, Kirsin, and Leviatan are involved. Then there are other algebraic formulations — but it seems to me that (1) the Arima–Iachello theory is soundly based physically, and (2) it behooves other theories to make successful contact with experiment. Ginocchio's Fermi model is special as it permits a check on the mapping procedure.

The second subsection has to do with the attempt to correct, perhaps to derive, IBM from shell model theory. An example is the mapping theory developed by Arima, Iachello, Otsuka and Talmi. In this case the S, D subset of shell model space is mapped onto s and d bosons by adjusting the latter description so that the matrix elements in the original $S - D$ space are matched by the matrix elements in the boson space. This is an effective recipe — but I emphasize the use of the word recipe. There are some concerns with this method, such as overcompleteness, but this is not the object of my concern. It is not shown why one should drop the rest of the shell model Hilbert Space. After all, as Talmi has told us, there are 10^{14}–10^{15} components to the actual wave function. What has happened to them? This remark applies to other mapping methods which were suggested during the meeting. A possibility lies in Leviatan partial dynamic symmetry.

I would like to make a suggestion in this regard. The idea comes from the derivation of the optical model. In that case, a high resolution study shows a tremendous amount of fluctuation in the cross-section — at low energy these are the compound nuclear resonances. These fluctuations have been suggested as a model for quantum chaos. Nevertheless this apparently chaotic behavior has a regularity which surfaces when an energy average is performed. Turning to the bound state system it can be shown that the mean field of the shell model can also be justified by a smoothing procedure which essentially averages over the fluctuations — but of course the average does affect the mean field.

In the present circumstances the 10^{14}–10^{15} shell model states are fluctuations. Smoothing over them in the region of interest leaves as a coherent residue the S and D states in this energy and mass domain. Of course this is a speculation, but it is capable of being demonstrated. In any event, it seems necessary to show that something of this sort occurs before one can be happy with IBA.

I have been following these developments for the last 20 years — three conferences at Erice, one at Dubrovnik and others — but I have not participated in any of the research, although I have enjoyed watching and in the meantime have entered into the good spirits associated with those involved in this enterprise.

REFERENCES

[1] H. Feshbach and F. Iachello, *Phys. Lett.* **45B**, 7 (1973 June); *Annals of Physics* **84**, 211 (1974 May).

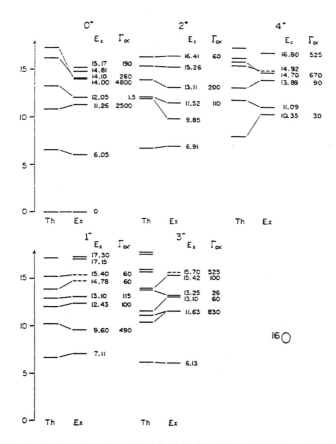

The calculated (TH) and experimental (EX) $J^\pi = 0^+, 2^+, 4^+, 1^-, 3^-$ $T = 0$ energy levels. The experimental excitation energy E_x (MeV) and the width for the (α, α) reaction Γ_α (keV) are also shown. Uncertain levels are shown in broken lines.

Contributed Abstracts

Mean Field description of the ground state of many boson systems relevant to nuclei

F. Aldabe, G. G. Dussel and H. M. Sofia

Abstract

In the present work we will give explicit expressions for the ground state wave function of a structureless boson systems, in different mean field approximations. We will initially make a review of the usual Hartree-Bose approximation and we will study its stability conditions that will yield the Particle-Hole Random Phase Approximation (PHRPA).

There exist a profound difference between the behaviour of boson and fermion systems already at the level of the mean field. In the fermionic case one may have or not the anomalous self-energy, and if it vanishes (or not), Hartree-Fock (or Hartree-Fock-Bogoliubov) provides an appropriate description of the system. In the bosonic case both types of terms must be necessarily taken into account. This difference is also related to the fact that for fermions the appearance of Cooper pairs marks a phase transition that can not be obtained in a perturbative way, but only after a deep change in the structure of the ground state. This phase transition can be understood in terms of the Bose condensation of the Cooper pairs. For bosons, once the Hartree-Bose approximation is used, its ground state is a Bose condensate and the fluctuations around this condensate will result in a ground state wave function that can be written wether as a condensate of pairs or as a coherent state formed by these pairs (this is the main conclusion of the present work). In any case the depletion factor will inform us about how much of the ground state is in the zero momentum state and will be an appropriate order parameter.

In all the cases we write explicitly the ground state wave function in such a way as to guarantee that the annihilation operators for the excitations of the system, acting over the ground state, give zero, showing that it can be always expressed in terms of a coherent mixture of pairs of bosons.

Interacting boson model for baryons

R. Bijker[1] and A. Leviatan[2]

[1] R.J. Van de Graaff Laboratory, University of Utrecht
P.O. Box 80000, 3508 TA Utrecht, The Netherlands

[2] Racah Institute of Physics, The Hebrew University
Jerusalem 91904, Israel

There is currently a great deal of interest in the spectroscopy and the electromagnetic properties of the nucleon. Experiments at new CW electron accelerators will provide a more complete understanding of the internal structure of the nucleon and its resonances as well as of QCD in the non-perturbative regime. Although the nucleon is the main constituent of nuclei, its structure in terms of a fundamental theory of strong interactions is not yet fully understood. Several models have been suggested to describe its properties. All these models share the same spin-flavor structure, but differ in their geometric structure.

Recently we have introduced a stringlike model of hadrons [1, 2] in which the nucleon and its resonances are treated as collective rotations and vibrations of the string. The string is idealized as a thin string along which the mass, charge and magnetization are distributed. A bosonic quantization of the string degrees of freedom suggests $U(7)$ as a suitable spectrum generating algebra. The resulting algebraic framework contains both single-particle and collective types of dynamics and allows one to calculate many properties of baryon resonances exactly (in some cases analytically). We have constructed a mass operator with the appropriate permutation symmetry among the baryon constituents. A fit of the mass spectrum of the nucleon and delta resonances is found in good overall agreement with the observed masses and is of comparable quality to that obtained in quark potential models [3], although the underlying dynamics is quite different.

Electromagnetic couplings are far more sensitive to wave functions and different models of hadronic structure. In particular, the distinguishing feature of hadronic models are their form factors. This holds for any extended object and is especially important for hadrons where the scale of the size (~ 1 fm) is comparable to that of the excitations (~ 300 MeV). In the collective stringlike model all form factors drop as powers of the momentum transfer (in agreement with the data) whereas in harmonic oscillator quark models they drop exponentially. In this contribution we discuss in detail the electric and magnetic form factors of the nucleon.

References

[1] R. Bijker and A. Leviatan, Proceedings of the 'XVI Oaxtepec Symposium on Nuclear Physics', Revista Mexicana de Fisica **39**, Suplemento 2, 7 (1993);
R. Bijker and A. Leviatan, in 'Symmetries in Science VII: Spectrum Generating Algebras and Dynamic Symmetries in Physics', ed. B. Gruber, Plenum Press, New York (1993), p. 87.

[2] R. Bijker, F. Iachello and A. Leviatan, Ann. Phys. (N.Y.), in press;
R. Bijker, F. Iachello and A. Leviatan, submitted to Phys. Rev. C;
R. Bijker, F. Iachello and A. Leviatan, submitted to Phys. Lett. B.

[3] N. Isgur and G. Karl, Phys. Rev. **18**, 4187 (1978); ibid. **D19**, 2653 (1979); ibid. **D20**, 1191 (1979); S. Capstick and N. Isgur, Phys. Rev. **D34**, 2809 (1986).

QUASIPARTICLE CALCULATIONS IN CORRELATED BASIS

J.Blomqvist[1], A.Insolia[2], R.J.Liotta[1] and N.Sandulescu[1,3]

[1] Royal Institute of Technology, Physics Department, S-10405 Stockholm

[2] Department of Physics, Univ. of Catania and INFN, I-95129 Catania

[3] Institute of Atomic Physics, P.O.Box MG-6, Bucharest

The current development in experimental techniques has provided complex high spin spectra for nuclei far from stability line. This has renewed the interest for a high quasiparticle excitations formalism, which goes beyond QRPA[1].

In this communication we present the alternative provided by the so-called QMSM (Quasiparticle Multistep Shell Model method)[2-6]. Within this framework first one calculates one and two quasiparticle (2qp) excitations by standard BCS and, respectively, QRPA. Then 3qp and 4qp excitations are calculated in a basis defined as the tensorial product af one-by-two and two-by-two quasiparticle states, respectively. The spurious state is controlled by simply rejecting from the tensorial basis the spurious 2qp state, which is decoupled at the QRPA level. This formulation is well suited for the truncation of the model space, which is performed in a self-consistent way by the overlap matrix between the basis vectors.

The method was applied in lead[2-4] and tin region[5,6]. The truncation procedure gave us the opportunity to obtain simple interpretations of the spectra. For instance we analysed from microscopical point of view the reliability of weak coupling model and the harmonic boson approximation.

REFERENCES

1. L.Zhao and A.Sustich, Ann. of Phys. 213(1992)378
2. C. Pomar, J. Blomqvist, R. J. Liotta and A. Insolia, Nucl.Phys.A515(1990)381
3. N.Sandulescu,A.Insolia,B.Fant,J.Blomqvist,R.J.Liotta, Phys.Lett.B288(1992)235
4. N.Sandulescu, A.Insolia, J.Blomqvist and R.J.Liotta, Phys.Rev.C47(1993)554
5. A.Insolia, N.Sandulescu, J.Blomqvist and R.J.Liotta, Nucl.Phys.A550(1992)34
6. N.Sandulescu,J.Blomqvist,R.J.Liotta, submitted to Nucl.Phys.A

IBFFM computations for odd-odd nuclei

S.Brant and V.Paar

Abstract

A survey of investigations of odd-odd nuclei in interacting boson-fermion-fermion model (IBFFM) during last ten years is presented. Computations of even-even, odd-even, even-odd and odd-odd members of the quadruplets of neighboring nuclei are discussed. An overview of IBFFM computations for nuclear structure of odd-odd K, Cu, Ga, As, Ag, In, Sb, I, Cs, Pr, Ir, Au and Tl nuclei is presented. Characteristic physical correlations of IBFFM are discussed. IBFFM band-pattern diagrams associated with dynamical symmetries of the core are surveyed.

Odd-A Xe nuclei described by the IBFM-1 model

Gh.Cata-Danil[1], D.Bucurescu[1], A.Gizon[2] and J.Gizon[2]

1 Institute of Atomic Physics, PO Box MG-6, Bucharest 76900, Romania
2 Institut des Sciences Nucléaires,IN2P3-CNRS/Université J. Fourier, Grenoble, France

Many experimental results on the neutron deficient Xe isotopes have been recently obtained both from beta decay and in-beam gamma ray spectroscopy, thus making them a very good ground for testing of nuclear models in a region of transition recognized for particular features due to its softness to gamma-deformation. In the present work we extend previous multishell IBFM-1 calculations [1], including isotopes from ^{131}Xe down to ^{119}Xe, and by comparison with the whole body of experimental data, we ascertain the possibilities of this model both for explaining complex sets of data and for prediction. The calculations are similar to earlier ones performed for the Ba isotopes [2].

The neutron orbitals taken in these calculations are $2d_{5/2}$, $1g_{7/2}$, $3s_{1/2}$, $2d_{3/2}$, $1h_{11/2}$, $2f_{7/2}$ and $1h_{9/2}$. The three strength parameters which define the usual simplified boson-fermion interaction have been determined by trying to describe the level schemes, static elecromagnetic moments and the electromagnetic decays for both positive and negative parity levels. Different parameter sets have been obtained for the two parities, one with little variation of the strengths for the positive parity, one with a more accentuated systematic variation for the negative parity. The following conclusions result from comparison with the experimental data:

- For the negative parity states: a good fit to the yrast states usually ensures a good description of non-yrast states as well. Low-lying levels are dominated by the $h_{11/2}$ orbital , the other two negative parity orbitals may be important in the higher levels. Both the beta and gamma bands of the core are found with notable contributions in the structure of certain bands identified in these odd-A Xe isotopes.

- For the positive parity states: the four orbitals are mixed in various degrees, the mixing increasing towards the lightest isotopes considered.For ^{121}Xe and ^{119}Xe,the calculations predict a strong mixing of the $d_{5/2}$ and $g_{7/2}$ orbitals but,due to the lack of relevant experimental data, it is difficult to assign the structure of the many low-lying $5/2^+$ and $7/2^+$ states occuring in these two isotopes.

[1] M.A. Cunningham, *Nucl.Phys.*A385(1982)204;221
[2] D. Bucurescu et al., *Phys.Rev.*C43(1991)2610

Nuclear Pair Shell Model

Jin-Quan Chen

An S-D collective pair shell model for proton-neutron systems is presented. A Hamiltonian with surface delta interaction (SDI) between identical nucleons and Q-Q interaction between neutrons and prontons is diagonalized exactly in a "realistic" S-D fermion subspce. The approach is applied to analyze the A=130 Xe-Ba region. Both the spectra and B(E2) value are in good agreement with experimental data. The effects of the single-particle energy splitting on the collectivity and pair structure are analyzed.

Phenomenological Model for Elastic and Inelastic Scattering of High Energy Hadrons from Nuclei

D.C.Choudhury and M.A.Scura
Department of Physics, Polytechnic University,
Brooklyn, New York, 11201

A phenomenological model within the framework of partial wave expansions in the adiabatic approximation is formulated to calculate the elastic and inelastic scattering of high energy nucleons, composite systems of nucleons, or mesons from nuclei. The present work is an extension and generalization of the earlier model which has been successful when applied to analyses[1,2] of the experimental data on the scattering of (1) 1.37GeV α particles from various Calcium isotopes obtained at Saclay and (2) 180MeV antiprotons from various nuclei obtained at CERN.

The motivation for the present work is to describe the elastic and inelastic scattering for a larger class of hadronic projectiles covering a greater range of energies than has been possible in the previous studies. The formulation utilized in the present investigation has been obtained by introducing a phenomenological approach involving complex angular momenta that is very much in the spirit of the approach to scattering theory first introduced by T. Regge[3]. We exploit the fact that the functional form for the S matrix used in previous studies[1,2], $S(\ell)$, where ℓ is the angular momentum quantum number associated with the ℓth partial wave, is in fact an analytic function of the complex variable $z=\ell+i\mu$. This property enables us to derive analytic closed form expressions for the differential cross sections for the elastic and inelastic scattering. The real parameter μ is interpreted to be connected with the real part of the nuclear phase shifts.

The details of the physical assumptions and mathematical formulation for the model and its application to pion-nucleus scattering for a wide range of energy[4,5] will be presented. We also discuss the validity of our formulation within the context of high energy elastic and inelastic scattering and draw conclusions regarding the structure of the low lying excited states for the target nuclei examined. The results of the present investigation tend to suggest that it is mostly the geometrical structure of the target nuclei which determines the differential cross sections for the elastic and inelastic high energy nuclear scattering.

[1] D.C.Choudhury, Phys. Rev. C **22**, 1848 (1980)
[2] D.C.Choudhury and T.Guo, Phys. Rev. C **39**, 1883 (1989)
[3] T.Regge, Nuovo Cimento **14**, 951 (1959)
[4] D.Marlow et al., Phys. Rev. C **30**, 1662 (1984)
[5] K.G.Boyer et al., Phys. Rev. C **29**, 182 (1984)

An Interacting Boson-Fermion Model for Weak Mixing Angles.

G.Dattoli[1] and G.Maino[2]

(1) ENEA, INN.FIS, Frascati, Roma (Italy).
(2) ENEA, INN.FIS, viale Ercolani 8, I-40138 Bologna (Italy).

March 24, 1994

We consider a simple model of weak quark mixing based on interactions involving both boson and fermion operators. We show that this model can be viewed as a standard three-level interaction where the Cabibbo - Kobayashi - Maskawa matrix is represented as the evolution operator of a three-level Hamiltonian. Finally, we prove that the number of parameters relevant to this problem can be reduced with respect to the usual approaches and that mixing angles and CP violating phase are obtained in terms of a single parameter linked to the Cabibbo angle. The possible extension to further quark generation is also discussed.

The interacting boson model with broken pairs: high spin states of ^{138}Nd.

G. de Angelis[1], M.A. Cardona[1], M. De Poli[1], S. Lunardi[2], D. Bazzacco[2], F. Brandolini[2], D. Vretenar[3], G. Bonsignori[4], M. Savoia[4], F. Terrasi[5], V. Roca[5]

[1] INFN, Laboratori Nazionali di Legnaro, Legnaro, Italy
[2] Dipartimento di Fisica and INFN Sezione di Padova, Padova, Italy
[3] Physics Department, University of Zagreb, Zagreb, Croatia
[4] Dipartimento di Fisica and INFN, Sezione di Bologna, Bologna, Italy
[5] Dipartimento di Scienze, and INFN Sezione di Napoli, Napoli, Italy

The interacting boson model has been recently successful in describing high spin states in nuclei by inclusion of non-collective two and four particle excitations. This approach applies well for transitional nuclei near closed shells. In fact, the high spin states of light Zr nuclei[1] as well as those of Hg isotopes[2] are nicely reproduced by the model. We have studied, through heavy ion reactions at the Tandem XTU of Legnaro, many nuclei of the transitional region with A$\approx$ 140 near N=82 and here we present the results for ^{138}Nd. The high-spin states structure in ^{138}Nd, populated by means of the ^{123}Sb(^{19}F,4n) reaction, is characterized by two $\Delta I = 2$ bands based on the two low-lying 10$^+$ states. According to the systematics of experimental data from neighboring nuclei, we expect the yrast 2qp band to be a two-neutron band $(\nu\ h_{11/2})^2$, and the band based on the 10$^+$ state at 3700 keV to have a two-proton configuration $(\pi\ h_{11/2})^2$. The results of the calculations for ^{138}Nd in the framework of the IBM with (neutron) broken pairs are shown in the figure. The calculations reproduce correctly the excitation energies of the ground state band and of the two neutron $(\nu\ h_{11/2})^2$ band.

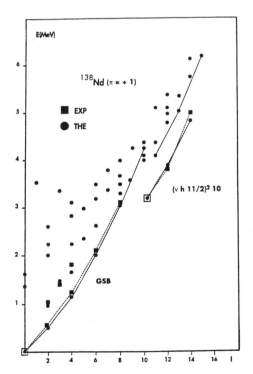

1) P. Chowdury et al., Phys. Rev. Lett. 67, 2950 (1991).

2) D. Vretenar, G. Bonsignori, and M. Savoia, Phys. Rev. C. 47, 2019 (1993).

Boson Mappings of the Fermion Dynamical Symmetry Model

J. Dobeš

Institute of Nuclear Physics, Czech Academy of Sciences,

CS 250 68 Řež, Czech Republic

P. Navrátil and H. B. Geyer

Institute of Theoretical Physics, University of Stellenbosch,

Stellenbosch 7600, South Africa

Boson mappings of the Fermion Dynamical Symmetry Model (FDSM) are introduced to investigate and clarify the relation between the FDSM and the proton-neutron Interacting Boson Model (IBM). Employing the Dyson boson mapping, exact FDSM results are obtained. Hermitizing the Dyson boson image of the FDSM Hamiltonian through the seniority similarity transformation or Belyaev-Zelevinsky mapping and retaining one- and two-body terms only, one obtains an IBM Hamiltonian. FDSM and boson mapped results are compared for a few typical cases. From the boson point of view, we re-examine recent statements about an effective SO(6) symmetry in the FDSM and the appearance of normal (with maximal F-spin) and exotic states in the model. Throughout possible spurious states which may appear as a result of an effective overcompleteness in the (linearly independent) boson basis, are properly identified. We discuss when and where these states appear in the spectra, and the possible implications these considerations may have for allowed representations in the IBM and FDSM.

Effective E1-operator in IBFM

D.N.Dojnikov* and V.M.Mikhajlov**

*Ioffe Physico-Tekhnical Institute, St.Petersburg, 194021 Russia
**Institute of Physics, St.Petersburg State University, 198904 Russia

The description of odd nuclei in the framework of IBM can be carried out on the ground of effective operators acting in the space of d-, s-bosons and ideal fermions within the valence shell. We have calculated these operators taking into account a number of intermediate states and the microscopical structure of the D-phonon (the image of the d-boson).

Since all matrix elements of the fermion E1-operators vanish within the valence shell, both in the intermediate states and in the D-phonon structure we consider fermion states in the neighbouring shells. Various processes leading to E1-transitions do include the alteration in the ds-space. Therefore the effective E1-operator can be represented as following

$$\hat{T}_{BF}^{E1} = \sum \left\{ (\alpha_j^+ \alpha_i)^{(k)} \left[a_{ij}^k (d^+ s + s^+ d) + b_{ij}^{kl} (d^+ d)^{(l)} \right] \right\}^{(1)}, \quad (1)$$

where α^+, α are ideal fermion operators. Parameters (a_{ij}^k, b_{ij}^{kl}) are calculated with a wide single-particle Saxon-Woods spectrum, factorized multiple forces and standard values of the effective E1-charges. In the table are shown experimental and calculated $B(E1)$ (10^{-5} W.u.) for ^{77}Se, ^{79}Kr. Our method of calculations gives also satisfactory results for energies and $B(E2)$, $B(M1)$ in this nuclear region [1].

^{77}Se				^{79}Kr			
L_i^π	L_f^π	$B(E1)_{exp}$	$B(E1)_{th}$	L_i^π	L_f^π	$B(E1)_{exp}$	$B(E1)_{th}$
$5/2_1^-$	$7/2_1^+$	1.74 ± 0.03	0.85	$5/2_1^+$	$5/2_1^-$	12_{-2}^{+3}	1.1
$7/2_1^-$	$5/2_1^+$	$0.54_{-0.20}^{+0.29}$	0.53	$7/2_1^-$	$5/2_1^+$	$2.8_{-0.5}^{+0.8}$	7.3
$7/2_1^-$	$9/2_1^+$	$0.79_{-0.23}^{+0.27}$	0.40	$7/2_1^-$	$7/2_1^+$	$5.9_{-1.0}^{+1.5}$	2.2
$7/2_1^-$	$7/2_1^+$	$2.2_{-0.6}^{+0.7}$	1.77	$7/2_1^-$	$9/2_1^+$	$2.1_{-0.4}^{+0.6}$	3.4
$9/2_1^-$	$9/2_1^+$	$2.5_{-1.3}^{+2.9}$	0.14	$9/2_1^-$	$7/2_1^+$	$2.6_{-0.6}^{+1.4}$	1.0
$9/2_1^-$	$7/2_1^+$	$5.2_{-1.9}^{+3.3}$	2.94	$11/2_1^-$	$9/2_1^+$	$3.1_{-0.6}^{+1.0}$	1.9
$5/2_2^-$	$7/2_1^+$	$0.54_{-0.06}^{+0.07}$	0.30	$13/2_1^-$	$11/2_1^+$	$5.2_{-2.6}^{+5.1}$	2.1
$7/2_2^-$	$9/2_1^+$	$0.13_{-0.06}^{+0.13}$	0.84	$15/2_1^-$	$13/2_1^+$	$4.0_{-0.7}^{+1.1}$	3.5
$5/2_3^-$	$5/2_1^+$	$5.1_{-2.0}^{+4.2}$	4.64	$17/2_1^-$	$15/2_1^+$	$6.2_{-1.3}^{+2.0}$	3.8
$5/2_3^-$	$7/2_1^+$	$5.0_{-1.8}^{+3.7}$	1.11	$19/2_1^-$	$17/2_1^+$	$4.8_{-1.1}^{+2.1}$	5.7
$13/2_1^-$	$11/2_1^+$	13_{-4}^{+5}	4.29				

1. D.N.Dojnikov et al, Bull.Acad.Sci. (phys.ser), 1993, v.57, N.2, p.80; N.3, p.180.

INTERACTION BETWEEN COLLECTIVE AND TWO QUASIPARTICLE STATES IN IBM

A.D.Efimov[*] and V.M.Mikhajlov[**]

[*]Ioffe Physico-Tekhnical Institute, St.Petersburg, 194021 Russia
[**]Institute of Physics, St.Petersburg State University, 198904 Russia

A coupling of collective and two–quasiparticle states (t.q.s.) in IBM was considered in [1], where the transition of two d-bosons into t.q.s. with $L = 4$ was taken into account (see term $V^{(a)}$ in (1)). However, our microscopic calculation show that the coupling of three d-boson configurations with t.q.s. can be not less important. Therefore corresponding terms have to be added in the Hamiltonian.

$$V_{int} = \frac{1}{\sqrt{1+\delta_{1,2}}} \frac{1}{\Omega\sqrt{6}} [V^{(a)}\sqrt{6\Omega}s^+(a_1^+ a_2^+)^{(L=4)}(dd)^{(L=4)} +$$

$$+ V^{(b)} s^+ s^+ (a_1^+ a_2^+)^{(L=4)}(ddd)^{(L=4)} + V^{(c)} s^+ s^+ (a_1^+ a_2^+)^{(L=6)}(ddd)^{(L=6)} + h..], \quad (1)$$

where a_i, d, s are quasiparticle, quadrupole and scalar boson operators respectively, Ω is the total number of bosons and t.q.s. We have calculated matrix elements in (1) for ^{102}Ru using factorized quadrupole forces and a realistic single-particle spectrum. The low energy quadrupole phonon of the modified RPA [2] is used as the microscopic image of the d-boson. Depending on a two–quasiparticle configuration the main contribution in (1) can be determined by either the term with $V^{(a)}$ or the one with $V^{(c)}$ (table 1). To illustrate the role of a new term with $V^{(c)}$ in (1) we have performed calculations of high-spin state energies ($I = 8\text{--}16$) in ^{102}Ru. Parameters of H_{IBM} have been found fitting energies of collective states with $I \leq 6$. We have adopted $V^{(c)}$ and the two quasiparticle state energy to be equal to -0.12 and 2.3 MeV respectively; $\Omega = 9$. E_c in table 2 is IBM energies. Table 2 shows that the results of our calculations (E_{th}) are in a reasonable agreement with experimental values (E_{exp}).

Table 1

Two particle configuration	$V^{(a)}$	$V^{(b)}$	$V^{(c)}$
	MeV		
$(g9/2)_\pi^2$	0.045	0.029	-0.19
$(d5/2)_\nu^2$	-0.048	-0.074	0
$(g7/2)_\nu^2$	0.24	-0.038	-0.056
$(h11/2)_\nu^2$	-0.40	-0.23	0.051

Table 2

L^π	E_c	E_{th}	E_{exp}
	MeV		
6^+	1.885	1.877	1.873
8^+	2.793	2.706	2.704
10^+	3.820	3.379	3.431
12^+	4.958	4.099	4.052
14^+	6.195	4.885	4.803
16^+	7.518	5.639	5.718

1. Yoshida N., Arima A., Otsuka T., Phys. Lett. 114B, 86(1982).
2. Efimov A.D., Mikhajlov V.M.,Izv.Akad.nauk SSSR, ser.fiz. 1992, 56, p.57.

SUPERDEFORMATION AND THE NEUTRON-PROTON INTERACTION

P. Federman and E. D. Kirchuk
Departamento de Física, Facultad de Ciencias Exactas y Naturales,
Universidad de Buenos Aires
Ciudad Universitaria, 1428 Buenos Aires, Argentina

The discovery of a superdeformed band in ^{152}Dy [1] originated intense experimental and theoretical efforts [2,3] that contributed to the understanding of rapidly rotating nuclei. By now superdeformed bands have been identified in many Gd, Tb, Dy, Hg and Pb isotopes, including cases of several such bands in one nucleus [2].

In this contribution we report results obtained in the framework of constrained Hartree-Fock-Bogoliubov calculations. The constraint on the quadrupole moment is instrumental to obtain excited superdeformed solutions.

Preliminary calculations indicate that, similarly to the case of normal deformation, the neutron-proton interaction plays a crucial role in the microscopic origin of superdeformation.

Superdeformation in ^{152}Dy is the result of the strong T=0 interaction between neutrons and protons that polarizes the protons into higher neutron orbits. The protons rise to orbits with strong spatial overlap favoured by the de-Shalit - Goldhaber rule, as in the case of normal deformation.

Thus, the origin of nuclear superdeformation can be traced to the same microscopic mechanism that originates normal deformation.

We wish to thank Stuart Pittel for enlightening discussions.

1. P. J. Twin et al., Phys. Rev. Lett. **57** (1986) 811.

2. R. V. F. Janssens,"Nuclear Shapes and Nuclear Structure at Low Excitation Energies", eds. M. Vergnes, J. Sauvage, P. H. Heenen and H .T . Duong, Plenum Press, New York, 1992, p.299.

3. W. Nazarewicz, "Recent Advances in Nuclear Structure", eds. D. Bucurescu, G. Cata-Danil and N. V. Zamfir, World Scientific, Singapore, 1991, p.175.

Symplectic Models as an Alternative to IBM

G.F.Filippov

Bogolyubov Institute for Theoretical Physics
252143 Kiev-143, Ukraine

The success of the IBM is due, apparently, mainly to a fortunate parametrization of the effective nuclear Hamiltonians expressed in terms of generators of some unitary or orthogonal group. Unfortunately, an interpretation of those Hamiltonians is not always clear and their derivation on the basis of a consistent microscopic approach meets almost insuperable difficulties. This raises hopes and optimism in those who study alternative ways of the development of the microscopic theory of nuclear spectra, in particular, within the framework of the so-called symplectic models. The history of the latter goes back to the well-known Elliott's work, where he proposed the microscopic model $SU(3)$ for the explanation of the p and s-d shell nuclei spectra. The merits of the Elliott model consist in the following. It is translational-invariant, takes into account precisely the Pauli principle and all basic conservation laws. It has enough resources to go out of its frames by taking into account those degrees of freedom that play a passive role in the original variant of this approach. If the basis of the model is widened by the most important $SU(3)$ multiplets of the open shells, it intrudes into the IBM "sphere of influence".

The symplectic models appeared as a generalization of the Elliott model. If the latter reproduces the dynamics of valence nucleons, the symplectic models include into the consideration the dynamics of the collective quadrupole modes. The Hamiltonian of the symplectic models takes an elegant form in the Fock–Bargmann space, the transitioin to which is performed by the generalized coherent states. In this space, the symplectic group generators, from which Hamiltonians are built, can be clearly interpreted since they contain constructions typical for the unified Bohr–Mottelson model. For example, the $Sp(6, R)$ generators Q_{rs} in the coordinate frame of the principal axes of the nucleus quadrupole deformation can be written as

$$Q_{rs} = A_{rs} + (f_3 + \frac{A-1}{2})\delta_{rs} -$$

$$- \sum_\nu (\mathbf{e}_r \mathbf{p}_\nu)(\mathbf{e}_s \mathbf{p}_\nu)[(1-\beta_\nu)c2\frac{\partial}{\partial \beta_\nu} + (A_{\nu\nu} + f_3 + \frac{A-1}{2})\beta_\nu] -$$

$$- \frac{1}{2i}\sum_\nu Fcrs_{\sigma\chi}[\frac{(1-\beta_\sigma)(1-\beta_\chi)}{\beta_\sigma - \beta_\chi}\hat{\tilde{M}}_\nu - \frac{1}{2}(\beta_\sigma - \beta_\chi)\hat{M}_\nu - i(\beta_\sigma + \beta_\chi)A_{\sigma\chi}].$$

Indices ν, σ, χ are the cyclic permutations of the integers 1,2,3; $\mathbf{e}_r$ are the basis vectors of the coordinate system where the components Q_{rs} and A_{rs} are defined; $\mathbf{p}_\nu$ are the basis vectors of the system referred to the principal axes of deformation; $A_{\sigma\chi}$ are the $U(3)$ generators in this system; M_ν and $\tilde{M}_\nu$ are the projections of the orbital angular momentum onto the deformation axes. Finally,

$$Fcrs_{\sigma\chi} = (\mathbf{e}_r \mathbf{p}_\sigma)(\mathbf{e}_s \mathbf{p}_\chi) + (\mathbf{e}_r \mathbf{p}_\chi)(\mathbf{e}_s \mathbf{p}_\sigma).$$

The $SU(3)$ multiplet $(f_1 - f_2, f_2 - f_3)$ is realized on the tensor product of the generator vectors of the vortex momentum.

Correlated Pair Operator Approach for Deformed Nuclei and its Relations to IBM Studies

Y K Gambhir, C R Sarma and N Bajpai

Department of Physics, Indian Institute of Technology,
Bombay-400 076, INDIA.

A practical scheme using the correlated pair operators (CPO) is developed for carrying out large spherical and / deformed shell-model configuration mixing calculations. Various approximate truncation schemes such as broken-pair approximation (BPA), equation of motion method emerge naturally from the present formalism. This method is useful in analysing explicitly the Hartree-Fock intrinsic states in terms of spherical shell-model configurations. For illustration some sample results for the nuclei in the 2p-1f region are presented. In addition this approach provides a natural starting point for the microscopic investigation of the n-p interacting boson model (IBM-2) and helps to describe the bosons of IBM in terms of fermions for deeper understanding.

Low-energy states fed by β/EC decay in odd-A Pr, Ce and La isotopes

J.Genevey[1], A.Gizon[1], J.Inchaouh[1], D.Barnéoud[1], R.Béraud[2], D.Bucurescu[3],
G.Cata-Danil[3], A.Emsallem[2], T. von Egidy[1,4], C.Foin[1], J.Gizon[1], C.F.Liang[5],
P.Paris[5], I.Penev[6], A.Plochocki[7], E.Ruchowska[7], C.Ur[3], B.Weiss[8]

1 ISN,IN2P3-CNRS/Univ. Joseph Fourier,F-38026 Grenoble, France
2 IPN,IN2P3-CNRS/Univ. Claude Bernard,F-69621 Villeurbanne, France
3 Institute of Atomic Physics,R-769000 Bucharest, Romania
4 Physik Department,Technische Universität Munchen,D-85748 Garching, Germany
5 CSNSM,IN2P3-CNRS,Bt 104,F-91405 Campus-Orsay, France
6 Institute of Nuclear Research and Nuclear Energy,B-1784 Sofia, Bulgaria
7 Institute of Experimental Physics,PL-00681 Warsav, Poland
8 Laboratoire de Radiochimie,Université,F-06034 Nice, France

Low spin states of odd-A neutron-deficient light La(Z=57), Ce(Z=58) and Pr(Z=59) isotopes have been investigated from the β/EC decays of radioactive chains with mass A=127,129 and 131. Neodynium to cerium isotope precursors were produced using 5-6 MeV/nucleon ^{40}Ca or ^{37}Cl beams from the SARA accelerator in Grenoble on thin isotopically enriched targets of ^{94}Mo and ^{96}Mo. The mass identifications were insured with the SARA on-line isotope separator working with a coupled He-jet ion-source system [1] or with an ion-guide [2]. Using the He-jet system, data have been obtained from delayed γ-X, γ-γ and γ-e$^-$ coincidences, singles multiscaling spectra and multipolarity assignments from internal conversion electron coefficients.

In addition to already known negative parity levels built on $\nu h_{11/2}$(high Ω) in even-Z nuclei and $\pi h_{11/2}$(low Ω) in odd-Z nuclei, various positive parity states have been observed [3].

For example, new $1/2^+$, $3/2^+$, $5/2^+$, $7/2^+$ level sequences have been identified in ^{127}Ce,^{129}Ce and ^{131}Ce. They are similar to those already established in odd-A Xe and Ba isotopes.

Complex sets of positive parity states with spins ranging from $3/2^+$ to $9/2^+$ or $11/2^+$ have been identified in odd-A La and Pr isotopes around A=130. They contain a $3/2^+$ quasirotational structure based upon $\pi d_{3/2}, \pi d_{5/2}$ configurations.

With the new level schemes, low-energy level systematics are proposed for odd-A Ce isotopes from A=127 to A=135.

As previously done for odd-A Ba and odd-A Xe nuclei [4], multishell Interacting Boson-Fermion model calculations have been performed for odd-A La,Ce and Pr isotopes.

[1] A. Plantier et al.,Nucl.Instr.Meth. in Phys.Res. B26(1987)314
[2] A. Astier et al.,4th Igisol Workshop,Sept.1992,Rydzyna,Poland
[3] J. Gizon, Invited Lecture, this Conference
[4] G. Cata-Danil et al., to appear in J.Phys.G. and this Conference

Solid Quark/Hadron Mixed Phase in Neutron Rich Matter

Norman K. Glendenning

Abstract

The high density nature of neutron rich matter is studied both in the absence of a phase transition to quark matter and in its presence. Even in the first case it cannot remain purely neutron, and a consistent picture constrained by nuclear properties, hypernuclei and the binding of the Λ in uniform matter is developed. Consequences for neutron stars are explored. We have shown recently that in all earlier work of the last two decades on the quark/hadron phase transition in neutron star matter, a degree of freedom was unintentionally frozen out which yielded a description of the transition as a *constant* pressure one [1, 2]. This had the explicit consequence of excluding the coexistence phase of hadrons and quarks from the star. The degree of freedom is the possibility of reaching the lowest energy state by rearranging electric charge between the regions of hadronic matter and quark matter in phase equilibrium. Because of this freedom, the pressure in the mixed phase varies as the proportions of the phases and is not a constant in the coexistence phase. It can therefore occupy a finite radial region in the interior of the star. The microphysics behind this preference for charge rearrangement is the charge-symmetric nuclear force which acts to relieve the high isospin asymmetry of neutron star matter as soon as it is in equilibrium with quark matter. This introduces a positive charge on the hadronic regions and a compensating negative charge on the quark matter. There is little need of leptons because charge neutrality can be achieved with lower energy among the *conserved* baryon number carrying baryons and quarks. The competition of the Coulomb and surface energies then establishes the shapes, sizes and spacings of the rarer phase in the background of the other so as to minimize the lattice energy. A Coulomb lattice structure of varying geometry is thus introduced to the interior of neutron stars. Many observable phenomena are likely to be associated with the presence of this solid, including all transport properties as well as pulsar glitches. Since the thickness, some kilometers, of mixed phase is a sensitive function of neutron star mass, or more particularly to its pressure profile, pulsars having such a region will be highly individualistic in their behavior according to fairly small variations in their mass. The early estimates that suggest that the phase transition occurs at a high density in neutron star matter are flawed by the fact that a very beta unstable description of the quark phase was used, $N_d = 2N_u$, which shifts the transition to a high density. We find both from qualitative arguments based on the finite size of nucleons and also from detailed models of hadronic and quark matter in beta equilibrium that the mixed phase is expected to begin at densities of about 3 times nuclear and is likely to exist in all neutron stars of mass $> M_\odot$.

[1] N. K. Glendenning, Nuclear Physics B (Proc. Suppl.) **24B** (1991) 110.

[2] N. K. Glendenning, Phys. Rev. D **46** (1992) 4161.

Decay Characteristics of Mixed-Symmetry States in ^{158}Gd

MK Harder, KT Tang
Department of Mathematical Sciences, University of Brighton,
Brighton, BN2 4GJ, UK

A Williams, S Judge
Insitut Laue-Langevin,38042 Grenoble, France.

ABSTRACT

The conversion electrons from the 157Gd(n,e-)158Gd reaction were detected with the high-resolution spectrometer BILL at the Institute Laue-Langevin (ILL), Grenoble. Multiscans were performed to try to resolve the subshells of transitions from the 1+ mixed-symmetry states at 3192 and 3201keV to the ground state and the first three 2+ states.

The ratio of the subshell intensities in the electron spectrum could yield mixing ratios of these decays, but the high density of states at these energies made positive identification difficult. More work has now been done with (n,g) studies at the ILL to identify the secondary gamma-transitions in the previously unmapped region of 2.7-3.8MeV. This gamma work is now completed and several new levels and placings have been made. These will be summarised at the conference.

Attention can now be turned back to the complex electron spectrum. With the gamma-transitions now known, it is expected that we will be able to present results of the mixing ratios of transitions depopulating the 1+ mixed-symmetry states. The predicted ratios for these transitions vary with different nuclear models. The preliminary results will be discussed.

PARTICLE-HOLE INTRUDER ANALOG MULTIPLETS

K.HEYDE[†], C.DE COSTER[†], P.VAN ISACKER[††] and J.L.WOOD[*]
[†] Institute for Theoretical Physics and Institute for Nuclear Physics, Proeftuinstraat 86, B-9000 Gent
[††] GANIL BP 5027, F-14023 Caen Cedex
[*] School of Physics, Georgia Institute of Technology, Atlanta, Georgia, U.S.A.

It has become clear, in recent years, that very near and even at the closed-shell configurations, excitations of particle-hole nature occur at very low energies. The study of such 'intruder' states in medium-heavy and heavy nuclei has by now been explored in a rather extensive way, in both odd-mass [1] and even-even [2] nuclei.

Given that these intruder excitations and the collective bands built on them are very similar in structure (level spacings, E2 transition rates, ...) to the low-lying collective bands in adjacent nuclei with the same number of active nucleons, an underlying symmetry leading to the introduction of intruder analog multiplets and the idea of intruder-analog spin, much like isospin, was presented [3]. Treating the multi-particle, multi-hole (mp-nh) excitations in terms of s and d bosons using the interacting boson model (IBM), the combined group $U_p(6) \otimes U_h(6)$ (where p stands for 'particle-like' bosons and h for 'hole-like' bosons) exhibits a rich structure.

Both the 'horizontal' or U(12) reduction and the 'vertical' non-compact $\tilde{U}(6,6)$ symmetries will be discussed and illustrated, in particular, in the Pb region encompassing the very neutron-dificient Pb, Hg, Pt, Os and W nuclei. We furthermore suggest a possible application to light nuclei ($1f_{7/2}$ shell) where together with the mp-nh excitations forming multiplets, within the IBM, the isospin lowering and raising operators may be included in the algebra resulting in an even larger group of multiplet configurations. Some suggestions for possible experiments to verify the latter idea will be made.

[1] K.Heyde, P.Van Isacker, M.Waroquier, J.L.Wood and R.A.Meyer, Phys. Repts. 102 (1983) 291
[2] J.L.Wood, K.Heyde, W.Nazarewicz, M.Huyse and P.Van Duppen, Phys. Repts. 215 (1992) 101
[3] K.Heyde, C.De Coster, J.Jolie and J.L.Wood, Phys. Rev. C46, (1992) 541

On the angular momentum dependence of the parity splitting in nuclei with octupole correlations

R.V.Jolos[a,b] and P.von Brentano[a]

[a] Institut für Kernphysik, Universität zu Köln, FRG
[b] Bogoliubov Theoretical Laboratory, JINR, Dubna, Russia

Experimental data [1,2] on the angular momentum dependence of parity splitting of yrast bands in different nuclei are analysed using a one-dimensional model of octupole motion with axial symmetry. A two parameter formula, based on a solution of the Schrödinger equation with a double-minimum potential,

$$\Delta\epsilon(I) \equiv \Delta E(I)/\Delta E(2) = exp[-\frac{I(I+1)/J_0(J_0+1)}{1+aI(I+1)} + \frac{6/J_0(J_0+1)}{1+6a}], \quad (1)$$

where $\Delta E(I)$ is the parity splitting, predicts that the parity splitting exponentially decrease with $I(I+1)$ and gives a good fit to data.

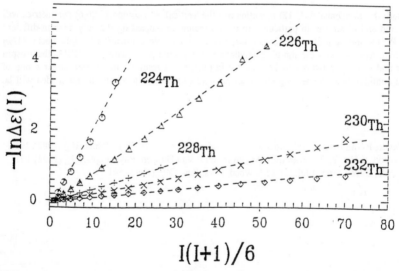

Fig.1. Experimental data for $-ln\Delta\epsilon(I)$ versus $I(I+1)/6$ for Th isotopes. The straight lines show the quality of the linear approximation.

References

[1]. Nucl.Data Sheets **42**,233(1984); **51**,241(1987); **63**, 17(1991).

[2]. P.C.Sood, D.M.Headly, and R.K.Sheline, At.Nucl.Data Tables **47**, 89(1991) and **51**,273(1992), and references therein.

SHAPE TRANSITION IN THE A≃70-80 REGIONS

E. D. Kirchuk

Departamento de Física (FCEN), Universidad de Buenos Aires
Ciudad Universitaria, 1428 Buenos Aires, Argentina

The phenomenon of shape coexistence has been observed recently in several new regions of the Periodic Table. In particular, nuclear deformation has been observed for several very neutron-deficient nuclei around A≃80, e.g. in the Sr, Zr and Mo nuclei with neutron numbers for which the Ge, Se and Kr isotopes, with A≃70, exhibit ground-state nuclear shape isomerism[1].

In a recent work[2], we proposed a common picture describing the onset of deformation for both the A≃80 and A≃100 regions within the framework of the Hartree-Fock-Bogoliubov method. In this presentation, our goal was to incorporate the A≃70 region in such a description.

The earlier calculations in the A≃80 region assumed a Z=28=N core, with the valence nucleons occupying the $1f_{5/2}$, $2p_{3/2}$, $2p_{1/2}$, and $1g_{9/2}$ orbits. The $N_p.N_n$ scheme[3] suggests that in the A≃70 region, the $1f_{7/2}$ orbit also plays an active role in producing collectivity. Thus, in these calculations, we modify the core accordingly. The role played by the $1f_{5/2}$ and $1g_{9/2}$ orbits is very important in this region. Our results compare reasonably well with other models, e.g. the IBM-Plus-Two-Fermion Model[4].

1. *Nuclear Structure of the Zirconium Region*, eds: J. Eberth, R. A. Meyer and K. Sistemich, (Springer-Verlag, Heidelberg, 1988).

2. E. D. Kirchuk, P. Federman and S. Pittel, *Phys. Rev.* **C47** (1993) 567.

3. R. F. Casten, *Phys. Rev. Lett.* **54** (1985) 1991.

4. S. T.Hsieh, H. C. Chiang and Der-Sam Chuu, *Phys. Rev.* **C46** (1992) 195.

Back-Angle Anomaly and Coupling between Seven Reaction Channels of ^{12}C+^{24}Mg using Algebraic Scattering Theory with Spin.

A. Lépine-Szily [1,2], M. M. Obuti [1,3], R. Lichtenthäler Filho [1],
J. M. Oliveira Jr.[1], and A. C. C. Villari [1,2]

1) IFUSP, DFN C.P. 20516, 01498 São Paulo, S.P. Brazil
2) GANIL, B.P. 5027, 14021 Caen Cedex, France.
3) SPring-8, JASRI, Hyogo 678-12, Japan.

March 2, 1994

The Algebraic Scattering Theory [1] was used to calculate the angular distributions for the seven reaction channels involving the ^{12}C+^{24}Mg system at E_{cm}=25.2 MeV. The final states considered are the elastic, the three inelastic channels leading to the first 2^+, 4^+ and $2'^+$ excited states in ^{24}Mg and the three transfer reaction channels leading to ^{16}O+^{20}Ne and its first excited states.

The Algebraic Scattering method allows the coupling of all seven states in a straight-forward way. The coupling between elastic, inelastic and α-transfer channels explains the backward anomaly observed in all angular distributions. The spin of the excited states is taken into account and spectroscopic informations are used, whenever possible. The results of calculations are compared to experimental measurements [2].

References:
1. Y. Alhassid and F. Iachello, Nucl. Phys. A501, 585 (1989)
2. A. Lépine-Szily et. al., Phys. Lett. B243, 23 (1990)

The Effect of S and D Pair Interaction on the Energy Level Statistics

Yu-xin Liu[a,b], En-guang Zhao[a,b], Gui-lu Long[c] and Hong-zhou Sun[a,b]

[a] CCAST(World Laboratory), P. O. Box 8730, Beijing 100080, China
[b] Institute of Theoretical Physics, Academia Sinica, P. O. Box 2735, Beijing 100080, China[1]
[c] Department of Physics, Tsinghua University, Beijing 100871, China

The energy level statistics of a quantum system has long been a significant topic in quantum mechanics. It has been pointed out that the fluctuations of energy levels around the average density depend only on the symmetry of the Hamiltonian. The Hamiltonian with a dynamical symmetry leads to a Poisson distribution. The breaking of the dynamical symmetry leads the distribution to Gaussion orthogonal ensemble (GOE). Allhassid and the collaborators' investigations on the chaotic properties of the Interacting Boson Model (IBM) show that near the $SU(3)$, $O(6)$ and $SU(5)$ symmetries of the IBM the fluctuation behavior of the states with angular momentum $L \geq 2$ is close to the Poisson statistics, changing gradually to the GOE statistics as the interaction moves away from these symmetries. It is known that the transition from the $O(6)$ symmetry to the $SU(3)$ symmetry is achieved by a continuous parameter χ in the quadrupole operator

$$\hat{Q}^{(2)}_\mu = (s^\dagger \tilde{d} + d^\dagger s)^{(2)}_\mu + \chi (d^\dagger \tilde{d})^{(2)}_\mu.$$

As χ changes from 0 to $\pm\frac{\sqrt{7}}{2}$, the system changes from the state holding $O(6)$ symmetry to the one possessing $SU(3)$ symmetry. In the framework of IBM, the collective motion of nucleus is generated by the coherent nucleon pairs $S(l=0)$ and $D(l=2)$. In the view of the OAI mapping, the parameter χ can be expressed as

$$\chi = \frac{\langle D|\hat{Q}^{(2)}\|D\rangle}{\langle D|\hat{Q}^{(2)}|S\rangle}.$$

It indicates that the energy level statistics of the nuclei in the transitional region from $O(6)$ symmetry to $SU(3)$ symmetry is determined by the competition between the quadrupole interaction between D pairs and that changing S pair to D pair.

---

[1] Mailing address

GAMMA EMISSION CALCULATIONS IN THE CASCADE-EXCITON MODEL

S.G. Mashnik
Bogolubov Theoretical Laboratory, Joint Institute for Nuclear Research
141980 Dubna, Moscow Region, Russia

Ye.S. Golubeva, A.S. Iljinov
Institute for Nuclear Research of the Russian Academy of Sciences
60-the October Anniversary prospect 7a, Moscow 117312, Russia

The Cascade-Exciton Model (CEM) of nuclear reaction [1] have been extended to describe the emission of energetic photons from intermediate energy nucleon-induced reactions. The CEM uses the Monte Carlo simulated method and assumes that the reactions occur in three stages: the intranuclear cascade, pre-equilibrium and the evaporative (+ fission) ones. We handled the emission of gamma rays at all these stages.

We assume that at the cascade stage of the reaction photons originate incoherently in single proton-neutron bremsstrahlung processes $np \to np\gamma$ and from elementary processes of radiative captures $np \to d\gamma$, and, following ref. [2], describe the late ones applying the detailed balance to the deuteron photoabsorption.

To describe photon emission at the second, pre-equilibrium stage of the reaction, we take into account both *Single-particle transition*, where a single nucleon changes its position (energy), and the amount of energy released is carried out by a γ [3], and *Two-particle transition*, also called as a *quasideuteron mechanism*, which assumes two correlated nucleons (a quasideuteron) to be responsible for the γ emission [4].

At last, at the third, evaporative stage of the reaction we handle the statistical mechanism of gamma emission using the detailed balance to the inverse process, photo-absorption, in a manner similar to ref. [5].

The first results obtained using the present approach are encouraging: we described satisfactorily data on nucleon-induced γ-emission in the energy range up to 200 MeV and analyzed the relative role of different photon emission mechanisms in these reactions.

References

[1] K.K. Gudima, S.G. Mashnik and V.D. Toneev, Nucl. Phys. **A401** (1983) 329.

[2] A.S. Iljinov, M.V. Kazarnovsky, B.V. Krippa, G.K. Matushko, Proc. 3rd All Union Seminar on Experiments at Moscow Meson Factory, (Moscow, 1983), p. 262.

[3] E. Běták, J. Kopecky, F. Cvelbar, Phys. Rev. **C46** (1992) 945.

[4] P. Obložinský, Phys. Rev. **C40** (1989) 1591.

[5] A.S. Iljinov, Yu.Ts. Oganessian, E.A. Cherepanov, Sov. J. Nucl. Phys. **33** (1981) 526.

Algebraic methods in molecular physics: recent developments of the Vibron Model

A. MENGONI[a,b] and T. SHIRAI[b]

[a]*ENEA, V.le G. B. Ercolani 8, 40138 Bologna, Italy*

[b]*JAERI, Tokai-mura, Ibaraki 319-11, Japan*

ABSTRACT

Recently, much attention has been gained by an algebraic model developed for the description of molecular rotational and vibrational degrees of freedom. The model, the Vibron Model (VM), has been originally [1] introduced for the description of rotational and vibrational motion of diatomic molecules. The VM has been subsequently extended for the treatment of polyatomic molecules. At the mean time, a number of other molecular properties have been derived in connection with the VM. Of those, the most relevant have been the calculation of the scattering amplitude for electron-molecule collision processes [2], the description of infrared transition intensities [3], the inclusion of electronic excitations [4] (with the inclusion of fermionic degrees of freedom in the boson Hamiltonian).

Most recently [5] the mean-field approximation (MFA) has been applied to the VM Hamiltonian in an attempt to provide a *geometrical* picture of an otherwise abstract algebraic model. This kind of approach has produced important results when applied to the Interacting Boson Model (IBM). In fact, in that case the three dynamic symmetries of the IBM have been shown to correspond to particular nuclear geometrical structures.

Recently we have derived a simple prescription which allows for a geometrical interpretation of the VM. More precisely, we have shown that there is a simple relation between the geometrical variables, e.g. the internuclear distance in a diatomic molecule, and the algebraic variables employed in the MFA. We have shown that there is a strong connection between the Morse potential and the MFA to the VM. This holds true not only for the energy spectrum but also for the potential energy surface. The latter is the ultimate goal for any model whose objective is the description of molecular properties.

Here we will present the latest developments of the works based on the VM. A review will be given on the recent applications of the VM to the calculation of vibrational spectra of polyatomic molecules. At the mean time an account will be given on the results obtained using the MFA to the VM with its implication on the interpretation of the algebraic Hamiltonian in terms of geometrical variables. The results of the calculation of the energy surfaces of diatomic molecules will be shown in comparison with the Morse potentials. Finally, the dipole moment function of a diatomic molecule will be shown as derived from the MFA to the VM.

[1] F. Iachello, *Chem. Phys. Lett.* **78**, 581 (1981).
[2] R. Bijker, R. D. Amado and D. A. Sparrow, *Phys. Rev. A* **33**, 871 (1986).
[3] F. Iachello, A. Leviatan, and A. Mengoni, *J. Chem. Phys.* **95**, 1449 (1991).
[4] A. Frank, R. Lemus, and F. Iachello, *J. Chem. Phys.* **91**, 29 (1989).
[5] A. Leviatan and M. K. Kirson, *Ann. Phys.* **188**, 142 (1989).

STRUCTURE ALTERATIONS IN ALIGNED MULTIPHONON BANDS

V.M.Mikhajlov*, A.K.Vlasnikov*, D.N.Dojnikov**

*Institute of Physics, St.Petersburg State University, 198904 Russia
**Ioffe Physico-Tekhnical Institute, St.Petersburg, 194021 Russia

We investigate the influence of accumulation of quasiparticles α^+ constituting multiphonon aligned states $|I> = (D_2^+)^{I/2}|0>$ on values of pairing gap Δ_I, chemical potential λ_I and phonon structure $\psi_{ik}^{(I)}$ ($D_2^+ = \sum \psi_{ik}^{(I)} [\alpha_i^+ \alpha_k^+]_2^{(2)}$; i,k are levels in spherical field). Our Hamiltonian $\hat{H}$ consists of a single–particle part, monopole pairing and quadrupole particle–hole interactions. In calculations we use a canonical basis (which allows to represent D_2^+ in the form $\sum_1 \rho_1 \alpha_{1+}^+ \alpha_{1-}^+$) and apply the symmetrical polynomial technique [1]. The alteration of Δ, λ and ψ with I increasing is governed by the particle number conservation on the average and the minimization of $<I|\hat{H}|I>$. The results of calculations show that with I growing (when phonon quasiparticles block single-particle levels) Δ_I and collectivity of D_2^+ decrease. The alteration of phonons and vacuum in the aligned band reveals itself in the reduction of $B(E2; I \rightarrow I-2)$ in comparison with the case of constant Δ, λ and ψ. The values of $\Delta_I/\Delta_{I=0}$ and $B(E2; I \rightarrow I-2)/B(E2; 2 \rightarrow 0)$ vrs I are presented in the figures for 20 valence particles distributed over six levels.

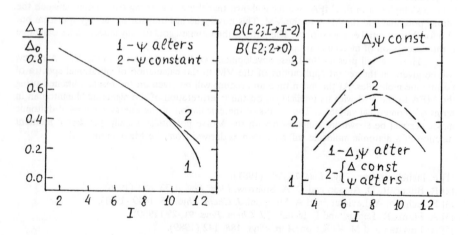

1. A.K.Vlasnikov, V.M.Mikhajlov. Bull.Acad.Sci.USSR (phys.ser), 1991, v.55, p.38.

INFLUENCE OF RESIDUAL INTERACTIONS ON AVERAGE FIELD AND PAIRING CORRELATIONS IN MULTIPHONON ALIGNED BANDS

V.M.Mikhajlov, A.K.Vlasnikov *(St.-Petersburg State University)*

The model [1] describing the multiphonon aligned states $|I> \sim (\pi_I D_\pi^+ + \nu_I D_\nu^+)^{I/2}|0>_I$ (where D_ρ^+ is the collective proton $\rho = \pi$ or neutron $\rho = \nu$ quadrupole operator with maximum spin z-projection and $|0>_I$ is the quasiparticle vacuum for the state with spin I) is extended to include the influence of residual forces on single–particle energies $\epsilon_i(\rho)$ and pairing gaps $\Delta(\rho)$. Retaining the most coherent contribution from the quadrupole particle-hole factorized interaction between unlike particles we obtain:

$$\epsilon_i(\rho) = \epsilon_i^{(0)}(\rho) + \kappa_{\pi\nu}\left\{2Q_0(\rho')\left[\sum_k{}^{(1)} q_{ik}L_{ik}L_{ii}^{-1}\tilde{\Omega}_i^{-1} <I|(\alpha_i^+\bar{\alpha}_k)_0^{(2)}|I>\right]_{(\rho)} - \right.$$
$$\left. - Q_2(\rho')\left[\sum_k{}^{(2)} q_{ik}M_{ik}L_{ii}^{-1}\tilde{\Omega}_i^{-1} <I|(\alpha_i^+\alpha_k^+)_2^{(2)}|I-2>\right]_{(\rho)}\right\};$$

$$\Delta(\rho) = G_0(\rho)\left[\sum_k L_{kk}\tilde{\Omega}_k\right]_{(\rho)} + \kappa_{\pi\nu}\left\{Q_2(\rho')\left[\sum_{ik}{}^{(1)}t_{ik}M_{ik} <I|(\alpha_i^+\alpha_k^+)_2^{(2)}|I-2>\right]_{(\rho)} - \right.$$
$$\left. -2Q_0(\rho')\left[\sum_{ik}{}^{(2)}t_{ik}L_{ik} <I|(\alpha_i^+\bar{\alpha}_k)_0^{(2)}|I>\right]_{(\rho)}\right\}\left[\sum_i \bar{\epsilon}_i^2 e_i^{-3}\tilde{\Omega}_i\right]_{(\rho)}^{-1};$$

where $\epsilon_i^{(0)}(\rho)$ is an unrenormalized single–particle energy, $\bar{\epsilon}_i(\rho) = (\epsilon_i - \lambda)_\rho$; $e_i(\rho) = (\bar{\epsilon}_i^2 + \Delta^2)^{1/2}_\rho$; $L_{ik}(\rho) = (u_i v_k + v_i u_k)_\rho/2$; $M_{ik}(\rho) = (u_i u_k - v_i v_k)_\rho$; $t_{ik}(\rho) = [q_{ik}(\bar{\epsilon}_i e_i^{-2} + \bar{\epsilon}_k e_k^{-2}]_\rho$;
$$\tilde{\Omega}_i(\rho) = (j_i + 1/2)_\rho - <I|(\sum \alpha_{im}^+ \alpha_{im})_\rho|I>;$$
$Q_0(\rho) = <I|[\sum_{ik} q_{ik}M_{ik}(\alpha_i^+\bar{\alpha}_k)_0^{(2)}]_\rho|I>$; $Q_2(\rho) = <I|[\sum_{ik} q_{ik}L_{ik}(\alpha_i^+\alpha_k^+)_2^{(2)}]_\rho|I-2>$;
$\sum_k^{(1)}$ means that levels k and i lie on the one side of the chemical potential λ, while λ is between levels i and k in $\sum_k^{(2)}$.

We have applied this method to the two–component single–j level model and compared results with [2] where the vacuum was determined separately for each state

$$(D_\pi^+)^K(D_\nu^+)^{I/2-K}|0>_{I,K}.$$

Both approaches give similar results for $I \leq \max(A_\pi, A_\nu)$, A_ρ is the particle number. For greater I the present method leads to smaller energies and greater $B(E2; I \to I-2)$. At $I > A_\pi + A_\nu$ when pairs of particles with spin greater then 2 play important role the method presented gives energies diminishing with angular momentum growing, thus structural changes of the trial wave function are demanded.

1. Vlasnikov A.K., Mikhajlov V.M. — Bull.Akad.Sci. phys.ser., 1993, v.57, n.9, p.168.
2. Vlasnikov A.K., Mikhajlov V.M. — ibid., 1992, v.56, n.11, p.98.

SUB-BARRIER FUSION IN CLASSICAL EQUATION OF MOTION APPROACH

R.C. Misra, Dept. of Physics, VSSD (PG) College, Kanpur - 208002 INDIA
(Kanpur University), Kanpur

We made some calculations for fusion crosssections of light mass heavy nuclei 12-C+12-C using classical equation of motion (CEOM) approach/1/. In CEOM approach each nucleon is considered as point particle without spin interacting through suitable two-body NN interaction/2/ given below,

$$V^N_{ij} = -V_o (1 - C/r_{ij}) \exp - (r_{ij}/r_o)$$

In the above relation V_o, C and r_o are depth parameter, repulsive core radius and range parameter respectively. These parameters are adjusted suitably to reproduce the binding energy and average size (rms radius) of the colliding nucleus very close to experimental values. These are given in Table below. The ground state properties of the colliding nuclei are calculated through computer code STATIC.

The colliding nuclei are placed in centre of mass frame far away from each other. These nuclei are bombarded with given energy and impact parameter in a given orientation. The critical impact parameter responsible for fusion is obtained. In this way we calculate impact parameters for various relative random orientations of colliding nuclei.

Results : The calculated values of fusion cross sections for the **chosen** system at Ecm = 7.43 MeV (Coulomb barrier Ec = 7.86 Mev) is 87 mb whereas experimental value is 286±20/3/. During collision deformation is produced in nuclear system and arrangement of nucleons changes. In this arrangement if the neutron splash is coming out, it produces fusion. In our case the fusion was persistent upto 50 NS (the time for which we made the calculations). We also calculated the shape deformation of colliding nuclei at different timings of the collision event. The results are very encouraging. In this study we found that clusters remain in the bound state till the end.

Acknowledgement : The author is thankful to Prof. Y.R. Waghmare, IIT/Kanpur for his kind cooperation and useful discussions in the above study. The author also acknowledges thanks to UGC, New Delhi for partial financial assistance in the above work.

References : /1/ R.C. Misra & Y.R. Waghmare, Proc. ICNRM, Calcutta 1989.
/2/ H.S. Koehler & Y.R. Waghmare, Nucl Phys 66 (1965) 261, Y.R. Waghmare, Phys Rev B136 (1964) 1261.
/3/ D G Kovar et al, Phys Rev C20 (1979) 1305.

V_o (MeV)	C (fm)	r_o (fm)	Calculated -BE (Mev)	Calculated rmsr (fm)	Experimental -BE (Mev)	Experimental rmsr (fm)
38.2×10^5	2.10	0.2	92.860	2.370	92.16	2.460

A TOY MODEL WHICH LEADS TO IDENTICAL BANDS

S.A. Moszkowski
Department of Physics, UCLA, Los Angeles, CA 90024
M. Mukerjee
Scientific American, New York, NY 10017

Abstract of contribution for IBM conference, Padua, June 1994

Consider identical particles in a two dimensional p shell. We suppose there are several species of such particles. For each species, there are four possible single particle states, $a\uparrow$, $a\downarrow$, $\overline{a}\uparrow$, and $\overline{a}\downarrow$. The interaction used in our toy model is essentially a δ interaction, for which all non-vanishing matrix elements are equal. It acts only between particles with oppositely directed spins, and leads to pairing in both $L = 0$ and $L = 2$ states. For s species, the energy of an $L = 0$ pair, which can be regarded as an S boson, is $2s$ in units of the interaction strength, while the energy of an $L = 2$ pair, a D boson, is s. In our model, the D-state pairing involves not only scattering of a pair from aa to aa and from aa to bb, but also from ab to ab and from ab to ba. Within this model, the wavefunctions and energies of the Yrast states (lowest states of given angular momentum) have a very simple form: For n particles and s species, the ground state, which is given as an antisymmetrized combination of $(n/2)$ S bosons, has energy ns.

The Yrast state with given L can be constructed by antisymmetrizing the product of $(n-L)/2 S$ bosons and $L/2 D$ bosons (each bosons being expressed as a nucleon pair). Its energy relative to the ground state is just $(L/2)s$, independent of the number of particles (provided, of course, that $n \geq L$). Thus the Yrast states form identical bands, and there is no interaction among the S and D bosons.

The non-Yrast states are more complicated, however. For example, the first excited 0^+ state, which can be regarded as a two phonon state with D and $\overline{D}$ boson, has excitation energy $2s - n/2 + O(1/s) + \cdots$ However, in the limit $s \gg n$, the interaction between D and $\overline{D}$ boson becomes insignificant compared to their single boson energies.

The effect of a spin-orbit splitting can also be studied. The identical band structure is only weakly affected for large number number of species.

On some properties of the valence bosons in relation with their rank.

G. Mouze
Laboratoire de Chimie Atomique, Faculté des Sciences
06108 Nice cedex 2, France

Useful for the description of the energetic structure of nuclear valence shells is the concept of *from-core* detachment energy (σ) of a valence nucleon or dinucleon. Thus, the total binding energy of the valence shells of an even-even nucleus is:

$$E_B(Z,N) - E_B(Z_0,N_0) = \sum_{i=1}^{N_\pi} \sigma_{2pi} + \sum_{j=1}^{N_\nu} \sigma_{2nj} + \sum_{i=1}^{N_\pi}\sum_{j=1}^{N_\nu} \varepsilon_{2pi\text{-}2nj},$$

where the σ' s, as well as the neutron-proton energies can be calculated from *experimental* mass data. The rule that detachment energies decrease as a function of the rank (i or j) of the nucleon or dinucleon admits almost no exception. The decrease follows a linear law. Even the detachment energy of the second fermion of a given boson enjoys these properties, and the slope of σ II may differ from that of σ I. Is this systematic decrease of the various detachment energies as a function of the rank due to a Coulomb effect, or might repulsive forces of nuclear origin play an important role ? An answer to this question can be given.

The resemblance of the variation of the σ's to that of some of the ε's as a function of the rank too needs an explanation. In the $1f_{7/2}$ valence shells, we find that the n-p interaction energies behave as linear functions of the difference of rank i - j of the partners as soon as i and j are different. For instance, the law of variation as a function of rank j is the same for $\varepsilon_{2p1\text{-}nj}$, and $\varepsilon_{p1\text{-}2nj}$, whereas $\varepsilon_{2p1\text{-}2nj}$ vary according to a very different law. But the variation of the detachment energy σ_{nj} is similar to that of $\varepsilon_{p1\text{-}nj}$ and $\varepsilon_{p1\text{-}2nj}$, whereas the variation of σ_{2nj} is similar to that of $\varepsilon_{2p1\text{-}2nj}$. In fact, these kinds of neutron-proton interaction energies could be considered as detachment energies from a proton, or diproton, taken as reference, instead of from the core, i.e. as new kinds of ionization potentials. However, for explaining the similarity of the variation of $\varepsilon_{p1\text{-}2nj}$ to that of $\varepsilon_{p1\text{-}nj}$, it seems necessary to assume that the constitutive fermions of the 2nj boson occupy diametrically opposed positions on the 2nj orbit.

The pairing energy, e.g. P_n, can be defined as $P_n = \sigma_{nII} - \sigma_{nI}$, instead of being a difference $S_{nII} - S_{nI}$ of separation energies. According to this new definition, P has to be considered as a constant in a given subshell, but only at rank one. In fact, P is a function of the rank in the subshell, but can be either an increasing one, or an decreasing one.

EFFECTIVE BOSON NUMBER AND ROLE OF PROTON-NEUTRON INTERACTION

H. Nakada

Center for Theoretical Physics, Yale University, New Haven, CT 06511, U. S. A.

In the OAI mapping with degenerate single-particle orbits[1], the s-boson degree of freedom and the boson number in the IBM has been connected with the factor $\sqrt{n-v}$ appearing through the seniority reduction formula. In the boson mapping[2] based on the so-called number-conserving quasiparticle method[3], a similar factor emerges also in non-degenerate systems, as an effect of the approximate number conservation. On this basis we define effective boson number as well as effective degeneracy. This effective boson number is required to get rid of a sizable N_d-dependence of a parameter, and has a microscopic origin connected with the number conservation.

This formulation of effective boson number and degeneracy is applied to $Z = 50-82$ and $N = 50-82$ shells. To clarify the effect of the proton-neutron interaction, we assume shell-model hamiltonian to be nucleus-independent. The hamiltonian is fixed so as to reproduce the energy levels of Sn-isotopes, $N = 82$-isotones and ^{132}Te.

In the $N = 50-82$ shell, the effective boson number is always close to the value normally adopted in the IBM. For $N = 82$ isotones (*i.e.*, the $Z = 50-82$ shell), the effective boson number is slightly quenched around the $Z = 64$ subshell. The bosons behave as hole-bosons at ^{144}Sm. On the other hand, once N departs from 82 and the proton-neutron interaction is set on, the proton bosons abruptly change their character, behaving as particle-bosons. Coincidentally, the effective boson number shows sizable increase.

Compared with Scholten's work[4], the present study will be characterized by the following points: (1) The origin of the effective boson number is directly connected with the number conservation. (2) The character change of bosons occurs even in spherical nuclei, not invoking deformed mean-field. (3) The role of the proton-neutron interaction in the effective boson number is manifestly shown.

There remains several problems; the $Z = 64$ subshell effect seems too weak compared with Ref.[5], and the present definition is not applicable at the point of shell closure.

References

[1] T. Otsuka, A. Arima and F. Iachello Nucl. Phys. **A309**(1978)1.
[2] H. Nakada and A. Arima, Nucl. Phys. **A524**(1991)1.
[3] C.-T. Li, Nucl. Phys. **A417**(1984)37.
[4] O. Scholten, Phys. Lett. **B127**(1983)144; in Proc. Workshop on Bosons in Nuclei, (World Scientific, 1984).
[5] R. F. Casten and D. D. Warner, Rev. Mod. Phys. **60**(1988)389.

Application of the Resonating Hartree-Fock Random Phase Approximation to the Lipkin Model

Seiya NISHIYAMA

Department of Physics, Kochi University, Kochi 780, JAPAN

To approach to topical many-body problems of Fermion systems with large quantum fluctuations, Fukutome and the present author have proposed theories called the resonating Hartree-Fock[1] (Res HF) and resonating Hartree-Bogoliubov[2] (Res HB) approximations by resonance of multiple mean field wave functions. The Res HF/HB approximations are based on an exact coherent state representation of Fermion systems on a unitary group and on a special orthogonal group, respectively. Each correlated ground state wave function is approximated by a superposition of non-orthogonal Slater determinants (S-dets) or HB wave functions with different correlation structures. Resonance of the wave functions occurs if the mean field energy functional has multiple local energy minima with near energies.

Very recently, in order to make clear essential features of the Res HF approximation and to show its superiority over the usual HF one, we have applied the approximation[3] to the exactly solvable Lipkin model[4] by making use of a newly developed orbital optimization algorithm.[5] For simplicity, we have assumed that a Res HF wave function is superposed by only two S-dets which give corresponding two local energy minima of monopole "deformation". The self-consistent Res HF calculation so as to minimize the energy functional including up to the second order variation completely determines the mixing coefficients and orbitals in each S-det and gives the excellent ground state correlation energy.

There are excitations due to small vibrational fluctuations of the orbitals and mixing coefficients around their stationary values that are generalization of and similar to collective excitations described by the usual random phase approximation (RPA). We have derived new approximations called the Res HF/HB RPAs in a manner analogous to the time dependent HF/HB derivations of the usual RPA.[6],[7]

The matrix of the above second order variation has the same structures as those of the matrix of the Res HF RPA[6] because the orbital variations are made in the full variation space. Then the quadratic steepest descent is considered to include surely the whole effect of RPA type fluctuations. If the convergence is achived after sweeping the whole variation space, the optimized orbitals should be supposed to contain both RPA type fluctuations up to higher orders and their mode-mode couplings since the Fermion nature of the wave function is kept all through the orbital optimization process. It is very important and interesting to apply the Res HF RPA to the Lipkin model and to prove such an argument by using the stationary solution of the Res HF approximation to the Lipkin model.[3]

References

1) H. Fukutome, Prog. Theor. Phys. **80** (1988), 417.
2) S. Nishiyama and H. Fukutome, Prog. Theor. Phys. **85** (1991), 1211; J. Phys. **G18** (1992), 317.
3) S. Nishiyama, Report of the Seminar and Workshop on *Large Amplitude Collective Motion* held at the National Institute for Nuclear Theory, Univ.of Washington, DOE/ER/40561-121-INT93-12-02.
4) H. J. Lipkin, N. Meshkov and A. J. Glick, Nucl. Phys. **62** (1965), 188.
5) A. Ikawa, S. Yamamoto and H. Fukutome, J. Phys. Soc. Jpn. **62** (1993), 1653.
6) H. Fukutome, Prog. Theor. Phys. **81** (1989), 342.
7) S. Nishiyama and H. Fukutome, Prog. Theor. Phys. **86** (1991), 371.

In-Beam Pure Spectra of ^{102}Cd

S. Rastikerdar
Physics Department, Isfahan University,
Isfahan, Iran

The neutron-deficient ^{102}Cd were produced and identified in-beam following the reaction ^{58}Ni+^{50}Cr at 195 MeV bombarding energy of the ^{58}Ni beams. Gamma rays from the reaction were detected using an array of 9 BGO shielded, intrinsic germanium detectors coupled to the Daresbury Recoil Separator[1]. After manipulation of the data to make the energy loss signal energy independent, a two dimensional spectrum of energy loss against gamma ray energy was produced and by careful analysis of the data the pure gamma ray spectrum associated with ^{102}Cd was generated for the first time. Comparison were made with the recently published data concerning the above isotope[2]. Good agreement was obtained. Since our spectra is pure, it not only helps to remove several ambiguities in the above data, but gives us several new gamma rays associated with ^{102}Cd for the first time. In this paper we report about the experiment, the data and its analysis together with the results in detail.

References:

(1) A. N. James et.al, Nucl. Instr. and Methods A267 (1988) 144.

(2) D. Alber et.al., Z. Phys. A-Hadrons and Nuclei 344 (1992) 1.

The SU(3) Generators at Superdeformation

K. Sugawara-Tanabe and A. Arima[†]

Otsuma Women's University, Tama, Tokyo 206 Japan
[†] *Riken, Wako, Saitama 351-01 Japan*

We have constructed the new $SU(3)$ group at superdeformation, so long as we can neglect spin-orbit force in a deformed axially symmetric Nilsson potential with a rational ratio between the two frequencies $\omega_\perp : \omega_z = 2 : 1$ [1]. The operators to describe this $SU(3)$ group are $\{a_x^\dagger, a_y^\dagger, b^\dagger, a_x, a_y, b\}$, where $\{a_k^\dagger, a_k \ ; \ k = x, y, z\}$ are the harmonic oscillator bosons and $\{b^\dagger, b\}$ are the new boson operators constructed from $\{a_z^\dagger, a_z\}$ using an analogy to Holstein-Primakoff transformation in spin operators and agree with the ladder operators [2]. Since we are constructing new bosons from the product of the original bosons, there is the degeneracy of the vacuum $|0>$ and $a_z^\dagger|0>$, where $|0>$ is the vacuum of a_z. Subsequently we have two kinds of bosons corresponding to N_{sh} = even and odd cases, and this is closely related to the dual nature of the new $SU(3)$ group. We found 8 generators that fulfill the commutation relation of $SU(3)$ from the byproduct of this set of boson operators in analogy to Elliot's $SU(3)$ model [3]. The new generators are available in the space neglecting $\Delta N_{sh} \neq 0$ in contrast to Elliot's case where $\Delta N \neq 0$ is neglected. From these generators we found the Casimir operator, and for the fixed value of the Casimir operator there correspond two superdeformed shells, N_{sh} and $N_{sh}+1$, or in other words new $SU(3)$ group is the product of $SU(3) \times SU(3)$. In the reduction of this $SU(3)$ group to $R_3 \times U_1$ subgroups, we find two quantum numbers to classify the group, i.e. $<\tilde{Q}_0>$ and $n_\perp/2$ where $n_\perp$ denotes the boson number in the axis $\perp$ perpendicular to the symmetry axis z. One N_{sh} contains both parity levels, and different parity levels correspond to the integer or half integer values of $n_\perp$, respectively. However these generators are not realistic except for $\tilde{L}_0$ operator, as the $\tilde{Q}_{\pm 1}$ and $\tilde{L}_\pm$ operators mix the different parity states. So we must take care of the difference between the realistic operators and these new generators.

[1] K. Sugawara-Tanabe and A. Arima, Nucl. Phys. A557(1973)157C.
[2] W. Nazarewitch and J. Dobaczewski, Phys. Rev. Lett. 68(1985)154.
[3] J. P. Elliot, Proc. Roy. Soc. A245(1958)128;562.

Angular anisotropy coefficients and Shape coexistence in Hg isotopes

K. Sugawara-Tanabe and K. Tanabe[†]

Otsuma Women's University, Tama, Tokyo 206 Japan
[†] Saitama University, Urawa, 338 Japan

We have calculated the angular distribution coefficient of γ-rays, $a_2(E_\gamma)$ from the giant-dipole-resonances for the high-spin and low-spin levels over Hg isotopes from $A = 190$ to 194. At first we solved the number- and angular momentum-constrained HFB equation self-consistently with only 4 interaction strength parameters which can reproduce the yrast energy level sequence, i.e. a total of 40 energy levels reasonably well. Then we have extended our formalism to include the temperature effect quantum mechanically. Here temperature is the indicator to express the excitation energy from the yrast band where $kT = 0$. Using the same 4 parameters we applied these thermal CHFB and thermal RPA equations for the thermal giant-dipole-resonances. The deformation can be estimated from the behaviour of the three strength functions S_x, S_y and S_z. We found the shape change occurs according to the change of angular momentum, mass number and temperature. For example in the yrast band of ^{192}Hg, the 0^+ state has oblate shape, but the 10^+ state spherical shape and the 30^+ state prolate shape. In ^{190}Hg the 30^+ state has spherical shape at $kT = 0$, but changes to oblate shape at $kT = 3$ MeV. In ^{194}Hg among the region within $kT = 0$ and 3 MeV, the 0^+ state belongs to oblate shape, while the 30^+ state to prolate shape. The resonance energy moves to the lower energy and the peak height of the absorption cross section decreases according with the increasing temperature or the increasing angular momentum. These characteristics are explained easily from the thermal energy weighted sum rule [1] and agrees with the experimental observation. The anisotropy coefficient $a_2(E_\gamma)$ is roughly approximated as proportional to the ratio of the strength function of x-axis S_x to the sum of three strength functions. When three components of the strength functions coincide each other, $a_2(E_\gamma)$ becomes 0. The critical energy E_γ^c giving $a_2(E_\gamma^c) = 0$, is lower for the higher temperature or lower angular momentum than the lower temperature or higher angular momentum, respectively. With increasing temperature, $a_2(E_\gamma)$ in the lower energy region is enhanced from the negative value where S_z dominates to the positive value where S_x or S_y dominates. On the other hand $a_2(E_\gamma)$ in the higher energy side is independent of temperature. These behaviours of $a_2(E_\gamma)$ are consistent with the experimental observations.

[1] K. Tanabe and K. Sugawara-Tanabe, Prog. Theor. Phys. 76(1986)356.

SCIENTIFIC PROGRAM

Monday morning, June 13, 1994

WELCOME

M. Mammi, Deputy-Rector of the University of Padova

A. Bettini, Vice-President of the Istituto Nazionale di Fisica Nucleare

---

CEREMONIAL SESSION

Chair: A. Bettini

F. Iachello	Open Problems in the Interacting Boson Model
I. Talmi	From Independent Nucleons to Interacting Bosons

Chair: D. M. Brink

R. F. Casten	Robust Predictions of the IBM: Implications for Nuclear Structure
T. Otsuka	IBM for O(6) Nuclei
J. Jolie	How Far Can a Dynamical Symmetry Approach be Pursued?

Monday afternoon, June 13, 1994

Chair: P. Lipas

A. Richter	The Scissors Mode and Related Modes Revisited in the IBM and Other Models
N. Lo Iudice	From Semiclassical to Microscopic Descriptions of Scissors Modes
G. Molnar	Mixed-Symmetry States in O(6) Nuclei
A. F. Diallo	Asymptotic Evaluation of F-Spin Content and M1 Sum Rule

Chair: P. G. Bizzeti

K. Heyde	Particle-Hole Excitations in the Interacting Boson Model
B. R. Barrett	Formulation and Application of Multiparticle and Multihole Configurations within the IBM-2
A. Leviatan	Partial Dynamical Symmetries in the Interacting Boson Model
S. Kuyucak	Applications of the 1/N Expansion to Nuclear Structure and Reaction Problems

Tuesday morning June 14, 1994

Chair: R. A. Ricci

A. Arima	From Seniority to Collectivity
P. von Brentano	Multiphonon Excitations in Nuclei
D. Rowe	Microscopic Theory and the Interacting Boson Model

Chair: P. Federman

H. B. Geyer	Boson Mappings and Phenomenological Boson Models
C. Johnson	Finite Boson Mappings of Fermion Systems
N. Yoshinaga	On Favored Pairs and Renormalization Effects from Non-Collective Pairs

Tuesday afternoon June 14, 1994

Chair: P. J. Brussaard

D. Vretenar	Broken Pairs and High-Spin States in the Interacting Boson Model
C. J. Lister	Testing the IBM Plus Broken Pair Model at High Spin in the A=80 Region
J. Dobes	Boson Mappings and Microscopy of the IBM

Chair: O. Scholten

P. Van Isacker	Symmetries in Odd-Mass Nuclei and Their Applications
J. Gizon	Low Energy States in Odd-A Xe, Ba, Ce Isotopes: Experiments and Description by the IBF Model
J. M. Arias	Intrinsic Frame Description of Composed Boson-Fermion Systems
J. Dukelsky	The Boson-Fermion Interaction
R. V. Jolos	Interacting Boson Fermion Model for Strongly Deformed Nuclei
V. G. Zelevinsky	Soft Mode in a Fermi System

Wednesday morning, June 15, 1994

Chair: M. Vergnes

A. Vitturi	Description of Octupole-Deformed Nuclei within the Interacting Boson and Interacting Boson-Fermion Models: An Intrinsic-Frame Approach
N. V. Zamfir	E1 and E3 Transition Rates in the sdf-IBA
Y. D. Devi	sdg Interacting Boson Model: Some Analytical and Numerical Aspects
M. Honma	Superdeformation and IBM
J. A. Cizewski	Supersymmetries in Superdeformed Nuclei

Chair: J. A. Evans

V. G. Soloviev	Contribution of the Two-Phonon Configurations to the Wave Function of Low-Lying States in Deformed Nuclei
D. D. Warner	New Challenges from Radioactive Beams

Wednesday afternoon, June 15, 1994

PARALLEL SESSIONS

SESSION A

Chair: A. Aprahamian

D. Bucurescu	Global Systematics of Unique Parity Quasibands in Odd-A Nuclei
A. Zilges	Electric Dipole Excitations in Rare Earth Nuclei
O. Vogel	IBA Calculation of the Average Gamma Deformation for the $A \approx 130$ Nuclei
M. K. Harder	Anomaly in O(6) ^{196}Pt
W. Andrejtscheff	Evolution of the Triaxial Asymmetry at Variation of the Symmetric Quadrupole Deformation
K. Sugawara-Tanabe	Revival of L-S Coupling Scheme at Superdeformation

SESSION B

Chair: G. Bonsignori

P. Navratil	On the Reconciliation of Microscopic and Fitted Boson g-factors
T. Mizusaki	IBM Approach to the Rotational Damping
A. Mengoni	Algebraic and Geometric Approaches to the Collective Enhancement Factor of Nuclear Level Densities
A. Gelberg	A Relation between F-spin Multiplets and the Strength of M1 Transitions
M. Sugita	Applications of IBM-3 to the $Z \sim N \sim 40$ Nuclei
C. De Coster	Sum Rules in the Proton Neutron Interacting Boson Model

SESSION A

Chair: G. de Angelis

H. Dejbakhsh	Exploring the Validity of Z=38 and Z=50 Proton Closed Shells in Even-Even Mo and Odd-A Tc Isotopes
S. M. Fischer	Supersymmetry in the Pt-Au Region
T. Borello-Lewin	New Experimental Systematics in the $A \sim 100$ Region
N. Yoshida	IBFM Calculation of Xe and Cs Isotopes
P. E. Garrett	Single-Nucleon Transfer Reactions and the IBFFM

SESSION B

Chair: S. A. Moszkowski

E. Lipparini	Surface Plasmons of Metal Clusters in the Interacting Boson Model
D. N. Dojnikov	Description of E1 Transitions in Fermion-Boson Systems
Zs. Dombradi	Structure of Odd-Odd Sb Nuclei
S. Swaminathan	Rotational Bands in the $Sp(6) \supset U(3)$ Algebra of the Fermion Dynamical Symmetry Model
A. V. Voinov	On Possible Equidistance of Some Groups of Levels in Deformed Nuclei at Excitation Energies up to 5 MeV

Thursday morning, June 16, 1994

Chair: G. Goldring

J. P. Elliott On the Number and Isospin Dependence of the IBM-3 Hamiltonian

J. N. Ginocchio The Effect of Isospin Conservation on Collective Motion for N~Z Nuclei

Chair: A. Lepine-Szily

R. Bijker Algebraic-Eikonal Approach to Proton-Nucleus and Electron-Molecule Scattering

R. Lichtenthaler Filho Algebraic Scattering Theory and Its Application to Low Energy Heavy-Ion Reactions

J. Cseh Interacting Boson Techniques in Cluster Studies

A. Frank Algebraic Methods in Molecular Physics

Friday morning, June 17, 1994, National Laboratory at Legnaro

Chair: V. Paar

Y. Ne'eman	QCD-derived Dynamical Foundation for the IBM
Yu. F. Smirnov	Q-Analogs of Some Nuclear Models
Y. Alhassid	Chaos in Nuclei
S. Pittel	Consistent Baryon Mapping of Quark Systems
G. Maino	Weak Interactions in the Interacting Boson-Fermion Model

Friday afternoon, June 17, 1994, National Laboratory at Legnaro

Chair: G. Lo Bianco

N. Benczer-Koller	Magnetic Moments: A Microscopic Look into the Nuclear Structure of Medium Weight and Heavy Nuclei
A. E. Stuchbery	Magnetic Moments in Transitional Nuclei
F. Brandolini	g-factor Measurements in Some Stable Even-Even Heavy Nuclei as a Severe Test of IBM and Other Theoretical Models
A. Wolf	Consistent Description of Magnetic Moments and M1 Transition Rates in the IBA

Chair: A. Covello

A. Ventura	Transitional Odd-Odd Nuclei in IBFM-2
T. Fenyes	Structure of Odd-Odd Ga and As Nuclei, Dynamical and Supersymmetries
N. Blasi	IBFFA Calculations on the Cs Isotopes

****

H. Feshbach	Conference Summary

PARTICIPANTS LIST

Y. Alhassid
Sloane Physics Laboratory
Yale University
New Haven, CT 06511
U.S.A.

C.E.A. Alonso
Depto. FAMN
Facultad de Fisica
Aptdo. 1065
41080 Seville
Spain

W. Andrejtscheff
Inst. for Nucl. Research & Nucl. Energy
Tzarigrad Choussee 72
1784 Sofia
Bulgaria

A. Aprahamian
Department of Physics
University of Notre Dame
Notre Dame, IN 46556
U.S.A.

J. M. Arias
Depto. FAMN
Facultad de Fisica
Aptdo. 1065
41080 Sevilla
Spain

A. Arima
RIKEN
Hirosawa, 2-1, Wako-shi
Saitama, 351-01
Japan

D. Balabauski
Faculty of Physics
Sofia University
Sofia
Bulgaria

B. R. Barrett
Department of Physics
Building 81
University of Arizona
Tucson, AZ 85721
U.S.A.

J. Bar-Touv
Ben-Gurion University
Beer-Sheva
Israel

D. Bazzacco
Dipartimento di Fisica
University of Padova
Via Marzolo 8
I-35131 Padova
Italy

T. Belgya
Institute of Isotopes
H-1525 Budapest
Hungary

R. Bijker
R.J. Van de Graaff Laboratory
P.O. Box 80000
3508 TA Utrecht
The Netherlands

P. G. Bizzeti
Dipartimento di Fisica
Universita di Firenze
Largo E. Fermi 2
50125 Firenze
Italy

N. Blasi
I.N.F.N.
Via Celoria 16
20133 Milano
Italy

G. Bonsignori
Dipartimento di Fisica
A. Righi Via Irnerio 46
40126 Bologna
Italy

T. Borello-Lewin
Instituto de Fisica
Universidad de Sao Paulo
C.P. 20516
Sao Paulo
Brasil

F. Brandolini
Dipartimento di Fisica
University of Padova
Via Marzolo 8
I-35131 Padova
Italy

P. von Brentano
Institut fur Kernphysik
University of Koln
Zulpicher Str. 77
50937 Koln
Germany

D. M. Brink
Dipartimento di Fisica
Universita degli Studi de Trento
Povo
Italy

P. J. Brussaard
Faculty of Physics and Astronomy
P.O. Box 80000
3508 TA Utrecht
The Netherlands

D. Bucurescu
Institute of Atomic Physics
Department of Theoretical Physics
Bucharest Magurele
P.O. Box MG-6
R-76900 Bucharest
Romania

R. F. Casten
Department of Physics
Brookhaven National Laboratory
Upton, New York 11973
U.S.A.

D. Chmielewska
Institute for Nuclear Studies
Dept. P-II
05-400 Swierk/Otwock
Poland

D. C. Choudhury
Polytechnic University
Physics Department
6 Metrotech Center
Brooklyn, NY 11201
U.S.A.

J. Cizewski
Department of Physics
Rutgers University
P.O. Box 849
Piscataway, NJ 08855-0849
U.S.A.

A. Covello
Universita di Napoli
Dipartimento di Scienze Fisiche
Mostra D'oltremare Pad. 19
I-80125 Napoli
Italy

J. Cseh
Institute of Nuclear Research
Debrecen Pf. 51
H-4001 Hungary

C. De Coster
Institute for Nuclear Physics
Proeftuinstraat 86
B-9000 Gent
Belgium

G. de Angelis
Laboratori Nazionali di Legnaro
Istituto Nazionali di Fisica Nucleare
Via Romea 4
I-35020 Legnaro (Padova)
Italy

H. Dejbakhsh
Cyclotron Institute
Texas A&M University
College Station, TX 77843
U.S.A.

Y. D. Devi
c/o V.K.B. Kota
Physical Research Laboratory
Navrangpura
Ahmedabad-380 009
India

A. F. Diallo
Apdo 308
Chitre, Herrera
Republic of Panama

J. Dobes
Institute of Nuclear Physics
Czech Academy of Sciences
CZ 25068 Rez near Prague
Czech Republic

D. Dojnikov
Cyclotron Laboratory
Ioffe PTI
St. Petersburg 194021
Russia

Z. Dombradi
Institute of Nuclear Research
Debrecen Pf. 51
H-4001 Hungary

J. Dukelsky
Grupe de Fisica Nuclear
Facultad de Ciencias
Universidadde Salamanca
37008 Salamanca
Spain

O. Dumitrescu
Institute of Atomic Physics
Department of Theoretical Physics
Bucharest Magurele
P.O. Box MG-6
R-76900 Bucharest
Romania

J. P. Elliott
MAPS
University of Sussex
Falmer, Brighton BN1 9QH
United Kingdom

J. A. Evans
MAPS
University of Sussex
Falmer, Brighton BN1 9QH
United Kingdom

P. Federman
Dep. de Fisica, F.C.E.N.
Ciudad Universitaria
1428 Capital Federal
Argentina

T. Fenyes
Institute of Nuclear Research
Debrecen Pf. 51
H-4001 Hungary

H. Feshbach
6-307 MIT
Cambridge, MA 02138
U.S.A.

S. Fischer
Department of Physics
University of Notre Dame
Notre Dame, IN 46556
U.S.A.

A. Frank
Instituto de Ciencias Nucleares
UNAM
Apdo. Postal 70-543
Mexico D.F. 04510
Mexico

P. Garrett
Universite de Fribourg
Institute de Physique
Perolles
CH-1700 Fribourg
Switzerland

A. Gelberg
Institut fur Kernphysik
University of Koln
Zulpicher Str. 77
50937 Koln
Germany

H. B. Geyer
Institute of Theoretical Physics
University of Stellenbosch
7600 Stellenbosch
South Africa

A. Giannatiempo
Dipartimento di Fisica
Universita di Firenze
Largo E. Fermi 2
50125 Firenze
Italy

J. Ginocchio
Los Alamos National Laboratory
MS B283
Los Alamos, New Mexico 87545
U.S.A.

J. Gizon
Institut des Sciences Nucleaires
53 Avenue des Martyrs
38026 Grenoble Cedex
France

G. Goldring
The Weizmann Institute of Science
Rehovot
Israel

M. Harder
Mathematical Sciences
Watts Building
University of Brighton
Lewes Rd., Brighton
E. Sussex BN2 4GJ
United Kingdom

K. Heyde
Institute for Nuclear Physics
Proeftuinstraat 86
B-9000 Gent
Belgium

M. Honma
Center for Mathematical Sciences
University of Aizu
Tsunuga, Iki-mach
Aizu-Wakamatsu
Fukushima 965
Japan

F. Iachello,
Sloane Laboratory-SPL 66
Yale University
New Haven, CT 06511
U.S.A.

C. Johnson
MS B283
Los Alamos National Laboratory
Los Alamos, NM 87505
U.S.A.

I. P. Johnstone
Department of Physics
Queen's University
Kingston, Ontario
Canada

J. Jolie
Universite de Fribourg
Institute de Physique
Perolles
CH-1700 Fribourg
Switzerland

R. Jolos
Bogoliubov Theoretical Laboratory
JINR
141980 Dubna
Russia

R. Kalinauskas
Institute of Physics
Nuclear Reasearch Laboratory
A.Gostauto 12
2600 VILNIUS
Lithuania

W. H. Klink
Department of Physics
University of Iowa
Iowa City, IA 52242
U.S.A.

N. Benczer-Koller
Department of Physics
Rutgers University
New Brunswick, NJ 08903
U.S.A.

S. Kuyucak
Department of Theoretical Physics
RSPS & Eng.
ANU
Canberra, ACT 0200
Australia

V. S. Lac
MAPS
University of Sussex
Falmer, Brighton BN1 9QH
United Kingdom

S. M. Lenzi
Dipartimento di Fisica
University of Padova
Via Marzolo 8
I-35131 Padova
Italy

A. Lepine-Szily
Instituto de Fisica
Universidad de Sao Paulo
C.P. 20516
Sao Paulo
Brasil

A. Leviatan
Racah Institute of Physics
The Hebrew University
Jerusalem 91904
Israel

R. Lichtenthaler
Instituto de Fisica
Universidad de Sao Paulo
C.P. 20516
Sao Paulo
Brasil

P. Lipas
Department of Physics
University of Jyvaskyla
P.O. Box 35
FIN-40351 Jyvaskyla
Finland

C. J. Lister
Physics Division, Building 203
Argonne National Laboratory
9700 South Cass Avenue
Argonne, IL 60302
U.S.A.

Y.-X. Liu
Institute of Theoretical Physics
Academia Sinica
P.O. Box 2735
Beijing 100080
China

G. Lo Bianco
Dipartimento di Fisica
Universita di Milano
Via Celoria 16
20133 Milano
Italy

N. Lo Iudice
Universita di Napoli
Dipartimento di Scienze Fisiche
Mostra D'oltremare Pad. 19
I-80125 Napoli
Italy

S. Lunardi
Dipartimento di Fisica
University of Padova
Via Marzolo 8
I-35131 Padova
Italy

G. Maino
ENEA Cre "E. Clementel"
Viale Ercolani 8
I-40138 Bologna
Italy

V. Manfredi
Dipartimento di Fisica
University of Padova
Via Marzolo 8
I-35131 Padova
Italy

S. Mashnik
Laboratory of Theoretical Physics
JINR
Dubna, Moscow Region
141980 Russia

A. Mengoni
ENEA
Vle G.B. Ercolani 8
40138 Bologna
Italy

T. Mizusaki
Department of Physics
University of Tokyo
Hongo, Bunkyo-ku
Tokyo 113
Japan

G. Molnar
Institute of Isotopes
H-1525 Budapest
Hungary

S. A. Moszkowski
Department of Physics
UCLA
405 Hilgard Avenue
Los Angeles, CA 90024
U.S.A.

M. A. Nagarajan
Laboratori Nazionali di Legnaro
Istituto Nazionali di Fisica Nucleare
Via Romea 4
I-35020 Legnaro (Padova)
Italy

H. Nakada
Department of Physics
Juntendo University
Inba-gun, Chiba 270-16
Japan

P. Navratil
Institute of Nuclear Physics
Czech Academy of Sciences
CZ 25068 Rez near Prague
Czech Republic

Y. Ne'man
Raymond & Beverly Sackler Faculty
of Exact Sciences
Tel-Aviv University
69978 Tel-Aviv
Israel

S. Nishiyama
Department of Physics
Kochi University
780 Kochi-City
Japan

S. Ohta
RIKEN
Hirosawa, 2-1, Wako-shi
Saitama, 351-01
Japan

T. Otsuka
Department of Physics
University of Tokyo
Hongo, Bunkyo-ku
Tokyo 113
Japan

V. Paar
Department of Physics
Bijenicka 32
University of Zagreb
41000 Zagreb
Croatia

M. P. Pato
Instituto de Fisica
Universidad de Sao Paulo
C.P. 20516
Sao Paulo
Brasil

A. Algora Pineda
Institute of Nuclear Research
Debrecen Pf. 51
H-4001 Hungary

S. Pittel
Bartol Research Institute
University of Delaware
Newark, DE 19716
U.S.A.

R. A. Ricci
Dipartimento di Fisica
University of Padova
Via Marzolo 8
I-35131 Padova
Italy

A. Richter
Institut fur Kernphysik
Technische Hochschule
Darmstadt
D-64289 Darmstadt
Germany

D. J. Rowe
Department of Physics
University of Toronto
Toronto
Ontario M55 1A7
Canada

L. Salasnich
Dipartimento di Fisica
Universita di Firenze
Largo E. Fermi 2
50125 Firenze
Italy

O. Scholten
KVI Zernihelaan 25
9747 AA Groningen
The Netherlands

M. Sferrazza
Dipartimento di Fisica
University of Padova
Via Marzolo 8
I-35131 Padova
Italy

S. Shlomo
Cyclotron Institute
Texas A&M University
College Station, TX 77843
U.S.A.

Y. F. Smirnov
Instituto de Fisica
Universidad Nacional
Autonoma de Mexico
Mexico D.F. 001
Mexico

H. M. Sofia
Departamento de Fisica
Comision Nacional de Energia
Av. del Libertador 8250
1429 Buenos Aires
Argentina

V. G. Soloviev
JINR
141980 Dubna
Russia

A. M. Sona-Bizzeti
Dipartimento di Fisica
Universita di Firenze
Largo E. Fermi 2
50125 Firenze
Italy

A. E. Stuchbery
Department of Nuclear Physics
RS Phys. SE
A.N.V.
Canberra ACT 0200
Australia

K. Sugawara-Tanabe,
Otsuma Women's University
Tama Tokyo 206
Japan

M. Sugita
Advanced Science Research Center
Tokai, Ibaraki 319-11
Japan

Z. Sujkowski
Institute for Nuclear Studies
05-400 Swierk
Poland

S. Swaminathan
50 Thiruvengadam Street
R. A. Puram, Madras 600028
India

I. Talmi
Physics Department
Weizmann Institute of Science
Rehovot 76100
Israel

P. Van Isacker
GANIL
BP 5027
F-14021 Caen Cedex
France

A. Ventura
ENEA Cre "E. Clementel"
Viale Ercolani 8
I-40138 Bologna
Italy

M. Vergnes
I.P.N.
91406 Orsay
France

A. Vitturi
Dipartimento di Fisica
University of Padova
Via Marzolo 8
I-35131 Padova
Italy

O. Vogel
Institut fur Kernphysik
University of Koln
Zulpicher Str. 77
50937 Koln
Germany

A. V. Voinov
Laboratory of Neutron Physics
JINR
141980 Dubna
Russia

D. Vretenar
Department of Physics
Bijenicka 32
University of Zagreb
41000 Zagreb
Croatia

D. D. Warner
DRAL Daresbury Laboratory
Daresbury
Warrington WA4 4AD
United Kingdom

A. Wolf
Nuclear Research Center Negev
P.O. Box 9001
Beer-Sheva 84190
Israel

N. Yoshida
Department of Informatics
Kansai University
Ryozenji-cho 2-1-1
Takatsuki-shi 569
Japan

N. Yoshinaga
Department of Physics
College of Liberal Arts
Saitama University
Saitama 338
Japan

N. V. Zamfir
Department of Physics
Brookhaven National Laboratory
Upton, NY 11973
U.S.A.

V. Zelevinsky
Nat. Superconducting Cyclotron Lab.
Michigan State University
East Lansing, MI 48824-1321
U.S.A.

A. Zilges
Institut fur Kernphysik
University of Koln
Zulpicher Str. 77
50937 Koln
Germany

L. Zuffi
Dipartimento di Fisica
Universita di Milano
Via Celoria 16
20133 Milano
Italy

Author Index

Algora, A. 673
Alhassid, Y. 591
Alonso, C.E. 283, 319
Alvarez, R.N. 575
Andrejtscheff, W. 405
Aprahamian, A. 443
Arias, J.M. 283, 319, 605
Arima, A. 157

Balantekin, A.B. 545
Barfield, A.F. 117
Barrett, B.R. 95, 117
Belgya, T. 87
Benczer-Koller, N. 631
Bijker, R. 507
Bizzeti-Sona, A.M. 665
Blasi, P. 665
Blasi, N. 683
Bonatsos, D. 575
Bonsignori, G. 225
Borello-Lewin, T. 449
Brandolini, F. 647
Brant, S. 469, 673
Bucurescu, D. 389
Burke, D.G. 457

Casten, R.F. 21
Castilho Alcaras, J.A. 257
Cata-Danil, G. 389
Chowdhury, P. 239, 631
Cizewski, J.A. 351
Cseh, J. 529

Daillo, A.F. 95
Davis, E.D. 413
Davis, D.E. 95
de Coster, C. 435
Devi, Y.D. 335
Diprete, D.P. 87
Dobaczewski, J. 189

Dobeš, J. 249
Dojnikov, D.N. 465
Dombrádi, Z.S. 469
Duarte, J.L.M. 449
Dukelsky, J. 291, 605

Elliott, J.P. 483
Evans, J.A. 483

Fazekas, B. 87
Fényes, T. 469, 673
Fischer, S.M. 443
Fleshbach, H. 693
Frank, A. 319, 537, 605

Gácsi, Z. 469
Garrett, P.E. 457
Gatenby, R.A. 87
Gelberg, A. 165, 299, 397, 425, 4
Geyer, H.B. 189
Ginocchio, J.N. 201, 495
Gizon, J. 275
Gomes, L.C. 449
Gulyás, J. 469

Harder, M.K. 401
Haydeh-Dejbakhsh 439
Hecht, K.T. 473
Herzberg, R.-D. 393
Heyde, K. 103, 435
Honma, M. 343
Horodynski-Matsushigue, L.B. 449

Iachello, F. 1
Iudice, N. Lo. 79
Ivascu, M. 389

Johnson, C.W. 201
Jolie, J. 45
Jolos, R.V. 299

Katkevičius, O. 257
Kim, K.-H. 33
Kneissl, U. 393
Kota, V.K.B. 335
Krasta, T. 257
Krusche, B. 401
Kumbartzki, G. 631
Kuyucak, S. 143

Lac, V.-S. 483
Lemus, R. 537
Lenzi, S.M. 319
Lévai, G. 529
Leviatan, A. 129, 495
Lichtenthäler Filho, R. 521
Lipparini, E. 461
Lister, C.L. 239, 631
Long, G.L. 483

Maino, G. 421, 617, 665
Margraf, J. 393
Matt, N. 631
Mengoni, A. 421
Metlay, M.P. 443
Mikhajlov, V.M. 465
Mizusaki, T. 33, 417
Molnár, G. 87
Mountford, A. 631

Nakajima, Y. 421
Navrátil, P. 189, 249, 413
Ne'eman, Y. 559

Otsuka, T. 33, 87, 343, 417, 425, 453

Paar, V. 469, 673
Petrov, P. 405
Pietralla, N. 165
Pittel, S. 291, 605
Pitz, H.H. 393
Podolyák, Z.S. 673

Richter, A. 59

Rowe, D.J. 177
Ruža, J. 257

Sagawa, H. 453
Saladin, J.X. 443
Satteson, M. 631
Savoia, M. 225
Scheid, W. 529
Shlomo, S. 439
Sijacki, D. 559
Šimičič, Lj. 673
Smirnov, Yu F. 575
Sofia, H.M. 319
Sohler, D. 673
Soloviev, V.G. 359
Stefanini, A.A. 665
Stroe, L. 389
Stuchbery, A.E. 639
Sugawara-Tanabe, K. 409
Sugita, M. 431
Swaminathan, S.R. 473

Talmi, I. 11
Tambergs, J. 257
Tanczyn, R. 631
Timár, J. 673

Ukita, G.M. 449
Ur, C.A. 389

van Isacker, P. 265, 397
Vass, T. 631
Ventura, A. 421, 461, 665
Vitturi, A. 283, 319
Vogel, O. 165, 397
Voinov, A.V. 477
von Brentano, P. 165, 393, 397, 417
Vretenar, D. 225, 239

Warner, D.D. 373
Wiedenhöver, I. 165
Wolf, A. 657
Wu, X. 443

Yates, S.W. 87
Yoshida, N. 453
Yoshinaga, N. 213

Zamfir, N.V. 327, 405
Zelevinsky, V.G. 307
Zilges, A. 393